WEATHER SATELLITES: SYSTEMS, DATA, AND ENVIRONMENTAL APPLICATIONS

WEATHER SATELLITES: SYSTEMS, DATA, AND ENVIRONMENTAL APPLICATIONS.

EDITED BY

P. KRISHNA RAO

SUSAN J. HOLMES

RALPH K. ANDERSON

JAY S. WINSTON

PAUL E. LEHR

American Meteorological Society
Boston
1990

Library of Congress Cataloging-in-Publication Data

Weather satellites.

Includes bibliographical references.
1. Meteorological satellites. 2. Astronautics in meteorology. I. Rao, P. Krishna, 1930– . II. American Meteorological Society.
QC879.5.W39 1990 551.6′354 90-333
ISBN 0-933876-66-1

ISBN 0-933876-66-1

Typeset and printed in the United States of America by Lancaster Press, Lancaster, PA.

Published by the American Meteorological Society, 45 Beacon Street, Boston, MA 02108.

Editorial support provided by Linda Esche, Laura Westlund, and Editorial Services of New England.

TABLE OF CONTENTS

PREFACE

This book has been compiled against the backdrop of varied and fast-developing progress in the use of meteorological satellites to observe the Earth and its environment. The purpose of this book is to provide students and practitioners of atmospheric sciences with a comprehensive overview of the field, and, by recounting a history of applied satellite meteorology, help them develop a perspective on the benefits to be derived from being able to observe Earth from space.

The idea was conceived by Dr. John H. McElroy in 1984, while he was the Assistant Administrator for Environmental Satellite, Data, and Information Services in NOAA, to commemorate the twenty-fifth anniversary of the launching of the first weather satellite. In 1985, with the departure of Dr. McElroy from NOAA, the current editors inherited responsibility for completing the book.

Initially, the editors and authors intended to produce a text that could be used at the advanced undergraduate or graduate level. As soon as we started receiving the manuscripts, we had to revise our original concept. Inputs received were voluminous and several text books on specialized topics could have been produced from the information. We decided to condense the information into one book and provide a comprehensive survey on environmental (primarily meteorological) remote sensing applications. During this process, we deleted a major section on Ocean Applications; it will be published as a separate book and confined to ocean remote sensing. We hope that this present format will be useful for both students and practicing scientists.

In this book, a number of references appear regarding company and product names. This is for identification purposes only and should not be interpreted as an endorsement by the authors.

A number of authors contributed to this project and they are recognized at the end of each section. In addition, major editorial assistance was provided by Gordon Vaeth and Donna Gicker. Extensive typing was done by Kathy LeFevre, Tina Cashman, and Donna Burns. We appreciate their constructive comments and help in shaping this book.

The Editors

ACKNOWLEDGMENTS

The editors and authors thank the following for their time, energy, patience, and contributions to this effort.

Computer Support
Thomas Babicki

Data Research Assistance
Al Chang
Phil Hovey
Tom Schmugge
Leroy Spayd
Roy Spencer

Editorial Assistance
Robert Dennis
Donna Gicker
Mark Holmes
Bill Replane
John Steger
Rita Terry
Gordon Vaeth

Draft Manuscripts
Bernard Fridovich
Sigmund Fritz
Christopher Hayden
James Sparkman
Vernon Suomi
David Wark
Harold Yates

Preliminary Section Editors
Abraham Schnapf
Harold Yates
Larry Heacock
William Callicott
Chuck Liddick
John Sherman III
Alan Strong
Paul McClain
Ed Epstein
Russell Koffler
Douglas MacCallum
Kenneth Hadeen

Review
Russell Ambroziak
Gary Davis
Pat DeLeonibus
Eric Flemming
Lyle Horn
Phil Hovey
Douglas MacCallum
Eileen Maturi
Paul McClain
Michael Nestlebush
Pasquale Nizza
James Purdom
Terry Schoeni
LeRoy Spayd
Albert Van Dyke
Thomas Vonderhaar
Linwood Whitney

Typing
Cindy Alston
Sandy Billings
Madelyn Bowman
Cindy Brown
Joy Burke
Donna Burns
Tina Cashman
Mildred Grissom
Gail Heddinghaus
Laura Kirschner
Tyeece Little
Kathy Lefevre
Olivia Smith
Emily Tidmore
Sallie Varner
Vickie Wells
Betty Wilson

Proofing
Donna Burns
Tina Cashman
Joan Dutton
Gail Heddinghaus
Mary Johnson
Monique Wilson

Illustrations
Paige Bridges
Gene Dunlop
Frank Dutton
Phil Golden
Ed King
Pasquale Nizza
John Shadid
Jerry Wright

Courtesy Illustrations

Academic Press
Air Force Geophysics Laboratory
American Geophysical Union
American Society of Photogrammetry and Remote Sensing
D. Reidel Publishing Company
European Space Agency
Elsevier Science Publishers
Institude of Electrical and Electroninc Engineers
Japan Meteorological Agency
Jet Propulsion Laboratory
John S. Wiley and Sons
National Academy Press
National Aeronautics and Space Administration
Pergamon Press
Royal Meteorological Society
World Meteorological Organization

CONTRIBUTORS

Abel, Peter G., NOAA/NESDIS, Washington, DC 20233
Alishouse, John C., NOAA/NESDIS, Washington, DC 20233
Anderson, Ralph K., NOAA/NESDIS, Washington, DC 20233
Arkin, Phillip A., NOAA/NWS, Wahington, DC 20233
Banks, Barbara A., NOAA/NESDIS, Washington, DC 20233
*Braun, John C., NOAA/NESDIS, Washington, DC 20233
*Braun, W. Charles, NOAA/NESDIS, Washington, DC 20233
Burdsall, Ellen L., NOAA/NESDIS, Washington, DC 20233
Callicott, William M., NOAA/NESDIS, Washington, DC 20233
Carlson, Marvin A., U.S. Geological Survey, Box 25046, Denver, CO 80225
Chalfant, Michael W., NOAA/NESDIS, Washington, DC 20233
Chang, Alfred T., NASA/Goddard Space Flight Center, Greenbelt, MD 20771
Clark, Dennis K., NOAA/NESDIS, Washington, DC 20233
Clark, J. Dane, NOAA/NESDIS, Washington, DC 20233
Clegg, Bruce R., Lawrence Livermore National Laboratory, Box 808, L-524, Livermore, CA 94550
Coakley, James A., Jr., Department of Atmospheric Sciences, Oregon State University, Corvallis, OR 97331
Coldren, Raymond L., NOAA/NESDIS, Washington, DC 20233
Cotter, Daniel J., NOAA/NESDIS, Washington, DC 20233
Cutchin, David L., University of California, San Diego, Scripps Institution of Oceanography, La Jolla, CA 92903
DeLeonibus, Pat S., NOAA/NESDIS, Washington, DC 20233
Derr, John S., U.S. Geological Survey, Box 25046, Denver, CO 80225
*Dvorak, Vernon F., NOAA/NESDIS, Washington, DC 20233
Ellickson, James K., NOAA/NESDIS, Washington, DC 20233
Ellrod, Gary P., NOAA/NESDIS, Washington, DC 20233
Fuller, Mark R., U.S. Fish and Wildlife Service, Patuxtent Wildlife Research Center,Laurel, MD 20708
Gelman, Melvyn E., NOAA/NWS, Wahington, DC 20233
Graul, Donald W., Ford Aerospace & Communications Corp., 3939 Fabian Way, Palo Alto, CA 94303-9981
Gray, Donald, NOAA/NESDIS, Washington, DC 20233
Greaves, James, NASA, Code EED, Washington, DC 20546
*Grid, Ronald S., NOAA/NESDIS, Washington, DC 20233

* Moved to a new position

Grody, Norman C., NOAA/NESDIS, Washington, DC 20233
Gruber, Arnold, NOAA/NESDIS, Washington, DC 20233
Hadeen, Kenneth D., Director, National Climate Data Center, NESDIS/NOAA, Federal Building, Asheville, NC 28801
*Hallgren, Richard E., NOAA/NWS, Silver Spring, MD 20910
*Hamilton, Gerald G., NOAA/NESDIS, Washington, DC 20233
Hasler, Arthur F., NASA/Goddard Space Flight Center, Greenbelt, MD 20771
Hawkins, Jamison S., NOAA/NESDIS, Washington, DC 20233
Hayden, Christopher, NOAA/NESDIS/SAL, University of Wisconsin,1225 West Dayton St., Madison, WI 53706
Hayes, Richard M., Office of the Oceanographer of the Navy, 34th and Massachusetts Ave. N.W., Washington, DC 20392-1800
Heacock, E. Larry, NOAA/NESDIS, Washington, DC 20233
Holmes, Susan J., NOAA/NESDIS, Washington, DC 20233
Honvault, Claude, ESOC, Pobert-Bosh-Strasse #5, 6100 Darmstadt, West Germany
Hornback, Charles E., NOAA/ERL, Space Environmental Laboratory, 325 Broadway,Boulder, CO 80303
*Hunolt, Gregory, NOAA/NESDIS, Washington, DC 20233
Jacobowitz, Herbert, NOAA/NESDIS, Washington, DC 20233
*Jager, Gilbert, NOAA/NESDIS, Washington, DC 20233
Johnson, Gary E., NOAA/NESDIS, Office of Research and Applications, Satellite Applications Laboratory, Climate Applications Branch, Federal Building, 608 East Cherry St., Columbia, MO 65201
Kahwajy, Faris T., NOAA/NESDIS, Washington, DC 20233
Kamiya, Yoshikazu, National Space Development Agency of Japan,300 South Grand Ave, Suite 2760, Los Angeles, CA 90071
Kniskern, Franklin E., NOS/Joint Ice Center, Washington, DC 20233
Koffler, Russell, NOAA/NESDIS, Washington, DC 20233
Lally, Vincent E., National Center for Atmospheric Research, P.O. Box 3000, Boulder, CO 80307-3000
LeDuc, Sharon K., NOAA/ERL, Atmospheric Science Modeling Division, Atmospheric Research & Exposure Assessment, EPA, MD 80, Research Triangle Park, Raleigh, NC 27711
Legeckis, Richard V., NOAA/NESDIS, Washington, DC 20233

Liddick, Charles E., NOAA/NESDIS, Washington, DC 20233
Lute, Dan, U.S. Department of the Interior Bureau of Reclamation, Northwest Region, P.O. Box 043, Boise, ID 83724
*MacCallum, Douglas H., NOAA/NESDIS, Washington, DC 20233
Mack, Elizabeth G., NOAA/NESDIS, Washington, DC 20233
Mairs, Robert L., NOAA/NESDIS, Washington, DC 20233
Masterson, John E., Universities Space Research Association (USRA), Boulder Program Office, P.O. Box 39, Boulder, CO 80306-0391
Matson, Michael, NOAA/NESDIS, Washington, DC 20233
Maturi, Eileen, NOAA/NESDIS, Washington, DC 20233
McClain, E. Paul, NOAA/NESDIS, Washington, DC 20233
*McClure, C. Hume, NOAA/NESDIS, Washington, DC 20233
McGinnis, David F., Jr., NOAA/NESDIS, Washington, DC 20233
McIntosh, Robert B., NOAA/NESDIS, Washington, DC 20233
McMillin, Larry M., NOAA/NESDIS, Washington, DC 20233
Members of The Japan Meteorological Agency, 3-235 Nakakiyoto, Kiyofe-shi, Tokyo, 204 Japan
Miller, Alvin J., NOAA/NWS, Wahington, DC 20233
*Moses, John F., NOAA/NESDIS, Washington, DC 20233
Nagatani, Ronald M., NOAA/NWS, Wahington, DC 20233
Needham, Bruce, NOAA/NESDIS, Washington, DC 20233
Nestlebush, Michael J., NOAA/NESDIS, Washington, DC 20233
Neuendorffer, Arthur C., NOAA/NESDIS, Washington, DC 20233
*Nizza, Pasquale, NOAA/NESDIS, Washington, DC 20233
Ohring, George, NOAA/NESDIS, Washington, DC 20233
Perkins, Marlin O., NOAA/NESDIS, Washington, DC 20233
Perry, Geoffrey E., 67 Trelawney Ave., Poughill, Bude, Cornwall EX23 9HB, England
Pichel, William G., NOAA/NESDIS, Washington, DC 20233
Pinker, Rachel T., University of Maryland, College Park, MD 20742
Planet, Walter G., NOAA/NESDIS, Washington, DC 20233
*Popham, Robert W., NOAA/NESDIS, Washington, DC 20233
Purdom, James F. W., NOAA/NESDIS, Department of Atmospheric Sciences, Colorado State University, Fort Collins, CO 80523
Rao, P. Krishna, NOAA/NESDIS, Washington, DC 20233
*Reinhardt, George, NOAA/NESDIS, Washington, DC 20233
Reynolds, Richard W., NOAA/NWS, Washington, DC 20233
Robison, William L., Lawrence Livermore National Laboratory, Box 808, L-524, Livermore, CA 94550
Ropelewski, Chester F., NOAA/NWS, Wahington, DC 20233
Roten, Ray J., Computer Graphics Center, Box 7106, North Carolina State University, Raleigh, NC 27695-7106
Schnapf, Abraham, Aerospace Systems Engineering, P.O. Box 160, Willingboro, NJ 08046
Schneider, Stanley R., NOAA/NESDIS, Washington, DC 20233
Schreitz, Richard, NOAA/NESDIS, Washington, DC 20233
Schwalb, Arthur, NOAA/NESDIS, Washington, DC 20233
Scofield, Roderick A., NOAA/NESDIS, Washington, DC 20233
Shaw, Archie E., Service Argos, 1808 McCormick Dr., Suite 10, Landover, MD 20785
Shellman, Marcus B., NOAA/NESDIS, Washington, DC 20233
Silva, H. James, NOAA/NESDIS, Washington, DC 20233
Smidt, George J., NWS, Gramax Building, Silver Spring, MD 20910
Smith, Brian G., NOAA/NWS, Silver Spring, MD 20910
Staton, Carl P., NOAA/NESDIS, Washington, DC 20233
Stearns, Charles R., NOAA/NESDIS, Washington, DC 20233
Stowe, Larry L., NOAA/NESDIS, Washington, DC 20233
Strong, Alan E., NOAA/NESDIS, Washington, DC 20233
Tarpley, J. Dan, NOAA/NESDIS, Washington, DC 20233
Tharps, Debbie T., NOAA/NESDIS, Washington, DC 20233
Thomasell, Albert, Jr., NOAA/NESDIS, Washington, DC 20233
Vaeth, J. Gordon, NOAA/NESDIS, Washington, DC 20233
Wark, Davis Q., NOAA/NESDIS, Washington, DC 20233
Watkins, Benjamin, NOAA/NESDIS, Washington, DC 20233
Weaks, Marcia L., NOAA/NESDIS, Washington, DC 20233
Weinreb, Michael P., NOAA/NESDIS, Washington, DC 20233
Welch, Melvin E., NOAA/NESDIS, Washington, DC 20233
*Whitney, Linwood F., Jr., NOAA/NESDIS, Washington, DC 20233
Wydick, James E., NOAA/NESDIS, Washington, DC 20233
Yates, Harold W., NOAA/NESDIS, Washington, DC 20233
Zbar, Fred S., NOAA/NWS, Silver Spring, MD 20910
Zegowitz, Vincent, Code W/OSO 141, Gramax, Rm 721, NOAA/NWS, Silver Spring, MD 20910

Section I

AN INTRODUCTION TO WEATHER SATELLITES

On April 1, 1960, the world's first meteorological satellite (TIROS-1) was launched by the United States. The spacecraft, the sensors, the communication links, the data, and the data uses of weather satellites today bear little resemblance to what they were at the time of that pioneering effort. In the three decades that have elapsed, satellite weight has grown from about 100 kg to nearly a metric ton, vidicon cameras have given way to scanning radiometers, hand-drawn analyses of the data have been replaced by computer-generated products, what was analog has become digital, and capabilities have expanded to include atmospheric profiling, ocean-sensing, and the collection and relay of environmental data recorded by remote reporting platforms. In addition, weather satellites now measure the space environment in which they operate.

Just as their technology and uses have evolved, so have weather satellites proliferated internationally. Meteorological spacecraft in polar orbit are operated by the United States and the Soviet Union; those in geostationary orbit by the United States, the European Space Agency, Japan, India, and imminently, the U.S.S.R. Weather satellites are planned or under development in the People's Republic of China and in Brazil.

Chapter I-1

OBSERVING SYSTEMS FOR WEATHER, CLIMATE, AND GLOBAL CHANGE—AN EVOLUTION FROM SURFACE TO SPACE[1]

I-1.1 INTRODUCTION

Today's atmospheric scientist would find it difficult to imagine the world without weather satellites, the first of which was launched in 1960. To put the use of weather satellites in proper perspective, we should look at the development of meteorology and how these satellites have affected the weather services of the world.

I-1.2 METEOROLOGY AND WEATHER PREDICTION BEFORE THE ADVENT OF SATELLITES

Atmospheric sciences were first described by Aristotle as early as the Third Century, B.C. Until the new technology that began in the mid-19th century, meteorology was described by unverifiable theories and philosophical discussions. Observations of the atmosphere were difficult to obtain. Systematic weather observations did not begin until shortly after the invention of the telegraph. The first near-synoptic meteorological observing network was organized in the 1840s by Joseph Henry, Secretary of the Smithsonian Institution. Weather reporting expanded nationwide in 1870 with the establishment of a federal weather service under the U.S. Army Signal Corps; in 1890 an independent U.S. Weather Bureau was established. The communication of weather and hydrologic information was greatly improved as inventions provided new technology: the radio early in the 20th century and the teletypewriter in 1928.

The first observational network reported surface conditions; however, scientists also needed data on conditions higher in the atmosphere. Professor Charles Marvin of the U.S. Weather Bureau invented the kite meteorograph in 1896 to record temperature, pressure, and humidity aloft. In 1898, the Weather Bureau opened several kite stations and began experimenting with daily synoptic observations. While the kite sounding program was faced with problems and severe limitations, it was the main means for obtaining upper air observations until airplanes were first used in 1931. Aircraft sounding, however, was both expensive and dangerous, so the Weather Bureau also tracked unmanned balloons to obtain wind directions and speeds aloft. Meteorographs and radio transmitters attached to these balloons were the first radiosonde instruments. The Weather Bureau began daily radiosonde observations in 1938. Today, the radiosonde is still an important tool for worldwide meteorological observations.

During the 1940s, two new methods for observing and investigating weather systems were developed. In 1943, an Army Air Force colonel flew a light, single engine airplane into the eye of a hurricane. This was the beginning of aircraft hurricane reconnaissance. By

[1] Primary Contributor: Richard E. Hallgren

the late 1940s, the technology of RAdio Detection And Ranging (radar) was being used for meteorological observations. Ever since, radar has been a highly effective observing tool. It is particularly useful for detecting and following severe storms and for tracking hurricanes as they approach coastal areas.

Revolutionary advances in large-scale weather prediction began in 1952 using the newly developed, high speed electronic computers to solve the equations that describe the atmosphere's behavior. By 1955, numerical weather predictions were being made operationally by U.S. Government agencies at the Joint Numerical Weather Prediction Center. Numerical Weather Prediction (NWP) models performed optimally with dense and uniform data input. Over most ocean areas, surface and upper air data coverage was almost nonexistent. Thus initial NWP analyses were incomplete, and forecasts were inadequate, particularly along the west coasts of continents.

I-1.3 DEVELOPMENT AND EVOLUTION OF WEATHER SATELLITES AND THEIR USE FOR WEATHER ANALYSIS AND PREDICTION

Possible solutions to the problems of inadequate weather observations became available in the 1940s. Rockets fitted with instruments and cameras collected data and took cloud pictures that were returned to Earth. The spectacular cloud photographs suggested that artificial satellites could be used as observing platforms in space. The successful launch of the TIROS-1 satellite in 1960 quickly confirmed these opinions.

I-1.3.1 Early Applications

The "silent" areas over the oceans were of immediate interest. If the satellite observations could be used, initial synoptic analyses would be more accurate and forecasts improved. The earliest investigations of satellite imagery focused on the identification of the cloud patterns of synoptic-scale circulation features and their general characteristics. Satellite pictures were used by the National Meteorological Center (NMC) in the preparation of surface analyses. During the mid-1960s, the polar orbiting Environmental Science Services Administration (ESSA) satellites provided global coverage that brought more widespread use of cloud imagery in synoptic analyses. Also, by that time, low resolution, real time data had become available worldwide through Automatic Picture Transmission (APT), a direct satellite-to-ground link between the satellite and local weather service users. Figure I-1-1 depicts the first type of global coverage by a satellite.

I-1.3.2 Weather Satellites Come of Age

By the early 1970s, the polar orbiting NOAA satellites provided global InfraRed (IR) imagery both day and night. IR images aided in distinguishing the heights of the cloud tops so that the use of satellite imagery in synoptic analysis was further improved. Atmospheric sounding sensors, first flown experimentally onboard Nimbus satellites of the 1960s, became operational on the NOAA satellites in 1972. These sounders produced vertical temperature profiles of the atmosphere in the cloudfree areas worldwide. Since 1978, polar orbiter sounders have produced temperature and moisture

Figure I-1-1. The first global coverage in a 24 hour period (TIROS-9), February 13, 1965.

profiles globally under most conditions of cloud cover. The data have been sent to NMC and meteorological centers worldwide.

The experimental geostationary satellites of the mid-1960s photographed the Earth once every 30 min. These photographs were made into time series films that showed the motion of clouds over the face of the Earth. These film loops were used as input to NWP from data sparse areas and to derive wind vectors from cloud motions.

Infrared sensors on Geostationary Operational Environmental Satellites (GOES) of the mid-1970s provided for real time monitoring of hemispheric weather conditions 24 hours a day. The IR data, with their ability to distinguish cloud top heights, allowed for improved determination of wind vectors at different levels in the atmosphere. In 1980, sounders were added to geostationary satellites; since 1982, these sounding data have been used for research and operations. More details about satellite sensors and their uses may be found in Chapter I-2, Sections III, IV, and VII through X.

I-1.3.3 Use of Satellite Information in Field Operations

Ever since the early days of TIROS-1, satellite imagery has been used to locate significant weather systems. With the arrival of the geostationary satellites in the mid-1960s, the National Environmental Satellite Center (NESC) produced the first time series "film loops" for distribution to selected field offices. Although these film loops were not available in real time, they were useful for visualizing the motions and characteristics of weather systems. The only satellite information available to the field stations in real time was that relayed by facsimile.

In the early 1970s, the National Environmental Satellite Service (NESS[2]—formerly NESC) established the first Satellite Field Services Station (SFSS) at Washington, D. C. Such stations were to supply interpretation of satellite imagery to National Weather Service (NWS) field stations. SFSS meteorologists prepared routine Satellite Interpretation Messages to advise field stations on current conditions and to alert them about quickly developing severe weather conditions. The number of SFSS was increased to four in 1974; the new SFSS were collocated with the Weather Service Forecast Office (WSFO) in San Francisco, the National Hurricane Center (NHC) in Miami, and the National Severe Storms Forecast Center (NSSFC) in Kansas City. Geostationary satellite data were used by the SFSS to supply data to many direct users, all WSFO nationwide, and many private and military customers. Three additional SFSS were later established in Anchorage, Honolulu, and Slidell, Louisiana.

In 1987, a Satellite Weather Information System (SWIS) was implemented which provides each WSFO with a direct, on-site capability to acquire and manipulate satellite imagery. Color enhanced imagery can be animated and electronically overlaid with numerical prediction products and conventional data. This has improved and hastened the interpretation of satellite imagery, permitting forecasters to more rapidly incorporate the results into the warning and forecast process.

During the 1970s, very active research quickly led to operational applications. These included identification of severe storm signatures, convective rainfall estimates for flash flood situations, and techniques for estimating tropical cyclone intensity. Since that time major advances have resulted in the ability to interpret satellite imagery and apply the results to weather and flood forecast problems. As forecasters became more knowledgeable about satellite imagery and convinced of its value as a forecast tool, studies continued to produce an abundance of new applications and techniques. Discussions of these operational applications appear in Section VII.

I-1.3.4 Use of Satellites in Environmental Monitoring

According to present observations and understanding, there is considerable evidence to show that changes are taking place in our global climate and natural environment. These changes are comparable to those since the last ice age. Changing concentrations of carbon dioxide, ozone, and other radiatively active trace gases are being noted. It is important to observe and document these changes and understand the complex interactive processes that affect the thermal balance of the atmosphere and the exchange of energy between ocean, land, and atmosphere.

The ocean circulation, biogeochemical cycles, and the total energy budget of the earth-atmosphere-ocean system are some of the key elements that must be monitored and evaluated to understand the global change. Only a well designed observational system of combined space-based observations and *in situ* measurements can provide the needed information.

The capabilities of satellites for ocean observations have been sufficiently demonstrated by Seasat, Geosat, Nimbus-7, and Advanced TIROS-N systems. Sea surface temperature distributions are being derived and temporal and spatial variations routinely monitored. These methods of observing global ocean parameters

[2] NESS became NESDIS, the current National Environmental Satellite, Data, and Information Service, December 1, 1982.

critical for weather and climate applications are possible only by deploying space-based systems. A description of sea surface temperature applications is contained in Section VIII.

Monitoring land surface conditions from satellites, important for weather and climate applications, is now feasible. The net radiation at the surface is the energy source for surface heating, and latent, sensible, and soil heat fluxes. The net heat flux is dependent on the type and condition of vegetation, albedo, surface temperature, and other parameters. These products play an important role in climate monitoring, diagnostic, and prediction studies. Again, satellites have a major role to play; a full description of various applications appears in Sections VIII through X.

I-1.3.5 Use of Satellites in Communications

The communications functions of weather satellites have evolved into a critically important capability. One of these functions is direct data readout, available in several forms. Since 1963, APT systems have permitted weather stations and individual observers worldwide to receive imagery from the polar orbiters, often with quite inexpensive equipment. Since the mid-1960s, equipment on geostationary satellites has transmitted imagery and soundings directly to users. Section VI gives a more complete picture of satellite data and product distribution. Since the middle and late 1970s, geostationary and polar orbiting satellites have carried subsystems to collect and relay data from remote *in situ* platforms. Section V describes such data collection.

I-1.3.6 Use of Satellites in International Programs

Spectacular advances in numerical forecasting and in data gathering by weather satellites have transformed global weather forecasting from an art to a science in the span of one generation. In 1961, the meteorological services of many nations of the world met to organize and initiate the World Weather Program under the sponsorship and support of the United Nations World Meteorological Organization (WMO) and the International Council of Scientific Unions (ICSU). The World Weather Program consisted of the World Weather Watch (WWW) and the Global Atmospheric Research Program (GARP). The WWW was the data gathering and processing part of the program; GARP was the research part. Geostationary and polar orbiting satellites furnished many of the observations for these programs. Over 140 nations participated in this program.

Chapter I-2 and Section III contain more detailed information on the uses of weather satellites for international programs.

I-1.4 FUTURE PLANS AND EXPECTATIONS

The meteorological successes of the past have led to great expectations for the future. Improvements in meteorological understanding and in the development of forecasting techniques have been substantial and rewarding. Information provided by satellite systems has contributed to steady improvement in the observation and forecasting of weather phenomena. Yet much remains to be done; weather satellites will be the key to solutions of some of the remaining problems of weather prediction.

The next generation of geostationary satellites will almost double the current spatial resolution. The imager will operate at much faster scan rates. Concurrent temperature and moisture soundings will be provided at frequent intervals. Data accuracy and increased resolution will also further our understanding of smaller scale processes.

Polar orbiting satellites of the future will continue to supply reliable sounding data for numerical forecasts. More accurate versions of current sensors will provide better accuracy and more detail in vertical profiles. Other sensors under development will permit: Doppler determination of winds; improved estimates of sea surface temperatures, wind speeds, and ice cover; measurement of ocean wave heights; detection of tides, currents and storms surges; and measurement of other parameters.

NWS interactive processing systems will handle large quantities of satellite and other data virtually automatically. The NWS Satellite Weather Information System (SWIS) will automatically acquire, store, display, and animate satellite imagery. The NWS Advanced Weather Interactive Processing System (AWIPS) for the 1990s will be the NWS operational system of the future. This system is being developed in anticipation of future observing systems and scientific developments; it will access many kinds of data and derived products for multiple uses. Interagency and international programs will benefit from new technologies. With time and more refined sensors, meteorological satellites will begin to furnish information that will lead to answers to many current climatological questions. Section IX contains a compilation of current climatological applications of satellite data.

In all of these considerations for the future, the focus is on continuation of current programs, refinements of present capabilities, and new techniques arising from current and currently planned satellite sensor data. Speculative or exotic technologies are not in the picture. Current plans for the future represent a maximal effort on the part of the United States to maintain its leadership role during the first 25 years of scientific progress through the use of meteorological satellites.

Chapter I-2

A HISTORY OF CIVILIAN WEATHER SATELLITES[1]

I-2.1 INTRODUCTION

A new era in meteorological observation and forecasting was ushered in with the launching of the Television and InfraRed Observation Satellite (TIROS-1) on April 1, 1960 (Figure I-2-1). TIROS, the world's first meteorological satellite, demonstrated the ability to acquire images of the Earth's cloud cover over much of the planet. Although TIROS-1 was experimental, the early results were welcomed by meteorologists who used them to supplement conventional observations. The impact of these initial space-acquired observations led to the evolution of a series of weather satellites that routinely monitor the Earth's atmosphere, land masses, and oceans.

I-2.2 EVENTS LEADING TO THE TIROS PROGRAM

Following World War II, captured V-2 and Viking rockets were used for aerological research. In 1954, the U. S. Navy Aerobee rocket returned high altitude motion pictures of the Earth and its cloud cover. Ensuing studies by the Rand Corporation and others in the mid-1950s led to the conceptual development of a satellite that could observe the Earth's cloud cover from space.

The initial development of the TIROS program was carried out by the Radio Corporation of America (RCA) under a contract let in 1956 by the Army Ballistic Missile Agency at Redstone Arsenal. It called for development of a concept for a rocket launched satellite carrying a TV system for weather reconnaissance. The program was designated JANUS; the resultant design was a slender, rodshaped, 9 kg spacecraft 0.13 m in diameter and 0.76 m long, equipped with a single TV camera. In 1958, the configuration was redesigned when a larger booster was made available. The new satellite weighed 39 kg, was 0.46 m in diameter and 0.76 m long.

In late 1958, the Thor-Able launch vehicle was made available and the spacecraft, now called TIROS-1, grew even larger. It now weighed 122 kg and was 1.1 m in diameter and 0.6 m high, not including antennas on the top and bottom of the spacecraft. The spacecraft consisted of two compact 1.3-cm vidicon television cameras (one wide angle and one narrow angle), two video recorders, a command and control electronic system, a communication system, solid rockets for spin control, power supply with rechargeable batteries, and a cylindrical solar array. During its development years, the program management shifted from the Army Ballistic Missile Agency to the Department of Defense Advanced Research Projects Agency to the U. S. Army Signal Corps, and finally, in 1959, to the newly created National Aeronautics and Space Administration (NASA) at the Goddard Space Flight Center.

[1] Primary Contributor: Abraham Schnapf

Figure I-2-1. The launching of TIROS-1, April 1, 1960 (courtesy of NASA).

I-2.3 THE EVOLUTION OF THE OPERATIONAL TIROS METEOROLOGICAL SATELLITE SYSTEM

I-2.3.1 TIROS Research and Development Satellites—TIROS-1 to 10 (1960 to 1965)

TIROS-1 was launched April 1, 1960, from Cape Canaveral, Florida, by a Thor-Able launch vehicle. Its primary objective was to demonstrate the feasibility of observing the Earth's cloud cover by means of slow scan miniature television cameras in an Earth orbiting, spin stabilized satellite. TIROS-1 was placed in a circular orbit 644 km above the Earth at a 48° inclination to the Equator. Television pictures were received on the first orbit. Figure I-2-2a shows one of the early TV pictures transmitted from TIROS-1 by its wide angle TV camera. Thus, a new and powerful meteorological tool became a reality.

TIROS-2, launched November 23, 1960, was used to demonstrate the utility of an experimental five channel

InfraRed (IR) and VISible (VIS) scanning radiometer and a two channel nonscanning instrument. The instruments measured the emitted and reflected thermal energy of the Earth's surface and its atmosphere. Measurements were used to make initial determinations of spatial and temporal variations in the heat balance of the Earth.

TIROS-3, 4, 5, 6, and 7 were launched between July 1961 and June 1963 to provide continuous observation of the Earth's cloud cover for limited operational use. With each of these satellites, particular care was taken to provide early warning of tropical storm development and movement. The satellites were also used experimentally to detect sea ice and snow cover.

TIROS-8, launched in December 1963, included a new 2.5-cm vidicon TV camera for Automatic Picture Transmission (APT); it had a very slow scan readout requiring 200 sec to transmit an 800 TV-line image. The 2-kHz bandwidth of the APT system permitted direct real time transmission of cloud pictures to relatively inexpensive APT ground stations around the world. Many nations could now receive two to four orbits of weather observations daily, covering a 1600 to 3200-km radius, directly from the spacecraft while the satellite was within radio range of the ground receiver.

TIROS-9, launched in January 1965, was reconfigured with its two television cameras aligned along the radius of the cylindrical satellite rather than parallel to its spin axis; thus the cameras could look straight down every time the satellite turned on its axis (at approximately 12 rpm). An improved attitude control system maintained a controlled rotation of the satellite in near-polar, Sun synchronous orbit. This "wheel" spacecraft could view the entire planet in a 24-hour period as it "rolled" along its orbit. Daily coverage capability of the satellite increased fourfold. TIROS-9 was the forerunner of the TIROS Operational System (TOS) satellites.

TIROS-10, the last of the research and development series, was launched in July 1965 to provide tropical storm observations. Table I-2-1 summarizes the TIROS and other U.S. meteorological satellites.

I-2.3.2 The TIROS Operational System (TOS) Satellites—ESSA-1 to 9 (1966-1969)

The world's first operational weather satellite system was placed into service with the launching of the Environmental Science Services Administration satellites, ESSA-1, on February 3, 1966, and ESSA-2, on February 28, 1966.[2]

The commitment to acquire worldwide observational data from satellites routinely and daily was fulfilled with the introduction of the TOS program. This system consisted of a pair of ESSA satellites in Sun synchronous orbit, each configured for a specific mission. The Advanced Vidicon Camera System (AVCS) obtained global imagery which were transmitted to the ESSA Command and Data Acquisition (CDA) stations at Wallops, Virginia, and Fairbanks, Alaska. The CDA relayed the data to the National Environmental Satellite Service at Suitland, Maryland, for processing and dissemination to forecasting centers of the U.S. and other nations. The odd numbered satellites (ESSA-1, 3, 5, 7, and 9), with redundant AVCS systems, were the global readout satellites. The even numbered satellites (ESSA-2, 4, 6, and 8) were equipped with redundant APT cameras. APT television pictures were transmitted directly to ground stations located around the world. All ESSA satellites operated in orbits of approximately 1450 km.

I-2.3.3 Improved TIROS Operational System (ITOS) Satellites—ITOS-1, NOAA-2 to 5 (1970-1978)

The second decade of meteorological satellites began with the launch of ITOS-1 on January 23, 1970. ITOS-1 combined the joint capabilities of two ESSA spacecraft; these were the direct readout APT system, and the global stored images of the AVCS system. In addition, ITOS-1 carried an operational two channel Scanning Radiometer (SR) providing day and night radiometric data for immediate transmission, as well as acquisition data stored for later playback to CDA stations. With the IR scanning radiometer, global observation of the Earth atmosphere and surface areas was available once every 12 hours from a single ITOS spacecraft. A second ITOS spacecraft, the National Oceanic and Atmospheric Administration satellite, NOAA-1, was launched on December 11, 1970.

The ITOS system evolved further with the development of the ITOS-D satellites. These had a new redundant sensor complement to provide day and night imaging by means of the Very High Resolution Radiometer (VHRR) and the medium resolution Scanning Radiometer (SR). The VHRR and SR systems replaced the AVCS and APT cameras of earlier systems. The Vertical Temperature Profile Radiometer (VTPR) for sounding the atmosphere and the Solar Proton

[2] Responsibility for the nation's operational civil meteorological satellites was assigned by Congress to the U. S. Weather Bureau in 1962. In 1965 the Bureau, renamed the National Weather Service, was made part of a newly created Environmental Science Services Administration (ESSA) which, in 1970, became the National Oceanic and Atmospheric Administration (NOAA). Satellite designators, such as NOAA-9, reflect these organizational name changes.

Table I-2-1 *Summary of U.S. civilian meteorological satellites*

NAME	LAUNCHED	PERIOD (MIN)	PERIGEE (KM)	APOGEE (KM)	INCLINATION (DEG)	REMARKS
TIROS-1	01 APR 60	99.2	796	867	48.3	1 TV-WA and 1 TV-NA
TIROS-2	23 NOV 60	98.3	717	837	48.5	1 TV-WA, 1 TV-NA, passive & active IR scan
TIROS-3	12 JUL 61	100.4	854	937	47.8	2 TV-WA, HB, IR, IRP
TIROS-4	08 FEB 62	100.4	817	972	48.3	1 TV-WA, IR, IRP, HB
TIROS-5	19 JUN 62	100.5	680	1119	58.1	1 TV-WA, 1 TV-MA
TIROS-6	18 SEP 62	98.7	783	822	58.2	1 TV-WA, 1 TV-MA
TIROS-7	19 JUN 63	97.4	713	743	58.2	2 TV-WA, IR, ion probe, HB
TIROS-8	21 DEC 63	99.3	796	878	58.5	1st APT TV direct readout & 1 TV-WA
Nimbus-1	28 AUG 64	98.3	487	1106	98.6	3 AVCS, 1 APT, HRIR "3-axis" stabilization
TIROS-9	22 JAN 65	119.2	806	2967	96.4	First "wheel", 2 TV-WA global coverage
TIROS-10	02 JUL 65	100.6	848	957	98.6	Sun synchronous, 2 TV-WA
ESSA-1	03 FEB 66	100.2	800	965	97.9	1st operational system, 2 TV-WA, FPR
ESSA-2	28 FEB 66	113.3	1561	1639	101.0	2 APT, global operational APT
Nimbus-2	15 MAY 66	108.1	1248	1354	100.3	3 AVCS, HRIR, MRIR
ESSA-3	02 OCT 66	114.5	1593	1709	101.0	2 AVCS, FPR
ATS-1	06 DEC 66	24 hr	41,257	42,447	0.2	Spin scan camera
ESSA-4	26 JAN 67	113.4	1522	1656	102.0	2 APT
ESSA-5	20 APR 67	113.5	1556	1635	101.9	2 AVCS, FPR
ATS-3	05 NOV 67	24 hr	41,166	41,222	0.4	Color spin scan camera
ESSA-6	10 NOV 67	114.8	1622	1713	102.1	2 APT TV
ESSA-7	16 AUG 68	114.9	1646	1691	101.7	2 AVCS, FPR, S-band
ESSA-8	15 DEC 68	114.7	1622	1682	101.8	2 APT TV
ESSA-9	26 FEB 69	115.3	1637	1730	101.9	2 AVCS, FPR, S-band
Nimbus-3	14 APR 69	107.3	1232	1302	101.1	SIRS A, IRIS, MRIR, IDCS, MUSE, IRLS
ITOS-1	23 JAN 70	115.1	1648	1700	102.0	2 APT, 2 AVCS, 2 SR, FPR, 3-axis stabilization
Nimbus-4	15 APR 70	107.1	1200	1280	99.9	SIRS B, IRIS, SCR, THIR, BUV, FWS, IDCS, IRLS, MUSE
NOAA-1	11 DEC 70	114.8	1422	1472	102.0	2 APT, 2 AVCS, 2 SR, FPR
NOAA-2	15 OCT 72	114.9	1451	1458	98.6	2 VHRR, 2 VTPR, 2 SR, SPM
Nimbus-5	11 DEC 72	107.1	1093	1105	99.9	SCMR, ITPR, NEMS, ESMR, THIR
NOAA-3	06 NOV 73	116.1	1502	1512	101.9	2 VHRR, 2 VTPR, 2 SR, SPM
SMS-1	17 MAY 74	1436.4	35,605	35,975	0.6	VISSR, DCS, WEFAX, SEM
NOAA-4	15 NOV 74	101.6	1447	1461	114.9	2 VHRR, 2 VTPR, 2 SR, SPM
SMS-2	06 FEB 75	1436.5	35,482	36,103	0.4	VISSR, DCS, WEFAX, SEM
Nimbus-6	12 JUN 75	107.4	1101	1115	99.9	ERB, ESMR, HIRS, LRIR, T&DR, SCAMS, TWERLE, PMR
GOES-1	16 OCT 75	1436.2	35,728	35,847	0.8	VISSR, DCS, WEFAX, SEM
NOAA-5	29 JUL 76	116.2	1504	1518	102.1	2 VHRR, 2 VTPR, 2 SR, SPM
GOES-2	16 JUN 77	1436.1	35,600	36,200	0.5	VISSR, DCS, WEFAX, SEM
GOES-3	15 JUN 78	1436.1	35,600	36,200	0.5	VISSR, DCS, WEFAX, SEM
TIROS-N	13 OCT 78	98.92	849	864	102.3	AVHRR, HIRS-2, SSU, MSU, HEPAD, MEPED
Nimbus-7	24 OCT 78	99.28	943	955	104.09	LIMS, SAMS, SAM-II, SBUV/TOMS, ERB, SMMR, THIR, CZCS
NOAA-6	27 JUN 79	101.26	807.5	823	98.74	AVHRR, HIRS-2, SSU, MSU, HEPAD, MEPED
GOES-4	09 SEP 80	1436.1	35,600	35,600	0.5	VAS, DCS, SEM, WEFAX
GOES-5	22 MAY 81	1436.1	35,600	35,600	0.5	VAS, DCS, SEM, WEFAX
NOAA-7	23 JUN 81	101.92	852	869	98.9	AVHRR, HIRS-2, SSU, MSU, HEPAD, MEPED
NOAA-8	28 MAR 83	101.2	826	801	98.2	AVHRR, HIRS-2, SSU, MSU, SEM, SAR
GOES-6	28 APR 83	1436.1	35,803	35,771	0.1	VAS, DCS, WEFAX, SEM
ERBS	05 OCT 84	96.8	608	393	5.7	ERBE, SAGE-II
NOAA-9	12 DEC 84	102.0	862	841	98.9	AVHRR, HIRS-2, SSU, MSU, SEM, SAR, SBUV, ERBE
NOAA-10	17 SEP 86	101.277	803	824	98.66	AVHRR, HIRS-2, SSU, MSU, SEM, SAR, SBUV, ERBE
GOES-7	26 FEB 87	1436.1	35,759	35,826	0.0493	VAS, DCS, WEFAX, SEM
NOAA-11	24 SEP 88	102.139	845	863	98.91	AVHRR, HIRS-2, SSU, MSU, SEM, SAR, SBUV, ERBE

Table I-2-1 *(Continued)*

APT	Automatic Picture Transmission TV	PMR	Pressure Modulated Radiometer
AVCS	Advanced Vidicon Camera System (1″ Vidicon)	SAM-II	Stratospheric Aerosol Measurement-II
AVHRR	Advanced Very High Resolution Radiometer	SAMS	Stratospheric and Mesospheric Sounder
BUV	Backscatter Ultraviolet Spectrometer	SAR	Search and Rescue
CZCS	Coastal Zone Color Scanner	SBUV	Solar Backscatter Ultraviolet Spectrometer
DCS	Data Collection System	SCAMS	Scanning Microwave Spectrometer
ERB	Earth Radiation Budget	SCMR	Surface Composition Mapping Radiometer
ERBE	Earth Radiation Budget Experiment	SCR	Selective Chopper Radiometer
ESMR	Electronic Scanning Microwave Radiometer	SEM	Solar Environmental Monitor
FPR	Flat Plate Radiometer	SIRS	Satellite Infrared Spectrometer
FWS	Filter Wedge Spectrometer	SMMR	Scanning Multichannel Microwave Radiometer
HB	Heat Budget Instrument	SPM	Solar Proton Monitor
HEPAD	High Energy Proton and Alpha Particle Detector	SR	Scanning Radiometer
HIRS	High Resolution Infrared Sounder	SSU	Stratospheric Sounding Unit
HRIR	High Resolution Infrared Radiometer	T&DR	Tracking and Data Relay
IDCS	Image Dissector Camera System	TED	Total Energy Detector
IR	Infrared 5 Channel Scanner	THIR	Temperature Humidity Infrared Radiometer
IRIS	Infrared Interferometer Spectrometer	TOMS	Total Ozone Mapping Spectrometer
IRLS	Interrogation, Recording and Location Subsystem	TV	Television Cameras (1/2″ Vidicon)
IRP	Infrared Passive		NA Narrow Angle-12°
ITPR	Infrared Temperature Profile Radiometer		MA Medium Angle-78°
LIMS	Limb Infrared Monitoring of the Stratosphere		WA Wide Angle-104°
LRIR	Limb Radiance Infrared Radiometer	TWERLE	Tropical Wind Energy Reference Equipment
MEPED	Medium Energy Proton and Electron Detector	VAS	VISSR and Atmospheric Sounder
MRIR	Medium Resolution Infrared Radiometer	VHRR	Very High Resolution Radiometer
MSU	Microwave Scanner Unit	VISSR	Visible Infrared Spin Scan Radiometer
MUSE	Monitor of Ultraviolet Solar Energy	VTPR	Vertical Temperature Profile Radiometer
NEMS	Nimbus E Microwave Spectrometer	WEFAX	Weather Facsimile

Monitor (SPM) for measurements of solar proton and electron flux in the vicinity of the satellite were added. The ITOS-1 series weighed 310 kg and the ITOS-D series weighed 340 kg. Four ITOS-D satellites were launched: NOAA-2, 3, 4, and 5 were placed in orbit in 1972, 1973, 1974, and 1975, respectively.

I-2.4 THE TIROS-N/NOAA OPERATIONAL SYSTEM

A third generation operational polar orbiting environmental satellite system was developed, and the first spacecraft, TIROS-N, was placed into operational service in 1978. This series of eleven spacecraft, TIROS-N/NOAA-A to J, was designed to provide operational service from 1978 to 1988. VIS imagery from TIROS-N are shown in Figure I-2-2b,c. There are two configurations in this series: five of the TIROS-N type, TIROS-N and NOAA-A to D, and six of the Advanced TIROS-N (ATN) type, NOAA-E to J. Figures I-2-3 and III-1-1 show the TIROS-N and ATN configurations.

I-2.4.1 The TIROS-N/NOAA-A to D System (1978-1981)

The TIROS-N system provided NOAA with the global meteorological and environmental data required for normal operations and for the experimental World Weather Watch Program. The spacecraft had a new complement of data gathering instruments. The Advanced Very High Resolution Radiometer (AVHRR) provided data not only for day and night imaging in the VIS and IR, but also for sea surface temperature determination, estimation of heat budget components, and identification of snow and sea ice. The TIROS Operational Vertical Sounder (TOVS) provided improved estimates of the vertical structure of the atmosphere. The Data Collection System (DCS) collected environmental data from fixed and moving platforms such as buoys, balloons, and other remote platforms; and transmitted the data to central stations for processing and relay to users. The Solar Environment Monitor (SEM) measured solar proton, electron, and alpha particle densities. For detailed descriptions of AVHRR, DCS, and SEM, see Chapter IV-2.

The data collected by the TIROS-N instruments, like those from all NOAA polar orbiting spacecraft, were stored onboard the satellite for transmission to the NOAA central processing facility at Suitland, Maryland, through the Wallops and Fairbanks CDA stations. Satellite data also were transmitted in real time direct readout at VHF and S-band frequencies to remote stations around the world in the APT and High Resolution Picture Transmission (HRPT) modes.

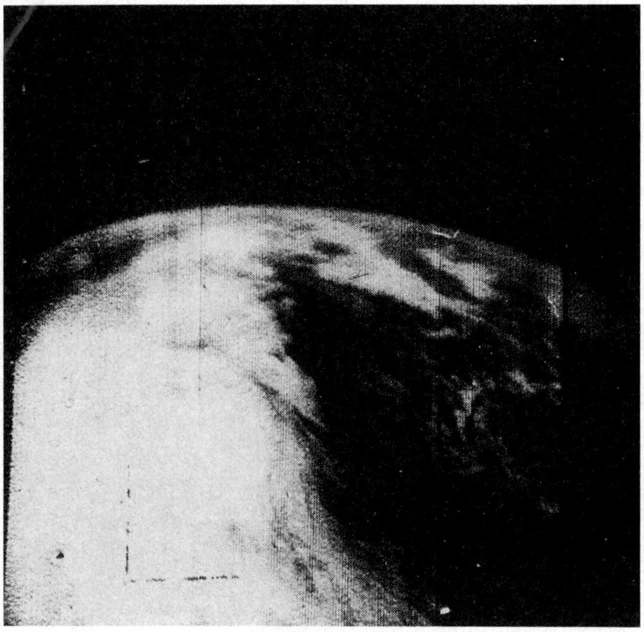

(a)

(b)

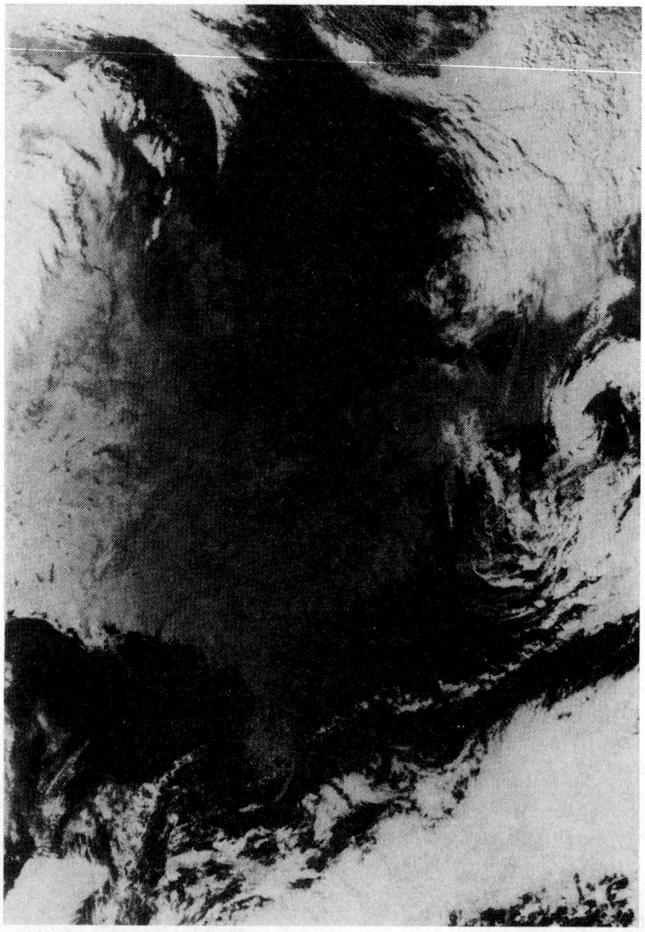

(c)

Figure I-2-2. Representative pictures from the TIROS series: (a) first TV picture from space by TIROS-1, (b) near-east scene of TIROS-N AVHRR VIS channel, and (c) eastern U.S. scene of TIROS-N VIS channel.

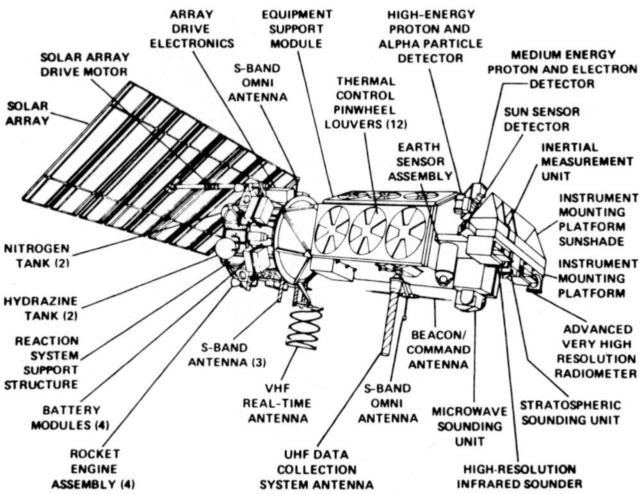

Figure I-2-3. The TIROS-N configuration (courtesy of RCA).

The TIROS-N satellites operated in a near polar circular Sun synchronous orbit with a nominal altitude of either 833 or 870 km. In the operational configuration, two satellites were positioned with a nominal orbit plane separation of 90°. Orbits for the TIROS-N, as for the later ATN series, were subject to the same constraints as earlier satellites, with operational requirements determining whether the spacecraft was to be launched in an afternoon (ascending) or a morning (descending) orbit. Launch of NOAA-D was superseded by the launch of the first Advanced TIROS-N and the spacecraft was placed in long-term storage for later use. For further details see Chapter III-1.

I-2.4.2 The Advanced TIROS-N (ATN)/NOAA-E to J System (1983 to present)

The last six TIROS-N/NOAA spacecraft, NOAA-E to J, were modified to add instrumental capacity (hence the Advanced TIROS-N designation). The spacecraft was lengthened approximately 0.5 m and the solar array was enlarged to provide additional power. Systems were added incrementally. Starting with NOAA-8 (NOAA-E before launch), the Search and Rescue (SAR) system was added. The Earth Radiation Budget Experiment (ERBE) was added to NOAA-9 (NOAA-F) and to NOAA-10 (NOAA-G). The Solar Backscatter UltraViolet (SBUV) radiometer was also added to NOAA-9 (NOAA-F). The instrument complement in each of the TIROS-N/NOAA-A to J series is shown in Table I-2-2.

As of October 1988, 32 TIROS/ESSA/ITOS/TIROS-N satellites had been successfully orbited. Three others failed to achieve orbit because of launch vehicle failure. Figure I-2-4 summarizes the orbital history of the TIROS satellites. Development, construction, and launch of the polar orbiting series were managed by the NASA Goddard Space Flight Center. Prototypes were funded by NASA; operational spacecraft by NOAA and its predecessor agencies.

Table I-2-2 *Polar satellite system launch schedules and spacecraft systems*

SATELLITE	LAUNCH DATE	TIME OF DAY	INSTRUMENT COMPLEMENT								
			AVHRR/2	HIRS/2	MSU	SSU	SEM[1]	DCS	SAR	SBUV[2]	ERBE
TIROS-N	10/13/78	PM	4-Channel	X	X	X	X	X			
NOAA-6(A)	6/27/79	AM	4-Channel	X	X	X	X	X			
NOAA-7(C)[3]	6/23/81	PM	5-Channel	X	X	X	X	X			
NOAA-8(E)	3/28/83	AM	4-Channel	X	X	X	X	X	X	D[4]	D
NOAA-9(F)	12/24/84	PM	5-Channel	X	X	X	X	X	X	X	X
						D					
NOAA-10(G)	9/17/86	AM	4-Channel	X	X	X	X	X	X		X
NOAA-H	9/24/88	PM	5-Channel	X	X	X		X	X	X	
NOAA-D	*12/89–6/90	AM	5-Channel	X	X			X			
NOAA-I	*9/90	PM	5-Channel	X	X	X		X	X	X	
NOAA-J	*12/91	AM	5-Channel	X	X			X	X		
NOAA-K	*4/93	PM	6-Channel	HIRS/3	AMSU			X	X		
NOAA-L	*8/94	AM	6-Channel	HIRS/3	AMSU			X	X		
NOAA-M	*11/95	PM	6-Channel	HIRS/3	AMSU			X	X		
EOS (Polar Orbiting Platform)	*6/96	A joint experimental and operational platform to be serviced in orbit									

[1] SEM will be included on future spacecraft as available.
[2] SBUV will fly only on PM satellites.
[3] NOAA-B failed to achieve useful orbit.
[4] D—Dummy unit
* Proposed launch date

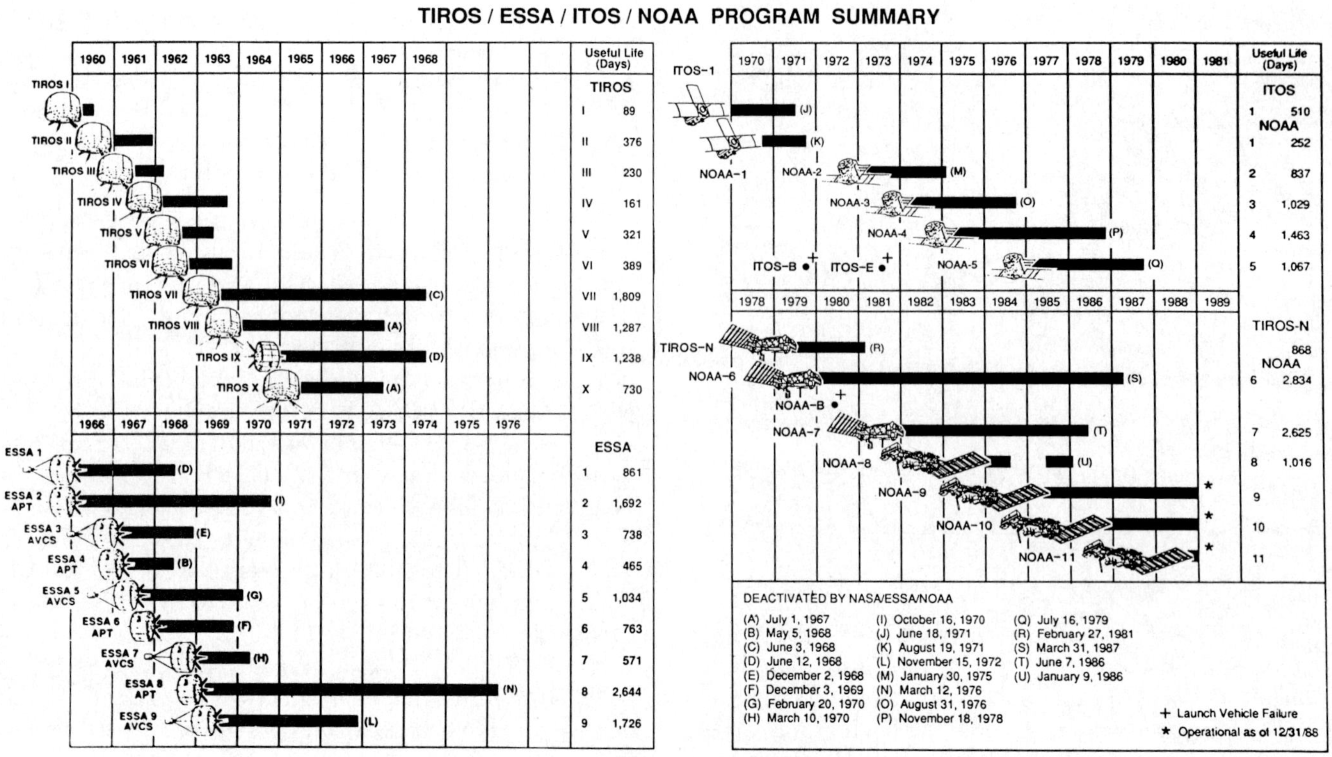

Figure I-2-4. The orbital history of the TIROS satellites.

I-2.5 THE NIMBUS TECHNOLOGY SATELLITES

The Nimbus Technology satellite program was initiated by NASA in the early 1960s to develop an observational system capable of meeting the research and development needs of the nation's atmospheric and Earth scientists. Initially intended as a replacement for TIROS, it eventually became the basic spacecraft for other space oriented experiments.

The general objectives of the program were to: (1) develop advanced passive radiometric and spectrometric sensors for surveillance of the atmosphere and oceans; (2) develop and evaluate new active and passive sensors for sounding the atmosphere and for mapping surface characteristics; (3) develop advanced space technology and ground data processing techniques for meteorological and scientific research; and (4) participate in global observation programs (World Weather Watch). The Nimbus system was used to test some of the advanced instruments used on operational TIROS satellites (see III-6.2.2). The Nimbus spacecraft was developed by NASA and built and integrated by General Electric. The instrument payload came from many universities and industrial groups such as RCA, Hughes, ITT, Ball, Texas Instruments, and Barnes.

A total of eight spacecraft were built: seven were orbited and one launch failed. All were launched between 1964 and 1978. Figure I-2-5 shows the launch dates and Table I-2-1 lists the payloads of the flights. The spacecraft configuration is shown in Figure I-2-6. Nimbus satellites were placed in Sun synchronous, near polar orbits and acquired global data twice every 24 hours. The spacecraft, 3 m tall, 1.5 m in diameter at its base, and 4 m across its solar cell panels, weighed about 950 kg.

I-2.6 THE APPLICATIONS TECHNOLOGY SATELLITE

I-2.6.1 ATS-1

The program that became the NOAA Geostationary Operational Environmental Satellite (GOES) program was a direct outgrowth of the NASA R&D Applications Technology Satellite (ATS) program. This program was initiated in 1966 to demonstrate communications technology by use of a satellite in a geostationary orbit. The excess capacity of the spacecraft suggested the inclusion of meteorological sensors for experimental observations of the Earth from orbital altitude (41,000 km). The Spin Scan Cloudcover Camera (SSCC), developed by Dr. V. Suomi of the University of Wisconsin and built by Santa Barbara Research Center, provided continuous images of the sunlit Earth disk every half hour. The spinning motion of the spacecraft, and a telescope and photomultiplier tube assembly, generated

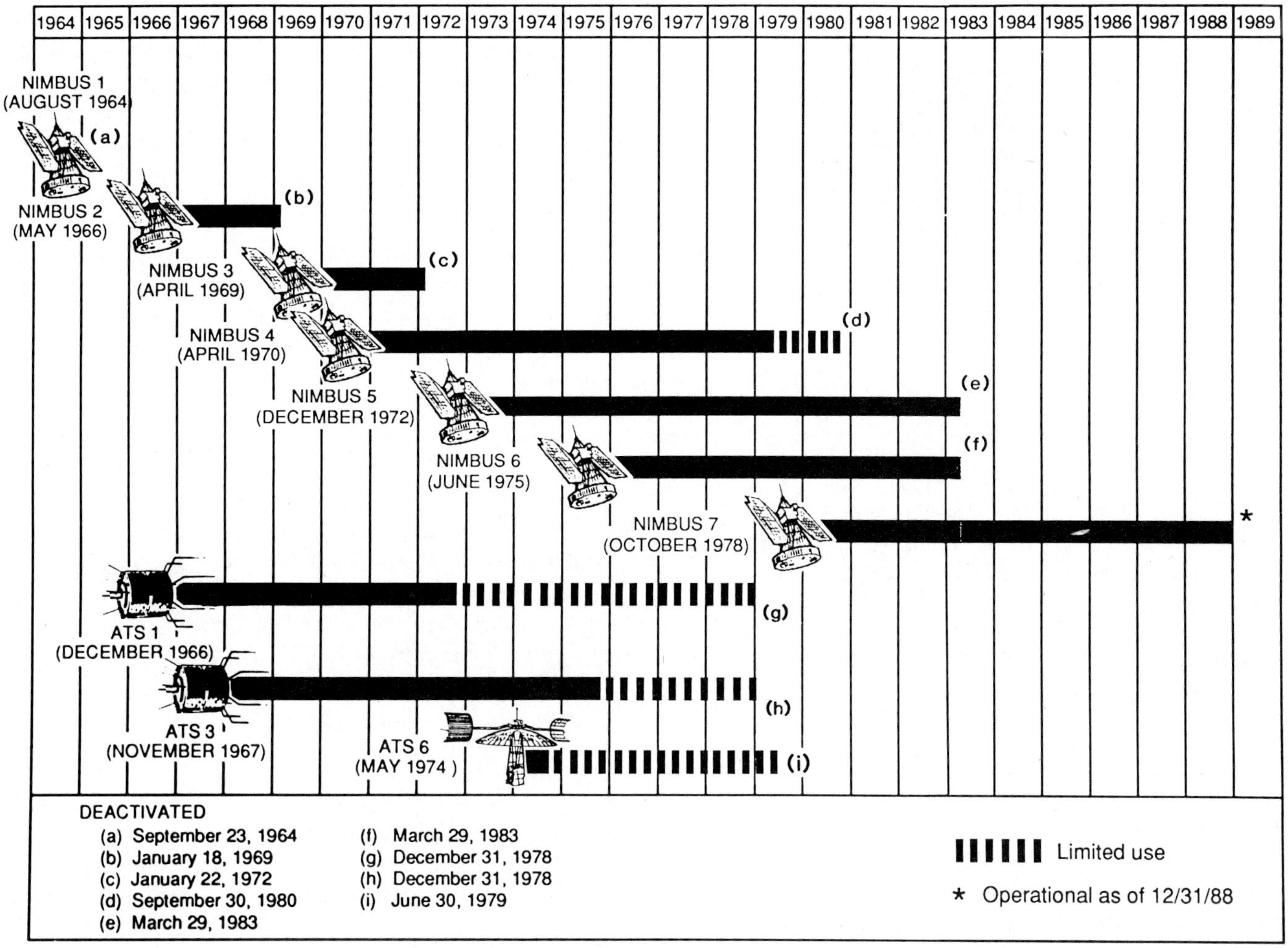

Figure I-2-5. The orbital history of the Nimbus spacecraft.

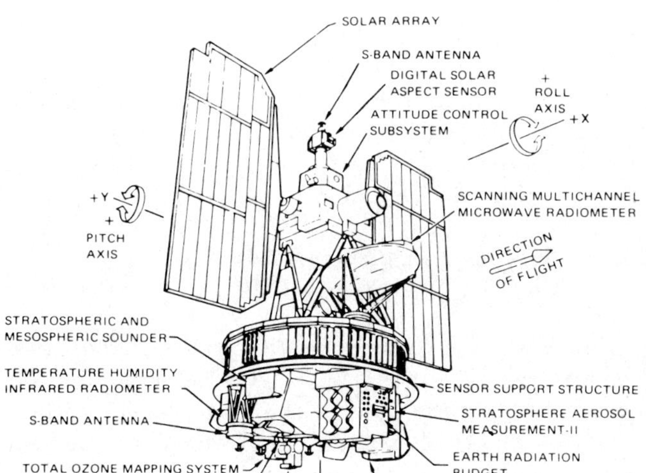

Figure I-2-6. The configuration of the Nimbus-7 spacecraft (courtesy of NASA).

line scans of the Earth and cloud cover. By incrementally moving the telescope with every spin of the satellite, an entire scene of the Earth disk was generated in 20 min and 10 min were required to reset the telescope for the next image. Spatial resolution at the subpoint was 3.2 km. ATS-1 included Weather Facsimile (WEFAX) data relay retransmission of weather data from a central NOAA ground station to ground stations within range of the satellite. The nearly continuous ATS-1 imagery gave convincing proof of the ability of geostationary spacecraft to monitor in real time the evolution of weather systems, particularly severe weather. The in-orbit weight of ATS-1 was 304 kg. The spacecraft was 1.5 m in diameter and 1.3 m long. Figure I-2-7 shows its external appearance.

I-2.6.2 ATS-3

ATS-3, launched November 1967, was slightly larger than ATS-1. It was put into stationary orbit over the

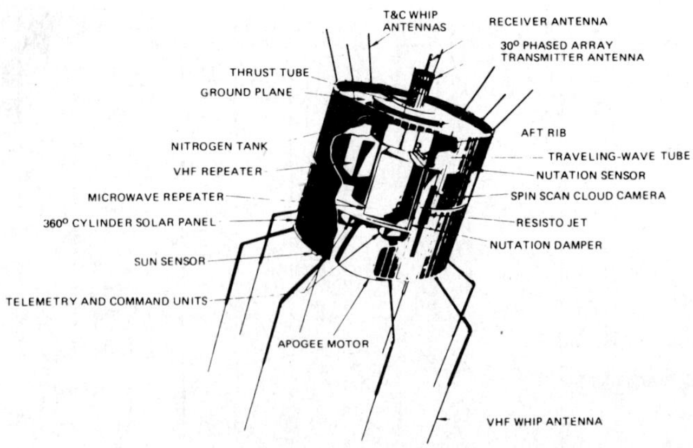

Figure I-2-7. An external view of the ATS-1 satellite (courtesy of NASA).

equator at 70°W to view the Western Hemisphere. The spin scan camera was modified to provide multicolor images.

I-2.6.3 ATS-6

ATS-6, launched May 1974, was a three-axis stabilized spacecraft intended primarily for communication experiments from geostationary orbit. It carried a meteorological sensor, the Geosynchronous Very High Resolution Radiometer (GVHRR), a two channel radiometer scanning in the VIS range (0.55 to 0.75 μm) with a 5.5-km spatial resolution, and in the IR range (10.5 to 12.5 μm) with an 11-km resolution. A full disk scene was scanned in 25 min.

I-2.7 THE OPERATIONAL GEOSTATIONARY SATELLITE SYSTEM

I-2.7.1 The Synchronous Meteorological Satellites/Geostationary Operational Environmental Satellites (SMS GOES) (1974-1978)

The success of the ATS-1 and 3 geostationary satellites led to development by NASA of an operational spacecraft designed specifically for meteorology. Five spacecraft were built for this series, two SMS and three GOES. NASA's prototype satellites, SMS-1 and 2, were launched in May 1974 and February 1975. Figure I-2-8 is a full disk SMS image.

These SMS GOES spacecraft provided continuous coverage from locations over the Equator. The principal instrument was the Visible Infrared Spin Scan Radiometer (VISSR). Also included was a Data Collection System (DCS) and a Space Environment Monitor (SEM). See Chapter III-1 for details.

Three additional satellites of the SMS design, GOES-1, 2, and 3, were launched in October 1975, June 1977, and June 1978, respectively. They were procured by NASA for NOAA and operated by NOAA after they had been checked out in orbit. For support of the operational mission, one GOES was placed over the equator at 75°W and a second at 135°W. Figure I-2-9 shows the orbital history of SMS GOES satellites.

I-2.7.2 GOES-D to H (1980-1987)

Five additional spacecraft, GOES-D to H, were procured by NOAA to support operational requirements through the 1980s. The GOES-D to H series were modified to carry the new Visible Infrared Spin Scan Radiometer Atmospheric Sounder (VAS). See Chapter III-1 for details.

Another significant change was the use of a despun S-band and UHF antennas for more effective communications with GOES ground stations. GOES-D, E, and F were launched in September 1980, May 1981, and April 1983. Figure III-1-2 shows the configuration of GOES-D to H. Launch of GOES-G failed in May 1986. The final spacecraft in this series, GOES-H, was orbited in February 1987.

I-2.7.3 GOES-I to M (1989 ...)

A new generation of geostationary operational satellites (GOES-I to M) is being procured from Ford-Aerospace. It is expected that the first will be launched in 1991. The spacecraft will be a three-axis stabilized system with a complement of new instruments. Additional details are given in Chapter III-1.

I-2.8 DEFENSE METEOROLOGICAL SATELLITE PROGRAM (DMSP)

Since the mid-1960s, the U.S. Air Force has operated polar orbiting spacecraft in the DMSP. The observations have emphasized imaging rather than atmospheric profiling although onboard instruments have been capable of both functions. Details of DMSP are available from the Department of Defense.

I-2.9 OTHER EXPERIMENTAL SATELLITE PROGRAMS

NASA has flown experimental satellites other than Nimbus and ATS for atmospheric research support. Two modular spacecraft for the Heat Capacity Mapping Mission (HCMM) and the Stratospheric Aerosol and Gas Experiment (SAGE) were launched by Scout launch vehicles in April 1978 and February 1979.

I-2.9.1 Heat Capacity Mapping Mission (HCMM)

The HCMM satellite, also known as the Applications Explorer Mission (AEM-A), provided day and night

Figure I-2-8. SMS picture of the whole Earth disk.

imagery of the Earth surface during its 2+ years of orbital operation. A two channel imaging radiometer measured reflected solar radiation and thermal radiation emitted by the Earth surface. HCMM operated from an orbit at 568- to 641-km altitude with a 97.6° inclination. The 1400 LST Equatorial crossing-time (ascending) was chosen so the radiometer could view the ground near both the hottest and the coldest parts of the day. The mission objective was to evaluate the utility of apparent thermal inertia information, derived from diurnal thermal surveys, for application in Earth science disciplines such as meteorology, agronomy, oceanography, and geology.

I-2.9.2 Stratospheric Aerosol and Gas Experiment (SAGE)

The Stratospheric Aerosol and Gas Experiment (SAGE) satellite, the second and last of the Small Applications Explorer Mission, was similar to the HCMM spacecraft. SAGE was placed into a 550- to 640-km orbit with an orbital inclination of 55°. It measured the extinction coefficients of solar irradiation wavelengths (at 0.38, 0.45, 0.6, and 1.0 μm) through the Earth atmosphere at sunset and sunrise on each orbit. SAGE accurately measured the vertical profiles of stratospheric aerosols and ozone. These measurements provided the first global aerosol data set ever obtained.

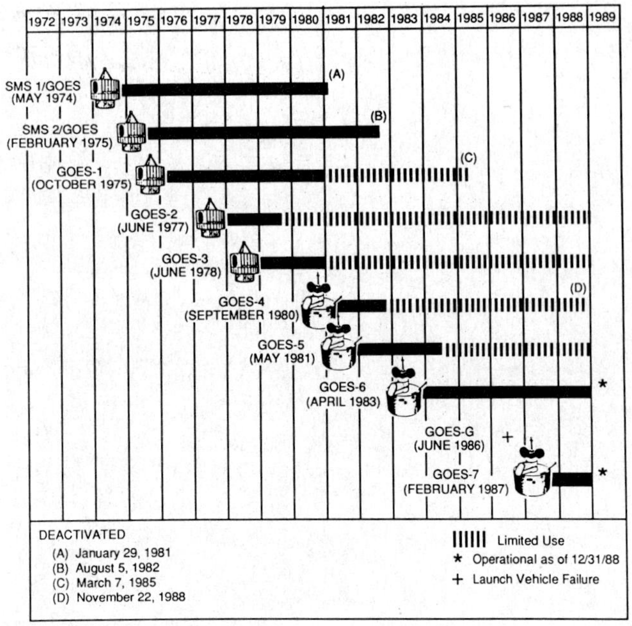

Figure I-2-9. The orbital history of the SMS GOES satellites.

I-2.9.3 Earth Radiation Budget Experiment (ERBE) Satellite Program

The ERBE Program began in 1984 with the successful launching of the ERBE spacecraft from the space shuttle in August 1984 and the launching of NOAA-9 in December 1984. Each carried an identical two instrument ERBE package. NOAA-G, which failed to reach orbit after launch in September 1986, also carried an ERBE. The ERBE spacecraft also included the SAGE/2 instrument for monitoring the vertical distribution of stratospheric aerosols, ozone, and nitrogen dioxide. The ERBE satellite was placed in 600-km orbit with 57° inclination to provide coverage at various times of the day as the orbit precessed.

The principal objectives of the ERBE Program were to: (1) determine, for a minimum of one year, the monthly average radiation budget at regional, zonal, and global scales and to determine the equator-to-pole energy transport; (2) determine the average diurnal variation in the radiation budget on regional and monthly scales; (3) initiate programs for the use of Earth radiation budget data to improve knowledge of regional and global climate processes; (4) provide stratospheric measurements of aerosols and their optical characteristics for use in computer models to assess climate change and to determine its predictability; and (5) map vertical profiles of stratospheric aerosols, NO_2, and O_3 globally from 70°S to 70°N and to establish a global data set. The ERBE Program is managed by NASA/GSFC and NASA Langley. Figure III-6-1 shows the ERBE satellite. Further details of the ERBE program are given in III-6.3.

I-2.9.4 Upper Atmosphere Research Satellite (UARS)

The UARS program has been under study by NASA, and launch of a spacecraft is planned for 1991. The UARS mission is to provide global data for the study of the physical processes acting within and on the stratosphere, the mesosphere, and the lower thermosphere. The specific areas to be studied are: energy input and loss, photochemistry, dynamics, and the connections between these processes and between atmospheric regions. The UARS is a single spacecraft carrying ten scientific instruments. The satellite orbit is planned for a 600-km altitude and an inclination of 57°.

I-2.10 INTERNATIONAL WEATHER SATELLITE PROGRAMS

Other countries have recognized the advantages of monitoring the Earth environment from space. The United States, the U.S.S.R., the European Space Agency, Japan, and India have developed and launched spacecraft for this purpose. Other nations (China, Brazil, and Argentina) are also developing experimental satellite programs. Many other nations can receive meteorological data directly from most of the satellites. Figure I-2-10 provides an overview of the U.S. and international satellite launches.

All nations have coordinated their efforts in satellite programs. In 1960 the World Meteorological Organization (WMO) organized the World Weather Watch (WWW). Under the WWW, world and regional centers were established for collecting and integrating satellite and conventional weather data, and disseminating the analyses and forecasts to WMO members. The WMO and the International Council of Scientific Unions (ICSU) developed the Global Atmospheric Research Program (GARP), a joint program for atmospheric research. GARP, a series of large-scale environmental data collection projects, has required extensive international cooperation. Significant quantities of meteorological satellite data were needed, especially over areas where conventional data are lacking. The primary GARP project was the Global Weather Experiment (GWE) in 1978-1979. Satellite data sets and products were coordinated to assure maximum compatibility among the operators of the Geostationary Meteorological Satellite Systems (the European Space Agency's Meteosat, Japan's Himawari, and the U.S. GOES). The U.S.S.R. participated in the GWE program using its polar orbiting Cosmos/Meteor system, in conjunction with the U.S. Nimbus and TIROS-N/NOAA satellites.

The Meteosat system is Europe's contribution to the WWW and GARP. The Meteosat system consists of

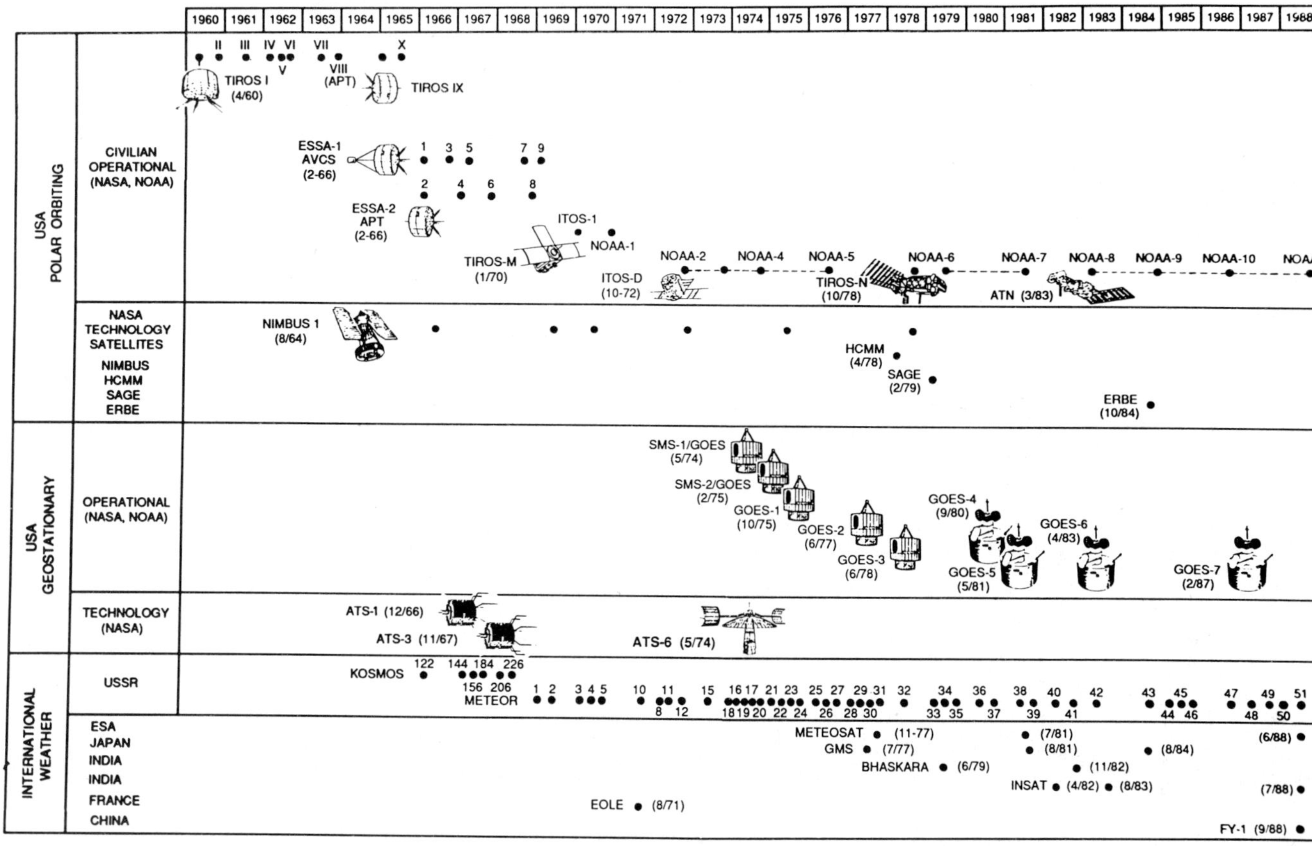

Figure I-2-10. Overview of meteorological satellites launched since 1960.

a spacecraft located in geostationary orbit at 0° longitude with the satellite control center at Darmstadt, West Germany. The Meteosat payload consists of a multispectral radiometer, a data collection system, and a WEFAX service. For details see Chapter III-3.

To participate in the GARP and WWW programs, Japan developed its own Geostationary Meteorological Satellite (GMS) program. In July 1977, Himawari-1, the first of these satellites, was launched into geostationary orbit by a Delta launch vehicle and located 36,000 km above the Equator at 140°E. The GMS program is managed by the National Space Development Agency of Japan (NASDA). For details see Chapter III-4.

The Soviet Union Meteorological Satellite Program has been extensive and to some degree parallels the U.S. polar orbiting programs. For details see Chapter III-2.

India's experimental satellite, Bhaskara-l, was launched by the Soviet Union June 1979 into a 550-km orbit at an inclination of 51°. It used passive MicroWave (MW) radiometry to measure atmospheric water vapor and liquid water content. The Indian National Satellite (Insat), a multipurpose geostationary spacecraft, carries a VHRR for meteorological observations and communications equipment. Insat-1B was launched in August 1983 and positioned at 74°E. For details see Chapter III-5.

France's space agency, Centre National d'Études Spatiales (CNES), developed the EOLE satellite that was launched in August 1971. EOLE collected data from free floating balloons in a cooperative U.S.–French experimental program.

The People's Republic of China is developing an experimental three-axis stabilized polar orbiting meteorological satellite. This spacecraft resembles the U.S. ITOS satellite and will be equipped with a two channel scanning radiometer. China is also planning an experimental geostationary meteorological satellite.

Contributors to Section I

Schnapf, Abraham	Aerospace Systems Engineering P.O. Box 160 Willingboro, NJ 08046
*Hallgren, Richard E. Smith, Brian G. Zbar, Fred S.	NOAA/NWS Silver Spring, MD 20910

* (Moved to a new position)

Section II

ELEMENTS OF REMOTE SENSING FROM SPACE

Remote sensing of the Earth from space is a relatively new science. Earth orbiting satellites gave geoscientists their first opportunity to see the broadscale patterns of land forms, vegetation, polar ice caps, and the atmosphere and its cloud patterns. Previously, these features had been studied and measured only from the surface of the Earth or from aircraft and rockets at relatively low altitudes, with very limited fields of view. The broad coverage available from satellites also brought problems. One was deciding what to look at or measure and how to do either or both. Instrumentation for remote sensing had to be designed to measure desired parameters, to be of sufficiently light weight to fly on a satellite, and to deliver data to the ground in a useful form.

It is interesting that the most effective use of the satellite for Earth studies has been in meteorology. To observe surface and atmospheric features from space, considerable effort and scientific ingenuity are required.

Chapter II-1 introduces the remote sensing system and the basic physics of such observations. Chapter II-2 presents the implications of remote sensing due to the characteristics of the earth-atmosphere system.

Chapter II-1

CHARACTERISTICS OF REMOTE SENSING

II-1.1 INTRODUCTION

All satellite remote sensing systems involve the measurement of ElectroMagnetic Radiation (EMR) which has been emitted, reflected, or scattered by the atmosphere or the surface. These EMR measurements allow the determination of actual physical values of the atmosphere and the surface. For example, temperature of the ocean surface is obtained by observing the infrared radiation emitted by it and then inferring the ocean temperature using Planck's law. Of course, the radiation reaching the sensor passes through the atmosphere which scatters and absorbs some of the surface radiation; these effects must be accounted for if the results are to be accurate. Remote sensing measurements provide a global coverage that is not possible with direct measurements, but they have difficulty achieving the accuracy of the more direct approach. The elements of remote sensing include:

- Sources of EMR, such as the Sun, the Earth and its atmosphere, or the satellite sensor itself.
- Interactions of EMR with the target being observed.
- Propagation of EMR from the target to the sensor.
- Interactions of EMR with the sensor.
- Processing of sensor data to infer a physical quantity.

II-1.2 ELECTROMAGNETIC RADIATION PROPERTIES AND INTERACTIONS

Although EMR has the properties of both waves and discrete particles, it is usually quantified according to its intensity (I) and wavelength (λ) in the IR, and intensity and frequency (ν) in the microwave region. The most frequently used units of wavelength are the meter (m) and the micrometer (μm, 10^{-6} m). Frequency is expressed in Hz. The characteristics of the EM spectrum are shown in Figure II-1-1. Although there are nominal regions of UltraViolet (UV), InfraRed (IR), MicroWave (MW), and Visible (VIS), no definite wavelength value divides one from the other. Remote sensing systems operate in various portions of the spectrum (VIS, IR, and MW).

The interaction of radiation with matter depends on many factors. Important factors are the state of the matter (solid, liquid, or gas) and the wavelength. These factors determine the form of the interaction, i.e., emission, scattering, absorption, or reflection. The types of interactions vary with the spectral region; the most significant are illustrated schematically in Figure II-1-2. For example, scattering is dominant over absorption for MW radiation less than 1 cm in length for precipitating clouds while absorption is more important for the IR.

II-1.2.1 Radiation Interactions with Solids and Liquids

Targets such as liquids and solids (land and sea at the surface; water droplets and ice crystals in the atmosphere) reflect, absorb, emit, and transmit radiant energy over a wide range of wavelengths. The emitted energy is described by a set of idealized blackbody radiation laws. According to the Stefan-Boltzmann law,

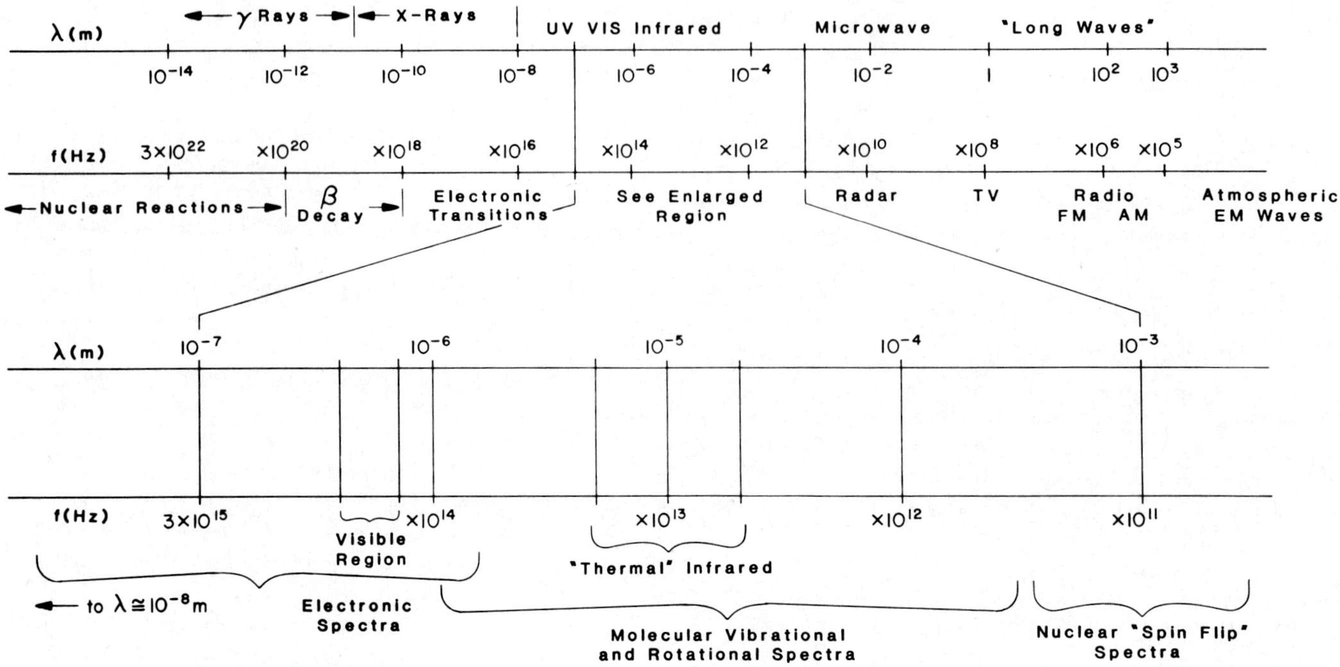

Figure II-1-1. A portion of the electromagnetic (EM) spectrum. The enlarged region includes most of the emitted solar radiation and that reemitted by the Earth surface. The human eye is sensitive to radiation around the peak of solar emission.

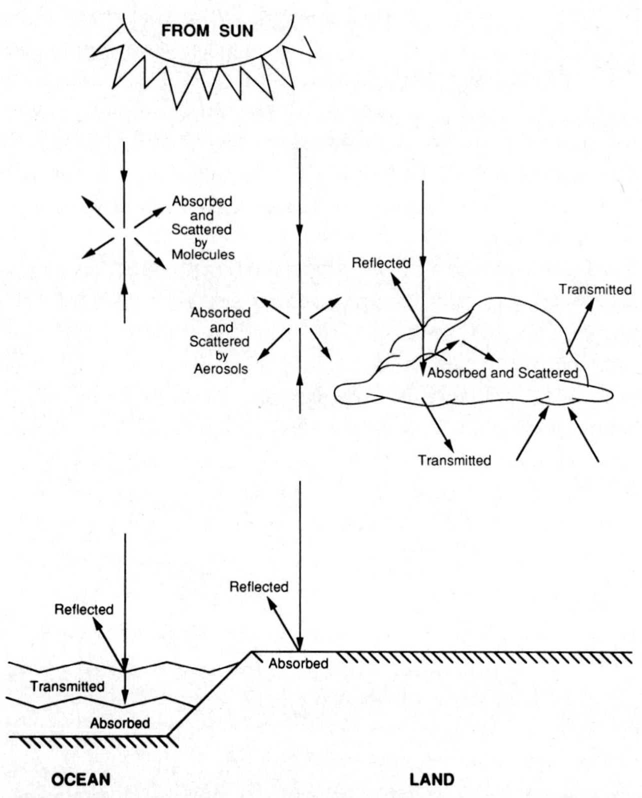

Figure II-1-2a. Possible interactions between VIS radiation and the surface, atmosphere, and clouds. Additional interactions are possible; the most significant are depicted. Radiation measured by the satellite is the net result of these many interactions contributing to the radiation for each field of view.

the amount of energy per unit area, E, emitted by a blackbody is given by

$$E = \sigma T^4, \qquad \text{(II-1-1)}$$

where σ is the Stefan-Boltzmann constant (5.67×10^{-8} W m^{-2} K^{-4}) and T is the temperature (K) of the radiator. The spectral distribution of blackbody radiation depends on the temperature of the blackbody. The wavelength at which the radiation is at a maximum (λmax in μm) is given by the Wien displacement law:

$$\lambda\text{max} = \frac{C_1}{T}, \qquad \text{(II-1-2)}$$

where C_1 is (2898 μm K).

Thus the Sun, whose radiative temperature is approximately 5900 K, has a peak at 0.5 μm, which is in the green region, the center of the VIS light spectrum (Figure II-1-3). The Earth, whose radiative temperature is approximately 300 K, has a peak near 10 to 12 μm in the IR region.

The Earth surface does not function as a perfect blackbody radiator since its efficiency as a radiator over the spectral range affects the total energy output. The radiance or intensity of the emitted radiation (I) at a given wavelength (λ) depends upon the surface temperature T, and the spectral emissivity of the surface ϵ_λ:

$$I_{(\lambda,T)} = \epsilon_\lambda B_{(\lambda,T)}. \qquad \text{(II-1-3)}$$

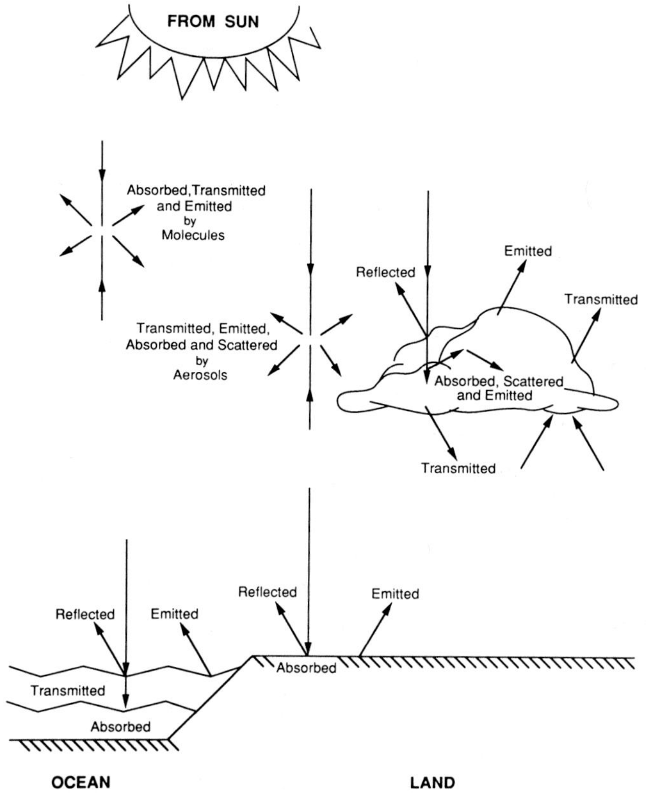

Figure II-1-2b. Same as Figure II-1-2a except interactions involving IR.

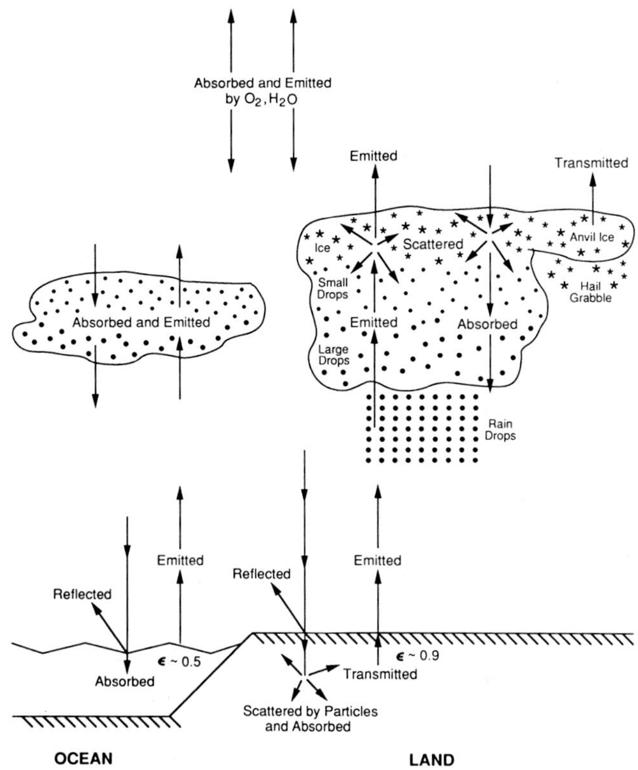

Figure II-1-2c. Same as Figure II-1-2b except interactions involving MW.

$B_{(\lambda,T)}$, the Planck function that describes blackbody radiation spectrally, is given by

$$B_{(\lambda,T)} = \frac{c_1}{\lambda^5[\exp(c_2/\lambda T) - 1]}, \qquad \text{(II-1-4)}$$

where c_1 and c_2 are constants.

Table II-1-1 contains emissivity values for various substances. Emissive properties related to specific wavelengths can change as surfaces change temperature. In the 8 to 14 μm range, where much sensing is done, emissivities can generally be considered constant (i.e., most materials can be treated as graybodies), but exceptions occur. For example, emission and absorption of substances, such as snow and vegetation, are strongly dependent on wavelength, and the substance is not a true graybody (Figure II-1-4). Snow is a relatively weak absorber in the VIS and near-IR regions, but is a strong absorber in the rest of the IR region.

At prevailing Earth temperatures, most radiation is emitted in the IR region. For this reason, it is sometimes called thermal, or longwave, radiation to distinguish it from the reflected shortwave radiation of the Sun. In addition to temperature, the measured thermal radiation depends on the reflection at the air-surface

Table II-1-1 *Typical emissivity values of various common materials (Hudson, 1969; Lillesand and Kiefer, 1979: ©John Wiley and Sons, Inc.)*

MATERIAL	TEMPERATURE (°C) OF SAMPLE ANALYZED	EMISSIVITY*
Human skin	32	0.98
Distilled water	20	0.96
Ice	−10	0.96
Carbon (candle soot)	20	0.95
Wet soil	20	0.95
Glass	20	0.94
Paint (average of 16 colors)	100	0.94
Brick	20	0.93
Dry soil	20	0.92
Concrete	20	0.92
Plaster	20	0.91
Sand	20	0.90
Wood	20	0.90
Snow	−10	0.85
Anodized aluminum	100	0.55
Buffed stainless steel	20	0.16
Highly polished gold	100	0.02

* Values measured normal to surface of object over all wavelengths. Emissivities measured at other angles and over discrete wavelength regions may vary somewhat.

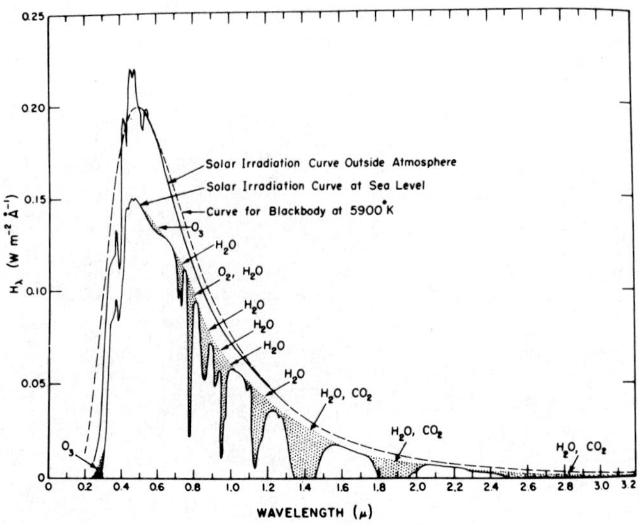

Figure II-1-3. Spectral curves of solar radiation incident at the top of the atmosphere and at the surface of the Earth for the Sun at zenith on a clear, cloudfree day (Air Force Cambridge Research Laboratories, 1965).

interface and the absorption and scattering properties within the material. For many practical purposes, ϵ_λ may be taken as unity, since the emissivity of the Earth surface in the IR is close to 1.0. Often the IR radiance measured by the satellite is entered into Planck's Equation (II-1-4) and the equation is solved for T to determine the temperature that a feature would have if it were in fact a blackbody radiator. This temperature is called the equivalent blackbody temperature.

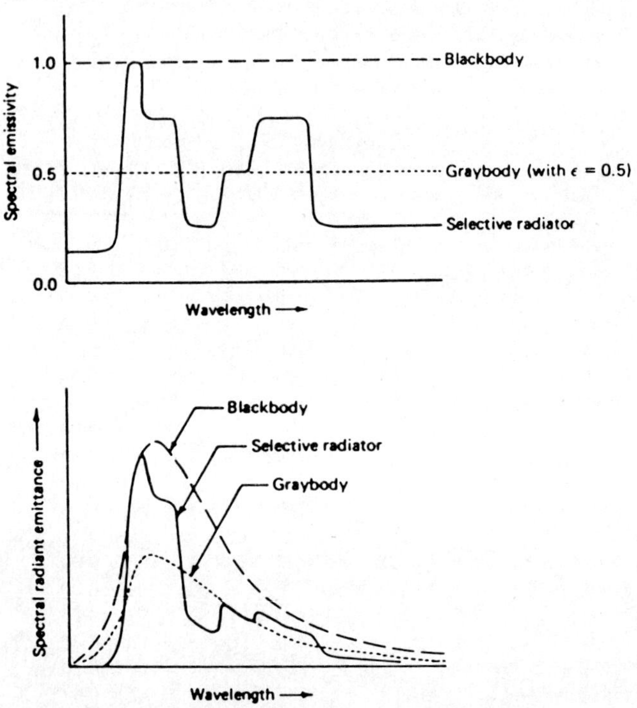

Figure II-1-4. Spectral emissivities and radiant emittances for a blackbody, a graybody, and a selective radiator (Hudson, 1969; Lillesand and Kiefer, 1979: © John Wiley and Sons, Inc.).

Microwave frequencies are used mainly to measure surface properties and the atmosphere (Ulaby, 1986). MW interactions with solids and liquids can be more complicated than VIS or IR interactions because changes in MW radiation result from the interactions with both the surface and subsurface. The interactions depend upon surface roughness, liquid water content, and scattering properties of the particles comprising the media. MW emissivities of a surface vary predominantly with liquid water content. Materials with a low water content, such as dry soil, snow and ice, have little MW absorption and the MW can penetrate deeply. Thus, MW emissivities are low and the resulting remotely observed MW intensities are low. The addition of a small amount of liquid water can result in emissivities approaching unity, as is the case for melting snow and moist soil. One would expect that a liquid surface such as the ocean would have an emissivity close to unity; however the water surface acts as an interface and reflects the radiation, thereby reducing the emissivity to approximately 0.5. For homogeneous surfaces, such as the ocean and soil, the emissivity is a function of the surface dielectric properties for a given frequency. This dielectric effect results in increasing emissivity and brightness temperature as MW sensor frequency increases.

Uniform materials which contain particles at about the size of MW wavelengths (e.g., ice crystals in snow or brine pockets in sea ice) can backscatter MW radiation within the medium. Radiation emitted from below the surface is backscattered downward, thereby reducing the effective emissivity at the surface. Scattering effects are strongly frequency dependent when particle size is about that of the MW wavelength. An increase in frequency results in an even lower brightness temperature. If liquid water is added to the medium containing scatterers, however, absorption by liquid water will predominate.

Effects caused by surface roughness are significant. The small-scale surface roughness depends upon the size and separation of the particles relative to the MW wavelength. When the MW wavelength is longer than the particle separation, the surface acts as if it were smooth. In the limiting case of an ocean surface, the water molecules are so closely bound that they act as a continuum, hence individual particle effects are not detected. When the wavelength approaches the particle separation, the surface behaves as a diffuse surface, such as occurs with pebbles on fine sand, and cloud droplets.

At viewing angles other than nadir, MW radiation is often polarized. As surface roughness increases, the polarization of the emitted or reflected radiation is rotated. The greater the roughness, the more spatially non-uniform is the polarization, i.e., depolarization occurs. This generally results in an increase of the radiation measured by the sensor. Conversely, a smooth or specular surface will reflect and emit polarized radiation at angles other than nadir, thereby reducing the amount of energy measured. The decrease in polarization due to roughness is slightly dependent upon frequency since the scale of the roughness will be related to the wavelength used. MW radiometers can be set to measure vertical or horizontal polarized radiation, or an average of the two. Measurements of both the vertical and horizontal polarized radiation can be used to obtain information about surface roughness.[1]

II-1.2.2 Radiation Interactions with Gases and Aerosols

Radiation interacts with gases and aerosols differently than with solids. While molecules tend to interact with radiation at specific wavelengths or wavelength bands more often than as a continuum, the reverse is true for solids and liquids. Because the energy absorbed or emitted by gases initially causes a change between discrete energy states allowed by the molecule, absorption occurs at discrete wavelengths that are determined by the composition of the molecule. Wavelengths where absorption occurs are known as absorption lines. The spectral regions of interactions for one of the atmospheric gases can be seen in Figure II-1-5.

The temperature of a gas is determined by the average speed of the molecules. Interaction between the molecules and radiation results in a transfer of energy between the radiation field and molecules that changes both. When absorption occurs, an atom is excited by the absorption of one or more photons. The molecule then converts the energy to heat that is transferred to other molecules through collisions as opposed to the scattering process that emits the radiation at the same wavelength at which it was absorbed. Eventually some of the molecules that are heated release the energy by emitting photons. Over a period of time, the energies of incoming and outgoing photons must balance. Usually a single high energy photon is absorbed and several photons of lower energy are emitted at longer wavelengths.

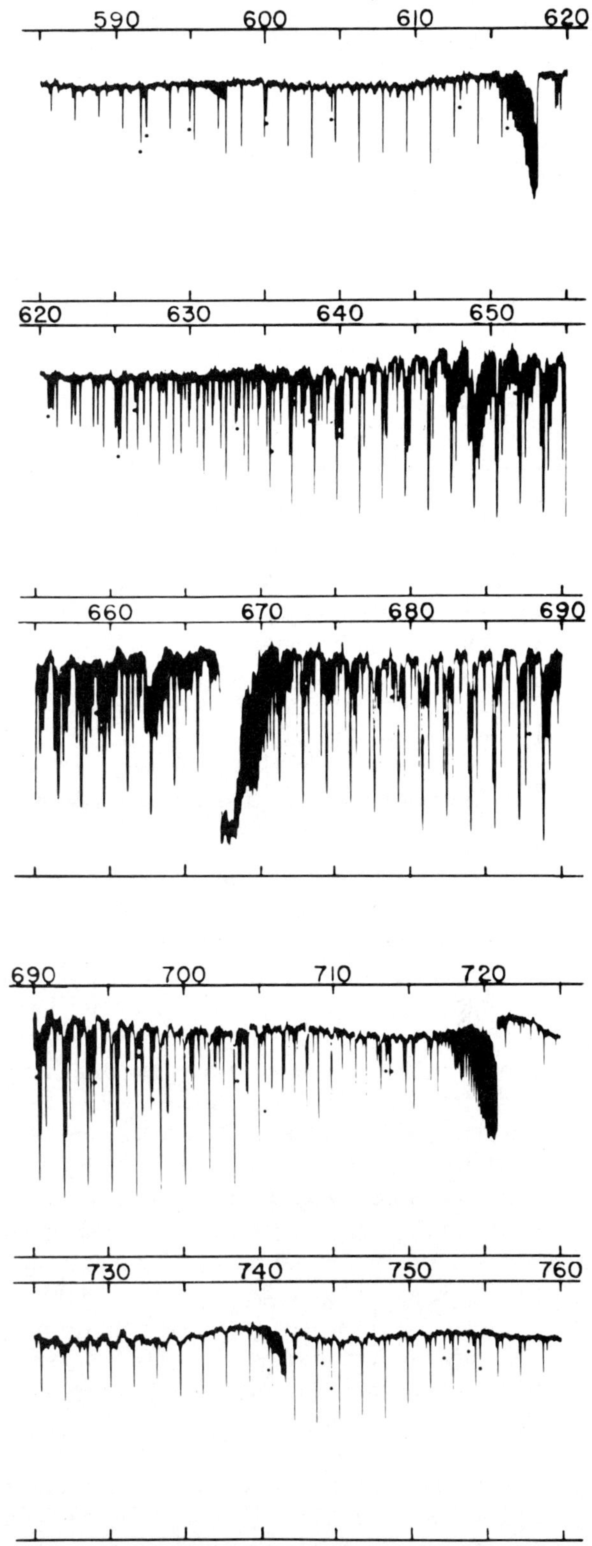

Figure II-1-5. Highly detailed spectral view of the 15 μm CO_2 "band" (Paso et al., 1980: © Academic Press). The abscissa shows the spectral range of the band in wavenumber (cm^{-1}) and the ordinate is the transmittance. Numerous weak isotopic bands (caused by isotopes of carbon and/or oxygen) and temperature sensitive "hot bands" complicate the fundamental band structure.

[1] Polarization is also significant in VIS measurements (e.g., the sea surface is highly polarized), but current sensors are not designed to measure polarization. The greatest amount of information is obtainable from spectral rather than polarization measurements.

Refraction occurs when EMR passes at an angle through a boundary between media with different refractive indices. Refraction effects are minimal when the atmosphere is stable. When turbulent conditions exist, waves are bent and this affects the angular distribution of the emerging radiation.

The interaction of EMR with aerosols is affected by the aerosol size and the wavelength of the radiation. The interactions can be described by polarization effects. In principle, the polarizability of an aerosol particle will determine exactly how it interacts with EMR. Unless the particle is spherical and homogeneous, computation of its EM properties is very difficult.

II-1.3 REMOTE SENSING SYSTEMS

The remote sensing systems used in environmental satellites have provided remarkably useful and accurate measurements. Yet the scientists and engineers who develop and construct remote sensing systems know that the capabilities have limits. There are numerous problems with the various components of the systems and many different techniques have been developed to compensate for these problems.

II-1.3.1 Limitations to Remote Sensing Systems

Remote sensing involves the detection of a variation in the radiation at the sensor due to the presence of the target. The best situations occur when the contrast between the radiation from the target and the background is greatest. Since many components in the earth-atmosphere system emit radiation at a relatively low level, a source of high energy output provides a signal at a level that exceeds the background. In the VIS, energy from the Sun exceeds the energy due to other sources and this situation is approached, even though scattering effects can be significant. IR energy emitted by components of the earth-atmosphere system is more difficult to isolate because all the components tend to emit the same type of energy. To measure atmospheric parameters accurately, it is necessary to know the radiation entering the atmosphere from the ground. To measure the surface parameters, it is necessary to know the changes made by the atmosphere.

More explicitly, both the earth and the atmosphere emit radiation over broad wavelength regions. Because of the broad overlap, it is difficult to find spectral regions where changes in the surface characteristics do not mask atmospheric signals and where the atmosphere does not mask surface signals. As an example, the radiation absorbed by the atmosphere is exactly replaced by radiation emitted by the atmosphere when the surface and atmosphere are at the same temperature. Under these conditions, little information can be gained about the atmosphere because no change occurs in the radiation passing through it. The problem is compounded when the sources of radiation and the targets change characteristics and spectral response with location and time. The quantity that must be known thus has an additional source of uncertainty. The interference between the various targets is explained more completely in Chapter II-2 and throughout the applications sections (VII through X).

Remote sensing measurements are also limited by the abilities of the sensors. Sensors are noisy at low energy levels, limited in their spectral characteristics, and subject to change with age and environmental conditions. Because of these limitations, instruments are limited in their spectral and spatial resolutions. Thus, any instrument is frequently a compromise between spatial and spectral resolutions.

Finally, remote sensing systems are limited by communication and processing capabilities. The volume of data produced and the need for ongoing comparisons with *in situ* data tax the capabilities that are available.

II-1.3.2 Improvements to Remote Sensing Systems

Because of the number of error sources in remote sensing systems, many techniques have been developed to improve results. Some sensors have been designed to take multiple views of the same target to reduce noise. Others obtain simultaneous views of the same target in a variety of wavelengths and use the multispectral information to achieve a greater ability to discriminate features. Other remote sensing systems use a multi-stage approach where features are concurrently measured by sensors flown by aircraft or by other satellites.

To improve vertical and spectral resolution, active sensors such as radar have been developed to send out signals that react with the target and then are reflected back to the source where they are measured. The greater vertical resolution is achieved by timing the signal rather than relying on the atmospheric characteristics as do passive systems. Greater spectral resolution is achieved by limiting the source to a known, very narrow spectral region. This allows the measurement of changes in wavelength, thus providing information about the character of the target. Finally, the energy can be increased to measure features not detectable with the lower levels available from natural radiation sources. Overall, active systems can provide more information than passive systems because the probing signal can be controlled. However, a disadvantage of active systems is their requirement for large amounts of power and massive components. Thus, their use from space has been restricted to large research craft such as Seasat, Skylab, and the Space Shuttle.

Operational spacecraft in use through the 1980s have used passive systems almost exclusively because they do not provide the space and power required for active remote sensor systems.

Lasers and masers are being developed for future active remote sensing systems on satellites. The laser (maser in the MW region) produces a continuous beam or pulse of EM radiation of nearly coherent, single frequency and of extremely high spectral power density. Because the wavelength of the emitted signal is known, Doppler shifts in the return signal can be used to measure atmospheric motions. Higher vertical resolution is also possible with these devices because, as with radar, the vertical resolution is determined by the time of the return pulse, rather than by the atmosphere.

II-1.3.3 Validation and Calibration

Approaches used to relate the sensor data to the actual quantity being measured require extensive data processing. The accuracies of quantitative data are dependent on the quality of the calibration and validation procedures. Calibration relates the output of an instrument to units of the parameter being measured (e.g., the conversion of voltages to units of radiance). Validation relates the values of physical parameters with those from a different instrument or from another technique of acceptable reliability.

For satellites, "ground truth" that serves as the alternative measurement is often more than just a check on the satellite instrument; ground truth data are generally used to develop a regression equation that relates a satellite measurement such as radiance to a physical parameter such as sea surface temperature. In such regression solutions, the satellite instrument is directly related to ground truth, and instrument calibration is less important. A limitation of regression is that if ground truth data are not available, regression coefficients cannot be adjusted to correct for instrumental changes through aging and the effects of the space environment. As will be seen in the following, monitoring these changes in space is also difficult.

It is preferable that means be provided to monitor the response and characteristics of the instruments. This could readily be accomplished, for example, by in-orbit servicing on manned space stations which are almost certain to become realities in the next two decades. However, operational spacecraft most likely will remain unmanned.

For non-regression solutions, continuous calibration of the spacecraft instrument is extremely important for relating output to physical units. Laboratory calibrations made prior to launch cannot be relied upon over long periods, especially when the instruments are subject to the environmental and chemical changes of space. Onboard calibration devices are subject to the same conditions that cause changes in the response of the instrument they calibrate. To compensate for these difficulties, a vicarious calibration is often provided by measuring a location on Earth with another instrument (e.g., White Sands, New Mexico, measured by an airborne radiometer) and using it as a known target for the spacecraft to measure. Such a target is used as an accessible calibration source that can be monitored for changes.

Chapter II-2

THE EARTH AS A TARGET FOR REMOTE SENSING

II-2.1 INTRODUCTION

The Earth is a complex target for remote sensing. The atmosphere alters ElectroMagnetic Radiation (EMR) passing through in a non-uniform way. Furthermore, atmospheric parameters vary over a wide range of time and space scales. Water vapor distributions, for example, may change within hours or days. Changes in ozone occur over periods of months and years, while changes in concentrations of atmospheric gases may span centuries. Accounting for these wide variations requires careful consideration in the design of remote sensing systems.

Many surface parameters display distinct signatures in several EMR wavelength intervals. Signatures of some surface features are not unique, however, so sophisticated interpretive techniques must be developed to deduce information about their variations.

II-2.2 THE EARTH SURFACE AS A TARGET

The intensity of radiation emitted from the surface is determined by its temperature and emissivity. If either is known, the other can be determined from the radiance measured just above the surface. If some additional information is known, such as the change in emissivity with wavelength, it is possible to use measurements at several wavelengths to determine both quantities but good accuracy is difficult to achieve with this approach. By observing in several wavelengths, extensive information can be obtained about surface features. Some wavelength ranges provide information from below the surface; for instance, VIS radiation penetrates water to significant depths and MW radiation penetrates snow and other surfaces. For all satellite measurements of the surface, however, corrections must be made for atmospheric scattering, absorption, and emission.

II-2.2.1 The Land Surface

A. Snow and Ice. Laboratory studies (O'Brien and Munis, 1975) show that the reflectance of snow (Figure II-2-1) varies from almost 100% in the VIS (0.6 to 0.7 μm) to near zero percent in the IR water absorption bands at 1.5 and 2.0 μm. At IR wavelengths greater than 0.8 μm, fresh snow reflects more strongly than aged snow, thus the age and condition of snow cover may be assessed. The 1.5 and 2.0 μm wavelengths, where snow is lowest in reflectivity and clouds are high in reflectivity, are ideal for separating snow from clouds (Hoffer et al., 1975).

In general, the underlying land surface reduces the reflectance of snow in the VIS portion of the spectrum, particularly for depths less than about 20 cm (Figure II-2-2). McGinnis et al. (1975) showed that snow depth was well related to satellite VIS brightness data. Satellite IR measurements are generally of little use for snow detection during winter when temperatures may

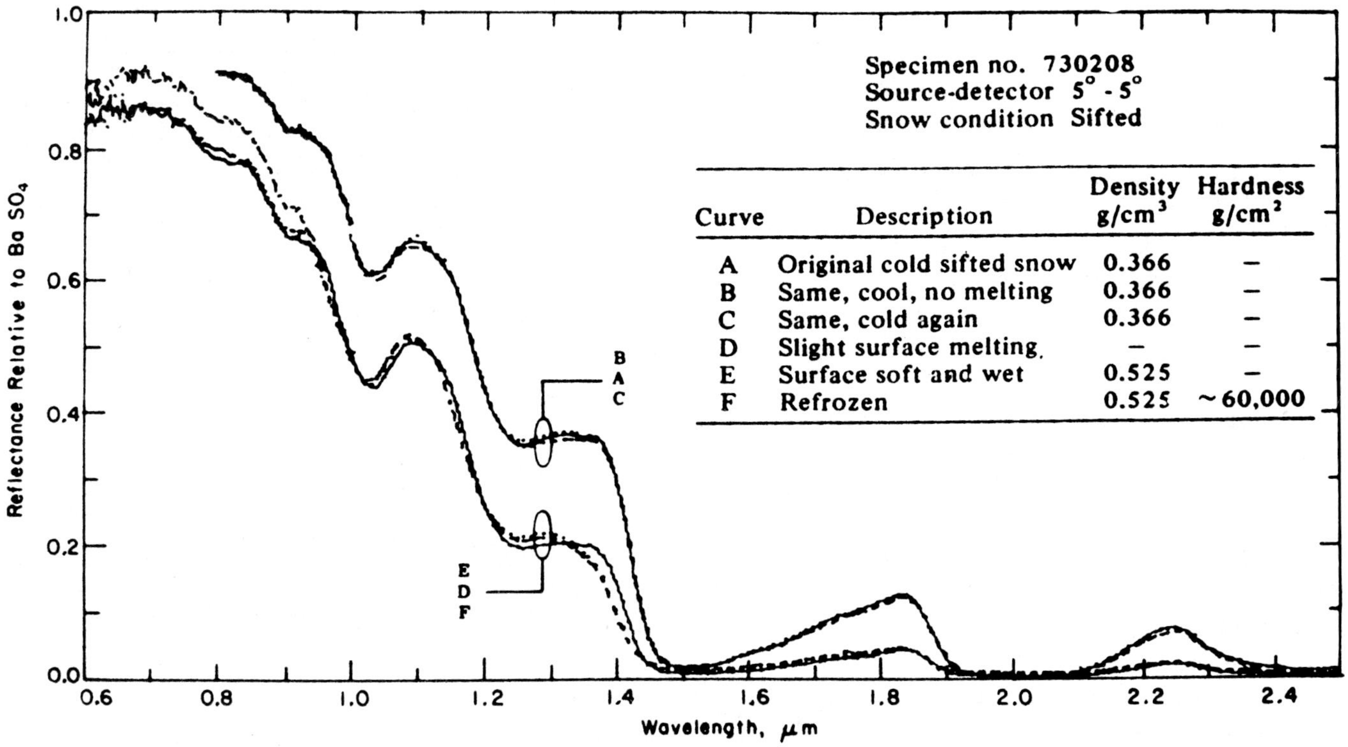

Figure II-2-1. Variations of sifted snow reflectance with wavelength for differing snow conditions (O'Brien and Munis, 1975).

be uniformly low over both snow and bare ground (Barnes and Bowley, 1972). When snow is melting, however, surface temperatures of the snow are substantially lower than adjacent snow-free surfaces, so IR measurements can readily detect the differences between snow and bare ground.

Snow has unique emissivity characteristics in the MW region. A snowpack contains a mixture of ice crystals or melting ice particles that depends on the snow temperature. The emissivity of snow depends on the absorption and scattering caused by the liquid water and the particle content. For dry snow the ice crystals scatter some of the upwelling radiation away from the sensor's field of view. The scattering effect increases with frequency so that the emissivity appears to decrease as the frequency increases, unlike most other natural surfaces. A classification technique has been developed which uses both the magnitude as well as the variation of emissivity with frequency to distinguish different surface types. These two quantities, the emissivity and its variation with frequency, can be inferred using dual frequency MW measurements. The brightness temperature "difference" is proportional to the change in emissivity with wavelength while the "average" of the two measurements is related to the emissivity itself.

Newly formed sea ice can be considered very smooth in its interaction with MW radiation. With age, bubbles, cracks, and other metamorphosing processes give an unsmooth texture and create scattering centers. Emissivities of heterogeneous substances (multiyear sea ice and dry snow) which contain scatterers have been computed by Poe et al. (1972), Hoffer and Matzler (1980), and Chang et al. (1976). Since the depth of MW penetration increases with decreasing frequency, multispectral measurements have been used for determining the subsurface characteristics of sea ice (Hollinger et al., 1984; Cavalieri et al., 1983). The analysis becomes more complex when stratification of different surfaces occurs, e.g., snow over ice, or ice over water (England, 1975). Identification of stratified ice and snow layers also may be possible from polarization measurements (Figures II-2-3 and II-2-4).

Snow and ice undergo a dramatic change (up to 50%) in MW emissivity with melting. Dry snow and ice absorb little and MW penetration can be many meters. They also contain many scatterers and the measured brightness temperature can be as low as 100 K. The brightness temperature decreases with increasing frequency. When melting begins, penetration decreases, the scattering terminates, and emissivity dramatically increases to more than 90%. Brightness temperature increases as frequency increases because of the water based effects.

B. Geological Features. Surface geology can be studied with VIS, near-IR, middle-IR, and MW data. For de-

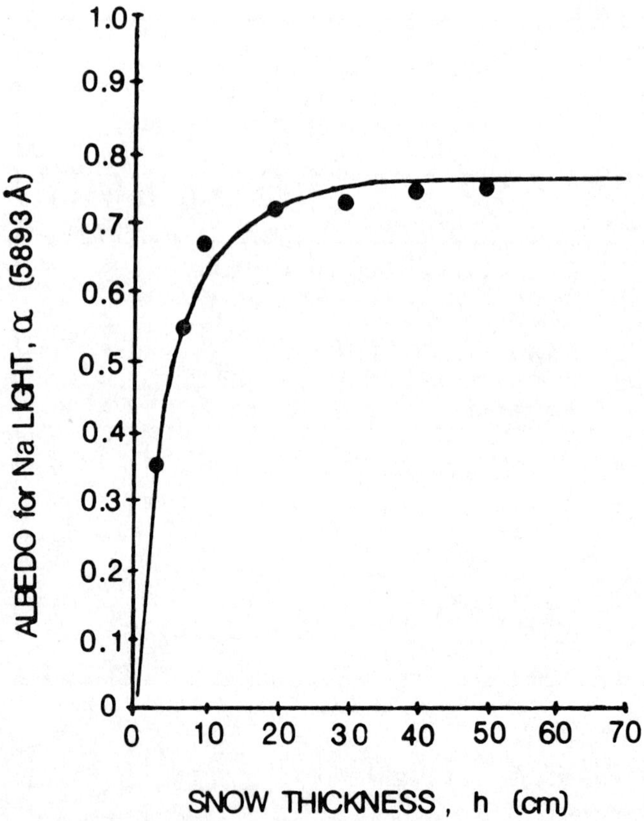

Figure II-2-2. Relation of albedo (sodium vapor light) of a snow surface to depth of snow, when snow is underlain by a black surface (Giddings and LaChapelle, 1961).

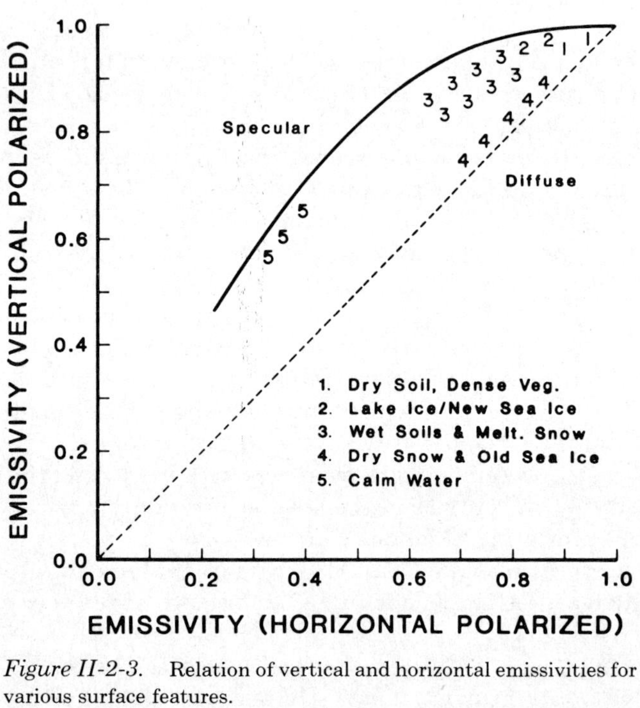

Figure II-2-3. Relation of vertical and horizontal emissivities for various surface features.

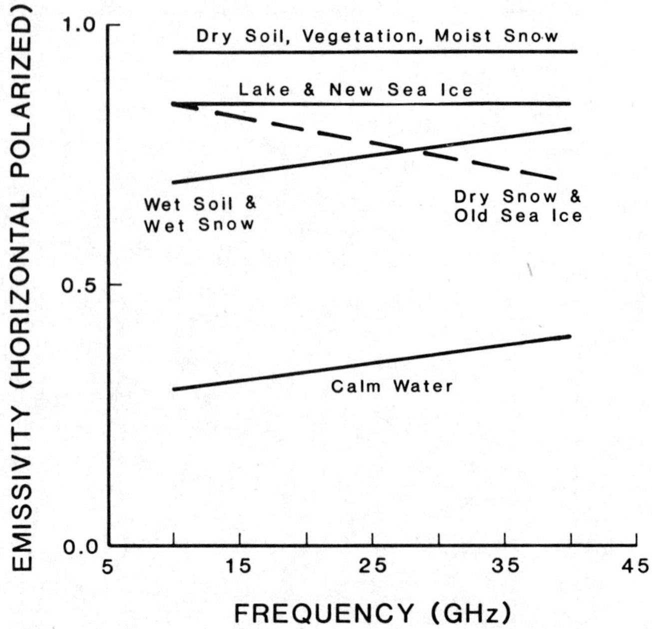

Figure II-2-4. Emissivity as a function of frequency for different surface types.

tails see the manual published by the American Society of Photogrammetry (Colwell et al., 1983). The reflectance spectra of minerals in the reflected IR show unique absorption patterns that can be detected remotely and used to identify them (Hovis, 1966). Middle-IR data are useful for identifying rocks and minerals because of their varied thermal responses in the 7 to 14 μm part of the spectrum. Since geologic structures are highly reflective in the MW, faults, joints, and folds in the surface; sand, gravel, and glacial deposits; and volcanic flows can all be detected by MW measurements. A major advantage of MW is that it penetrates overlying vegetation to reveal the underlying geologic features.

C. Vegetation and Soils. Vegetation has marked differences in the VIS and in the near-IR. The characteristic spectral reflectance of a green maple leaf is shown in Figure II-2-5. Absorption by leaf pigments predominates in the VIS region from 0.38 to 0.70 μm, creating minima in reflectance at 0.47 and 0.68 μm. The familiar green reflectance peak occurs at 0.55 μm. Between 0.70 and 1.3 μm (near-IR), there is high reflectance and minimal absorption. These reflectance properties can be used to determine the physical state of vegetated areas. A marked increase in reflectance occurs when leaf moisture content becomes very low, e.g., in the case of senescence or during extreme drought. Altered leaf structure caused by disease can also affect near reflectance. VIS and near-IR data are sensitive indicators of the presence of green vegetation

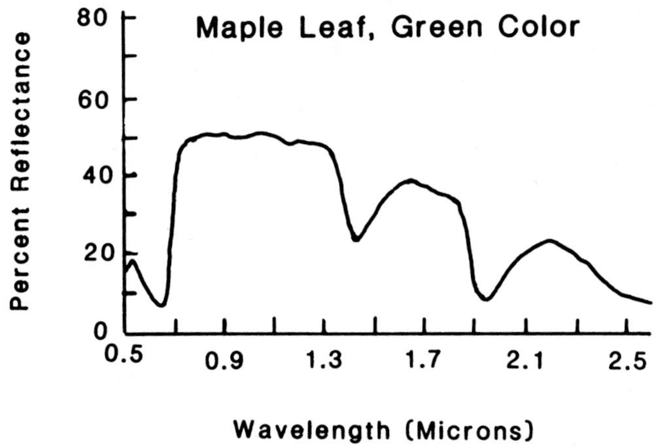

Figure II-2-5. The effect of pigmentation on leaf reflectance (Hoffer and Johannsen, 1969).

and can be used to construct a vegetation index (Tucker, 1978). Also, observations near 3.8 μm can be used to detect fires or deforestation (Matson and Dozier, 1981).

The spectral response of soil in the VIS and near-IR is affected by organic matter, iron oxide, composition, surface roughness, and moisture. In the near and middle-IR, large reductions in reflectance are noted as soil moisture increases (Merritt and Hall, 1973). At MW frequencies, smooth soil acts as a smooth surface because the particles are much smaller than the MW wavelength, but when tilled, soil becomes a rough surface. Like dry snow, dry soils can be penetrated greatly, but the addition of water retards penetration and increases emissivity. MW studies of subsurface characteristics were conducted by Schmugge (1983) and Wang et al. (1983) using thermal emission. Choudhury et al. (1979) introduced geometric factors in emissivity calculations using electromagnetic and radiative transfer theory for a homogeneous but unsmooth soil surface. Many natural materials behave as a continuum in MW and can be approximated by their average dielectric properties. The emissivity for a smooth, isothermal surface of a homogeneous material is given by the Fresnel transmission coefficients for a dielectric interface (Chen and Peake, 1961). Table II-2-1 lists the dielectric constants for selected surfaces and the calculated emissivities for nadir viewing. The real (refractive) and imaginary (loss) components of the dielectric constants correspond to a temperature of 20°C for water and soil, a salinity of 35% for sea water (Klein and Swift, 1977), and a soil moisture of 0.4 cm^3/cm^3 with an average wilting point of 0.2 cm^3/cm^3 (Wang and Schmugge, 1980). Also tabulated is the skin depth, which corresponds to the top layer thickness from which 63% of the radiation emanates.

Compared with the other surfaces listed in Table II-2-1, water has the largest dielectric constant and the smallest skin depth and emissivity. This is in contrast to ice and dry soil which have the lowest dielectric constants and largest emissivity values. Wet soil has properties between those of water and dry land.

D. Urban Areas. The albedos of concrete and asphalt of urban areas do not contrast sharply with those of vegetation in surrounding rural areas (Table II-2-2). Materials used in urban areas, however, do produce a different middle-IR response than vegetated rural areas. Thus, urban areas appear as heat islands in the satellite middle-IR imagery (Matson et al., 1978 and Chapter VIII-1).

II-2.2.2 The Oceans

Several techniques have been used to measure features of interest in oceanography. Oceans reflect, scatter, and absorb incident sunlight, and reflect and emit radiation in the IR and MW regions. For example, the composite "roughness" of the sea surface can be observed by MW radars or radiometers. Knowledge of roughness can be used to estimate wave height, wind stress, and wind velocity (Wright, 1966). Polarization information appears particularly useful for determi-

Table II-2-1 *Dielectric constant and emissivity for selected materials*

MATERIAL	FREQ. (GHz)	DIELECTRIC REAL PART	CONSTANT IMAG. PART	SKIN DEPTH (MM)	EMISSIVITY
Water (Sea)	5	73 (66)	22 (36)	4 (2)	0.36
Wet Soil	5	20.3	5.4	8.0	0.58
Water (Sea)	20	34 (33)	36 (37)	0.4	0.40
Wet Soil	20	10.8	8.7	1.0	0.64
Water (Sea)	35	18	28	0.2	0.45
Wet Soil	35	7.0	6.7	0.6	0.70
Dry Soil	—	2.8	0	—	0.94
Ice	—	3.2	<.01	—	0.90

Table II-2-2 *Albedos for the shortwave (wavelengths less than 4.0 μm) portion of the electromagnetic spectrum (Sellers, 1965)*

SURFACE FEATURE	ALBEDO %
A. Bare Areas and Soils	
Snow, fresh fallen	75–95
Snow, several days old	40–70
Sand dune, dry	35–45
Sand dune, wet	20–30
Soil, dark	5–15
Soil, moist gray	10–20
Soil, dry clay or gray	20–35
Soil, dry light sand	25–45
Concrete, dry	17–27
Road, blacktop	5–10
B. Natural Surfaces	
Desert	25–30
Savanna, dry season	25–30
Savanna, wet season	15–20
Chaparral	15–20
Meadows, green	10–20
Forest, deciduous	10–20
Forest, coniferous	5–15
Tundra	15–20
Crops	15–25

nation of surface roughness and for estimating wind speed over oceans (Webster et al., 1976).

Ocean current features such as the Gulf Stream boundaries, warm and cold Gulf Stream rings and eddies, and internal waves along continental shelves can be sensed by MW imaging radars because of their influence on ocean surface roughness (Beal et al., 1981; Fu and Holt, 1982). MW radars measure the strength and shape of the backscattered return signal to determine large-scale currents (Huang, 1979), surface layer winds, and waves. Large features of the ocean bottom topography, sea mounts and deep trenches, are manifest on the surface because they produce local variations in the gravity vector and so distort the surface.

Satellite IR radiometers can measure the sea surface temperature (SST) in cloudfree areas with accuracies of 0.5 to 1.0°C (Robinson et al., 1984). Accuracy of absolute SST measurements is limited by instrument noise, errors in correction for atmospheric absorption of the emitted radiation, instrument calibration, and ground processing. The IR radiation from water represents a very thin (0.1 mm) layer at the surface.

Information about the ocean can be obtained from ocean color. A portion of the direct solar radiation and diffuse sky light that illuminates the oceans penetrates the upper layers of water and is scattered upwards. In the absence of a strong reflectance from the ocean surface, this scattered light can be observed. Some light in the 400 to 600-nm range may penetrate to depths of tens of meters. Backscattered light appears as a variety of colors ranging from brilliant purple-blues for purest water, blue-greens for biologically enriched water, to yellow-greens and yellows for water approaching eutrophication. For clear water, absorption is strongest at 600 to 700 nm, decreases to a minimum at around 450 nm, and begins to increase rapidly in the violet and ultraviolet regions (Figure II-2-6). A comprehensive review of the spectral absorption and scattering coefficients of sea water is contained in Smith and Baker (1981). Other background material is contained in Morel and Prieur (1977), who discuss effects of organic and inorganic suspensions on radiance transmission, and in Gordon and McCluney (1975), Duntley et al. (1974), Kiefer and Austin (1974), Morel (1980), Gordon and Clark (1980a, 1980b), and Clark et al. (1980). For water with suspended material, the color shifts. An example of the shift in color in the observed Munsell colors produced by varying amounts of chlorophyll *a* (CHL-A), the pigment of phytoplankton, is shown in Figure II-2-7. Here the spectral distribution and approximate visual colors are shown for four conditions ranging from fairly pure (clear) water, to water rich in suspended phytoplankton. The complete radiative transfer processes connected with optics are de-

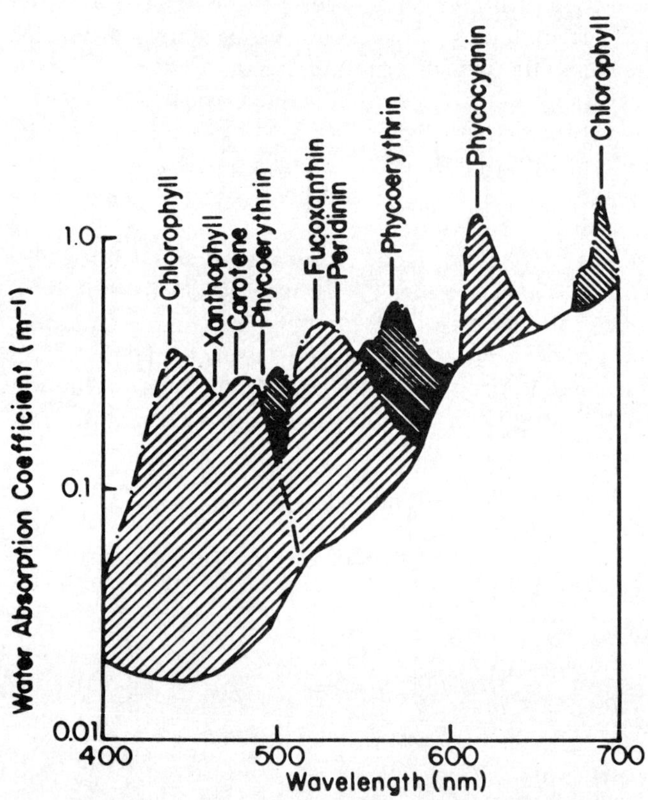

Figure II-2-6. A clear water absorption curve for the VIS region, with the relative absorption bands for the major pigments present in marine phytoplankton (Clark and Sherman, 1986).

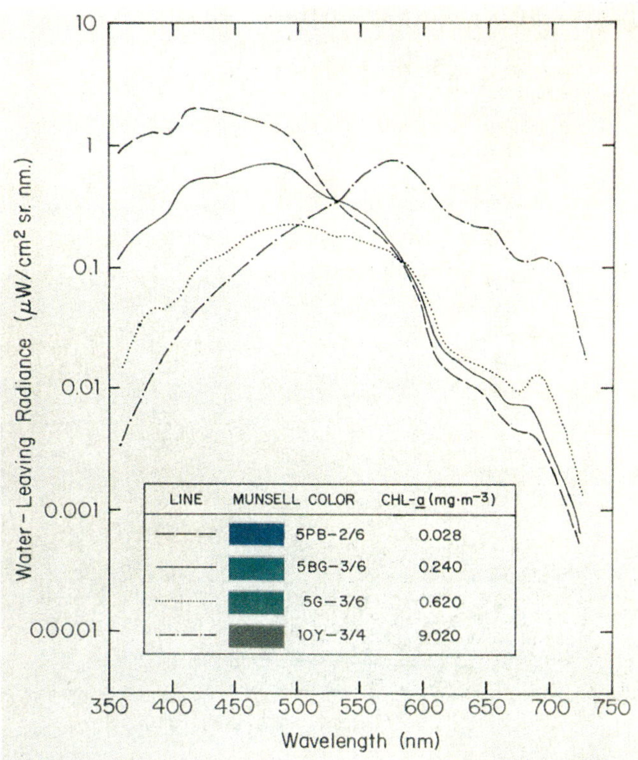

Figure II-2-7. Backscattered spectral radiance $L_w(\lambda)$ distributions for several chlorophyll *a* concentrations in surface water. Corresponding Munsell colors are indicated.

scribed by Preisendorfer (1976), Jerlov and Nielsen (1974), and Jerlov (1968).

Solar radiation reflected from the surface interferes with measurements of ocean color. When the sea surface is flat, its reflectivity ranges from 2% at normal incidence to 100% for grazing incidence. The most striking example of grazing incidence is the image of the Sun at the specular reflection point. As the surface is roughened by wind waves and swell in an area, the solar image becomes distorted by reflection and refraction; glitter or glint patterns of varying sizes are produced. Although glitter masks the ocean color, glitter patterns can be used to estimate sea surface roughness, to detect oil slicks, and to visualize fluid flow.

II-2.3 THE ATMOSPHERE AS AN ERROR SOURCE

Surface radiation passing through the atmosphere is modified by clouds, aerosols, and atmospheric gases. Atmospheric absorption, scattering, and refraction alter the intensity, frequency, and direction of EMR. VIS and IR radiation are particularly impacted by these atmospheric interactions. In the VIS, scattering by atmospheric gases is much more important than absorption in determining the spectral distribution of solar energy. In the IR, however, absorption by atmospheric gases and clouds is much more important than scattering; IR windows are mainly used for surface applications to minimize the atmospheric attenuation. UV is absorbed and scattered. In the MW, effects by clouds and precipitation vary depending upon the frequency; window channels are often used to minimize the atmospheric effects. It is most useful to start with the simple case of a clear atmosphere without aerosols, clouds, or precipitation.

II-2.3.1 The Clear Atmosphere

In a clear atmosphere, some of the radiation passing through a layer is absorbed by molecules in the layer and some radiation is added to the beam by other molecules in the layer. The change in a beam of radiation passing through a layer is determined by the concentration and temperature of the absorbing gas. Figure II-2-8 depicts the O_2 attenuation for different altitudes which represent different temperature and pressure situations.

The atmospheric gases determine the spectral absorbency of the atmosphere. The major atmospheric gases, oxygen and nitrogen, are uniformly mixed except where intense combustion (e.g., petroleum fuel and forest fires) substantially changes the local oxygen content of the air. The normal composition in the lower troposphere of an atmosphere free of water vapor and contaminants is shown in Table II-2-3 and the vertical distribution of three of these in Figure II-2-9.

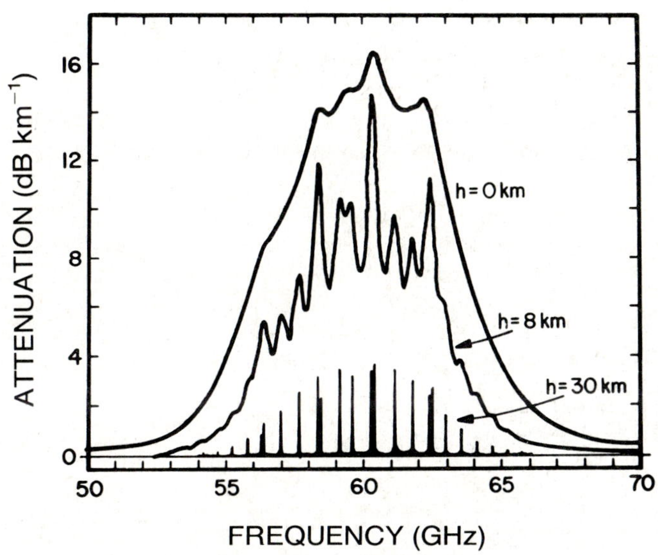

Figure II-2-8. Computed absorption coefficients for oxygen at three altitudes. The figure shows that the individual lines completely overlap at sea level, partly overlap at 8 km, and are resolved at 30 km (Meeks and Lilley, 1963).

Table II-2-3 *Normal composition of clean, dry air near sea level* (U.S. Standard Atmosphere, 1962)*

CONSTITUENT GAS	GAS SYMBOL	CONTENT (% BY VOLUME)	MOLECULAR WEIGHT
Nitrogen	N_2	78.084	28.0134
Oxygen	O_2	20.9476	31.9988
Argon	Ar	0.934	39.948
†Carbon dioxide	CO_2	0.0314	44.00995
Neon	Ne	0.001818	20.183
Helium	He	0.000524	4.0026
Krypton	Kr	0.000114	83.80
Xenon	Xe	0.0000087	131.30
Hydrogen	H_2	0.00005	2.01594
†Methane	CH_4	0.0002	16.04303
Nitrous Oxide	N_2O	0.00005	44.0128
†Ozone	O_3	Summer: 0 to 0.000007	47.9982
		Winter: 0 to 0.000002	47.9982
†Sulfur dioxide	SO_2	0 to 0.0001	64.0628
†Nitrogen dioxide	NO_2	0 to 0.000002	46.0055
†Ammonia	NH_3	0 to trace	17.03061
†Carbon monoxide	CO	0 to trace	28.01055
†Iodine	I_2	0 to 0.000001	253.8088

* On basis of C^{12} equals 12.000.

† The content of these gases may undergo significant variations from time to time or from place to place relative to the normal.

Above 90 km, molecular oxygen is dissociated by solar UV radiation and becomes less uniformly distributed aloft. Above 100 km, diffusion becomes important and the lighter atmospheric gases tend to accumulate. At 2000 km the principal constituents are ionized helium and hydrogen, and electrons. Thus, the mean molecular weight of the atmosphere decreases from about 29 at 100 km, to 20 at 400 km, and 16 at 700 km. Nonuniformly distributed constituents such as H_2O, CO_2, O_3, CH_4, and N_2O are active in the IR. The distributions of these and other constituents are seen in Figure II-2-10. The concentrations of many of these constituents are too low for them to have significant radiative effects.

The principal absorbers of EMR are carbon dioxide, water vapor, and ozone. Figure II-1-5 shows the absorption lines in the 15 μm band of CO_2. Ozone is active in the UV with absorption bands centered below a wavelength of 0.3 μm and emission of radiation principally in an IR band centered near 9.6 μm. At the

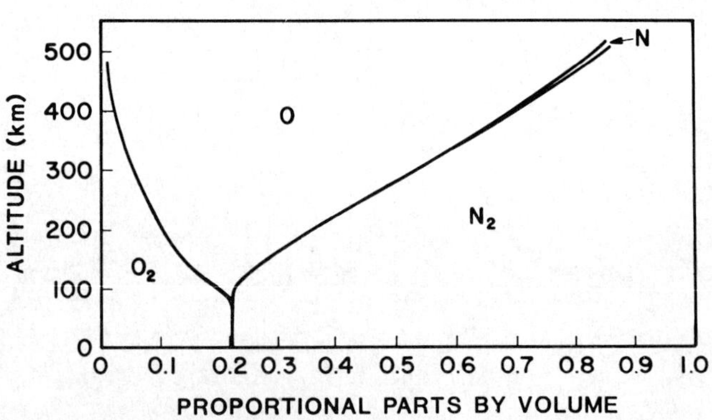

Figure II-2-9. Relative proportions of the principal constituents at altitudes up to 500 km (Air Force Cambridge Research Laboratories, 1965).

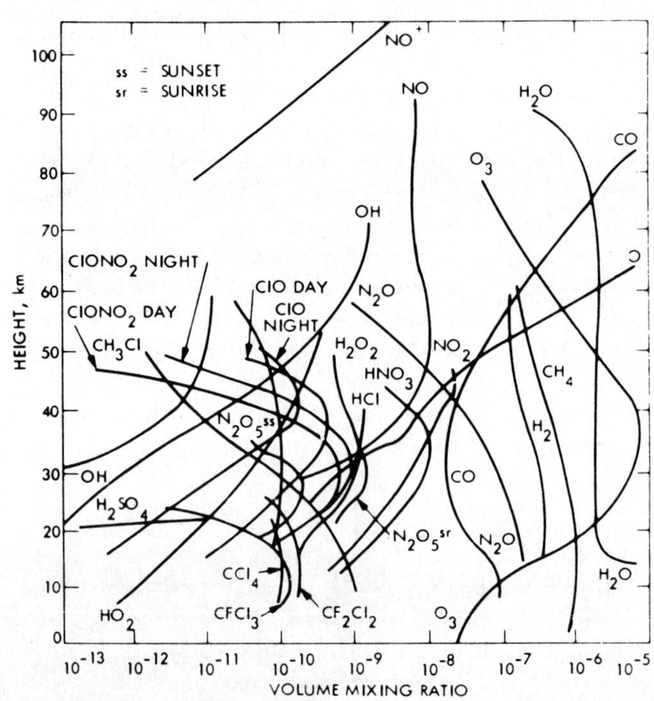

Figure II-2-10. Distribution of some chemically active constituents in the stratosphere, mesosphere, and lower thermosphere (Huntress, 1978; courtesy of JPL, California Institute of Technology, Pasadena, CA).

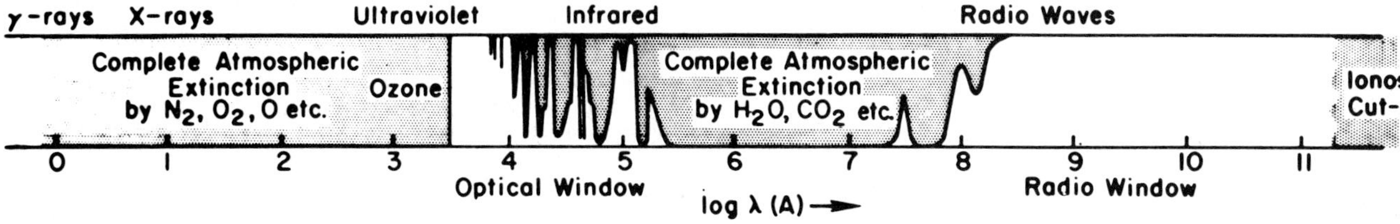

Figure II-2-11. The atmosphere transmittance spectrum from gamma rays to radio waves (Aller, 1963: © John Wiley and Sons, Inc.). The wavelength is in Angstrom units.

centers of strong absorption lines, only radiation emitted at the top of the atmosphere reaches the satellite. Radiation emitted from below the top of the atmosphere (usually by the surface or clouds) is absorbed within these wavelengths. Gases in the atmosphere absorb across the entire spectrum. There are some regions in the IR, however, where the atmosphere is nearly transparent between absorption lines. At the most transparent regions between lines, radiation from the ground reaches the satellite with little attenuation. Regions of low atmospheric absorption are called atmospheric windows (Figure II-2-11). It is through these windows that much remote sensing of the Earth takes place. Since these atmospheric windows are not totally free from atmospheric absorption, gases and suspended particles in the atmosphere absorb and emit radiation, thus changing the radiation measured at the sensor. These effects vary greatly in space and time and often necessitate corrections of remotely sensed data.

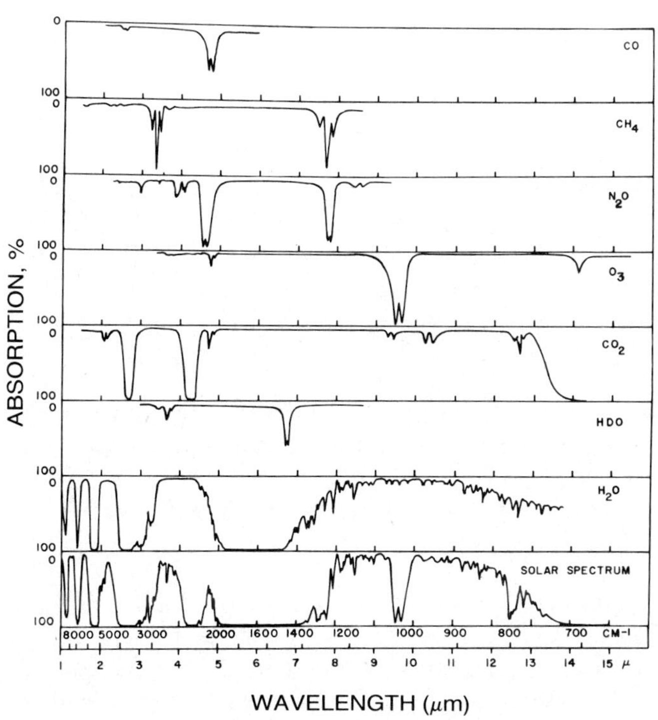

Figure II-2-12. The near-IR transmittance spectra of the atmosphere. This is an expansion of the 4.0–5.2 log (λ) region in Figure II-2-11 and includes separation of the gaseous components (J. H. Shaw, in Howard et al., 1955).

Since current operational instruments use filters that are several absorption lines in width, it is difficult to obtain a measurement that is not affected by the atmosphere to some extent. Figure II-2-12 depicts the detailed individual constituent absorptions within the near-IR window. Unfortunately the windows do not always lie in spectral regions suitable for surface monitoring.

Although energy is smaller than in the VIS and IR regions, the MW region between 1 cm and 10 m offers an extensive window for remote sensing of the surface (Figure II-2-11). At MW frequencies, the principal molecular absorption in the atmosphere is attributable to oxygen and water vapor. Water vapor resonance centers are at 22.235 GHz with a much stronger line at 183.31 GHz. Oxygen has about 45 lines centered near 60 ±10 GHz, and an isolated line at 118.75 GHz. Numerous lines also exist for ozone, chlorine, nitrogen, and hydrogen compounds; O_2, H_2O, O_3, N_2O, CO, SO_2 and the ClO radical are active in the MW region. The two curves of Figure II-2-13, computed with the inclusion of O_3, show the integrated absorption spectrum with and without the presence of water vapor.

II-2.3.2 Clouds and Precipitation

Clouds strongly influence radiation entering or leaving the atmosphere. Clouds only a few meters thick can be seen in VIS and IR images.

At VIS wavelengths, radiation transmitted and reflected by clouds consists predominantly of that which has been scattered many times between entering and leaving the cloud layer. For a spherical water droplet, the distribution of scattered energy can be computed with the theory developed by Mie (1908). Figure II-2-14 is an example of such a distribution. Note the large scattering that typically occurs in a small range of angles about the forward direction. Distributions are similar for other droplet sizes. To describe fully the passage of radiation through a cloud layer, one must

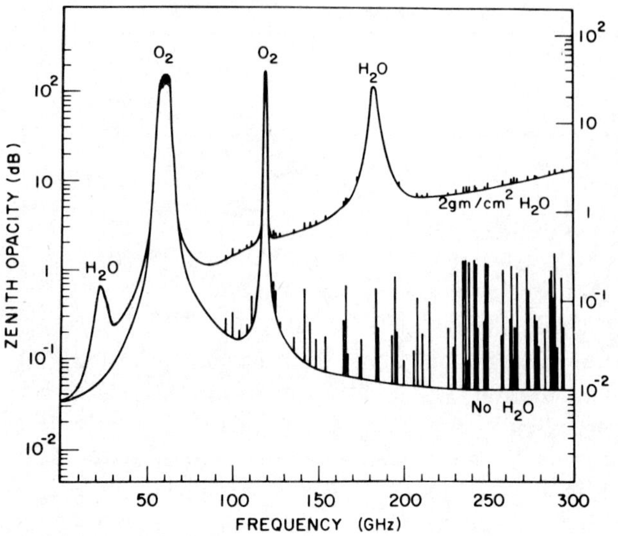

Figure II-2-13. Integrated absorption (minus log of transmittance) for an atmosphere consisting of oxygen, water vapor, and ozone. The effects are greatly reduced when water vapor is included (Waters, 1976).

also specify its optical thickness (Figures II-2-15a and II-2-15b).

The net spectral distribution of VIS radiation is altered little by absorption and scattering, so clouds usually appear white or gray. A summary compiled by Lumb (1964), Figure II-2-16, demonstrates the effects of various cloud types on total irradiance.

In the thermal IR, most clouds emit like full radiators at the temperatures of the water droplets or ice crystals at their tops. The integrated effects of the clouds interacting with radiation result in the sensor not being able to detect radiation from the surface and atmosphere below the cloud. Instead, it measures almost as if the top of the cloud were a surface radiator with IR absorption/emission occurring over a wide range of wavelengths. The range in temperatures of the clouds causes the signatures in the IR window to vary widely, thus providing information about the cloud height. This is further discussed in Chapter VII-1. Cloud contamination in the IR hinders the use of many automated techniques for determining surface and atmospheric conditions.

Microwave interactions with clouds are very different from VIS and IR interactions. Clouds of water droplets smaller than the radiation wavelength do not scatter the radiation. They act as absorbers/emitters, but their effect is minimal because the amount of liquid involved is small. A thunderstorm cloud, however, may affect a MW measurement, but it would have to be horizontally extensive (5-10 km) in relation to the size of the field of view (FOV) to produce significant degradation in the signal from the surface. It is difficult, however, both

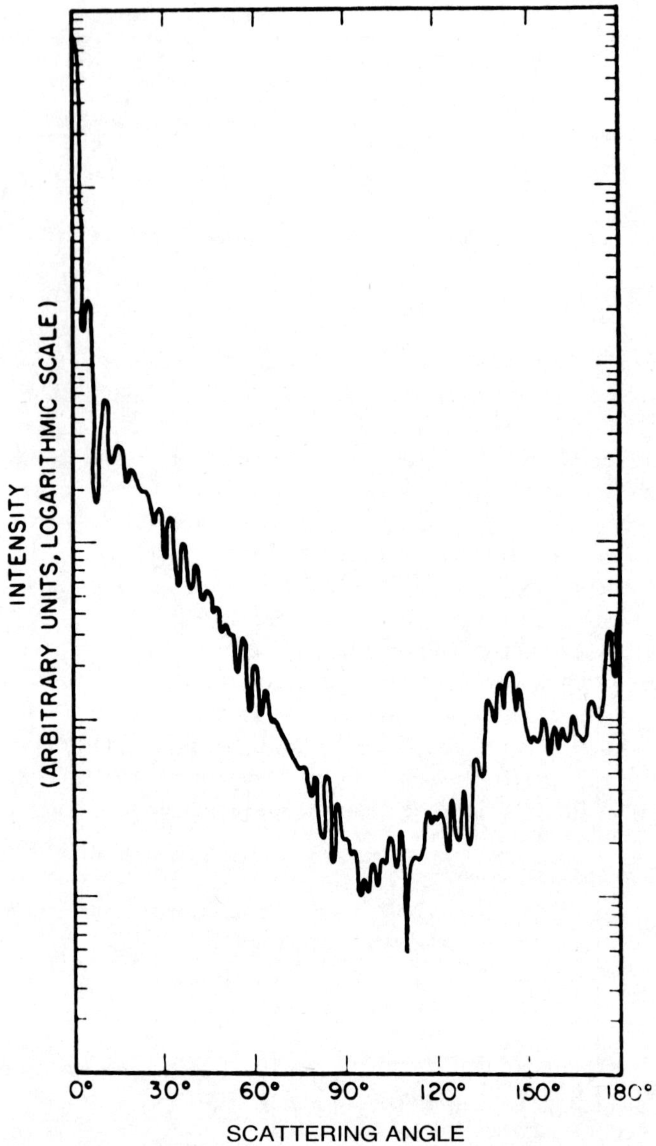

Figure II-2-14. Mie scattering function for a cloud drop size distribution with mean radius 6 μm and a standard deviation of 1 μm.

to estimate the extent of such degradation and to correct for it. Rain has a scattering effect, as shown in Figure II-2-17, where single particle albedo is plotted as a function of liquid water density at 10, 20, and 30 GHz. Frozen precipitation, because of its scattering characteristics, results in some of the lowest MW brightness temperatures observed over land. The large ice particles (graupel or hail) also result in polarization due to organized orientation. The effects of clouds and liquid precipitation over land are generally difficult to observe at frequencies less than 50 GHz. Conversely, the surface emission over oceans exhibits a large contrast in brightness temperature when compared with radiation from clouds. It is possible to use multiple

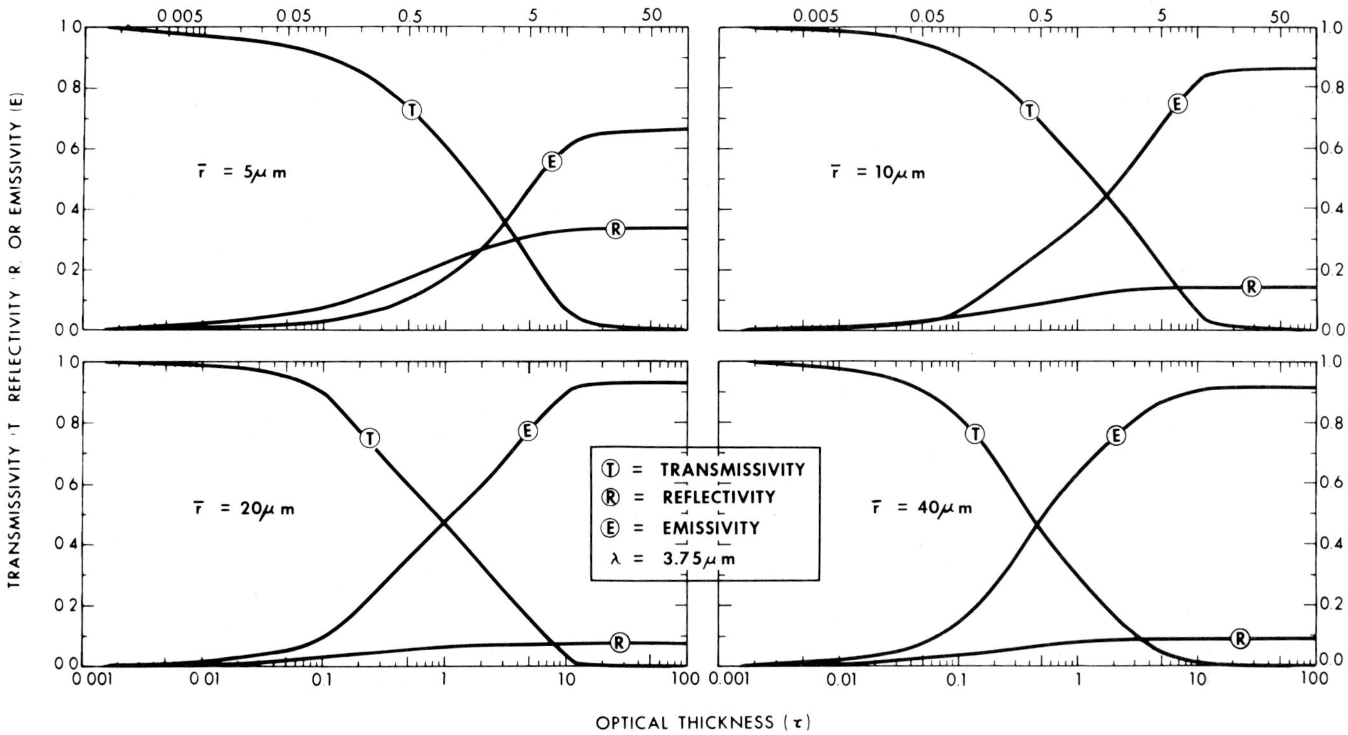

Figure II-2-15a. Transmissivity (T), reflectivity (R), and emissivity (E) versus cloud optical thickness (τ) for four different sizes of spherical particles (r) at $\lambda = 3.75\ \mu$m.

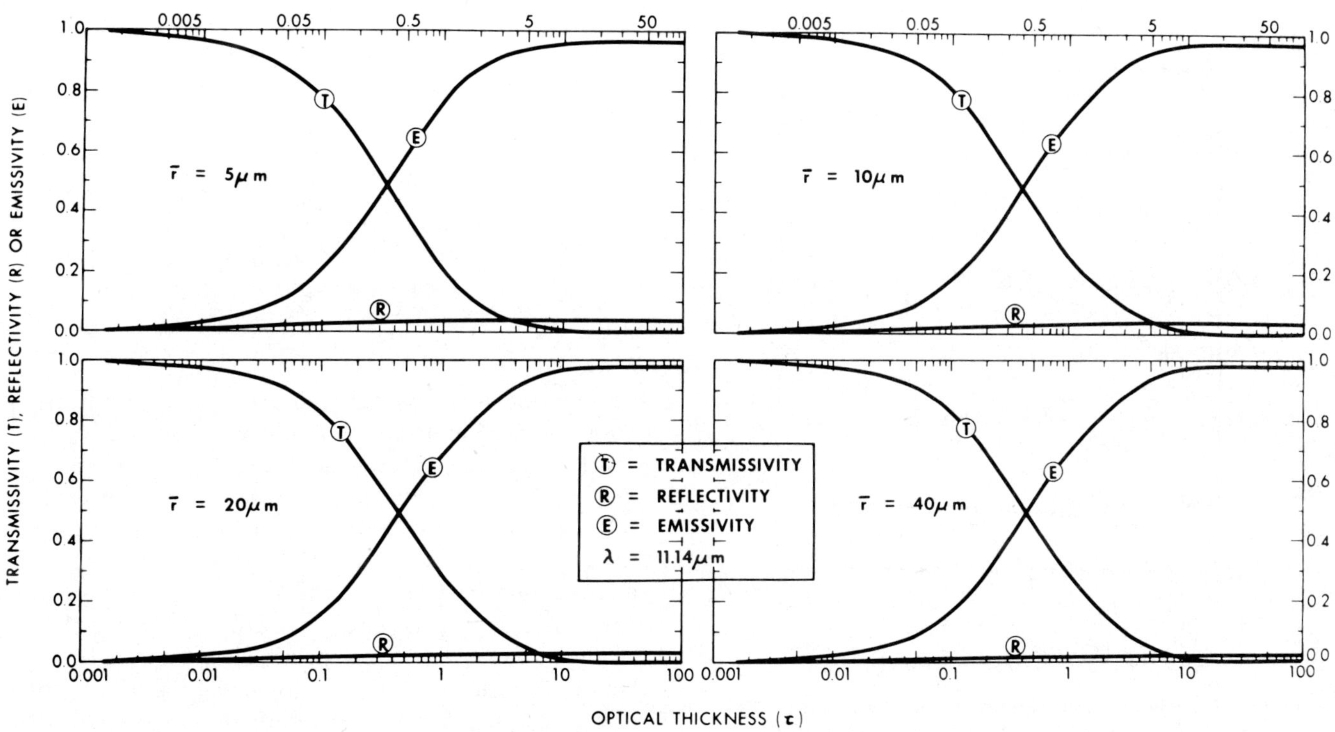

Figure II-2-15b. Transmissivity (T), reflectivity (R), and emissivity (E) versus cloud optical thickness (τ) for four different sizes of spherical particles (r) at $\lambda = 11.14\ \mu$m.

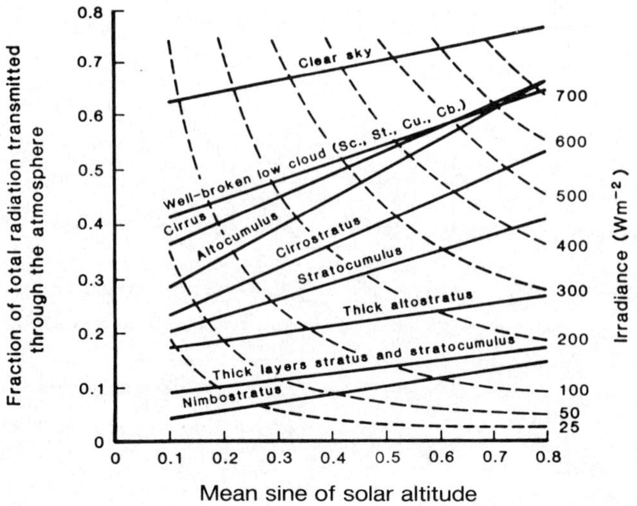

Figure II-2-16. The straight lines represent the empirical relations between solar radiation and solar angle for different cloud types from measurements in the North Atlantic (52°N, 20°W). The curves are isopleths of irradiance (courtesy Royal Meteorological Society; Lumb, 1964; Monteith, 1973). Sc-stratocumulus, St-stratus, Cu-cumulus, Cb-cumulonimbus.

MW frequencies to measure atmospheric liquid water content over the oceans and so correct for cloud effects.

II-2.3.3 Aerosols

Over most of the solar spectrum, from 0.3 to 0.9 μm, an aerosol-free atmosphere is almost non-absorbing.

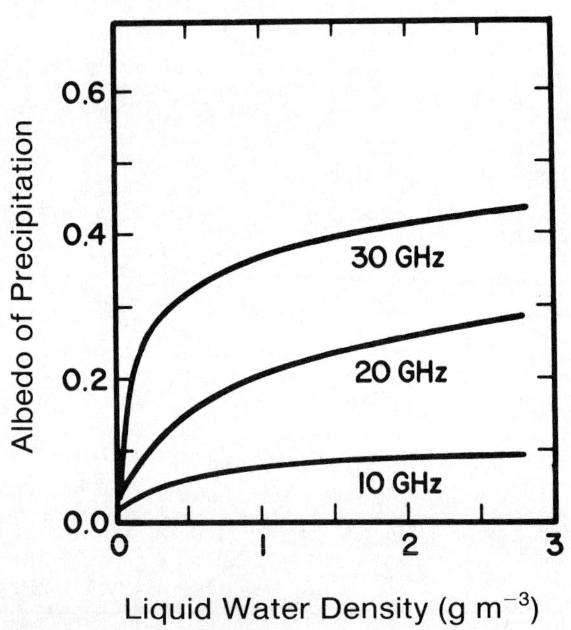

Figure II-2-17. Single scattering albedo as a function of liquid water density at 10, 20, and 30 GHz. Calculations were made for a temperature of 0°C and are based upon a Marshall-Palmer drop size distribution.

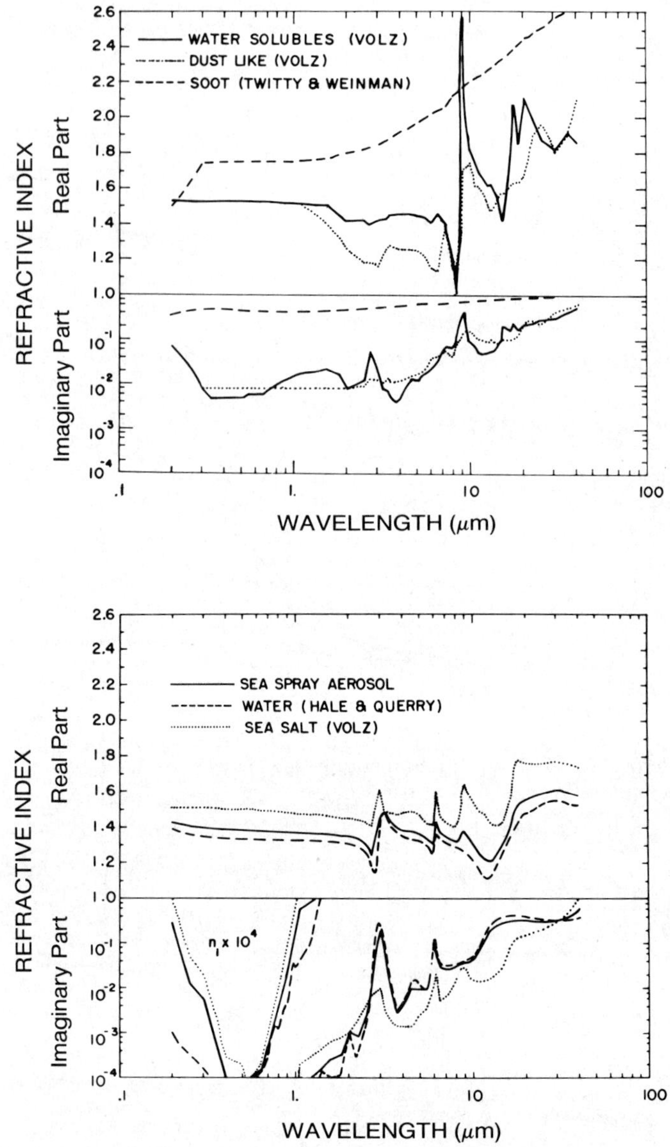

Figure II-2-18a. Refractive index versus wavelength for indicated aerosol types (Shettle and Fenn, 1979).

When aerosols are present, absorption and scattering of solar radiation occur, but for small amounts of aerosols there is little effect on satellite VIS measurements. For large amounts of aerosols, such as extensive dust or large loadings of volcanic dust in the stratosphere, both VIS and IR estimates of surface features like sea surface temperature and ocean color need correction. The MW is not affected by aerosols.

Dust and other airborne particles approximately the size of the radiation wavelength make the relation between scatter and wavelength difficult (Figure II-2-18a, b). Observation of a dust-loaded atmosphere, such as that found with a subtropical anticyclone in the VIS,

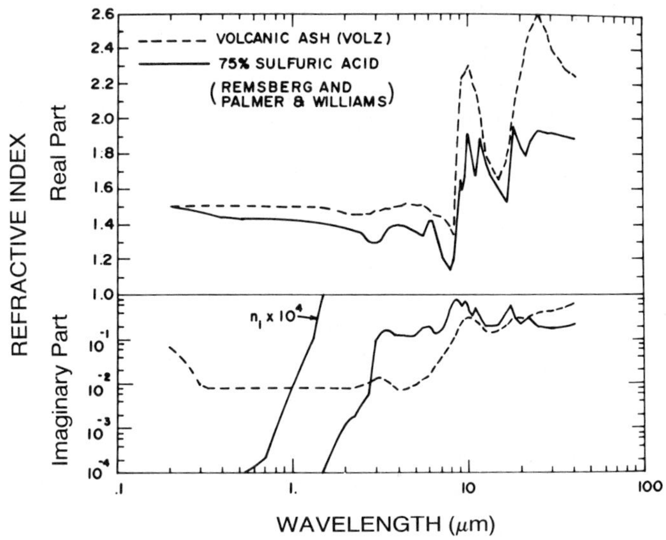

Figure II-2-18b. Refractive index versus wavelength for indicated aerosol types (courtesy of E. P. Shettle).

finds scattering to be independent of wavelength. See Chapter IX-2 for further discussion of clouds and aerosols.

II-2.4 THE ATMOSPHERE AS A TARGET

Ironically, the very interactions between the atmosphere and EMR which create error sources and limit the usefulness of remote sensing of the surface of the Earth make possible the estimation of atmospheric parameters. Satellite measurements have been successfully used to estimate temperature, moisture, cloud amount and height, wind through the tracing of cloud motion, precipitation, aerosol content, and ozone. Details of the methods for deriving these parameters will be discussed in Sections VI, VII, and IX. A few general concepts concerning remote vertical sounding of the atmosphere are given here.

II-2.4.1 Atmospheric Sounding and Constituent Estimation

Vertical sounding of the atmosphere is achieved by looking at spectral regions with a range of absorption characteristics, obtaining radiances emitted at different atmospheric heights, and then determining a vertical profile of the parameter that fits the observed radiative emittances. For example, if temperature is being measured, a spectral region is selected where the absorption is caused by a gas with a uniform mixing ratio in the atmosphere so that the concentration of the gas is a known factor and the changes in radiance indicate only changes in temperature of the absorbing gas. A radiance at a given wavelength (channel) represents a weighted average temperature over a certain range of heights where the weighting is determined by the derivative of the atmospheric transmittance with respect to height. This quantity has become known as the "weighting function." The weighting function is dependent upon the concentration and vertical distribution of the gas, and its temperature and pressure. At present, three wavelength regions involving two gases are commonly used to determine atmospheric temperature: the 15 μm and 4.3 μm IR regions using CO_2, and the 50 GHz MW region using oxygen. Atmospheric sounding is further discussed in Chapters VI-2 and VII-5.

One of the limitations of remote sensing is that the range of heights represented by the sounder radiance of a single channel is fairly large. This difficulty is compounded by the fact that it is not always possible to resolve spectral regions narrower than a few lines in width (Figure II-2-19). Vertical resolution varies with instrument, wavelength, and the parameter being estimated. The major goal in developing new remote sounding instrumentation is to achieve the sharpest possible weighting functions so that vertical resolution can be improved. Figure II-2-20 shows typical weighting

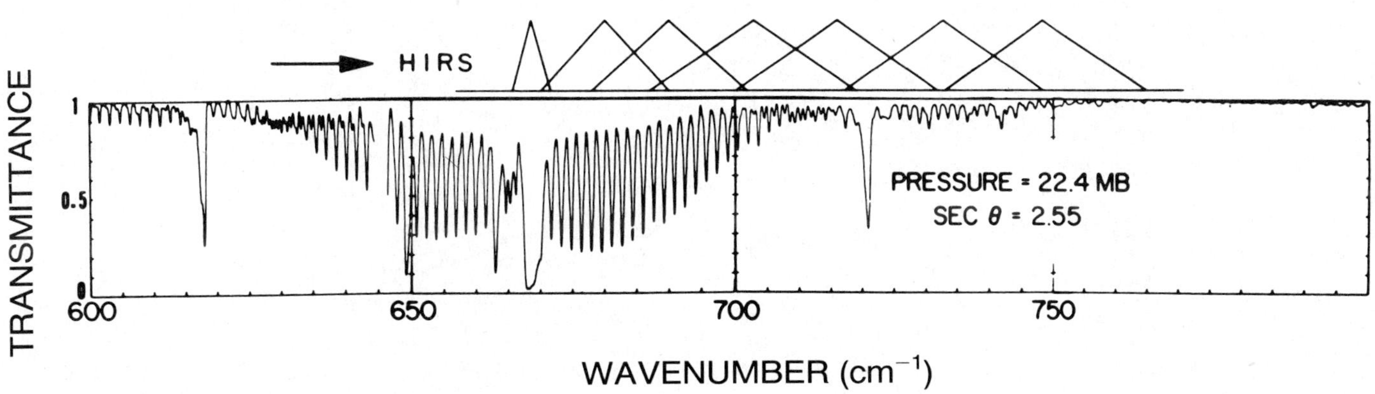

Figure II-2-19. Locations of idealized spectral intervals of HIRS Channels 1-7 compared with atmospheric spectrum measured from balloon at altitude of 22.4 mb (Weinreb et al., 1981). Ordinate is in units of transmittance.

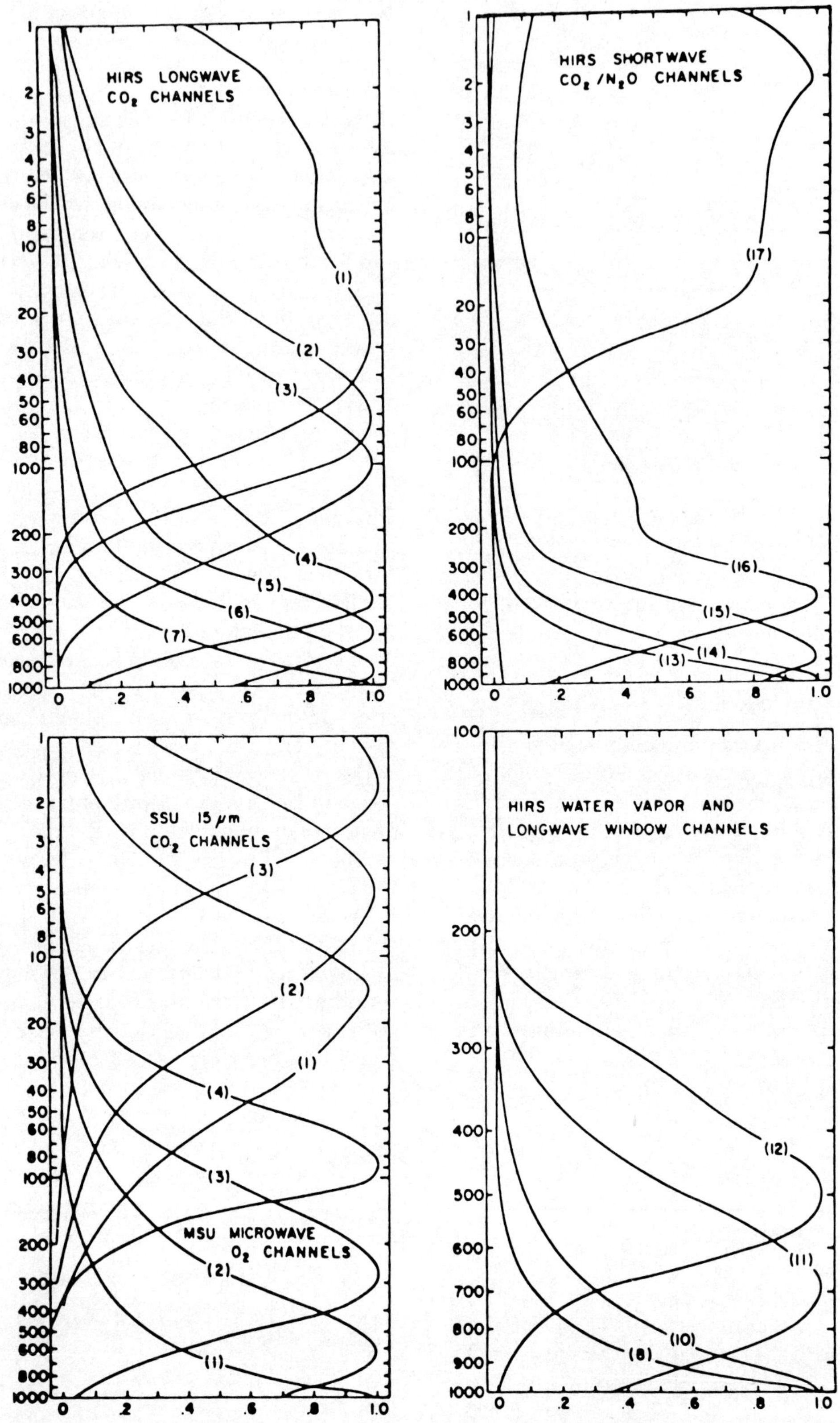

Figure II-2-20. TOVS weighting functions (normalized). Ordinate is pressure (mb) and abscissa is the normalized weight (Smith et al., 1979).

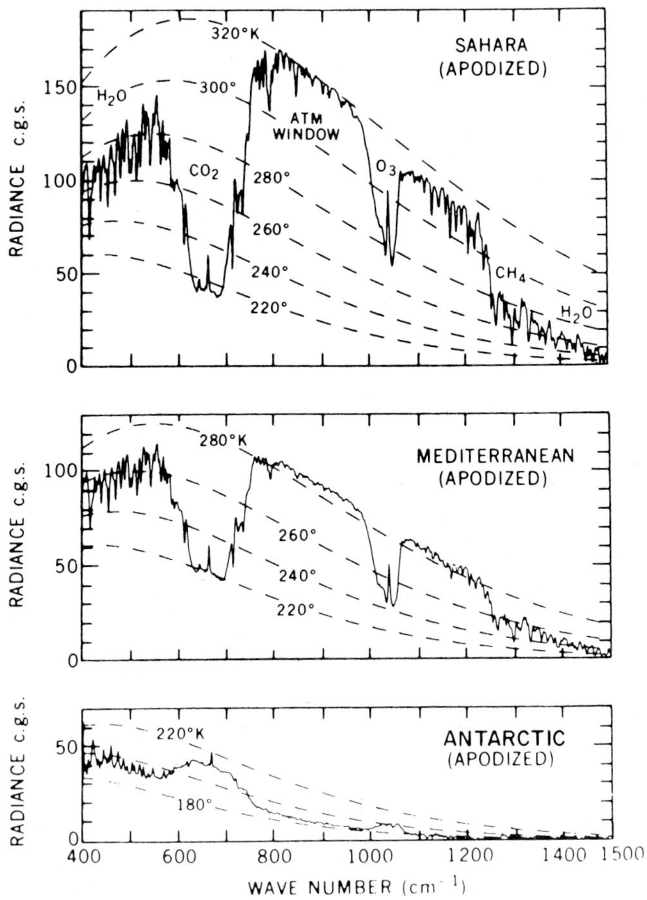

Figure II-2-21. Spectra of the radiation emitted by the earth-atmosphere system as seen by the Nimbus satellite above three different points on the Earth. Radiance units (cgs) are $erg \cdot s^{-1} \cdot cm^{-2} \cdot sr^{-1}/cm^{-1}$ (Hanel et al., 1971).

functions for temperature and moisture sounding instruments.

To determine variable gas concentrations in the atmosphere, some satellite measurements use instruments that do not look directly at the Earth surface. Some designs use limb scanning with either cold space or the Sun as a background, others use solar radiation backscattered from the Sun or the atmosphere, while some proposed designs use lidar and radar as active sources of radiation. An advantage of using solar radiation is that its intensity is so much greater than that emitted by atmospheric sources, hence the emitted radiation can be ignored and only absorption has to be considered. Concentrations of water vapor and ozone are routinely obtained by operational satellites. Details are given in Sections VI, VII, and IX.

The appearance of the earth-atmosphere system as a target is illustrated by Figure II-2-21, which shows some typical IR emission spectra of the earth-atmosphere system. In the spectral region labelled atmospheric window, the blackbody emission from the surface is transmitted through the atmosphere to the satellite with relatively little attenuation; the observed brightness temperature is then approximately equal to the actual surface temperature. At these wavelengths, the observed brightness temperatures (dashed lines) range from 320 K over the hot Sahara Desert, to 280 K over the cooler Mediterranean Sea, to less than 200 K over the cold Antarctic continent.

II-2.4.2 Clouds, Precipitation, and Aerosols as Targets

A variety of VIS and IR methods have been developed for observing cloud properties. The window IR is particularly useful for estimating cloud top temperature and height. Sequential VIS and IR imagery are used extensively for analysis of cloud type, height, cloud motions and associated winds, precipitation, and changes in weather systems. Automated techniques to classify cloud type and height, and to determine cloud cover, have been developed using IR and have been improved by using multispectral techniques.

MW studies show promise in providing additional information about clouds and precipitation, particularly over the oceans where the background brightness temperatures are nearly constant and the low emissivity brightness temperatures are distinct from the brightness temperatures of clouds and precipitation. Brightness temperature measurements in the window regions have been used to estimate cloud liquid water and water vapor content (Westwater and Guiraud, 1980; Grody et al., 1980). Due to the more direct relationship between radiation and precipitation, passive MW techniques show even greater promise than IR techniques. Rain penetration at low frequencies was considered in an experiment by Jones et al. (1981). They flew a 4.5 to 7.2 GHz MW radiometer onboard an airplane to obtain estimates of the sea surface winds and precipitation of Hurricane David. This experiment showed the ability of low frequency MW to penetrate rain.

Techniques to detect the presence of aerosols are being refined using UV, VIS, and near-IR. Multispectral techniques are used to study dust from deserts, air pollution, related haze, and smoke. When intense concentrations of aerosols are evident in the window IR imagery, atmospheric motions are studied using the aerosols as tracers. See Sections VII and IX for additional information.

BIBLIOGRAPHY AND REFERENCES FOR SECTION II

Air Force Cambridge Research Laboratories, 1965: *Handbook of Geophysics and Space Environments*. McGraw-Hill Book Company, Inc., New York, NY, 6-1, 16-2.

Aller, L. H., 1963: *The Atmosphere of the Sun and Stars*. The Ronald Press Co., New York, NY, 412 pp.

Austin, R. W., 1974: The remote sensing of spectral radiance from below the ocean surface. N. G. Jerlov and E. S. Nielsen (eds.), *Optical Aspects of Oceanography*, Academic Press, New York, NY, 317-344.

Barnes, J. C. and C. J. Bowley, 1972: Snow studies using thermal infrared measurements from Earth satellites. *Final Report, Subcontract No. 19334 to ARA Inc., NOAA/NESS Contract No. 135350*. Environmental Research and Technology, Inc., Concord, MA, 112 pp.

Barnes, R. B., 1968: Diagnostic thermography. *Appl. Opt., 7*, 1673-1685.

Beal, R. C., P. S. DeLeonibus, and I. Katz (eds.), 1981: *Spaceborne Synthetic Aperture Radar*. Johns Hopkins Press, Baltimore, MD, 215 pp.

Berg, C. P., D. R. Wiesnet, and M. Matson, 1981: Assessing the Red River of the North 1978 flooding from NOAA satellite data. *Satellite Hydrology*, M. Deutsch, D. R. Wiesnet and A. Rango (eds.), Amer. Water Res. Assoc., Minneapolis, MN, 309-315.

Born, M. and E. Wolf, 1970: *Principles of Optics*. 4th ed., Pergamon Press, Oxford, 808 pp.

Brown, O. B. and R. E. Cheney, 1983: Advances in satellite oceanography. *Rev. of Geophys. and Space Physics, 21*, 1216-1230.

Cavalieri, D. J., S. Martin, and P. Gloerson, 1983: Nimbus-7 SMMR observations of the Bering Sea ice cover during March 1979. *J. Geophys. Res., 88*, 2743-2754.

Chang, A. T. C., P. Gloerson, T. Schmugge, T. T. Wilheit, and H. J. Zwally, 1976: Microwave emission from snow and glacier ice. *J. Glaciology, 16*, 23-39.

Chen, S. N. C. and W. H. Peake, 1961: Apparent temperature of smooth and rough terrain. *IEEE Trans. Ant. Prop., AP-9*, 567-572.

Choudhury, B. J., T. J. Schmugge, A. Chang, and R. W. Newton, 1979: Effect of surface roughness on the microwave emission from soils. *J. Geophys. Res., 84*, 5699-5706.

Cicerone, C., C. Delwiche, R. Harriss, and R. Dickinson, 1984: Critical processes affecting the distribution of chemical species: Biological and surface sources. *Global Tropospheric Chemistry - A Plan for Action*. National Academy Press, Washington, DC, 55-68.

Clark, D. K. and J. W. Sherman, III, 1986: Nimbus-7 coastal zone color scanner: Ocean color applications. *Marine Technology Soc. Journal, 20*, 43-5C.

———, E. T. Baker, and A. E. Strong, 1980: Upwelled spectral radiance distribution in relation to particulate matter in sea water. *Bound. Layer Meteor., 18*, 287-298.

Colwell, R. N., J. E. Estes, and G. A. Thorley, 1983: Geological applications. *Manual of Remote Sensing, 2*, Amer. Soc. of Photogram., Falls Church, VA, 1667-1954.

Cox, C. S., 1974: Refraction and reflection of light at the sea surface. N. G. Jerlov and E. S. Nielson (eds.), *Optical Aspects of Oceanography*, Academic Press, London, England, 51-75.

——— and W. Munk, 1954: Measurements of the roughness of the sea-surface from photographs of the sun glitter. *J. Opt. Soc. Am., 44*, 838-850.

Craig, R. A., 1965: The upper atmosphere, meteorology and physics. *International Geophysics Series, 8*, Academic Press, New York, NY, 509 pp.

Curran, P. J., 1985: *Principles of Remote Sensing*. Longman Inc., New York, NY, 271 pp.

Ditchburn, R. W., 1959: *Light*. 2nd ed., Interscience Publishers, New York, NY, 833 pp.

Donner, L. and V. Ramanathan, 1980: Methane and nitrous oxide: Their effect on the terrestrial climate. *J. Atmos. Sci., 37*, 119-124.

Duntley, S. Q., R. W. Austin, W. H. Wilson, C. F. Edgerton, and S. E. Moran, 1974: Ocean color analysis. *Technical Report SIO 74-10*, Scripps Inst. of Oceanography, La Jolla, CA, 67 pp.

England, A. W., 1975: Thermal microwave emission from a scattering layer. *J. Geophys. Res., 80*, 4484-4496.

Fishman, J., V. Ramanathan, P. Crutzen, and S. Liu, 1979: Tropospheric ozone and climate. *Nature, 282*, 818-820.

Ford, J. P., J. B. Cimino, and C. Elachi, 1982: The SIR-B science plan: Imaging radar science working group. *JPL Publication 82-78*, Natl. Aero. and Space Adm., JPL, Pasadena, CA, XIX-67, 67 pp.

Foster, J. L., D. K. Hall, and A. T. C. Chang, 1984: An overview of passive microwave snow research and results. *Geophys. Space. Phys., 22*, 195-207.

Fu, L. L. and B. Holt, 1982: Seasat views oceans and sea ice with synthetic aperture radar. *Jet Propulsion Laboratory Pub. 81-120*, Pasadena, CA, 200 pp.

Giddings, J. C. and E. LaChapelle, 1961: Diffusion theory applied to radiant energy distribution and albedo of snow. *J. Geophys. Res., 66*, 181-189.

Gorden, R. G., 1967: On the pressure broadening of molecular multiplet spectra. *J. Chem. Phys., 46*, 448-455.

Gordon, H. R. and D. K. Clark, 1980a: Atmospheric effects in the remote sensing of phytoplankton pigments. *Bound. Layer Meteor., 18*, 300-313.

——— and D. K. Clark, 1980b: Remote sensing optical properties of a stratified ocean: An improved interpretation. *Appl. Optics, 18*, 3428-3430.

——— and W. R. McCluney, 1975: Estimation of the depth of sunlight penetration in the sea for remote sensing. *Appl. Optics, 14*, 413-416.

Grody, N. C., D. G. Gray, C. S. Novak, J. S. Prasad, M. Piepgras, and C. A. Dean, 1985: Temperature soundings from the DMSP microwave sounder. *Advances in Remote Sensing Retrievals*, A. Deepak, Hampton, VA, 249-265.

———, A. Gruber, and W. C. Chen, 1980: Atmospheric water content over the tropical Pacific derived from the Nimbus-6 scanning microwave spectrometer. *J. Appl. Meteor., 19*, 986-996.

———, 1976: Remote sensing of atmospheric water content from satellites using microwave radiometry. *IEEE Trans. Ant. Prop., AP-24*, 155-162.

Hameed, S., R. Cess, and J. Hogan, 1980: Response of the global climate to changes in atmospheric composition due to fossil fuel burning. *J. Geophys. Res., 85*, 7537-7545.

Hanel, R. A., B. Schlachman, D. Rogers, and D. Vanous, 1971: Nimbus-4 Michelson interferometer. *Appl. Opt., 10*, 1376-1382.

Hoffer, R. M. and C. Matzler, 1980: Investigations of snow parameters by radiometry in the 3- to 60-mm wavelength region. *J. Geophys. Res., 85*, 453-460.

———, M. D. Fleming, L. A. Bartolucci, T. E. Anuta, S. G. Luther,

D. W. Levandowski, R. Borger, P. Krebs, and P. Spencer, 1975: Computer- aided analysis of SKYLAB multispectral scanner data in mountainous terrain for land use, forestry, water resource and geologic applications. *Final Report, NASA Contract No. NAS9-13380*, SKYLAB EREP Project 398, *LARS Information Note 121275*, Lab. for Applications of Remote Sensing, Purdue Univ., West Lafayette, IN, 381 pp.

——— and C. J. Johannsen, 1969: Ecological potentials in spectral signature analysis. *Remote Sensing in Ecology*, Philip L. Johnson (ed.), Univ. of Georgia Press, Athens, GA, 1-16.

Hogg, D. C., M. T. Decker, F. O. Guiraud, K. B. Earnshaw, D. A. Meritt, K. P. Moran, W. B. Sweezy, R. G. Strauch, E. R. Westwater, and C. G. Little, 1983: An automatic profiler of the temperature, wind and humidity in the troposphere. *J. Appl. Meteor., 22*, 807-831.

Hollinger, J. P., B. E. Troy, R. O. Ramseier, K. W. Asmus, M. F. Hartman, and C. A. Luther, 1984: Microwave emission from high Arctic sea ice during freeze-up. *J. Geophys. Res., 89*, 8104-8122.

Holton, J. R., 1975: The dynamic meteorology of the stratosphere and mesosphere. *Meteor. Monogr., 15*, Amer. Meteor. Soc., 216 pp.

Houghton, J. T., F. W. Taylor, and C. D. Rodgers, 1984: *Remote Sounding of Atmospheres*. Cambridge Univ. Press, New York, NY, 6, 210.

Hovis, W. A., 1966: Infrared spectral reflectance of some common minerals. *Appl. Opt., 5*, 245-248.

Howard, J. N., D. E. Burch, and D. Williams, 1955: Near infrared transmission through synthetic atmospheres. *Geophys. Res. Paper No. 40,* Air Force Cambridge Research Center, Bedford, MA.

Huang, N. E., 1979: New developments in satellite oceanography and current measurements. *Rev. of Geophys. and Space Phys., 17*, 1558-1568.

Hudson, R. D., Jr., 1969: *Infrared System Engineering*. Wiley, New York, NY.

Hummel, J. R. and R. A. Reck, 1981: The direct-thermal effect of $CHClF_2$, CH_3CCl_3 and CH_2Cl_2 on atmospheric surface temperatures. *Atmos. Environ., 15*, 379-382.

Huntress, W. T., 1978: Upper atmosphere research satellite program. *Final Report, Jet Propulsion Laboratory Publication, 78-54*, Pasadena, CA, 62 pp.

Idso, S. B., T. J. Schmugge, R. D. Jackson, and R. Reginato, 1975: The utility of surface temperature measurements for the remote sensing of surface soil water status. *J. Geophys. Res., 80*, 3044-3050.

Jerlov, N. G. and E. S. Nielsen (eds.), 1974: *Optical Aspects of Oceanography*. Academic Press, London, England, 494 pp.

Jerlov, N. G., 1968: *Optical Oceanography*. Elsevier, New York, NY, 231 pp.

Jones, W., P. G. Black, V. E. Delnore, and C. T. Swift, 1981: Airborne microwave remote-sensing measurements of Hurricane Allen. *Science, 214*, 274-280.

Kakar, R. K. and B. H. Lambrigtsen, 1983: A statistical correlation method for the retrieval of atmospheric moisture profiles by microwave radiometry. *J. Clim. Appl. Meteor., 23*, 1110-1114.

Kiefer, D. A. and R. W. Austin, 1974: The effect of varying phytoplankton concentration on the submarine light transmission in the Gulf of California. *Limnol. Oceanogr., 19*, 55-64.

Kinsman, B., 1965: *Wind Waves*. Prentice-Hall, Inc., Englewood Cliffs, NJ, 676 pp.

Klein, L. A. and C. T. Swift, 1977: An improved model for the dielectric constant of sea water at microwave frequencies. *IEEE Trans. Ant. Prop., AP-25*, 104-114.

Krueger, A. J., B. Guenther, A. J. Fleig, D. F. Heath, E. Hilsenvath, R. McPeters, and C. Prabhakara, 1980: Satellite ozone measurements. *Phil Trans. Roy. Soc. London, A296*, 191-204.

Kunzi, K., S. Patil, and H. Rott, 1984: Snow-cover parameters retrieved from Nimbus-7 scanning multichannel microwave radiometer (SMMR) data. *Rev. Geophys. Space Phys., 22*, 452-467.

Kyle, T. G. and A. Goldman, 1975: Atlas of computed infrared atmospheric absorption spectra. *NCAR Tech. Note TN-STR112*, National Center for Atmospheric Research, Boulder, Colorado.

Legeckis, R. and W. Pichel, 1984: Monitoring of long waves in the eastern equatorial Pacific 1981-83 using multi-channel SST charts. *NOAA Technical Report NESDIS 8*, Dept. of Commerce, Washington, DC, 19 pp.

Liebe, H. J., G. G. Gimmestad, and J. D. Hopponen, 1977: Atmospheric oxygen microwave spectrum experiment versus theory. *IEEE Trans. Ant. Prop., AP-25*, 327-335.

Lillesand, T. M. and R. W. Kiefer, 1979: *Remote Sensing and Image Interpretation*. John Wiley and Sons, New York, NY, 1-34, 387-389.

London, J. and J. K. Angell, 1982: The observed distribution of ozone and its variations. *Stratospheric Ozone and Man, Vol. 1*, F. A. Bower and R. B. Ward (eds.), CRC Press, Boca Raton, FL, 8-42.

———, 1978: Distribution of atmospheric ozone and how it is measured. *Air Quality Meteorology and Atmospheric Ozone*. ASPM STP 653, A. L. Morris and R. C. Barras (eds.), Amer. Soc. for Testing and Materials, Philadelphia, PA, 339 pp.

Lumb, F. E., 1964: The influence of cloud on hourly amounts of total solar radiation at the sea surface. *Q. Jn. R. Meteor. Soc., 90*, 43.

Margulis, L. and J. E. Lovelock, 1978: The biota as ancient and modern modulator of the earth's atmosphere. *Pure Appl. Geophys., 116*, 239-243.

Matson, M. and J. Dozier, 1981: Identification of subresolution high temperature sources using a thermal IR sensor. *Photogram. Eng. and Remote Sensing, 47*, 1311-1318.

———, E. P. McClain, D. F. McGinnis, Jr., and J. A. Pritchard, 1978: Satellite detection of urban heat islands. *Mon. Wea. Rev., 106*, 1725-1734.

McGinnis, D. F., Jr., D. R. Wiesnet, and J. A. Pritchard, 1975: Determination of snow depth and snow extent from NOAA-2 satellite very high resolution data. *Water Resour. Res., 11*, 897-902.

Meeks, M. L. and A. E. Lilley, 1963: The microwave spectrum of oxygen in the earth's atmosphere. *J. Geophys. Res., 68*, 1683-1703.

Merritt, E. S. and C. Hall, 1973: Soil moisture estimation applications of Nimbus-3 HRIR-D (0.7-1.3 μm) observations. *Final Report, NOAA/NESS Contract No. 2-37098*, p. 19 and addend.

Mie, G., 1908: Beitragezur Optik Truber Medien, speziell Kolloidaler Metolosungeir. *Ann. der Physik., 25*, 377-445.

Monteith, J. L., 1973: *Principles of Environmental Physics*. E. Arnold Pub. Ltd., London, 28.

Morel, A., 1980: In-water and remote measurements of ocean color. *Bound. Layer Meteor., 18*, 177-201.

——— and L. Prieur, 1977: Analysis of variations in ocean color. *Limnol. Oceanogr., 22*, 709-722.

National Research Council, 1983: Carbon Dioxide Assessment Committee, *Changing Climate*. National Academy Press, Washington, DC, 1-4, 496.

Neiburger, M., 1949: Reflection, absorption, and transmission of insolation by stratus cloud. *J. Meteor., 6*, 98-104.

Nickerson, E. C. (ed.), 1986: Geophysical Monitoring for Climatic Change (GMCC) Summary Report through 1984. *Report #13*, Air Resources Laboratory, ERL, NOAA, Rockville, MD, 59 pp.

O'Brien, H. W. and R. H. Munis, 1975: Red and near-infrared spectral reflectance of snow. *Research Report 332*, Cold Regions Research and Engineering Lab., U.S. Army Corps of Engineers, Hanover, NH, 18 pp.

Olsen, R. L., D. V. Rogers, and D. B. Hodge, 1978: The aR^b relation in the calculation of rain attenuation. *IEEE Trans. Ant. Prop., AP-26*, 318-329.

Oort, A. H., 1983: Global atmospheric circulation statistics, 1958-73. *NOAA Professional Paper 14*, Dept. of Commerce, Washington, DC, 180 pp.

Palmen, E. and C. W. Newton, 1969: Atmospheric circulation systems. *Int. Geophy. Series, 13*, Academic Press, New York, NY, 603 pp.

Parrish, A., R. L. De Zafra, P. M. Solomon, J. W. Barrett, and E. R. Carlson, 1981: Chlorine oxide in the stratospheric ozone layer; ground-based detection and measurement. *Science, 211*, 1158-1160.

Paso, R., J. Kauppinen, and R. Anttila, 1980: Infrared spectrum of CO_2 in the region of the bending fundamental $_2$. *J. of Mol. Spect., 79*, 237.

Petterssen, S., 1956: *Weather Analysis and Forecasting*. Vols. I and II, McGraw-Hill Book Co., Inc., New York, NY, 266, 428 pp.

Phillips, O. M., 1977: *The Dynamics of the Upper Ocean*. Cambridge Univ. Press, New York, NY, 336 pp.

Poe, G., A. Stogryn and A. T. Edgerton, 1972: Microwave emission characteristics of sea ice. Aerojet Electrosystems Co., USDOC Contract No. 2-35340, Final Tech. Rep. No. 1749R-2. *NOAA Professional Paper l4*, 180 pp.

Preisendorfer, R. W., 1976: *Hydrologic Optics*. Dept. of Commerce, NOAA, Washington, DC, 7 vols.

Ramanathan, V., 1975: Greenhouse effect due to chlorofluorocarbons: Climatic implications. *Science, 190*, 50-52.

Rasmussen, R. A. and M. A. K. Khalil, 1985: Atmospheric Ozone 1985. WMO, Global Ozone Research and Monitoring Project, Report No. 16, Geneva, Switzerland, 1986.

Robinson, I. S., N. C. Wells, and H. Charnock, 1984: The sea surface thermal boundary layer and its relevance to the measurement of sea surface temperature by airborne and spaceborne radiometers. *Int. J. Remote Sensing, 5*, 19-45.

Rosenkranz, P. W., M. J. Komichak, and D. H. Staelin, 1982: A method for estimation of water vapor profiles by microwave radiometery. *J. Appl. Meteor., 21*, 1364-1370.

———, 1975: Shape of the 5mm oxygen band in the atmosphere. *IEEE Trans. Ant. Prop., AP-23*, 498-506.

Ryan, P. R., 1984: El Niño. *Oceans, 27*, 1-84.

Schanda, E., C. Matzler, and K. Kunzi, 1983: Microwave remote sensing of snow cover. *Int. J. Remote Sensing, 4*, 149-158.

Schawlow, A. L. and C. H. Townes, 1958: Infrared and optical masers. *Physical Review, 112*, 1940.

Schmugge, T. J., 1983: Remote sensing of soil moisture: recent advances. *IEEE Trans. Geos. Remote Sensing, GE-21*, 336-344.

———, 1980: Effect of soil texture on the microwave emission from soil. *IEEE Trans. Geos. Remote Sensing, GE-18*, 353-361.

Sellers, W. D., 1965: *Physical Climatology*. Univ. of Chicago Press, Chicago, IL, 21 pp.

Shettle, E. P. and R. W. Fenn, 1979: Models for the Aerosols of the Lower Atmosphere and the Effects of Humidity Variations on Their Optical Properties. *Environmental Research Papers, 676*, Air Force Geophysics Lab., Hanscom AFB, MA.

Shimabukuro, F. I., P. L. Smith, and W. J. Wilson, 1977: Estimation of the daytime and nighttime distribution of atmospheric ozone from ground-based millimeter wavelength measurements. *J. Appl. Meteor., 16*, 929-934.

Smith, R. C. and K. S. Baker, 198l: Optical properties of the clearest natural waters (200-800). *Appl. Opt., 20*, 177-l84.

——— and K. S. Baker, 1977: The bio-optical state of ocean waters and remote sensing. *Limnol. Oceanogr., 23*, 247-259.

———, 1974: Solar radiation in the upper layers of the Sea. N. G. Jerlov and E. S. Nielsen (eds.), *Optical Aspects of Oceanography*, Academic Press, New York, 95-119.

Smith, W. L., H. M. Woolf, C. M. Hayden, D. Q. Wark, and L. M. McMillin, 1979: The TIROS-N operational vertical sounder. *Bull. Amer. Meteor. Soc., 60*, 1177.

Spencer, R. W., W. S. Olson, W. Rongzhang, D. Martin, J. A. Weinman, and D. A. Santek, 1983: Heavy thunderstorms observed over land by the Nimbus-7 scanning multichannel microwave radiometer. *J. Clim. Appl. Meteor., 22*, 1041-1046.

Stogryn, A., 1967: The apparent temperature of the sea at microwave frequencies. *IEEE Trans. Ant. Prop., AP-15*, 278-286.

Thacker, D. L., G. J. Gibbins, P. R. Schwartz, and R. M. Devilacqua, 1981: Ground-based microwave observations of mesospheric H_2O in January, April, July, and September 1980. *Geophys. Res. Lett., 8*, 1059-1062.

Tucker, C. J., 1978: A comparison of satellite sensor bands for vegetation monitoring. *Photogram. Engin. and Remote Sensing, 44*, 1369-1380.

Ulaby, F. T., R. K. Moore, and A. K. Fung, 1986: *Microwave Remote Sensing—Active and Passive*. Volume III from Theory to Applications. Artech House, Dedham, MA, 2162 pp.

U.S. Committee for the Global Atmospheric Research Program, 1975: *Understanding Climatic Change: A Program for Action*. National Academy Press, Washington, DC, 239 pp.

U.S. Standard Atmosphere, 1976: U.S. Government Printing Office, Washington, DC, 227 pp.

U.S. Standard Atmosphere, 1962: U.S. Government Printing Office, Washington, DC, 278 pp.

Van de Hulst, H. C., 1957: *Light Scattering by Small Particles*. John Wiley and Sons Inc., New York, NY, 470 pp.

Van Vleck, J. H. and V. F. Weissopf, 1945: On the shape of collision broadened lines. *Rev. Mod. Phys., 17*, 227.

Wang, J. R., P. E. O'Neill, T. J. Jackson, and E. T. Engman, 1983: Multifrequency measurements of the effects of soil moisture, soil texture, and surface roughness. *IEEE Trans. Geo. Remote Sensing, GE-21*, 44-51.

——— and T. T. Schmugge, 1980: An empirical model for the complex dielectric permittivity of soils as a function of water content. *IEEE Trans. Geo. Remote Sensing, GE-18*, 288-295.

Wang, W. C., J. P. Pinto, and Y. Yung, 1980: Climatic effects due to halogenated compounds in the Earth's atmosphere. *J. Atmos. Sci., 37*, 333-338.

———, Y. Yung, A. Lacis, T. Mo, and J. Hansen, 1976: Greenhouse effects due to man-made perturbation of trace gases. *Science, 194*, 685-690.

Waters, J. W., 1976: Absorption and emission by atmospheric gases. *Methods of Experimental Physics: Vol. 12: Astrophysics, Part B*, M. L. Meeks (ed.), Academic Press, New York, NY, QB-173.

———, W. J. Wilson, and F. I. Shimabukuro, 1976: Microwave measurement of mesospheric carbon monoxide. *Science, 191*, 1174-1175.

———, K. F. Kunzi, R. L. Pettyjohn, and D. H. Staelin, 1975: Remote sensing of atmospheric temperature profiles with the Nimbus-5 microwave spectrometer. *J. Atmos. Sci., 32*, 1953-1969.

Webster, W., T. T. Wilheit, D. B. Ross, and P. Gloerson, 1976: Spectral characteristics of the microwave emission from a wind-driven foam-covered sea. *J. Geophys. Res., 81*, 3095-3099.

Weinreb, M. P., H. E. Fleming, L. M. McMillin, and A. C. Neuendorffer, 1981: Transmittances for the TIROS Operational Vertical Sounder. *NOAA Tech. Rep. NESS 85*, Dept. of Commerce, Washington, DC.

Westwater, E. R. and F. O. Guiraud, 1980: Ground-based microwave radiometric retrieval of precipitable water vapor in the presence of clouds with high liquid water content. *Radio Sci., 15*, 947-957.

World Climate Report 55, Report of WMO (CAS)/Radiation Commission of IAMAP Meeting of Experts on Aerosols and their Climatic Effects. Williamsburg, VA, March 1983, 28-30.

Wright, J. W., 1966: Backscattering from capillary waves with application to sea clutter. *IEEE Trans. Ant. Prop., AP-l4*, 749-754.

Yariv, A., 1975: *Quantum Electronics*. J. Wiley and Sons, New York, NY, 570 pp.

Contributors to Section II

Abel, Peter G.
*Braun, W. Charles
Clark, Dennis K
DeLeonibus, Pat S.
Grody, Norman C.
Jacobowitz, Herbert
Legeckis, Richard V.
McGinnis, David F. Jr.
McMillin, Larry M.
Neuendorffer, Arthur C.
*Reinhardt, George
Strong, Alan E.
Thomasell, Albert Jr.
Yates, Harold W.

NOAA/NESDIS
Washington, D.C. 20233

* (Moved to a new position)

Section III

NATIONAL OBSERVING CAPABILITIES

Weather satellites are now operated by the United States, the Soviet Union, the European Space Agency, Japan, and India. This section contains descriptions of the spacecraft and observing systems of these nations, and the observing capabilities of NASA's research and development satellites. Further details on the U.S. satellite systems are contained in Sections IV, V, and VI.

The polar orbiting satellite activities are coordinated through the International Polar Orbiting Meteorological Satellites (IPOMS) Committee created in 1985. Membership is open to all interested nations.

The equivalent coordinating mechanism for geosynchronous spacecraft is the Coordination of Geostationary Meteorological Satellites (CGMS). The U.S., European Space Agency, Japan, and the U.S.S.R. are members; the respective satellite views can be seen in Figure III-1. Although the Indian National Satellite (Insat), a multi-use, national purpose spacecraft, is not part of the CGMS it is described herein because it represents a major observing capability.

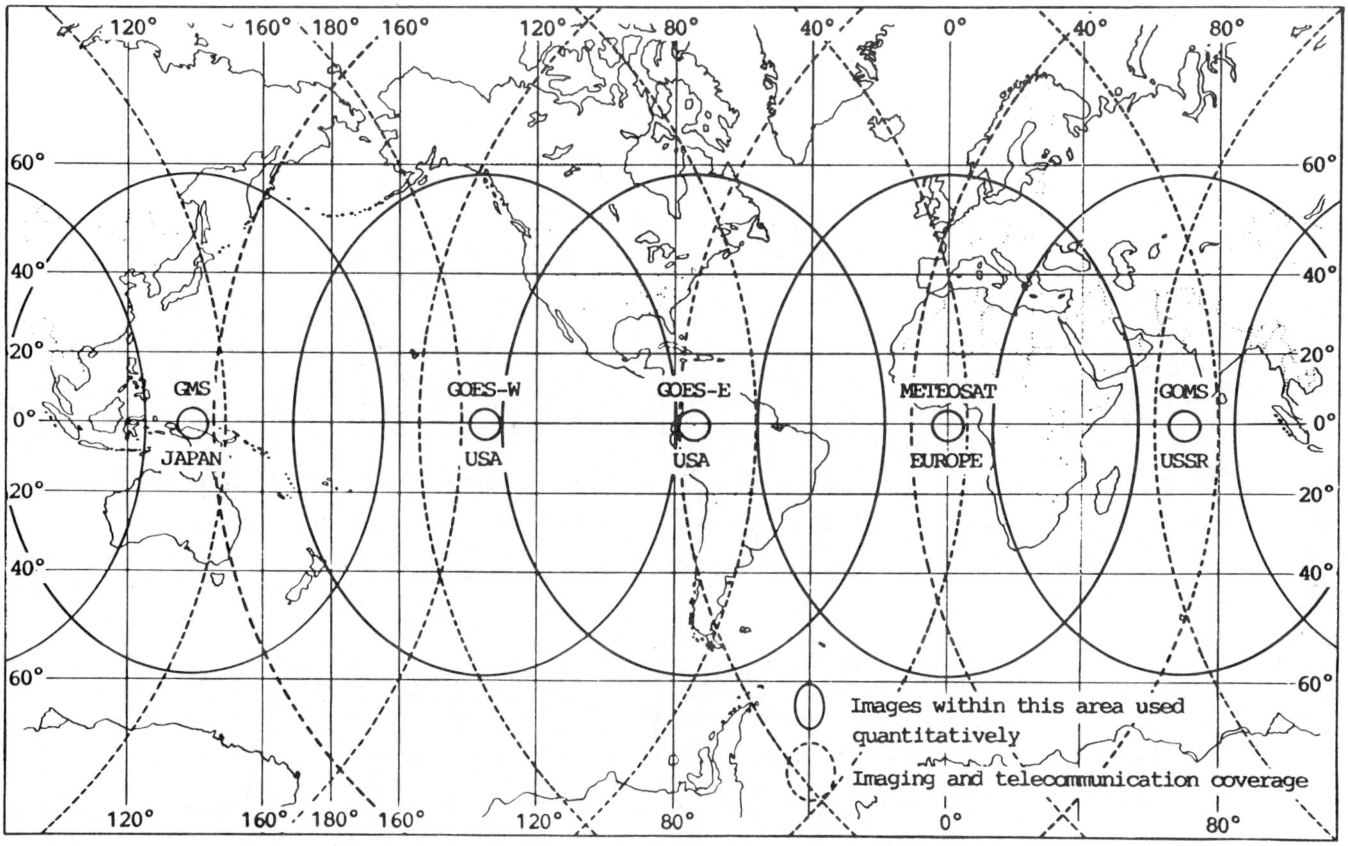

Figure III-1. Coverage of the geosynchronous satellites associated with the Coordination of Geostationary Meteorological Satellites (courtesy of CGMS).

Chapter III-1

THE UNITED STATES OPERATIONAL POLAR AND GEOSTATIONARY SATELLITES[1]

III-1.1 INTRODUCTION

The United States maintains a civil program to provide meteorological observations from spacecraft in both polar and geostationary orbits. The program is managed by the National Oceanic and Atmospheric Administration (NOAA) of the Department of Commerce. The National Environmental Satellite, Data, and Information Service (NESDIS), a subunit of NOAA, plans and operates the satellite systems, produces and distributes weather products based on satellite data, and archives the data.

III-1.2 THE POLAR ORBITING OPERATIONAL ENVIRONMENTAL SATELLITES (POES)

The successful launch of TIROS-1 on April 1, 1960, led to the establishment of the U.S. operational weather satellite system. The POES system, initiated with the February 3, 1966, launch of ESSA-1, is the direct descendant of the earlier polar spacecraft. See Chapter I-2 for a historical review of the polar orbiting satellites. NOAA POES spacecraft are designated NOAA-1, 2, 3, etc. Spacecraft in prelaunch state are designated NOAA-A, B, C, etc. After establishment in orbit, the alphabetical designators are replaced by numerals. For example, NOAA-F, launched on December 12, 1984, became NOAA-9 after being declared operational.

III-1.2.1 POES Mission and Orbit

The primary POES mission is to provide daily global observations of weather patterns and environmental conditions in the form of quantitative data usable for numerical weather analysis and prediction. POES spacecraft are used to observe and derive cloud cover, ice and snow coverage, surface temperatures, vertical temperature and humidity profiles, and other variables. The system also is equipped to detect and relay Search And Rescue (SAR) signals from downed aircraft and ships in distress; measure solar particle flux along the orbit; relay data from fixed or moving environmental observation platforms; determine the Earth locations of moving platforms; and provide continuous data broadcasts for ground receiving stations.

POES orbits are selected to provide maximum global coverage within the limitations imposed by the communications and data handling facilities and the timeliness needs of data users. The orbit is Sun synchronous

[1] Primary Contributor: Daniel J. Cotter

and circular at a mean 850 km altitude. Two nominal altitudes, 833 and 870 km, have been chosen to permit concurrent operation of two satellites. At the 833 km altitude, the orbital inclination is 98.739°, the nodal period 101.58 min, and the number of orbits per day 14.18. At the 870-km altitude, these parameters are 98.899°, 102.37 min, and 14.07 orbits per day. The current POES mode of operation requires two operational satellites in orbit; a replacement is injected into the orbit of a failed satellite about 4 months after its failure. Under normal conditions, one satellite will be in an orbit with a southbound Equator crossing time at about 0730 Local Solar Time (LST), the other with a northbound Equator crossing at about 1430 LST. The orbit-by-orbit longitude of Equator crossing varies with the rotation of the Earth.

III-1.2.2 The Advanced TIROS-N (ATN) Spacecraft

The current POES spacecraft is the Advanced TIROS-N (ATN). ATN was first used for the NOAA-8 mission (March 1983) and for every POES mission since. Except for NOAA-D, which will be taken out of long-term storage and scheduled for a 1989 launch, all POES will be ATN through the early 1990s.

The ATN is 4.2 m long, 1.9 m in diameter, and carries a 2.4 × 4.9 m solar array for power. Except for expendable equipment, the spacecraft weighs about 1010 kg and carries a payload of about 375 kg. The solar array is rated at a minimum of 515 W at the worst solar angle; this provides some 40 W of reserve for an expected full operational power requirement of about 475 W. Attitude control can be maintained to within 0.2° about each axis; operating attitude can be determined to within 0.14°. An Atlas booster is used to orbit the ATN. Each ATN is expected to operate for at least 2 years.

Figure III-1-1 shows the major components of the ATN and the spacecraft locations of primary sensors. As the sensor complement is not identical on every POES flight, this figure shows a nominal, rather than an actual, configuration.

III-1.2.3 Primary POES Sensors

The POES sensor payload varies from mission to mission, based on in-orbit experience and changing requirements. Each has an atmospheric sounding capability and a high resolution imaging capability. The sensors described below are (or will be) used on satellites of the POES system currently (and for the foreseeable future). Details on current sensors are given in Chapter IV-2.

The TIROS Operational Vertical Sounder (TOVS). TOVS is a three instrument system consisting of:

- The High-resolution Infrared Radiation Sounder (HIRS/2). The HIRS/2 is a 20 channel instrument for taking atmospheric measurements, primarily in the IR region. The data acquired can be used to compute atmospheric temperatures from the Earth surface to 50 mb, water vapor content in three atmospheric layers, and the total ozone content of the atmospheric column. HIRS/2 will be replaced by an improved version onboard NOAA-K, L, and M in the 1990s.

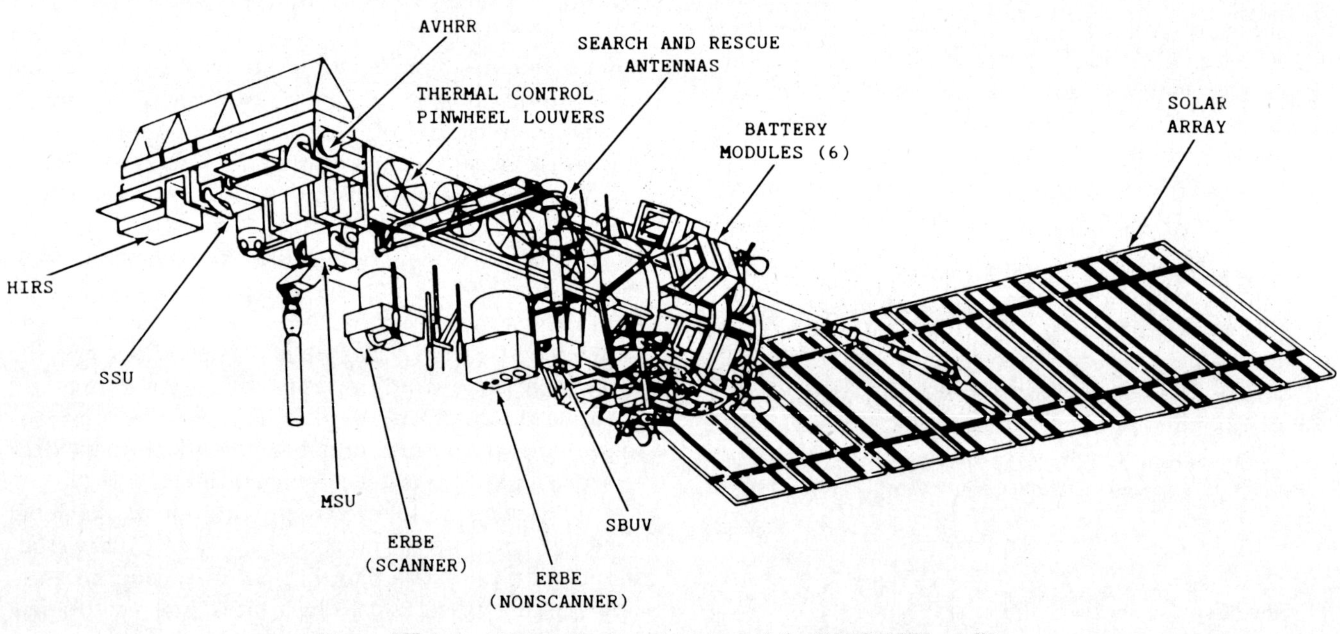

Figure III-1-1. The nominal configuration of the POES ATN satellites.

- The Stratospheric Sounding Unit (SSU). The SSU is a three channel instrument, provided by the United Kingdom, that uses the selective absorption technique. The pressure in a carbon dioxide gas cell in the optical path determines the spectral characteristics of each channel, and the mass of CO_2 in each cell determines the atmospheric level at which the weighting function of each channel peaks. It will be replaced by an advanced passive MW sensor in the 1990s. (Not all POES carry this instrument.)
- The Microwave Sounding Unit (MSU). This four channel Dicke radiometer makes passive MW measurements in the 5.5-mm O_2 band. Unlike the IR instruments of TOVS, the MSU is little influenced by clouds in the field of view. An MSU is onboard NOAA-11, launched in 1988, and will be onboard NOAA-D in 1990, NOAA-I in 1991, and NOAA-J in 1992. After that, the MSU will be replaced (on "K", "L", and "M") by the Advanced Microwave Sounding Unit (AMSU).

The Advanced Very High Resolution Radiometer (AVHRR). The AVHRR five channel scanning radiometer with 1.1-km resolution is sensitive in the VIS, the near-IR, and the IR "window" regions. This instrument will be carried through NOAA-J. NOAA-K, L, and M will have a similar instrument with six channels and other improvements. AVHRR data are tape-recorded onboard the spacecraft for readout at the Fairbanks and Wallops Command and Data Acquisition stations. These data can be recorded at 1.1-km resolution (the basic resolution of the AVHRR instrument) or at 4 km resolution. The stored high resolution (1.1-km) imagery is known as Local Area Coverage (LAC). Owing to the large number of data bits, only about 11+ min of LAC can be accommodated on a single recorder. In contrast, 115 min of the lower resolution (4-km) imagery, called Global Area Coverage (GAC), can be stored on a recorder, enough to cover an entire 102 min orbit of data.

The Space Environment Monitor (SEM). The SEM consists of two separate instruments and a data processing unit. The Total Energy Detector (TED) measures a broad range of energetic particles in 11 bands that range from 0.3 keV to 20.0 keV. The Medium Energy Proton and Electron Detector (MEPED) senses protons, electrons, and ions with energies from 30 keV to several tens of MeV. The SEM is scheduled for POES flights with northbound afternoon Equator crossing times.

III-1.2.4 Other POES Sensors

The Solar Backscatter Ultraviolet Radiometer (SBUV). The SBUV is a non-scanning, nadir viewing instrument designed to measure scene radiance in the spectral region from 160 to 400 nm. SBUV data are used to determine the vertical distribution of and total ozone in the atmosphere, and solar spectral irradiance. The instrument is normally flown on POES missions with afternoon northbound Equator crossing times.

The Earth Radiation Budget Experiment (ERBE). The ERBE is a NASA research instrument, two of which have been flown (on NOAA-9 and NOAA-10). The ERBE consists of two parts. The first is a fixed field of view instrument with four Earth viewing radiometers and a Sun viewing radiometer equipped with shutters. Two of the Earth viewing radiometers monitor broad-band outgoing Earth radiation from 0.2 to 50 μm; the other two monitor outgoing shortwave radiation from 0.2 to 5.0 μm. The second part is a three channel scanner with an instantaneous field of view of about 3°. The three channels are: 0.2 to 50 μm; 0.2 to 5.0 μm; and 5 to 50 μm. ERBE data contribute to understanding the total and seasonal planetary albedo and Earth radiation balances, zone by zone. This information is used for recognizing and interpreting seasonal and annual climate variations and contributes to long-term climate monitoring, research, and prediction.

III-1.2.5 Search and Rescue Satellite Aided Tracking (SARSAT) System

SARSAT is part of an international program to save lives. SARSAT equipment on POES is provided by Canada and France. Similar Soviet Union equipment, called COSPAS (Space System for the Search of Distressed Vessels), is carried on the polar orbiting spacecraft of their satellite navigation system.

SARSAT and COSPAS spacecraft relay emergency radio signals from aviators, mariners, and land travelers in distress to ground stations, or Local User Terminals (LUT), where the location of the distress signal transmitter is determined by Doppler shift techniques. Information on the nature and location of the emergency is then passed to a Mission Control Center (MCC) that alerts the Rescue Coordination Center (RCC) closest to the emergency. In the United States, the Air Force operates the USMCC; actual search and rescue is conducted by the Air Force or the Coast Guard, depending on the location of the emergency. NOAA will take over the USMCC by 1990.

SARSAT and COSPAS monitor a common frequency of 121.5 MHz. SARSAT also receives the military distress frequency of 243.0 MHz. An experimental frequency of 406.025 MHz is being evaluated for its ability to acquire location accuracies of 2 to 5 km, compared to the 8 to 16 km of the current system.

III-1.2.6 The Argos Data Collection and Location System

The Centre National d'Études Spatiales (CNES) of France builds and furnishes the data collection and location system known as Argos. Argos systems on POES spacecraft provide a random access system by means of which environmental data sensed at remote platforms can be collected by POES and retransmitted to ground processing stations. The remote platforms may be fixed or moving; the location of a moving platform is computed by using differential Doppler shift techniques.

Data from these platforms, mixed with other satellite data, are received at NESDIS where the platform messages are separated and relayed to the Service Argos U.S. Processing Center, Lanham, MD, for further processing, determination of platform location, and relay to users. About two-thirds of the remote platform data are received by users within 3 hours of their acquisition by the satellite. These data are also distributed over the Global Telecommunications System (GTS) of the WMO. POES will carry Argos well into the 1990s.

III-1.2.7 The Direct Broadcast System (DBS)

Data from the TOVS, AVHRR, and SEM instruments onboard POES satellites are broadcast continuously and can be received by any ground station within radio range. There are three categories of direct broadcast service.

Automatic Picture Transmission (APT). APT transmits data from two channels of the AVHRR at a reduced resolution of 4 km. The analog APT transmission consists of 120 lines of data per minute at 137.50 or 137.62 MHz. The VIS channel is used for daytime data and one IR channel for day and night data; a second IR channel is available for additional night coverage, and, in general, channels can be changed to meet requirements. Data from each channel are blocked into half lines, so an unaltered image of the transmission shows the two in side-by-side frames. Relatively simple and inexpensive ground facilities are adequate for receipt and use of APT transmissions. More than 1200 APT ground stations are known to be in operation globally.

High Resolution Picture Transmission (HRPT). Data from all five channels of the AVHRR are provided at full 1.1-km resolution by S-band HRPT at 1698.0 and 1707.0 MHz. These digital transmissions also contain TOVS, SEM, Data Collection System, and spacecraft telemetry information. Receiving stations, some 100 worldwide, use the AVHRR data provided by HRPT for imagery and for quantitative applications.

Direct Sounder Broadcasts (DSB). Ground stations can receive real time sounding data acquired by TOVS through two separate POES communications links. TOVS data are available in the HRPT S-band data stream and in the POES VHF beacon transmissions at 136.77 or 137.77 MHz. TOVS digital data are identical in both transmissions. The VHF beacon transmission permits users without S-band equipment to obtain TOVS data at a rate of 8 kbps, rather than the 665.4-kbps rate of HRPT.

III-1.2.8 POES Communication Links

POES communication links include uplinks for commanding the satellite, receiving data collection platform messages, and receiving SAR distress signals. Downlinks include the APT, the HRPT, and the VHF beacon links of the DSB. Additional downlinks are for the SAR and for the HRPT transmissions to NOAA data acquisition ground stations.

Global data tape recorded on POES spacecraft are played back as the satellites pass over the NOAA ground stations at Fairbanks, Alaska, and Wallops, Virginia, and over the CNES ground station at Lannion, France. The Lannion station receives stored data from orbits that cannot be acquired by the Fairbanks and Wallops stations. These data and the accompanying spacecraft telemetry are then retransmitted through commercial communications satellites to the NESDIS data processing center at Suitland, Maryland.

III-1.2.9 Planned POES Improvements and Changes

Early in the 1990s, NOAA plans to replace the SSU and MSU instruments of the TOVS with the AMSU. This two component MW instrument will be carried on NOAA-K, L, and M. AMSU/A, a 15-channel instrument covering the 20 to 90 GHz range, will provide most of the data for temperature soundings. The United Kingdom is to provide AMSU/B, a five channel instrument covering the 90 to 184 GHz range, to obtain data for atmospheric humidity profiles, delineations between sea ice and open water, and information on snow thickness and soil moisture. The TOVS HIRS/2 will be retained, possibly with some channel changes. The AMSU will improve the POES sounding capability with more reliable data in cloudy areas.

Together with the AMSU, NOAA plans to add a sixth channel to the AVHRR and possibly make other modifications. HRPT and direct sounder transmission will be modified to include these new data, but APT will remain essentially unchanged.

Addition of the AMSU will require a change in the telemetry bands used for direct-to-user broadcasts. One of the beacon frequencies now used (136.77 MHz) will be reallocated for commercial use in 1990. Various options are being considered to assure sufficient data flow, and minimal effects on spacecraft and ground station telemetry hardware, and on global direct broadcast data users.

Beginning with NOAA-11, launched late in September 1988, the time for Equatorial crossings changed from 1430 to 1340 LST. The change required a redesign of spacecraft thermal controls and instrument shades because the spacecraft sees the Sun constantly at higher viewing angles during all sunlit flight times. The change coordinates Eastern Pacific data collection with a U.S. National Weather Service rescheduling of the start up time of its numerical prediction runs.

III-1.3 THE GEOSTATIONARY OPERATIONAL ENVIRONMENTAL SATELLITE (GOES)

The first geostationary weather satellite was the NASA ATS-1, launched in December 1966. Today's operational geostationary satellites are part of the series that started with the May 1974 launch of NASA SMS-1. The first NOAA geostationary operational weather satellite, the GOES-1, was launched on October 16, 1974.

III-1.3.1 GOES Mission and Orbit

The GOES mission is to provide the nearly continuous, repetitive observations that are needed to predict, detect, and track severe weather. GOES spacecraft are equipped to observe the Earth disk in the VIS and IR regions of the spectrum. GOES spacecraft observe and measure cloud cover, surface conditions, snow and ice cover, surface temperatures, and the vertical distributions of atmospheric temperature and humidity. They are also instrumented to measure solar X-rays and other energetics, collect and relay environmental data from platforms, and broadcast instrument data and environmental information products to ground stations.

GOES spacecraft operate at the geosynchronous altitude of 35,800 km above the surface. Two[2] GOES provide overlapping coverage that centers on the United States and extends over the eastern Atlantic Ocean and the western Pacific Ocean (Figure III-1). GOES-East is located over the Equator at 75°W; GOES-West is at 135°W.

III-1.3.2 The GOES Spacecraft

The current GOES spacecraft (Figure III-1-2) consists of a vertical cylinder covered with solar cells, containing most of the instrumentation and an antenna assembly that remains directed toward Earth. The spacecraft is 6.86 m long, including the antenna assembly; the outside diameter of the main cylinder is

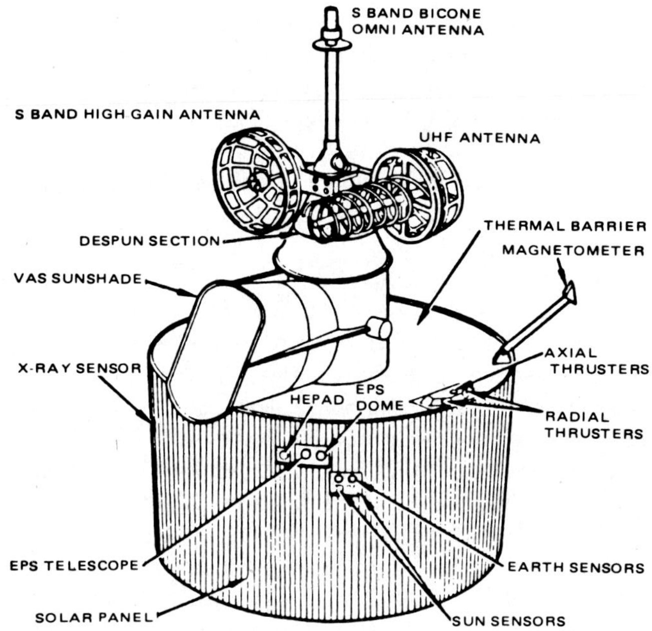

Figure III-1-2. Sensor configuration of the GOES-D to H spacecraft.

3.33 m. The spacecraft weighs 346 kg and has a design lifetime of seven years. The solar panels provide 330 W of power. The GOES spacecraft spins at 100 rpm about its vertical axis, which is oriented parallel to the Earth's axis.

III-1.3.3 GOES Sensors

The Visible and Infrared Spin Scan Radiometer (VISSR) Atmospheric Sounder (VAS). The VAS is a VIS and IR radiometer that provides image (VISSR mode) and dwell sounding (VAS mode) data. In the imaging mode, west to east scan lines are produced as the spin of the spacecraft views the Earth through the instrument's scan mirror. North to south scanning is done by stepping the scan mirror between Earth views. Atmospheric sounding data are obtained in the VAS mode by repeatedly scanning the same Earth swath; from 1 to 46 repeated views of an atmospheric column are needed to achieve a signal to noise ratio sufficiently large for sounding computations. The dwell time demands of the sounding mode do not permit simultaneous imaging and sounding with VAS. In the multispectral imaging mode, it is possible to collect and transmit VIS data for every scan line, together with views of the scene in up to three different IR bands.

The VAS instrument contains a Ritchey-Chretien optical system with two focal planes. Visible spectrum signals are obtained at the principal focus and relayed by fiber optics to photomultiplier tubes. IR signals are relayed to solid state detectors in the cooled focal plane

[2] In the event of a satellite failure, the remaining satellite is moved to a central U.S. longitude to provide maximum coverage.

of the instrument. A filter wheel positioned in the optical path permits selection from any of 12 IR spectral bands with central wavelengths between 3.9 and 15.0 μm. The instrument resolution is 1 km in the VIS range and 7 km or 14 km in the IR ranges.

During normal imaging operations, each GOES provides full Earth disk image data every 30 min (20 min for scan, 10 min for retrace) with 1-km VIS data resolution during daylight and 7-km IR data resolution during daylight and darkness. For observing unusual weather situations, more frequent scanning is possible, but for reduced geographic areas.

The Space Environment Monitor (SEM). The SEM has four subsystems. The Solar X-ray Sensor detects activity in 0.5 to 3.0 A and 1 to 8 A bands. Charged particle flux is measured by the Energetic Particle Sensor (EPS), which detects protons in the 0.8 to 500 MeV region over 7 log ranges, alpha particles over 6 log ranges in the 3.2 to 400 MeV region, and electrons with energies ≥2 MeV. The High Energy Proton and Alpha Detector (HEPAD) detects protons and alpha particles with energies ≥850 MeV. The magnetometer measures the magnitude of the ambient magnetic field parallel and transverse to the spacecraft spin axis.

III-1.3.4 The GOES Data Collection System (DCS)

The GOES Data Collection System (DCS) gathers and relays environmental data transmissions that originate at remote automatic data collection platforms. These platforms can be set to transmit data in four modes: at fixed times; in response to DCS interrogation; in an emergency mode, during which the platform signals for a DCS interrogation within 1 min; and in an adaptive random reporting mode. In this last mode, platform transmissions are made at random times but are subject to transmission conflicts. To compensate for conflicts, random reporting platforms transmit data more often than would otherwise be necessary. This mode is "adaptive" in that the reporting rate can be keyed to the level of environmental threat observed at the platform. This flexibility and the relatively lower cost of random reporting equipment make this newest mode the most frequent choice when equipping or modifying platforms.

Platforms measure and report on atmospheric state, ocean state, river levels, soil moisture, and seismological events. Platform data are relayed through the DCS to the Wallops, Virginia, Command and Data Acquisition Station (CDAS) and then to the Central Data Distribution Facility (CDDF) in near-real time. At the CDDF, the data are sorted and distributed to the organization that funds the platform and to other users. NOAA users who operate large numbers of platforms usually rely on their own ground stations for immediate access to data from their platforms. Beginning in 1988, platform data distribution will be centralized at the Wallops facility for delivery to all users.

III-1.3.5 GOES Direct Broadcast Services

The two types of direct broadcast services provided by GOES can be intercepted by any receiving station within range. Both are S-band. One is for sensor and spacecraft data. The other transmits environmental information in WEFAX format.

Stretched Sensor Data. VISSR VAS data are acquired and transmitted to the Wallops ground station while the Earth is within view of the instrument (roughly 20° of spacecraft rotation). During the remainder of each rotation, the ground station puts these data in order, appends calibration and Earth location information, and retransmits them through the satellite at a bit rate reduced from about 28 to 2.1 Mbps. These "stretched" sensor data are retransmitted at 1687.1 MHz.

The data are stretched so that ground stations can receive and process GOES sensor data with less complex and expensive equipment than would otherwise be needed. This transmission is for internal use by NOAA and, while available to any user, is not considered a public service.

WEFAX. The weather facsimile (WEFAX) service broadcasts processed environmental information at 1691.0 MHz. A daily operational message confirms schedules and contents. Standard meteorological charts and processed GOES imagery at 8-km resolution and POES imagery at resolutions ranging from 8 to 12 km are broadcast by WEFAX.

III-1.3.6 GOES Communications Links

The GOES communications subsystem sends commands to the spacecraft and provides uplinks for WEFAX and data collection platform transmissions. Downlinks from the spacecraft are for raw sensor and telemetry data, stretched sensor data, data collection platform data, data collection platform interrogation, and WEFAX.

III-1.3.7 The GOES-I to M Series

GOES-H, the last in the current NOAA GOES series, was launched in February 1987. In 1985, the Ford Aerospace and Communications Corporation was chosen to design and build the follow-on series, GOES-I to M, for use during the 1990s. The first is to be ready for launch in 1991.

GOES-I to M will have a sensing capability for simultaneous imaging and sounding. Additional channels, with sharper responses in some sounding chan-

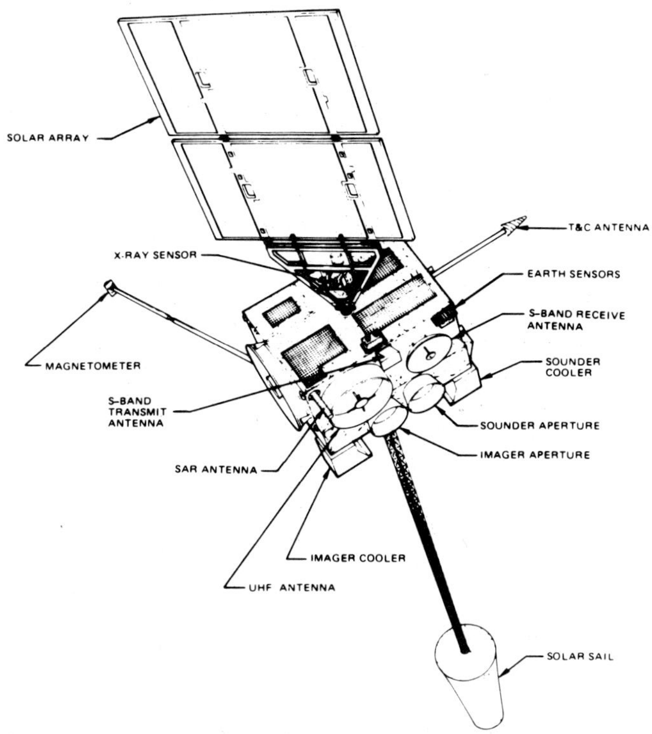

Figure III-1-3. Configuration of the GOES-I to M satellites (a proportionally longer solar array will be included on flight models of these spacecraft).

Table III-1-1 *Spectral coverage and resolution, GOES-I to M imager*

SPECTRAL BANDS (μm)	SUBPOINT RESOLUTION (km)
0.55–0.75	1
3.80–4.00	4
6.50–7.00	8
10.20–11.20	4
11.50–12.50	4

nels, also will be provided. Three-axis stabilization will permit maintenance of unobstructed, continuous views of Earth for the separate imager and sounder instruments.

Each satellite consists of a six sided body (Figure III-1-3) to carry the instrument payload, a continuous drive solar array, a solar sail to offset solar pressure torque, an omnidirectional telemetry and command antenna, and the Space Environment Monitor (SEM)

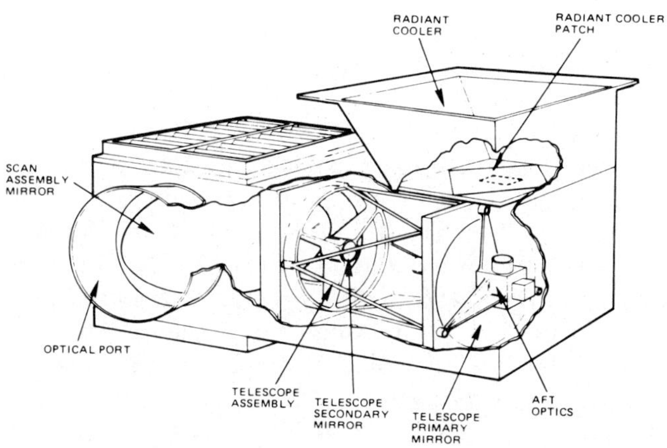

Optical FOV	Square		
Channels Detectors and IGFOV	Visible	Silicon	1 km
	Shortwave	InSb	4 km
	Moisture	HgCdTe	8 km
	Longwave 1	HgCdTe	4 km
	Longwave 2	HgCdTe	4 km
FOV defining element	Detector		
Channel-to-channel alignment	14 μr (0.5 km)		
Radiometric calibration	Space and 290K IR blackbody (varies with housing temperature)		
Signal quantizing	10 bits, all channels		
Scan capability	Full Earth, sector, area		

Figure III-1-4. GOES-I to M imaging instrument.

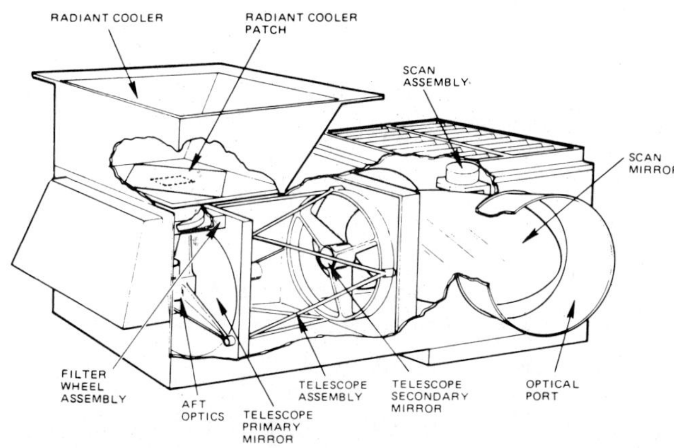

FOV defining element	Field stop
Telescope aperture	30.5 cm (12 in) diameter
Channel definition	Interference filters
Radiometric calibration	Space and 290K - IR blackbody
Field sampling	4 areas N-S on 10 km centers
Nominal IGFOV	215 μrad (all channels)
Scan step angle	280 μrad (110 km nadir)
Step and dwell time	0.1 seconds
Scan capability	Full Earth and space
Sounding areas	10 km x 40 km to 60° N-S and 60° E-W
Signal quantizing	12 bits all channels
Output data rate	29920 bits per second

Channels		Detector Type	Nominal Square IGFOV (μrad)
1-7	(Longwave IR)	HgCdTe	215
8-12	(Mid-wave IR)	HgCdTe	215
13-18	(Shortwave IR)	InSb	215
19	(Visible)	Silicon	215

Figure III-1-5. GOES-I to M sounder instrument.

Table III-1-2 *Spectral characteristics, GOES-I to M sounder*

CHANNEL	METEOROLOGICAL SIGNIFICANCE	CENTRAL WAVENUMBER (CM^{-1})	HALF POWER BANDWIDTH (CM^{-1})
Longwave (LW)			
1	Temperature Soundings	679.0	10
2		691.0	12
3		700.0	10
4		710.0	10
5		735.0	13
6		748.0	16
7		832.0	20
Midwave (MW)	Surface Temp. & Cloud Detection		
8		907.0	50
9	Tot. Ozone Amt.	1035.0	35
10	Water Vapor Sounding	1225.0	60
11		1365.0	50
12		1467.0	140
Shortwave (SW)			
13	Temperature Soundings	2190.0	35
14		2213.0	35
15		2240.0	35
16		2275.0	35
17	Surface Temp. & Cloud Detection	2520.0	100
18		2671.0	100
Visible (VIS)			
19	Cloud Detection	14367.0	1000

magnetometer. Thermal control will be by use of passive techniques, heaters, and louvers. The solar array, a two panel assembly, is attached to a continuously rotating drive assembly that also functions as a Sun tracking platform for the X-ray sensor of the SEM.

In the GOES-I to M design, two independent instruments will be used; one for radiometric imaging, the other for radiometric sounding. Their common components—telescopes, scanners, and coolers—will be identical. This feature and design improvements (such as advanced bearings, lubricants, and anticontamination techniques) will reduce risks. Redundant electronics and other life stretching measures will be used to achieve the specified 5 year operational lifetime.

Direct readout ground stations will receive broadcasts from GOES-I to M at the same radio frequencies, data rates, and in the same data formats as those now used.

The GOES-I to M imager (Figure III-1-4) will be a calibrated, five channel scanning radiometer, with one VIS channel at 1-km resolution and four IR channels, three at 4-km resolution and one at 8-km resolution. Scan can be controlled to provide coverage ranging from full Earth disk, with space background in 30 min, to small area scans of 1,000 × 500-km segments in 20 sec. The imager will include star sensing for precise Earth location of data. The spectral coverage of the imager is shown on Table III-1-1 and Figure III-1-4.

The GOES-I to M sounder (Figure III-1-5) will be a 19 channel (18 IR, 1 VIS) scanning radiometer with a filter wheel for channel selection. The resolution will be a 10 × 40-km FOV area between 60°N and 60°S, although the scan capability will include the full Earth disk and space. Area and frequency of scan will be fully programmable. The scanning mirror will step in 10-km Earth increments, with a step and dwell time of 0.1 sec. The sounder and imager will also include star sensing. The spectral characteristics of the sounder are shown in Table III-1-2 and Figure III-1-5.

GOES-I to M will continue the SEM DCS service of the current GOES series. Users of the DCS will be able to receive information directly by intercepting GOES rebroadcasts, or from NOAA through scheduled distributions. WEFAX services also will be continued.

The SAR capability of GOES-H will be continued on GOES-I to M. The search and rescue transponders of GOES-I to M will permit the instantaneous detection of distress signals from the Earth disk below and greatly speed the alerting of rescue forces. The polar SARSAT system will still be required to determine location.

GOES-I to M will deal with pixel registration by use of the spacecraft microprocessor system to compensate for variations in mirror motions and other pointing disturbances. This will dispense with the need to dampen disturbances by mechanical means.

Solution of the Earth location problem is also computational. On the spacecraft, the scene in the instrument IFOV will be located relative to the instrument optical axis. These azimuth and elevation coordinates then will be transformed to the orbital axis of the spacecraft. Accuracy of the spacecraft axis position will be maintained by using radio ranging, landmark, and star positions processed at the NOAA CDAS. These will be converted to spacecraft ephemeris data and transmitted to the spacecraft processor. Scene Earth location will be established by combining these sets of reference.

The GOES-I to M design will have space and power for an additional sensor package, such as a solar X-ray imager.

Chapter III-2

THE RUSSIAN SATELLITES[1]

III-2.1 INTRODUCTION

In April 1967, the Congress of the World Meteorological Organization (WMO) met in Geneva, Switzerland, to discuss a world weather service. Although the meeting coincided with the flights of the Cosmos-144 and 156 meteorological satellites, the U.S.S.R. did not announce until two months later that these satellites were part of the new "Meteor" system of weather reporting (TASS, 1967).

The Meteor system included weather satellites, stations to receive and process the data, a service to control the operations of the satellites, and a regular prediction service. The data were recorded and stored for transmission to Earth by radio when the satellites were over Soviet territory. Later satellites carried Automatic Picture Transmission (APT) equipment making real time pictures available anywhere within receiving range.

Three centers receive satellite information: Moscow, Novosibirsk, and Khabarovsk. Individual satellite pictures are made into a large photo-mosaic of the area under study. Satellite data used by the Soviet Hydrometeorological Service are instantly relayed to the World Meteorological Organization (TASS, 1970).

During the years 1962–66, 14 Cosmos satellites were launched, primarily as experimental spacecraft. Following the 1967 launches of Cosmos-144 and 156, six basically meteorological Cosmos were launched. Table III-2-1 gives details of those satellites known to have carried meteorological equipment.

The Soviet Union signaled the completely operational status of a weather satellite system in 1969 when it began naming certain satellites Meteor.

III-2.2 METEOR

The day the first Meteor was launched, it was identified as a weather satellite reporting on cloudiness, snow and ice cover, and Earth and atmosphere radiation both day and night (TASS, 1969; see also Figure III-2-1).

The original satellites of the system were named Meteor-1, 2, 3, etc. This series continued through Meteor-31 in 1981 (Table III-2-2). Since the early days of the Meteor program there have been a number of modifications to its payloads. Meteor-8 contained spectrometric apparatus to determine the vertical temperature profile of the atmosphere to approximately 35 km.

Although no mention of an APT system was made in the TASS announcement of the launch of Meteor-10, its APT signals were recorded in Kettering, England, on February 3, 1972. Meteor-10 was also used to test ion-plasma electrojet engines. These orbital correction engines were developed to guarantee stabilizing capability for long duration flights (Andronov, 1972). With these engines, it was possible to change the altitude of the satellite by 16.9 km (Artsimovich et al., 1974).

[1] Primary Contributor: Geoffrey E. Perry, M. B. E., The Kettering Group. Original manuscript submitted: December 1984.

Table III-2-1 *Cosmos satellites contributing to the development of the Meteor system*

SATELLITE	LAUNCH SITE	LAUNCH VEHICLE	LAUNCH DATE	INCLINATION (DEG)	ORBITAL PERIOD (MIN)	APOGEE (KM)	PERIGEE (KM)	DECAY DATE	REMARKS
Cosmos-4	T	A-1	26 Apr 62	65.0	90.6	330	298	29 Apr 62	First recoverable Cosmos and first cloudcover television.
Cosmos-7	T	A-1	28 Jul 62	65.0	90.1	369	210	1 Aug 62	Recovered.
Cosmos-9	T	A-1	27 Sep 62	65.0	90.9	353	301	1 Oct 62	Recovered.
Cosmos-14	K	B-1	13 Apr 63	49.0	92.1	512	256	29 Aug 63	Test bed for solar batteries and attitude stabilization system.
Cosmos-15	T	A-1	22 Apr 63	65.0	89.8	371	173	27 Apr 63	Recovered.
Cosmos-23	K	B-1	13 Dec 63	49.0	92.9	613	240	27 Mar 64	Similar to Cosmos-14.
Cosmos-44	T	A-1	29 Aug 64	65.0	99.5	860	618	—	Possibly the first test of a complete system.
Cosmos-45	T	A-2	13 Sep 64	64.9	89.7	327	206	18 Sep 64	UV and IR measurements. Recovered.
Cosmos-58	T	A-1	26 Feb 65	65.0	96.8	659	581	—	Similar to Cosmos-44.
Cosmos-65	T	A-2	17 Apr 65	65.0	89.8	342	210	25 Apr 65	Recovered. Similar to Cosmos-45.
Cosmos-92	T	A-2	16 Oct 65	65.0	89.9	353	212	24 Oct 65	Recovered. Similar to Cosmos-45.
Cosmos-100	T	A-1	17 Dec 65	65.0	97.7	650	650	—	Similar to Cosmos-44.
Cosmos-118	T	A-1	11 May 66	65.0	97.1	640	640	—	Similar to Cosmos-44.
Cosmos-122	T	A-1	25 Jun 66	65.0	97.1	625	625	—	Developmental satellite with operational infrared sensors.
Cosmos-144	P	A-1	28 Feb 67	81.2	96.9	625	625	14 Sep 82	First operational weather satellite from Plesetsk.
Cosmos-149	K	B-1	21 Mar 67	48.4	89.8	297	248	7 Apr 67	Aerodynamic stabilization.
Cosmos-156	P	A-1	27 Apr 67	81.2	97.0	630	630	—	Similar to Cosmos-144.
Cosmos-184	P	A-1	27 Oct 67	81.2	97.1	635	635	—	Used ion detectors as orientation aid.
Cosmos-206	P	A-1	14 Mar 68	81.0	97.0	630	630	—	Similar to Cosmos-144.
Cosmos-226	P	A-1	12 Jun 68	81.2	96.9	650	603	18 Oct 83	Similar to Cosmos-144.
Cosmos-243	T	A-2	23 Sep 68	71.3	89.6	319	210	4 Oct 68	Supplemental meteorological payload. Recovered.
Cosmos-320	K	B-1	16 Jan 70	48.5	90.0	342	240	10 Feb 70	Aerodynamic stabilization.
Cosmos-384	P	A-2	10 Dec 70	72.9	89.5	314	212	22 Dec 70	Similar to Cosmos-243. Recovered.

Notes: 1. Orbital data given in the original *TASS* announcement of the launch. 2. Launch sites denoted by T = Tyuratam (45.6°N, 63.4°E); K = Kapustin Yar (48.4°N, 45.8°E); and P = Plesetsk (62.8°N, 40.1°E). 3. Launch vehicles indicated by alpha-numeric designators due to Charles S. Sheldon II in *TRW Space Log,* Vol. 8, No. 4, pp. 2–23, Winter 1968–69. 4. Decay/recovery dates taken from the NASA Satellite Situation Report, Vol. 24, No. 1, Goddard Space Flight Center, February 29, 1984. 5. Prepared by Geoffrey E. Perry, MBE, The Kettering Group.

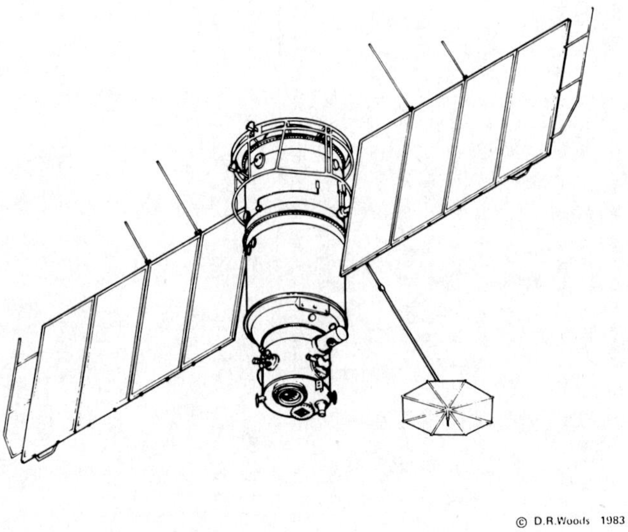

Figure III-2-1. The Meteor series spacecraft.

III-2.3 METEOR-2

The improved operational Meteor was the second generation Meteor-2, the first of which was launched on July 11, 1975 (Table III-2-2). After that launch, Pravda reported that Meteor-2 (Figure III-2-2) carried improved equipment: an optical–mechanical VIS television scanner, an optical–mechanical IR television scanner, and radiometric apparatus for continuous observations of radiation in near-Earth space. The satellite also carried a precision orientation system, an electrical solar cell system with independent solar tracking, radio telemetry for transmitting housekeeping data to Earth, and radios for precise measurement of orbital parameters and for transmitting data to Earth (Pravda, 1975).

Full details of the APT signals for the Meteor-2 series and experimental meteorological satellites are given in Table III-2-3. Figure III-2-3 is an example of APT imagery from the seventh Meteor-2.

Infrared APT signals have been observed only during the first few weeks after launch of a Meteor-2 satellite. These images produce very low resolution cloud cover imagery. Few examples have been published. One of the earliest, from the first Meteor-2, was recorded by Virgil Neher (Winkler, 1976). Figure III-2-4 is a rare example of the transition from VIS to IR APT (recorded in June 1982 by Alan Salmen, of Lower Hutt, New Zealand).

III-2.4 METEOR OPERATIONS

Collection of Meteor data was intended to be worldwide, with twice daily transmissions at 0000 and 1200 UTC. This system is switched off when a satellite passes over the Northern Hemisphere during hours of darkness. The switching on and off is reported routinely in operations messages.

The U.S.S.R. planned for the launch of one or two satellites annually through 1985. Information on future plans was unavailable at this writing.

The National Paper to the Second United Nations Conference on the Exploration and Peaceful Uses of Outer Space (National Paper: U.S.S.R., 1981) describes the handling of data in the satellite, and its transmission to ground receiving stations. This report went on to list trends envisaged for the system up to 1985. These were: increase the orbital altitude to ensure complete coverage of the Earth surface in the equatorial zone; use torque motors to synchronize the system and increase its efficiency; use improved, higher resolution IR equipment with direct transmission for local coverage (APT); and experiment with SHF MW equipment to obtain all-weather observations of ice and snow cover and to determine cloud moisture content.

III-2.5 METEOR-PRIRODA SATELLITES

Some of the Meteor satellites were identified in the WMO publication as "experimental." The launch announcement of Meteor-18 stated that it was "chiefly programmed to collect meteorological information," implying an additional unspecified function, and that the scientific information was being received at the State Center for Research of Environment and Natural Resources as well as at the Hydrometeorological Center (TASS, 1974). Later, the mission of the 25th satellite of the Meteor-2 series, also referred to as Meteor-Priroda (Nature), was defined to:

- obtain multizonal pictures of clouds and of the underlying surface over limited areas.
- obtain data relating to the spatial distribution of areas of precipitation, the total water content of cloud, the position of the limit of ice cover and the compactness of the ice.
- obtain data on the exchange of moisture content in the atmosphere.
- obtain data on the temperature of the underlying surface.
- measure reflected radiation and the polarized component of the radiation in order to determine the phase composition of the cloud.
- measure the intensity of fluxes of corpuscular radiation.
- measure radiant heat in the upper atmosphere.

Observations of the Kettering Group and other amateurs in the United Kingdom show that the transmitter is switched off during passes to the west of the European land mass. From time to time, however, Len Maxim has received imagery from these satellites, oc-

Table III-2-2 *Meteor satellites*

SATELLITE	LAUNCH DATE	INCLINATION (DEG)	ORBITAL PERIOD (MIN)	APOGEE (KM)	PERIGEE (KM)	REMARKS
Meteor 1	27 Mar 69	81.2	97.9	713	644	First named Meteor.
Meteor 2	6 Oct 69	81.2	97.7	690	630	
Meteor 3	17 Mar 70	81.2	96.4	643	555	Decayed 18 Nov 83.
Meteor 4	28 Apr 70	81.2	98.1	736	637	
Meteor 5	23 Jun 70	81.2	102.0	906	863	First at this height.
Meteor 6	15 Oct 70	81.2	97.5	674	633	
Meteor 7	20 Jan 71	81.2	97.6	679	630	
Meteor 8	17 Apr 71	81.2	97.2	646	620	
Meteor 9	16 Jul 71	81.2	97.3	650	618	
Meteor 10	29 Dec 71	81.2	102.7	905	880	First with APT.
Meteor 11	30 Mar 72	81.2	102.6	903	878	
Meteor 12	*30 Jun 72	81.2	103.0	929	897	
Meteor 13	*26 Oct 72	81.2	102.6	904	893	
Meteor 14	20 Mar 73	81.2	102.6	903	882	
Meteor 15	29 May 73	81.2	102.5	909	867	
Meteor 16	5 Mar 74	81.2	102.2	906	853	
Meteor 17	24 Apr 74	81.2	102.6	907	877	
Meteor 18	9 Jul 74	81.2	102.6	905	877	APT on 137.15 MHz announced.
Meteor 19	*28 Oct 74	81.2	102.5	917	855	
Meteor 20	17 Dec 74	81.2	102.4	910	861	
Meteor 21	1 Apr 75	81.2	102.6	906	877	
Meteor-2 1	11 Jul 75	81.3	102.5	903	872	First Meteor-2. APT on 137.3 MHz announced.
Meteor 22	18 Sep 75	81.2	102.3	918	867	
Meteor 23	25 Dec 75	81.3	102.4	913	857	
Meteor 24	7 Apr 76	81.2	102.3	906	863	
Meteor 25	15 May 76	81.2	102.4	908	866	Carried German Democratic Republic spectrometer-interferometer.
Meteor 26	*15 Oct 76	81.3	102.5	904	871	
Meteor-2 2	*7 Jan 77	81.3	103.0	932	830	
Meteor 27	5 Apr 77	81.2	102.5	909	869	
Meteor 28	29 Jun 77	98.0	97.5	685	602	First in retrograde Sun synchronous orbit.
Meteor-2 3	14 Dec 77	81.2	102.5	906	872	
Cosmos 1066	23 Dec 78	81.2	102.2	908	848	Possible Meteor-2 failure.
Meteor 29	25 Jan 79	98.0	97.4	656	628	Carried GDR interferometer-spectrometer.
Meteor-2 4	1 Mar 79	81.2	102.3	908	857	
Meteor-2 5	31 Oct 79	81.2	102.6	904	877	
Meteor 30	18 Jun 80	98.0	97.3	678	589	
Meteor-2 6	9 Sep 80	81.2	102.4	906	868	
Meteor-2 7	*14 May 81	81.3	102.5	904	868	
Meteor 31	10 Jul 81	97.9	97.6	688	611	Designated Meteor-Priroda. Carried Bulgaria-1300 experiment. Iskra amateur satellite launched with this.
Meteor-2 8	25 Mar 82	82.5	104.2	976	954	First at this inclination.
Meteor-2 9	*14 Dec 82	81.3	102.0	904	836	
Meteor-2 10	28 Oct 83	81.2	101.0	901	780	
Meteor-2 11	5 Jul 84	82.5	104.0	974	954	
Meteor-2 12	7 Feb 85	82.5	104.0	975	950	

Notes: 1. Orbital data given in the original *TASS* announcement of the launch whenever available. Otherwise from *Soviet Space Programs: 1976–80,* Part 1, Appendix III, U.S. Government Printing Office, Washington, D.C. December 1982. 2. With the exception of the four retrograde launches into Sun synchronous orbits from Tyuratam, all launches have been from Plesetsk. 3. With the exception of the three F-2 launches at 82.5°, all satellites were launched by the A-1 launch vehicle. 4. Only one satellite, Meteor 3, has decayed. The decay date was taken from the NASA Satellite Situation Report, Vol. 24, No. 1, Goddard Space Flight Center, February 29, 1984. 5. Launch dates denoted by * are determined in UTC and differ from the *TASS* dates by being one day earlier in each case as determined by Moscow Standard Time. The time difference does not account for the discrepancy in the case of Meteor 19 which was launched at approximately October 28.5 and announced as October 29 by *TASS.* 6. Cosmos-1484, launched on July 24, 1983, into a Sun synchronous orbit similar to that of the retrograde Meteors was announced as an Earth resources payload. Although considered a Meteor failure by some western analysts, signals were received in Stockholm from this satellite. 7. Prepared by Geoffrey E. Perry, MBE, The Kettering Group.

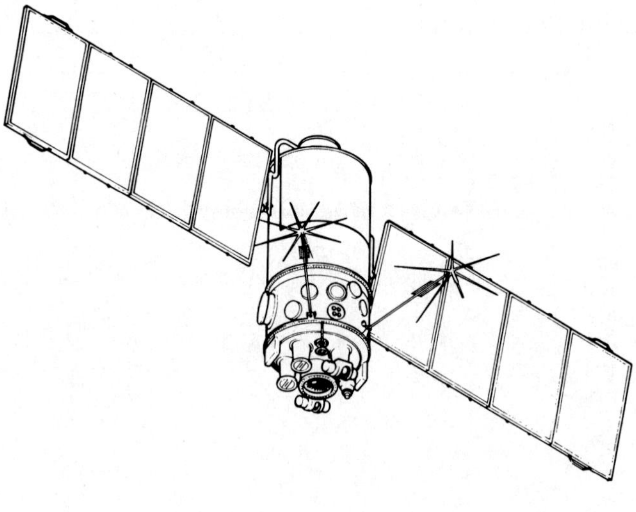

Figure III-2-2. The Meteor-2 series spacecraft.

casionally during the hours of darkness when only a totally black image with the usual numerical edge code is received.

Meteor-18 and 25 were identified as experimental, as was the first Sun synchronous spacecraft, Meteor-28. One launch during the operational period of the First GARP Global Experiment (FGGE) was planned but no mention was made of further launchings. Additional Sun synchronous satellites were launched in 1979, 1980, and 1981.

Meteor-31 transmitted APT at 240 lines per min and carried the small Iskra (Spark) amateur radio satellite "piggyback" (Johnson, 1981). Both Meteor-30 and 31 transmitted during 1984; Meteor-30 transmissions were received by the Kettering Group more often than those from Meteor-31.

Cosmos-1484, launched in 1983 in the Sun synchronous orbit characteristic of the Meteor-Priroda satellites, was designed to have an Earth resources mission. Interception of its transmissions in Sweden showed it to be a success.

A 1981 issue of *Issledovaniye Zemli iz Kosmosa* (1981b), *Earth Research from Space*, was devoted entirely to remote sensing. Most of the work was based on data provided by the Meteor-30 "Fragment" instrumentation unit. The Fragment complex was developed with the assistance of specialists from Karl Zeiss-Jena in the German Democratic Republic (GDR). In addition to the Fragment complex, Meteor-30 carried a BIK-E experimental onboard information complex consisting of:

- a medium resolution multizonal scanning unit with a tapered optical–mechanical scanner (MSU-SK) operating in four spectral bands.
- a high resolution multizonal scanning unit with electronic scanning using charge coupled devices (MSU-E) operating in three spectral bands.
- an information conversion and multiplexing unit.
- a digital radio transmission unit operating at 460 to 470 MHz with a resolution of 240 m.

Meteor-28, the third experimental Meteor satellite, was launched into a retrograde orbit on June 29, 1977. Satellites in such orbits pass over a given point on the Earth surface at the same local solar time throughout the year. This is important for Earth resources surveys, photographic reconnaissance, and meteorology.

III-2.6 THE FANAS MESSAGE

On March 8, 1978, the Secretary-General of the WMO, at the request of the U.S.S.R., wrote to inform WMO members that, starting on January 15, 1978, the Hydrometeorological Service of the U.S.S.R. would transmit orbital data and predictions for Meteor-2 satellites in the national code form, Forecast for Ascending Node for Automatic Satellites (FANAS), over the Global Telecommunications System.

The message is in five sections, the first giving information on the orbital period and longitudinal increment and daily data on the ascending node. Time is in hours, minutes, and seconds, UTC; longitude is in degrees and decimals of a degree. First and last ascending nodes are given for odd numbered days and only first ascending nodes for even numbered days.

The second section is in plain language (English). It gives information on the current operating mode of the onboard TV system and the radio frequency of APT transmissions. The third section gives the time in minutes and seconds, measured from the ascending node epoch, at which the satellite crosses circles of latitude at 10° intervals for the part of the orbit over which the TV transmitter operates.

The fourth section, rarely used, carries an explanation of the content of the FANAS message and definitions of the digital groups of the transmitted message. The fifth and final section provides information on latitude in degrees and decimals of a degree within the specified octant, and height above the Earth surface in tenths of a kilometer for each even minute measured from the ascending node epoch. This section may also contain nonscheduled information on the operation of the instruments, changes of operating modes, recommendations, requests, and the time of transmissions from geosynchronous satellites.

III-2.7 GEOSYNCHRONOUS METEOROLOGICAL SATELLITE (GOMS)

In May 1966, the Soviet Union adopted the geosynchronous orbit for meteorological purposes by placing

Table III-2-3 *Characteristics of automatic picture transmission (APT) signals from Soviet meteorological satellites*

CHARACTERISTICS	METEOR-2 IMPROVED OPERATIONAL SYSTEM	METEOR EXPERIMENTAL SYSTEM
Characteristics of radio signals		
Carrier frequency	137.3 MHz*	137.15 MHz
Modulation	Frequency	Amplitude-Frequency
Deviation	±15 kHz	9.6 kHz
Output power of radio transmitter	Not less than 5 W	Not less than 5 W
Characteristics of low-frequency signals for television pictures		
Sub-carrier frequency	2400 Hz ± 10 percent	2400 Hz
Sub-carrier modulation	Amplitude with coefficient of modulation up to 90 percent, level "white" corresponds to maximum amplitude	Amplitude, level "white" corresponds to maximum amplitude
Automatic control signals	The line phasing signal is transmitted by amplitude modulation of the sub-carrier frequency 256 Hz with relative instability not worse than 5 parts per million, intensity of modulation approximately 100 percent, length of phasing signal 90–100 microseconds	
Characteristics of recording equipment for television pictures		
Index of co-operation	264	528
Scanning speed	120 line/min	240 line/min
Scanning density	3.8 line/mm	7.6 line/mm
Dimension of picture	195 × 290 mm	195 × 290 mm
Number of half-tones	Not less than 12	Not less than 12
Direction of scanning	right/left	—
Characteristics of low-frequency signals for infrared pictures		
Sub-carrier frequency	400 Hz	
Sub-carrier modulation	Amplitude with coefficient of modulation up to 90 percent, level "white" corresponds to maximum amplitude	
Automatic control signals	Line phasing signal transmitted by amplitude modulation of sub-carrier frequency 512 Hz with relative instability not worse than 5 parts per million, intensity of modulation approximately 100 percent, length of phasing signal 0.15–0.2 seconds	
Characteristics of recording equipment for infrared pictures		
Scanning speed	20 line/min	
Scanning density	1 line/mm	
Direction of scanning	right/left	
Number of half-tones	Not less than 9	

* Frequencies of 137.4 and 137.85 MHz are also used to avoid interference when more than one satellite is above the horizon of Soviet ground stations.

television cameras on the third and fourth Molniya-1 communications satellites (in highly elliptical orbits with apogee at approximately 40,000 km above the Northern Hemisphere). The first picture was taken on May 18, 1966 (Izvestiya, 1966c).

Molniya-1 No. 4, had a television camera that was

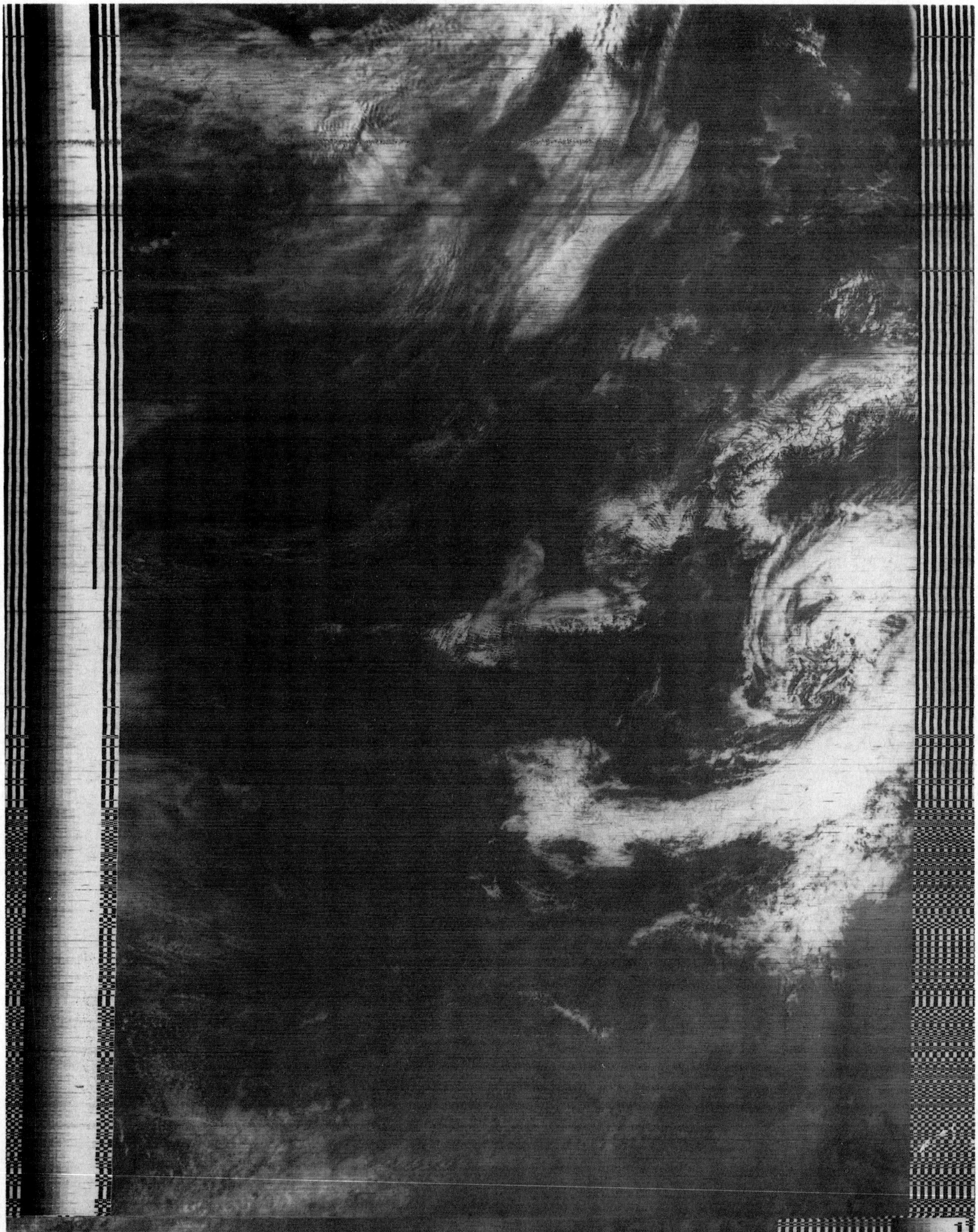

Figure III-2-3. Visible APT from the seventh Meteor-2, 0825 UTC, April 25, 1982. Note the poor contrast between land and sea areas. The black lines immediately to the left of the VIS image decode to give binary 43 at the top, decreasing by unit steps to 40 at the bottom of the picture. Lateral displacement of the synchronization bars and binary edge code in the lower part of the picture are presumed to carry encrypted information (courtesy of L. Currington).

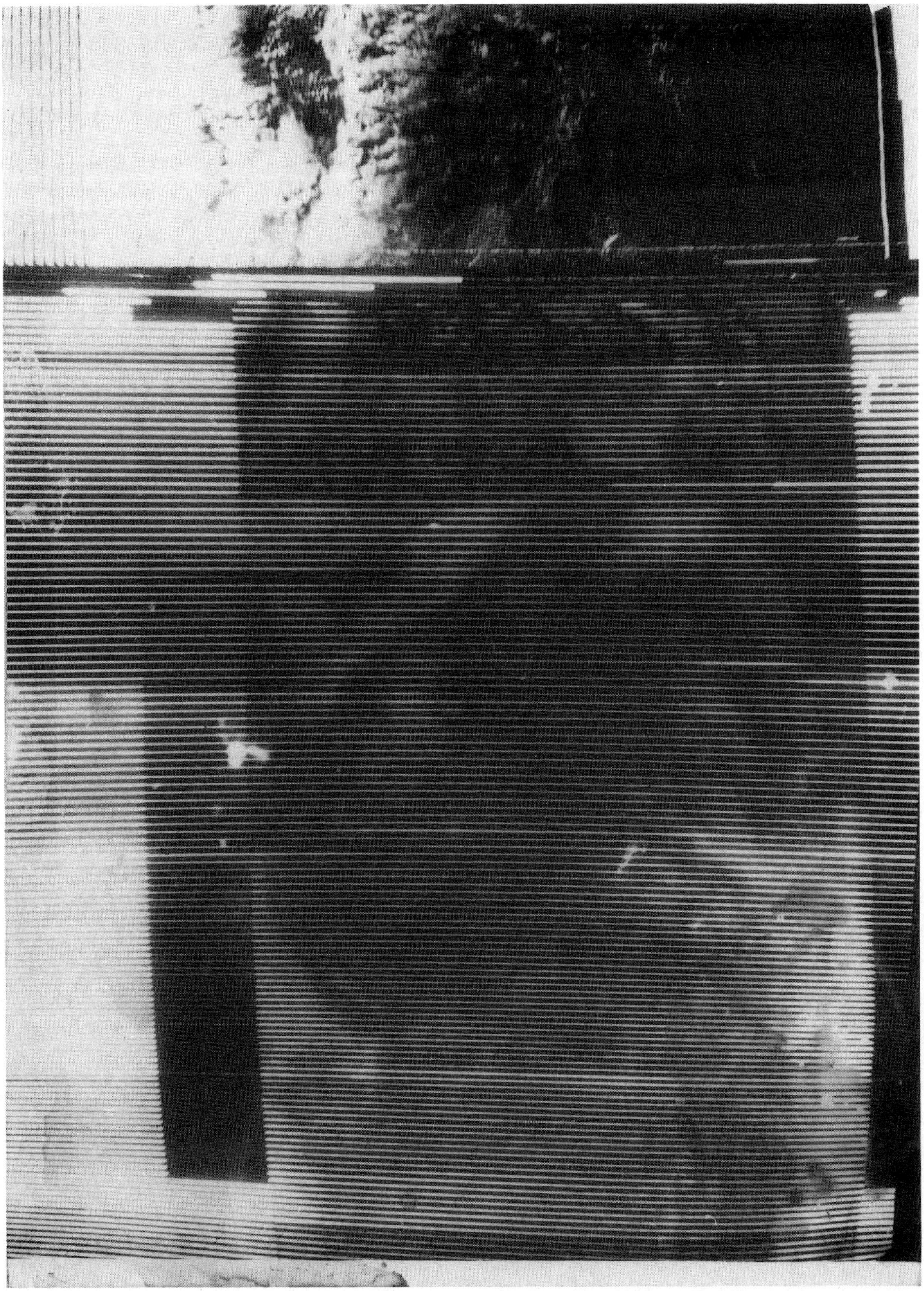

Figure III-2-4. Example of the transition from VIS APT at 2 lines per sec to IR APT at 20 lines per min recorded in New Zealand from Meteor-2, June 1982 (courtesy of A. Salmen).

steerable from Earth and included both wide-angle and narrow-angle lenses and various filters, so that several kinds of observations could be made using remote controls on the ground (Krasnaya, Zvezda, 1966).

With the successful launches and uses of geostationary satellites by other nations, plans were made to have a chain of four or five such satellites spaced equally around the Equator to permit wind estimates and convection surveys in the tropics. It was eventually agreed that to provide the desired global coverage in the tropics, two geostationary satellites would be furnished by the United States and one each by Japan, the European Space Agency (ESA), and the Soviet Union (National Aeronautics and Space Administration, 1979).

Although a Soviet satellite was to have been launched within a year or two, none had been launched by 1988. In late 1977, the United States was told that the U.S.S.R. could not provide the Indian Ocean satellite. Since the lack of this satellite would have resulted in a gap in data coverage for the Global Weather Experiment, the United States and ESA collaborated to provide and operate a GOES satellite as a substitute. GOES-1 was moved to 55° to 60°E by the end of 1978 and turned over to ESA for operation (National Aeronautics and Space Administration, 1979; ESA Bulletin, 1978). Plans for GOMS were understood to be still active in 1987.

III-2.8 OCEANOGRAPHIC COSMOS SATELLITES

Cosmos-1076, 1151, 1500 and 1602 were all announced, at time of launch, as having oceanographic missions to investigate sea, ice, and wind conditions. Cosmos-1500 was immediately effective in providing imagery which enabled the freeing of Soviet merchant ships trapped in the ice by a sudden freeze in the Arctic Ocean. Subsequently it was revealed that the satellite was equipped with side looking radar (TASS, 1984).

As early as November 13, 1983, amateurs were receiving APT at four lines per second on 137.4 MHz. The pictures had the numerical edge code characteristic of the Meteor-Priroda satellite transmissions, but consisted of side-by-side images, one of which was "cloud-free." An image recorded in Welwyn Garden City, England, April 10, 1984, revealed the VIS APT section was adjacent to the edge code and had a width some 6.6 times that of the edge code. Next to this was the cloudfree section, some 3.4 times the width of the edge code, a narrow vertical gray scale, and a totally black area some 4.4 times the width of the edge code. A series of observations from Italy, at 3-day intervals during August 1984, permitted identification of Cosmos-1500 as the source of these transmissions.

Papers presented at the 35th Congress of the International Astronautical Federation in Lausanne described the instrumentation and modes of operation of Cosmos-1500 (Afans'ev et al., 1984). The side looking radar operates at 3.15 cm, with 1.5 to 2-km resolution over a viewing field of 460 km. A multichannel, low resolution scanner provides imagery in four spectral bands. The field of view from 650 km is 1930 km with a ground resolution of 1.5 km. The side looking radar and multispectral scanner can function simultaneously; each has the same line scanning pattern at the Earth surface. The satellite information system can either transmit directly or record and subsequently play back within radio range of receiving centers. The memory can store 6.5 min of data in image strips 2750 km long and 1930 km wide. This capacity furnished data for the production of maps of the Arctic and Antarctic ice caps (Burtzev et al., 1984).

Data are received and processed by the main and regional centers at Moscow, Novosibirsk, and Khabarovsk and are sent to stations at Leningrad, Petropavlovsk-Kamchatka, Yuzhno-Sakhalinsk, four locations along the northern seaboard of the Soviet Union, and shipboard receiving stations.

Reception of similar imagery at 137.4 MHz has also been reported from Cosmos-1602. Signals have been recorded in Europe and New Zealand. The storage capability has been retained. Peter Wakelin received a VIS image of Japan at Sunningdale, U.K., on November 29, 1984, during a transmission between 0939 and 0945 UTC. The numerical edge code gave a minute count between 376 and 380 min after 2100 UTC, some 6 hours earlier, when Cosmos-1602 was southbound down the east coast of the People's Republic of China.

Chapter III-3

THE EUROPEAN METEOSAT[1]

III-3.1 INTRODUCTION

In 1972, eight member states of the European Space Research Organization (ESRO) decided to work on a program for the development of pre-operational geostationary meteorological satellites (Meteosat). Execution of the program was entrusted to the organization now known as the European Space Agency (ESA). Subsequently, ESA was charged with operation of the resulting spacecraft, of which two models (F1 and F2) were launched (in 1977 and 1981). However, ESRO still intended to establish a community responsible for setting up a system of meteorological spacecraft derived from the original Meteosat. In Geneva on May 24, 1983, 12 countries signed a convention creating the international organization known as EUMETSAT to establish, maintain, and operate systems of operational satellites. The first was to be a continuation of the Meteosat pre-operational program. Four other countries signed later. EUMETSAT will rely on ESA to carry out the program. The Meteosat Operational Program is composed of three satellites, to be launched in early 1989, early 1990, and 1992. The program will fulfill five missions.

III-3.2 MISSIONS OF THE METEOSAT OPERATIONAL PROGRAM

III-3.2.1 Earth Imaging

The radiance of the Earth surface and its cloud cover are remotely sensed simultaneously in three spectral bands (Table III-3-1):

- The VIS image, sensed in the 0.4 to 1.1 μm region, consists of 5000 lines, each containing 5000 picture elements.
- The thermal IR image, sensed in the 10.5 to 12.5 μm region, consists of 2500 lines, each containing 2500 picture elements.
- The WV absorption band image, sensed in the 5.7 to 7.1 μm region, consists of 2500 lines, each containing 2500 picture elements.

For each spectral channel, the radiometric information is encoded into 256 gray levels. An example of each type of image is shown in Figure III-3-1.

III-3.2.2 Dissemination

Preprocessed images and other meteorological data are relayed to user stations. The transmissions are in both digital and analog formats (WEFAX pictures).

Images are processed in two stages, preprocessing and image processing proper. Preprocessing consists of demultiplexing the data stream and compensating for radiometer and imaging chain imperfections, including amplitude equalization of both VIS channels. The processing of images provides for:

- The computation of a deformation model to specify the difference between actual location of points in the images and the location they would have if the satellite position and orientation were absolutely fixed and ideal.
- The conversion of the raw image to one of standard aspect, i.e., as if it were taken from the fixed and ideal position with an absolute accuracy better than an rms error of two IR pixels and a consecutive image accuracy better than an rms error of 0.5 IR pixels.

[1] Primary Contributor: Claude Honvault

Table III-3-1 *Characteristics of the Meteosat radiometer channels (courtesy of ESA)*

PARAMETERS	VISIBLE 0.4–1.1 μm	INFRARED 10.5–12.5 μm	WATER VAPOR 5.7–7.1 μm
Number of channels	2 simultaneous + 2 redundant	1 + 1 redundant	1 + 1 redundant
Field of view (mrad)	0.065	0.14	0.14
Resolution at subsatellite point (km)	2.5	5	5
Number of lines/Images of ±9°	5000	2500	2500
Detectors	Si Photodiodes	HgCdTe	HgCdTe
Performance	S/N > 200 for 25% albedo	NEΔT < 0.4 K at 290 K	NEΔT < 1.0 K at 260 K

- Absolute radiometric calibration information for the IR and WV channels provided for every image.

The main characteristics of the digital transmission are listed in Table III-3-2. The recommended gain per temperature (G/T) for the acquisition station (Primary Data User Station, PDUS) is 11.5 dB · K^{-1}. The image handling subsystem of the station varies according to the user specifications. Major options are: tape recorder, color image display, hard copy, and extended computer equipment.

The main characteristics of the analog transmission are listed in Table III-3-3. The recommended gain per temperature (G/T) for the acquisition station (Secondary Data User Station, SDUS) is 2.5 dB · K^{-1}.

The Meteosat Operational Program (MOP) satellites will have the capability to relay eight additional channels, with a capacity of 2400 bps, for Meteorological Data Dissemination (MDD). The characteristics of the MDD ground stations are listed in Table III-3-4.

III-3.2.3 Collection of Environmental Data

Environmental data gathered by various types of fixed and mobile Data Collection Platforms (DCP) are collected by as many as 66 channels on the satellite. This mission supports DCP that transmit messages on an agreed upon schedule (self-timed platforms), and those that transmit only in response to specific environmental criteria (alert platforms). The main characteristics of the transmission of messages from platform to satellite are presented in Table III-3-5.

The formats of messages have been coordinated with those of Japan and the U.S. through the group for Coordination of Geostationary Meteorological Satellites (CGMS).

The DCP data are retransmitted to the SDUS with the WEFAX images, and can be acquired provided the equipment is compatible with the modulation specifications given in Table III-3-6.

III-3.2.4 Extraction of Meteorological Parameters

The actual extraction of products is performed automatically for the region within a 55° great circle arc from the subsatellite point. Before products are distributed, they are quality controlled by experienced meteorologists. Products which have a 32 × 32 IR pixel resolution, except for Cloud Top Height charts which have a 4 × 4 IR pixel resolution, are as follows:

- Cloud Motion Vectors (CMV): extracted twice per day at 0000 UTC and 1200 UTC and distributed by 0230 and 1430 UTC. About 600 vectors per run are produced with a target accuracy ranging from 5 m s^{-1} rms at low levels to 10 m s^{-1} at high levels.
- Sea Surface Temperatures (SST): extracted daily at 1200 UTC, based on a 3 hour composite, and are distributed by 1430 UTC. Approximately 800 pieces of data are processed daily with a target accuracy of $T \leqslant 1.5$°C rms.
- Cloud Analysis (CA): provides cloud cover in up to three layers together with cloud top temperature. They are extracted at 0000 UTC and 1200 UTC and distributed by 0230 UTC and 1430 UTC.
- Upper Tropospheric Humidity (UTH): provides average relative humidity between 700 and 300 mb. Values are coordinated with the WV channel measurements. UTH are extracted at 0000 and 1200 UTC and distributed by 0230 and 1430 UTC. The target accuracy is ⩽20% relative humidity rms.
- Cloud Top Height (CTH): provides a WEFAX map of cloud tops at 1500-m intervals between 4.5 and 12 km. Extraction is at 0300, 0900, 1200, and 1500 UTC; data are broadcast within 1 hour through the WEFAX channel of Meteosat.

CMV, SST, CA and UTH, encoded into WMO SATOB Bulletins, are disseminated through the Global Telecommunications System (GTS) via the Offenbach Regional Telecommunications Hub. CTH maps are broadcast as WEFAX pictures via the satellite.

III-3.2.5 Archiving of Digital Data and Image Negatives

All available images are regularly archived in digital form together with the extracted meteorological products (except CTH maps). Two sets of three images per day are recorded on photographic film. A comprehensive catalogue is maintained on all archived data.

Figure III-3-1a. Meteosat VIS image (courtesy of ESA).

III-3.3 DESCRIPTION OF THE METEOSAT SPACECRAFT

The basic design of the operational satellite is that of the (pre-operational) Meteosat-F2 with two major improvements to the mission capability:

- the mission performance transponder provides eight additional channels (2400 bits/sec) for meteorological data dissemination.
- the WV absorption channel (5.7 to 7.1 μm) is available in parallel with VIS and IR channels.

Figure III-3-1b. Meteosat IR image (courtesy of ESA).

A number of minor modifications based on F1 and F2 experience and technology improvements are incorporated for increased reliability.

Figure III-3-2 shows the spacecraft and its main features. The main body (a cylinder) is covered with solar cells for supplying electrical energy. Most of the subsystems, including the radiometer, are located in this cylinder. The second cylinder carries:

- the Electronically Despun Antenna (EDA), which is an array of radiating dipoles electronically fed so they simulate an S-band antenna that is artificially and permanently oriented towards the Earth.

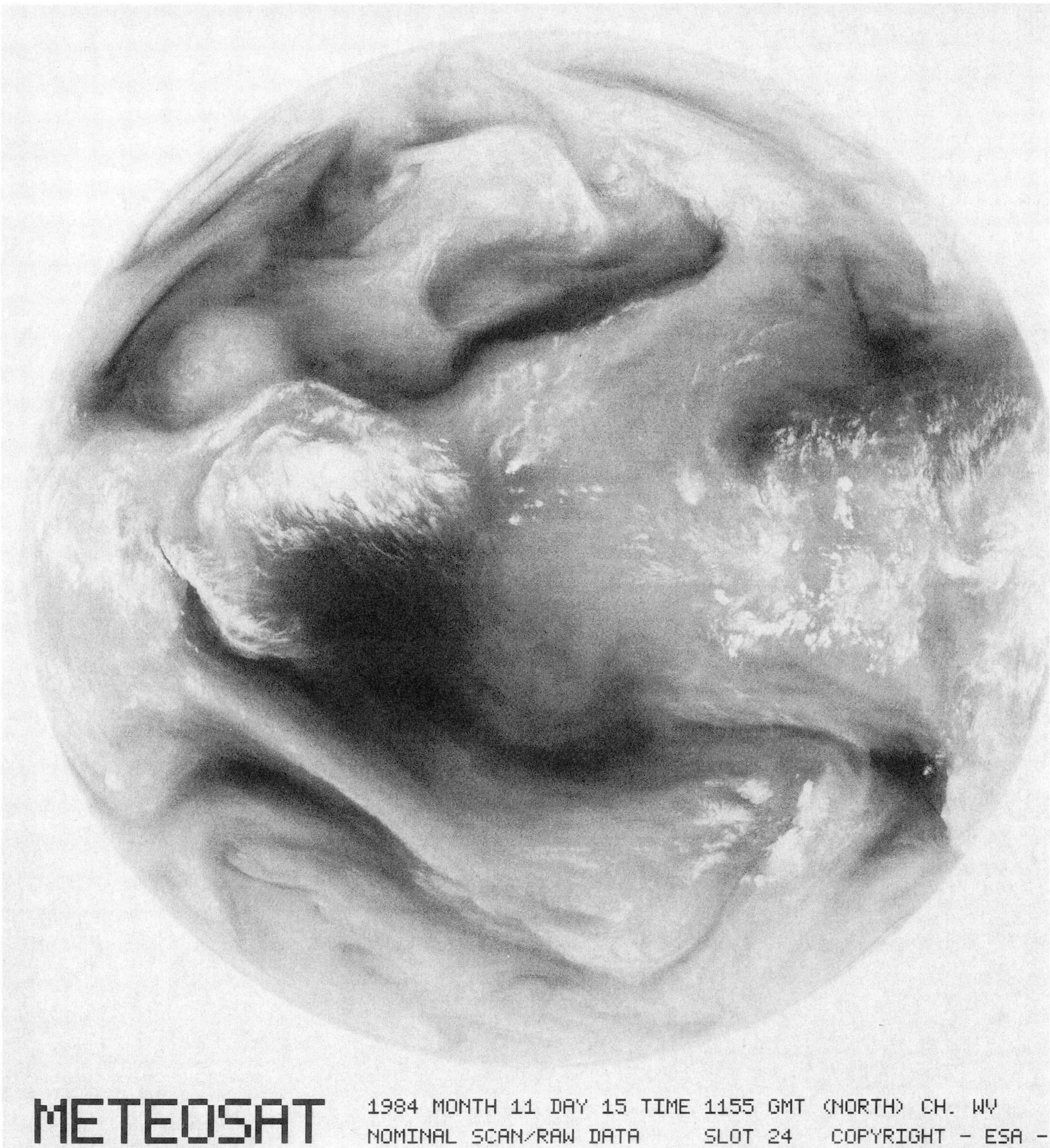

Figure III-3-1c. Meteosat WV image (courtesy of ESA).

- most of the telecommunications equipment.
- additional antennas.

The two cylinders mounted on top of the drum are toroidal pattern, S-band and UHF-band antennas.

The payload consists of a high resolution radiometer and a data transmission system. The radiometer is an electro-optical instrument which includes a large telescope with a 3650-mm focal length. A set of 250 $\times$ 250 μm detectors (silicon photodiodes in the VIS spectral band, and HgCdTe 70 $\times$ 70 μm detectors for WV and IR) located in the focal plane of the telescope

Table III-3-2 *Characteristics of Meteosat digital transmissions (courtesy of ESA)*

PARAMETERS	VALUE
Frequency	1694.5 MHz or 1691 MHz
Long-term frequency stability (3 years)	Better than 1.2×10^{-5}
Modulation	PCM (SP-L) PM
Index	1.2 rad ± 5%
Bit rate	166 kbps for Meteosat 1 and 2 333 kbps for the Meteosat Operational Program (MOP) 1, 2, 3

Table III-3-3 *Characteristics of Meteosat analog transmissions (courtesy of ESA)*

PARAMETERS	VALUE
Frequency	1691 MHz (or 1694.5 MHz)
Modulation	AM/FM
Frequency deviation	9 kHz peak
Subcarrier frequency	2400 Hz
Line rate	4 lines/sec

measure the radiance of the Earth and its cloud cover in the VIS, thermal-IR, and WV absorption spectral bands. By virtue of the 100-rpm spin motion of the satellite, the radiometer scans the Earth along the east-west axis; the scan along the south-north axis is achieved by tilting the optical telescope axis at the end of each east-west scan. A set of three images, one in each spectral band, is produced once every 30 min.

During each spacecraft spin period, the radiometer "sees" the Earth for only 1/20th of a revolution. A memory onboard allows one line of image to be stored and transmitted to the ground during the remaining 19/20 of the revolution. The transmission speed is 166 kbps. Should the memory fail, it can be bypassed and the transmission speed becomes 2.7 Mbps. The transmission speed for MOP 1, 2, and 3 will be 333 kbps.

Table III-3-4 *Characteristics of meteorological data dissemination ground stations (courtesy of ESA)*

PARAMETERS	VALUE
Transmitter	
Frequency band	2105 ± 1.2 MHz
Frequency stability (0 to 50°C)	4×10^{-9}
Maximum acceptable drift	2 kHz/month
Receiver	
Frequency band	1694.5 ± 1.2 mHz
G/T	$6.0\ \mathrm{dB} \cdot \mathrm{K}^{-1}$
Modulation scheme and code	PCM (SP-L)/PM
Modulation index	1.2 rad

Table III-3-5 *Characteristics of transmission by DCP (courtesy of ESA)*

PARAMETERS	VALUE	
	INTERNATIONAL DCP	REGIONAL DCP
Frequency bands allocated	402.0 to 402.1 MHz	402.1 to 402.2 MHz
Number of channels	33	29 (+4 reserved)
Width of channels	3 kHz	
Types of modulation	PCM/bi-phase Manchester ± 60°	
Bit rate	100 bps	
EIRP limits of one platform	13 to 22 dBW	
Limits of power flux density per channel	-145 ± 6 dBW m^{-2}	

Table III-3-6 *Operational modulation specifications (courtesy of ESA)*

PARAMETERS	VALUE
Modulation	PCM (NRZ-L) PSK/FM
Frequency deviation	12.5 kHz peak
Bit rate	12.5 kbps

The communication package consists of a transponder and its antenna subsystem. The transponder operates at 1675-2015 MHz and 402 MHz (the S- and UHF-bands) and ensures the transmission of satellite images (WEFAX and high resolution) and data collection platform messages to user stations, the collection of data transmitted from platforms, and the exchange of meteorological data between users.

The attitude measurement subsystem includes two accelerometers, two pairs of Sun sensors and two Earth IR sensors. The data from this system permit an attitude reconstitution on the ground to within 0.1 degree

Table III-3-7 *Performance characteristics of the attitude and orbit control system (AOCS; courtesy of ESA)*

PARAMETERS	VALUE
Spin rate	100 rpm ± 1%
Spin rate control	Possibility of adjustment better than ±0.1 rpm
Nutation	<5 arc seconds after damping
Longitude station keeping (east-west)	±1°
Inclination (north-south)	<0.8°
Eccentricity of the orbit	<0.002
Maximum amount of hydrazine at beginning of life	29 kg

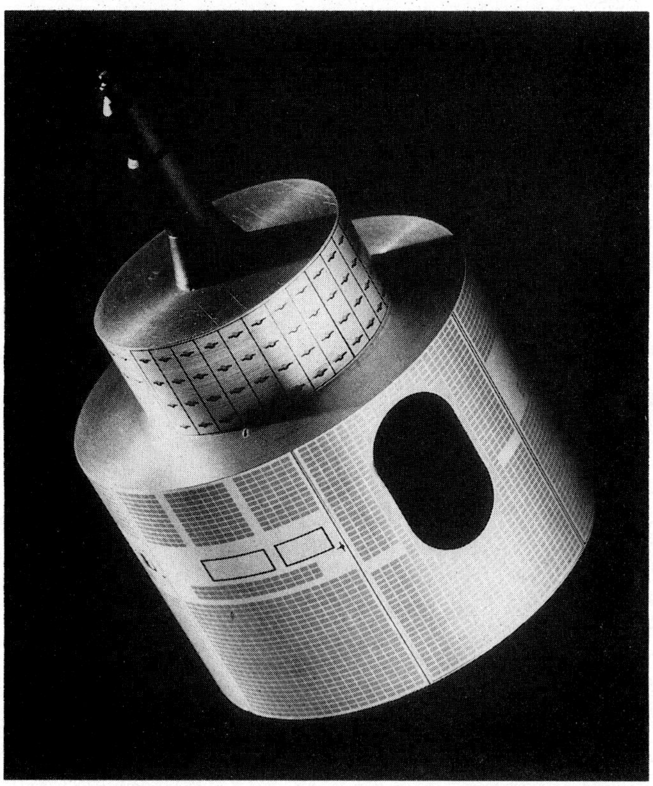

Figure III-3-2. The Meteosat spacecraft (courtesy of ESA).

after three hours. The Attitude and Orbit Control Subsystem (AOCS) includes three hydrazine tanks, six thrusters, and two passive dampers. It permits the execution of the following main tasks:

- orbital correction maneuvers
- spin axis control and correction maneuvers
- nominal spin rate acquisition and control
- passive nutation damping.

The performance characteristics of the AOCS are listed in Table III-3-7.

III-3.4 DESCRIPTION OF THE METEOSAT GROUND SEGMENT

The ground segment (Figure III-3-3) is split into two parts: ground facilities to carry out the Meteosat missions and user stations. The ground facilities consist of four elements (the first three elements are located in the European Satellite Operations Center (ESOC), Darmstadt, West Germany):

- The Meteosat Operations Control Centre (MOCC) is responsible for operational management of the satellite and the tracking facilities.

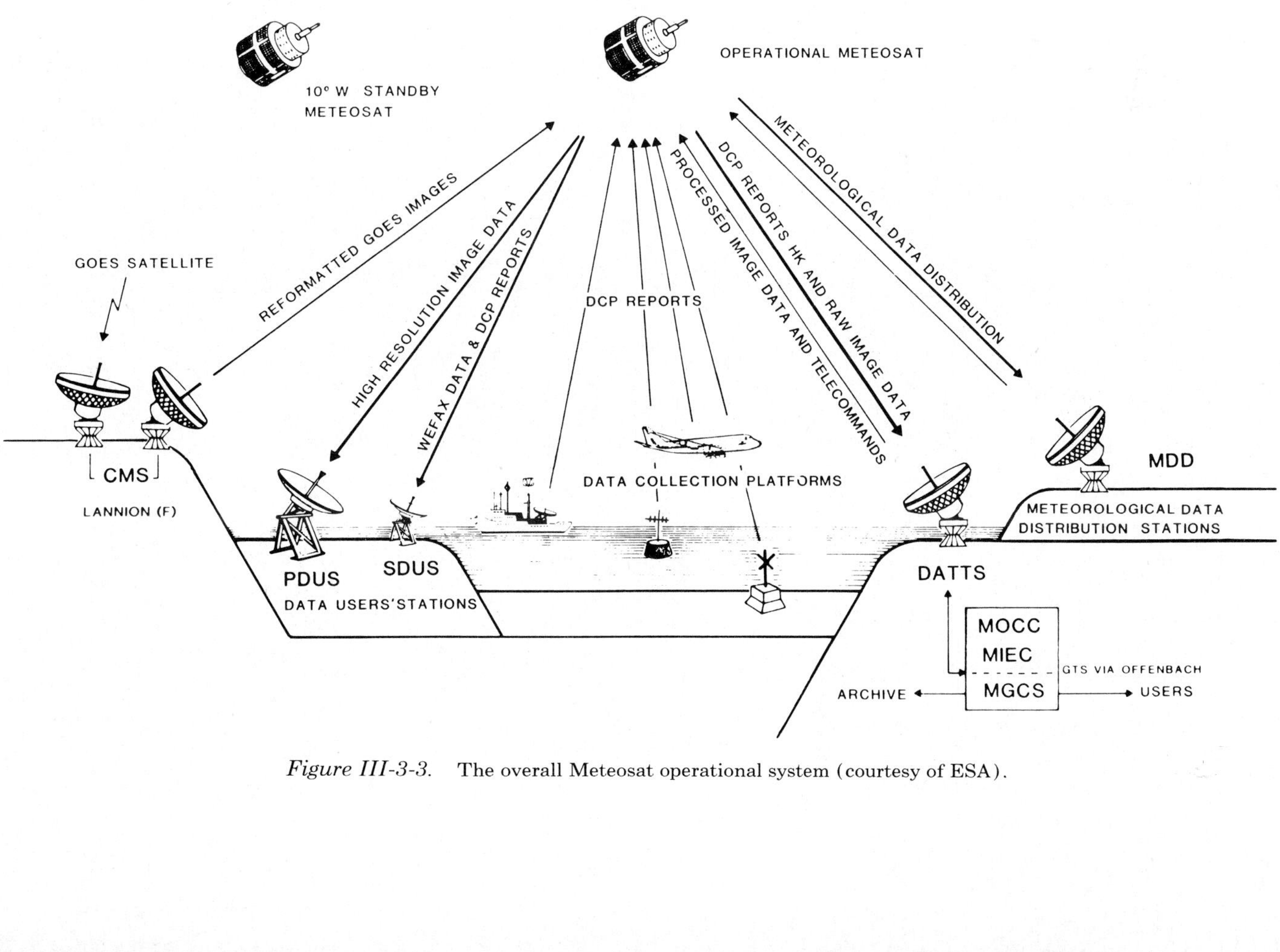

Figure III-3-3. The overall Meteosat operational system (courtesy of ESA).

- The Data Referencing and Conditioning Centre (DRCC) is mainly responsible for processing and formatting the image data.
- The Meteorological Information Extraction Centre (MIEC) extracts from the processed images specific meteorological information, such as wind fields, sea surface temperature charts, cloud system analyses, and radiation balances. The tasks of MIEC and DRCC are performed in the Meteosat Ground Computer System (MGCS).
- The Data Acquisition Telecommand and Tracking Station (DATTS), located at Michelstadt, near ESOC, is responsible for the acquisition of the satellite's radiometric and housekeeping data and messages from the data collection platforms. It transmits telecommands to the satellite, and meteorological data or images to user stations. Finally, in association with a land-based transponder at Kourou (in French Guiana), it carries out ranging measurements to locate the satellite precisely.

The different user stations are:

- The Primary Data User Stations (PDUS), which receive rectified Earth images transmitted in digital form through the satellite.
- The Secondary Data User Stations (SDUS), which receive image data in a standard (WEFAX) format and messages from the Data Collection Platforms (DCP); all are transmitted through the satellite.
- The DCP, which take measurements of the local environment and transmit them to the satellite. DCPs can be installed on the ground, on buoys, on ships, and on aircraft.
- The Meteorological Data Dissemination Stations (MDDS), which have receiving and transmission facilities and can exchange meteorological information with national meteorological services.

There are more than 500 SDUS and about 10 PDUS within telecommunication range of Meteosat. These stations are used operationally every day for weather forecasting and as an aid to airlines.

Chapter III-4

THE JAPANESE GMS (HIMAWARI)[1]

III-4.1 OBJECTIVES AND DEVELOPMENT HISTORY

The Geostationary Meteorological Satellite (GMS) program was initiated by the Japan Meteorological Agency (JMA); development was started by their National Space Development Agency (NASDA) in July 1973, and the spacecraft was manufactured by Nippon Electric Company in association with Hughes Aircraft Company of the United States. The first GMS, named "Himawari" (Sunflower), was launched from the United States on July 14, 1977.

The successors to the initial GMS, GMS-2 and GMS-3, were launched from Japan on August 11, 1981, and August 3, 1984, respectively. There are four major missions in the GMS project:

A. Weather Watch by VISSR
- Imaging Earth surface and cloud distribution, and observation of meteorological phenomena such as typhoons, cyclones, fronts, etc.
- Meteorological parameter extraction such as temperature on both Earth surface and cloud top, cloud top height, cloud amount, cloud motion winds, etc.

B. Collection of Meteorological Observation Data
- Collection of meteorological data from Data Collection Platforms (DCP) installed on ship, buoy, aircraft and other miscellaneous land stations.

C. Direct Broadcast of Cloud Images
- Dissemination of digital image data (stretched VISSR data) via satellite simultaneously with imaging to users of the Medium-scale Data Utilization Station (MDUS).
- Dissemination of processed analog image data (WEFAX data) via satellite to users of the Small-scale Data Utilization Station (SDUS).

D. Monitoring of Solar Particles (Space Environment Monitoring)
- Monitoring solar particles; protons (0.8-100 MeV), alpha particles (8-370 MeV) and electrons (≤2 MeV).

III-4.2 SPACECRAFT CONFIGURATION AND SALIENT CHARACTERISTICS

The current operational satellite, GMS-3, is a spin stabilized, geostationary meteorological satellite with mechanically despun antennas. The configuration and block diagram of GMS-3 are shown in Figures III-4-1 and III-4-2 and salient characteristics (size, mass, etc.) are listed in Tables III-4-1 and III-4-2.

The spacecraft consists of a spinning section rotating at 100 rpm and an Earth oriented antenna assembly with despun S-band and UHF antennas for high gain communications links with ground stations. Sun and Earth sensors on the spinning section supply reference signals to control the despin assembly. The spinning section also carries the VISSR and supporting subsystems.

[1] Primary Contributors: Dr. Yoshikazu Kamiya, NASDA, and Members of JMA

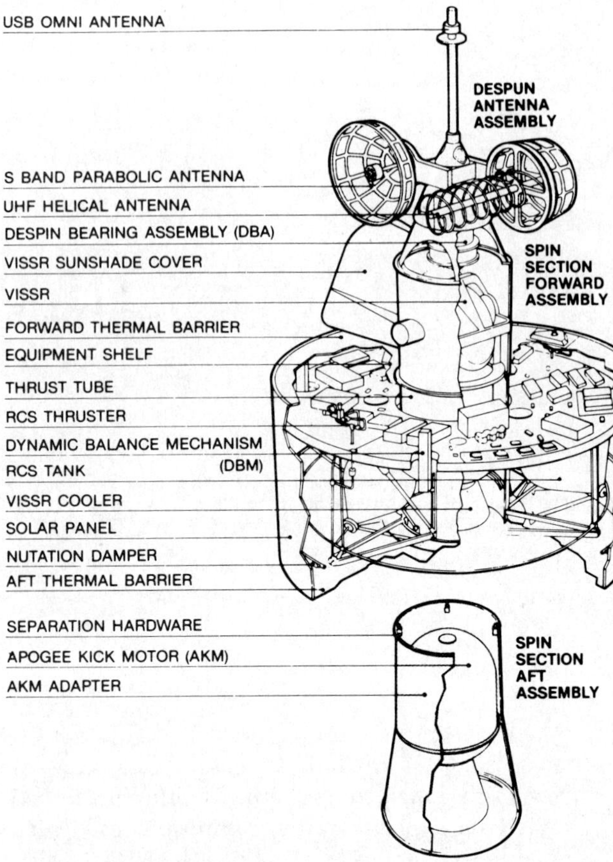

Figure III-4-1. Configuration of the GMS-3 spacecraft.

The satellite design life is limited to 5 years by the amount of hydrazine fuel onboard. However, solar panel power of 263 W includes approximately 16 W margin at the end of 5 years. Therefore, the satellites may be operated for some time after the onboard fuel has been consumed.

III-4.2.1 Communications Subsystem

The GMS-3 communications subsystem consists of primary and redundant units which can be switched on or off by ground command except for the Unified S-Band (USB) receivers which are always on. This subsystem provides for:

- S-band transmission of VISSR signals to Command and Data Acquisition Stations (CDAS).
- S-band relay of image data from CDAS to MDUS and SDUS.
- S-band relay of trilateration ranging signals to or from CDAS and two Turn Around Ranging Stations (TARS).
- S-band transmission of telemetry signals to, and S-band command reception from, CDAS and Tracking and Control Stations (TACS).
- S-band to UHF relay of DCP interrogation signals from CDAS (at S-band) to the DCP (at UHF).
- UHF to S-band relay of report signals from DCP (at UHF) to CDAS (at S-band).
- USB tracking, telemetry, and command capability as the primary link during transfer orbit and as backup during synchronous orbit (see also III-4.3.1).

Communications subsystem performance is summarized in Table III-4-3.

III-4.2.2 Space Environment Monitor (SEM)

The SEM measures solar protons, alpha particles, and electrons with five silicon detectors. SEM characteristics are listed in Table III-4-4.

III-4.2.3 VISSR and VISSR Digital Multiplexer

The VISSR is used to obtain VIS and IR images of the Earth and its cloud cover using an optical telescope and detector system. A complete $20° \times 20°$ scan of the full Earth disk can be accomplished every 30 min using the 100-rpm spinning motion of the satellite for the west to east scan and 2500 140 μrad steps of the scanning mirror for the north to south scan. This requires 25 min for image acquisition, 2.5 min for retrace, and 2.5 min for stabilization. The function of the VISSR telescope is shown in Figure III-4-3, the spin scan geometry in Figure III-4-4, and VISSR characteristics in Table III-4-5.

The VISSR Digital Multiplexer (VDM) is a high speed analog to digital converter and Pulse Code Modulation (PCM) encoder. It converts four channels of VIS and one channel of IR VISSR signals into digital form, multiplexes them with image and message synchronization data, and differentially encodes them. VDM characteristics and Earth scan timing are shown in Table III-4-6 and Figure III-4-5.

III-4.2.4 VISSR Data Characteristics

The characteristics of GMS VISSR data are essentially the same as those of the GOES VISSR of the United States (Chapter III-1).

III-4.3 GEOSTATIONARY METEOROLOGICAL SATELLITE SYSTEM (GMSS) OF JAPAN

The GMS satellite and ground facilities constitute the GMS system (Figure III-4-6). The primary ground facilities include the CDAS, the Data Processing Center (DPC), the TARS, and the Tracking and Control

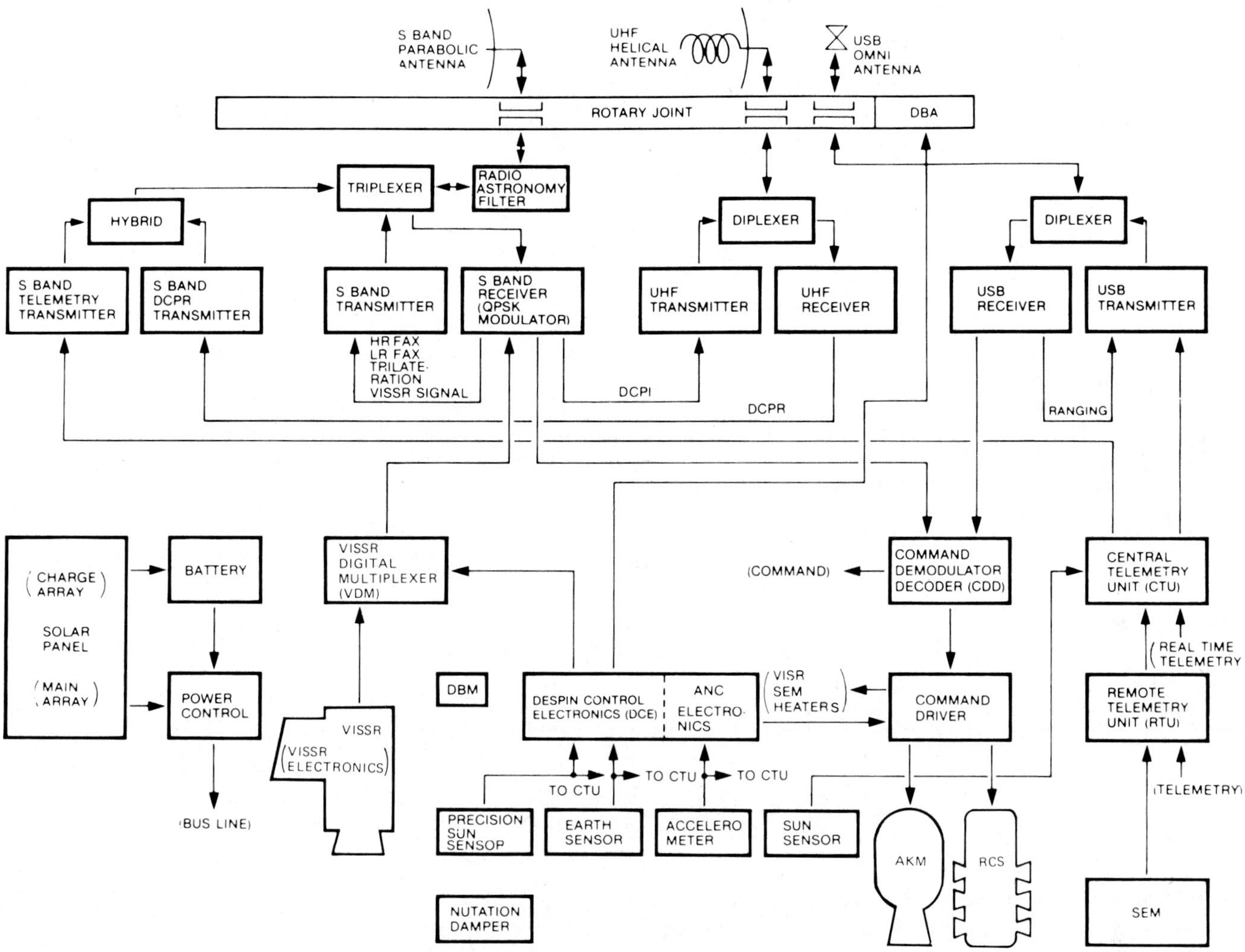

Figure III-4-2. The GMS spacecraft system.

Center (TACC). The TACC and TACS were established by the NASDA, TARS-2 was established and operated by the Australian Bureau of Meteorology, and the other facilities were established and are operated by the JMA.

III-4.3.1 Roles of the Ground Facilities

A. The CDAS[2] is the window to the GMS spacecraft. The commands to operate the spacecraft are transmitted through CDAS antenna. Observed image data, meteorological data from DCP and telemetry data from the spacecraft are received at CDAS and relayed to DPC. Direct broadcast signals such as digital Stretched VISSR (S-VISSR) data produced at CDAS and WEFAX data produced DPC are transmitted via the spacecraft to the data utilization station (MDUS and SDUS) users.

B. DPC is the central data processing center of the GMSS. A large computer system is installed to perform:

- Image data processing to produce facsimile chart and meteorological parameters
- Transmission of these products
- The operation and control of the GMS spacecraft.

C. The TACC shares the housekeeping and station keeping operations of the GMS and serves as backup for MSC ground facilities. Its large capacity computer analyzes telemetry data and determines and predicts satellite orbit and attitude information for satellite maneuvers. Commands for station

[2] The CDAS and DPC form the Meteorological Satellite Center (MSC).

Table III-4-1 *Prominent characteristics of GMS-3*

DIMENSIONS	
Diameter	2.1 m
Height	4.4 m (with AKM) 3.5 m (without AKM)
MASS	
Separated weight	680.9 kg
On-station	
BOL (incl. margin)	303.2 kg
EOL (incl. margin)	286.7 kg
LIFE	
Mission life	4–5 yr
Design life	5 yr
RELIABILITY	
Five years	0.656
ORBIT	
Geosynchronous	140°E
Station keeping:	
E-W	0.5°
N-S	1.0°
LAUNCH	
Vehicle	N-11
Site	Tanegashima

Table III-4-2 *GMS-3 subsystem characteristics*

VISSR and VDM	VISIBLE	INFRARED
Spectral band	0.50–0.75 μm	10.5–12.5 μm
Resolution	1.25 km	5 km

SEM	
Measurement range	0.8–100 MeV (protons) 8–370 MeV (alpha particles) ≥2 MeV (electrons)

COMMUNICATIONS	S-BAND	UHF	USB
Uplink	2026–2035 MHz	402 MHz	2100 MHz
Downlink	1682–1695 MHz	468 MHz	2280 MHz
Antenna	Parabola	Helical	Bicone
Coverage	Global	Global	Omni
G/T	-22 dB · K^{-1}	-23 dB · K^{-1}	-40 dB · K^{-1}
EIRP	40 + 1.5 dBm* 34 ± 1.5 dBm** 57 ± 1.5 dBm***	46 +2/−1.5 dBm	27.5 ± 1.5 dBm

TELEMETRY AND COMMAND	TELEMETRY	COMMAND
Capacity	256 × 2 channels	254 (pulse) +1 (serial)
Modulation	****(PCM/PSK/PM ****(FM/PM, FSK/PK *(PCM/PSK *(FM/PM, FSK/PK	PCM/FSK-AM/PM
Bit rate	250 bps	128 bps

CONTROLS	
Spin-stabilized (spin rate 100 = 1 rpm)	
Antenna pointing	±0.5° (N–S) ±0.39° (E–W, Sun reference)

ELECTRICAL POWER	
Solar array power	263 W (EOL S/S)
Batteries	5.5 A-hr × 2 set

RCS	
Propellant	Hydrazine
3 tanks, 2 axial and 4 radial thrusters	

AKM
Thiokol TE-M-616

THERMAL CONTROL
Passive

* S-band telemetry, **DCPR, ***VISSR data, ****USB telemetry

keeping and system maintenance and housekeeping are generated here.

D. The TACS receive telemetry signals from the satellite and transfer them to the TACC. TACC commands are also transmitted to the satellite by TACS.

E. There are two types of user stations: the Medium-scale Data Utilization Station (MDUS) and the Small-scale Data Utilization Station (SDUS). MDUS receives the S-VISSR image data from the GMS spacecraft and SDUS receives the WEFAX image data.

F. DCP have been installed on aircraft, ships, buoys, and remote islands to collect weather and environmental data.

III-4.3.2 Mission Operations

GMS data received by the CDAS are transmitted by microwave to the DPC. The DPC processes the data to generate a variety of image products and products for extraction of meteorological parameters, such as wind fields. The products produced by the MSC are used by JMA and other organizations to prepare forecasts and issue warnings of severe weather conditions. The primary roles of the Meteorological Satellite Center (MSC) are:

- Management of GMS operations.
- Production of full disk, partial disk, sectorized, and polar stereographic pictures from the VISSR images.

Table III-4-3 *Communication subsystem performance summary*

S-BAND TRANSPONDER	
Design	All solid state hybrid circuitry; channelized and fully redundant; dual conversion with limiting signal multiplexing/demultiplexing, and interface with UHF at IF
RF power	20 W at transmitter output for VISSR and multifunction (Low power 2 W)
	200 mW at transmitter output for DCPR
	1.25 W at transmitter output for S-band telemetry
Noise figure	3.8 dB at receiver input
EIRP	57 ± 1.5 dBm for VISSR and multifunction (Low power 47 = 1.5 dBm)
	34 ± 1.5 dBm for DCPR
	40 ± 1.5 dBm S-band telemetry
G/T	-22 dB·K^{-1}
Antenna	Vertically polarized earth coverage parabola, mechanically despun and pointed at earth
UHF TRANSPONDER	
Design	All solid state hybrid circuitry; fully redundant; dual conversion; interfaces with S-band through UHF if processor and AGC at IF
RF Power	4 W at transmitter output
Noise figure	2.5 dB at receiver input
EIRP	46 +2/−1.5 dBm
G/T	-23 dB·K^{-1}
Antenna	Right-hand circular polarized bifilar helix, mechanically despun and pointed at earth
USB TRANSPONDER	
Design	All solid state hybrid circuitry; fully redundant
RF power	4.0 W at transmitter output
Noise figure	3.1 dB at receiver input
EIRP	27.5 ± 1.5 dBm
G/T	-40 dB·K^{-1}
Antenna	Vertical linear polarized ± 50° coverage bicone, mounted coincident with spin axis on despun section

- Extraction of meteorological parameters.
- Collection and editing of SEM data.
- Determination and prediction of GMS orbit and attitude.

Table III-4-4 *SEM characteristics*

Functions	Measures quantity of protons, electrons, and alpha particles within specified energy bands
Design	Five detectors for multimode particle counting, submultiplexer, and 8 bit parallel readout
Measurement range	Protons 0.8 to 100 MeV Alpha particles 8 to 370 MeV Electrons ≥2 MeV

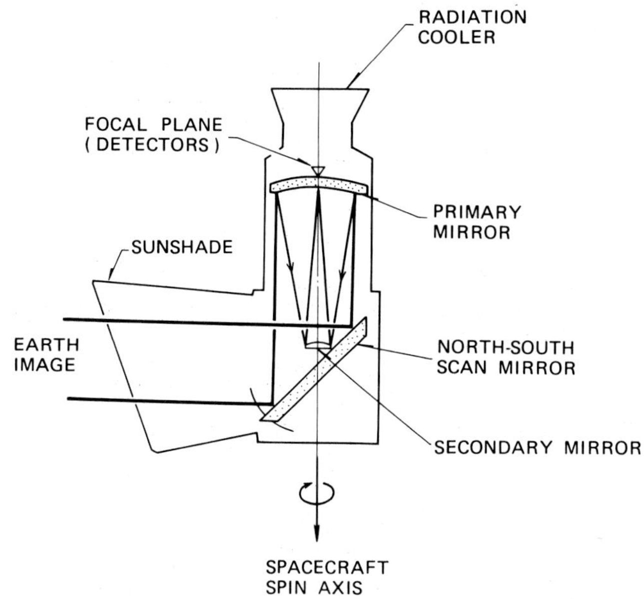

Figure III-4-3. The GMS VISSR telescope optics.

- Transmission of command signals to operate the satellite subsystems, and analysis of housekeeping data to monitor their status.
- Collection, editing, and transmission on the GTS of data from the various DCP.
- Archiving of all data and data products derived from GMSS.

III-4.3.3 Data Products of GMSS

Various meteorological parameters are extracted from the VISSR data:

A. Cloud Motion Vectors (CMV). GMS cloud motion winds are useful sources of upper air winds in the data sparse area over the ocean. Cloud motion winds are calculated by tracking target clouds using time sequential images taken successively at 30-min intervals. They are derived automatically by using specific algorithms and also manually by an analyst using image display.

The extracted cloud motion winds are transmitted to the Forecast Department of JMA via Automatic Data Editing and Switching System (ADESS) and to worldwide users via GTS. These winds are used as the initial data for numerical weather prediction and as the basic data for analysis of typhoons.

B. Satellite Cloud Information Chart (SCIC). Cloud area distribution information is useful for the short-range weather forecasting and nowcasting.

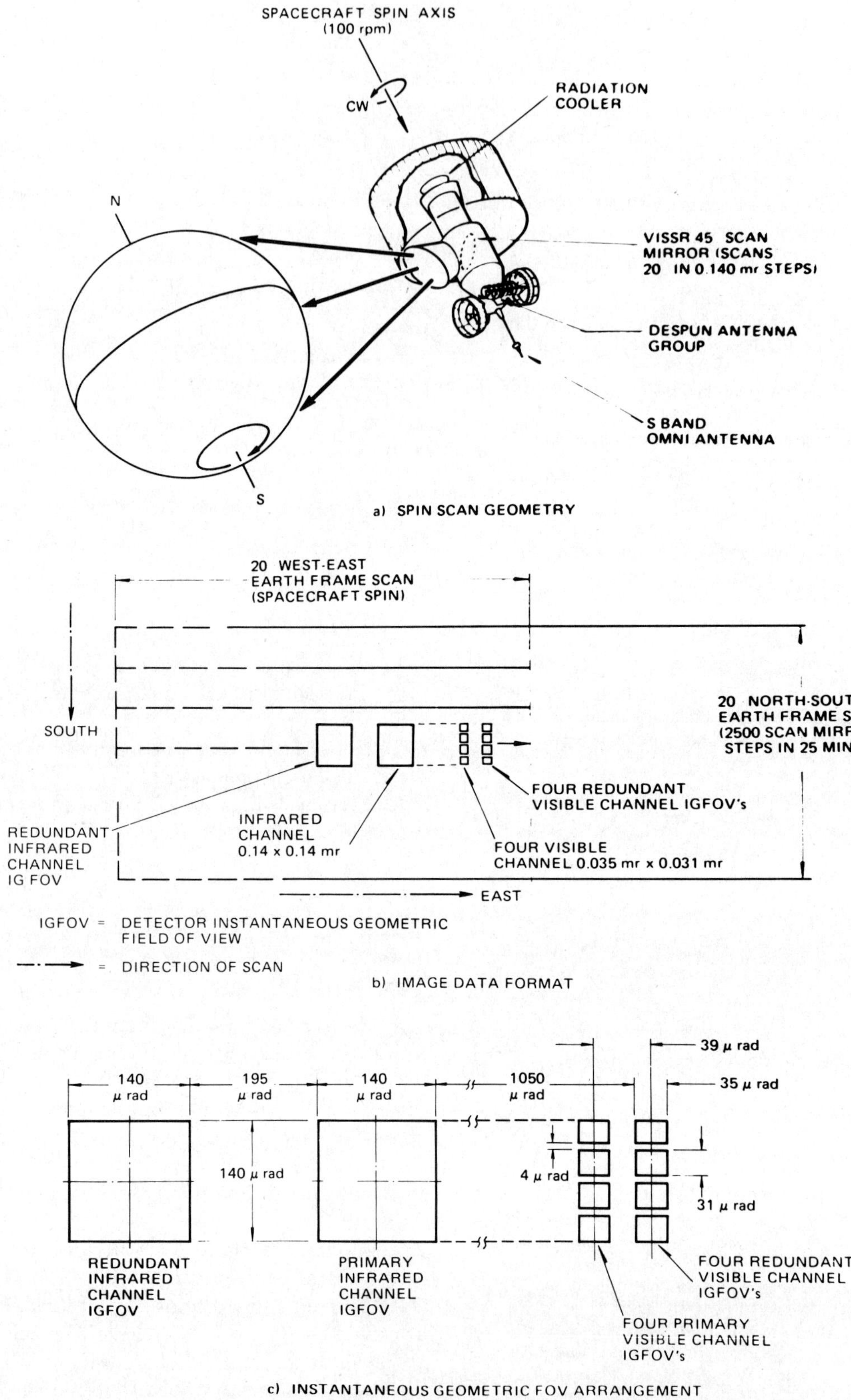

Figure III-4-4. GMS VISSR spacecraft characteristics: (a) spin scan geometry, (b) image data format, and (c) instantaneous geometric field of view arrangement.

Table III-4-5 *VISSR characteristics*

Functions	Provides VIS and IR spectrum mapping of Earth and its cloud cover.	
Design	Spin scan radiometer includes north-south scan mirror, 40.64 cm diameter 291.3 cm focal length Ritchey-Chretien optical system, PMT and HgCdTe detectors, −95 K radiation cooler with servo temperature control, and beryllium housing.	
	VIS CHANNELS	IR CHANNELS
Number	4 (+4 redundant)	1 (+1 redundant)
Instantaneous geometrical field of view (GFOV)	35 × 31 μrad	140 × 140 μrad
Band	0.50 to 0.75 μm	10.5 to 12.5 μm
Resolution	1.25 km	5.0 km
Scanning lines/frame	2500 × 4	2500
Scan step repeatability, (1 μm)	±1.8 μrad	±1.8 μrad
Noise performance	S/N ≥ 45 (albedo = 100%)	NEΔT ≤ 0.5 K (300 K)
	S/N ≥ 6.5 (albedo = 2.5%)	NEΔT ≤ 1.5 K (220 K)

SCIC, mainly produced through automatic procedures, contains blackbody temperature (TBB) contour lines and four types of categorized cloud area information derived from the IR histogram data. The analyst switches back and forth between the VIS and IR pictures on the image display, and adds onto the machine made chart the cloud type, movement, and the change of activity. The SCIC is disseminated as part of the meteorological observations over the JMA eight times a day.

C. Cloud Amount. A continuous observation of global cloud distribution is necessary to monitor the Earth climate system. Cloud amount derived from GMS IR data shows the cloud distribution in the area from 60°N to 60°S and from 80°E to 160°W. The area covers the

Table III-4-6 *VDM characteristics*

	VIS CHANNELS	IR CHANNELS
Number	4	1
Quantization	64 (square root)	256 (linear)
Sample rate	437.5 K samples/sec	109.4 K samples/sec
Bandwidth	150 kHz	37.5 kHz
Output format differentially encoded into two 2-bit bytes		

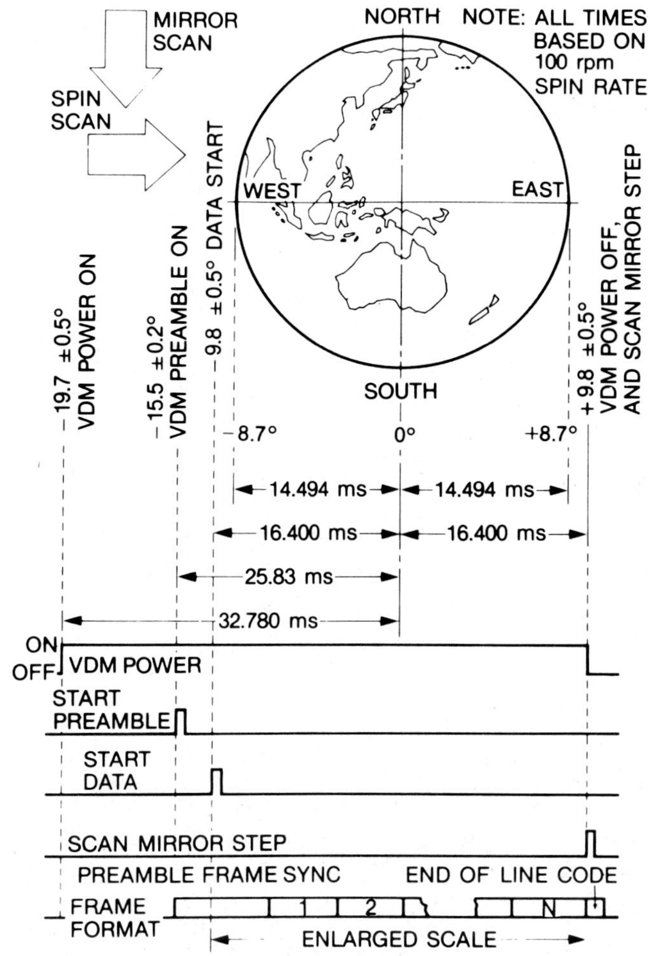

Figure III-4-5. The VISSR data multiplexer Earth scan timing.

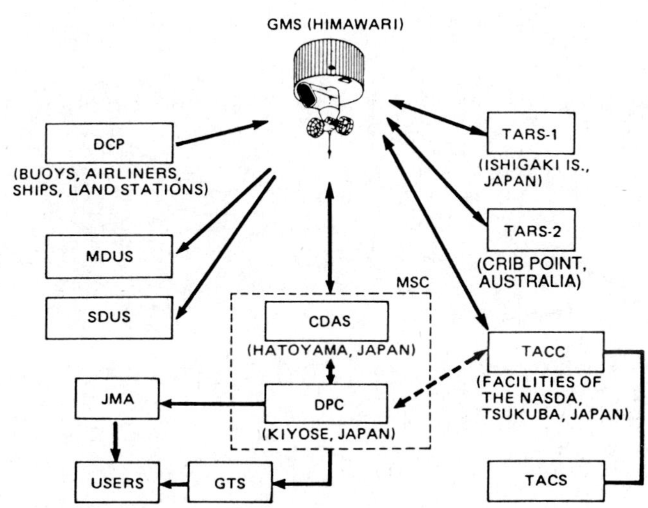

Figure III-4-6. Schematic of the overall GMS system.

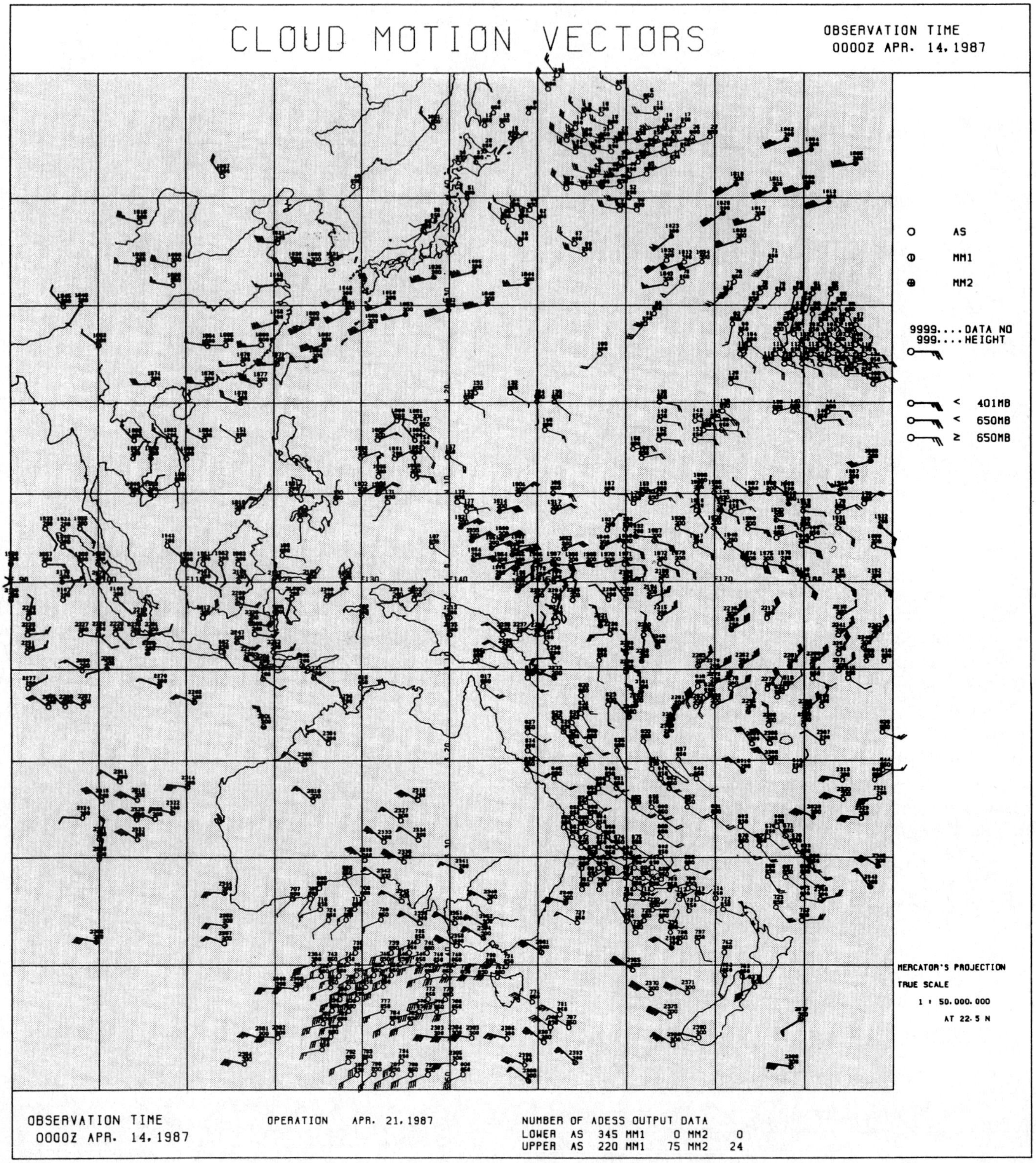

Figure III-4-7. The cloud motion vectors derived from GMS data.

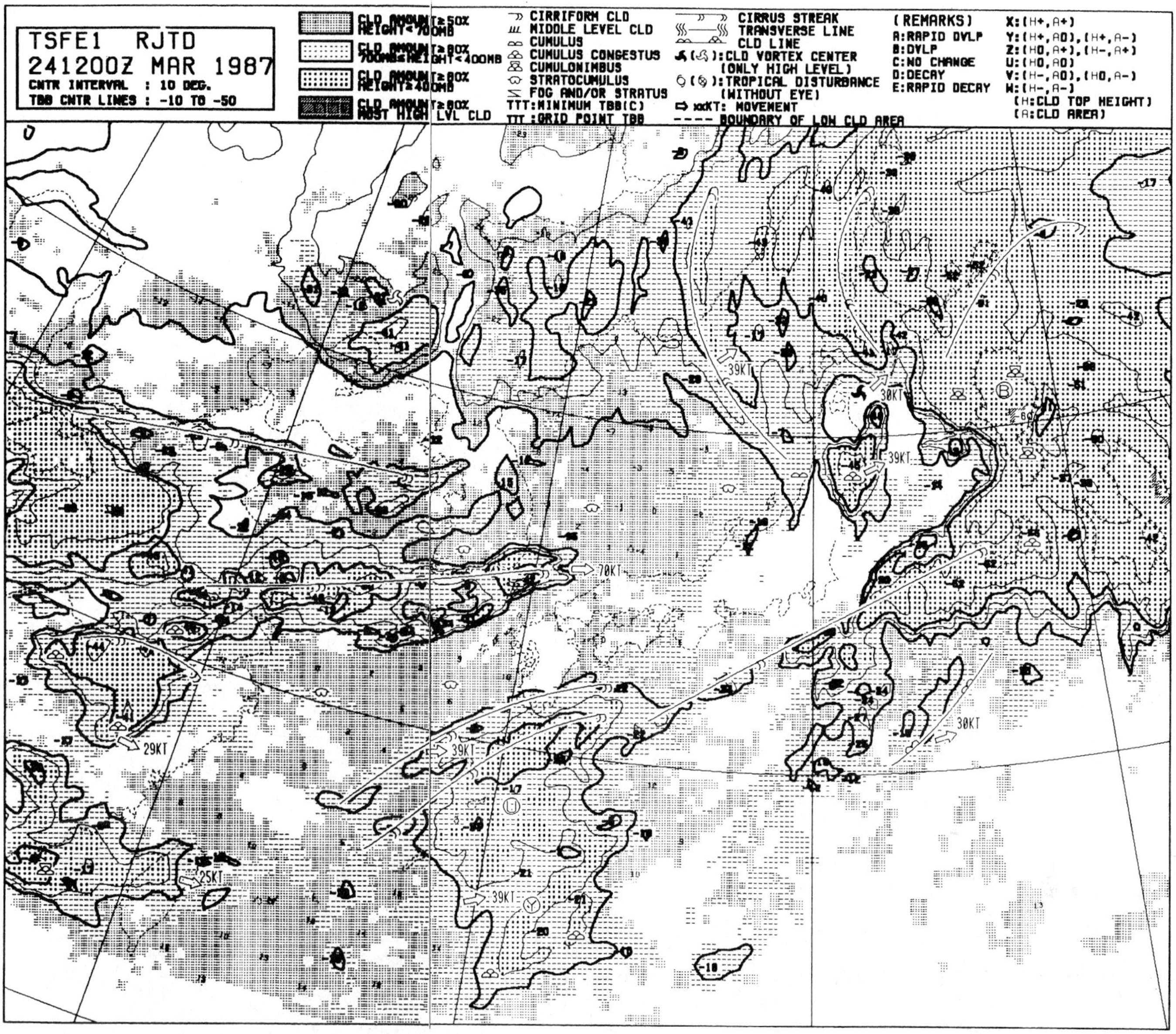

Figure III-4-8. The Satellite Cloud Information Chart (SCIC) derived from GMS data.

Intertropical Convergence Zone (ITCZ) whose tendency is closely related to the global climate system.

Figures III-4-7, III-4-8, and III-4-9 show examples of various satellite derived data.

III-4.3.4 Direct Broadcasts of the GMS Images

GMS provides two kinds of direct broadcast services: real-time digital cloud images (S-VISSR) and analog images (WEFAX). These services provide meteorological agencies in the coverage area of GMS spacecraft with useful information for understanding the latest weather situation and issuing weather advisories, warnings, and daily weather forecasting.

A. Stretched VISSR data are the digital image data disseminated to users simultaneously with VISSR observation. The S-VISSR data can be processed not only with a high grade computer system but also with a personal computer system.

Some of the meteorological products extracted from VISSR are shown in the illustrations. Utilizing S-VISSR, users can extract their own meteorological parameters necessary for the regional requirements. The routine broadcast of the S-VISSR data began in March 1988.

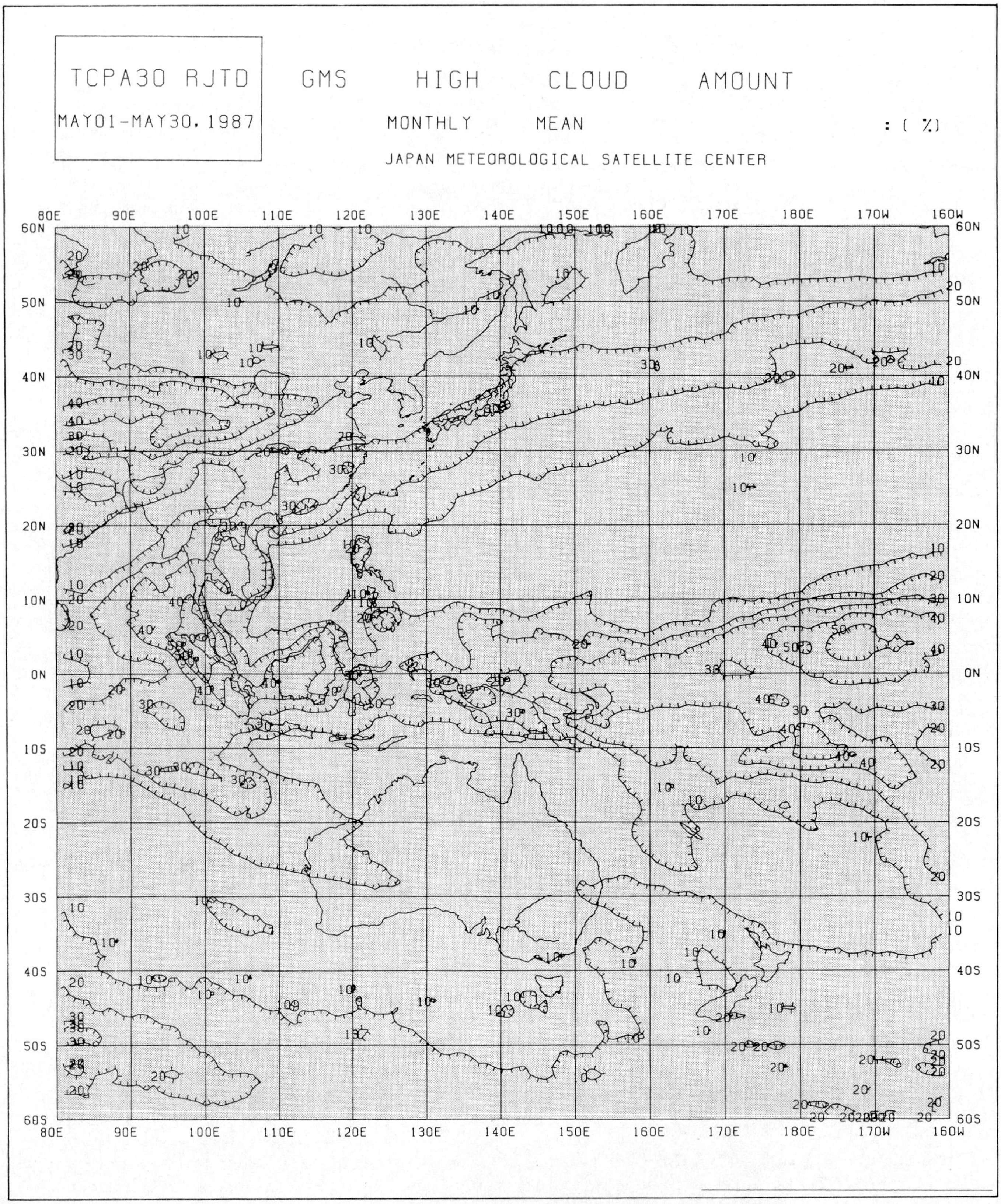

Figure III-4-9. The monthly mean cloud amount chart derived from GMS.

Figure III-4-10. The WEFAX image data transmitted via the GMS spacecraft over the WEFAX service.

A–D:	Infrared	four-sectorized Earth disk pictures
H:	Infrared	polar-stereographic projection picture
I:	Visible	polar-stereographic projection picture
J:	Enhanced infrared	polar-stereographic projection picture

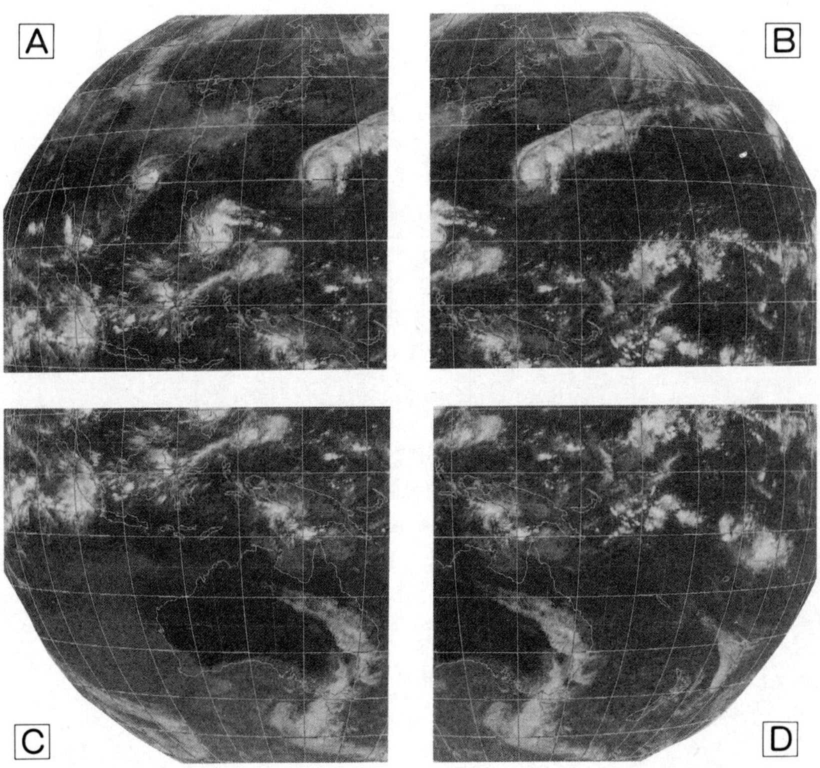

Table III-4-7 *Specification of DCP*

Frequency bands allocated	402.0–402.1 MHz (IDCP) 402.1–402.4 MHz (RDCP)
Bit rate	100 bps
Power	No more than 52 dBm
Type of modulation	PCM-PSK (±60°)

B. WEFAX is produced by the computer system using VISSR image data processed by: sampling, brightness conversion, superimposing grid and coastal lines, and adding annotation. There are two types of WEFAX pictures: four-sectorized Earth disk pictures (Figure III-4-10 A-D) and polar-stereographic pictures (Figure III-4-10 H-J).

III-4.3.5 Data Collection System (DCS)

DCS is the spacecraft system that collects the meteorological data observed by automated remote weather stations (Data Collection Platforms) and distributes the data to users through GTS. The GMS DCS is a part of the International DCS (IDCS) coordinated with other satellite operators. DCP specifications are listed in Table III-4-7. The DCP data are collected through GMS and converted to the WMO codes at MSC and distributed through GTS.

A. International DCP (IDCP) support mobile DCP on ships, aircraft and balloons which move from the coverage of one geostationary meteorological satellite to another. Thirty-three channels are provided for IDCP.

B. Regional DCP (RDCP). The GMS DCS supports regional DCP which are covered by one geostationary satellite. These are stationary DCP on moored buoys, isolated islands, mountains, etc., and regionally mobile DCP. The GMS provides 100 channels for RDCP.

There are two types of DCP: self-timed DCP which transmit the observation data automatically on scheduled time, and interrogation DCP which transmit the observation data by the interrogation command from MSC. Only self-timed DCP can use IDCS.

Chapter III-5

THE INDIAN INSAT[1]

III-5.1 OVERVIEW

The Indian National Satellite System (Insat-1) is a multipurpose system performing three distinct functions from a single space platform. The overall system is planned to consist of (Insat, 1983):

- Two identical geostationary satellites located at 74°E and 94°E and a Master Control Facility (MCF) for command, control, and monitoring of the satellites.
- A telecommunications ground network of 31 Earth stations and a network operations control center.
- A Meteorological Data Utilization Center (MDUC), 20 secondary data utilization centers, 100 land-based and 10 ocean-based automatic unattended Data Collection Platforms (DCP), and a Disaster Warning System.
- A television broadcast system consisting of a direct broadcast of TV signals to numerous community receivers on the ground, and a terrestrial rebroadcasting system that uses the satellite to distribute TV signals from a few central facilities to numerous terrestrial transmitters.
- A radio network with uplink facilities and a large number of five channel receive only terminals for satellite radio reception and rebroadcast.

The Insat-1 system concept is diagrammed in Figure III-5-1.

The telecommunications, meteorological, and broadcast functions are provided by four different systems on the Insat-1 spacecraft. The telecommunications payload consists of 12 transponders, each with 36 MHz bandwidth, operating at 5935 to 6425 MHz (Earth to satellite) and at 3710 to 4200 MHz (satellite to Earth). Meteorological observations are made by a Very High Resolution Radiometer (VHRR) with VIS and IR channels with resolutions of 2.75 km and 11 km, respectively. The VHRR can provide full Earth coverage every 30 min and sector scans in 5 min.

A data relay transponder receives the data transmitted by unattended DCP at 402.75 MHz and retransmits the data to Earth in the C-band. The TV broadcast and radio networking functions are provided by two high power S-band transponders operating at 5855 to 5935 MHz (Earth to satellite) and at 2555 to 2635 MHz (satellite to Earth). Each transponder can handle one TV broadcast channel and five injected carrier channels for radio network broadcasts, disaster warnings, and similar functions.

The Insat-1 operations are carried out from the MCF located at Hassan, India. The MCF is equipped for telemetry data reception and processing, ranging and orbit determination, and command generation and transmission. Also available are equipment and systems for testing all satellite payloads. A quick look facility is provided for monitoring VHRR imaging during regular operations.

Meteorological data transmitted from VHRR are received and processed in the MDUC at New Delhi. DCP data also are collected, processed, and retransmitted at MDUC.

---

[1] Primary Contributor: Donald W. Graul
Original manuscript submitted: January 1985.

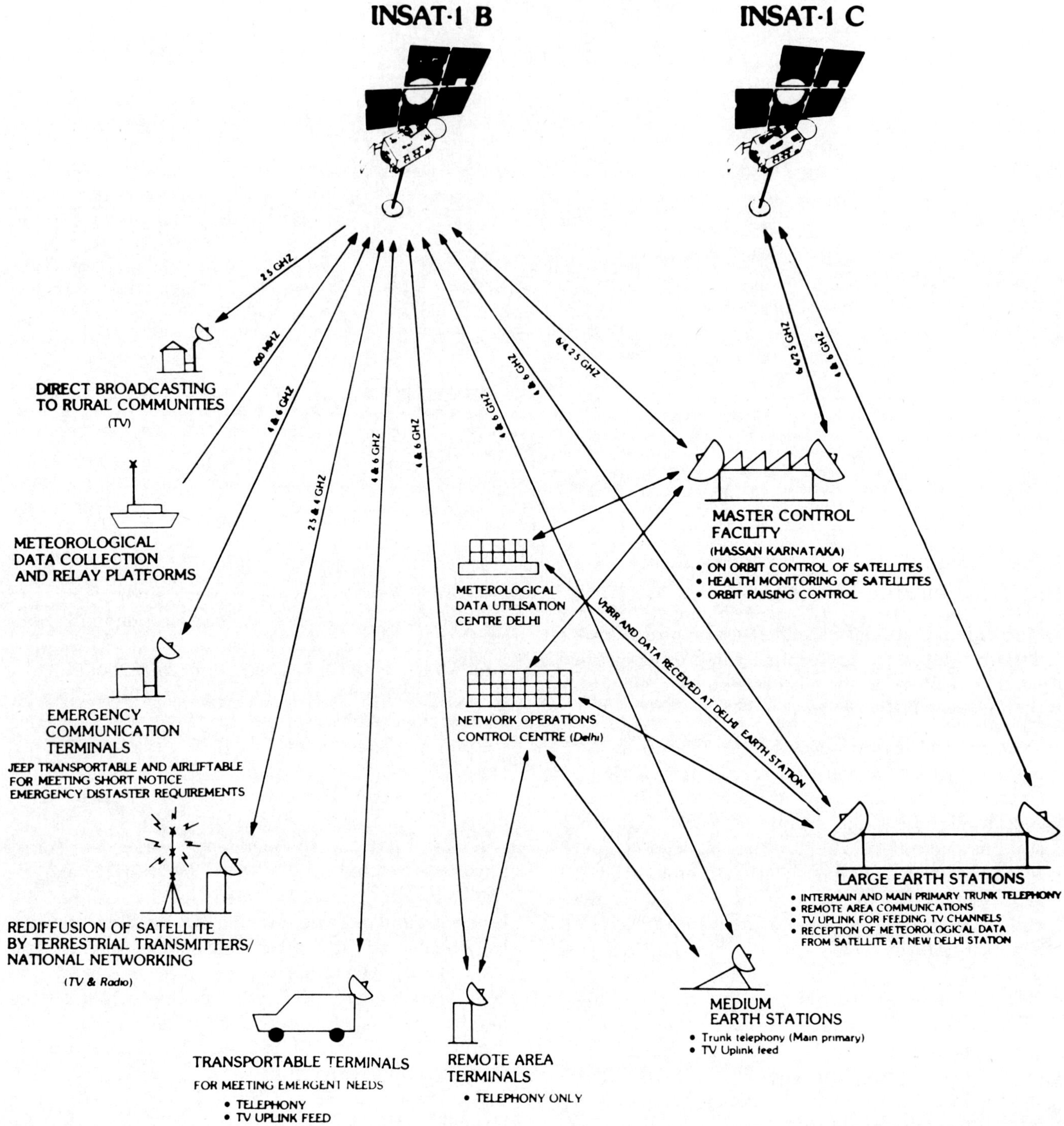

Figure III-5-1. Insat system concept schematic.

III-5.2 INSTRUMENTATION

III-5.2.1 Very High Resolution Radiometer (VHRR)

The VHRR is a two channel instrument that measures direct and reflected radiation in the VIS (0.55 to 0.75 μm) and the IR (10.5 to 12.5 μm) spectral bands. The Field Of View (FOV) of the optics is a square element with 2.75 × 2.75 km resolution in the VIS and 11 × 11 km in the IR from geosynchronous altitude. These instantaneous FOV are scanned across the Earth surface by a controlled mirror. The bidirectional scan

Table III-5-1 *Insat-1 VHRR performance characteristics*

PARAMETERS	VALUE
Scanned Field	
E/W	20°
N/S	19.96°
IR Channel Wavelength	10.5 to 12.5 μm
Visible Channel Wavelength	0.55 to 0.75 μm
IR Field of View, max.	
E/W	243 μrad
N/S	326 μrad
Visible FOV, max.	
E/W	78 μrad
N/S	67 μrad
IR Channel NEΔT	0.19 @ 300 K
Visible Channel S/N (min.) (@ 2.5 albedo)	13/1
Detector Registration (max.)	
E/W	30 μrad
N/S	38 μrad
Scan Repeatability (max.)	22 μrad
Scan Linearity (max.)	
E/W	85 μrad
N/S	24 μrad
Frame Time (w/calibration and reset)	23.2 min
Digital Resolution	10 bits
Modulation Transfer Function (min.) @ 25°C	
Visible	
E/W	0.26
N/S	0.49
IR	
E/W	0.27
N/S	0.37

E/W signifies east-west and/or west-east direction.
N/S signifies north-south and/or south-north direction.

takes one second to cross 20° from east to west and 0.2 sec to step 307 μrad from north to south. In all, 22 min and 44 sec are required to scan a 20° by 20° lat./long. image of the Earth. In addition to the normal 20° by 20° scan, the VHRR can produce a 20° (east-west) and 5° (north-south) sector image. The start of each sector can be selected in 1° north-south increments so that small portions of the Earth's disk can be scanned every 6 min. In-orbit calibration of VIS and IR data channels is accomplished electronically using deep space and an internal blackbody as temperature reference points. The VHRR is made up of two units:

- A module containing the scan assembly, the telescope and optics assembly, and the radiant cooler assembly.
- An electronics unit containing the power converter, the timing system, the signal processing system, data block and data line formatting control logic, scan control electronics, and interface circuitry.

The VHRR transmitter output is multiplexed with the C-band telecommunications channels for transmission to ground. Table III-5-1 lists Insat-1 VHRR measured performance characteristics.

III-5.2.2 Data Relay Transponder (DRT)

The DRT consists of a UHF antenna, a down converter, an upconverter, and a data relay amplifier. The slot array UHF antenna receives the 402.75-MHz DCP signal and relays it to the ground at 4038.1 MHz.

III-5.3 SATELLITE DESCRIPTION

The Insat-1 satellite is designed and manufactured by Ford Aerospace & Communications Corporation for the Government of India, with specifications provided by the Indian Department of Space. Insat-1 is a biased momentum three axis stabilized satellite with a 7-year design life. A precision attitude control system provides the stability needed for meteorological imaging and antenna pointing accuracy. The satellite uses an asymmetrical solar array to provide a clear view of

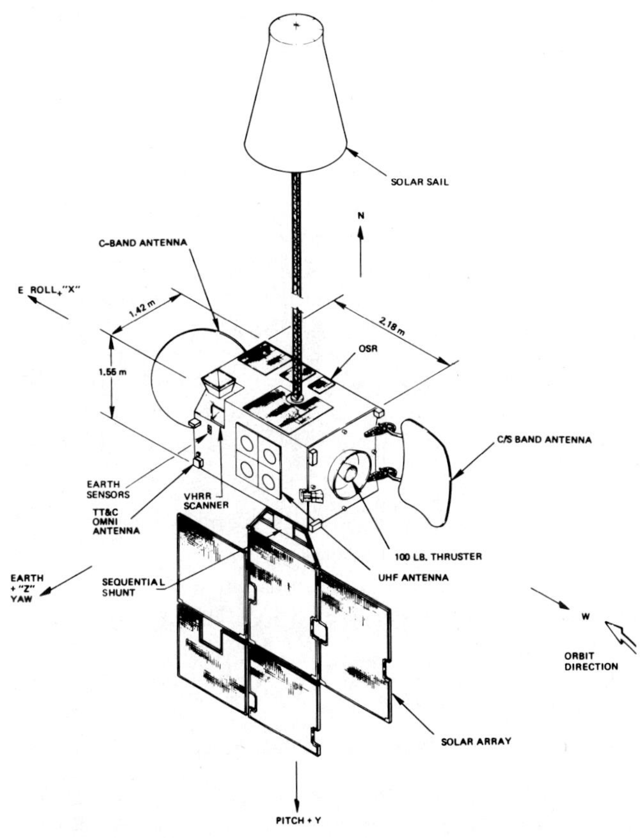

Figure III-5-2. Insat in-orbit configuration schematic.

Figure III-5-3. Insat data collection platform.

space for the radiation cooler of the VHRR. A solar sail counterbalances solar radiation pressure on the solar array. A magnetic torquer is used for fine control. Figure III-5-2 shows the Insat-1 configuration.

III-5.4 MASTER CONTROL FACILITY (MCF)

The MCF, located in Hassan, near Bangalore, is the major ground system element used to support Insat-1 satellite operations. Functions provided by the MCF include:

- Telemetry reception and processing for satellite hardware monitoring and control
- Command generation and transmission
- Reception and processing of VHRR data for quick look evaluation
- Ranging and orbit determination for orbital analysis and control
- Instrumentation test and evaluation.

The MCF also is the focal point for overall Insat-1 system operations coordination. It has dedicated voice and teleprinter links with the Network Operations Control Center and MDUC at New Delhi.

The MCF is equipped with two independent Satellite Control Earth Stations, each with a fully steerable 14-m diameter antenna and a Satellite Control Center with telemetry, ranging and command equipment, computer systems, and recording and test equipment.

III-5.5 METEOROLOGICAL DATA UTILIZATION CENTER (MDUC)

MDUC, the key meteorological ground facility, is located in New Delhi. Satellite data are received at the Insat Earth station located about 60 km from New Delhi and are transmitted over a MW ground link to the MDUC. There the data are demodulated and processed to produce cloud cover imagery and other weather system data.

The 400-kbps VHRR data stream is first ingested by a preprocessing computer. On completion of an image scan, the data are processed to produce both VIS and IR cloud cover imagery. Also computed are sea

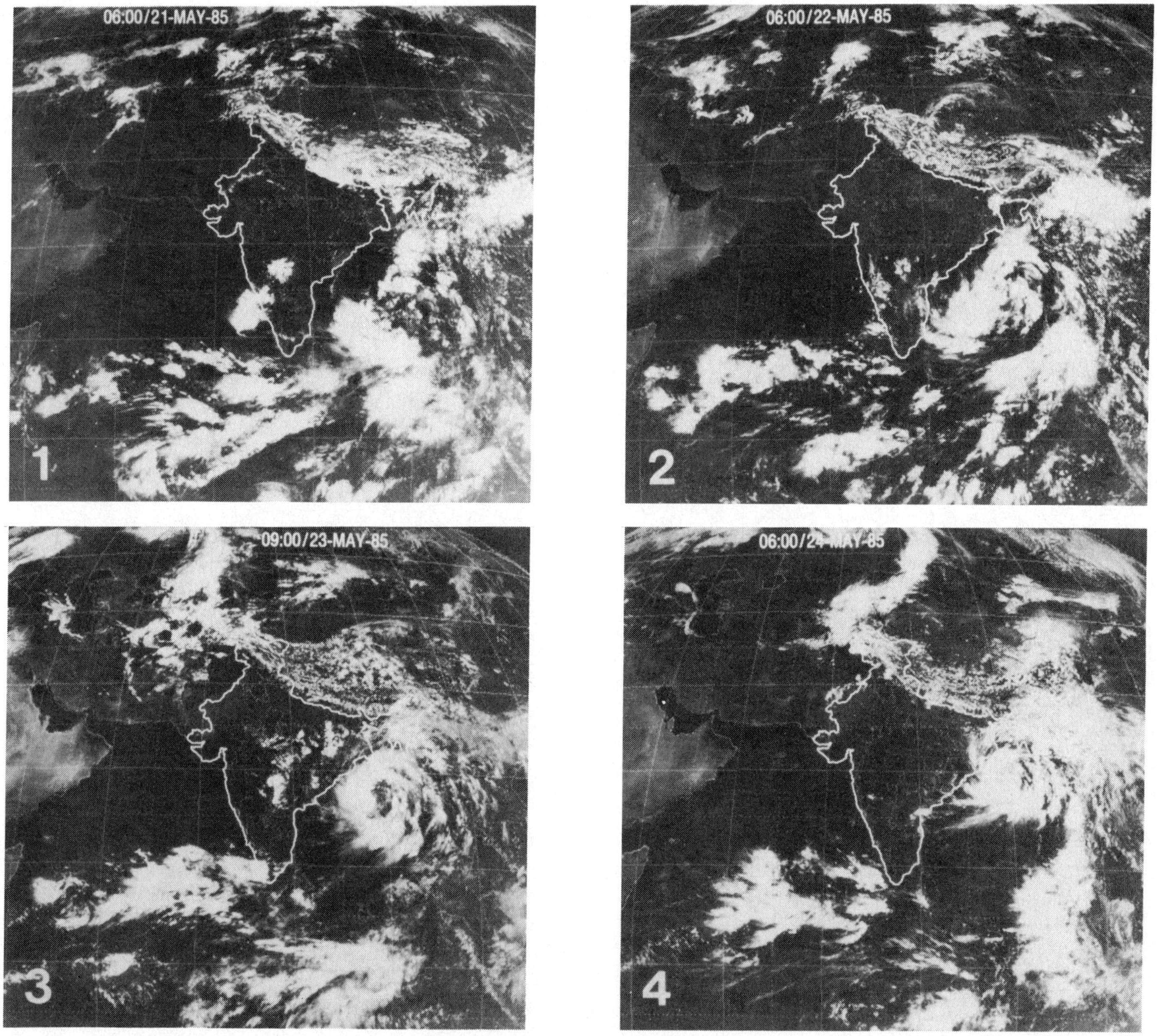

Figure III-5-4. Insat VHRR images of May 1985 Bangladesh cyclone.

surface temperatures, cloud heights, and cloud velocity information.

In addition to VHRR data, MDUC also receives and processes data from the Insat DCP. Meteorological data from the DCP are used in local weather forecasting.

The Disaster Warning System, when operational, will use satellite data to forecast severe local weather conditions and to issue warnings to affected areas over a satellite S-band channel. Warning messages, addressed to specific locations, will be used to issue warnings to the nearby communities.

III-5.6 DATA COLLECTION PLATFORM (DCP)

The DCP is an unattended, battery powered set of instruments that automatically collects local meteorological, hydrological, or oceanographic data and transmits them to the satellite. Ten to 13 parameters are sensed and converted to a digital data stream at 4.8 kbps. This data stream, containing a station identification code, is transmitted to the satellite at 402.75 MHz. Each DCP transmits the data in a burst mode at random times during a designated 10-min interval each hour. About 100 land-based and 10 ocean-based DCP are planned for the Insat-1 system. Figure III-5-3 shows a typical DCP.

III-5.7 IN-ORBIT DATA RESULTS

Insat-1B, launched on August 30, 1983, was positioned at 74°E. Initial orbital checkout of satellite systems and payloads was completed by the end of September and regular operational use of the satellite started on October 15. A number of cyclonic developments over the Indian Ocean and Bay of Bengal were tracked and precautionary warnings issued to minimize damage. Figure III-5-4 shows a series of VHRR images, taken by Insat-1B, of the Bangladesh cyclone of May 1985.

Chapter III-6

THE UNITED STATES RESEARCH SATELLITES[1]

III-6.1 INTRODUCTION

For more than 25 years, NOAA and NASA have been working in close partnership to fulfill complementary missions in the civil national space program: NASA at the forefront of new space technology and its applications, and NOAA to provide space data services to both national and international users.

NOAA/NASA working procedures with respect to NOAA's operational polar orbiting and geostationary satellites were first defined in a basic agreement signed in 1964. Under this agreement, NOAA determined the requirements of its user community and specified the geophysical parameters to be measured. In response, NASA prepared the plan of approach, specified and procured the necessary flight hardware, conducted launch operations, and carried out an in-orbit checkout of the spacecraft. Data acquisition, processing, distribution, and archiving were carried out by NOAA. The roles and responsibilities defined in this agreement are still basically in force. NOAA is now responsible for spacecraft command and control after "handover" by NASA upon completion of in-orbit checkout.

The first generations of low Earth orbiting meteorological satellites (the TIROS series) and geostationary satellites (the ATS/SMS series) were developed under NASA funding. For many years NASA budgeted for the prototype unit of each major system change, with NOAA funding later units. Although NOAA is now responsible for the funding of improvements to operational systems, NASA continues to evaluate and develop new systems as each is proposed.

[1] Primary Contributor: James Greaves

III-6.2 HISTORY

Three satellite programs initiated by NASA have had a profound impact on the nation's meteorological satellite capabilities. Beginning with the TIROS series, NASA has over the years improved the design and development of new generations of operational polar satellites. The Nimbus program served as a test bed for new instruments and techniques. The experimental geosynchronous Applications Technology Satellites (ATS) evolved into the operational GOES series.

III-6.2.1 TIROS

The current generation of Polar Orbiting Environmental Satellites (POES) evolved over a quarter of a century from TIROS-1. Since the initial series of ten TIROS research and development satellites, four generations of POES (TOS, ITOS, TIROS-N, Advanced TIROS-N) have evolved.

III-6.2.2 Nimbus

The Nimbus project, initiated by NASA in the early 1960s, eventually became the nation's principal satellite platform for remote sensing research.

Table III-6-1 *Transfer of Nimbus technology to operational spacecraft*

NIMBUS INSTRUMENT		OPERATIONAL INSTRUMENT	
APT	Automatic Picture Transmission System	APT	Automatic Picture Transmission System
AVCS	Advanced Vidicon Camera System	AVCS	Advanced Vidicon Camera System
SIRS	Satellite Infrared Spectrometer	ITPR	Infrared Temperature Profile Radiometer
IRLS	Interrogation, Recording and Location System	DCS	Data Collection Systems
HIRS	High Resolution Infrared Sounder	HIRS	High Resolution Infrared Sounder
THIR	Temperature-Humidity Infrared Radiometer	AVHRR	Advanced Very High Resolution Radiometer
MRIR	Medium Resolution Infrared Radiometer	SR	Scanning Radiometer
NEMS	Nimbus Experiment Microwave Spectrometer	AMSU	Advanced Microwave Sounding Unit
		MSU	Microwave Sounding Unit
PMR	Pressure Modulated Radiometer	SSU	Stratospheric Sounding Unit
BUV	Backscatter Ultraviolet	SBUV	Solar Backscatter Ultraviolet Spectrometer
SBUV	Solar Backscatter Ultraviolet Spectrometer		

As applications and research vehicles, the seven Nimbus spacecrafts were used to test a variety of new and advanced meteorological and geophysical remote sensing instruments. Many of these NASA experimental sensors became operational sensors for NOAA's satellite program (Table III-6-1).

Nimbus satellites launched between 1964 and 1978 carried out 48 experiments to survey the atmosphere, map land and water characteristics, and observe weather and climate patterns. Nimbus launch dates are given in Figure I-2-5 and payloads are listed in Table I-2-1.

The Nimbus spacecraft were placed in Sun synchronous, near polar orbits and acquired global data twice every 24 hours. The basic spacecraft weighs about 950 kg and is similar in shape to an ocean buoy. It is approximately 3 m tall, 1.5 m in diameter at the base, and 4 m wide with the solar panels fully extended. Figure I-2-6 is a line drawing of Nimbus-7. A circular structure at the base of the satellite houses the major subsystems and experiments. A three axis attitude control keeps the base of the spacecraft pointing toward nadir throughout orbit.

Nimbus-7 (last in the series) has proven to be one of the single most significant sources of experimental space data relating to atmospheric and oceanic processes. Nimbus-7 was launched with eight experiments to acquire these data which are validated and archived by NOAA.

III-6.2.3 SMS/GOES

During the middle 1960s, a NASA research program based on the use of geostationary satellites was initiated; the spacecraft were the Applications Technology Satellite (ATS) series. Although the primary ATS mission was to improve communications satellite technology, the spacecraft structure had sufficient excess capacity to permit mounting meteorological sensors.

On December 7, 1966, the ATS-1 was launched and placed into a geostationary orbit 35,900 km above the Pacific Ocean. For the first time, continuous observations were made of nearly one-third of the Earth surface. ATS-1 showed that a picture of the Western Hemisphere (from approximately 55°N to 55°S) could be taken every half hour with a spin scan camera. ATS-3, launched in November 1967, took the first color pictures of the full Earth disk. Sequential pictures of the same area facilitate the early detection of severe storms and provide real time data on cloud and frontal movements. See I-2.6 for further details of ATS satellites.

The ability to acquire atmospheric observations from geostationary altitudes led to NASA's development of a satellite designed specifically for meteorological uses. NASA's prototype Synchronous Meteorological Satellite, SMS-1, was placed in orbit in May 1974 and positioned above the Equator at 45°W. Then came SMS-2 in February 1975, followed in October 1975 by the first operational version for NOAA, GOES-1 (Geostationary Operational Environmental Satellite). Six GOES have since been launched, the last in the current series GOES-H (GOES-7 upon becoming operational) on February 26, 1987. A major advance occurred September 9, 1980, when GOES-4 went up carrying the first atmospheric sounder used at geostationary altitude. This Visible Infrared Spin Scan Radiometer (VISSR) Atmospheric Sounder (VAS) was also carried on GOES-5, 6, and 7. See I-2.7 for other details of SMS GOES.

III-6.3 EARTH RADIATION BUDGET SATELLITE (ERBS)

The average temperature of the Earth has fluctuated dramatically throughout history. Changes in the Earth's orbit, variations in solar activity, the greenhouse effect, cloud cover, volcanic debris, and human activities are some of the variables that may influence global temperature. The study of the Earth radiation budget is an integral part of any program to increase understanding of the processes and the predictability of climate. NASA's Earth Radiation Budget Experiment (ERBE) was created to gather radiation budget data for use in climate studies.

ERBE was designed to provide observations and measurements of incoming and outgoing radiation to determine, for a minimum of 1 year: 1) monthly average radiation regionally, zonally, and globally; 2) Equator-to-pole energy transport gradient; and 3) monthly average diurnal variation in regional radiation budget. ERBE observations can be of use in validating elements of climate and long-range weather prediction models. ERBE observations can also provide a baseline against which to compare subsequent radiation budget observations.

III-6.3.1 The Spacecraft

While satellites are excellent observation platforms, the requirement for measuring thermal and solar radiation over the entire Earth at all times of day is extremely demanding. To make such measurements, it was arranged for ERBE instruments to be flown on two NOAA operational satellites (NOAA-9 and NOAA-10), and also on the NASA Earth Radiation Budget Satellite (ERBS).

The ERBS (Figure III-6-1) was placed in a 57° inclination orbit, the spacecraft passing over nearly all

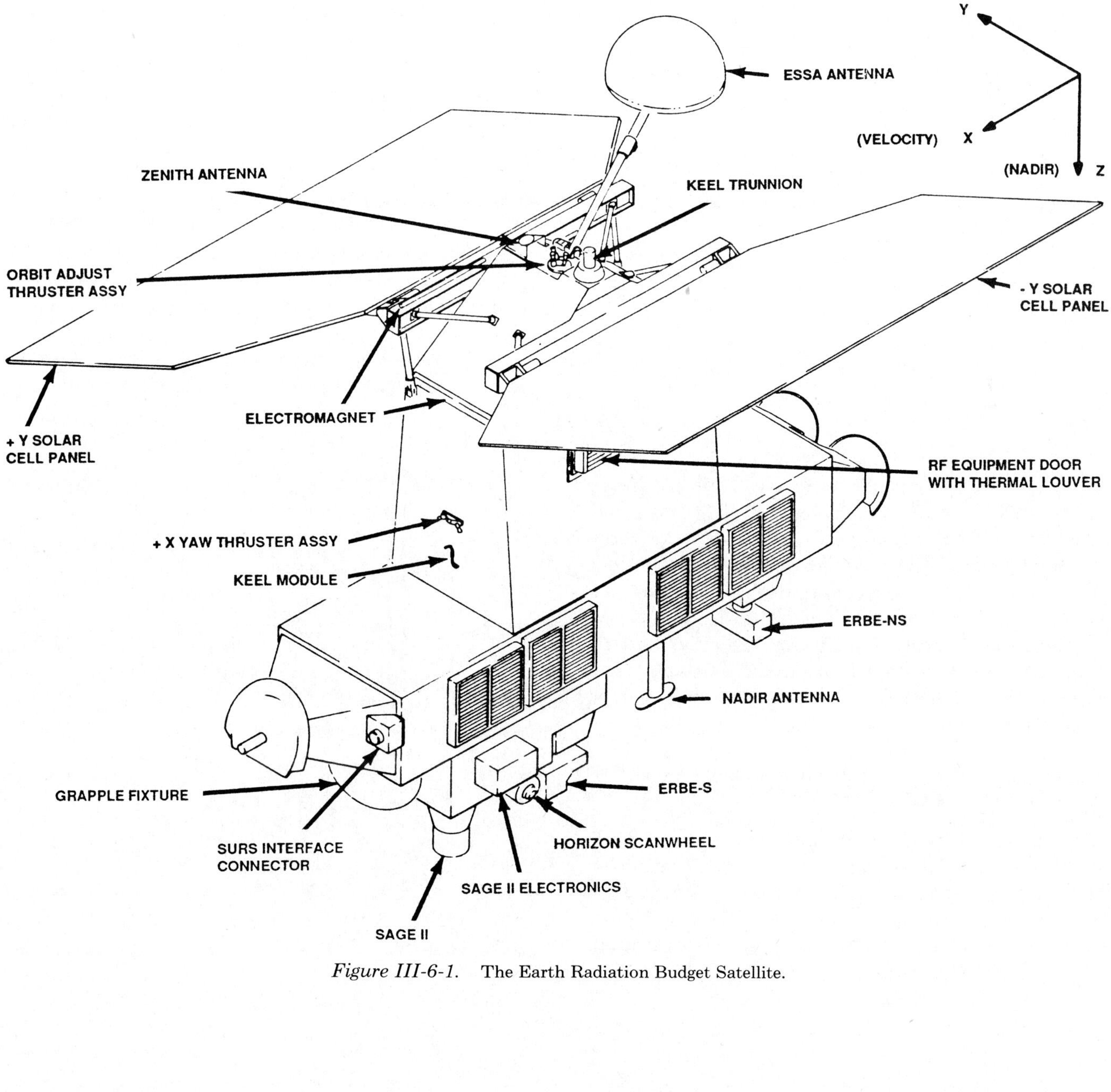

Figure III-6-1. The Earth Radiation Budget Satellite.

parts of the Earth from 57°S to 57°N at all times of day during a 2-month period. In October 1984, ERBS was placed in a preliminary orbit by the Shuttle and then boosted to its final orbit of 610 km by an onboard gas jet propulsion system which could also be used for later orbit adjustments.

III-6.3.2 The Instruments

To meet the scientific objectives of ERBE, the experiment includes a Scanner instrument (ERBE-S) and a Non-Scanner instrument (ERBE-NS).

ERBE-S is a scanning radiometer with three channels: longwave, shortwave, and total. The shortwave channel measures reflected solar radiation, the longwave channel measures emitted radiation, and the total channel serves as a check on the other two. The telescopes of three channels are mounted on a continuously moving scanhead assembly that scans cross track every 4 sec. Each channel makes 74 radiometric measurements per scan. The Instantaneous Field Of View (FOV) of each channel is about 40 km of the Earth surface at nadir.

ERBE-NS has five sensors, each using cavity radiometer detectors. Four point directly at the Earth: two wide FOV sensors view the entire disk of Earth; two medium FOV sensors view a 10° region. The fifth, a solar monitor, measures the total solar radiant output. One wide and one medium FOV sensor measure in the shortwave; the other two measure total radiation. These sensors are among the most accurate radiometers ever used to observe Earth from space.

III-6.3.3 The Science Team

To get the maximum benefits from the ERBE, a team of 60 scientists was assembled to advise on planning the experiment and to analyze and use the data. Included are experts in satellite instrumentation and measurement methods, data analysis, Earth radiation budget, and weather and climate processes. Results will be used to further understanding of the time and space variations of cloudiness and Earth radiation budget parameters, and to verify mathematical–physical models of the circulation of the earth-atmosphere system for climate studies.

III-6.4 THE UPPER ATMOSPHERE RESEARCH SATELLITE (UARS)

Data gathered by the UARS, scheduled for a 1991 launch, are to be used to improve understanding of the chemistry, dynamics, and energetics of the stratosphere, mesosphere, and lower thermosphere with particular focus on long-term changes in the concentration and distribution of ozone and other atmospheric constituents. The UARS program was initiated in response to growing concerns about possible stratospheric ozone depletion.

Although the basic concepts regarding the catalytic destruction of ozone by reactive gases remain unchanged, a global scale data base for further research is currently lacking. UARS will provide that needed data base.

UARS will be the largest atmospheric research satellite ever flown (Figure III-6-2). The payload will consist of nine experiments specially selected to complement each other and ensure success of the overall science mission. A tenth instrument will be flown to continue a NASA program of long-term total solar irradiance measurements. In addition to nine experimental groups, the UARS program includes ten theoretical groups with specific responsibilities for data analysis and interpretation. Results from the Solar Backscatter Ultraviolet (SBUV) spectrometer on NOAA operational satellites will be included in the UARS science data base. This science data base will reside in the UARS Central Data Handling Facility (CDHF) located at Goddard Space Flight Center. UARS investigators will have near real time access to all UARS data via electronic link.

The UARS will be delivered by the Shuttle orbiter into a 600-km circular orbit at an inclination of 57°. With the exception of an energetic particle measurement instrument, the instruments are passive remote sensors. UARS will make the first direct measurements of upper atmospheric winds through high resolution spectroscopic techniques, and will gather data for com-

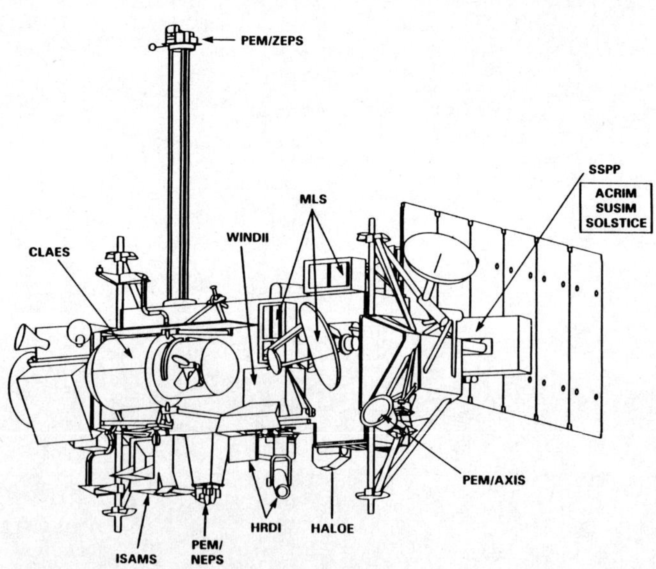

Figure III-6-2. The Upper Atmosphere Research Satellite (UARS).

parison with the predictions of 3-dimensional models of the upper atmosphere.

The nine UARS instruments consist of three groups: (a) three energy input experiments; (b) four chemical species and temperature experiments; and (c) two wind experiments. See Table III-6-2 for a brief description of each.

III-6.5 THE FUTURE IN EARTH SCIENCE REMOTE SENSING

Great strides have been made in developing means for observing the Earth. These techniques have focused mainly on such elements of the Earth system as properties of the atmosphere, the oceans, and the land surface. The Landsat and meteorological satellite programs are successful examples of this approach. More recently, issues have emerged that necessitate a long-term interdisciplinary approach. Key examples are the carbon dioxide greenhouse effect, the human consequences of the anticipated depletion of the stratospheric ozone layer, the El Niño-related modifications to weather patterns, and acid precipitation. To study these issues requires cooperation among atmospheric and soil scientists, foresters, agronomers, biological oceanographers, and others.

The key to progress on these and other interdisciplinary issues in Earth science during the 1990s will be to approach the Earth as an integrated system. The fundamental processes that govern and integrate the Earth system are the hydrologic cycle, the biogeochemical cycles, and climatic processes. The comprehensive, integrated study of these processes has been termed "Earth system science," with a strong focus on developing the capability to predict the evolution of the Earth system on time scales of decades to a century.

The first major new interdisciplinary mission for Earth science in the 1990s will be the Earth Observing System (EOS) (Figure III-6-3). EOS is the first interdisciplinary mission to be planned. The goal of EOS is to advance the understanding of the Earth system on a global scale through developing a deeper understanding of the components of that system, the interactions among them, and how the Earth system is changing. In order to quantify changes in the Earth system, EOS will be a long-term mission providing systematic, continuing observations from low Earth orbit. Crucial to the successful achievement of the mission's objectives will be the development of the EOS Data and Information System (EOSDIS) which will provide access to the data from EOS instruments and other relevant data sets and to the scientific results of research using these data.

The EOS Program will include space and ground based research, serviceable spacecraft carrying remote sensing instruments, and a distributed information system and global data base. The EOS Program includes participants from the United States, the European Space Agency (ESA), Japan, and other nations. U. S. participants include NASA, NOAA, USGS, and others.

EOS is to be an evolutionary program with an orbital observing capability that grows over time and is maintained for 15 years. This 15 year mission lifetime will be accomplished via instrument and platform redun-

Table III-6-2 *UARS experimental investigations*

INSTRUMENT	DESCRIPTION	PRIMARY MEASUREMENTS
SOLSTICE—Solar-stellar intercomparison experiment	Full disk solar irradiance spectrometer incorporating stellar comparison	Solar spectral irradiance: 115–430 nm
SUSIM—Solar ultraviolet spectral irradiance monitor	Full disk solar irradiance spectrometer incorporating onboard calibration	Solar spectral irradiance: 120–400 nm
PEM—Particle environment monitor	X-ray proton, and electron spectrometers	In situ energetic electrons and protons; remote sensing of electron energy deposition
CLAES—Cryogenic limb array etalon spectrometer	Neon and carbon dioxide cooled spectrometer sensing atmospheric infrared emissions	T, CF_2Cl_2, $CFCl_3$, $ClONO_2$, CH_4, O_3, NO_2, N_2O, HNO_3, and H_2O
ISAMS—Improved stratospheric and mesospheric sounder	Mechanically cooled spectrometer sensing atmospheric infrared emissions	T, O_3, NO, NO_2, N_2O, HNO_3, H_2O, CH_4, and CO
MLS—Microwave limb sounder	Microwave radiometer sensing atmospheric emissions	ClO, H_2O, O_3 and atmospheric pressure
HALOE—Halogen occultation experiment	Gas filter/radiometer sensing sunlight occulted by the atmosphere	HF, HCl, CH_4, CO_2, O_3, and H_2O
HRDI—High resolution Doppler interferometer	Fabry-Perot spectrometer sensing atmospheric emission and scattering	Two-component wind: 10–110 km
WINDII—Wind Imaging Interferometer	Michelson interferometer sensing atmospheric emission and scattering	Two-component wind: 80–110 km

Figure III-6-3. Artist's concept of the Earth Observing System (EOS).

dancy, and orbital replacement and servicing. Current plans call for the launch of four instrumented polar orbiting platforms. In conjunction with the international Space Station Program, NASA plans to launch one polar platform in 1996 and the European Space Agency plans to launch one polar platform in 1997. The NASA platform will assume an 824-km, Sun synchronous orbit with a 1330 LST Equator crossing time. The ESA platform will assume an 824-km, Sun synchronous orbit with a 1000 LST Equator crossing time. In addition, Earth observing instruments are planned for flight as attached payloads on the manned Space Station base, which will assume a 335-460-km, 28.5° inclination orbit. A second NASA polar platform is planned for launch in 1998. A Japanese polar platform is also planned for launch in 1998.

The first and second NASA polar orbiting platforms (NPOP-1 and NPOP-2) will be designed for orbital servicing, to allow repair, replacement, or updating of the instrumentation and spacecraft to fulfill a broad range of scientific requirements over an extended lifetime. It is planned that both NASA platforms will be launched by Titan IVs, and that Delta IIs or Atlas Centaurs will be used for servicing missions. The servicing missions will require a matching orbital altitude (824 km). It is anticipated that the in-orbit rendezvous will be accomplished via an active carrier vehicle which will perform final maneuvers.

The planned instrument suite for the EOS polar platforms is international in scope and includes the operational instrument set sponsored by NOAA. The current planning scenario for the payload (subject to change upon completion of final principal investigator instrument selection and completion of Phase B definition and design studies) is:

A. NASA Research Facility Instruments
   - Moderate Resolution Imaging Spectrometer (MODIS), including MODIS-Tilt and MODIS-Nadir

- High Resolution Imaging Spectrometer (HIRIS)
- Synthetic Aperture Radar (SAR)
- Geodynamics Laser Ranging System (GLRS)
- Laser Atmospheric Wind Sounder (LAWS)
- Atmospheric Infrared Sounder (AIRS)
- Wide Band Data Collection System (WBDCS)

B. ESA Research Facility Instruments
- Medium Resolution Imaging Spectrometer (MERIS)
- High Resolution Imaging Spectrometer (HIRIS)
- Atmospheric Lidar (ATLID)
- Scatterometer (SCATT/2)
- Altimeter (ALT/2)
- Synthetic Aperture Radar C-Band (SAR-C)

C. Japanese Research Facility Instruments
- Intermediate Thermal Infrared Radiometer (ITIR)
- Advanced Microwave Scanning Radiometer (AMSR)

D. Principal Investigator (PI) Instruments
Principal Investigator instruments involve the provision of a scientific instrument for flight on the polar platforms or as an attached payload on the manned Space Station, and the reduction, analysis, and interpretation of data from that instrument by a group of scientists. The following is a partial list of potential PI instruments.
- Magnetometer (MAG)
- Correlation Radiometer (CR)
- Magnetospheric Particle Detector (MPD)
- Precision Positioning System–Precision Orbit Determination System (PPS-PPD)

E. Operational Facility Instruments
- Advanced Medium Resolution Imaging Radiometer (AMRIR)
- Global Ozone Monitoring Radiometer
- Space Environment Monitor (SEM)
- Altimeter (ALT/1)
- Scatterometer (SCATT/1)
- Earth Radiation Budget Instrument (ERBI)

F. Research and operational support facilities are expected to be provided by NASA and NOAA.

The EOSDIS is a cornerstone of the NASA EOS Program. The EOSDIS will be comprised of both centralized facilities and distributed elements intended to improve access to Earth science data. The EOSDIS will provide for the command and control of the NASA polar platforms and instruments, payload and platform data processing facilities, active short-term archives, data transfer to permanent archives, and exchange of information with other EOS participants.

The information system supporting EOS will include elements provided by the program's international participants, in addition to U.S. facilities. All EOS data will be available immediately to researchers and other users without exclusion. International collaboration provided by EOS will give scientists and researchers access to integrated global data bases for the study of Earth system science.

BIBLIOGRAPHY AND REFERENCES FOR SECTION III

Afans'ev, Yu. A., et al., 1984: The experimental oceanographic satellite Cosmos 1500. *Proc. of the 35th Congress of the Intl. Astronautical Fed., Lausanne, Switzerland, October 7-13, 1984*, American Institute of Aeronautics and Astronautics, New York, NY.

Andronov, I., 1972: *The Meteors Are on Watch.* Pravda, Moscow, May 25, 1972, p. 3.

Artsimovich, L. A., et al., 1974: *Development of a Stationary Plasma Engine (SPE) and its testing on the meteor artificial Earth satellite. Kosmicheskiye Issledovniya, 12* (3), 451-468.

Aviatsiya i Kosmonavtika, 1967: No. 9, 30-35.

Burtzev, A. I., et al., 1984: Monitoring of the Arctic and Antarctic ice cover with Cosmos 1500 satellite radar images. *Proc. of the 35th Congress of the Intl. Astronautical Fed., Lausanne, Switzerland, October 7-13, 1984*, American Institute of Aeronautics and Astronautics, New York, NY.

Coordination on Geostationary Meteorological Satellites, 1987: *Consolidated Report of C.G.M.S. Activities*, EUMETSAT, Darmstadt-Eberstadt, West Germany, 255 pp.

ESA Bulletin, 1978: No. 15, August, European Space Agency, Paris, France, 52-59.

———, 1975: No. 3, October, European Space Agency, Paris, France, 35.

Filatova, N. I., et al., 1980: *Geotektonika, 5*,105-118.

Gerasminov, L. M. and V. Yu Luskina, 1980: *Novosibirsk Tektonika Sibiri, 8*, 102-120.

Heacock, E. L. (ed.), 1985: NOAA ENVIROSAT-2000 Report: Comparison of the Defense Meteorological Satellite Program (DMSP) and the NOAA Polar-orbiting Operational Environmental Satellite (POES) program. Dept. of Commerce, Washington, DC, 422 pp.

Insat - The Indian National Satellite System, 1983: Dept. of Space, Government of India, New Delhi, India.

Issledovaniye Zemli iz Kosmosa, 1981a: No. 1, January-February, Editorial, 5-6.

———, 1981b: No. 5, September-October.

Izvestiya, 1968: Moscow, March 17.

———, 1967: Moscow, March 17, p. 5.

———, 1966a: Moscow, August 21, p. 5.

———, 1966b: Moscow, August 19, p. 4.

———, 1966c: Moscow, May 20, p. 6.

Johnson, N. I., 1981: *The Soviet Year in Space: 1981.* Teledyne Brown Engineering, p. 18.

Krasnaya Zvezda, 1966: Moscow, October 22, p. 1.

Marchenko, M., 1979: *Tekhnika Molodezhi, 9*, 23.

Maxim, L., personal communication to G. E. Perry, November 20, 1979.

National Aeronautics and Space Administration, 1979: *Aeronautics and Space Report of the President, 1978 Activities*. Natl. Aero. and Space Adm., Washington, DC, 59-60.

National Paper: *USSR, United Nations Conf. on Exploration and Peaceful Uses of Outer Space, September 2, 1981*, 17-29.

Perry, G. E., 1981: *Proc. of the 1979 Australian Astronautics Convention*, W. A. Perth, Vol. 1, 75-83.

Pravda, 1975: Moscow, *Meteor-2 in Flight*, July 13, p. 3.

———, 1969: Moscow, January 21, p. 3.

———, 1968: Moscow, March 9.

———, 1967: Moscow, October 26, p. 3.

Schwalb, A. and W. J. Hussey, 1985: NOAA ENVIROSAT-2000 Report, GOES-Next Overview. Dept. of Commerce, Washington, DC, 67 pp.

———, 1982: Modified version of the TIROS-N/NOAA A-G satellite series (NOAA E-J) - Advanced TIROS-N (ATN). *NOAA Technical Memorandum NESS 116*, Dept. of Commerce, Washington, DC, 23 pp.

———, 1978/87: The TIROS-N/NOAA A-G satellite series. *NOAA Technical Memorandum NESS 95*, Dept. of Commerce, Washington, DC, 75 pp.

Space Exploration and Applications, 1968: United Nations, Austria, A/CONF. 34/2, Vol. 1.

TASS (in English), 1984: Moscow, January 23, 1136 UTC.

———, 1974: Moscow, October 10, 1529 UTC.

———, 1970: Moscow, April 29.

———, 1969: Moscow, March 27, 1625 UTC.

———, 1967: Moscow, June 1, 1525 UTC.

Trifonov, Yu. V., 1981a: *Meteor-Series Satellites Intended for Studying the Earth from Space. Issledovaniye Zemli iz Kosmosa,* No. 5, September-October, 8-20.

———, 1981b: *Technical Equipment Complex for Experiment in Remote Sensing of Earth from Space. Issledovniye Zemli iz Kosmosa,* No. 5, September-October, 21-27.

Trofimov, D. M. and B. I. Dmitriyeva, 1981: *Issledovaniye Zemli iz Kosmosa,* No. 4, July-August, 39-44.

TRUD, 1968: Moscow, January 24, p. 4.

U.S. Geological Survey: *Landsat Data User's Handbook* (revised edition), 4-6.

Vetlov, I. P., 1980: *Issledovaniye Zemli iz Kosmosa,* No. 2, 11-27, Table I.

Winkler, L. R., 1976: *CQ, 32* (10), 51.

World Meteorological Organization Reference W/MS/SB.2, March 8, 1978.

World Meteorological Organization, 1975: *World Weather Watch Global Observing System - Satellite Subsystem,* WMO No. 411, Geneva, with supplements.

Contributors to Section III

Cotter, Daniel J.	NOAA/NESDIS Washington, DC 20233
Graul, Donald W.	Ford Aerospace & Communications Corp. 3939 Fabian Way Palo Alto, CA 94303-9981
Greaves, James	National Aeronautics and Space Administration Code EED Washington, DC 20546
Heacock, E. Larry	NOAA/NESDIS Washington, DC 20233
Honvault, Claude	ESOC Robert-Bosh-Strasse #5 6100 Darmstadt West Germany
Kamiya, Yoshikazu	National Space Development Agency of Japan 300 South Grand Avenue, Suite 2760 Los Angeles, CA 90071
Members of	The Japan Meteorological Agency 3-235 Nakakiyoto Kiyofe-shi Tokyo, 204 Japan
Perry, Geoffrey E.	67 Trelawney Avenue Poughill Bude Cornwall EX23 9HB England
Vaeth, J. Gordon	NOAA/NESDIS Washington, DC 20233

Section IV

SATELLITE SENSORS AND THEIR DATA

Satellite sensors are the most important part of the remote sensing system. They provide the basic measurements essential for understanding and monitoring the Earth, atmosphere, and oceans. Chapter IV-1 describes the physical principles of various sensors and their capabilities. Chapter IV-2 describes current U.S. operational satellite sensors and onboard data processing. The characteristics of experimental and international sensors were introduced in Section III.

Chapter IV-1

REMOTE SENSING INSTRUMENTATION

IV-1.1 CLASSES OF INSTRUMENTS AND THEIR BASIC COMPONENTS

Remote sensing instruments measure the intensity of electromagnetic radiation emitted, reflected, or transmitted by the target of interest. Instruments that receive natural emitted, transmitted, or reflected radiation from the target are passive; most instruments fall in this category. Instruments that emit a pulse or stream of radiation and measure the echo from the target are active, such as radar and lidar.

Figure IV-1-1 shows the component elements of a generalized remote sensing instrument. Any instrument may have any or all of the following individual components.

A. The pointing, tracking, or spatial scanning component is required for data location information. This capability is provided by mechanically scanning mirrors in most UV, VIS, and IR instruments, with spacecraft motion providing one dimension of the scan and a rotating mirror the other. The Advanced Very High Resolution Radiometer (AVHRR), High Resolution Infrared Radiation Sounder (HIRS), and Stratospheric Sounding Unit (SSU) carried by the Advanced TIROS-N (ATN) spacecraft are examples of sensors using mechanically scanning mirrors as is the Visible and Infrared Spin Scan Radiometer (VISSR) Atmospheric Sounder (VAS) on the GOES spacecraft.

Passive MW instruments, e.g., the ATN Microwave Sounding Unit (MSU), may also use mechanical scanning (in this case with two scanning reflective antenna systems). Some sensors, such as radar, also use electronic scanning. In electronic scanning, the antenna consists of a number of individual elements. Their signals are combined electronically to scan and focus without physical motion.

In the case of other instruments that can view the entire scene instantaneously, this component is not needed. Examples are video (television) cameras with wide field optical elements and multi-element detector arrays.

B. The radiation collector consists of lenses, mirrors, or MW antennae which focus the target scene onto the image plane bearing the detectors. Special equipment, such as filters, spectrometers, gas correlation cells, polarization optics, interferometers, and time coding modulators may also be used to process incoming radiation.

C. The detector receives the processed radiation and converts it into a more manageable form, usually an electrical signal. Detectors have one or several million individual elements, depending upon the objective of the instrument and the wavelength in which it operates.

D. The electronic processing component converts the signal from the detector into data representing a measured value of the desired parameter.

E. The recording and storage component makes a permanent record of the measurement. Tape recorders and solid state Random Access Memories (RAM) are most commonly used for spacecraft instruments.

F. The target illumination component of an active sensor irradiates the target. In radar systems, coherent

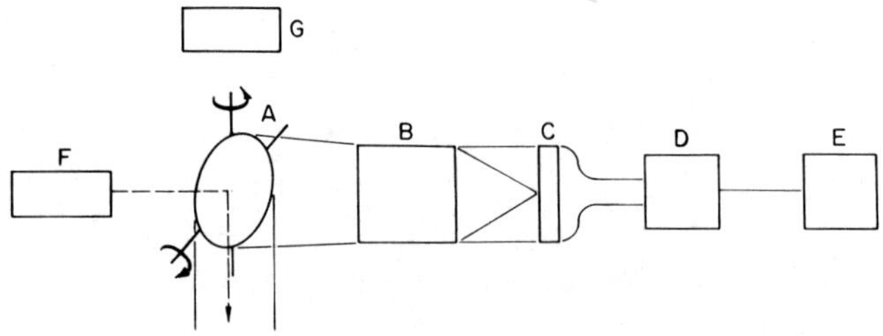

Figure IV-1-1. Schematic of a remote sensing instrument.

microwaves are generated; in lidar the radiation is emitted by a laser. The target illumination component usually emits pulses to derive range, but in some cases a continuous beam with frequency modulation is used.

*G. **The calibration source*** is an onboard source of radiation used periodically by the instrument to determine the instrumental sensitivity or response, or to determine if it has changed since launch.

With passive systems, it is difficult to obtain a uniform response across a large detector; therefore the quantitative information is not reliable. When accurate quantitative information is required, individual detectors or detector arrays coupled with scanning systems are used almost exclusively. For research, broad spectral coverage is needed to allow evaluation of various measurements. For operational purposes, the required spectral input is fixed and the selected measurements are usually provided by some form of filter. An excellent review of passive instruments on both research and operational spacecraft can be found in Houghton et al. (1984).

Active systems provide the option of selecting wavelengths that will minimize interference from natural radiation or make use of unique spectral characteristics of the target. Active sensors are not currently flown onboard ATN and GOES spacecraft, but are expected to play an important role in future operational spacecraft.

IV-1.2 ELECTROMAGNETIC RADIATION DETECTORS

Electromagnetic radiation is detected by its effects on a material body. The detector can be thought of as a device that links the electromagnetic energy of a radiation field to some kind of display apparatus by converting electromagnetic energy into electrical or mechanical energy. In most detectors, the detected energy is converted into electrical signals.

Detectors fall primarily into two categories: quantum counters and thermal detectors. Quantum counters respond nearly uniformly to the number of quanta incident on the detector, provided the energy of each quantum is greater than some threshold value characteristic of the detector. Thermal detectors respond to the total amount of energy incident upon them; thus they respond in the same way to a small number of high energy quanta as to a large number of lower energy quanta. The important characteristics of detectors are:

- the useful spectral range.
- the time constant, which determines the information rate capacity of the detector.
- the noise equivalent power, or the detectivity, which determines the lowest signal level detectable.
- the practical physical size of the detector, which determines the ultimate, overall sensitivity of the optical system in which the detector is used.

Other parameters, such as electrical resistance (impedance), linearity, and saturation level, are important but generally do not determine system performance in an optimally designed system.

Detectors can be grouped into three categories: single element, multi-element array, and image devices. Single element detectors in imaging systems use 2-dimensional scanning to provide an image. Multi-element arrays cover one dimension of the scene and their associated optical system mechanically covers the second dimension. Image devices such as vidicons, image orthicons, or image dissectors need no scanning mechanisms.

In the UV, VIS, and very near-IR regions of the spectrum, there are two commonly used types of detectors—photoelectric emission devices and solid state devices. In the IR region, quantum and thermal solid state devices predominate.

A photoelectric emission device has a cathode made of a metallic or composite surface. In a vacuum this surface emits electrons for each quantum of incident radiation whose energy meets or exceeds the "work function" characteristic of the surface. Another electrode in the vacuum collects the emitted electrons and produces a current proportional to the quanta received per second. The lower the work function, the greater the spectral range of the detector. The lowest work functions are those of the alkaline earth metals—Li, Na, K, Rb, Cs, and Fr. Frequently a photomultiplier tube is used. This tube has additional electrodes that collect the emitted electrons and amplify the current by secondary emission—one high energy electron striking an electrode can cause the release of up to ten secondary electrons.

Solid state detectors are made from a variety of materials that are photoconductive or photovoltaic. When subjected to electromagnetic radiation, photoconductors show a reduction in electrical resistance and photovoltaics show an electrical voltage across their surfaces. Photoconductors are the most versatile and most used detectors. Each photoconductive material has a characteristic "band edge" or minimum energy state needed to excite a bound electron into a conducting state; it will respond to all photons with an energy greater than the band edge energy. Frequently the materials used in solid state detectors are "doped" with n- and p-type impurity atoms. These doped materials are then used to form a diode from which the signal may be more easily extracted. In the VIS and near-IR regions, Si and Ge are frequently used materials for solid state detectors. In the IR region, more complex materials such as HgCdTe, InSb, PbS, and Ge doped with various metals are usually used.

Figure IV-1-2 shows the detectivity (D^*) for a number of the most used photoconductive detectors and their normal operating temperatures. This is for the case where the detector is used to measure radiation with brightness temperature at or about 300 K. The energy available at this temperature limits the achievable D^*. The curve at the top of this figure shows this limit for a detector looking at a 300 K scene with a full hemisphere (2π sr) field of view (FOV). Note that D^* is a minimum at about 10 μm where the blackbody curve of a 300 K object peaks and that the curve is roughly the inverse of the Planck curve. Somewhat greater sensitivity can be achieved by limiting the FOV of the detector with a cold mask. Also shown are two thermal detectors, an immersed thermistor, and a Ge bolometer (Strong, 1954) which have a flat response since they respond linearly to energy. Responses of the photoconductors show a gradually declining D^* toward shorter wavelengths and sharp cutoffs toward longer wavelengths.

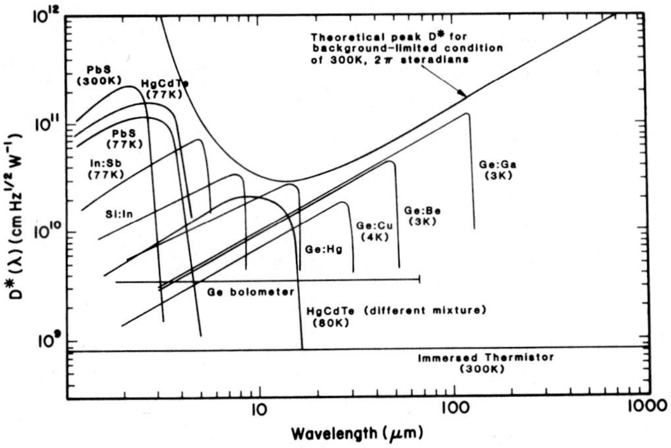

Figure IV-1-2. The spectral detectivity (D^*) as a function of wavelength (λ) of a selection of IR detectors.

IV-1.3 TECHNIQUES OF SPATIAL SCANNING

When single element or linear array detectors are used to cover large areas, the instrument must scan the scene in one or two dimensions. For a vidicon, the entire FOV of the camera is projected onto a sensitive surface. Electronic spatial scanning of the sensitive surface reads out each pixel, or single element, of the image. For example, the Return Beam Vidicon (RBV) camera on the early Landsat spacecraft imaged an area (186 km)^2 onto a (2.54 cm)^2 image plane with 2325 $\times$ 2325 pixels. The surface resolution is about 80 m. Assume the RBV is exposed to the scene for 1/250 sec. A typical polar orbiter moves at 7 km sec^{-1} and so has moved 28 m. With this rate of motion each pixel receives energy from a surface about 80 $\times$ 108 m. The spacecraft orbits 186 km every 26 sec, hence time is available for overlapping coverage.

In the IR, where sensitive materials for vidicons are not available, a surface area is projected onto one or a few detectors, and the optical axis is moved across the desired area. The FOV is moved along the orbit in one direction and usually is scanned perpendicular to the satellite track in the other dimension. Many methods can be used to accomplish this. A 45° mirror rotated about the optical axis is the simplest method. This was used on the following instruments (nadir resolution given in parentheses): MRIR (30 km), HRIR (7 km), AVHRR (1.1 km), CZCS (0.8 km), and SCMR (0.5 km). The relationships between resolution, FOV, satellite orbital velocity, and mirror speed must be considered. Higher resolution demands higher scan mirror rotation speed, so the radiation sampling frequency also must be set to give similar resolution in the mirror scan direction. Since the spacecraft is moving and the mirror is rotating, the ground target track is the resultant of two motions. The Earth's spherical shape also contributes to the geometry of what is included in the FOV. Figure IV-1-3 depicts the scan geometry from a polar orbiter.

A variation on the 45° rotating mirror is the conical scan, used on the Scanning Multichannel Microwave Radiometer (SMMR). In the conical scan technique, the mirror rotates about a vertical axis and produces a scan on the ground that is the intersection of a cone and the spherical Earth. The main advantage of this technique is that the target is viewed at a constant angle from the target zenith. The ground position is a series of arcs perpendicular to the satellite ground track. Another spatial scanning device is the rocking

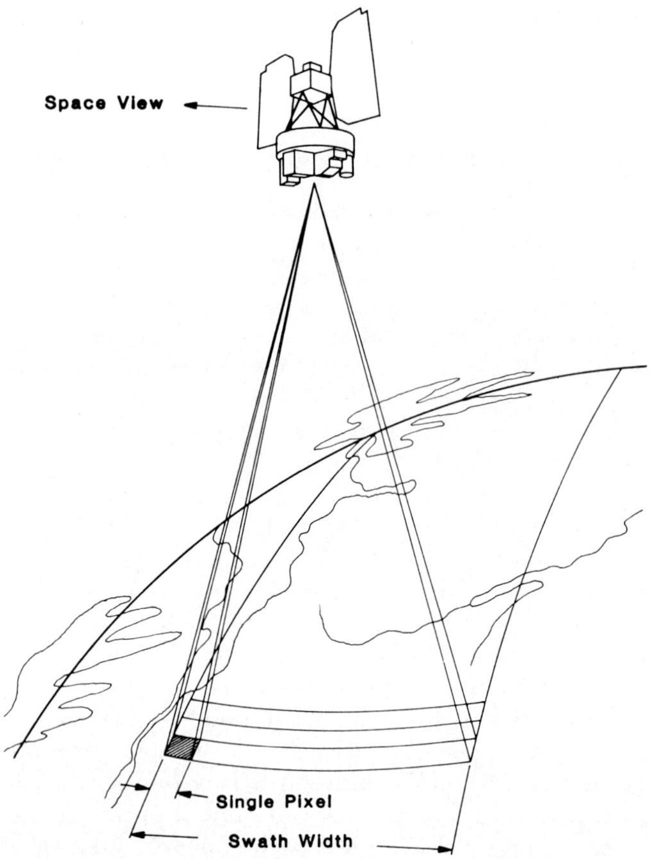

Figure IV-1-3. Imaging from low orbit. The instrument scans across-track. Spacecraft motion provides along-track coverage.

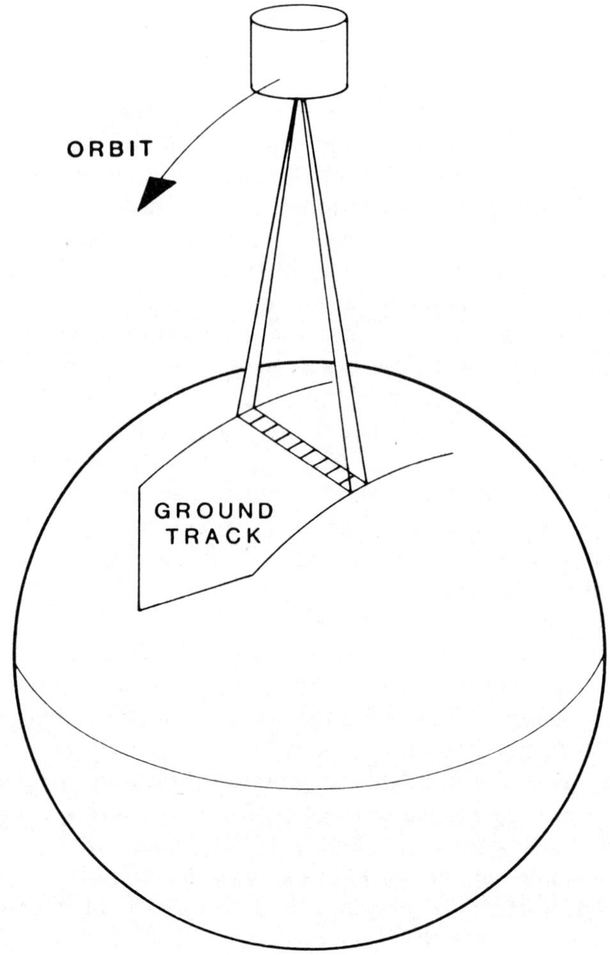

Figure IV-1-4. "Pushbroom" scanning (vertical scale exaggerated).

mirror used in both the MultiSpectral Scanner (MSS) and Thematic Mapper (TM) instruments on Landsat. The rocking mirror on the MSS scans six FOV imaged on six detectors simultaneously, perpendicular to the satellite track. The mirror rocks from −5.6° to +5.6° from nadir, with data taken in one direction only. With a ground resolution of 80 m for each detector, each scan covers 480 m and the mirror cycle repeats 14.6 times per second. The TM, with 30-m resolution, simultaneously scans 16 ground lines perpendicular to the satellite track and collects data as the mirror moves both ways. Because the satellite is orbiting, the ground pattern is a zigzag.

Another method of spatial scanning is the "pushbroom" mode which is used on the French SPOT satellite instrument. In this mode, a wide linear array of individual ground areas is imaged onto a linear array of detectors. The satellite motion "pushes" this pattern across the surface (Figure IV-1-4).

For geostationary satellites, two methods have been used depending on spacecraft stabilization. The GOES satellites are spin stabilized, rotating at 100 rpm about an axis perpendicular to that of the Earth. This spinning motion sweeps the FOV across the Earth; a 45° mirror is stepped one line per revolution to provide a full-Earth image of 1800 lines (Figure IV-1-5). The VISSR instrument sweeps eight FOV simultaneously. For a three axis stabilized spacecraft, 2-dimensional scanning must be provided. The instrument on the NASA ATS-6 and on the Indian Insat uses a mirror which scans in both dimensions. This is advantageous in that the Earth is viewed continuously. The instrument on a spin stabilized geostationary satellite views the Earth less than 5% of the time.

IV-1.4 TECHNIQUES FOR SPECTRAL SELECTION

There are specific wavelength requirements for measurements to be used for various studies. The study of the radiation balance of the Earth (see Chapter IX-1) calls for measurement of all radiation wavelengths, from the shortest to the longest. Solar "blind" studies, in which the radiometer is used to measure the

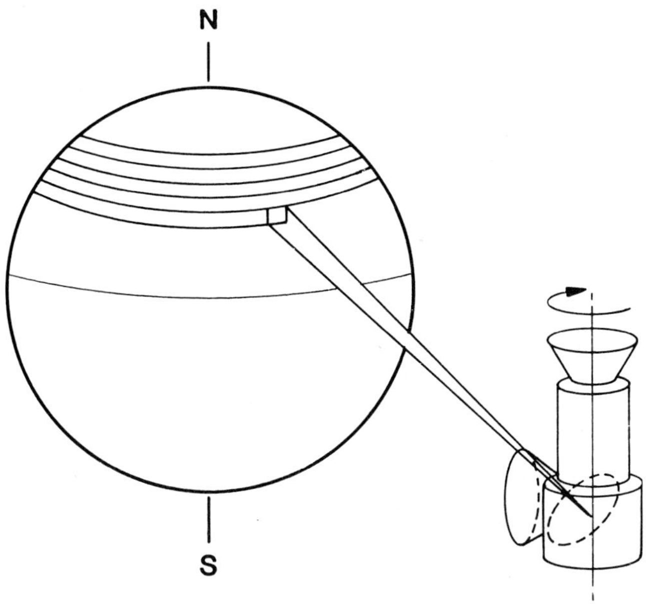

Figure IV-1-5. Spatial scanning from a geostationary spinning spacecraft.

very center of a dark Fraunhofer line, require filtering to less than one Angstrom. Still others require fairly broad coverage with good spectral resolution throughout.

Absorption, reflection, dispersion, interference, and transmission are all useful for achieving the desired spectral separation of incident radiation. Interference is the method most commonly used. If two waves are 180° out of phase, the result is destructive interference. If two waves are in phase, there is constructive interference. When a broad region is to be measured with good spectral resolution, interferometers, either of the Michelson or Fabry-Perot type, are used almost exclusively.

The devices most commonly used for achieving spectral separation in the UV, VIS, and IR parts of the spectrum are filters, particularly interference filters. Other devices such as diffraction gratings, gas absorption cells, and interferometers are also used.

Filters are the simplest devices used for spectral separation in satellite instruments. These are either "body" filters, a flat slab of transparent material into which a selective absorber has been dissolved; or interference filters, fixed interferometers consisting of multiple layers of different indices of refraction. The design and fabrication of interference filters are quite specialized (Born and Wolf, 1970; Driscoll and Vaughan, 1978).

The diffraction grating (Strong, 1954) is a convenient, although somewhat inefficient, device for spectral selection. Good efficiency over a limited spectral range can be achieved by cutting the grooves of the grating at an angle equal to the angle of diffraction of the desired wavelength.

IV-1.4.1 Interferometers

Because of their high efficiency and versatility, interferometers are rapidly becoming some of the predominant remote sensing instruments for broad spectral coverage. Interferometers vary the path length between two or more beams by different amounts to obtain alternate intervals of constructive and destructive interference. The mechanism by which this is done differs between types of interferometers. Although the Michelson interferometer is the most popular, many others are available. Born and Wolf (1970) describe 14 different kinds. Two Michelson interferometers, called the IRIS (InfraRed Interferometric Spectrometer), were flown on the Nimbus-2 and 4 satellites.

IV-1.4.2 Gas Correlation Spectroscopy

Infrared radiation emerging through the top of the atmosphere can be related to its altitude of origin depending upon its wavelength. For example, radiation at spectral line centers, where absorption coefficients are strongest, originates from high altitudes. Radiation in spectral regions between lines, where absorption coefficients are weak, tends to originate from low atmospheric levels. Conventional IR spectrometers and filter radiometers do not have adequate resolution to observe and measure the lines of atmospheric gases. The radiation detected by such instruments includes spectral regions with both strong and weak absorption, and therefore originates in a relatively broad layer of the atmosphere.

For optimal vertical resolution in sounding the atmosphere, special techniques are needed for resolving parts of spectral lines. One, known as gas correlation spectroscopy, was first used by scientists of the United Kingdom for measuring temperature in the 15 μm bands of carbon dioxide (Abel et al., 1970). Gas cells filled with samples of carbon dioxide are used in this technique, which can operate by means of selective absorption or selective chopping.

In selective absorption, the gas cell is located in a line with a conventional filter isolating the 15 μm region. Since the gas cell blocks radiation at and near centers of the strong spectral lines, the combined system is especially sensitive to radiation emitted from the upper troposphere and lower stratosphere, as illustrated by the weighting functions in Figure IV-1-6.

Selective chopping is used for sounding above 30 km. The incident radiation is observed alternately through two cells, one filled with CO_2, the other either empty

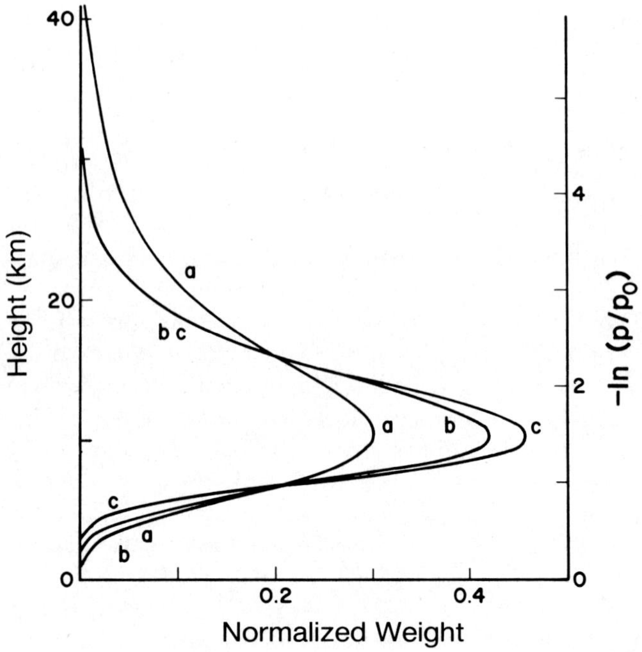

Figure IV-1-6. Demonstration of the effect of selective absorption. Curve a is a weighting function for a 5-cm^{-1} interval near 690 cm^{-1}; b is a weighting function for the same interval as for curve a, but with a short path of CO_2 included; c is an ideal weighting function for a monochromatic frequency in the wing of a spectral line (Abel et al., 1970).

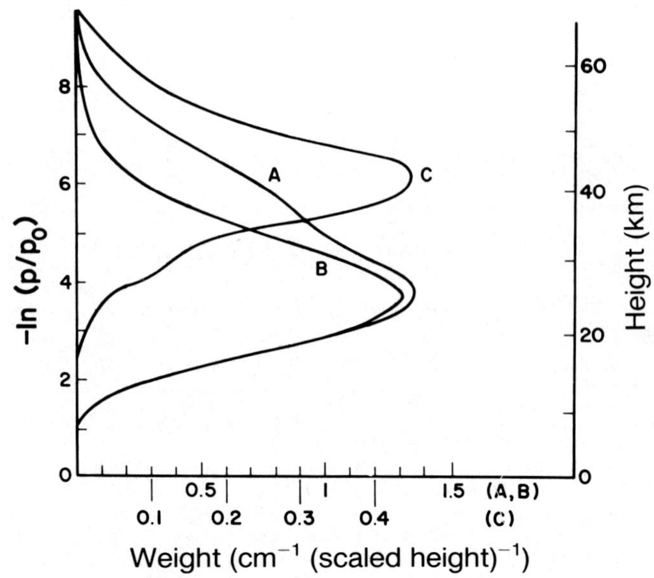

Figure IV-1-7. Demonstration of the effect of selective chopping. Curve A is a weighting function for a 5-cm^{-1} interval centered on 668 cm^{-1}; B is a weighting function for the same interval as A but including a 1 cm path of CO_2 at 0.05 atmospheric pressure; C is curve A minus curve B, or the weighting function for selective chopping between A and B (Abel et al., 1970).

or with CO_2 at a relatively low pressure. An alternating signal is produced by radiation at and near the centers of the spectral lines. This radiation is emitted by the upper atmosphere, and the resultant weighting function peaks at levels higher than can be obtained with conventional spectrometry. The height of the weighting function is determined by the choice of the cell length and the amount and pressure of the gas in the filled cell. It is, in fact, inversely related to the gas pressure in the cell. Figure IV-1-7 illustrates the effects of selective chopping. The principles of selective absorption and selective chopping were used to design the Selective Chopper Radiometers (SCR) carried by the Nimbus-4 (Abel et al., 1970) and Nimbus-5 (Ellis et al., 1973) satellites for obtaining atmospheric temperature profiles.

Another instrument, the Pressure Modulated Radiometer (PMR), uses the selective chopping principle but avoids the difficult and sensitive optical balancing required in double-beam systems such as the SCR. The PMR consists of a single cell of gas in which the pressure is modulated at a low frequency (~15 to 40 Hz) by an oscillating piston. This modulates the transmissions of the cell, principally at or near the centers of the absorption lines of the gas within. As with the SCR, the alternating signal from the PMR is produced by radiation emitted from the upper atmosphere. The height of the weighting function depends on the choice of the mean cell pressure—the higher the mean cell pressure, the lower the altitude of the weighting function. Figure IV-1-8 shows the correspondence between weighting functions and mean cell pressures. One can also choose the altitude of the emitting region by selecting the scan angle so that it introduces a desired component of the satellite's orbital velocity along the line of sight. This introduces a Doppler shift of the

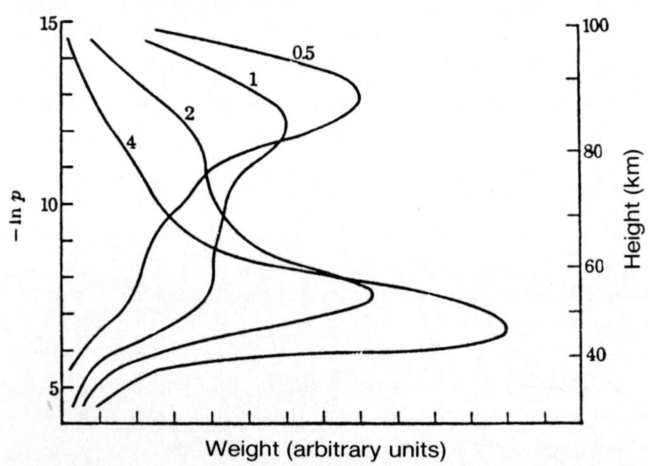

Figure IV-1-8. Weighting functions for a pressure modulated radiometer with cell pressures (numbers on curves) from 0.5 to 4.0 mb (Curtis et al., 1974).

spectral lines of the gas in the cell relative to the atmospheric emission lines. The effect of changing scan angle on the weighting function is shown in Figure IV-1-9. The Nimbus-6 PMR (Curtis et al., 1974) operated in the 15 μm carbon dioxide bands and was used for temperature sounding. The Nimbus-7 carried the Stratospheric And Mesospheric Sounder (SAMS) (Drummond et al., 1980), which consists of several PMR for temperature sounding (at 15 μm) and for measuring concentrations of CO, NO, CH_4, H_2O, and N_2O. Each of the TIROS-N series also carries a Stratospheric Sounding Unit (SSU), made up of three PMR, for sounding upper atmospheric temperatures and related meteorological properties. Figure IV-1-10 illustrates the weighting functions for the SSU.

IV-1.4.3 Microwave Techniques

The Microwave Sounding Unit (MSU), a part of the TIROS-N Operational Vertical Sounder (TOVS), measures atmospheric emissions in the 50 to 60 GHz range for profiling temperature in the troposphere. Four signal processing channels, at the MSU front end, each sense a narrow spectral sample of the emissions from the Earth. Figure IV-1-11 shows a schematic of the MSU.

Spectral separation is achieved using double side-band, superheterodyne reception techniques prior to phase detection. These reception techniques involve "mixing" the signal input with the coherent output of a local oscillator to convert it to an Intermediate Frequency (IF) and then filtering and amplifying the IF signal. Each of the four channels has a relatively narrow, double sideband, bandpass spectrum defined by the local oscillator, the mixer, and the IF components.

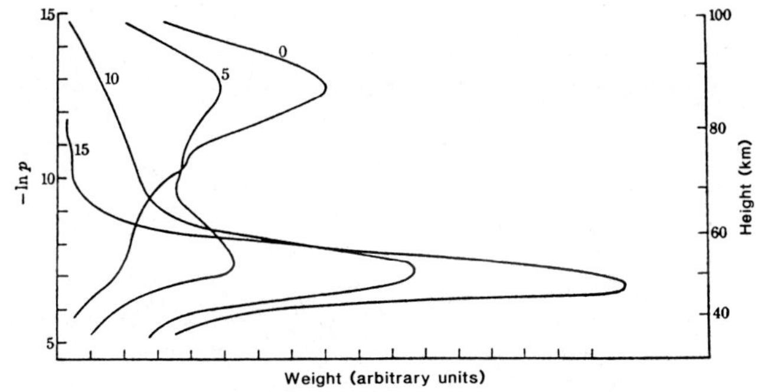

Figure IV-1-9. Weighting functions for a pressure modulated radiometer viewing at different angles to the nadir (numbers on curves equal degrees off nadir). At different angles the emission lines from the atmospheric CO_2 shift across the absorbing lines of the CO_2 in the cell due to the relative motion of the spacecraft and the atmosphere (Curtis et al., 1974).

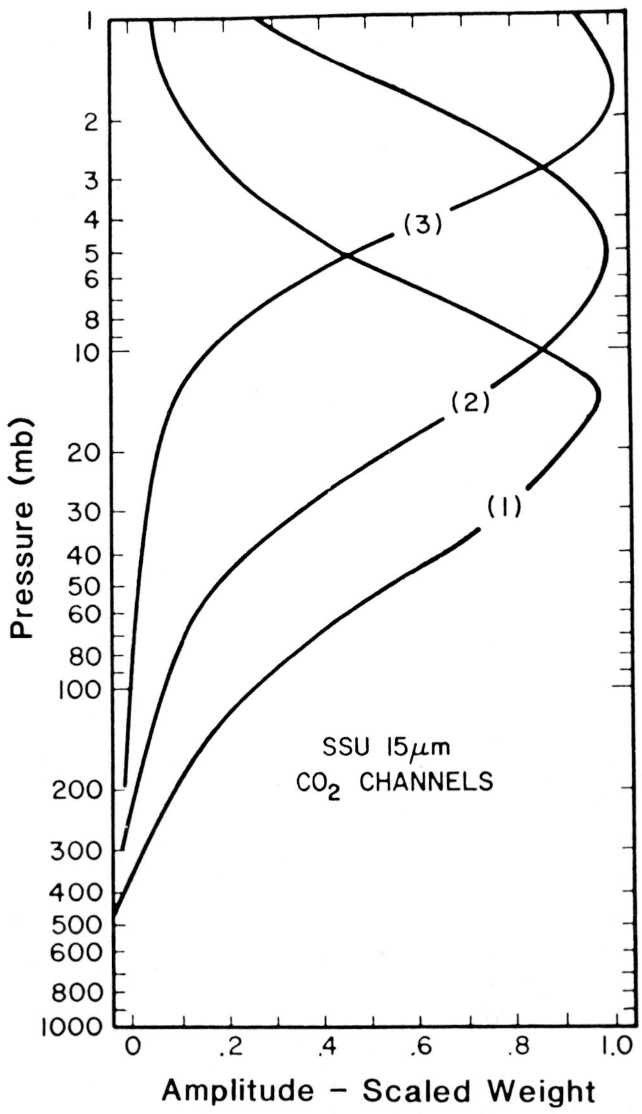

Figure IV-1-10. Weighting functions (scaled to make the maximum unity) for the Stratospheric Sounding Unit (SSU) carried as part of the sounding system on the TIROS operational satellites (Smith et al., 1979).

IV-1.4.4 Summary of Techniques

Table IV-1-1 contains a summary of some of the advantages and disadvantages of the commonly used methods of spectral scanning. If only a few channels of moderate spectral bandpass are required, filters may offer a cost advantage over other techniques.

Table IV-1-2 lists the techniques used for spectral scanning or separation by most of the instruments that have flown on weather satellites. Most VIS and IR instruments utilize filters of some type to define their spectral bandpasses, while MW instruments utilize heterodyne techniques to achieve the desired center frequency and bandpass.

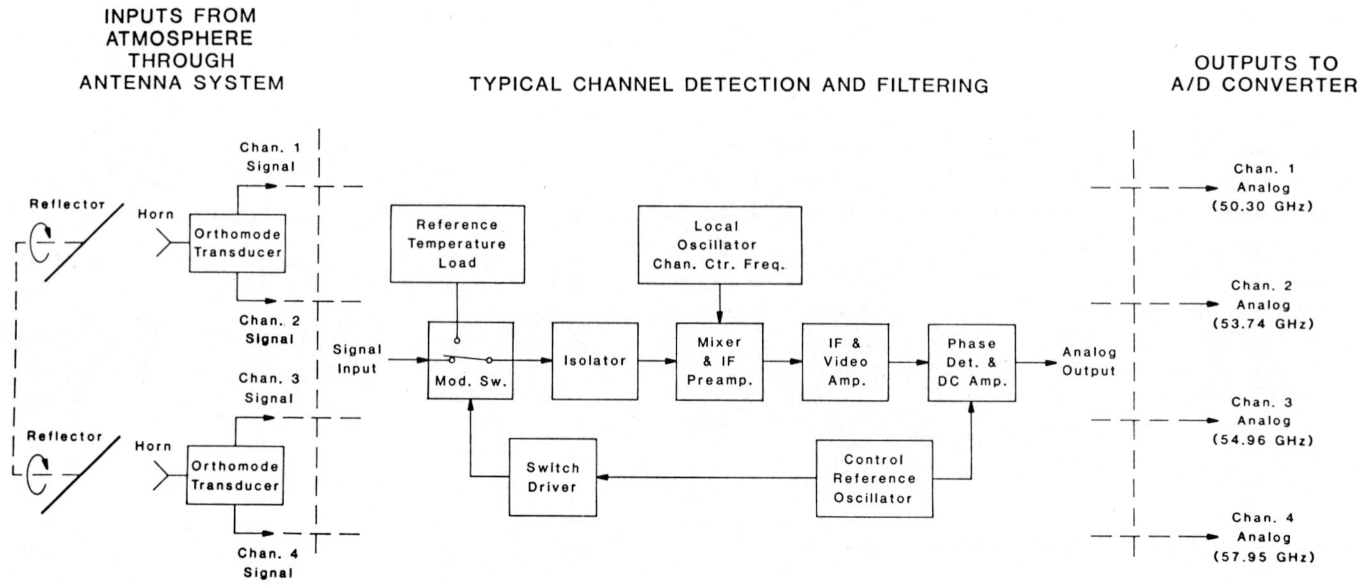

Figure IV-1-11. Schematic of the Microwave Sounding Unit (MSU) carried as part of the sounding system on the TIROS operational satellites.

IV-1.5 ACTIVE SYSTEMS

In active systems, a signal is sent out, allowed to interact with the target, and then observed and analyzed for characteristics of the target. Such systems have several advantages over passive systems:

- They make accurate measurements of the distance between the satellite and the target.
- Through the selection of signal frequencies and the use of multiple frequencies, they furnish more information on target properties.
- They avoid extraneous interference by careful selection of signal frequencies.

Active systems are now becoming feasible for operational use, mainly because recent developments have resulted in practical systems and larger modern spacecraft can now accommodate the power, size, and weight demands of active systems. Active systems can be divided into two categories: (1) radar and (2) lidar. They differ mainly in the wavelengths used; the principles of operation are almost identical.

IV-1.5.1 Radar

A. Physical Principles. Because most atmospheric scatterers are small relative to MW wavelengths, the magnitude of the backscatter coefficient for atmospheric constituents other than large raindrops is small. Thus, the atmosphere is essentially transparent at radar frequencies. As a result, spaceborne radars receive energy scattered from the Earth surface rather than from the atmosphere. This backscatter or radar cross-section is a function of many variables, but primarily of the dielectric properties of the scattering surface. Surface roughness also has an effect, as do features such as soil type, moisture content, and vegetative cover of land surfaces. In scattering from the ocean, rough-

Table IV-1-1 *Summary of spectral scanning techniques*

TECHNIQUE	ADVANTAGES	DISADVANTAGES
Gratings	reproducible spectral response, do many channels simultaneously, slit function usually triangular, mechanically rugged, stable	low throughput, requires many detectors or precise mechanical movement of grating, polarization sensitivity, order sorting
Interference Filters	high throughput, wider bandwidths, potential cost advantage	irregular spectral shape, temperature dependence, angular dependence, transmit higher orders
Interferometer	high throughput, many channels available, cover wide spectral range	requires precise mechanical motion, very sensitive to noise spikes, slow spatial scanning due to movement of the interferometer mirror
Gas Correlation Spectroscopy	narrow bandwidths, bandwidth can be varied	limited spectral range, high S/N difficult to achieve

Table IV-1-2 *Spectral scanning characteristics of satellite instruments*

INSTRUMENT	PURPOSE	METHOD(S) OF SPECTRAL SCANNING
Advanced Very High Resolution Radiometer (AVHRR)	VIS and IR images, sea surface temperature	filters and spectral response of detectors, one filter and one detector per channel
High Resolution Infrared Spectrometer (HIRS)	vertical temperature & water vapor profiles, and total ozone	interference filters in a rotating wheel, two IR and 1 VIS detectors
Microwave Sounding Unit (MSU)	vertical temperature profiles in cloudy areas	heterodyne
Stratospheric Sounding Unit (SSU)	extend vertical temperature profiles into stratosphere	gas (CO_2) pressure cell & filters
Vertical Temperature Profile Radiometer (VTPR)	vertical temperature profiles	interference filters in a rotating filter wheel, one detector
Scanning Radiometer (SR)	VIS & IR cloud images, some sea surface temperatures	filters
Very High Resolution Radiometer (VHRR)	VIS & IR cloud images	filters
Visible Infrared Spin Scan Radiometer (VISSR)	VIS & IR cloud images from geosynchronous altitude	filters
VISSR Atmospheric Sounder (VAS)	vertical temperature and humidity profiles from geosynchronous altitude	filters
Solar Backscatter UltraViolet (SBUV)	infer ozone distribution from solar backscattered UV radiation	stepped grating, photomultiplier tube response
Earth Radiation Budget Experiment (ERBE)	measure solar constant & outgoing radiation from Earth	filters & detector response
Coastal Zone Color Scanner (CZCS)	sea surface temperature and ocean color	grating & filter
Thematic Mapper (TM)	high resolution images	filters
Satellite Infrared Spectrometer (SIRS)	temperature & humidity profiles	grating
MultiSpectral Scanner (MSSR)	multispectral Earth images	filters
Infrared Interferometer Spectrometer (IRIS)	Earth's infrared spectrum	Michelson interferometer
Temperature-Humidity Infrared Radiometer (THIR)	surface or cloud top temperature & humidity measurements (6.3 μm H_2O band)	filters
Nimbus E Microwave Spectrometer (NEMS)	vertical temperature profiles, H_2O information, & cloud water	heterodyne
Scanning Microwave Spectrometer (SCAMS)	vertical temperature profiles & H_2O information	heterodyne
Selective Chopper Radiometer (SCR)	vertical temperature profiles	gas absorption cells
Pressure Modulated Radiometer (PMR)	vertical temperature profiles	gas absorption cells
Stratospheric and Mesospheric Sounder (SAMS)	temperature profiles & gas concentrations	gas absorption cells
Limb Infrared Monitor of the Stratosphere (LIMS) & Limb Radiance Infrared Radiometer (LRIR)	temperature profile & gas concentrations from limb radiances	filters
Scanning Multichannel Microwave Radiometer (SMMR)	sea surface temperature, winds and water vapor, cloud liquid water, precipitation, sea ice concentration, snow coverage	heterodyne
Special Sensor Microwave Temperature (SSMT) & Special Sensor Microwave Imager (SSMI)	vertical temperature profiles ocean surface wind speed, snow & ice coverage, precipitation, cloud liquid water, water vapor, surface identification, soil moisture, surface temperature	heterodyne and filters

ness is the dominant factor (Brown, 1978), especially for non-nadir viewing. The dominant mechanism in these radar measurements is a resonant scattering caused by variations in incidence angle of the radar beams, the radar wavelength, and the wavelength of the ocean waves.

The desired spatial resolution from a satellite radar system is obtained by selecting one or more of three variables: the length of the emitted pulse; the size of the transmitting and receiving antenna; and the Doppler shift of the returned signal. Pulse limited operation is shown in Figure IV-1-12. For a beam limited system, the spatial resolution is determined by the diffraction pattern of the system antenna. For a Doppler system, the rate of change of the radius vector between the satellite and a target on Earth produces a Doppler shift

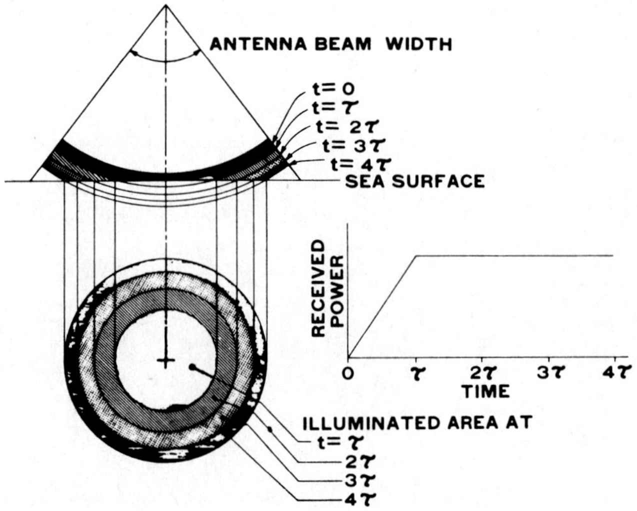

Figure IV-1-12. The geometry of a pulse of duration τ incident on a calm sea, illustrating the pulse-limited spatial resolution and the shape of the returned pulse (Walsh et al., 1978).

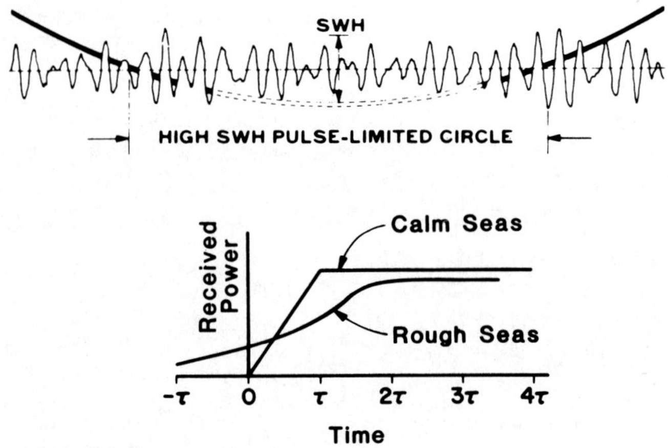

Figure IV-1-13. A narrow pulse of duration incident on a rough sea (i.e., the significant wave height (SWH) is greater than $c\tau/2$). The variations in wave height cause a spreading of the leading edge of the returned pulse (Walsh et al., 1978).

in the returned signal that can be used for locating the target.

B. Three Types of Radar have been used from spaceborne platforms to make measurements of geophysical parameters: radar altimeters, scatterometers, and synthetic aperture radars.

The radar altimeter is a short pulse, nadir viewing radar designed to measure precisely the distance from the satellite to the Earth surface. The instrument can also measure the shape of the returned pulse and the backscattering cross-section, σ_o. The latter two measurements permit interference of significant wave height and wind speed. The significant wave height can be inferred from the slope of the leading edge of the returned pulse and the wind speed can be inferred from the σ_0 measurement. Higher waves tend to decrease the slope of the leading edge of the pulse and also the backscattering, thus leading to the inference of the presence of higher winds. The major component of information derived from a radar altimeter pertains to the geoid and small-scale anomalies. These anomalies may be caused by sea mounts, trenches, ocean currents, and fronts.

Satellite radar altimeters operate in the pulse limited mode. Figure IV-1-13 illustrates this principle by showing a pulse of duration τ incident on a rough sea surface, a sea surface where the significant wave height is much greater than the pulse length, and the shape of the returned pulses.

Three radar altimeter experiments (McGoogan et al., 1974; Stanley, 1979; MacArthur, 1976) done in space yielded significant data. Table IV-1-3 compares the characteristics of these experiments.

To achieve a precision of less than 10 cm, many corrections (Tapley et al., 1982) are required. Pulse travel is slowed by charged particles in the ionosphere and by water vapor in the troposphere. The pulse shape (particularly the leading edge) is spread by high waves, so that a sea state correction is required. Uncertainties in spacecraft altitude and attitude are also potential sources of error. More precise altimeters are included in the Topex and Geosat missions.

The scatterometer is a radar system designed to infer wind speed and direction by precisely measuring the backscattering cross-section or the normalized radar cross-section. It is a side-looking radar with a large swath width. Figure IV-1-14 shows the swath of the Seasat scatterometer. Individual pixel sizes are determined by the beamwidth in the azimuthal direction and by the Doppler shift in the range direction. The operational characteristics of the Seasat scatterometer are given in Table IV-1-4.

The accuracy of scatterometer measurement (Johnson et al., 1980) is influenced by spacecraft attitude

Table IV-1-3 *Characteristics of satellite altimeters (MacArthur, 1976: © IEEE)*

	SKYLAB	GEOS-3	SEASAT-1
Mean Altitude (km)	435	840	800
Frequency (GHz)	13.9	13.9	13.5
Antenna Beamwidth (deg)	1.5	2.6	1.6
Peak RF Power (W)	2000	2000	2000
Average RF Power (W)	0.05	0.29	6.5
Pulsewidth (uncompressed)	130 ns	1 s	3.2 s
Pulsewidth (ns) (compressed)	10	12.5	3.125
Repetition Frequency (Hz)	250	100	1020
Altitude Precision (rms)	1 m	50 cm	10 cm

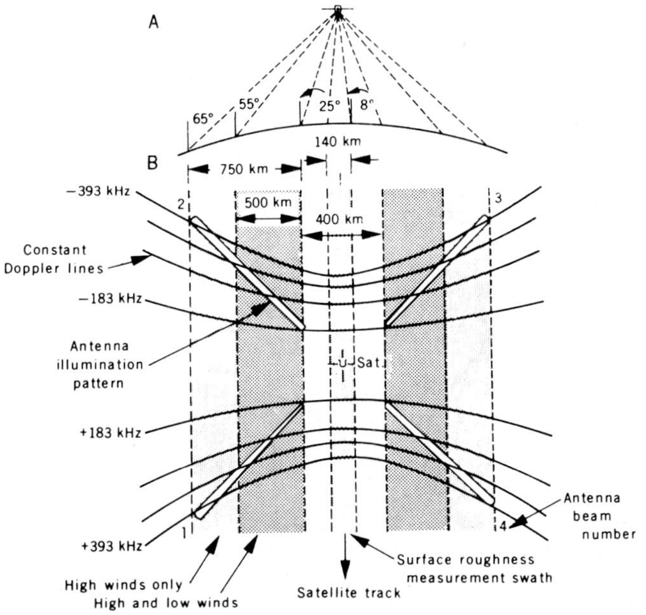

Figure IV-1-14. The measurement geometry of the Seasat scatterometer (Jones et al., 1979: © AAAS).

uncertainties, communication noise, uncertainties in transmitted and received power, and atmospheric attenuation. Fixed, or bias, errors may be corrected by careful comparison with accurate surface measurements of wind speed and direction. Random errors, unfortunately, cannot be corrected.

The Synthetic Aperture Radar (SAR) (Harger, 1970) is unique; it forms very high resolution 2-dimensional images. It is a side-looking radar that uses two methods (one for each dimension) to achieve resolution. A SAR typically generates a very short pulse and a series of range gates to resolve range. The motion of the satellite relative to a point on the Earth produces a Doppler shift in the returned signal, permitting azimuthal resolution. The azimuth resolution can be one half the azimuthal dimension of the transmitting antenna and is independent of the radar wavelength or the satellite altitude (Harger, 1970). The price for such high resolution is a very high data rate and complicated data processing on the ground.

Table IV-1-4 *Characteristics of Seasat scatterometer (Grantham et al., 1977)*

Frequency	14.6 GHz
Bandwidth	±500 kHz
Pulse Length	4.8 ms
Pulse Repetition Frequency	34 Hz
Beam Size	0.5° × 25°
Antenna Boresight Angle	47°
Data Rate	540 bps
Peak Transmitted Power	100 W
DC Input Power	136 W
Weight	102 kg

Table IV-1-5 *Characteristics of some SAR systems (Jordan, 1980; 1982)*

PARAMETER	SEASAT SAR	SIR-A	SIR-B
Orbital altitude (km)	800	260	235
Orbital inclination (deg)	108 (72)	38	57
Frequency (GHz)	1.28	1.28	1.28
Polarization	H H	H H	H H
Look angle (deg)	20.5	47	15 to 60
Swath width (km)	100	50	20 to 50
Peak power (kW)	1	1	1
Antenna dimensions (m)	10.7 × 2.16	9.4 × 2.16	10.7 × 2.16
Antenna gain (dB)	35	33.6	33.0
Bandwidth (MHz)	19	6	12
Range resolution (m)	25	40	58 to 17
Azimuth resolution (m)	25	40	25

Imaging radar experiments such as the Seasat SAR and the Shuttle Imaging Radar (SIR) have yielded images with information content of interest to the study of hydrology, oceanography, geography, geology, and agriculture. The characteristics of the Seasat SAR and two of the SIR experiments, given in Table IV-1-5, are described by Ulaby et al. (1986).

IV-1.5.2 Lidar

A. General. Lidar uses the reflected signal from a laser beam to sense atmospheric properties such as temperature, concentrations of water vapor and other gaseous atmospheric components, and wind velocity. Discussions of the physical principles of lidar and the design of lidar experiments are given by Measures (1984) and Zuev (1982). Although lidar has not as yet been used on weather satellites, its prospective use from spacecraft has been advocated by many investigators (e.g., Remsberg and Gordley, 1978).

A typical lidar system for studying the atmosphere consists of a single unit containing a laser, a collecting telescope, and a detector. The laser emits a short pulse of radiation of a definite length and frequency (Figure IV-1-15).

Atmospheric properties can be measured with a spatial resolution limited only by the length of the laser pulse, which can be as short as a few meters. The accuracy with which they can be determined is limited by detector noise, fluctuations in the flux of the returning photons, and fluctuations in the number of photons from sunlight scattered into and detected by the receiver.

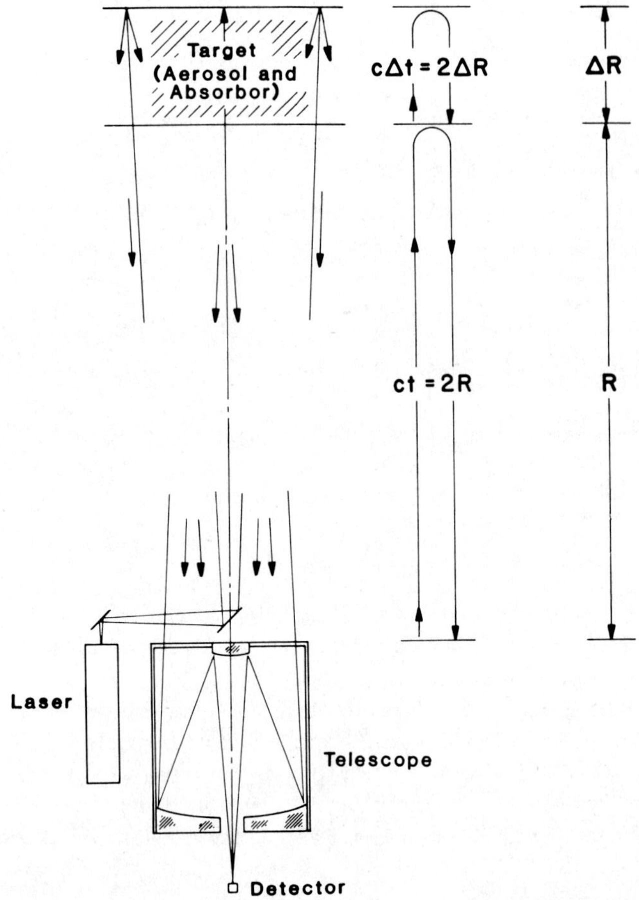

Figure IV-1-15. Schematic diagram of a lidar system. A short monochromatic pulse is emitted by the laser and sent out along the telescope axis in the system shown. The pulse slowly spreads out laterally as it propagates outward. Aerosols at a distance R will cause some of the laser radiation to be backscattered into the telescope at a time $t = 2R/c$. The intensity of the backscattered signal as a function of t yields information on the distribution of aerosols along R assuming that the scattering cross-section is independent of R.

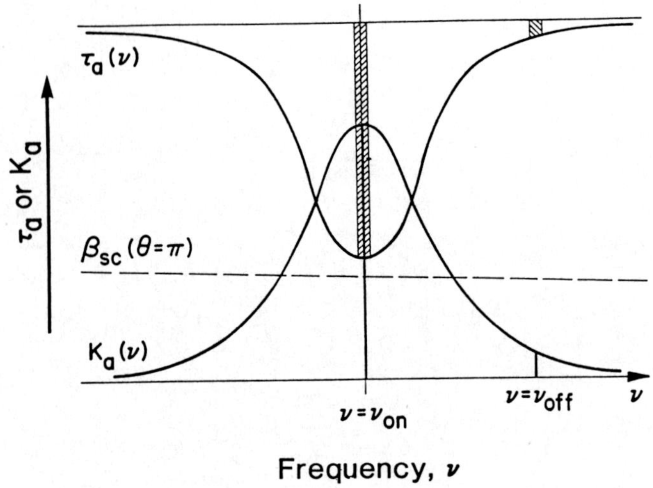

Figure IV-1-16. The principle of differential absorption lidar (DIAL). A lidar system simultaneously emits two short pulses of energy. One is at the frequency ν_{on} at the center of an absorption line as indicated by the minimum in the transmission curve K_a. The other is at the frequency ν_{off} in the far wing of the same absorption line. Energy from each pulse is backscattered into the telescope. The ratios of the return signals give information on the density of the absorbers as a function of distance from the lidar system.

Three techniques studied theoretically have been used experimentally to measure concentration profiles of atmospheric constituents and atmospheric motion. These are the DIfferential Absorption Lidar (DIAL), the Raman scattering technique, and Doppler shifted lidar. Recent theoretical and experimental progress has been reviewed by Zuev (1982), Carswell (1983), and Kostko (1976). The types of lasers adaptable to lidar systems are described by Yariv (1975), Schafer (1973), Rhodes (1979), and Tang (1979).

B. DIAL Methods. The DIAL technique uses measurements at two different frequencies. One is chosen to be at the center of an absorption line of the constituent to be studied. The other is chosen to be far enough away so that absorption by the constituent is negligible, but close enough so that backscattering and extinction by other molecules and aerosols are nearly the same at both wavelengths (Figure IV-1-16). The ratio of the return signals is related to the amount of the constituent. The ratio analyzed as a function of detector time gives the concentration as a function of range.

Use of DIAL techniques to measure atmospheric temperatures was first suggested by Mason (1975), who proposed three frequencies to determine simulta-

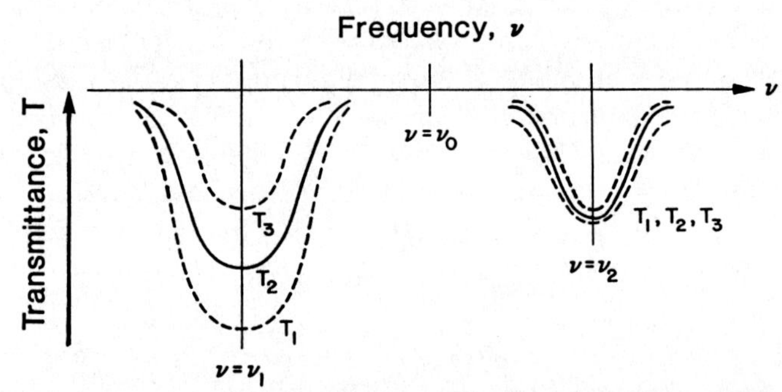

Figure IV-1-17. A three frequency method of determining absorber density and temperature profiles. A lidar system simultaneously emits three short laser pulses at frequencies ν_o, ν_1, and ν_2. An increase in attenuation corresponds to a decrease in transmittance. The pulse at ν_o is not attenuated by the absorber, and the attenuation of ν_2 is independent of temperature. The pulse at ν_1 has an attenuation which is highly temperature dependent. The ratios of the signals can be used to determine simultaneously the absorber concentration and its temperature as a function of the distance from the laser transmitter.

neously the concentration of water vapor and temperature along a path in the atmosphere. This method is illustrated in Figure IV-1-17.

Several ground-based experiments have been conducted to measure corrected atmospheric water vapor profiles using DIAL techniques (Cahen et al., 1982). Profiles of other atmospheric constituents have also been measured with ground-based equipment (Altmann et al., 1980; Lundquist et al., 1981). Aircraft DIAL measurements of atmospheric constituents were made by Browell (1982).

C. Raman Techniques. Another method of measuring water vapor concentration is by observing the intensity of Raman backscattering. A review of the principles and applications of this technique for remote sensing is given by Inaba (1976). Raman scattering has also been studied for making atmospheric temperature profile measurements (Cooney and Pina, 1976). Because the scattering cross-sections for the DIAL technique are much larger than for the Raman technique, DIAL is probably more suitable for use on satellites.

D. Doppler Techniques. Several Doppler lidar techniques appear to have much potential for sensing atmospheric wind fields. Among these new techniques are the Doppler heterodyne lidar, the Doppler incoherent lidar, and the Doppler dual-frequency lidar. An overview of Doppler lidar research and development work is described by Bilbro and Vaughan (1978), Bilbro (1980), and Benedetti-Michelangeli et al. (1972). A comparison between wind measurements by lidar and classical methods has been made by several authors (Post et al., 1978). Satellite-based lidar systems appear to be potentially useful for global wind measurement since they could provide good space and time resolution.

Chapter IV-2

ATN AND GOES SENSORS[1]

IV-2.1 INTRODUCTION

The technology and observing capabilities of the sensors aboard the world's environmental satellites are a complex subject. Instead of trying to cover them all, this chapter describes the instruments carried by the Advanced TIROS-N (ATN) and Geostationary Operational Environmental Satellite (GOES) as typical and representative of the state of the art.[2]

IV-2.2 ATN SENSORS

There are six meteorological sensors onboard the polar orbiting ATN family:

- High resolution InfraRed Sounder (HIRS/2)
- Stratospheric Sounding Unit (SSU)
- Microwave Sounding Unit (MSU)
- Advanced Very High Resolution Radiometer (AVHRR/2)
- Solar Backscatter Ultraviolet Radiometer (SBUV/2)
- Earth Radiation Budget Experiment (ERBE)

A seventh, the Space Environment Monitor (SEM), measures energetic particles, protons, electrons, and alpha particles in the satellite's orbit.

The HIRS/2, SSU, and MSU are three separate and independent instruments that as a package make up the TIROS Operational Vertical Sounder (TOVS). TOVS makes measurements that permit calculation of (1) temperature profiles from the surface to 10 mb, (2) water vapor content at three levels, and (3) total ozone.

IV-2.2.1 High Resolution Infrared Radiation Sounder (HIRS/2)

HIRS/2 (Figure IV-2-1), an adaptation of the HIRS/1 instrument flown on Nimbus-6, measures incident radiation in 19 IR spectral regions, including both longwave (15 μm) and shortwave (4.3 μm) regions, and one VIS (0.69 μm) wavelength.

HIRS/2 utilizes a 15-cm diameter optical system to gather energy emitted from the atmosphere. The Instantaneous Fields of View (IFOV) of all the channels are stepped across the satellite track by a rotating mirror. This cross-track scan, combined with the satellite's motion in orbit, provides coverage of a major portion of the Earth surface.

The energy received by the telescope is separated by a dichroic beamsplitter into longwave (above 6.4 μm) and shortwave (below 6.4 μm) energy, controlled by field stops, and passed through bandpass filters and relay optics to the detectors. In the shortwave path, a second dichroic beamsplitter transmits the VIS channel to its detector. Essential parameters of the instrument are shown in Table IV-2-1.

A. HIRS/2 Operation–Scan System. In orbit, the instrument output is locked to the spacecraft clock. The scan mirror synchronizes its stepping to this clock,

[1] Primary Contributor: Arthur Schwalb

[2] Ed. Note: The material in this chapter is an abridgment and adaptation of Schwalb, 1987; 1982; and NOAA/NASA/Hughes, 1980.

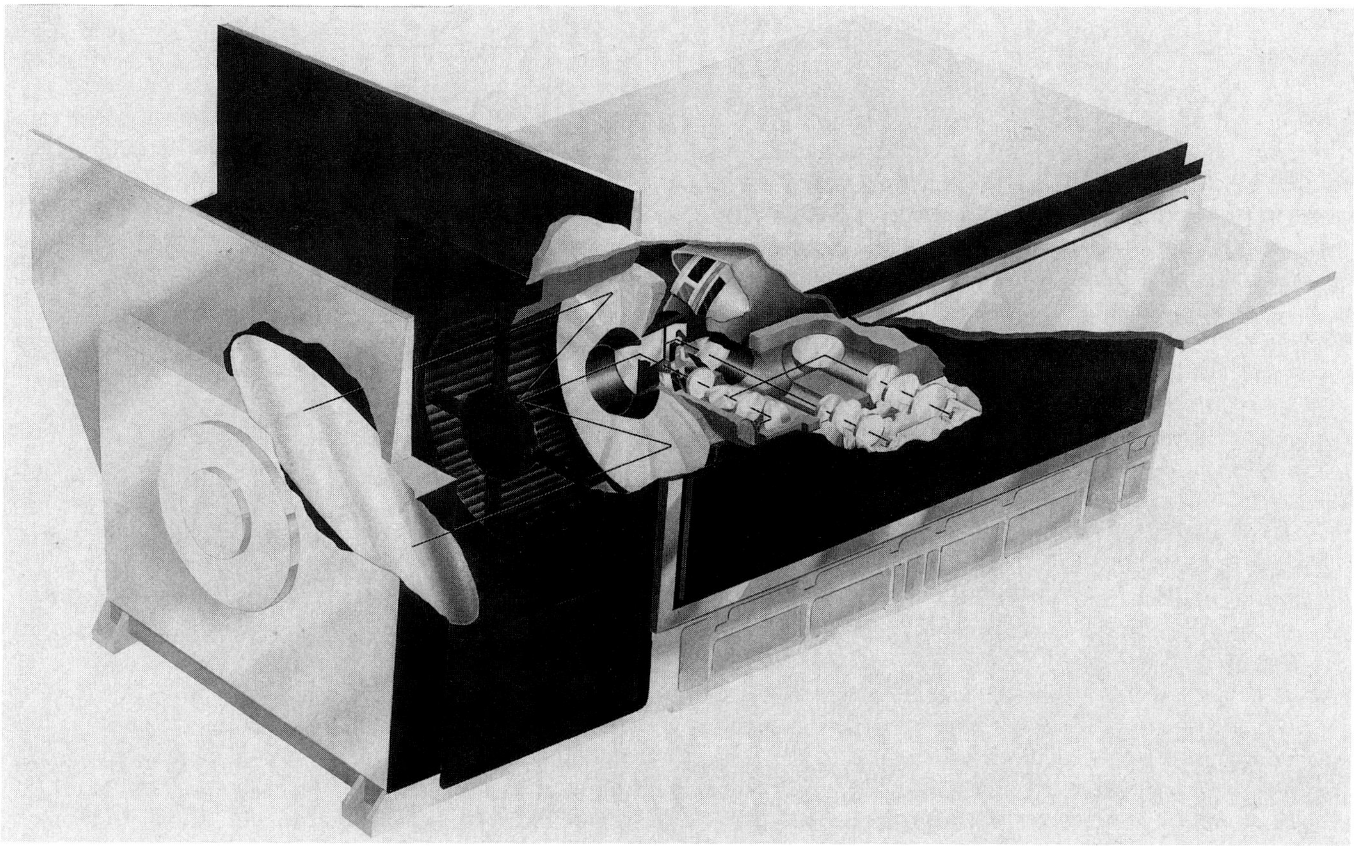

Figure IV-2-1. High Resolution Infrared Radiation Sounder (HIRS/2; courtesy of ITT/Aerospace Optical Division).

starting a new scan line in conjunction with the other TOVS components upon receipt of a major frame pulse from the spacecraft. The mirror steps from its initial (home) position in 55, 1.8° steps (56 data points) over its 99° swath (measured from the spacecraft). At completion of data acquisition at the last position (#56), the mirror returns rapidly to the first (home) position and repeats the Earth scan pattern. Each scan line requires 6.4 sec. Synchronization with both the other instruments occurs every 128 sec (every 20 scan lines).

Table IV-2-1 *HIRS/2 system characteristics*

CHARACTERISTIC	VALUE
Calibration	Stable blackbodies (2) and space background
Cross-track scan	±49.5° (±1120 km)
Scan time	6.4 s
Number of steps	56
Optical FOV	1.25°
Step angle	1.8°
Step time	100 ms
Ground IFOV (nadir)	17.4 km diameter
Ground IFOV (end of scan)	58.5 km cross-track by 29.9 km along-track
Distance between IFOVs	42 km along-track
Data rate	2880 bps

The instrument can be commanded automatically to enter a calibration mode every 256 sec. In this calibration mode, at the beginning of a scan line, the mirror swings rapidly around (equivalent to the time for eight scan elements) to a space view where it stops for the length of time necessary to complete one line (equivalent to 48 scan elements). All channels of the instrument are sampled during this period. The mirror is next moved to a position where it views a cold calibration target. Data are taken for the equivalent of 56 scan steps at which time the scan mirror is stepped to view the internal warm target. After another 56 samples, the mirror continues its motion to the start of scan (home) position, where it begins normal Earth scan. The total calibration sequence is equivalent in time to three scan lines. No Earth data are obtained during this period.

B. The HIRS/2 Optical System is based largely on the design used for HIRS/1. Small changes in the longwave design were instituted primarily to reduce vignetting (fading near the limits of the FOV). The optical path (port) to the cooler has been kept as small as possible to reduce the heat loading effects. The effective FOV of the instrument has been defined by the field stops.

The first dichroic beamsplitter transmits the longwave and reflects the shortwave and VIS channels. Therefore, two field stops define the FOV, one for longwave channels, the other for shortwave channels. Immediately behind the field stops, the bandpass defining filters are grouped on a wheel which rotates so that the energy reaching the detectors is defined by each filter in turn. The shortwave filters are located along an outer circumference, while the longwave filters are arrayed along an inner circumference. A chopper tooth is rigidly attached to the wheel and rotates with it; signal integration is confined to the time interval over which an optical filter is viewed. Length of the filter (and therefore the length of integration) has been chosen to provide an adequate signal-to-noise ratio.

The relay lens system is used to focus the received energy on the detectors. It reduces vignetting by ensuring that a minimum of energy from beyond the FOV reaches the detectors.

C. HIRS/2 Spectral Channel Characteristics. HIRS/2 makes measurements in 20 spectral regions, the specifications for which are given in Table IV-2-2. The individual bandpass filters are positioned on the filter wheel in a manner consistent with the requirement to ensure registration with the long and shortwave window channels. Longwave to shortwave registration is achieved by adjusting the two detectors relative to each other. Response outside of bands is held to the minimum level consistent with state-of-the-art filter design.

D. HIRS Radiant Cooler and Detectors. HIRS/2 uses two solid state IR detectors which operate most efficiently near 105 K. As is the case for the AVHRR, the longwave detector is mercury cadmium telluride (HgCdTe), the shortwave indium antimonide (InSb). A silicon detector operating at ambient temperatures is used for the VIS channel.

Table IV-2-2 *HIRS/2 channels (nadir resolution 17.4 km)*

CHANNELS	WAVELENGTHS (μm)	PRIMARY USES
1–5	14.95–13.97	Temperature profiles, clouds
6–7	13.64–13.35	Carbon dioxide and water vapor
8	11.11	Surface temperature, clouds
9	9.71	Total ozone concentration
10–12	8.16–6.72	Humidity profiles, detection of thin cirrus clouds
13–17	4.57–4.24	Temperature profiles
18–20	4.00–0.69	Clouds, surface temperatures under partly cloudy skies

The radiant cooler, functionally the same as that used for the AVHRR, can reach an operating temperature of about 97 K if permitted to run uncontrolled. This provides an adequate reserve margin to ensure continuous operation at the 105 K control point.

To ensure that outgassing products do not condense on the cooler during the initial time in orbit, the cooler is heated to 303 K for approximately 2 weeks following launch. The same heaters are also available to drive off contaminants in the event cooler performance degrades during the 2-year lifetime. No data are obtained from the instrument IR channels during any period of decontamination.

E. HIRS/2 Scan System. A scan housing assembly, which can be detached from the baseplate, contains all electronic components of the scan system. The system has been designed so that during Earth scan stepping, the mirror steps 1.8° to each new position with minimum overshoot and settles to within 0.1° in 35 ms. A mirror position encoder provides a positive indication of the scan position of the mirror.

F. HIRS/2 Electronics and Data Handling. The instrument output consists of digitally converted data levels for each spectral interval at sufficient dynamic range and quantizing resolution to allow extraction of all radiometric information. The range of signal from each channel is adjusted to conform to the range of input temperature expected in that spectral interval. Each channel signal is offset to make full use of the 13-bit digital system.

IV-2.2.2 The Stratospheric Sounding Unit (SSU)

The SSU (Figure IV-2-2), supplied by the United Kingdom Meteorological Office, employs a selective absorption technique to make measurements in three channels. The principles of its operation are based on the Selective Chopper Radiometer (SCR) flown on Nimbus-4 and 5, and the Pressure Modulation Radiometer (PMR) flown on Nimbus-6. Basic characteristics are shown in Table IV-2-3.

The SSU makes use of the pressure modulation technique to measure radiation emitted from carbon dioxide at the top of the atmosphere. A cell of CO_2 gas in the instrument's optical path has its pressure changed in a cyclic manner (at a rate of about 40 Hz). The spectral characteristics of the channel and, therefore, the height of the weighting function are then determined by the pressure in the cell during the period of integration. By using three cells filled at different pressures, weighting functions peaking at three different heights can be obtained. The primary objective of the instrument is to obtain data from which stratospheric (25 to 50-km) temperature profiles can be determined. This instrument is used in conjunction with

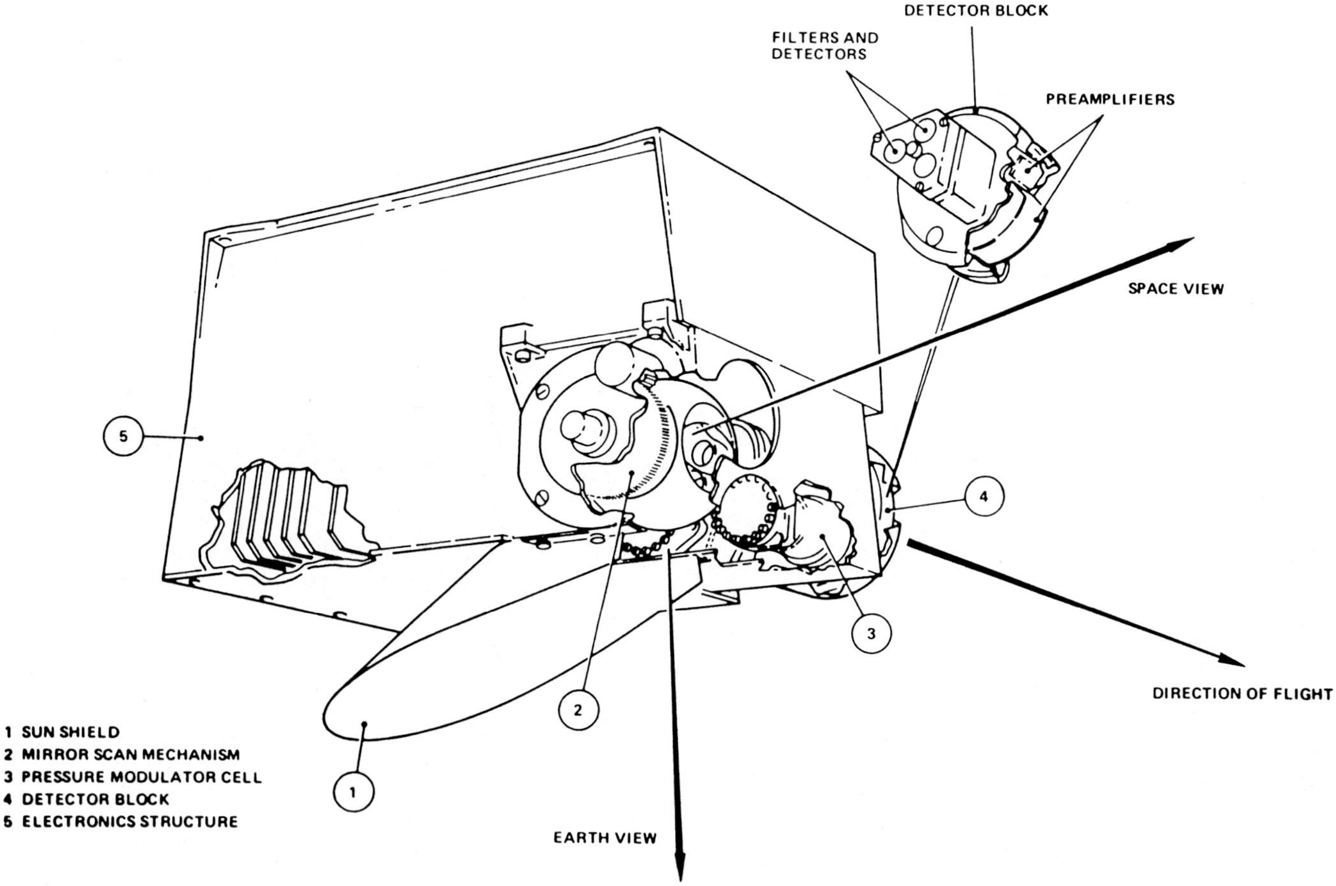

Figure IV-2-2. Stratospheric Sounding Unit (SSU).

the HIRS/2 and MSU to determine temperature profiles from the surface to the 50-km level.

A. SSU Operation. The single primary telescope with its 10° IFOV is step-scanned perpendicular to the subpoint track. Each scan line is composed of eight individual 4 sec steps and requires a total of 32 sec, including mirror retrace.

Table IV-2-3 *SSU characteristics (nadir resolution 147.3 km)*

CHANNEL NUMBER	CENTRAL WAVE NO. (CM^{-1})	CELL PRESSURE (MB)	PRESSURE OF WEIGHTING FUNCTION PEAK (MB)	(KM)
1	668	100	15	29
2	668	35	5	37
3	668	10	1.5	45

Calibration	Stable blackbody and space
Angular field-of-view	10°
Number of Earth views/line	8
Time interval between steps	4 s
Total scan angle	±40° from nadir
Scan time	32 s
Data rate	480 bps

The SSU uncooled pyroelectric detectors integrate the radiance in each channel for 3.6 sec during each step. The integrated output signal level is sampled eight times during this period. Quantization is to 12-bit precision.

B. SSU Scan System. A single 8-cm diameter scan mirror serves the three instrument channels. To reduce any possible bias errors, separation of the optical axes of the three channels is kept to a minimum. The mirror and the various apertures have dimensions that allow a clear path for rays up to 3° outside the nominal FOV. A fiberglass shield has been included to prevent reflected sunlight from contaminating the measurements.

C. SSU Optics and Electronics. The SSU detector is a flake of Tri-Glycine Sulphate (TGS). The flake is attached to the end of a conical, gold plated, nickel, light pipe. The 1.1 mm diameter exit aperture of the pipe defines the illuminated area on the flake, and the 6-mm diameter input end of the pipe with the objective lens defines the FOV.

The three detectors are mounted in a common block which also provides an integral housing for the preamplifiers. In-orbit calibration checks are confined to two points near the extreme of the radiometer's input range. It is important, therefore, that the response of each channel be linear. The electronics were designed with linearity to within 0.1% as a requirement. Monitoring of the operating frequencies of the cells, which depend on their internal pressures, provides the only direct means of verifying that the weighting functions are unchanged.

An interference filter is used to reject radiation at more than 50 cm^{-1} from the center of the Q-branch of the 15 μm CO_2 band. This filter is mounted between the absorption cell and the field lens.

D. The SSU Pressure Modulated Cell (PMC) consists of a sealed CO_2 cell and a means of modulating it. The cell provides a 1-cm CO_2 path with germanium optics coated to provide high transmission at 15 m. The cell is mounted on a cylinder within which is mounted a piston. The available modulation for a peak-to-peak piston amplitude of 3 mm is about 28%.

The mean pressure in the cell is about seven times the pressure at the peak of the weighting function. Fluctuations in depth of modulation are reflected in radiometer sensitivity changes. To ensure stable operation in orbit, the amplitude of piston motion is kept constant by an electronic servo system.

E. SSU In-Flight Calibration. A calibration sequence is initiated every 256 sec (8 scans) during which the radiometer is first stepped to a position to view unobstructed space and then to an internal blackbody at a known temperature. This calibration mode is synchronized with the HIRS/2 instrument.

Figure IV-2-3. Microwave Sounding Unit (MSU; courtesy of Jet Propulsion Laboratory).

IV-2.2.3 The Microwave Sounding Unit (MSU)

Figure IV-2-3 shows the MSU which is an adaptation of the Scanning Microwave Spectrometer experiment flown on Nimbus-6. The instrument is a four channel Dicke radiometer making passive measurements in four regions of the 5.5-mm oxygen region. The frequencies are listed in Table IV-2-4 which also gives the instrument parameters.

The instrument has two scanning reflector antenna systems, orthomode transducers, four Dicke superheterodyne receivers, a data programmer, and power supplies. The antennas scan ±47.4° either side of nadir in 11 steps. The beamwidth of the antennas is 7.5° (half power point), resulting in a ground resolution at the subpoint of 109 km.

Microwave energy received by each antenna is separated into vertical and horizontal polarization components by an orthomode transducer. Each of the four resulting signals is fed to one of the radiometer channels. The incoming noise temperature is modulated at a rate of 1 kHz by a Dicke switch so that a constant comparison is made between the ambient temperature reference load and the incoming signal. A two-point calibration is accomplished by a cold space view and a housing view once each scan period. Each radiometer channel is sensitive to inputs originating from temperatures ranging from 0 to 350 K. Each radiometer channel gain is then approximately 35 $mV \cdot K^{-1}$; digitization is to 12-bit precision.

A. MSU Operation. This instrument uses the spacecraft clock system to maintain scan synchronization. The rotating reflectors, synchronized to the clock, start a new scan line in conjunction with the other TOVS instruments upon receipt of a major frame pulse. The reflector steps from its initial home position in ten 9.45° steps (11 Earth views) over its 102° swath (measured

Table IV-2-4 *MSU characteristics (nadir resolution 105 km)*

CHARACTERISTIC	VALUE CH1	CH2	CH3	CH4	TOLERANCE
Frequency (GHz)	50.3	53.74	54.96	57.95	±20 MHz
RF bandwidth (MHz)	220	220	220	220	Maximum
NEΔT (K)	0.3	0.3	0.3	0.3	Maximum
Antenna beam efficiency*	>90%	>90%	>90%	>90%	
Dynamic range (K)	0–350	0–350	0–350	0–350	
Calibration	Hot reference body and space background each scan cycle				
Cross-track scan angle	±47.35°				
Scan time	25.6 s				
Number of steps	11				
Step angle	9.47°				
Step time	1.84 s				
Angular resolution	7.5° (3 dB)				
Data rate	320 bps				

* Specified. Instruments as built have measured beam efficiencies of >95%.

from the satellite). From the last Earth view position, the reflector rapidly moves four steps to view space, ten additional steps to view the housing, and then returns to the home position to begin another scan line. Each scan line requires 25.6 sec so that synchronization with the other two TOVS instruments occurs every 128 sec (five scan lines). Unlike the HIRS/2 and SSU, there is no special calibration sequence that interrupts normal scanning.

B. The MSU Scan System consists of the antenna subsystem, antenna drive, position measuring subsystem, and structural support. The antenna system consists of two rotating reflectors with fixed corrugated horns. The reflectors are attached to bearings rotated by highly accurate pulley drives attached to a 90° stepper motor and a set of miniature drive belts. The antenna positions are monitored by a potentiometer and a disk-type encoder. The encoder generates angular positions with a resolution of 256 counts (gray code) per revolution.

The aluminum antenna support structure has a MW blackbody calibration load where the antenna dwells for 1.9 sec during each scan period (25.6 sec). The calibration load consists of parallel rows of knife edge material made by casting an iron-filled epoxy onto an aluminum backing plate. These essentially isothermal targets, with embedded temperature monitors, provide one of two calibration points for each antenna. The second calibration point is the cold space view.

C. MSU Elements. The instrument electronics consist of three chassis—the RF, IF/video, and data. These three components are mechanically connected by support brackets, radiator plates, and the antenna subsystem baseplate.

D. The MSU Data Unit consists of a multiplexer and an Analog to Digital (A/D) converter. The A/D has 12-bit precision with a relative accuracy of +0.05%. The multiplexer accepts the analog data, and monitors signals from the four channels and commutates them in sequence for the A/D converter. The sequencing and synchronizing signals for the multiplexer, the A/D converter, and the scan system are provided by a digital programmer that interfaces with the spacecraft clock and synchronization signals.

IV-2.2.4 The Advanced Very High Resolution Radiometer (AVHRR/2)

The AVHRR (Figure IV-2-4) is a five channel scanning instrument sensitive to VIS, near-IR, and IR portions of the spectrum. The channels were selected to permit multispectral analyses of meteorological, hydrological, and oceanographic parameters. The channels and their purposes are listed in Table IV-2-5.

A. AVHRR Modules. The instrument is comprised of five modules assembled to form a single instrument.

1. The AVHRR Scanner Module includes the 80-pole hysteresis synchronous motor, the motor housing, and the scan mirror. The scan motor continuously rotates the mirror at 360 rpm to produce cross-track scanning in orbit. The scan mirror size (20.96 cm across the minor axis; 29.46 cm across the major axis) is adequate to fill the FOV of the 20.32-cm diameter telescope. The IFOV was chosen so that the satellite motion along its orbit would cause successive scan lines to be contiguous at the subpoint.
2. The AVHRR Electronics Module is bolted to the instrument inboard side panel. Electronic functions including data processing, temperature control, telem-

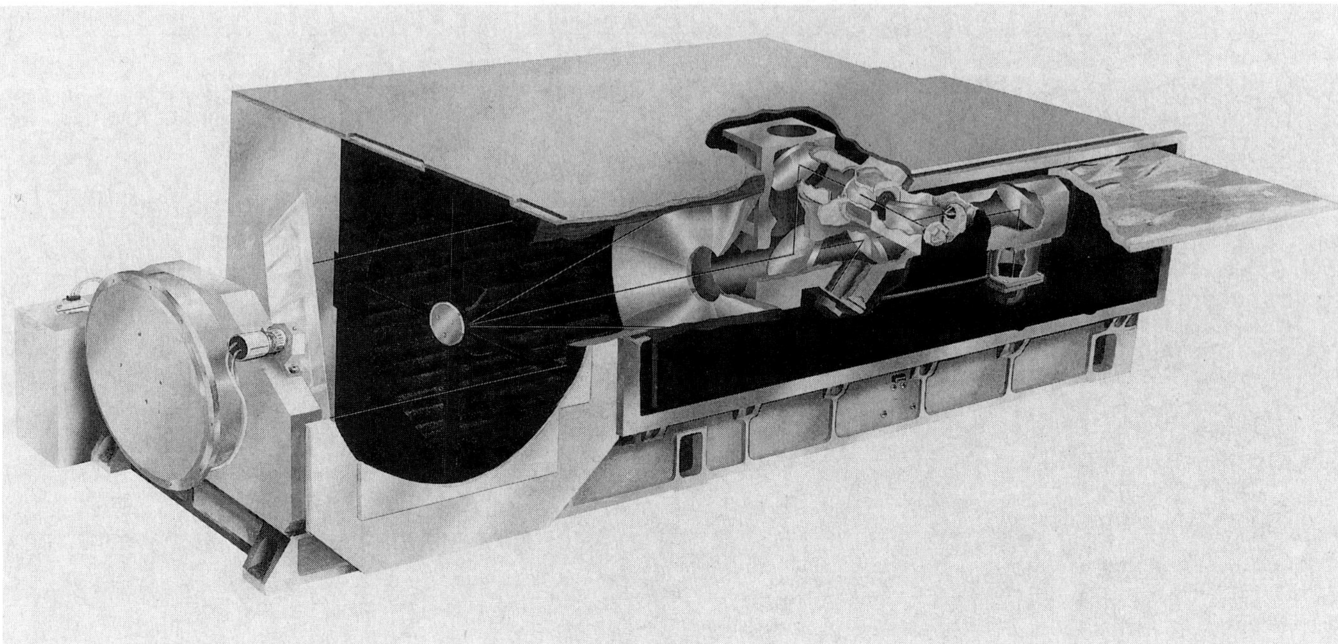

Figure IV-2-4. Advanced Very High Resolution Radiometer (AVHRR/2; courtesy of ITT Aerospace/Optical Division).

etry generation, scan, and motor logic are centered within this part of the instrument.

3. The AVHRR Radiant Cooler consists of a cooler housing, a first-stage radiator, a second-stage radiator (referred to as the "patch"), and a cooler cover. The cooler cover shades most of the radiator surface from input radiation from the Earth. The first and second stage have unobstructed views of space and radiate sufficient energy to bring the patch to its desired operating temperature of 105 K. A proportional heater provides sufficient energy to the patch to keep it from falling below this temperature. Should cooler performance unexpectedly degrade in orbit, it would be possible to restabilize the patch temperature at 107 K and continue normal operation.

Table IV-2-5 *AVHRR channels (nadir resolution 1.1 km; >2600 km swath width)*

CHANNELS	WAVELENGTH (μm)	PRIMARY USES
1	0.58–0.68	Daytime cloud/surface mapping
2	0.725–1.10	Surface water delineation, ice and snow melt
3	3.55–3.93	Sea surface temperature, nighttime cloud mapping
4	10.30–11.30	Sea surface temperature, day and night cloud mapping
5	11.50–12.50	Sea surface temperature, day and night cloud mapping

4. The AVHRR Optical System consists of a 20.3-cm aperture telescope combined with secondary optics that separate the radiant energy into the five discrete spectral bands listed in Table IV-2-5. The IFOV for all channels is specified to be 1.3 ± 0.1 mrad. Polarization effects have been minimized to the extent practical by orientation of internal optical components. The instrument has been designed so that the IFOV of the five channels can be made coincident within ±0.1 mrad (8%). Resolution is 1.1 km at the subpoint.

5. The AVHRR Baseplate is the instrument structure upon which all other modules are secured. A cover plate over the electronics modules and the cooler housing serve as a radiator for thermal control.

B. AVHRR Operating Characteristics. VIS and near-IR channels use silicon detectors to measure incident radiation. Both are square, 0.254 cm on a side. The defining field stops are 0.06 cm squares; analysis indicates that 99% of the rays passed through the field stop are collected by the detectors. The instrument easily meets the specified 3:1 signal-to-noise (at 0.5% albedo) requirement.

The IR channels use detectors cooled to 105 K. The detector for the 3.8 μm channel is indium antimonide (InSb), while that for the 11 μm channels is mercury cadmium telluride (HgCdTe). The InSb has a 0.0173-cm square active area and is mounted with an aplanatic lens forming a hermetic seal. A Noise Equivalent Temperature Difference (NEΔT) better than 0.12 K (for a 300 K scene) is achieved from this channel. The

HgCdTe detector is optimized for best sensitivity between 10.5 and 11.4 μm. The detector is 0.0173 cm square and is bonded to the aplanatic lens. The NEΔT is better than the specified 0.12 K.

The zero point for detected energy is restored to a preset zero level once each scan while viewing space. The output of the radiometer during the remainder of the scan is equal to the difference in detected energy between space and the radiating surface. Figure IV-2-5 is a simplified diagram showing the movement of the electronic signals from the detectors to the point where the outputs are sampled by the Manipulated Information Rate Processor (MIRP). The role of the MIRP is described in IV-2.3.

The instrument has an internally generated ramp to permit routine validation of the linearity of the electronics. In-orbit calibration of the IR channels is possible because the instrument output is linear with input energy. During every scan line, the instrument views cold space (zero radiance) and its housing (approximately 290 K). The housing portion of the instrument is a blackbody target. Platinum Resistance Thermometers (PRT), whose output values are included in the data stream, are embedded in the housing and monitor the temperature of the target. By determining the instrument output while viewing the known warm target and cold space (also in the data format), it is possible to ascertain the instrument response curve.

C. AVHRR Direction of Scan. The AVHRR scans across the Earth from the space view side of the spacecraft toward the Sun. This direction of scan ensures that picture data, as displayed on a normal facsimile machine, will maintain correct geographic orientation for the chosen orbits. This scan geometry also permits the instrument to restore its zero level at the beginning of each scan line while viewing space.

The TOVS instruments, in contrast, scan in the opposite direction, i.e., from Sun side to space side. This scan geometry is optimum for a temperature profile radiometer where space view calibration is accomplished at the end of a scan line.

Looking in the direction of the spacecraft velocity vector, the AVHRR scans from right to left, the TOVS instruments from left to right. Since temperature profile radiometers are not generally used to produce picture displays, this scanning mode presents few problems for users.

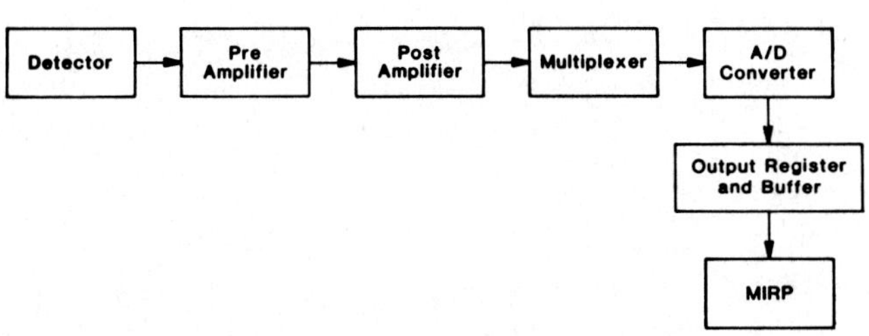

Figure IV-2-5. AVHRR data flow.

IV-2.2.5 The Solar Backscatter Ultraviolet Radiometer (SBUV/2)

The SBUV/2 radiometer design is based upon the technology developed for the Total Ozone Mapping Spectrometer (TOMS) flown on Nimbus-7. It provides data for computing the vertical distribution of ozone.

To collect this information, two separate measurements in the 160 to 400-nm spectral range are made by the SBUV/2 instrument. These are: (1) the spectral radiance of the solar UV radiation backscattered from the strong ozone absorption band in the atmosphere and (2) the direct solar spectral irradiance.

The SBUV/2 instrument consists of two components, an electronics and logic module and a sensor/detector module. The Earth viewing sensors are mounted on the exterior surface of the equipment support module of the ATN satellite (Figure III-1-1) while the electronics/logic module is located within. The basic components of the sensor module are:

- A scanning double monochromator
- A cloud cover radiometer
- A diffuser plate
- Detectors.

A. SBUV/2 Characteristics. The SBUV/2 monochromator measures backscattered solar radiation in an 11.3° FOV in the nadir direction at 12 discrete, 1.1 nm wide wavelength bands between 252.0 and 339.8 nm. The solar irradiance is determined at the same 12 wavelength bands by deploying, upon command, a diffuser which reflects sunlight into the instrument FOV. The atmospheric radiance measurement, relative to the solar irradiance, is the significant factor being determined.

The SBUV/2 monochromator can also measure the solar irradiance or the atmospheric radiance with a continuous spectral scan from 160 to 400 nm in increments nominally 0.148 nm. These measurements provide data on photochemical processes in the atmosphere.

A separate narrowband filter photometer channel, called the Cloud Cover Radiometer (CCR), continuously measures the Earth surface brightness at 380 nm, i.e., outside the ozone absorption band. The CCR is located in the same structure as the monochromator. The CCR FOV is the same size as, and is co-aligned with, the monochromator's FOV.

B. SBUV/2 Operating Modes. The instrument operates in five distinct modes:

- Discrete Mode. The instrument sequentially measures scene radiance and solar spectral irradiance in 12 discrete spectral bands.
- Sweep Mode. The instrument senses input energy as the spectral band pass is "swept" from 160 to 400 nm continuously. Each measurement has an equal integration time and is equally spaced across the spectral band. Either the scene spectral radiance or solar spectral irradiance may be measured in this mode.
- Wavelength Calibration Mode. This is equivalent to the discrete mode, but the spectral wavelengths scanned are those of the onboard calibration lamp.
- Monochromator Stop Mode. The spectral scan is interrupted with the grating fixed in position at the time of command receipt.
- Monochromator Caged Mode. The monochromator is caged for launch at a predetermined position.

The sequencing for each monochromator mode is controlled from either a fixed, ground-programmable memory or from a random access memory programmable by command. The desired memory is selected by command.

Each monochromator mode defines a unique wavelength sequence and a data sampling sequence. The wavelengths are always scanned from the long wavelengths to the short wavelengths. This is followed by a rapid retrace to the long wavelength and a wait for the start of the next 32-sec TIROS Information Processor (TIP) frame. See IV-2.3 for information about TIP. During this time, the preamplifiers are switched out, and up to ten precision voltage levels are inserted into the electronics for calibration of the analog electronics and the voltage to frequency converters.

Beginning at the start of the first major frame following a "Discrete Mode" command, the gratings sequentially move to and dwell at the 12 discrete wavelengths. The signal at each wavelength is integrated for 1.25 sec. An additional 0.75 sec is allowed for moving to and settling at the next wavelength. Thus, the 12 discrete wavelengths are covered in 24 sec. This allows 8 sec for returning to the first discrete wavelength, electronic calibration, and waiting for the start of the next major frame.

Beginning at the start of the first major frame following receipt of a "Sweep Mode" command, the wavelength range from 400 nm to 160 nm is scanned in nominally 0.074-nm steps. The monochromator signal is integrated for 0.1 sec resulting in nominally 0.148 nm sample increments. Therefore, approximately 1631 spectral measurements are made between 400 and 160 nm. This takes about 164 sec. The grating drive then retraces to 400 nm and waits for the start of the next major frame. Thus, the total cycle time for the Sweep Mode is 192 sec. Electronic calibration takes place during the period between 164 and 192 sec.

The "Wavelength Calibration Mode" is functionally similar to the "Discrete Mode." Beginning at the start of a major frame, it moves to and dwells at 12 separate wavelengths, each separated by nominally 0.296 nm, around any desired line source. At each wavelength, data are integrated for 1.25 sec.

IV-2.2.6 Earth Radiation Budget Experiment (ERBE)

The purpose of the ERBE is to determine the monthly average radiation budget on regional, zonal, and global scales. A second objective is to determine average diurnal variations in the radiation budget. These instruments have been flown experimentally on NOAA-9 and 10 and are not considered to be a component of the operations payload.

There are two ERBE instrument packages, the Non-Scanner and the Scanner. The Non-Scanner contains wide and medium FOV channels viewing in a fixed mode. The Scanner consists of narrow FOV channels viewing in a scanning mode. Figure III-1-1 shows their location on the ATN spacecraft.

A. The ERBE Non-Scanner instrument consists of five channels, four of which view the Earth. Two of these are wide FOV (i.e., view the total Earth disk beneath the satellite) and two are medium FOV (31.8° FOV, equivalent to a 10° Earth central angle beneath the satellite). These four channels are mounted on a single-axis gimbal which, when activated by command, allows these channels to view the Sun for periodic calibration. The fifth channel is not gimbaled but is located to view the Sun periodically and provide a measurement of the solar constant. The spatial and spectral scales of the five sensors and their measurements are listed in Table IV-2-6.

Table IV-2-6 *ERBE non-scanner characteristics*

CHANNEL	SPECTRAL INTERVAL (μm)	FILTER	FIELD OF VIEW
1 (Wide FOV)	0.2–50+	None	Limb-to-Limb
2 (Wide FOV)	0.2–5	Suprasil-W Dome	Limb-to-Limb
3 (Medium FOV)	0.2–50+	None	10° ECA*
4 (Medium FOV)	0.2–5	Suprasil-W Dome	10° ECA*
5 (Solar)	0.2–50+	None	18° Conical

* Earth central angle

The detectors are essentially cavity radiometers that have a broadband spectral response from about 0.2 to 50 μm. The detectors of one wide FOV and one medium FOV channel are placed beneath a Suprasil-W hemispheric dome filter which provides spectral isolation since the filter does not transmit above 5 μm. Hence, one of each size channel makes broadband or total radiation measurements and the other measurements of the shortwave spectral band characterized by the filter. For both sets of channels, the Earth-emitted longwave radiation component is determined by subtracting the shortwave (Earth-reflected radiation) channel measurements from the total measurement. The solar channel is a narrow FOV cavity radiometer measuring the total solar spectral range.

B. The ERBE Scanner views an essentially continuous pattern scanning from horizon to horizon in a direction normal to the orbit plane. The IFOV is 3° by 4.5°. The mechanical system is configured to provide the same capability for viewing Earth and Sun as does the Non-Scanner instrument.

Table IV-2-7 shows the three wavelength regions of operation for the scanning channels: a long wavelength band (5 to 50 μm), a short wavelength band (0.2 to 5 μm), and a total band (0.2 to 50 μm).

The scanner assembly is a lightweight, compact, three-bearing, mechanical configuration which scans in sawtooth fashion across the ground track of the spacecraft. The scanner instrument package weighs about 29 kg and requires 25 W average orbit power. The data output is 960 bps.

NASA Langley Research Center procured the ERBE instruments flown for NASA by NOAA on NOAA-9 and NOAA-10. NASA Langley is responsible for managing and processing the data from these instruments.

IV-2.2.7 The Space Environment Monitor (SEM)

Making *in situ,* non-meteorological measurements, the SEM provides an important capability through its measurement of solar proton, alpha particle, and electron flux density, energy spectrum, and total particulate energy disposition at satellite altitude. Its two components are the Total Energy Detector (TED) and the Medium Energy Proton and Electron Detector (MEPED).

A. The SEM Total Energy Detector (TED) uses a curved plate analyzer and channeltron detector to determine the intensity of particles in the energy bands from 0.3 to 20 KeV. Four curved plate analyzers (two measuring electrons, two protons) measure incoming particles reaching the instrument. Outputs from the analyzers are sent to the detectors and then to the Data Processing Unit (DPU) for multiplexing into the final output data stream.

B. The SEM Medium Energy Proton and Electron Detector (MEPED) senses protons, electrons, and ions with energies from 30 KeV to more than 60 MeV. Components of this instrument are four directional, solid state, detector telescopes and one omnidirectional sensor. All five components use solid state nuclear detectors. Outputs from the detectors are connected to a signal analyzer which senses and logically selects events that exceed specific threshold values. These data are fed to the DPU and are included with the instrument output.

Table IV-2-7 *ERBE scanner characteristics*

CHANNEL	SPECTRAL INTERVAL (μm)	FILTER	INSTANTANEOUS FIELD OF VIEW
6	0.2–5	Suprasil-W	3° × 4.5°
7	5–50	Diamond plus shortwave cut-off	3° × 4.5°
8	0.2–50	None	3° × 4.5°

IV-2.3 THE ATN TIROS INFORMATION PROCESSOR (TIP) AND MANIPULATED INFORMATION RATE PROCESSOR (MIRP)

Data sensed by the instruments described above are processed onboard the satellite for real time direct readout or for recording on tape for readout at the Fairbanks and Wallops stations. The TIROS TIP formats all low bit rate instrument data (i.e., from the HIRS/2, SSU, MSU, SBUV/2, ERBE, and SEM) and the spacecraft telemetry, for transmission to the ground. It also controls the data outputs from these sources. TIP adds synchronization, identification, and a time code before simultaneously transferring the data to the beacon transmitter (which operates at 136.77 or 137.77 MHz depending on whether it is a "morning" or "afternoon" spacecraft), to the tape recorder interface, and to the MIRP.

The AVHRR is the ATN high data rate sensor and MIRP the high data rate processor. Using the data from AVHRR, the MIRP provides separate outputs for:

- High Resolution Picture Transmission (HRPT) at 1.1 km in real time
- Automatic Picture Transmission (APT) at a reduced resolution of 4 km in real time
- tape recording of Local Area Coverage (LAC) at 1.1-km resolution of selected areas for storage and readout
- tape recording of Global Area Coverage (GAC) at 4-km resolution for storage and readout.

In addition to formatting, MIRP adds synchronization, identification, telemetry, time code, and the output of TIP to the AVHRR output (except for APT).

IV-2.3.1 AVHRR/HRPT Processing

The AVHRR scan mirror rotates at 360 rpm. Once during each revolution of the mirror, a synchronization pulse is generated as the mirror reaches a precise (pre-Earth) position. This pulse is used as the reference for all subsequent event timing, which is controlled by the MIRP. Upon receipt of the synchronization pulse from AVHRR, the MIRP begins a data sampling process which takes an appropriate number of samples at specified times during the scan line. Data are stored in a memory bank and read out for the HRPT data format. Full resolution data from the AVHRR with TIP information added are output for the HRPT and LAC transmissions.

IV-2.3.2 AVHRR/GAC Processing

MIRP uses one of three AVHRR lines for GAC data. Data volume and resolution are further reduced by averaging four adjacent samples along the line and skipping the following sample. This pattern is continued for the length of the scan line. The averaged values are inserted in the GAC format for eventual relay to the central processing facility. The TIP output is also inserted in the GAC format.

IV-2.3.3 AVHRR/APT Processing

Any two AVHRR channels can be chosen by ground command for output to the APT transmitter. The data processing algorithm was designed so that data from every third line from each of the two channels of the original high resolution output are formatted for the APT system. The APT data output, an amplitude modulated signal, is significantly different from the digital outputs. Two channels of AVHRR data are time-division multiplexed into an output data stream that has been preprocessed to achieve both bandwidth reduction and geometric correction. The bandwidth reduction algorithm is similar to that used for globally stored data (GAC) in that only one line of three from the original AVHRR output is processed. The algorithm is different in that nearly equal geometric resolution is maintained along the scan line. This linearization is accomplished by using a separate resolution reduction in each of five regions either side of nadir.

IV-2.4 GEOSTATIONARY (GOES) SENSORS

The weather observation sensor on the GOES series of NOAA operated spacecraft is the Visible Infrared Spin Scan Radiometer Atmospheric Sounder (VAS). First flown on GOES-4 in 1980, it is an outgrowth of the spin scan radiometer carried by earlier GOES satellites.

The other environmental sensor on GOES is its Space Environment Monitor (SEM), similar in purpose to the SEM of the polar orbiting ATN series except that it gathers data at geostationary altitude rather than at the lower altitude of the ATN.

IV-2.4.1 The VISSR Atmospheric Sounder (VAS)

The VAS (Figure IV-2-6 and Table IV-2-8) is a more advanced version of the Visible Infrared Spin Scan Radiometer (VISSR) carried by earlier GOES satellites through GOES-C (launched June 1978). Like VISSR, it scans the Earth by means of an optical reflecting telescope.

Radiation enters the telescope (Figure IV-2-7) at right angles to its optical axis via the optically flat scan mirror, which is provided with an angular-positioned stepping mechanism to facilitate scanning the terrestrial scene. The scanned pattern resembles a conventional television image. West-to-east scan lines are formed by rotation of the spinning spacecraft. The lines are moved by stepping the mirror to scan north to south. Synchronizing signals are also sent through the spacecraft so that imaging signals can be assembled on the ground.

Visible (0.55 to 0.75 μm) data are detected and converted to electrical signals by an array of eight photomultiplier tubes. Each tube views an eighth of the scan line width, so that eight lines are scanned on each spin of the spacecraft. Resolution at the subsatellite point is about 0.9 km.

For IR observations, VAS has six detectors. There are two small mercury cadmium telluride (HgCdTe) long wavelength detectors, used mostly for imaging, with an approximate resolution of 6.9 km. There are also four larger IR detectors, two of HgCdTe and two of indium antimonide (InSb) with a resolution of 13.8 km; these are used for soundings. These detectors, cooled to 94 K by a radiative cooler, convert the IR energy to electrical signals for transmission.

For multispectral scan imaging, VAS can be programmed to transmit VIS data on every scan line and to transmit IR signals in pairs. By the selective use of filters (through an onboard commandable filter wheel), detectors, and by interweaving the scans of the large and small detectors (as facilitated by their north-south spacing in the IR focal plane), a full frame of scans can image the Earth in as many as three IR channels.

For atmospheric sounding, VAS includes a filter wheel that covers seven detection bands in the region of CO_2 absorption, three in areas of water vapor ab-

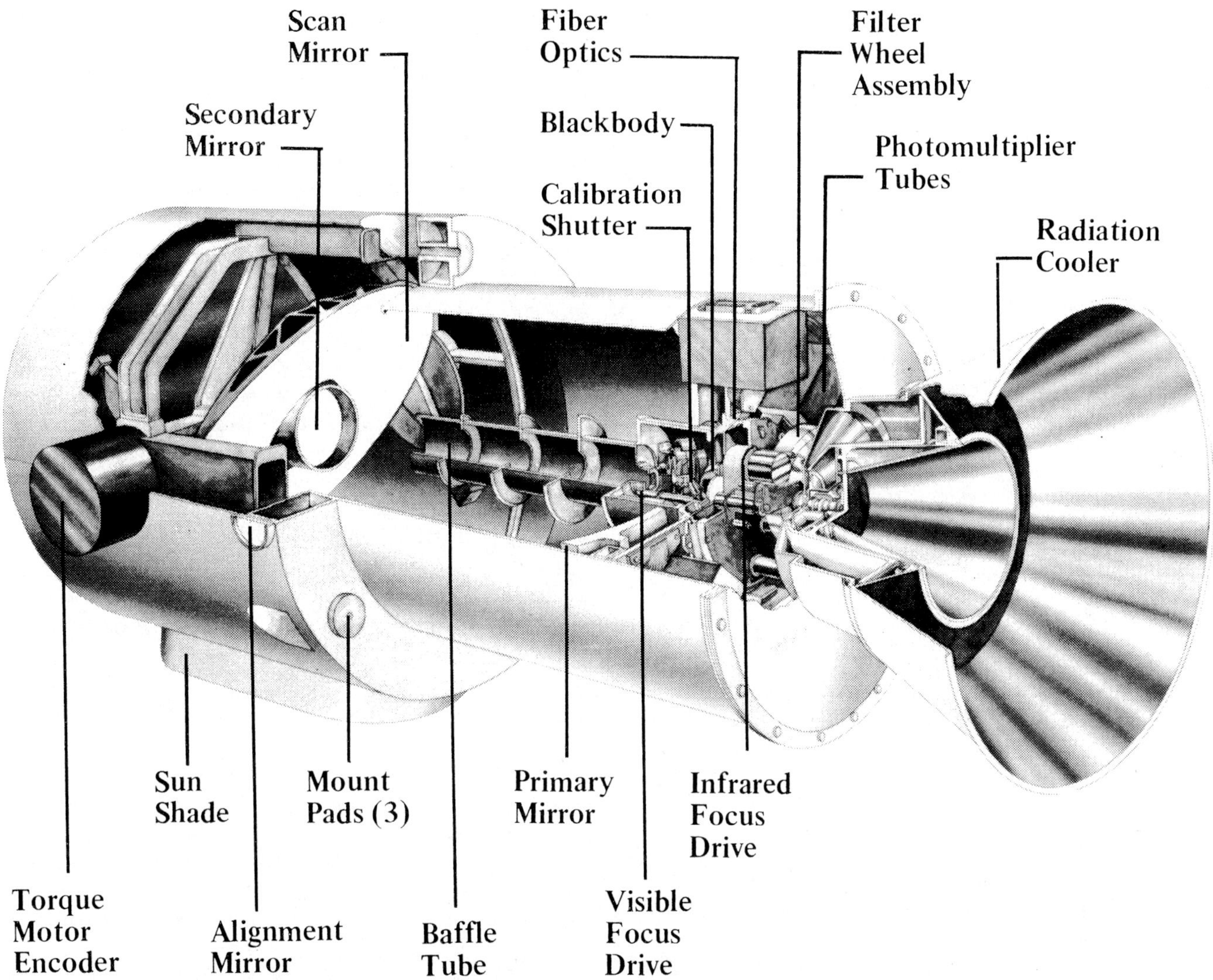

Figure IV-2-6. Visible Infrared Spin Scan Radiometer Vertical Atmospheric Sounder (VAS).

sorption, and two "window" bands, one spectrally nearer the longwave absorption band detectors and one nearer the shortwave detectors. Narrow detection bands are created by combinations of detectors sensitive over relatively wide bands in the spectral regions of interest with narrowband filters that are mechanically inserted into the optical path by the servo-positioned wheel.

Table IV-2-9 lists the spectral bands that can be inserted into the optical path by the filter wheel. The numbers in the "Spin Budget" column are the number of times that each line must be scanned to produce (by reinforcement) a satisfactory signal to noise ratio for a usable sounding.

The cylindrical shape of VAS permits it to fit inside the thrust tube of a classically designed spinning spacecraft structure, thereby minimizing the effects on spacecraft weight, balance, and moment of inertia. Design requirements on imaging capability dictate a telescope of relatively large optics. To avoid the weight of large precision optics made from conventional material such as glass mirrors mounted in steel structures, the entire VAS structure and principal optics are made of beryllium. North to south VAS scanning is accomplished by stepping the scan mirror. An angle position encoder integral with the mirror stepping mechanism converts the position information to electrical signals, which are sent to the CDA stations to aid in reassembly of the Earth scene. Mirror motion is divided into 1820 steps, each representing 192 μrad optically.

A Ritchey-Chretien optical system is used to provide a large field at the focal plane. This large field accommodates the several detector elements required for electrical conversion of the focused image.

Table IV-2-8 *VAS characteristics*

Size	
Length	~1.5 m
Radial dimension	~0.65 m
Weight	
Scanner	64.3 kg
Electronics module	10.5 kg
Structural material	Beryllium (structure and mirrors)
Optics	Ritchey-Chretien system 40.64-cm diameter primary mirror 292.1-cm focal length Optically flat scan mirror
Detectors	
Visible spectrum	8 photomultiplier tubes coupled to focal plane by optical fibers which define IGFOV: 21 × 25 μrad
Infrared spectrum	Solid state detectors cooled to 95 K 2 HgCdTe; IGFOV: 192 μrad 2 HgCdTe; IGFOV: 384 μrad 2 InSb; IGFOV: 384 μrad
Spectral bands	By filters inserted in IR optical path; 12 filters selectable from rotating wheel
Radiative cooler	Servoed operating temperature ~95 K Heat rejection capacity ~16 mW Minimum temperature ~82 K
In-orbit calibration	Radiometric IR: Temperature monitored blackbody, cyclically reflected into optical axis VIS: Optical aperture providing sun signal of 50% Earth albedo Electronic IR and VIS: Stairstep voltage into amplifier chain
Onboard programmer	Reprogrammable by command link Selects and controls scan sector size and center position Controls filter wheel and selects transmitting detector pair

At the image plane, a relatively large FOV is available. Each detector element is dimensioned to define the FOV that its signal is intended to represent. For example, the smallest IR field is 192 μrad, defined by a square detector 80 μm on each side. At synchronous altitude, 192 μrad is equivalent to 6.9 km along the Earth surface at the satellite's suborbital point.

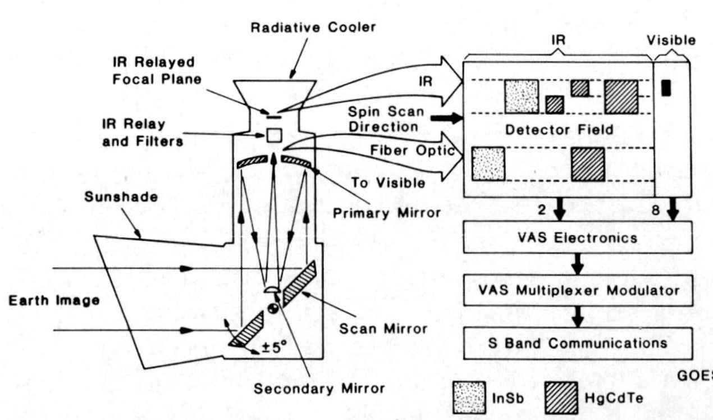

Figure IV-2-7. VAS telescope optics.

Two focal planes are used in VAS. VIS signals are obtained at the principal focus. An optical fiber for each of the eight FOV defines the field to be measured and conveys the impinging light within that FOV to a photomultiplier tube, which converts the light intensity to a proportional electrical current.

Infrared radiation must be sensed by solid state detectors cooled to a low temperature to reduce their intrinsic electrical noise to a level below the electrical equivalent of the least intense radiation to be measured. This cooling is provided by a cooler which radiates excess heat into space. Because of spacecraft design constraints, the cooler must be located away from the prime focal plane. This requirement is accommodated by relay of the IR signals to another, conveniently placed, focal plane. The relay optics provide an appropriate location for an IR focusing mechanism and filter assembly out of the VIS light path. Filters are inserted in the IR path only, and used in the multispectral imaging and sounding modes.

IV-2.4.2 The GOES Space Environment Monitor (SEM)

The SEM performs three functions: measurement of magnetic field strength and direction in the vicinity of the spacecraft; assessment of solar X-ray flux; and sensing of energetic particles of the solar wind and radiation belts around the Earth. Although the subsystems are independent, the data produced by the three functions are interrelated.

The magnetic field is sensed by two detectors: one with sensitive axis parallel to the spacecraft spin axis and one with sensitive axis perpendicular to the spin axis (radial). The data from the axial detector measure the field without further conversion. The output of the radial sensor must be processed by a phase sensitive detector to determine its magnitude and vector direction angle relative to spacecraft indices. When the spacecraft is in its normal on-station orientation in synchronous equatorial orbit, the axial sensor measures the north to south component of the magnetic field and the radial sensor measures the radial component indexed to the Earth–satellite line of sight.

The flux of electrons, protons, and alpha particles in the vicinity of the spacecraft is measured in numerous energy ranges. The measurement is made by counting, for specified periods, the numbers of impinging particles detected by solid state devices. The High Energy Proton and Alpha Detector (HEPAD) and an Energetic Particle Sensor are the satellite instruments used for these measurements. HEPAD

Table IV-2-9 *Spectral bands of VAS filter wheel*

SPECTRAL BAND	PRESSURE LEVEL (MB)*	WAVELENGTH (μm)	BAND	DETECTOR TYPE	SPIN BUDGET
1	65	14.73	CO_2	HgCdTe	6
2	100	14.48	CO_2	HgCdTe	14
3	325	14.25	CO_2	HgCdTe	11
4	450	14.01	CO_2	HgCdTe	7
5	Surface	13.33	CO_2	HgCdTe	7
6	700	4.525	CO_2	InSb	35
7	Surface	12.66	H_2O	HgCdTe	7
8	Surface	11.17	Window	HgCdTe	1
9	375	7.261	H_2O	HgCdTe	16
10	330	6.725	H_2O	HgCdTe	3
11	280	4.444	CO_2	InSb	46
12	Surface	3.945	Window	InSb	4

* Level of maximum weighting function for the spectral band.

senses protons and alpha particles from about 370 to more than 850 MeV. It is essentially a Cerenkov detector. The Cerenkov crystal is installed within a telescope in association with two solid state detectors; the telescope is shielded to establish the instrument FOV.

Solar X-ray flux emitted from the fully visible hemisphere of the Sun is measured in two spectral ranges by ion chamber detectors which view the Sun via a collimator on each spacecraft. The collimator excludes energetic electrons and attenuates bremsstrahlung and other undesired radiation. It tracks the Sun's apparent north to south motion, which varies seasonally. A single-axis positioner, stepped at proper time intervals, provides the Sun tracking.

BIBLIOGRAPHY AND REFERENCES FOR SECTION IV

Abel, P. G., P. J. Ellis, J. T. Houghton, G. Peckham, C. D. Rodgers, S. D. Smith, and E. J. Williamson, 1970: Remote sounding of atmospheric temperature from satellites II. The selective chopper radiometer for Nimbus-D. *Proc. Roy. Soc. London, A320*, 35-55.

Altmann, J., W. Lahmann, and C. Weitkamp, 1980: Remote measurement of atmospheric N_2O with a DF laser lidar. *Appl. Opt., 19*, 3453-7.

Bendetti-Michelangeli, G. F., F. Congeduti, and G. Fiocco, 1972: Measurement of aerosol motion and wind velocity in the lower troposphere by Doppler optical radar. *J. Atmos. Sci., 29*, 906.

Bilbro, J. W., 1980: Atmospheric laser Doppler velocimetry: An overview. *Opt. Eng., 19*, 533.

——— and W. W. Vaughan, 1978: Wind field measurement in the nonprecipitous regions surrounding severe storms by an airborne pulsed Doppler lidar system. *Bull. Amer. Meteor. Soc., 59*, 1095.

Born, M. and E. Wolf, 1970: *Principles of Optics*. Pergamon Press, Oxford, England, 808 pp.

Boyle, W. S. and G. E. Smith, 1970: Charge coupled semiconductor devices. *Bell System Technical Journal, 49*, Amer. Telephone and Telegraph, New York, NY.

Browell, E. V., A. G. Carter, and S. T. Shipley, 1983: NASA multipurpose airborne DIAL system and measurements of ozone and aerosol profiles. *Appl. Opt., 22*, 522-534.

———, 1982: Lidar measurements of tropospheric gases. *Opt. Eng., 21*, 128-132.

Brown, G. S., 1978: Backscattering from Gaussian distributed perfectly conducting rough surfaces. *IEEE Trans. Antennas Propag., AP-26*, 472-482.

Cahen, C., G. Megie, and P. Flament, 1982: Lidar monitoring of the water vapor cycle in the troposphere. *J. Appl. Meteor., 21*, 1506-1515.

Carswell, A. I., 1983: Lidar measurements of the atmosphere. *Can. J. Phys., 61*, 378-395.

Chen, H. S., 1985: *Space Remote Sensing Systems*. Academic Press, Inc., New York, NY, 257 pp.

Claasen, J. P., H. S. Tung, R. K. Moore, and W. J. Pierson, Jr., 1972: Radar sea return and the RADSCAT satellite anemometer. *Proc. of the IEEE Intl. Conf. Engineering in the Ocean Environment, Newport, RI*, IEEE Pub. 72 CHO 660-1 OCC.

Cooney, J. and M. Pina, 1976: Laser radar measurements of atmospheric temperature profiles by use of Raman rotational backscatter. *Appl. Opt., 15*, 602.

Curtis, P. D., J. T. Houghton, G. D. Peskett, and C. D. Rodgers, 1974: Remote sounding of atmospheric temperature from satellites V. The pressure modulator radiometer for Nimbus-F. *Proc. Roy. Soc. London, A337*, 135-150.

Driscoll, W. G. and W. Vaughan (eds.), 1978: *Handbook of Optics*. Optical Society of America, McGraw-Hill, New York, NY, 1172 pp.

Drummond, J. R., J. T. Houghton, G. D. Peskett, C. D. Rodgers, M. J. Wale, J. Whitney, and E. J. Williamson, 1980: The stratospheric and mesospheric sounder on Nimbus-7. *Phil. Trans. Roy. Soc. London, A296*, 219-241.

Ellis, P., G. Holah, J. T. Houghton, T. S. Jones, G. Peckham, G. D. Peskett, D. R. Pick, C. D. Rodgers, H. Roscoe, R. Sandwell, S. D. Smith, and E. J. Williamson, 1973: Remote sounding of atmospheric temperature from satellites IV. The selective chopper radiometer for Nimbus-5. *Proc. Roy. Soc. London, A334*, 149-170.

Grantham, W. L., E. M. Bracalente, W. L. Jones, and J. W. Johnson, 1977: The Seasat-A satellite scatterometer. *IEEE J. Ocean Eng., OE-2*, 200-206.

Harger, R. O., 1970: *Synthetic Aperture Radar Systems: Theory and Design*. Academic Press, Inc., New York, NY, 240 pp.

Houghton, J. T., F. W. Taylor, and C. D. Rodgers, 1984: *Remote Sounding of Atmospheres*. Cambridge Univ. Press, England, 210 pp.

Inaba, H., 1976: Detection of atoms and molecules by Raman scattering and resonance fluorescence. Chapter Laser Monitoring of the Atmosphere, E. D. Hinkley (ed.), *Topics in Applied Physics, V-14*, Springer-Verlag, New York, NY, 380 pp.

Johnson, J. W., L. A. Williams, Jr., E. M. Bracalente, F. B. Beck, and W. L. Grantham, 1980: Seasat-A satellite scatterometer instrument evaluation. *IEEE J. Ocean Eng., 5*, 138-144.

Jones, W. L., P. G. Black, D. M. Boggs, E. M. Bracalente, R. A. Brown, G. Dome, J. A. Ernst, I. M. Halberstam, J. E. Overland, S. Peteherych, W. J. Pierson, F. J. Wentz, P. M. Woiceshyn, and M. G. Wurtele, 1979: Seasat scatterometer: Results of the Gulf of Alaska workshop. *Science, 204*, 1413-1415.

Jordan, R. L., 1982: The SIR-B plan: Imaging radar science working group. *JPL Publication 82-78*, NASA, JPL, Pasadena, CA, XIX67.

———, 1980: The Seasat-A synthetic aperture radar system. *IEEE J. Ocean Eng., 5*, 154-164.

Kostko, O. K., 1976: Use of laser radar in atmospheric investigations. *Sov. J. Quant. Electron., 5*, 1161-1177.

Lundquist, S., C. O. Falt, U. Persson, B. Marthinson, and S. Eng, 1981: Air pollution monitoring with a Q switched laser lidar using heterodyne detection. *Appl. Opt., 20*, 2534-8.

MacArthur, J. L., 1976: Design of the Seasat-A radar altimeter. *Proc. Oceans 76*, IEEE 76, CH1118-9 OEC, 10B-1 to 10B-8.

Mason, J. B., 1975: Lidar measurement of temperature: A new approach. *Appl. Opt., 14*, 76-78.

McGoogan, J. T., L. S. Miller, G. S. Brown, and G. S. Hayne, 1974: The S-193 radar altimeter experiment. *Proc. IEEE, 62*, 793-803.

Measures, R. M., 1984: *Laser Remote Sensing*. J. Wiley and Sons, New York, NY, 510 pp.

NOAA/NASA/Hughes Aircraft Company/Space and Communications Group, 1980: *GOES D, E, F Data Book*, Natl. Aero. and Space Adm., GSFC, Greenbelt, MD, 100 pp.

Post, M. J., R. L. Schwiesow, R. E. Capp, D. A. Haugen, and J. T. Newman, 1978: A comparison of anemometer- and lidar-sensed wind velocity data. *J. Appl. Meteor., 17*, 1179-1181.

Remsberg, E. E. and L. L. Gordley, 1978: Analysis of differential absorption lidar from the space shuttle. *Appl. Opt., 17*, 624-630.

Rhodes, C. K., 1979: Excimer lasers. *Topics in Applied Physics*, Vol. 30, Springer-Verlag, New York, NY, 194 pp.

Schafer, F. P. (Ed.), 1973: Dye lasers. *Topics in Applied Physics*, Vol. 1, Springer-Verlag, New York, NY, 285 pp.

Schwalb, A., 1982: Modified version of the TIROS-N/NOAA A-G satellite series (NOAA E-J) - Advanced TIROS-N (ATN). *NOAA Technical Memorandum NESS 116*, Dept. of Commerce, Washington, DC, 29 pp.

———, 1987, 1978: The TIROS-N/NOAA A-G satellite series (NOAA E-J) - Advanced TIROS-N (ATN). *NOAA Technical Memorandum NESS 95*, Dept. of Commerce, Washington, DC, 75 pp.

Smith, W. L., H. M. Woolf, C. M. Hayden, D. Q. Wark, and L. M. McMillin, 1979: The TIROS-N operational vertical sounder. *Bull. Amer. Meteor. Soc., 60*, 1177-1187.

Stanley, H. R., 1979: The Geos-3 Project. *J. Geophys. Res., 84*, 3779-3783.

Strong, J., 1954: *Concepts of Classical Optics*. W. H. Freeman, San Francisco, CA, 692 pp.

Tang, C. L. (ed.), 1979: Quantum Electronics. *Methods of Experimental Physics*. Vol. 15B, Academic Press, Inc., New York, NY, 342 pp.

Tapley, B. D., G. H. Born, and M. E. Parke, 1982: The Seasat altimeter data and its accuracy assessment. *J. Geophys. Res., 87*, 3179-3188.

Ulaby, F. T., R. K. Moore, and A. K. Fung, 1986: *Microwave Remote Sensing—Active and Passive*. Volume III from Theory to Applications, Artech House, Dedham, MA, 2162 pp.

Walsh, E. J., E. A. Uliana, and B. S. Yaplee, 1978: Ocean wave heights measured by a high resolution pulse-limited radar altimeter. *Bound. Layer Meteor., 13*, 263-276.

Yariv, A., 1975: *Quantum Electronics*. John Wiley and Sons, Inc., New York, NY, 570 pp.

Zuev, V. E. C., 1982: *Laser Beams in the Atmosphere*. Trans. by J. S. Wood, Consultants Bureau, New York, NY, 504 pp.

Contributors to Section IV

Alishouse, John C.
*Braun, W. Charles
*Reinhardt, George
Schwalb, Arthur
Vaeth, J. Gordon
Weinreb, Michael P.
* (Moved to a new position)

NOAA/NESDIS
Washington, DC 20233

Section V

SATELLITE COMMAND AND DATA RECEPTION

An operational satellite system consists of spacecraft and ground systems. After launch of a spacecraft, extensive communication continues between the ground and the spacecraft. The ground system receives data, sends instructions (commands) to the spacecraft, and controls the instruments. The scanning modes of instruments are altered to acquire alternate forms of data; recorders on the spacecraft are directed to store or to play back data; and the satellite is commanded to move, adjust its orbit, or change its orientation. The functioning of spacecraft instruments, temperature levels, power status, fuel levels, etc., are monitored. Information obtained from the spacecraft and the ground systems on the "health" of the satellite is called "housekeeping data." The ground system coordinates these various functions through schedules sent to the satellite.

The satellites are also used to record or relay messages from ground transmitters or remotely located sensors. The data collected are relayed to ground stations for processing and extraction of information. The satellites collect data from various stationary and moving platforms. A full description of these systems is given in Chapters V-3 and V-4.

The process of receiving data from the satellite is called "data acquisition." There are basically two methods of data acquisition: by command from a central facility and via direct readout. Only a command facility such as a Command and Data Acquisition (CDA) Station can acquire data stored onboard. However, any properly equipped ground station can acquire data transmitted continuously or near-continuously via direct readout. Central processing of data into products and their distribution are described in Section VI.

The NOAA/NESDIS polar and geostationary satellite systems complement each other by providing various types of data at various resolutions, different areas of coverage, and times of data availability. The command and data acquisition requirements of the two systems differ significantly. Each operates independently, but each complements the operational mission of the other.

Chapter V-1

SATELLITE LAUNCH, ORBIT, COMMAND AND CONTROL, AND DATA ACQUISITION

V-1.1 LAUNCH INTO POLAR ORBIT

The NOAA polar orbiting spacecraft are launched from Vandenberg Air Force Base, California, by a modified Atlas-E ballistic missile. The Atlas is 28.7 m long and 3.05 m in diameter. The fairing is 7.42 m long and 2.13 m in diameter. At liftoff it carries 69 kl of liquid oxygen and 43 kl of RP-1 fuel, a highly refined kerosene.

From liftoff to +60 sec, an airborne autopilot programmer in the launch vehicle flight-control system provides steering and discrete commands. The General Electric Radio Tracking System (GERTS) ground station and guidance computer then take over the guidance through the launch vehicle pulse beacon system.

The vehicle is powered by one sustainer, two boosters, and two vernier engines using liquid oxygen and liquid hydrocarbon propellants. A 0.97-m diameter fitting fastens the spacecraft to the launch vehicle. The fairing attached to the forward face of the launch vehicle protects the spacecraft during the boost flight. Figure V-1-1 shows the Atlas-E launch vehicle and Figure V-1-2 the launch-to-orbit injection sequence.

V-1.2 LAUNCH INTO GEOSTATIONARY ORBIT

GOES spacecraft are launched from the U.S. Air Force Eastern Space and Missile Center, Cape Kennedy, Florida, by three-stage Delta rockets. With its fairing, the vehicle (Figure V-1-3) has an overall length of about 35 m and a maximum diameter of 2.4 m.

The first stage contains the main engine, a modified Thor booster powered by liquid oxygen and liquid hydrocarbon propellants. There are also nine strap-on solid fuel rocket motors. Two liquid propellant vernier engines provide roll control during liftoff. At liftoff, the main engine and five strap-on rockets are ignited. About one minute into the flight, these strap-ons burn out and are jettisoned. The remaining strap-ons are then ignited; they burn out and are jettisoned about one minute later. The main engine continues its thrust until almost four minutes after liftoff; then it, too, is jettisoned.

The second stage contains a liquid fuel, pressure-fed, engine that ignites just after main engine cutoff. Shortly afterward, the fairing, which covers the third stage and the spacecraft, is jettisoned. This stage places the vehicle into a parking orbit at about 167 to 185 km where spin-up and second stage separation occur. Roll control throughout this portion of the flight is provided by a nitrogen gas system with eight fixed nozzles. Orbit inclination at this point is about 28.7°, which results from using a launch site near that latitude.

The third stage is a spin stabilized, solid-propellant motor. Following the second stage events, the vehicle is reoriented and at the first Equatorial crossing the third stage ignites, injecting the combination into an

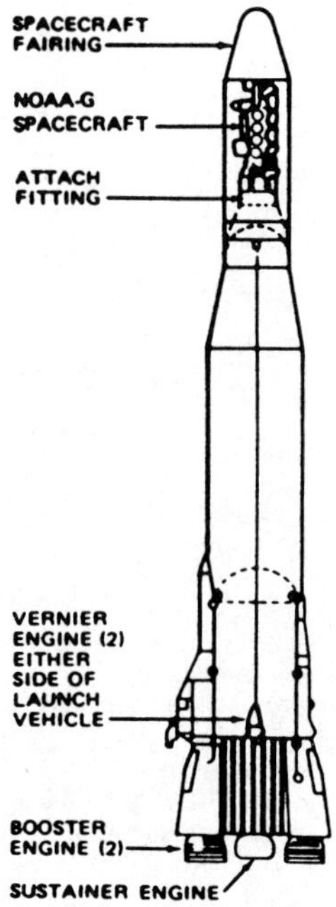

Figure V-1-1. Atlas-E launch vehicle.

elliptical transfer orbit with its apogee at or near geostationary altitude and its perigee at parking orbit altitude. The inclination at this point has been reduced to about 24°.

After burnout of the third stage, the spacecraft, with apogee boost motor attached, is separated. The Delta Inertial Guidance System (DIGS), consisting of an inertial sensor package and a digital guidance computer has guided the launch vehicle from liftoff to this point.

To prepare for apogee injection, spacecraft circuits take over nutation control and the spin axis is reoriented by ground command. Selection of the apogee height for the transfer orbit is based on whether the satellite is to be in a drift orbit or on station after the next and final launch event, the ground commanded Apogee Motor Firing (AMF). With AMF at an apogee of geostationary altitude, the spacecraft will achieve geosynchronous orbit. If the desired station is east of the AMF point, an apogee of less than geostationary altitude is programmed for the transfer orbit. An apogee at greater than geostationary altitude for the AMF will result in westward drift. Once the satellite has drifted to its desired station, its altitude is adjusted so that the orbit is geosynchronous. Figure V-1-4 shows the details of a geostationary launch.

V-1.3 THE POLAR OPERATIONAL ENVIRONMENTAL SATELLITE (POES) ORBIT

The POES system normally consists of two spacecraft in circular Sun synchronous orbits at altitudes of 833 km and 870 km. These altitudes are staggered to minimize observing and acquisition conflicts. With their orbital planes about 90 degrees apart, one has a southbound Equator crossing time of about 0730 LST and the other a northbound Equator crossing time of 1340 LST.

These Sun synchronous orbits ensure that these Equator crossings always occur at the same LST. Sun synchronism is achieved by placing the satellite at an inclination of 98.7° to 98.8° retrograde so that the orbital plane will precess, like a gyroscope, eastward at the same rate (0.986° per day) at which the Earth moves about the Sun. Sun synchronism makes for consistent scene illumination. The circularity of the orbit helps to ensure uniform observations by the satellite and efficient control of it by ground stations. The orbital period is about 101 min. Because the entire surface of the Earth passes beneath each satellite and can be viewed by it, global coverage is obtained. Figure V-1-5 shows a typical POES orbit.

V-1.4 THE GEOSTATIONARY ORBIT

The geostationary, or geosynchronous, orbit (at 35,800 km above the Equator) makes possible continuous sensing and communications between the satellite and the area of the Earth effectively within its view. From a command, control, and data acquisition standpoint, the geostationary orbit greatly simplifies ground-based antenna systems since they are relieved of the fast-tracking needed to follow the polar orbiters. Instead, the antenna systems basically need only be "locked" onto the spacecraft's fixed position in the sky.

Two satellites normally make up the fully operating GOES system; GOES-East at 75°W and GOES-West at 135°W. The inclinations, attitudes, and longitudinal positions of GOES spacecraft can be changed by onboard gas jets controlled by ground command. Location along the orbit is maintained to within 0.5° and inclination to within 0.1°. Attitude deviation is kept to within 0.2° of orbit normal. At the end of the useful lifetime of a GOES, any gas remaining onboard is used to boost the satellite to a higher altitude and thus remove it as a derelict from the crowded geostationary belt.

In addition to GOES-East and GOES-West, a third GOES is operated at 65°W. This is a GOES spacecraft that has lost its imaging capability but has retained its ability to transpond and communicate. It is used for broadcasting east WEFAX transmissions. Because imaging is not involved, station keeping, inclination, and attitude requirements are not so stringent for this GOES as for GOES-East and West.

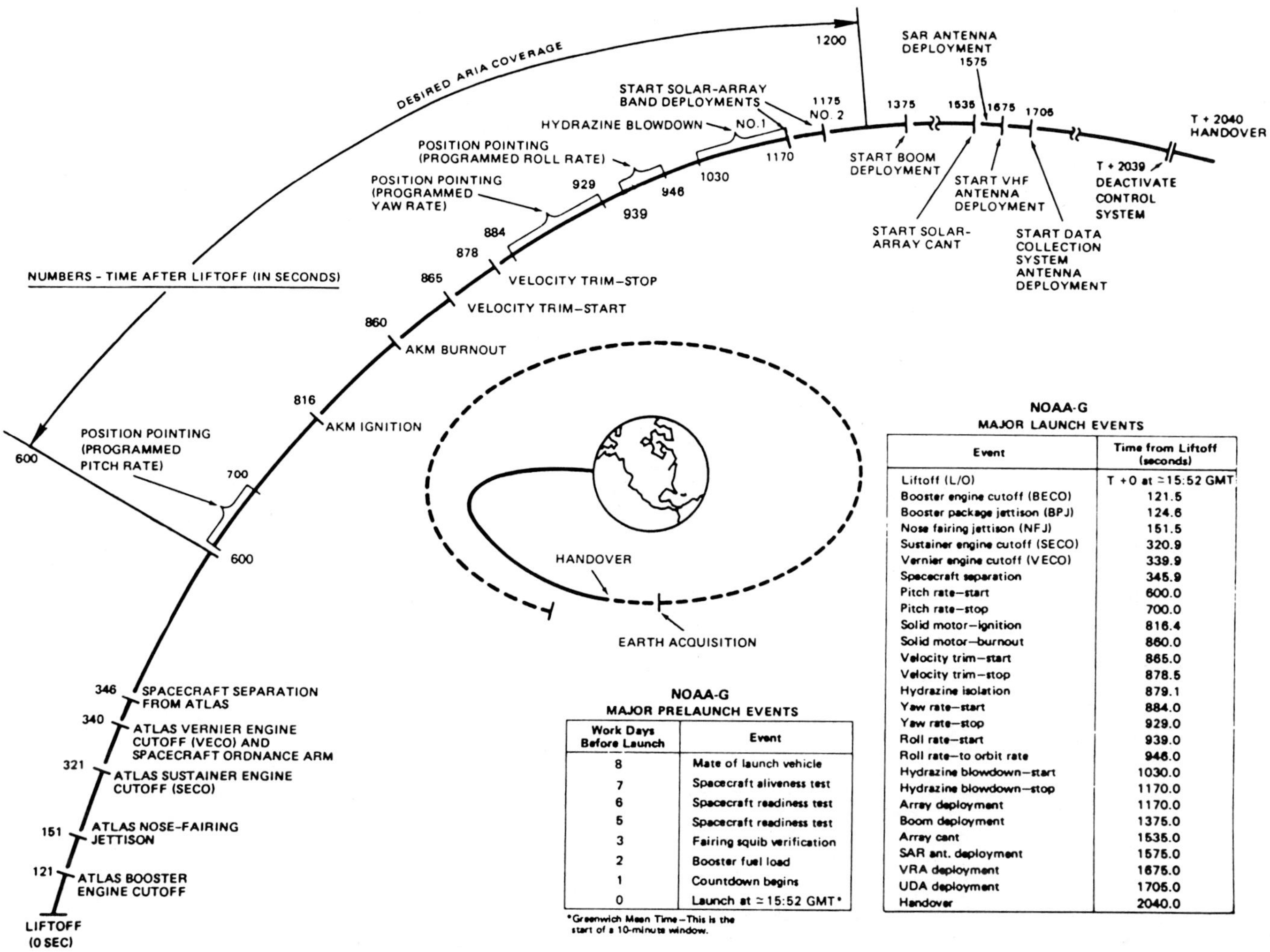

NOAA-G
MAJOR PRELAUNCH EVENTS

Work Days Before Launch	Event
8	Mate of launch vehicle
7	Spacecraft aliveness test
6	Spacecraft readiness test
5	Spacecraft readiness test
3	Fairing squib verification
2	Booster fuel load
1	Countdown begins
0	Launch at ≃15:52 GMT*

*Greenwich Mean Time—This is the start of a 10-minute window.

NOAA-G
MAJOR LAUNCH EVENTS

Event	Time from Liftoff (seconds)
Liftoff (L/O)	T +0 at ≃15:52 GMT
Booster engine cutoff (BECO)	121.5
Booster package jettison (BPJ)	124.6
Nose fairing jettison (NFJ)	151.5
Sustainer engine cutoff (SECO)	320.9
Vernier engine cutoff (VECO)	339.9
Spacecraft separation	345.9
Pitch rate—start	600.0
Pitch rate—stop	700.0
Solid motor—ignition	816.4
Solid motor—burnout	860.0
Velocity trim—start	865.0
Velocity trim—stop	878.5
Hydrazine isolation	879.1
Yaw rate—start	884.0
Yaw rate—stop	929.0
Roll rate—start	939.0
Roll rate—to orbit rate	946.0
Hydrazine blowdown—start	1030.0
Hydrazine blowdown—stop	1170.0
Array deployment	1170.0
Boom deployment	1375.0
Array cant	1535.0
SAR ant. deployment	1575.0
VRA deployment	1675.0
UDA deployment	1705.0
Handover	2040.0

Figure V-1-2. POES launch-to-orbit injection sequence.

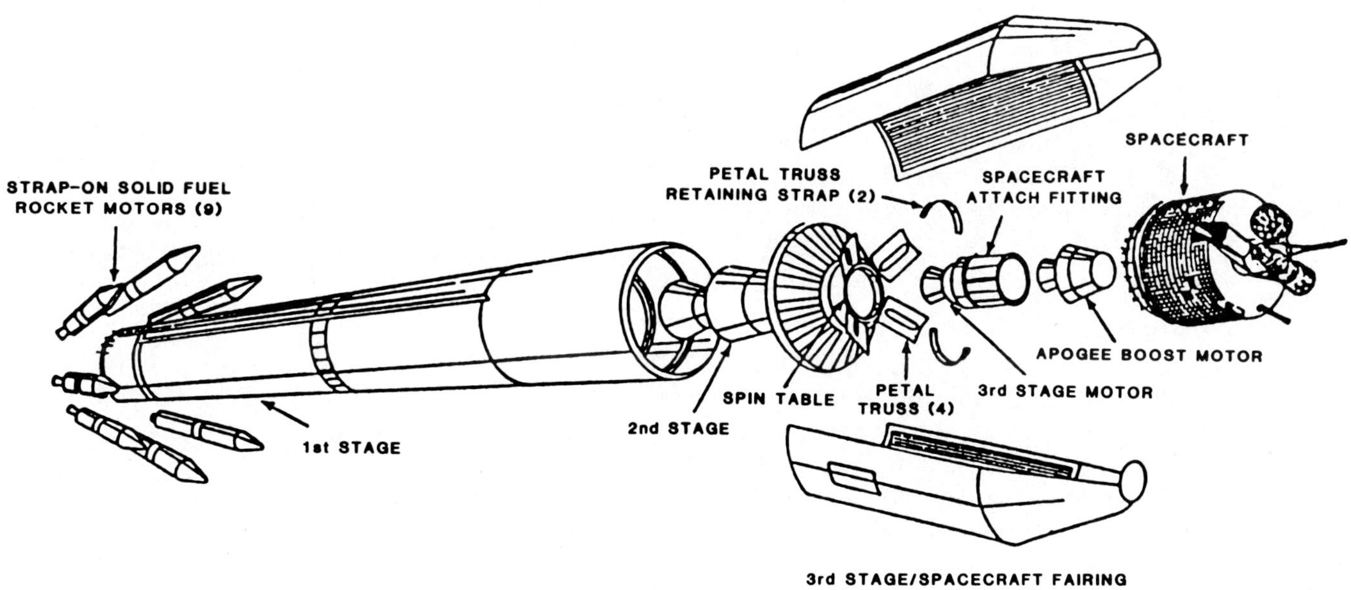

Figure V-1-3. Delta launch vehicle and GOES separation schematic.

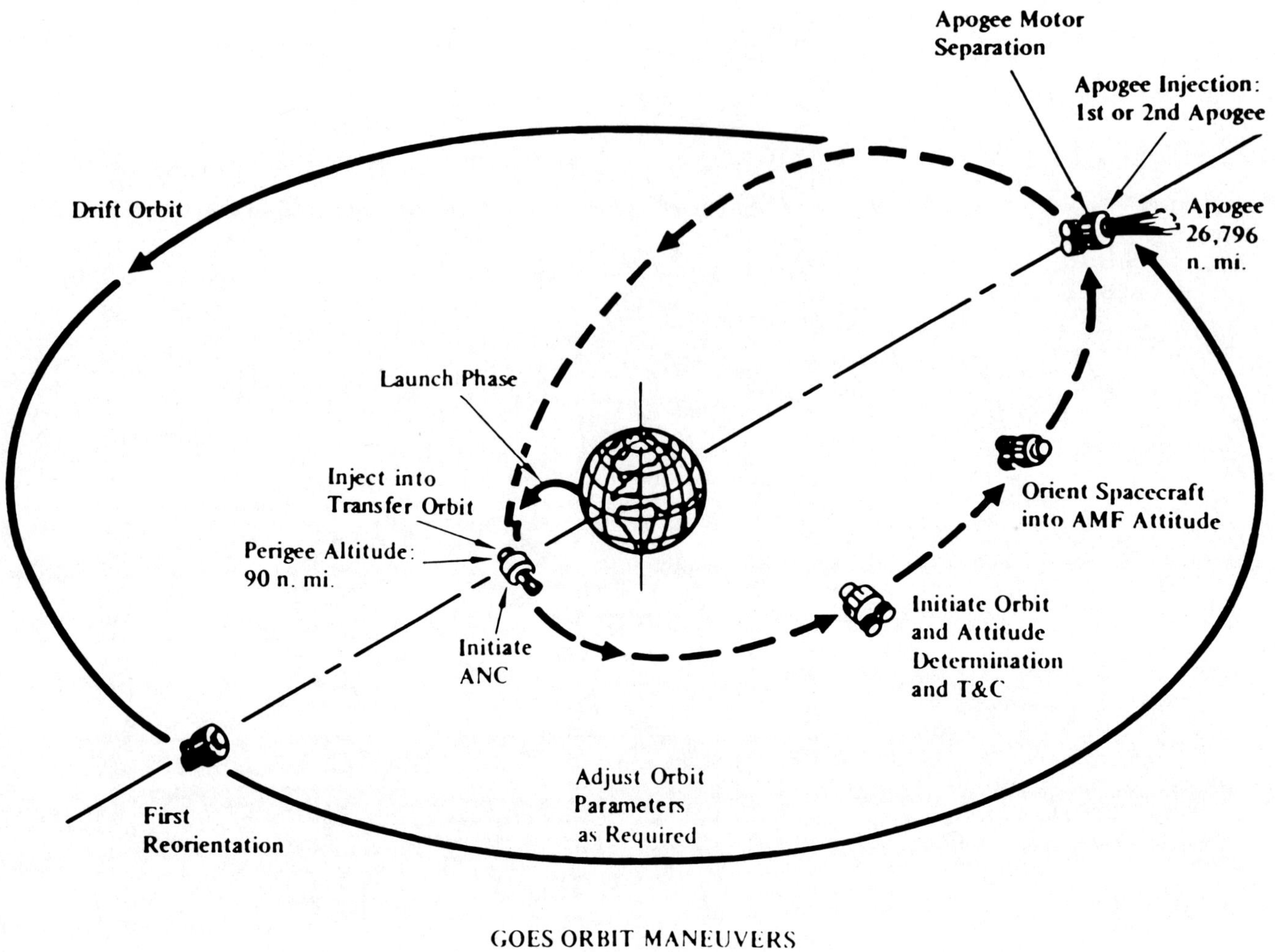

Figure V-1-4. GOES launch sequence.

V-1.5 COMMAND, CONTROL, AND DATA ACQUISITION

The POES and GOES satellite systems are owned and operated by NOAA/NESDIS of the U.S. Department of Commerce. The spacecraft are procured and launched for NOAA/NESDIS by NASA. After being placed in orbit and their performance verified, they are turned over to NESDIS by NASA for full operational control.

This control is exercised for NESDIS by its Satellite Operations Control Center (SOCC) located in Federal Office Building #4 at Suitland, Maryland, near Washington, DC. SOCC, the "brain center" of the operation, is responsible 24 hours a day for the performance of the satellites and of the NESDIS Command and Data Acquisition (CDA) stations that support them. SOCC monitors spacecraft health and safety and effects remedial efforts when anomalies or malfunctions occur. Based on the requirements of the data users, SOCC also schedules the spacecraft observations.

Transmission to the spacecraft of the decisions and commands originated by SOCC is carried out through the CDA stations at Wallops, Virginia, and Fairbanks, Alaska (Figures IV-1-6 and IV-1-7). These stations provide the radio links between satellites and ground by which commands are transmitted to the satellite and data are received at the ground. Both Wallops and Fairbanks "work" the polar orbiters, but because of geographic location, only Wallops can "see" and operate GOES. The nominal maximum acquisition time during which each station "sees" the POES spacecraft is 15 min per pass.

V-1.5.1 POES Command, Control, and Data Acquisition

POES imagery and data are read out at Wallops by a 26-m, S-band, tracking antenna and at Fairbanks by both 26-m and 13-m S-band tracking antennae. Because Wallops has no backup antenna for its 26-m sys-

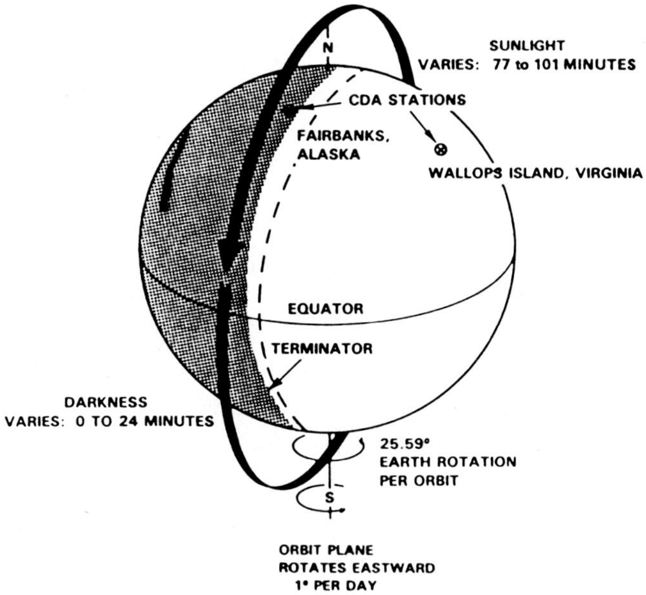

ORBITAL CHARACTERISTICS
Apogee – 833 - km (450 nmi)
Perigee – 833 - km (450 nmi)
Minutes per orbit – 101.35
Degrees inclination – 98.7

NOAA-G Orbit

Figure V-1-5. Typical POES orbit.

tem, Fairbanks acts as backup when a Wallops outage occurs.

Fairbanks, by reason of its high latitude location, can normally acquire eight POES passes per spacecraft per day. Wallops acquires four. The number of orbits per day per spacecraft is 14. Thus, every day two orbits are missed by the two NESDIS CDA stations. The data from these "missed" orbits are stored onboard for subsequent readout or "dumping" when the spacecraft is later within reach of Fairbanks or Wallops. Temperature profile data over the Western Pacific and Eastern Atlantic, being of special importance to the National Meteorological Center, cannot be delayed. The data missed by the U.S. stations are therefore read out for NOAA by the Lannion Centre de Météorologi Spatiale in France and are relayed to Wallops through GOES-East.

The key to understanding how and why POES imagery is planned and scheduled by SOCC lies in understanding the dual role played by the POES AVHRR. In one mode it produces 4-km VIS and IR imagery, in the other 1.1-km imagery. Onboard tape recorders—five sets of electronics and ten tape drives—are used to store the imagery during orbit. The low resolution data are known as Global Area Coverage (GAC). An entire orbit of GAC (115 min) can be stored by a single recorder. Only 11.5 min of high resolution imagery (LAC or Local Area Coverage) can be similarly stored. LAC must, therefore, be scheduled and timed carefully to meet the data needs of users with recording capability.

Advanced planning for POES missions involves 17 days of lead time for global (GAC) and high resolution (LAC) data. The scheduling system is flexible enough to accommodate short-term changes within 8 to 48 hours, depending on the complexity of the change and the station acquisition pattern of the spacecraft involved. Long lead schedules are done on large NOAA NAS 9070 mainframe computers. Day to day schedules are generated with four day lead times on minicomputers; these schedules control the spacecraft onboard computers for up to 30 hours.

Some of the observing systems carried by POES (described in Section III) require in-flight calibration, planning for which is another SOCC task.

The SOCC hardware consists of Data General System 230 computers. These S230 computers function as the heart of the POES command and control system, performing command and telemetry operations by direct link with CDA station computers.

The two CDA sites are also staffed for continuous real time operations. Their primary functions are to encode and transmit spacecraft commands, and to receive, demodulate, and transfer or process payload data. Commands and data transfers are controlled by computer-to-computer links with SOCC. Commanding can be done at SOCC, or independently at either acquisition site.

Communication between the CDA stations and SOCC is through a leased commercial satellite, except that narrow bandwidth information (commands, command verification, etc.) is transmitted over terrestrial links between Wallops and SOCC. Observational data, received from Fairbanks and Wallops through the commercial satellite terminal at Suitland, are routed to the NESDIS Office of Satellite Data and Product Distribution (OSDPD) for processing and forwarding to users.

The following listing of the communication links (arbitrarily numbered 1 through 8) between spacecraft and ground shows the major command, control, and data acquisition functions performed by and with POES:

Link 1—148.56 MHz

Command uplink—used by the Fairbanks and Wallops CDA stations to send all commands to the spacecraft.

Link 2—137.77 MHz and 136.77 MHz

Beacon transmission—provides direct full-time broadcast of the low data rate instru-

Figure V-1-6. Wallops Station, Virginia.

ments onboard the satellites. A major use of these transmissions is to provide direct readout of temperature sounding information. One frequency is used on the "morning" satellite, the other on the "afternoon" satellite. Use of 136.77 MHz will be discontinued about 1989 in accordance with Consultative Committee on International Radio (CCIR) Recommendation 595.

Link 3—137.50 MHz and 137.62 MHz

VHF real time downlink—provides 4-km APT analog imagery to global users, consisting of one VIS and one IR channel from the AVHRR. One frequency is used on the "morning" satellite, the other on the "afternoon" satellite.

Link 4—1698 MHz and 1707 MHz

S-band real time downlink—provides high-resolution AVHRR imagery via HRPT. Data are real time only, *not* stored, and contain the output of all the low data rate instruments (plus the high resolution data from all the AVHRR channels). One frequency is used on the "morning" satellite, the other on the "afternoon" satellite.

Link 5—1698 MHz, 1702.5 MHz, and 1707 MHz

S-band playback.

Link 6—401.64 MHz

Data Collection System (DCS) uplink—used by buoys, balloons, and other environmental sensing platforms to communicate their observations to POES for recording and pro-

Figure V-1-7. Fairbanks command and data acquisition facility.

cessing (including position-locating) after stored readout.

Link 7—121.5 MHz, 243.0 MHz, and 406.025 MHz
Search and Rescue uplink—transmits distress signals to spacecraft from parties in distress on the ground for relay to local user terminals and initiation of rescue efforts.

Link 8—1544.5 MHz
Search and Rescue downlink—transmits distress messages from the satellite to ground-based local user terminals.

V-1.5.2 GOES Command, Control, and Data Acquisition

For GOES, Wallops has ten antenna systems; the prime ones are two 18-m S-band transmit–receive "dishes" with a 13-m S-band backup. A 7-m S-band antenna transmits WEFAX products to the GOES-Central spacecraft. Other smaller antennas operate with the GOES Data Collection System (DCS).

As with POES, scheduling, control, and monitoring of GOES satellite health and safety are the responsibility of SOCC. SOCC receives data product requirements for mission fulfillment via the OSDPD and converts them into control schedules for the command of satellite subsystems and control of ground system configuration. Schedule summaries are widely distributed to users to assist them in data acquisition activities.

The functions of GOES command, primary data acquisition, and preliminary processing are performed at the Wallops CDA station. In addition to receiving and preprocessing telemetry, the CDA receives, processes, and retransmits GOES VAS data, provides part of the DCS ground-to-spacecraft-to-ground communication link, and provides the ground-to-spacecraft link for WEFAX broadcasts.

The GOES system, its spacecraft, sensors, data collection system, and direct broadcast services (including "stretching" the imagery data for retransmission) have been described previously in Chapter III-1.

Summarizing the capabilities represented by that system, SOCC and the Wallops CDA station have four major GOES missions to perform:

- VAS Imaging and Sounding
- Data Collection System Operation
- WEFAX Broadcasts
- Space Environmental Monitoring.

Imagery (VIS and IR) and soundings are acquired at the Wallops CDA station for preliminary processing which reformats, applies calibration, implants Earth location gridding, and reduces the data rate for retransmission through the satellite for user acquisition. The lower resolution IR imagery is simultaneously transmitted in analog form via landline to selected users. This preprocessing and distribution system significantly reduces the complexity and cost of user equipment.

The GOES DCS consists of several thousand user-owned data collection platforms (river gages, tide gages, buoys, balloons, aircraft, etc.), data acquisition equipment at Wallops, and data distribution equipment at the NESDIS Central Data Distribution Facility (CDDF) in the World Weather Building, Camp Springs, MD. The remotely located platforms transmit reports through the satellite as described in Chapter III-1. When received at the CDA station, the reports are formatted and forwarded to the CDDF via landline for distribution to users, also by landline. Dedicated and call-in lines connecting users directly with Wallops are also used.

The WEFAX identifies both a service and communications subsystem. Products for broadcast in this medium are created primarily by acquiring and processing both geostationary and polar satellite images into sections or chips. Chips containing meteorological charts and operational messages defining product content and broadcast schedules are also prepared and broadcast.

These products are routed from NESDIS at Suitland to the Wallops CDA station in analog form on telephone lines. Wallops transmits them to one or more of the GOES satellites supporting this service. The spacecraft, in turn, rebroadcasts them to users within communications range.

Two WEFAX broadcast services are in operation: GOES-East and GOES-West. Receiving equipment, low in complexity and cost, is readily available to those who want this service.

SEM data are transmitted from GOES through the satellite's telemetry stream. They are acquired directly by NOAA's Space Environmental Laboratory at Boulder, Colorado, where they are separated from the telemetry, processed, and distributed.

GOES communication links (arbitrarily numbered 1 through 14) are described below:

Link 1—2034.2 MHz
Command link from ground to spacecraft.

Link 2—1681.6 MHz
Raw VAS downlink over which VAS is received in 28 megabit bursts at Wallops for "stretching" and retransmission (over Link 3) to the spacecraft for broadcast.

Link 3—2029.1 MHz
Stretched VAS uplink via which VAS imagery, preprocessed at Wallops and "stretched" from 28 to 2.1 Mbps, is retransmitted to GOES for broadcast.

Link 4—1687.1 MHz
Stretched VAS broadcast frequency.

Link 5—2033.0 MHz
WEFAX uplink for transmission of meteorological products to GOES for facsimile broadcast.

Link 6—1691.0 MHz
WEFAX broadcast frequency.

Link 7—1694 MHz
Telemetry downlink to Wallops.

Link 8—2214.0 MHz
Redundant telemetry downlink to Wallops and NASA Satellite Tracking Data Network stations.

Link 9—2034.9 MHz
DCS uplink for transmitting to GOES interrogations and commands intended for participating platforms and National Bureau of Standards time intended for broadcast.

Link 10—468.85 MHz
Downlink over which GOES retransmits DCS platform interrogations and commands and the NBS time code.

Link 11—401.9 MHz
Uplink over which DCS platforms convey their observations to GOES which then transponds them over Link 12.

Link 12—1694.5 MHz
DCS GOES-to-Wallops downlink.

Link 13—406.05 MHz
Search and Rescue uplink.

Link 14—1698.65 MHz
Search and Rescue downlink.

Chapter V-2

SATELLITE DIRECT READOUT[1]

V-2.1 INTRODUCTION

Satellite direct broadcasting, more commonly called direct readout service, is used to distribute "raw" or minimally preprocessed satellite data to anyone anywhere in the world in real or near-real time. Remotely sensed data are transmitted directly from polar orbiting or geostationary satellites to ground stations within signal range of the satellite. The services are Automatic Picture Transmission (APT), High Resolution Picture Transmission (HRPT), and Direct Sounder Broadcasts (DSB) from POES and Weather Facsimile (WEFAX), Visible Infrared Spin Scan Radiometer (VISSR), and VISSR Atmospheric Sounder (VAS) data transmissions from GOES. Satellite collection of data from remote platforms is treated separately.

Until the early 1970s, direct readout services were a relatively small part of the satellite data distribution network. Most users were first-order meteorological forecast offices that received data generated by NOAA central processing facilities and distributed by landline, radio-teletype, or facsimile. All direct readouts were on VHF frequencies; the only services available were daytime APT images and a few processed GOES images sent via WEFAX.

Then, about 1975, changes in the user community began to take place, mainly because of the availability of data from new satellite sensors. Multichannel high resolution and atmospheric sounding data became available from POES in 1972. In 1974, high resolution GOES transmissions were scheduled and scores of processed satellite images and weather charts were added to the WEFAX service. Products from all instruments also were made available to the direct readout users.

The quantity, quality, and timeliness of satellite data were greatly improved, as was accessibility of satellite data to the general public. Additional APT and WEFAX stations and HRPT, DSB, and VISSR stations came into existence.

Today, nearly 90% of the world's users of satellite observations acquire them through direct readout. More than 120 countries have more than 5000 such stations worldwide. The data they receive often are relayed to scores of other sites.

NOAA/NESDIS is the focal point in the U.S. Government for the direct readout program. The NESDIS Office of External Relations currently coordinates direct readout activities. It provides a liaison service between NESDIS offices and the user community, provides information on constructing or purchasing ground stations, distributes publications, supports conferences, and issues bulletins and information notes on the services. It maintains records of the locations of readout stations and their use of the data, and acts as an international clearinghouse for technology and information exchange.

[1] Primary Contributor: Robert W. Popham

V-2.2 POLAR SATELLITE DIRECT READOUT SERVICES

Polar satellites provide APT, HRPT, and DSB services. APT consists of low resolution analog signals on VHF frequencies, receivable on simple ground equipment. The data are used extensively by U.S. and foreign government and private groups. HRPT is in the form of digital data products frequently converted to images. Ground resolution is four times higher than APT data. HRPT signals are provided on S-band frequencies; reception and processing requires relatively expensive ground equipment. Primary users are government weather and land use agencies; several universities also receive HRPT data. The DSB provides atmospheric data on both S-band and VHF; because of the complicated mathematical processing required, only a few government meteorological agencies use DSB data. Satellite APT, HRPT, and DSB systems operate and transmit continuously; some of the data are recorded for later transmission to a NOAA CDA station.

V-2.2.1 Automatic Picture Transmission (APT) Services

APT services and subsystems were initially designed to broadcast direct readout satellite imagery to low cost ground receiving equipment. This low equipment cost has accounted for the proliferation of APT stations around the world.

APT services have been available almost continuously since 1963. NASA and RCA (now a part of General Electric Corporation) originally developed APT data transmission techniques to meet U.S. Weather Bureau data distribution requirements. Seven stations in three countries received APT data from TIROS-8 in 1963; today, more than 5000 APT stations, established in at least 123 countries, receive APT data from U.S. and U.S.S.R. satellites.

APT readout has the lowest cost and is the most generally useful of all the direct readout services. Full capability systems can cost as much as $70,000, but radio amateurs and students have constructed stations for as little as $200. APT stations are portable and can be installed almost anywhere.

APT signals from U.S. satellites are transmitted on 137.50 or 137.62 MHz, while Soviet satellites transmit between 137.30 and 137.85 MHz. U.S. satellites transmit 4-km VIS and IR images. Basic receivers consist of a low cost steerable helix, yagi, or omni-directional antenna, VHF receivers, and a display device such as a personal computer, a facsimile receiver, a photographic device, or a CRT.

APT products and Global Area Coverage (GAC) products appear to be similar; both are derived from the operation of the AVHRR. The difference is in the way they are acquired. APT data are acquired whenever the satellite passes within range of a ground station, usually four times in a 24-hour period. Each station receives only while the spacecraft is within range (for about 15 min). The same observations are simultaneously recorded onboard the spacecraft as GAC data. Later, the GAC data from each orbit are transmitted to a CDA station, processed, redistributed via WEFAX, and archived.

User applications of APT data are mainly in the fields of meteorology and oceanography. Obviously most of these applications are to real time operational and experimental activities. APT data reception is also useful in many educational programs. Construction of a station teaches electronics, communications, computer usage, and photography; data interpretation and applications involve many disciplines and fields. The images arouse student interest in Earth sciences. APT programs in schools have contributed to both regular course work and extracurricular activities.

V-2.2.2 High Resolution Picture Transmission (HRPT)

HRPT systems were initially designed to provide weather observations to meteorologists. In 1972, NOAA meteorologists in San Francisco began receiving and using HRPT data as an aid to forecasting weather for the West Coast. Then, few scientists other than weather personnel were aware that HRPT services existed. Today, HRPT signals are received and used by fishermen, shipping, oil, and power companies, water management personnel, foresters, and a host of other interests.

There are now more than 130 HRPT stations in existence and more are planned. Stations have been established in 44 countries, in the Antarctic, and on some military and commercial vessels. Nearly half the HRPT stations in existence today were formerly APT stations whose operators gained their expertise by first receiving APT data, then upgrading their stations.

Like APT, HRPT services also are derived from the AVHRR instrument. Ground resolution at the satellite subpoint is 1.1 km, rather than the 4 km of APT. Also, five channels of information are sent as digital signals.

The cost of HRPT ground equipment ranges from about $50,000 to as much as $1,300,000 for a station designed to meet specific user processing requirements. Three universities have constructed ground receiving equipment sufficient to meet their needs for under $100,000, and radio amateurs have designed and built stations for far less.

V-2.2.3 Direct Sounder Broadcasts (DSB)

DSB consist of a direct transmission of TOVS data as they are being recorded on the spacecraft. Since these are the raw TOVS data, those receiving DSB require facilities to extensively process the data into temperature and moisture soundings.

As of January 1986, 20 stations in 16 countries received and processed DSB direct readout data. The reception of such data, however, is of such importance that several other countries are working to establish receiver facilities.

V-2.3 GEOSTATIONARY SATELLITE DIRECT READOUT SERVICES

GOES direct readout data distribution consists of WEFAX, VISSR, and VISSR/VAS services. WEFAX is used to broadcast processed polar and geostationary satellite images, selected National Weather Service weather charts, and specific text messages at scheduled times, on S-band frequencies, to low cost stations.

Complete VISSR images require about 20 min per transmission and are sent every half hour. Both the viewed area and the frequency of transmission can be varied, however. As a service, VISSR transmissions are sent directly in real time to any station within receiving range. Transmissions are on S-band frequencies; reception and processing requires rather expensive ground equipment. VISSR/VAS transmission provides geostationary sounding data as well as VISSR imagery. Modified ground stations can receive VISSR/VAS data.

As of January 1, 1988, approximately 300 government, military, and private sector WEFAX stations receive broadcasts from the GOES, and 200 to 300 stations receive WEFAX broadcasts from Meteosat and GMS. VISSR transmissions are received in 23 stations in seven countries.

WEFAX receiving stations are operated by government, military, scientific, commercial and academic agencies, and by amateurs. WEFAX transmissions from GOES, Meteosat, and GMS are all at the same frequency, 1691.0 MHz. Meteosat WEFAX also is transmitted on a special frequency, 1694.5 MHz. With minor exceptions, WEFAX data from any of the satellites can be received on the same type of ground equipment.

With the 1976 launch of GOES-1, a new ground network was established to send VISSR products to U.S. weather offices. Called GOES-TAP, this network carries high resolution sectorized images. GOES-TAP largely replaced WEFAX as the primary source of satellite imagery for National Weather Service forecast offices. But with a large international community dependent upon WEFAX for satellite coverage, WEFAX service was continued and other products were added. Broadcasts of image products were increased from eight to 36 daily; NWS weather charts were added later. Since 1978, nearly 200 separate products have been added to the GOES WEFAX schedule. Today, WEFAX broadcasts to a large international user community, to a few sea-going vessels, and to several U.S. private and academic users.

International coordination of WEFAX products, services, transmitting equipment, and frequencies is through an informal body, the Coordination on Geostationary Meteorological Satellites (CGMS). Each of the nations or consortiums that operate geostationary satellites has representatives on this committee.

Chapter V-3

THE GOES DATA COLLECTION SYSTEM (DCS)

V-3.1 INTRODUCTION

The GOES DCS relays data from remote sites located at or near the Earth surface via GOES spacecraft to receiving stations within radio range. The two operational GOES located over the Equator at 75°W (GOES-East) and 135°W (GOES-West) are within range of most of the Earth from 0° westward to 165°E. Poleward coverage is limited to about latitude 77° and is further restricted to lower latitudes in either direction away from satellite subpoint longitudes.

The third GOES (GOES-Central), located approximately midway (113°W) between the two operational satellites, is used during eclipse periods to prevent loss of data through either operational spacecraft. If one of the regular GOES fails, this backup spacecraft can be moved in to replace it.

V-3.2 SYSTEM DESCRIPTION

The GOES DCS uses the spacecraft transponder to relay UHF transmissions from Data Collection Platforms (DCP) to ground receivers by S-band (1694.5 MHz). Also, S-band transmissions from the NESDIS ground system can be relayed through the spacecraft to receivers within radio range. Each spacecraft transponder can handle up to 233 reply channels. The 200 regional or domestic channels (401.7 to 402.0 MHz) use 1.5 kHz channel separation. The 33 international channels (402.0 to 402.1 MHz) have 3.0 kHz channel separation in common with both Meteosat and GMS spacecraft.

The DCS can handle at least 12,000 DCP messages per hour. With advances in stability and accuracy of timing oscillators and the introduction of a random reporting mode, the number of messages per hour could increase greatly. However, the transmission rate for all operational modes is 100 bps.

The four functional subsystems of the GOES DCS (see Figure V-3-1) are:

- Deployed Data Collection Platforms (DCP)
- Two operational GOES spacecraft
- Command and Data Acquisition Station (CDA) at Wallops Station, VA
- Central Data Dissemination Facility (CDDF).

All DCP used on the regional channels of the GOES DCS must be certified by NESDIS. DCP used on the international channels can be certified by other organizations that operate geostationary meteorological satellites. Self-timed DCP contain a transmitter and timing device preprogrammed to transmit information from attached sensors. Interrogate DCP contain both a transmitter and receiver, and normally transmit in response to a signal received via the spacecraft. Random reporting DCP contain a transmitter and timing device but normally transmit in response to an attached sensor input.

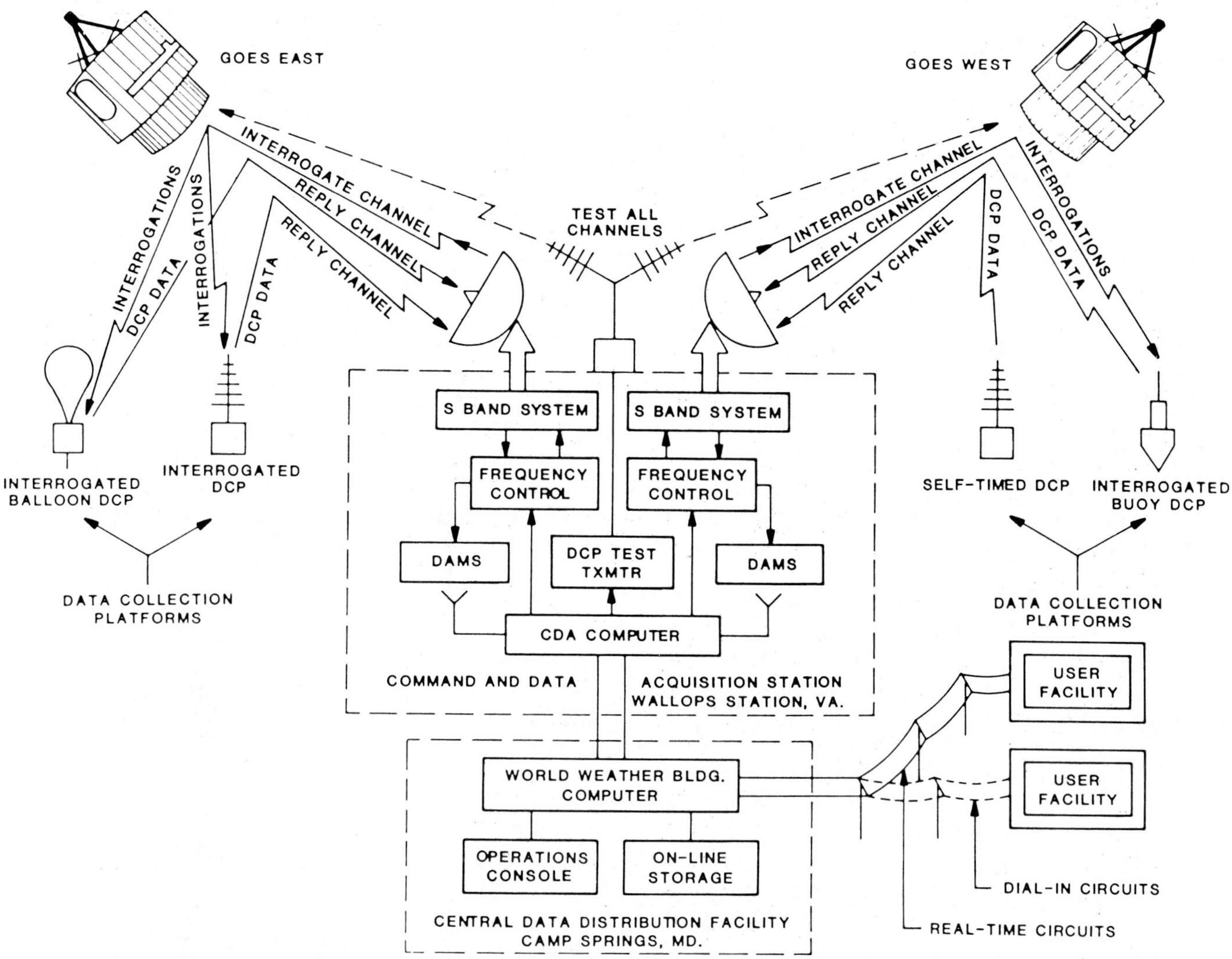

Figure V-3-1. The GOES DCS subsystem (MacCallum and Nestlebush, 1983).

V-3.3 LIMITATIONS AND ELIGIBILITY

DCP collect a variety of environmental data at unmanned remote sites, on ocean buoys, and on free floating balloons. The types of data include the usual atmospheric data collected at manned stations, sea surface temperatures, river heights and rates of flow, and upper air temperatures and humidities. International agreements limit GOES DCS frequencies to the relay of environmental data.

All data acquired through the GOES DCS are considered to be in the public domain, i.e., available to anyone who needs them. Certain private users may request proprietary treatment of collected data if special conditions are met. Private or non-U.S. users who want to collect environmental data needed for federal, state, or local government programs may have direct access to the GOES DCS if they have a sponsor and meet all other prerequisites. Only a U.S. federal, state, or local government agency can act as a sponsor. All users of the GOES DCS must also have a written agreement with NESDIS. Any user may have a data collection direct readout facility and receive the S-band data from GOES, but the facility must operate in a receive mode only.

V-3.4 APPLICATIONS

In general, the uses made of DCS data are the same as the uses made of routinely collected data. A few specific examples of DCS data usages follow.

V-3.4.1 Hydrology

The headquarters of the Pacific Northwest Region of the Bureau of Reclamation, in Boise, Idaho, operates a "real time" hydrometeorological GOES DCS. There are 213 remote stations in the current network, a Direct Readout Ground Station, and a Central Computer Facility (CCF). Normal reporting from the remote stations is automatic every four hours; all stations also can report significant changes as they occur via the random reporting channel.

The system was designed for use in forecasting conditions within complex reservoir and river systems of the region. Managers use it to keep track of conditions affecting flood control, irrigation, water quality, recreation, power, fish and wildlife, etc. Recently the system began to monitor dissolved gas content in rivers and to collect data for computer scheduling of irrigation water. Future plans include adding seismic sensors to DCP.

All data received by the CCF are immediately processed and stored in the short-term data file where they are available to users through time share terminals. At 0500 every day, the CCF uses the data from the previous day to compute the long-term file. Long-term file data, which include midnight reservoir elevations and contents, maximum and minimum temperatures, mean daily flows, etc., are also available through time share terminals. Graphs of data files are extended daily to determine trends.

Data from random reporting platforms are available within 15 min, a feature that permits operators to make decisions quickly to adjust flows to meet water storage or release requirements.

The CCF is also used as a 24 hour per day "watch dog" for the system. If the CCF determines an event to be significant, it will immediately and automatically notify management offices via telephone.

All short-term file data are automatically transferred hourly to the Columbia River Operational Hydromet Management System (CROHMS) data bank in Portland, Oregon, where they are then available to all state and federal agencies.

V-3.4.2 Meteorology/Oceanography

The NOAA Data Buoy Center (NDBC) operates moored buoys in coastal and offshore U.S. waters and fixed platforms at headland stations along the U.S. coasts. The DCP reports are used by the National Weather Service (NWS), military weather services, and private forecast organizations for marine weather and oceanographic analysis and prediction. Coastal communities, marine transportation interests, fisheries, recreational boating, and offshore construction enterprises benefit most directly. Data also are used for initial analysis in numerical forecasting models, for ground truth, for validation of theoretical models, and for climate monitoring studies.

These platforms measure wind speed and direction, wind gusts, air temperature, barometric pressure, significant wave height, wave period, wave spectra, and sea surface temperature. Most platforms have redundant wind and pressure sensors. The normal buoys have self-timed DCP which can be interrogated when necessary. The headland stations also use self-timed DCP.

The acquisition network for these platforms consists of NESDIS, National Meteorological Center (NMC), NDBC, and NWS communications. GOES DCS provides the communication link between platforms and NESDIS. NESDIS provides ground support facilities for acquisition and initial distribution of data. NMC receives data from NESDIS and processes it twice an hour. The processing prepares data reports into packets for distribution to users over NWS communication circuits. NDBC manages the platform network; this includes remote control of DCP, maintenance of data bases for information processing, and quality control of the data.

Formats used in report preparation include the WMO codes FM 12-VII for land surface observations and FM 13-VII for marine surface observations. Wave data are disseminated in U.S. code only. Approximately 95% of all data are disseminated by NMC within 30 min of observation time.

To alleviate some of the problems of conventional systems and to satisfy the need for timely data, NOAA/National Ocean Service uses the GOES DCS for expeditious delivery of data from the Shipboard Environmental (Data) Acquisition System (SEAS), moored buoys, tide gages, and remote, unmanned Coastal Marine Automated Network (C-MAN) stations. These units report marine meteorological and oceanographic parameters, including automatic bathythermograph data and, in the case of SEAS, manual entry of data by shipboard personnel.

The existing oceanographic observation program is scheduled for considerable expansion by use of the GOES DCS. The SEAS program will use domestic and international channels so that the near global coverage of other nations' geostationary meteorological satellites can be included.

V-3.4.3 Aircraft to Satellite Data Relay (ASDAR)

ASDAR is carried aboard wide-bodied jets such as the Boeing 747, McDonnell-Douglas DC-10, and the Lockheed L-1011 aircraft. It is interconnected with existing onboard avionics systems to obtain 3-dimensional position, temperature, wind, and turbulence information. These observations are made every 7 min during cruise or level flight. ASDAR is a self-timed platform equipped with its own high accuracy clock;

its observations are transmitted once per hour using international channels of the U.S., European Space Agency, and Japanese geostationary meteorological satellites to provide virtually complete global coverage. The major purpose of ASDAR is to increase observations for numerical prediction models and to significantly increase the observation set from data-sparse areas. Also, since ASDAR automates aircraft observation reporting, the error rate for ASDAR is significantly lower than for aircraft using the conventional voice relay system.

ASDAR also takes observations at a fast rate to compile vertical profiles during aircraft ascent and descent. These vertical profile observations add significantly to data from the existing radiosonde network. Observations are spaced 10 mb apart in the boundary layer, i.e., in the first 100 mb above the surface, and 50 mb apart between the boundary layer and cruise level.

A unique aspect of ASDAR as a DCP is its susceptibility to Doppler shift of the transmitted signal as the aircraft flies at speeds of up to 600 knots. Doppler shift is greatest at the horizon of the satellite's communication range, i.e., about 80 degrees from the satellite subpoint. The maximum shift, in the neighborhood of 350 Hz, occurs as the aircraft moves directly toward or away from the satellite at the horizon. This shift, in effect, can reduce the satellite footprint for ASDAR reception if some additional drift occurs in the ASDAR transmitter.

V-3.4.4 Seismology

Seismic data are valuable if processed immediately after an event. This is especially true along the west coasts of Central and South America which are highly seismic and where even small earthquakes can cause extensive damage. Currently it takes many hours and, in some cases, days before data are available from Central and South American seismic stations. New technology and software developments by the U.S. Geological Survey (USGS) make possible the use of a small, inexpensive microcomputer that can monitor a network of DCP seismic stations. Such networks operate unattended 24 hours a day and transmit data that permit the rapid, accurate, and automatic location of seismic events. Existing flood alert systems using GOES DCS could be added to establish a comprehensive hazard alert and regional disaster response system (Figure V-3-2).

The USGS, cooperating with the Agency for International Development, has developed a prototype earthquake event detector that uses the GOES DCS (Carlson et al., 1983). The prototype system consists of a short period vertical seismometer, a microcomputer to monitor its output, and a DCP to transmit the output through the satellite. The microprocessor monitors the seismic signals, selects the seismic events, and stores the associated waveform data for DCP transmission to the satellite. Using a ten sample per second digitizing rate, it is possible to transmit about 50 sec of waveform data for each seismic event.

Algorithms are used to pick the first arrival, or "P" time, the period, and the amplitude (Murdock and Hutt, 1983). However, the error rate for "P Picker" algorithms is large, so without waveform data, the errors cannot be corrected or sometimes cannot even be detected.

During times of heavy earthquake activity, a DCP that transmits waveform data requires more channels than can be allocated to any one seismic station. To avoid this problem, an interim system was set up in which a trained seismologist is on location to interpret the seismograms. Instead of using normal communications channels, the analyst key codes seismic data into a DCP microcomputer. This is so efficient that most earthquake data can be transmitted in a time allocation of one minute per day. Because this is point-to-point communication not permitted on the GOES DCS, it has been limited to one development and test operation. It has been refined, however, using data compression techniques and presently is operational on the Argos system from an observatory on the Ivory Coast.

V-3.4.5 Space Environment

The joint NOAA/Air Force Space Environment Services Center (SESC) in Boulder, Colorado, provides a forecast and alert service to users whose operations depend on knowledge of geophysical conditions near the Earth. Measurements of the Earth's magnetic field are necessary for this service. To be useful, these measurements must be received within minutes of a magnetic event.

The Remote Geophysical Observing Network (RGON) consists of 18 GOES DCS platforms that store and report one minute data values from three-axis magnetometers on a 12-min schedule. The platforms are located across North America from Sachs Harbour, Canada, to San Juan, Puerto Rico, and from Narssarssuaq, Greenland, to Honolulu, Hawaii. Two minutes from the time data are telemetered from the RGON through the GOES DCS, they are available for use at the SESC. The main users of information derived from these data are the DOD, NASA, and operators of powerline, pipeline, and high frequency communication networks affected by magnetic field disturbances.

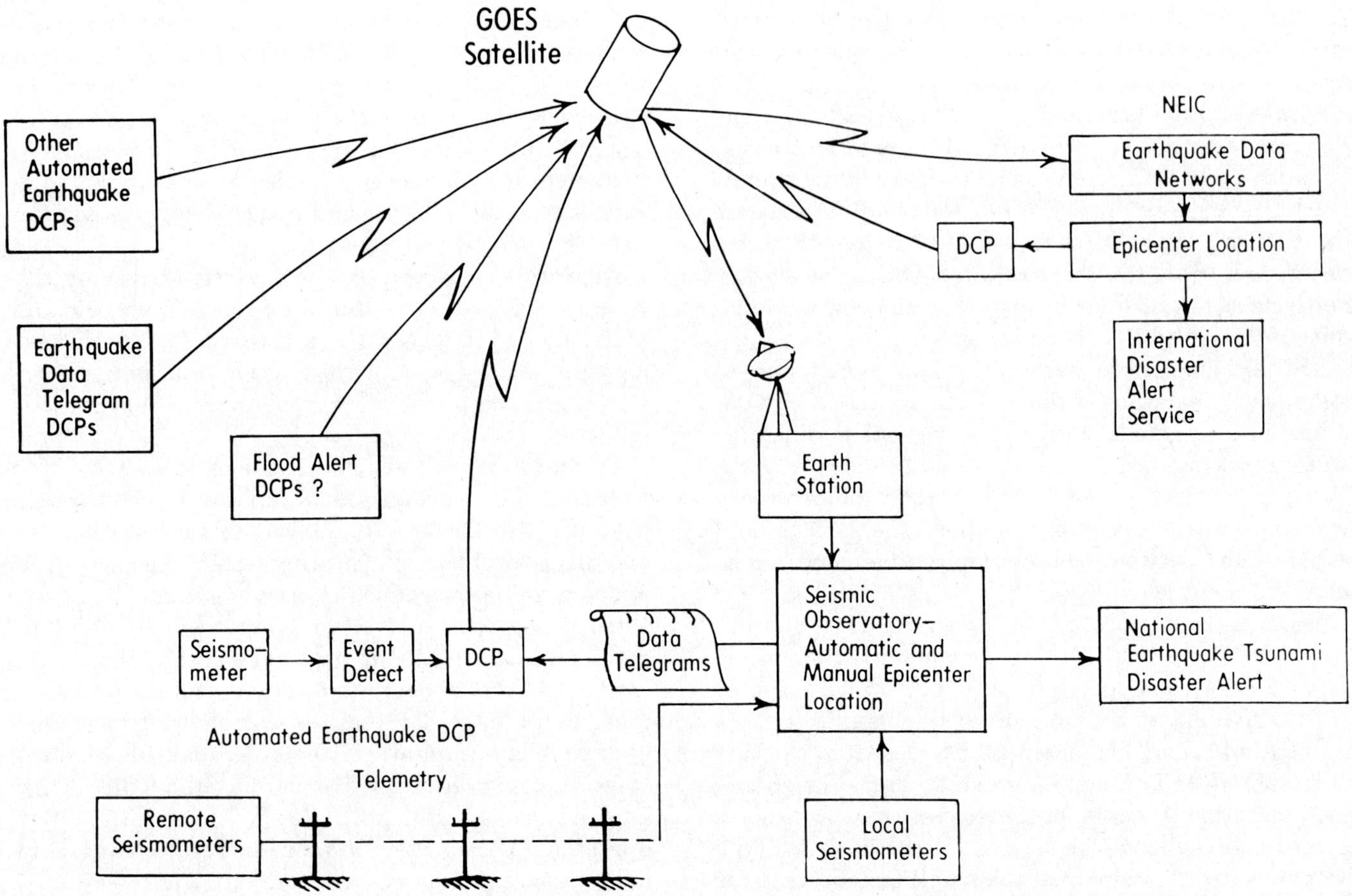

Figure V-3-2. Regional disaster response system (Carlson et al., 1983).

V-3.4.6 Radiation

The Lawrence Livermore National Laboratory is using GOES DCS to help trace soil borne radionuclides in the Marshall Islands. The DCS transmits environmental data continuously from field sites at Enewetak and Bikini Atolls to the Laboratory.

Because of the on-site difficulties in collecting the various data, the GOES DCS was selected as the best means for remotely recording the microclimate and soil water dynamics by a land-based station located on the atolls (Office of Systems Engineering, 1979; Blumenstock and Rex, 1969).

There are three sites in the northern Marshall Islands instrumented with micrometeorological and soil moisture sensors. The Micrometeorological Sensor Processor (MSP), designed at the Laboratory, collects data at 20-min intervals; these data are transmitted every three hours by the self-timed DCP. Modest, multi-use instrumentation computers retrieve these data from the GOES DCS CDDF in Maryland and process it in the Laboratory.

Chapter V-4

ARGOS DATA COLLECTION AND LOCATION SYSTEM (DCLS)

V-4.1 INTRODUCTION

The Argos DCLS provides an operational satellite-based means to collect, locate, and disseminate environmental data. Development of the system was achieved through a cooperative effort among Centre National d'Études Spatiales (CNES), NOAA, and NASA, and officially began with an interagency agreement signed in 1974. The system became operational in October 1978 with the successful launch of TIROS-N and was used during the Global Weather Experiment (GWE) in 1979. The initial agreement was modified in 1986 to provide for the establishment of the new U.S. Argos Processing Center. The Argos DCLS is scheduled to be onboard all NOAA satellites through NOAA-M (to around 1997).

V-4.2 ARGOS DCLS ADMINISTRATION

The Argos DCLS is administered by Service Argos, Inc., within the United States, and by CLS/Service Argos, S. A., in France. Argos system usage and performance is monitored by the Argos Operations Committee (AOC), a bilateral oversight organization that meets once a year. AOC, comprised of members from CNES, NOAA, and NASA, is responsible for ensuring that the system is managed according to the agreement.

The Argos system is reserved for programs involved with the collection of environmental data, defined as measurements of physical, chemical, or biological properties of solid Earth, waters (rivers, lakes, and oceans), and the atmosphere (including space). Programs not directly involved with environmental monitoring may be approved by AOC, provided that the program lasts for only a limited amount of time.

V-4.3 THE ARGOS SUBSYSTEMS

The Argos DCLS has four subsystems:

- The Platform Transmitter Terminals (PTT), used on buoys, balloons, animals, ships, and fixed land stations.
- The Data Recovery Unit (DRU) package aboard the NOAA TIROS satellites.
- The telemetry receiving stations, data ingest and decommutation system, and communications links to the data processing centers.
- Two data processing centers, located at Landover, Maryland, and Toulouse, France. Each center processes platform location and sensor data and disseminates the data to users through communications networks.

V-4.3.1 Platform Transmitter Terminals (PTT)

There are two basic PTT categories: the locator type, equipped with good quality oscillators used for both location and data collection; and the data collection-only type, equipped with lower quality oscillators.

The characteristics of Argos PTT are listed in Table V-4-1.

The accuracy of PTT positions obtained from the Argos DCLS depends on many factors, including the stability of the transmitter oscillator, but is usually better than 0.5 km.

V-4.3.2 Onboard Equipment

The space segment of the Argos DCLS consists of two polar orbiting NOAA satellites. Onboard DRU receive messages transmitted by PTT within the satellite's area of coverage (a swath approximately 2500 km either side of the subsatellite track). Processing by the onboard equipment identifies the PTT, records the time of data reception contained in the message, and measures the Doppler shift of the incoming signal. Since there are normally two satellites in operation, a minimum of four positions per day are obtained from a PTT in Equatorial regions and up to 28 positions per day in polar areas. In temperate latitudes, eight positions per day are typical.

The Argos DCLS data are available from the satellites in three modes:

- S-band, which plays back recorded TIROS Information Processor (TIP) data on 1698, 1702.5 or 1707 MHz at 2.66 Mbps.
- Real time S-band on 1698 or 1707 MHz at 665.4 kbps.
- Real time VHF beacon on 136.77 or 137.77 MHz at 8.32 kbps.

V-4.3.3 Ground Reception and Processing

Each time a satellite passes over one of the three telemetry receiving stations (Wallops Island, Virginia; Fairbanks, Alaska; and Lannion, France) all recorded TIP data are transmitted. All Argos data received by these stations are sent via the NOAA/NESDIS facility at Suitland, Maryland, to the Argos Data Processing Centers at Toulouse and Landover where the following operations are performed:

- Decoding of PTT messages and conversion of data into physical units.
- Very accurate computation of satellite orbits.
- Computation of PTT positions from orbital data and Doppler shifts.
- Computer storage of all results for user access.

Users requiring real time data can also acquire these data directly from the satellite through the High Resolution Picture Transmission (HRPT) S-band downlink or VHF beacon. Since each transmission is rebroadcast by the satellite as soon as it is received, anyone with a downlink receiver within range of the satellite can receive data from PTT in view of the satellite at the same time. Although positions can be determined from the information received by a LUT, the accuracies so far are on the order of kilometers, mainly because of software limitations.

Table V-4-1 *Argos platform characteristics (Service Argos, 1988)*

Carrier frequency	401.650 MHz		
Aging (during life)	±2 kHz		
Short-term stability (100 ms)	$\leq 2.10^{-9}$ (All platform types) $\leq 10^{-9}$ (Location-type platforms)		
	SLOPE OF LEAST SQUARES FIT (HZ/MIN)	RESIDUAL DEVIATION ABOUT SLOPE	ACCURACY
Medium-term stability (20 min)	≤0.20	≤2.10–9	a few hundred m
	≤1.00	≤5.10–9	a few km
	<4.00	<10–8	a few tens of km
Long term (2 hours)	±400 Hz		
Radiated power (according to antenna)	≤1 dB (Power variation) ≤5 ms (Rise time)		
Antenna	Vertical polarization and circular right-hand		
Message length	360–920 ms		
Repetition period for message	40–90 s (Requiring location) 90–200 s (Not requiring location)		
Data sensors	4–32 eight-bit sensors for environmental data		

V-4.3.4 Data Distribution

Data collected by the Argos DCLS may be obtained from Service Argos. They are available on computer files accessible by telephone, telex, or communication networks, generally within 4 hours after the receipt of data. Processed data are available on Computer Compatible Tapes (CCT), print outs, or floppy disks every 2 to 4 weeks. Formatted data are distributed over the Global Telecommunication System (GTS). The user can access three types of files.

- The "AJOUR" file, containing the past 24 hours of location and sensor data for each PTT.
- The "TELEX" file, containing, in chronological order for the past 100 hours, one sensor message and corresponding location per satellite pass for each PTT.
- The "DISPOSE" file containing, in chronological order for the past 100 hours, all location data and all sensor messages for each satellite pass and each PTT.

V-4.4 ARGOS SYSTEM APPLICATIONS

Argos data are used for studies of the sea, land, and air, as well as for biological research. Some of the more extensive applications are described below.

V-4.4.1 Drifting Buoys

The 1979 GWE included an array of more than 300 drifting buoys in the Southern Hemisphere oceans that were monitored by the Argos system (Fleming et al., 1979). Data randomly transmitted from the drifting buoys were acquired by the Argos system and processed into atmospheric pressure and sea surface temperature information; buoy positions were obtained from processed data. This drifting buoy experiment provided *in situ* operational meteorological data to the national weather services of the Southern Hemisphere and crucial data for the Global Atmospheric Research Program (GARP). The Tropical Ocean and Global Atmospheric (TOGA) and World Ocean Circulation Experiment (WOCE) are two other examples of global oceanographic programs relying on buoy data derived through the Argos system.

V-4.4.2 International Ice Patrol

The Coast Guard has used DCLS for operations and research from the Nimbus-6 Random Access Measurement System to the present Argos DCLS. The Coast Guard International Ice Patrol first placed a satellite tracked drifting buoy in the Labrador Current in April 1976 to obtain surface current velocities for predicting iceberg drift trajectories (Weir, 1978). Since then the International Ice Patrol has incorporated this method of measurement, using the Argos DCLS, into its operations. These operational applications have included iceberg tracking (Robe et al., 1980), oil spill monitoring (Mountain et al., 1980), the search and rescue datum marker buoy project (Hayes, 1981), and the foreign fishing vessel transmitting terminal project (Hayes and Murray, 1982).

V-4.4.3 Ship Observations

Scripps Institution of Oceanography (SIO) and the French Office de la Recherche Scientifique et Technique Outre-Mer (ORSTOM) jointly use a network of commercial ships that voluntarily make oceanographic observations in the Pacific. The principal instrument used in this program is the Expendable Bathy-Thermograph (XBT). About half of the ships in the SIO/ORSTOM network are equipped with microcomputers that digitize the XBT temperature versus depth "trace" and record it on tape. The XBT data processing and recording system used is the one designed by the Technical Planning and Development Group at Oregon State University (OSU) (Mesecar and Wagner, 1979).

Shipboard microcomputers also prepare and display a summary of the trace in the standard WMO "bathymessage" format. Officers of each of the volunteer ships manually transcribe the bathymessage from the screen of the microcomputer for transmission by the ship radio to designated shore stations, as is done with conventional marine weather observations. This process for getting XBT data into the GTS is inconvenient and subject to errors. To solve this problem, OSU incorporated an Argos PTT into the XBT system (Mesecar and Wagner, 1982). In the spring of 1982, the first of these was placed aboard the M/V Lillooet which operates between Australia and the U.S. west coast. The system has since been automatically transmitting XBT bathymessages and ship position information. More recently Scripps has tested the prototype of a Polar Research Laboratories (PRL) Argos transmitter which accepts bathymessages from the computer via a serial, asynchronous RS-232 connection. The PRL unit stores the message and retransmits it automatically at 60 sec intervals.

V-4.4.4 Tracking Superpressure Balloons

The Argos DCLS has been used to locate and collect atmospheric data from superpressure balloons since 1978. During the GWE 313 balloons were launched from tropical islands and circled the Earth at an altitude of 14 km. Their paths and instruments provided information on winds and temperatures at the top of the tropical troposphere. These data were not otherwise available since the balloons were drifting above aircraft altitudes. The Argos DCLS acquired an average of four locations per day from each of the 313 balloons. With an average lifetime of 60 days, the balloons were the source of otherwise unavailable critical data during the GWE.

V-4.4.5 Surface Weather Observations in Antarctica

The Automatic Weather Stations (AWS) used for surface weather observations in Antarctica measure wind speed and direction and atmospheric temperature and pressure at a nominal height of 3 m. Data transmitted every 200 sec to the Argos DCLS include five values for each item with updating every 10 min. Four AWS units are equipped to measure relative humidity and vertical temperatures.

Raw data are obtained monthly from Service Argos with the processing carried out at the University of Wisconsin. These data are available in the form of a monthly summary of 3-hourly values for each AWS unit and the complete data set is available on magnetic tape.

Argos DCLS real time data are transmitted on the VHF beacon and over the HRPT S-band downlink of the polar orbiting satellites. The U.S. Navy at McMurdo, Ross Island, Antarctica, has two systems for direct readout. One uses the HRPT and the other uses the beacon transmission. These data are used in weather forecasting for aircraft operations.

At present there are 20 AWS units in the Antarctic between the Antarctic Peninsula and Dumont d'Urville; 16 are operational. Maintenance and new installations are done during the summer.

V-4.4.6 Animal Tracking

It has often been difficult for biologists to study the behavior of animals because the presence of the field worker can disrupt normal behavior; also many species occur in such remote areas or move so rapidly and widely that they cannot be observed regularly. In the early 1960s, the use of radio telemetry techniques to solve a wide variety of wildlife and biological problems began. Some examples of early work as well as papers about more recent application of biotelemetry can be found in Amlaner and MacDonald (1980). This volume also contains papers describing the tracking of polar bears with satellites (Kolz et al., 1980). Some pioneering work was conducted with free ranging elk carrying relatively heavy transmitters (Craighead et al., 1972). Later, several attempts were made to develop smaller satellite-compatible transmitters that could be carried by a variety of animal species (e.g., Jennings and Gandy, 1980). The technology for emergency locator transmitters and for locator packages for monitoring locations of weather balloons and buoys led to packages usable for tracking polar bears (Kolz et al., 1980; Schweinsburg and Lee, 1982) and sea turtles (Timko and Kolz, 1982; Service Argos, 1984a). These total package sizes remained large (5 kg). Major problems connected with the design of smaller satellite-compatible transmitters are described by Kuechle et al. (1979).

By 1982, lighter PTT compatible with the Argos DCLS were produced. A 500-g transmitter was packaged specifically for use on marine mammals such as porpoises and dolphins. In 1983, PTT were successfully tested on whales (Service Argos, 1984a) and in 1984 on caribou (Fancy et al., 1988). The Argos Newsletter of April 1984 reported that zoologists from the University of Aberdeen had used PTT to track basking sharks.

The Applied Physics Laboratory (APL) of Johns Hopkins University designed packages powered by a combination of solar cells and rechargeable nickel cadmium batteries to be carried by birds (Strikwerda et al., 1985; 1986). Service Argos certified this circuitry in 1983 and the U.S. Fish and Wildlife Service Patuxent Wildlife Research Center tested packages on captive and wild birds (Fuller et al., 1984). The prototype PTT have provided reliable locations for several months from eagles in the Mid-Atlantic States, from swans migrating from Alaska across Canada, and from giant petrels along the Antarctic Peninsula (Strikwerda et al., 1986).

V-4.5 ARGOS SYSTEM ENHANCEMENTS

V-4.5.1 United States Processing Center

The new U.S. Argos Processing Center (USAPC), located in Landover, Maryland, became fully operational in April 1987. Shorter delays from the recorded global TIP data have been realized due to both a faster data transmission time between NOAA and the Argos processing centers, and new and improved software within both NOAA and the two redundant Argos processing centers.

V-4.5.2 Improved Orbitography

In April 1987, two software algorithm modifications were made to reduce systematic location errors:

- All the orbitography platforms, which are used as fixed precise reference points for location calculations, were referenced to the same geocentric system.
- The complete Goddard Earth Model 10 (GEM 10) was incorporated to take certain terms into account, in particular resonance coefficients.

V-4.5.3 Real Time Data

In mid-1987, Service Argos began acquiring the data contained in the real time S-band data received at the Fairbanks, Alaska, and Wallops Island, Virginia, CDA stations. These data are processed just as if they were the recorded data. This allows data turn-around within 20 min to those users' platforms within the satellite footprint of these two stations. Argos system location accuracies are retained, as opposed to the degraded accuracies obtained through currently available LUT. These data are also presented to the users in the same Argos format to which they have become accustomed.

V-4.5.4 Very Accurate Locations

A study by Service Argos, Argos—Profile, to determine the maximum achievable location accuracy was conducted during early 1987. The objective of this experiment was to analyze very accurate position determinations for 60 points on the Kerguelen Archipelago, located in the southern Indian Ocean. Fifteen standard PTT were used for the experiment. Each PTT was left at the same position for at least six days so that sufficient measurements for accuracy could be obtained. As a control point, one of the PTT was sited close to the Kerguelen orbitography beacon. Elevation data were supplied by helicopter altimeter, as location can be supplied only if elevation is known. Some of the processing features used by the experimental system were:

- use of interpolated orbital data
- selection of Doppler measurements (from satellite passes higher than 8° above the horizon)
- correction for oscillator frequency drift
- calculation of mean position.

The experiment showed that with a few enhancements and retrospective processing, the Argos system could achieve absolute location to within 50 m and relative location to within 30 m. The accurate location system was implemented at the end of 1987.

V-4.5.5 Expendable Bathythermograph (XBT) System

In conjunction with the TOGA experiment, Service Argos developed an end-to-end XBT system. This system enables the results of an XBT sounding to be processed by the Argos system and automatically sent to the GTS. The data are sent as messages in the required WMO code format BATHY. The data acquired from the XBT probes are also made available for Argos system users from the Argos data dissemination computer files AJOUR, DISPOSE, and TELEX.

The XBT microcomputer-based system includes: the launcher; the microcomputer interface and computer; the acquisition, bathymessage, and encoding software; and the GTS data processing software resident in the Argos data processing system. This XBT system was acceptance tested in January 1988, and sea-trial tested in March and April. The Institut Francais de Recherche pour l'Exploitation de la Mer (IFREMER—the French oceanographic agency) has ordered 15 complete XBT shipboard units which will be progressively implemented by the TOGA network. Forty World Weather Watch (WWW) ships of opportunity will be equipped with these units by 1989.

V-4.5.6 Technical Advancement

User platforms, with their sensors and transmitters, are the creatively designed part of the system, allowing innovation on the part of the user or manufacturer. Advances in buoy platform technology are proceeding at a swift pace, and inexpensive "expendable" units are available in the marketplace. The increasing demand for more intelligent and sophisticated sensor-transmitter configurations, to a large extent fueled by the requirements of animal tracking applications, has dramatically affected technical design in terms of solar power, battery life, miniaturization (weight), and other factors. The incorporation of microprocessor technology into PTT can be of tremendous value, allowing platforms to be customized to meet specific requirements of the users.

V-4.5.7 New Satellite Instrumentation

For the NOAA satellite series H through J, Service Argos, in cooperation with NOAA and NASA, will increase the data rate between the four Argos DRU and the TIROS Information Processor. This modification will increase by 20% the number of PTT able to be received simultaneously onboard the satellite. NOAA satellite series K through M will see yet another increase in the DRU to TIP data rate and the DRU complement will be doubled. These latter modifications will further increase the onboard PTT reception capacity fourfold.

BIBLIOGRAPHY AND REFERENCES FOR SECTION V

Amlaner, C. J. and D. W. MacDonald (eds.), 1980: *A Handbook on Biotelemetry and Radio-Tracking*. Pergamon Press, Oxford, England, 826 pp.

Blumenstock, D. I. and D. F. Rex, 1969: Microclimate observations at Eniwetok Atoll. *Atoll Research Bulletin, 71*.

Bouyoucos, G. J. and A. H. Mick, 1940: An electrical resistance method for the continuous measurement of soil moisture under field conditions. *Michigan Agriculture Experimental Station Technical Bulletin 172*.

Bumpus, D. F. and L. M. Lauzier, 1965: *Serial Atlas of Marine Environment*. American Geographical Society, New York, NY.

Carlson, M. A., S. T. Algermissen, J. N. Jordan, D. A. Taff, and

R. N. Hunter, 1983: A prototype seismic event detector and satellite based telemetry system. *San Francisco 1983 Natl. Conf. on Resources Management Applications: Energy and Environment*.

Craighead, F. C. Jr., J. J. Craighead, C. E. Cole, and H. K. Buechner, 1972: Satellite and ground based radio tracking of elk. *Animal Orientation and Navigation: A Symp.*, S. R. Galler, K. Schmidt-Koenig, G. J. Jacobs and R. E. Belleville (eds.), NASA SP 262, 99-111.

Fancy, S. G., L. F. Pank, D. C. Douglas, C. H. Curby, G. W. Garner, S. C. Amstrup, and W. L. Regelin, 1988: Satellite telemetry: A new tool for wildlife research and management. *Fish and Wildlife Service Res. Pub., 172*, U.S. Dept. of Interior, 54 pp.

Fleming, R., T. Kaneshige, W. McGovern, and T. Bryan, 1979: The Global Weather Experiment II: The second special observing period. *Bull. Amer. Meteor. Soc., 60*, 1316-1322.

Fuller, M. R., N. Levanon, T. E. Strikwerda, W. S. Seegar, J. Wall, H. D. Black, F. P. Ward, P. W. Howey, and J. Partelow, 1984: Feasibility of a bird-borne transmitter for tracking via satellite. *Proc. of the 8th International Symp. on Biotelemetry, Nijmegen, Netherlands*, H. R. Kimmich and H. J. Klewe (eds.), Intl. Society on Biotelemetry, 375-378.

Hayes, R. M. and J. J. Murray, 1982: Applications of a satellite-tracked fishing vessel transmitting terminal to fisheries management and science. *Science Council Studies Number 4, Special Session on Remote Sensing*. NW Atlantic Fisheries Organization, Dartmouth, Nova Scotia, 77-83.

———, 1981: Surface drift in the Gulf Stream from a small, undrogued, air- deployed satellite-tracked Lagrangian drifter. Abstract. *Mini-Symp. on Applications of Aerospace Remote Sensing in Marine Research, ICES Statutory Meeting, October 6-10, Woods Hole, MA*.

Heinmiller, R. J., J. E. Masterson, and J. C. McWilliams, 1984: A report on the Drifters Program. *Argos Users Conference, May 21-23, Seattle, WA*, Service Argos, Landover, MD.

Hughes Aircraft Company/Space and Communications Group, NOAA/NASA, 1986: *GOES G/H Data Book*. 104 pp.

Jennings, J. G. and W. F. Gandy, 1980: Tracking pelagic dolphins by satellite. *A Handbook on Biotelemetry and Radio-Tracking*, C. J. Amlaner, Jr. and D. W. MacDonald (eds.), Pergamon Press, Oxford, England, 753-755.

Kolz, A. L., J. Lentifer, and H. G. Fallek, 1980: Satellite radio tracking of polar bears instrumented in Alaska. *Handbook on Biotelemetry and Radio-Tracking*, C. J. Amlaner, Jr. and D. W. MacDonald (eds.), Pergamon Press, Oxford, England, 743-752.

Kuechle, V. B., D. P. DeMaster, and D. B. Siniff, 1979: State of the art and needs of the earth platform. *Proc. of the 13th International Symp. on Remote Sensing of the Environment, April 23-27*, Environ. Res. Inst. of MI, Ann Arbor, MI.

Large, W. G., J. C. McWilliams, and P. P. Niiler, 1983: Progressive development of the CASID Buoys. *Natl. Data Buoy Office—Marine Technology Society Buoy Symp., April 27-29, New Orleans, LA*, NDBO, AMSF, MI, 318-332.

MacCallum, D. H. and M. J. Nestlebush, 1983: Geostationary operational environmental satellite/data collection system. *NOAA Technical Memorandum NESDIS 2*, Dept. of Commerce, Washington, DC, 50 pp.

McWilliams, J. C. and J. E. Masterson, 1982: *Plan for the Development and Utilization of Ocean Surface Drifting Buoys.* Univ. Corporation for Applied Research, Boulder, CO, 37 pp.

Mesecar, R. and J. Wagner, 1982: Satellite transmission of XBT bathymessages. *Exposure, 10,* 6 pp.

——— and J. Wagner, 1979: An XBT digital recording and display system. *Proc. of the Oceans 1979 Conf., San Diego, CA*, IEEE, New York, NY.

Mountain, D. G., J. J. Murray, and K. A. Mooney, 1980: Operational use of satellite tracked buoys during the Ixtoc l oil spill. *Record of the 16th Annual Conf. of Marine Technology Society, October 6-8, Washington, DC*, 36-41.

Murdock, J. N. and C. R. Hutt, 1983: A new event detector designed for the Seismic Research Observatories. *U.S. Geological Survey Open-File Report 83-785*.

National Environmental Satellite, Data, and Information Service, 1985: NOAA Satellite Programs Briefing. Dept. of Commerce, Washington, DC, 264 pp.

Office of Systems Engineering, 1979: Geostationary operational environmental satellite/data collection system. *NOAA Technical Report NESS 78*, Dept. of Commerce, Washington, DC, 80 pp.

Robe, R. Q., D. C. Maier, and W. Russel, 1980: Long-term drift of icebergs in Baffin Bay and Labrador Sea. *Cold Regions Science and Technology*, *1*, 183-193.

Schwalb, A., 1987, 1979, 1978: The TIROS-N/NOAA A-G satellite series (NOAA E-J) - Advanced TIROS-N (ATN). *NOAA Technical Memorandum NESS 95*, Dept. of Commerce, Washington, DC, 75 pp.

Schweinsburg, R. E. and L. J. Lee, 1982: Movement of four satellite-monitored polar bears in Lancaster Sound, Northwest Territories. *Arctic, 35*, 504-511.

Service Argos, 1988: *Argos Platform Transmitter Terminals General Specifications and Certification, Ver. 2*. Service Argos, Landover, MD.

———, 1984a: *Argos Newsletter*. April 1984, Number 19.

———, 1984b: *Location and Data Collection Satellite System User's Guide*. May 1984, Toulouse, France, 36 pp.

Strikwerda, T. E., M. R. Fuller, W. S. Seegar, P. W. Howey, and H. D. Black, 1986: Bird-borne satellite transmitter and location program. *Applied Physics Laboratory Technical Digest, 7*, Johns Hopkins Univ., Baltimore, MD, 203-208.

———, H. D. Black, N. Levanon, and P. Howey, 1985: The bird-borne transmitter. *Applied Physics Laboratory Technical Digest, 6*, Johns Hopkins Univ., Baltimore, MD, 60-67.

Summy, A. D. and I. Anderson, 1983: Operational uses of TIROS oceanographic drifters by the International Ice Patrol (1978-1982). *Proc. of the 1983 Symp. on Buoy Technology, New Orleans, LA,* Marine Technology Society, Gulf Region, 46-250.

Timko, R. E. and A. L. Kolz, 1982: Satellite sea turtle-tracking. *Marine Fisheries Review, 44*, 19-24.

Weir, C. R., 1978: Operational use of the free-drifting, satellite-tracked buoys. *Report of the International Ice Patrol Service in the North Atlantic Ocean, Season of 1976* (Appendix D). Bulletin Number 62, CG-188-31, Washington, DC, 73-76.

Contributors to Section V

Carlson, Marvin A.	United States Geological Survey Box 25046 Denver, CO 80225
Clegg, Bruce R.	Lawrence Livermore National Laboratory Box 808, L-524 Livermore, CA 94550
Cutchin, David L.	University of California, San Diego Scripps Institution of Oceanography La Jolla, CA 92903

Derr, John S.	United States Geological Survey Box 25046 Denver, CO 80225
Fuller, Mark R.	U.S. Fish and Wildlife Service Patuxent Wildlife Research Center Laurel, MD 20708
Hayes, Richard M.	Office of the Oceanographer of the Navy 34th and Massachusetts Avenue, N. W. Washington, DC 20392-1800
Hornback, Charles E.	NOAA/ERL Space Environment Laboratory 325 S. Broadway Boulder, CO 80303
Lally, Vincent E.	National Center for Atmospheric Research P. O. Box 3000 Boulder, CO 80307
Lute, Dan	U. S. Dept. of the Interior Bureau of Reclamation, Northwest Region P. O. Box 043 Boise, ID 83724
Masterson, John E.	Universities Space Research Association (USRA) Boulder Program Office P. O. Box 391 Boulder, CO 80306-0391
Robison, William L.	Lawrence Livermore National Laboratory Box 808, L-524 Livermore, CA 94550
Roten, Ray J.	Computer Graphics Center Box 7106 North Carolina State University Raleigh, NC 27695-7106
Shaw, Archie E.	Service Argos 1808 McCormick Dr., Suite 10 Landover, MD 20785
Smidt, George J.	National Weather Service Gramax Building Silver Spring, MD 20910
Zegowitz, Vincent	Code W/OSO141 Gramax, Room 721 NOAA/NWS Silver Spring, MD 20910
Liddick, Charles E. *MacCallum, Douglas H. McIntosh, Robert B. Nestlebush, Michael J. *Nizza, Pasquale *Popham, Robert W. Stearns, Charles R. Vaeith, J. Gordon	NOAA/NESDIS Washington, DC 20233

* (Moved to a new position)

Section VI

CENTRAL DATA PROCESSING AND DISTRIBUTION

Data processing and distribution are important elements of the satellite program. The important parts of processing and distribution are timeliness, repeatability of data content, and uninterrupted delivery. The other key elements are data presentation and the retention of data records.

The volume of data generated in a day from both the polar and geostationary satellites is very large; approximately 10^{11} bits. The data cover the entire globe and data processing is complex. The satellite orbital geometry, Sun elevation angle, instrument calibration, and other factors to convert the measurements to geophysical units must be considered. Final granules are provided to the users in various analog, digital, and graphical formats. To manage and produce a complex data system like this, a massive computer system is necessary. The chapters in this section describe in detail the central data processing, operational processing of data into products, distribution of centrally processed data and products, and the archival of satellite data for retrospective use. Many of the products are discussed further in the Applications Sections, particularly Section VII.

Chapter VI-1

PREPROCESSING SATELLITE DATA[1]

VI-1.1 INTRODUCTION

Data received from a satellite require special handling before they can be used. Data from different instruments are sampled at different rates, multiplexed (combined) into a single bit-stream, and recorded or transmitted to a receiving station in a compact format. Such unprocessed "raw" data must be transformed into a more manageable and useful format with the addition of important ancillary data, such as calibration and Earth location parameters. The result is known as "conditioned" data. This transformation from raw to conditioned data, which is called "preprocessing," includes the following ordered steps:

- Data broadcast from the satellite are "ingested" or received into the preprocessing computer.
- Different types of data are separated into individual data sets according to the originating satellite and instrument.
- Instrument data are "calibrated" or described in forms that can be related to physical quantities.
- "Earth location" information is added so that each measurement can be directly related to a location on the Earth.
- The performance of the instrument and datatransmission are regularly monitored.
- The final conditioned data are stored in a data base accessible by the product processing programs and in a format compatible for archiving.

It is important that these preprocessing steps be executed without error at a rate equal to that at which the data are broadcast. The usefulness of meteorological data for current operations is quite short-lived. Therefore, the processing systems require almost immediate input of these data with all errors and irregularities flagged. If preprocessing systems cannot reliably detect errors, the product systems must perform their own quality checking. Given the time constraints involved, this redundancy should be avoided.

The material in this section pertaining to the NOAA series of polar orbiting satellites is a synopsis of topics taken from Schwalb (1979) and Lauritson et al. (1979). The material on the GOES series is taken from Hughes (1980), NOAA et al. (1980), and Mazur (1983).

VI-1.2 INGEST COMPUTER SOFTWARE CONSIDERATIONS

When received by the ingest computer, the satellite data are separated by source instrument and spacecraft, quality checked, and stored as "raw" data. The next functions, appending calibration and Earth location information, are described below. It is important to note the volume of data to be preprocessed, e.g., high resolution and multichannel observations constitute an enormous amount of data. The following two examples are indicative of the processing involved.

VI-1.2.1 The AVHRR Data Stream

The AVHRR instrument currently carried on the NOAA polar orbiting satellites observes the Earth's

[1] Primary contributor: James K. Ellickson

radiation in five wavelengths, two in the VIS spectrum and three in the IR (Chapters III-1 and IV-2). The ground resolution of each sample is about 1.1 km directly below the spacecraft. Earth-scan data are retained from a swath extending 55.4° to either side of nadir (2048 samples per scan per channel). Quantification of sensor output requires that each sample be recorded in ten-bit words. Thus, for all five channels there are 102,400 bits per scan. Since the satellite is traveling about 7 km each second, the instrument must scan the Earth six times per second to achieve the 1.1 km resolution. A nominal orbit is about 100 min. Thus the rate at which observational data are produced by this one instrument is 614,400 bits per second, or 3,686,400,000 bits per orbit.

The volume of raw data is expanded by adding time tags, calibration, and Earth location information. Conditioned, full resolution AVHRR data are produced at a rate of 118,400 bits per scan. At this rate, only 7 min of data can be stored on a standard 9-track, 1600-bits per inch (bpi), 2200 ft magnetic tape. Transmission of 10 min of such data across a typical 9600 baud (bps) telephone line would require about 12.5 hours.

VI-1.2.2 The GOES VISSR Atmospheric Sounding (VAS) Data Stream

About 1600 scans produce a picture of the Earth disk. This amounts to almost 12×10^9 bits of VIS data alone. Since these pictures are taken at least every half hour during the day, an enormous amount of data is generated.

The NESDIS Command and Data Acquisition (CDA) Station at Wallops Island, Virginia, preprocesses all GOES data and retransmits them—formatted, calibrated, and Earth located in real time. The bit rate of GOES retransmitted data is 2.11136×10^6 bps. These conditioned data are retransmitted from the CDA through the spacecraft to major readout facilities.

VI-1.3 INSTRUMENT CALIBRATION

In general, calibration involves exposing a sensor to an external source that has been calibrated against a primary or secondary standard from one of the national laboratories, and then establishing a relation between the electrical output of the sensor and the amount of electromagnetic radiation (radiance) entering the sensor (Williamson, 1977).

For example, all the radiometers flown on the NOAA ATN series undergo extensive prelaunch calibrations at the instrument manufacturer's facilities to establish their stability, linearity of response, and sensitivities. Instruments operating in the IR and MW regions of the spectrum are calibrated against precision blackbody sources whose calibrations are, in general, traceable to a National Bureau of Standards (NBS) reference. Instruments operating in the VIS and near-IR regions are calibrated against lamps whose output is viewed through the aperture of an integrating sphere. Calibrations for these lamps are also traceable to NBS.

Prelaunch IR calibrations are performed in a thermal-vacuum environment to simulate the environment of space. During the thermal-vacuum exposure, calibrations are done at several instrument–operating temperatures to provide a measure of deviation of instrument response as a function of temperature. VIS and near-IR calibrations are performed at ambient temperatures.

VI-1.3.1 Infrared/Microwave Instrument Calibration

During a typical IR or MW calibration, a radiometer is exposed to a blackbody source whose temperature is varied, in discrete steps, over the entire dynamic range of interest. In the VIS, the intensity of the source is varied. Data recorded during calibrations form a baseline from which the accuracy and precision of the radiometer measurements can be determined. Analyses of prelaunch calibration data from each of the primary radiometers on the NOAA ATN satellites are performed at NESDIS. Information on the accuracy or precision of calibration is available to data users at the time of each launch.

Even with extensive prelaunch calibration, one cannot rely on such static data to achieve the accuracy desired from remote sensing devices. Radiometer performance is subject to degradation in orbit. Unless some method is provided to assess such degradation, prelaunch calibrations may quickly become useless or questionable. Therefore, the NOAA ATN radiometers are designed for routine, in-orbit self-calibration in the IR and MW regions during scan sequences.

In-orbit calibration is done by programming the radiometer scan mirror to view scenes of known radiance: (1) space (at near-absolute zero) and (2) part of the instrument housing, which is designed as an on-board blackbody with its temperature maintained at approximately the operating temperature of the radiometer (288 K). The radiometer temperature sensor outputs are multiplexed into its telemetry. Thus, a two-point, in-flight calibration is provided.

Calibration of the temperature sensors of the on-board blackbody is generally done prior to launch. The method preferred by NESDIS is to use the instrument itself as a transfer standard. Since the radiometer is calibrated at several different temperatures, calibration curves using radiometrically derived temperatures can be generated for each temperature sensor in the on-board blackbody.

In the GOES program a similar procedure is followed, with one major difference: while NOAA polar instruments view the blackbody through the telescope optics, the GOES radiometer (VAS) does not have an external thermal target. The effects of the VAS optics must be accounted for separately.

When the the VAS scanner is not in thermal equilibrium, viewing the internal blackbody will not produce the same output voltage level in the IR channels as that produced by a blackbody of the same temperature located outside the VAS. During orbital flight, the components of the scanner are not in thermal equilibrium so the "effective" temperature of the internal source must be calculated.

VI-1.3.2 Visible Channel Calibration

Visible channel calibration is needed to produce data directly related to the Earth's albedo. The two-target approach described above for the IR and MW channels is impractical. A space-view of 0% albedo is easily obtainable, but finding a reliable "second point on the curve" has not been successful for NOAA polar or geostationary spacecraft. No operational NOAA spacecraft carries a calibration target for its VIS channels. Earth features (e.g., White Sands, New Mexico) are not reliable targets because of atmospheric effects. The moon may be viewed (by GOES) within one day of full illumination, but only at certain times of the year. The VAS can view the Sun indirectly through an energy reducing prism, which gives the equivalent of an Earth albedo response of about 50%. But these target data are not sufficiently accurate for use in calibration. Thus, modification of the VIS channels is done by a channel-to-channel normalization procedure that eliminates effects of different sensor sensitivities.

VI-1.3.3 Visible Channel Normalization

The output of the VIS channels is normalized to remove "striping" in the imagery. This striping, in which some lines appear brighter than others, is caused by sensitivity and offset differences among the eight VIS sensors. Normalization is done as follows:

- Raw full-resolution VIS data are collected.
- Extended targets with uniform albedos are selected statistically from these data. Ideally, the signal level from the eight channels should be identical for each target. When striping is evident, however, they are not. Therefore, an empirical analysis is done to determine the adjustments required to make all channels respond identically to a common, uniform target.
- The adjustments for each channel are incorporated into corrections made to the VIS data as they are preprocessed at the NESDIS ground station.

VI-1.4 EARTH LOCATION OF SATELLITE DATA

Once the satellite instrument data have been separated, identified according to source, and calibrated, they then must be directly associated with locations on the Earth. This process is called "Earth location." There are many ways to do this, but usually three procedures suffice. They are:

- Imagery data which will be used qualitatively may be gridded. Certain samples (for an image these are often "pixels") are altered to show the latitude or longitude lines, or geographic or political boundaries. This often is vital when dense clouds obscure recognizable land features.
- Data used quantitatively may require that the Earth location of any or all data samples be known. This is often done by calculating the latitude and longitude positions of selected samples and finding the Earth location of the others by interpolation.
- All of the satellite data may be mapped into a well-defined projection, such as Mercator, polar stereographic, or a nominal satellite projection. This is a computationally more demanding process.

Two considerations that require exercising caution in Earth location procedures are:

- Care should always be taken to handle time as accurately as possible. This is especially true for the polar satellites which move at speeds of over 7-kms^{-1}, i. e., a 1 sec error translates directly into a 7-km Earth location error. This is a real problem because the on-board clock may drift as much as 1 sec over a period of months, at which time it is reset by ground command to synchronize with UTC.
- It is worth comparing pointing requirements between geostationary and polar satellites. NOAA polar satellites orbit at altitudes of about 850 km. This means a kilometer on the Earth subtends about 4 arc–min in the satellite's field of view. The GOES satellites, however, orbit at altitudes of about 35,800 km. Since a kilometer at the Equator subtends less than 5.8 arc-sec from geostationary altitude, a much greater pointing accuracy is required of the VAS to obtain the same Earth location accuracy as is available from polar satellite altitude.

VI-1.4.1 Earth Location Algorithm Development

All of these methods require similar algorithm development. A mathematical transformation must be used either to match latitude and longitude to a given sample, or to match a sample with a given latitude and longitude. To do this, the orbit and orientation of the spacecraft, and the geometry of sensor scanning, must be known.

VI-1.4.2 Earth Location of GOES Data

Because GOES is relatively motionless with respect to the Earth, the gravitational force between the Earth and the satellite is effectively constant. Only the gravitational effects of the Sun and the moon, and the force of solar radiation have any effect on the orbit. To make a one- or two-day prediction of satellite position, a simple "two-body" solution of satellite motion is adequate. This type of model is described in most orbit determination texts (e.g., Escobal, 1976).

Since GOES spins like a precise gyroscope, its orientation is determined by its axis of spin, which is usually parallel to that of the Earth. GOES altitude can be determined in a number of ways, but all make use of the VAS scan mirror position when viewing other objects in space. The horizons of the Earth are automatically sensed by the ground preprocessing system and become part of the data stream. Thus, if the position of the satellite (with respect to the Earth) is known, horizon data give information as to the direction in which the Earth appears with respect to the spin axis. Measuring the scan line locations of Earth landmarks can also be useful.

VI-1.4.3 Earth Location of Polar Satellite Data

In contrast to GOES, the polar satellites use spinning momentum wheels to maintain the spacecraft orientation with respect to the Earth. Any errors detected in this "three-axis stabilized" system are recorded as part of the telemetry and can be used to adjust a "nominal" attitude model. To date this has not been necessary operationally at NESDIS.

A rotating mirror scans the Earth at 90° angles to the path of travel and sensor sampling is performed so that the midscan point is aligned with the nadir direction. When the orbital position is known precisely, the altitude-control system makes Earth location relatively straightforward.

VI-1.5 QUALITY CONTROL CONSIDERATIONS IN BUILDING SATELLITE INSTRUMENT DATABASES

Quality control is used routinely at NESDIS to examine data, statistically or as imagery, and to find processing or instrument problems. Because of timeliness requirements and the enormous volume of data, real-time error correction is rarely attempted. However, action can often be taken to correct certain errors for future data intake, as the following examples indicate.

GOES imagery is routinely displayed on an interactive image display system, the VISSR Image Registration and Gridding System (VIRGS), used primarily for image navigation and gridding. When image quality problems are noted, the Satellite Operations Control Center (SOCC) is notified and total system troubleshooting begins. In the upgraded Wallops GOES ground system of 1986, a subsystem (the Display Monitor) is used for this purpose. This system displays GOES data during retransmission to users; the display is examined to identify problems and isolate their sources.

Some statistical analysis is made of polar satellite data during ingestion and calibration; e.g., the values of space-view and calibration target data are assumed to be uniform and consistent. These data are discarded by the calibration software when they exceed certain bounds. Also their standard deviations are estimated and reported as statistics on instrument noise.

In summary, the ability to view imagery and to analytically monitor instrument data, especially calibration data, is vital for the production of quality satellite data products.

VI-1.6 DATABASE CONSIDERATIONS

The design and construction of an instrument database is determined by user requirements. A few of the factors most often considered are:

- Is the database local or global? A global database is usually mapped onto a latitude and longitude oriented grid. A local data base may be usable in satellite scan oriented format.
- Will the data be used in real time, accumulated, or archived? Polar satellite data can be mapped as intermediate datasets before they are processed into a global database. GOES data, however, are often needed immediately after receipt (e.g., for severe storm monitoring) and must be available from the database as soon as received.
- Should multichannel instrument data be separated by channel or should the values from each channel be formatted together for each scanner position? If imagery is the major product, separating the channels during preprocessing may simplify product formation. If quantitative comparisons of channel data are to be made, such as for multichannel sea surface

temperature calculations, the data are more useful if formatted together.

- What data accuracy must be maintained? If the data are mapped, some interpolations are made. The errors that can be introduced during these additional calculations must be controlled.
- What data packing is appropriate? Efficient use of storage must be weighed against the computational overhead of retrieving the data.
- What physical constraints must be considered? Storage media characteristics may influence the size of a database; a major consideration is the most efficient use of space.

In summary, the most efficiently constructed satellite instrument database is characterized by its timeliness, utility, quality, and reliability. These four characteristics must be properly balanced to meet the needs of each product system. The six major product systems in use at NESDIS are described in Chapter VI-2.

Chapter VI-2

SATELLITE DATA PRODUCT PROCESSING

VI-2.1 INTRODUCTION

After raw satellite data have been processed into their more manageable conditioned form, with Earth location and calibration appended, further processing is needed to generate products. These can generally be grouped into two categories: imagery and numerical products. Both require large computer resources, particularly for global use. The primary NESDIS numerical products include atmospheric soundings, cloud motion wind estimates, sea surface temperatures, radiation budget values, and ozone amounts. Image products are produced centrally for VIS, IR, and WV channels. These are sent to users for display either on hard copy facsimile machines or as animated video images. Image data can also be accessed digitally for interactive analysis. Some of the products, such as the manually produced sea surface temperature analysis, are derived from imagery and numerical data.

VI-2.2 ATMOSPHERIC SOUNDING DATA FROM SATELLITE RADIOMETERS

VI-2.2.1 Introduction

Data acquired from the TIROS Operational Vertical Sounder (TOVS) are used to produce atmospheric soundings. The TOVS software transforms upwelling IR and MW radiant energy data into vertical temperature and water vapor profiles. See Chapter VII-5 for details of sounding derivation. A variety of other parameters, such as total ozone, cloud height and amount, and tropopause temperature and pressure, are also derived. Soundings are produced from data acquired by two polar orbiting satellites with ascending Equator crossing times of 0730 and 1430 LST. Approximately 50,000 soundings, spaced 80 to 300 km apart, are produced daily with data from each satellite. Objective analysis and noise screening techniques that eliminate bad, redundant, and questionable data reduce this number to about 10,000 soundings per day per satellite that can be transmitted.

VI-2.2.2 TOVS Sounding Processing

A flow diagram of the sounding processing is shown in Figure VI-2-1. NOAA satellite data are acquired through the Data Access and Command Subsystem (DACS). From DACS, the data are ingested by the Data Processing SubSystem (DPSS) which reformats and divides the data into various product-oriented streams. Parameters for Earth location and in-flight thermal calibration are appended to the TOVS data. This is the 1B dataset that is archived. The File Transfer (FT) moves the TOVS 1B data to the Central Computer Facility (CCF) for sounding product processing.

Two subsystems, the On-Line TOVS (orbit-by-orbit) Processing (Fleming et al., 1988; Werbowetzki, 1981) and the off-line Coefficient Update, are used to process the sounding data. A step-by-step functional outline of each module follows:[1]

[1] At the end of 1988, NESDIS was in the process of converting to a physical retrieval algorithm.

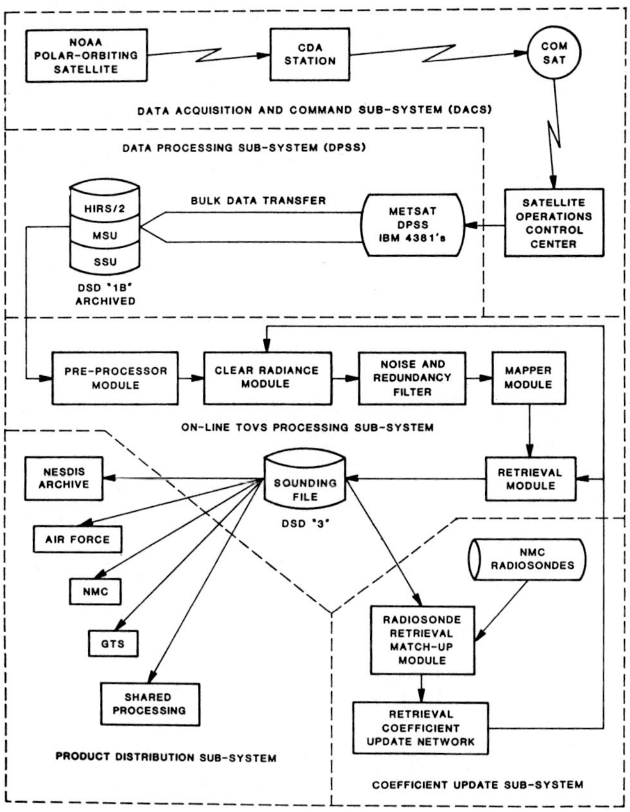

Figure VI-2-1. TOVS dataflow diagram.

On-Line TOVS Processing Subsystem

A. Preprocessor Module
- Filter obviously bad data.
- Limb correct MSU data.
- Filter MSU data for rain contamination (Nappi et al., 1986).
- Co-locate MSU data to HIRS/2 fields of view.
- Limb correct HIRS/2 data.
- Correct HIRS/2 window channels for water vapor.

B. Clear Radiance Module (McMillin, 1978)
- Determine cloud contamination (clear, partly cloudy, or cloudy).
- Calculate equivalent clear column radiances for HIRS/2 in partly cloudy areas.
- Calculate cloud height and amount.
- Select radiance data for calculation of soundings.

C. Noise and Redundancy Filter (Reale et al., 1986)
- Use selected channel temperatures to determine sounding quality.
- Use selected channel temperatures to identify redundant soundings.

D. Mapper Module
- Update surface temperature fields.
- Calculate and map albedo field.
- Limb correct SSU data.
- Map SSU fields.

E. Retrieval Module
- Calculate vertical temperature profiles (40 levels).
- Calculate vertical water vapor profiles (15 levels).
- Calculate tropopause temperature and pressure.
- Calculate total ozone.

Coefficient Update Subsystem

A. Radiosonde/Retrieval Matchup Module
- Match retrieval and radiosondes in time and space.
- Filter retrievals and radiosondes for bad samples.
- Interpolate 00 UTC and 12 UTC radiosondes to the time of the closest matched retrieval.
- Select matched radiosondes and retrievals for global distribution.

B. Coefficient Update Network
- Compute updated retrieval regression coefficients from matchups every week.
- Verify updated coefficients by using an independent sample of matchups.

VI-2.2.3 TOVS Sounding Product Distribution Subsystem

The TOVS product distribution subsystem (Figure VI-2-1) routinely sends 10,000 soundings per day per satellite (Figure VII-5-2b) to the National Meteorological Center, Air Force Global Weather Central, the European Centre for Medium-Range Weather Forecasts, and the NESDIS archive. Users served by the Global Telecommunications System (GTS) receive 2,500 soundings per day (spaced 500 km apart). A detailed list of sounding parameters is shown in Table VI-2-1.

VI-2.3 WIND FIELDS FROM GEOSTATIONARY SATELLITE DATA

VI-2.3.1 Introduction

Winds measured from cloud motions are derived routinely for synoptic analysis times of 00, 06, 12, and 18 UTC, using 8 × 8-km resolution IR data from GOES-East and GOES-West. Geographical coverage extends from 45°N to 45°S and from 175°E to 25°W. Operational winds are produced by two methods: automated picture-pair cloud tracking and manual cloud tracking using an interactive computer system. Specific details of these methods are described in Chapter VII-4. Figure VII-4-1 shows the distribution of satellite cloud track wind products for a 24-hour period.

Table VI-2-1 *NOAA polar environmental satellite sounding products*

PRODUCT DESCRIPTION	ACCURACY GOALS	SPATIAL RESOLUTION/ GEOGRAPHICAL COVERAGE	SCHEDULE	USERS (OUTPUT MEDIA)
Layer-mean temperatures (K) Sfc-850 mb 300–200 mb 30–10 mb 850–700 mb 200–100 mb 10–05 mb 700–500 mb 100–70 mb 05–02 mb 500–400 mb 70–50 mb 02–01 mb 400–300 mb 50–30 mb 01–4.4 mb	Sfc-850 mb ± 2.5 K 850-tropopause ± 2.25 K Tropopause-2.0 mb ± 2.5 K	20–50 km resolution/global with 80–300 km spacing depending upon the extent of the meteorological activity	Orbit-by-orbit	NMC (Disk) EDIS (CCT)
Layer precipitable water (mm) Sfc-700 mb 700–500 mb 500–300 mb	±30% of the true value (or 1 mm, whichever is larger)			
Tropopause pressure (mb) and experimental temperature (K)	Pressure: ±50 mb Temperature: ±2.5			
Experimental total ozone (Dobson)	±15% tropical, ±50% polar			
Equivalent blackbody temperatures (K) for 20 HIRS/2 stratospheric channels, 4 MSU channels, 3 SSU channels	±2 K			
Experimental cloud cover	±20%			
Thickness (m) between selected standard pressure levels	Equivalent to accuracy goals for layer-mean temperatures (in meters)	Same as above	Orbit-by-orbit	All customers served by GTS teletype
Precipitable water (mm) between selected standard pressure levels	±30%			
Tropopause pressure (mb) and temperature (K)	Pressure: ±50 mb Temperature: 2.5 K			
Cloud cover	±20%			
Clear radiances mW (m^2 sr cm^{-1})$^{-1}$	±2%	Same as above	Orbit-by-orbit	All customers served by GTS teletype
Earth located, calibrated radiances from SSU plus selected HIRS/2 and MSU channels	Instrument specifications	MSU/SSU individual field of view; HIRS reduced to MSU resolution	Orbit-by-orbit	British Meteorological Office, Bracknell, U.K. (Telecommunication lines)

VI-2.3.2 Automated Cloud Tracking

The automatic picture-pair program which produces low level cloud drift wind estimates consists of the following primary functions:

A. Image Selection—the best two images in a three-hour window immediately prior to synoptic observation time are selected based upon:
   - Latitudinal coverage
   - Timeliness
   - Absence of line dropouts and scan count errors

B. Relative Registration of Sequential Images
   - Preselected landmarks (usually coastlines) are compared by the computer to determine any offset between the two images.
   - Translational corrections are made so the landmarks in the two images exactly coincide. Image registration error is usually less than one-eighth of an image element.

C. Cloud Motion Measurement and Wind Computation
   - Target Selection—16 × 16 arrays (8 km IR) at preselected grid points in the first image are examined for numbers of samples corresponding to low level clouds (warmer than the NMC 700-mb temperature). Grid points with insufficient numbers of samples are skipped.
   - Motion Sensing—Using cross-correlation, the motion of samples within the temperature range is computed. Satellite orbit and attitude information is used to convert cloud displacement to standard meteorological winds with direction in degrees and speed in knots.

D. Editing
   - The number of candidate winds is examined and for totals less than a minimum, processing is reinitiated using a new pair of images.
   - Using the previous picture-pair winds and the latest NMC 850-mb analysis as a first guess, an objective analysis of the computed winds is performed and inconsistent winds are deleted.

E. Distribution
   - Global Analysis Centers
   - Global Telecommunication Systems
   - Archives

VI-2.3.3 Interactive Cloud Tracking

Satellite wind estimates from the upper atmosphere are made using the WIND Computation Operation (WINDCO) program (Smith and Phillips, 1972) and the NESDIS interactive computer facility. WINDCO is used to obtain wind vectors between 1000 and 150 mb. The wind vector is derived by tracking a cloud element through a set of three images to determine its speed and direction. The tracking system works in two ways: pattern matching, where the pattern of the clouds within a small area is matched and the displacement of the pattern between the first and last images is measured; or single pixel tracking, where single cloud elements are individually followed by an analyst through viewing a sequence of images in animation. Interactive winds are processed as follows:

- A series of three images taken at 30-min intervals is selected and obtained in digital form for machine processing.
- The cloud field is examined visually to identify traceable cloud elements, such as cirrus flow. This process is necessary because some clouds associated with thunderstorms or jet streams are not suitable as wind tracers.
- Cloud motion vectors are obtained over as much of the area of imagery as possible.
- Cloud top temperatures are established by using the VISSR IR values; a temperature is assigned to each cloud vector.
- Cloud top temperatures are converted to cloud top height. Temperature profiles of the atmosphere at satellite wind data points are constructed with the aid of temperature fields provided by NMC. Height values for each cloud vector temperature are extracted from the vertical temperature profiles and are assigned to the cloud vectors.
- Satellite cloud motion vectors are sorted according to atmospheric height, edited on the interactive computer, and made available for distribution.

VI-2.4 SATELLITE DERIVED SEA SURFACE TEMPERATURES (SST)

VI-2.4.1 Introduction

A technique for producing SST based on single-window IR measurements was used from 1979 until 1981. This technique, called the Global Operational Sea Surface Temperature Computation (GOSSTCOMP), did not permit adequate cloud filtering and water vapor corrections (McClain, 1979; Strong and Pritchard, 1980).

In 1981, the Multichannel Sea Surface Temperature (MCSST) technique became operational. This technique included improved cloud filtering and water vapor corrections. The additional information provided by multichannel data, lower noise levels of the data, and temperature dependent bias corrections incorporated into this technique improved the results significantly (Figure VI-2-2). This technique is discussed more fully in Chapter VIII-4. The derived MCSST are stored in databases archived for users, and from which products,

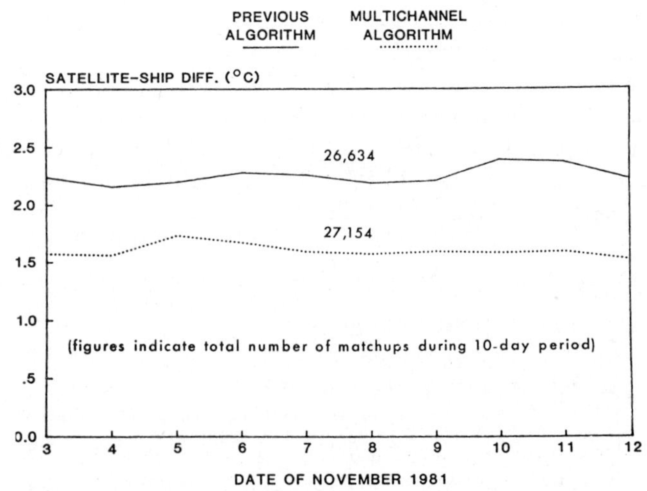

Figure VI-2-2. Daily global RMS temperature differences (satellite-ship) for a 10-day period in November 1981. Results using the new MCSST algorithm and the previous operational TIROS-N (GOSSTCOMP) algorithm are shown.

such as MCSST fields at various resolutions, can be produced.

VI-2.4.2 Multichannel Sea Surface Temperature Processing

MCSST are calculated using 4-km resolution Global Area Coverage (GAC) data from the AVHRR. These data, recorded on the satellite, provide nearly day and night coverage twice every 24 hours. For the calculation of MCSST, the GAC data are processed in 11 × 11 scan-line scan-spot arrays or targets. A target ranges in size from 36 × 43 km to 36 × 81 km, depending on the viewing angle of the sensor. Target centers are spaced approximately 25 km apart to provide significant overlap between adjacent targets. Some 840,000 targets are processed daily; approximately 320,000 pass all preprocessing checks, of which about 140,000 are daytime and 180,000 are nighttime. Those sufficiently cloud-free are used to compute 140,000 to 150,000 MCSST every 24 hours.

MCSST are derived from daytime targets by examining 2 × 2 subunits (unit arrays) of each 11 × 11 target. Cloud tests are made on unit arrays, and MCSST are calculated from one or more cloudfree unit arrays from each target. Thus, the daytime MCSST density can be varied from one every 25 km to one every 8 km. Regardless of density, however, each MCSST is derived from a 2 × 2 unit array of GAC data and so represents an area about 8 km on a side. With one observation per usable GAC target, 100,000 to 115,000 SST can be derived daily from the 140,000 daytime targets. As the nighttime multichannel algorithm cannot use VIS data, the IR channels alone must be used. As in the daytime algorithm, MCSST are calculated from 2 × 2 unit arrays, but only one is calculated from each nighttime target. Typically, some 30,000 to 35,000 SST are derived daily from the 180,000 nighttime targets:

The following is a step-by-step outline of the MCSST processing.

A. Target Processing
- Check conditioned database quality indicators for each scan line.
- Check magnitude and consistency checks of AVHRR calibration coefficients.
- Do gross land/sea test over a 1/2° latitude-longitude grid.
- Determine whether the target is in daylight or darkness.

B. Daytime MCSST Processing
- Reject unless more than 20 of 121 GAC target spots have a bi-directional reflectance (Ch 2) of less than 20% (Gross Cloud Test).
- Reject unless difference between maximum and minimum Ch 2 counts within the array is ≤3 counts (VIS Channel Uniformity Test).
- Reject unless Ch 2 is less than a predetermined threshold that is dependent on the solar and satellite zenith angles and the relative sun-satellite azimuth angle (VIS Cloud Test).
- Calculate simple average count value for each of the five radiometer channels.
- Convert VIS channel counts to bidirectional reflectance and IR channel counts to brightness temperature.
- Correct for atmospheric attenuation and calculate MCSST (in °C) using brightness temperatures (in K) from Ch 4 and Ch 5 using equations such as those discussed in Chapter VIII-4.
- Accept only MCSST values between −2°C and +35°C and within 7°C of climatology.
- Compute latitude and longitude of the unit array center.

C. Nighttime MCSST Retrieval Algorithm
- Check that at least 30 of the 121 elements of the target have Ch 3 (3.7 μm) temperatures higher than −3°C (Gross Cloud IR Test).
- Locate warmest Ch 4 target array element, using the one nearest the target center.
- Taking each of the four 2 × 2 unit arrays containing the warmest element in common, retain only those in which the difference between the maximum and minimum count is less than 10 counts in Ch 3 and 2 counts in Ch 4 (Thermal IR Uniformity Test).
- Calculate a simple average count value for each IR channel.

- Convert IR channel counts to brightness temperatures.
- Predict Ch 3 from Ch 4, and Ch 4 from Ch 5, and reject if observed values depart by more than a few degrees from the predicted (cloud-free) values (Thermal IR Cloud Test).
- Reject the unit array if Ch 4 or Ch 5 exceeds Ch 3 by more than a few tenths of a degree (Uniform Low Stratus Test).
- Correct for atmospheric attenuation and calculate three MCSST estimates—one with a dual window equation (Ch 3 and Ch 4), one with a split window (Ch 4 and Ch 5), and one with a triple window (all three IR channels).
- Compare the three MCSST from the step above and reject the unit array if the range is more than 1°C—use the triple window value for arrays accepted (SST Intercomparison Test).
- Accept MCSST if it differs by 7°C or less from the climatological SST and lies between −2°C and +35°C.
- Calculate the latitude and longitude of the unit array center.

VI-2.4.3 The Sea Surface Temperature Database

SST values are derived from AVHRR data with the MCSST algorithm (McClain, 1981; McClain et al., 1983). Some 100,000 to 115,000 daylight and 30,000 to 35,000 darkness 8-km resolution observations are calculated daily. Observations are spaced every 8 km in U.S. coastal and selected research areas, every 15 km in the eastern North Pacific and western North Atlantic, and every 25 km elsewhere. Every six hours, all SST observations are entered into a geographic data base that can be interrogated via Remote Job Entry (RJE) or Time Sharing Option (TSO) terminals at any time. One observation from every 2½° lat./long. square is transmitted twice daily on the Global Telecommunications System (GTS). The primary users are NWS and the Navy Fleet Numerical Oceanography Center. Users worldwide can obtain these data electronically via GTS or by mail from the Satellite Data Services Division of NESDIS.

VI-2.4.4 Satellite SST Analyzed Fields

Satellite MCSST are analyzed at a number of grid point scales. The same basic analysis technique is used at all grid scales for every interval:

- Define a search area whose size is a function of the surface temperature gradient for each gridpoint.
- Weigh the observations in the search area inversely with the square of the distance from the grid point.
- Update the grid point MCSST that was calculated in the previous field analysis by replacing it with a weighted grid point MCSST average and the current one.

A global analysis is produced daily on a 1° lat./long. grid (Figure VI-2-3). Analyses for regions nearer to the U.S. are produced weekly on a ½° lat./long. grid (Figure VI-2-4), and for the coastal areas of the U.S. twice a week on a ⅛° lat./long. grid. All are in the form of isotherm contour charts; all are available at computer terminals at any time. These charts are used by Weather Service Forecast Offices (WSFO) as guidance in producing oceanographic products. Charts may be obtained by subscription from the Satellite Data Services Division of NESDIS.

VI-2.4.5 Current Combined Satellite and Conventional Data Products

The NWS combines conventional ship, buoy, and Expendable Bathythermograph (XBT) data with

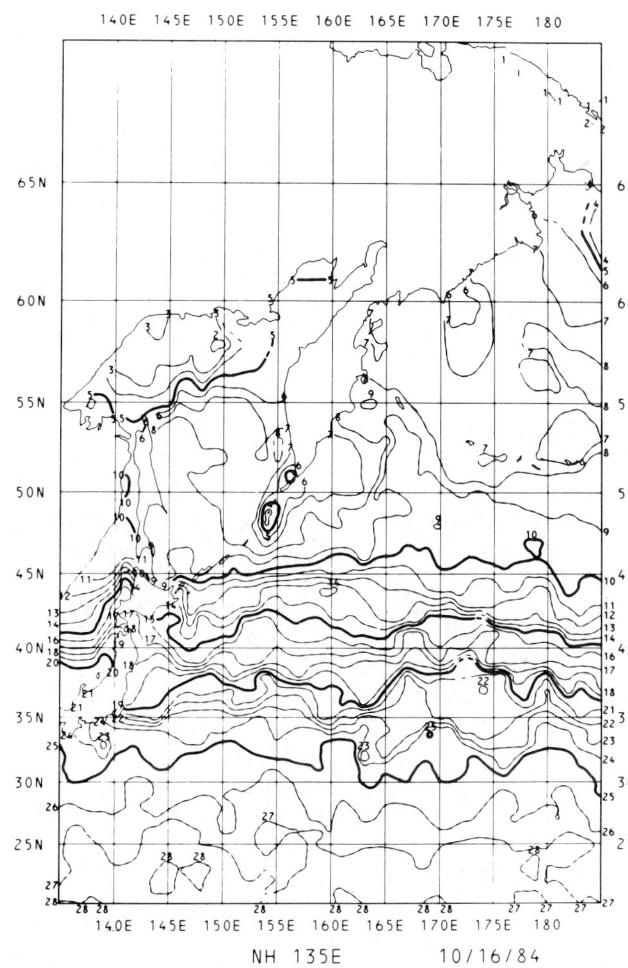

Figure VI-2-3. Sea surface temperature analyzed from a 1.0° lat./long. grid.

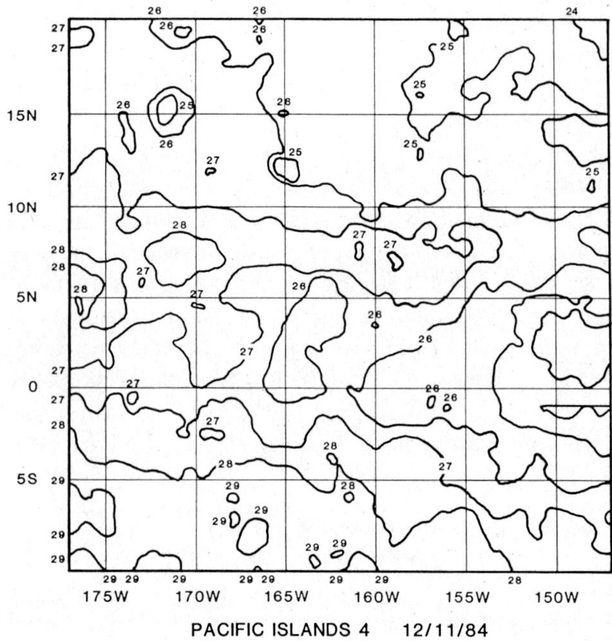

Figure VI-2-4. Sea surface temperature analyzed from a 0.5° lat./long. grid.

MCSST into a single global SST analysis (Bell and Clark, 1984). Running 15-day global blended analyses are updated daily on a 2° grid and sent out on GTS at 5° resolution. Monthly mean blended 2° lat./long. SST and anomaly charts for use in climate research and monitoring (Chapter IX-4) are available by mail from NWS.

VI-2.4.6 Current Manual SST Products

Local, high-resolution ocean products are produced for U.S. coastal areas by regional NOAA oceanographic facilities. Oceanographers there use satellite images, satellite MCSST charts, conventional data, and blended products to produce SST and ocean-feature charts manually. Ocean-feature charts use mostly VIS imagery data; however, conventional ship and buoy temperatures, and sometimes XBT data, are used to locate eddies having insignificant surface thermal signatures. Some SST products (e.g., Figure VI-2-5) are generated from MCSST analyses with frontal features determined from satellite images. Such products are

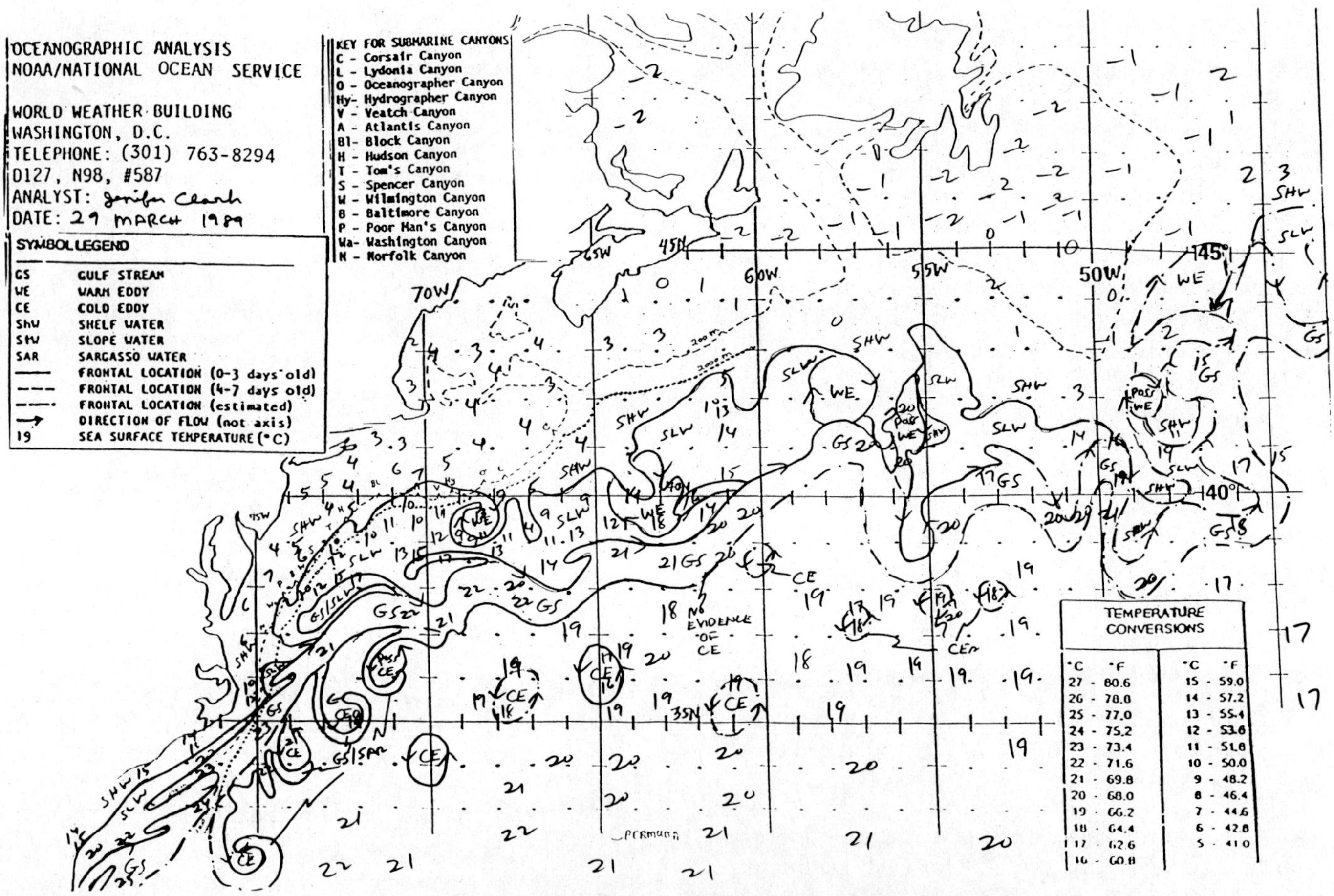

Figure VI-2-5. Manually produced SST product.

available to fishermen and commercial sea-transport companies by autotelecopier, radio facsimile, and mail.

VI-2.5 OPERATIONAL EARTH RADIATION BUDGET

VI-2.5.1 Introduction

Earth-atmosphere radiation budget products have been generated since June 1974 (except for March-December 1978 when data were unavailable). From 1974 to 1979, observations obtained from the two-channel Scanning Radiometer (SR) on-board NOAA operational satellites were used as a data source (Gruber, 1977). These two channels observed reflected energy in the VIS (0.5 to 0.7 μm) and emitted longwave radiation in the atmospheric window (10.5 to 12.5 μm). Emphasis was placed on the broad and temporally averaged results, such as mean monthly zonal and global averages, and maps of radiation budget parameters such as albedo, outgoing longwave radiation, absorbed radiation, and net radiation.

With the implementation in 1979 of the TIROS-N series of satellites, which carried AVHRR in place of SR, the processing system that produces the planetary radiation budget estimates was redesigned (Gruber et al., 1983). The new procedures operate directly on the incoming data stream. Previously the processing was done on a mapped database in which some processing of the raw data had already been accomplished.

VI-2.5.2 Earth Radiation Budget Procedures

The basic data element used by the radiation budget processing system is an 11 × 11 matrix of contiguous AVHRR Global Area Coverage (GAC) spots, which is termed a GAC target (sometimes also referred to as a 50-km area). The width of a target along the orbital track is 36 to 37 km, while in the scan direction it varies from 42 to 196 km.

The current processing system uses two of the four or five AVHRR channels. Absorbed solar energy or albedo is determined from VIS reflectivity values of Channel 1 (0.58 to 0.68 μm). The Outgoing Long-wave Radiation (OLR) is obtained from Channel 5 (11.5 to 12.5 μm) on five channel AVHRR models; and from Channel 4 (10.5 to 11.5 μm) on four channel models. The latter channel is somewhat more transparent to atmospheric moisture.

Each individual GAC target array is examined to ascertain the number of individual observations present. If there are less than 60 (roughly half of the possible number), nothing further is done and the target is labeled as missing. If there are at least 60 observations present, the digital count values for each channel are averaged to obtain representative values for the target matrix. The proper calibration equations are then applied to these average values, and a visible albedo and/or a long-wave radiance is calculated. The final flux values, as described below, are produced from these data.

VI-2.5.3 Earth Radiation Budget Products

A. Outgoing Long-wave Radiation. The mean long-wave radiance is normalized to a central satellite zenith value using an empirically derived limb-darkening coefficient (Abel and Gruber, 1979). The use of a central satellite zenith angle introduces negligible error in the final value, since the angle varies little over a GAC target. The normalized radiance is converted to a brightness temperature using the Planck radiation equation. A flux equivalent temperature is then obtained by applying empirical coefficients to the brightness temperature. Finally, an outgoing flux is calculated by applying the Stefan-Boltzman equation to the flux equivalent temperature.

B. Albedo. The calibration equation that is applied to Channel 1 transforms digital counts to albedo for that spectral range at the top of the atmosphere, normal to the incoming solar energy of the mean Sun-Earth distance. The solar constant used is based on Thekaekara (1971). This percent reflected, or "albedo," must be normalized. It is divided by the cosine of the solar zenith angle of the target center and corrected for the Sun-Earth distance at the time of observation. It is assumed that this value of normalized spectral albedo is equivalent to the broadband albedo and that it remains constant over the day.

C. Product Organization. Routinely produced radiation budget products fall into three general classes:

- Nighttime OLR—derived from emitted radiation sensed by AVHRR Channel 5 (or Channel 4) during the mostly nighttime satellite passes.
- Daytime OLR—derived from emitted radiation sensed by AVHRR Channel 5 (or Channel 4) during the mostly daytime satellite passes.
- Absorbed Solar Radiation—derived from the reflected solar radiation measured by AVHRR Channel 1 and from the computed daily available solar radiation (the mean amount of solar radiation striking the top of the atmosphere over the target on the day of observation) during the northbound daylight satellite passes.

Each dataset is organized into average values in a square grid (resolution ~175 km at 45° lat. on polar stereographic projections for the Northern and Southern Hemispheres and in a 2.5° lat. × 2.5° long. global grid which is used for mapping on a Mercator projection).

VI-2.5.4 Supplementary Earth Radiation Budget Products

In addition to the output products discussed above, there are other products that are generated in an "off-line" mode. These may be considered as examples of the first level of operational product applications. They consist of the calculation and archival of monthly and seasonal means using operational daily products such as nighttime and daytime long-wave radiation, and absorbed solar radiation (Gruber et al., 1986). These data products are currently being used in climate diagnosis and research by the Climate Analysis Center, National Center for Atmospheric Research, and others. Chapters IX-1 and IX-4 provide examples, analysis, and application of the Earth radiation budget products.

VI-2.5.5 Earth Radiation Budget Product Improvement

A. Albedo. Efforts are ongoing to enhance the product quality. Two areas under active study are (1) the spectral range of reflected energy observations and (2) the angular pattern of reflected energy. The current albedo algorithm takes the measurement from a single narrow-band channel to represent the entire reflected solar envelope. A new method being tested utilizes two narrow-band channels (Wydick et al., 1987) as input to a regressional algorithm to estimate spectral broadband albedo. The current albedo algorithm also assumes isotropy for the surface reflectance as observed from space. Angular models for both bidirectional and solar zenith angle dependencies developed from analysis of Nimbus-7 Earth Radiation Budget (ERB) studies (Taylor and Stowe, 1984) are being adapted to AVHRR processing. This will allow for more accurate albedo estimates, particularly for low solar elevations where reflecting surfaces are strongly anisotropic.

Better diurnal sampling would also lead to improved albedo estimates. Processing of both morning and afternoon AVHRR data helps somewhat, but the early morning observations require more major angular adjustments. Future use of geostationary VIS observations to calculate albedo would observe the diurnal variations directly, especially of clouds. However, geostationary satellites cannot provide complete global coverage.

B. Outgoing Long-Wave Radiation. Efforts are also being made to improve the estimates of OLR. The possibility of using IR channels from TOVS to estimate OLR is being pursued. The advantage of this method is that these channels have information about the temperature, water vapor, ozone, and carbon dioxide structure of the atmosphere, information lacking for the most part in the AVHRR window channels. Experiments using linear combinations of these sounding channels show improvements over the current single channel algorithm, making this method an excellent candidate for replacing the current one channel method.

VI-2.6 SATELLITE DERIVED ATMOSPHERIC OZONE DATA

VI-2.6.1 Introduction

The Solar Backscattered Ultraviolet Radiometer, Version 2 (SBUV/2), an operational, remote-sensing instrument (Chapter IV-2), is used to produce global maps of total ozone concentration and vertical distribution in the atmosphere.

The solar UV radiation measured by the satellite is primarily that which has penetrated the stratosphere and has been reflected back by the troposphere and the surface. Since ozone attenuates UV, the amount of ozone in the atmosphere above the reflecting surfaces can be estimated by determining how much of the reflected UV has been absorbed.

Several variables must be considered in computing the intensity of the Backscattered UltraViolet (BUV) light measured at the 12 wavelengths of the SBUV sensor. Pressure and reflectivity at the reflecting surface in the Instantaneous Field Of View (IFOV) must be determined. The solar zenith angle determines the path of the UV light.

The ratio between the total backscattered radiance and the incident solar flux is used to determine total ozone and ozone profiles. See Chapter IV-2 for description of the SBUV/2 radiometer. The total ozone algorithm uses the four longest wavelength bands (312.5, 317.5, 331.2, and 339.8 nm); the profiling algorithm uses the shorter wavelengths.

VI-2.6.2 Steps in Processing Data to Produce Ozone Measurements

A. Inputs to Processing. The input data listed below are used in producing total and profile ozone measurements. The first three listed vary daily; the next three are reference data; and the last is updated daily.

- SBUV/2 raw data—sensor, in-flight calibration, and housekeeping.
- Ephemeris data—satellite position and velocity vectors at constant time intervals.
- Meteorological data—temperature profiles, cloud amounts, and cloud top heights.
- Prelaunch calibration coefficients—SBUV/2 radiance and irradiance counts-to-engineering units conversion constants.
- Surface pressure data—global grid of surface pressures derived from NMC terrain height data.

- Ancillary data—constants required to initialize the ozone algorithm.
- Snow and ice—daily global grid of snow and ice conditions.

B. Data Collection. SBUV/2 raw data and meteorological data from a satellite or ground network are entered into the processing system. Ephemeris data are appended to the SBUV/2 raw data prior to entry into the ozone processing system. Prelaunch calibration coefficients, surface pressure data, and ancillary data are initially included. Prelaunch calibration coefficients are used in the computation of normalized radiances. The ancillary dataset contains the 12 band-center wavelengths, their ozone absorption and Rayleigh scattering coefficients, total ozone and multiple scattering correction lookup tables, and *a priori* profiling information. These data are used to initialize the ozone algorithm.

Snow and ice data are used in determining whether high reflectivity is from clouds or the Earth's surface.

VI-2.6.3 Processing Algorithm

The SBUV/2 algorithm consists of two parts, one for a total ozone and the other for the ozone profile. For details on the algorithm, see Klenk et al. (1982).

A. Total Ozone Algorithm. The procedure for deriving total ozone from backscattered radiances involves using the solar zenith angle, the surface reflectivity, and the surface pressure to obtain the columnar amount of total ozone by means of lookup and interpolation from a table of precomputed radiances.

B. Profile Algorithm. Determination of the vertical ozone profile is done by use of an optimum statistical inversion technique (Rodgers, 1976). The use of this technique is described by Schneider et al. (1981). The wavelengths from 251.9 to 305.8 nm are used for profiling. Also, at solar zenith angles greater than 75.5°, the 312.5-nm radiance is also used. The optimum method yields the ozone layer amounts and their standard deviations in 12 atmospheric pressure layers. Mixing ratios at 16 standard meteorological levels, between 0.3 and 40 mb, are calculated from layer amounts.

VI-2.6.4 Products

Four operational products based on SBUV/2 data are produced:

A. The Conditioned (1B) Dataset—contains (1) all SBUV/2 sensor data and support data for deriving atmospheric ozone and solar flux; (2) instrument inflight calibration data and housekeeping functions for monitoring post-launch instrument changes; (3) prelaunch calibration factors, and (4) computed current day instrument calibration and albedo correction factors. The SBUV/2 data consist of atmospheric radiance and solar irradiance measurements. The support data include cloud and temperature data from TOVS, ancillary data to initialize the algorithm, surface pressure data, and snow and ice data.

B. Product Master File (PMF)—contains the derived ozone information located in space and time, meteorological information that supports the computation, parameters on the validity of individual ozone retrievals, and radiance information from SBUV/2 measurements.

C. Product User File (PUF)—contains a subset of the information contained on the Product Master File; this file is used to produce a reduced size ozone information dataset.

D. Historical Instrument File (HIF)—data files, created by the Instrument Support Subsystem, that characterize instrument performance and albedo correction over time.

VI-2.6.5 Distribution and Use of Ozone Products

The 1B datasets, PMF, PUF, and HIF are archived at the National Climatic Data Center (NCDC) in Washington, D. C. Except for the HIF, which is an internal dataset, they are available for distribution.

Ozone data from SBUV/2 are used by the CAC of the National Weather Service to monitor and document long-term trends of stratospheric ozone, and by the international scientific community for studying the photochemistry of ozone.

VI-2.7 PROCESSING IMAGERY FROM THE POLAR SATELLITE

VI-2.7.1 Introduction

Satellite imagery is processed geometrically and radiometrically. Geometric processing is the positioning of the raw images. An example is plotting the imagery on a predefined map array.

Radiometric processing is controlled changing of the intensity or count values of the imagery. Also, intensity values of the cloud images are often changed to highlight or suppress certain features depending upon the intended use. For example, display of imagery usually requires an alteration of image counts to fit the gray scale range of the display device.

VI-2.7.2 Global Mapped Data Bases

NOAA polar satellites record both VIS and IR images of the entire globe. Cloud images, displayed on polar stereographic or Mercator projections, are used to observe, analyze, and forecast meteorological events.

High-speed computers plot the satellite data onto standard map projections. These mapped data are distributed to users electronically in real time and archived in digital form.

A. Polar Stereographic Projection Mapping is done on a 1024 × 1024 square grid mesh for each hemisphere (Figure VI-2-6). The resolution is 14.8 km at the equator and 29.6 km at the poles.

B. Mercator Projection Mapping (Figure VI-2-7), NESDIS uses a coordinate system with 4050 grid points per row and 984 rows. Resolution between latitudinal grid points is 9.8 km at the Equator and 7.5 km at the 40°N and 40°S limits.

VI-2.7.3 Mapping Process

A. Preprocessing of the Imagery. Before mapping, the data are corrected for effects caused by the sensor, the atmosphere, and the Sun. IR image data are processed through a limb-darkening algorithm to correct for the atmospheric attenuation effects. VIS image data are processed through a Sun normalizing algorithm to correct for varying effects of reflected sunlight at different angles and times of the day.

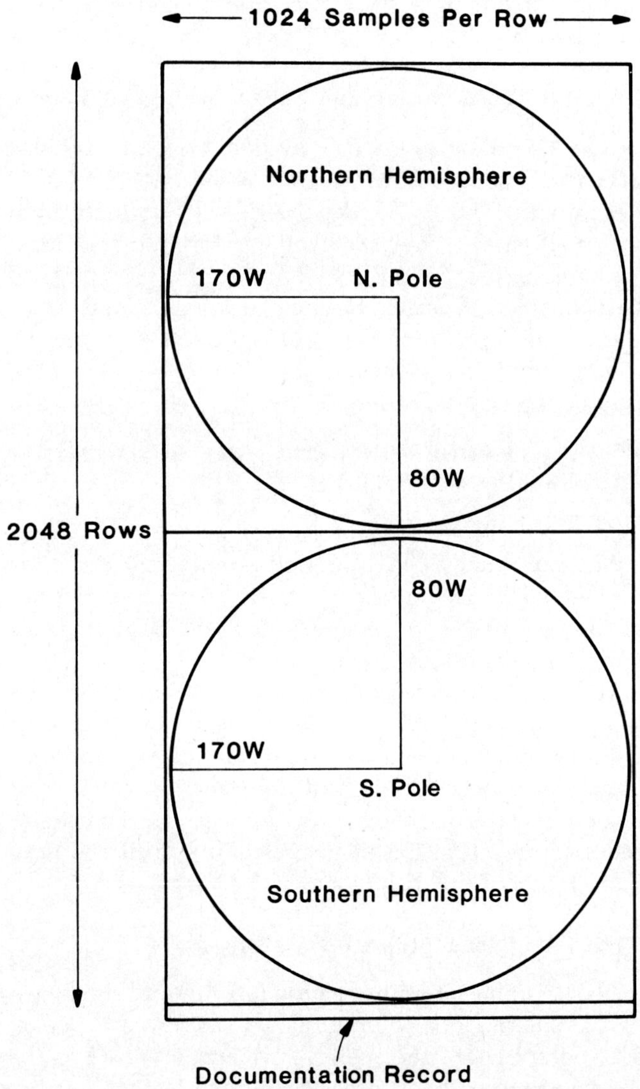

Figure VI-2-6. Polar stereographic projection map files.

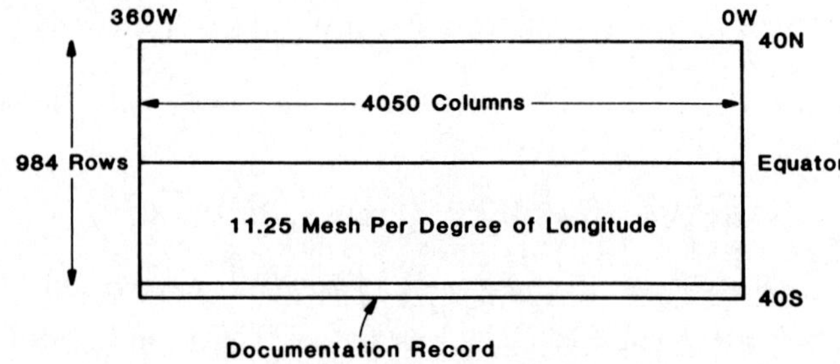

Figure VI-2-7. Mercator projection map files.

B. Generation of Mapped Mosaics. Latitude and longitude values are assigned to the raw image data at every eighth sample along each scan of data. These Earth location pairs are transformed at NESDIS into I, J map coordinates. The raw image data at each I, J benchmark are presented in a column, scan (X, Y) array. Interpolation can then be used to plot data values for intermediate points.

C. Output Products. To deliver mapped data quickly, partial sectors of the mosaics are extracted and sent to users by telephone facsimile circuits or by WEFAX broadcasts. At the end of each processing day, the global composite mapped databases (Figures VI-2-8 and VI-2-9) are archived in digital form.

VI-2.7.4 Stretched-Gridded Imagery

Image strips of polar satellite data, produced one pass at a time, are stretched to eliminate most of the distortion caused by the Earth's curvature and the scanning geometry of the radiometer. Latitude, longitude, and political boundaries are also superimposed on the image. This stretched-gridded product is produced as follows:

A. The Predicted Grid Dataset. A predicted Earth-locator grid containing scan line column coordinates is produced using orbital elements and other spacecraft parameters.

B. Output Product. The stretched-gridded product (Figure VI-2-10) is produced in photographic form or in facsimile format for the users. This product is not archived in digital form.

VI-2.8 IMAGE SECTORS FROM GOES DATA

Full disk GOES data, VIS, 11.5 μm IR and 6.7 μm WV channels are processed by computer into sectors distributed to NOAA field offices and other users (Chapter VI-3).

The sectorizer computer in the CDDF extracts sec-

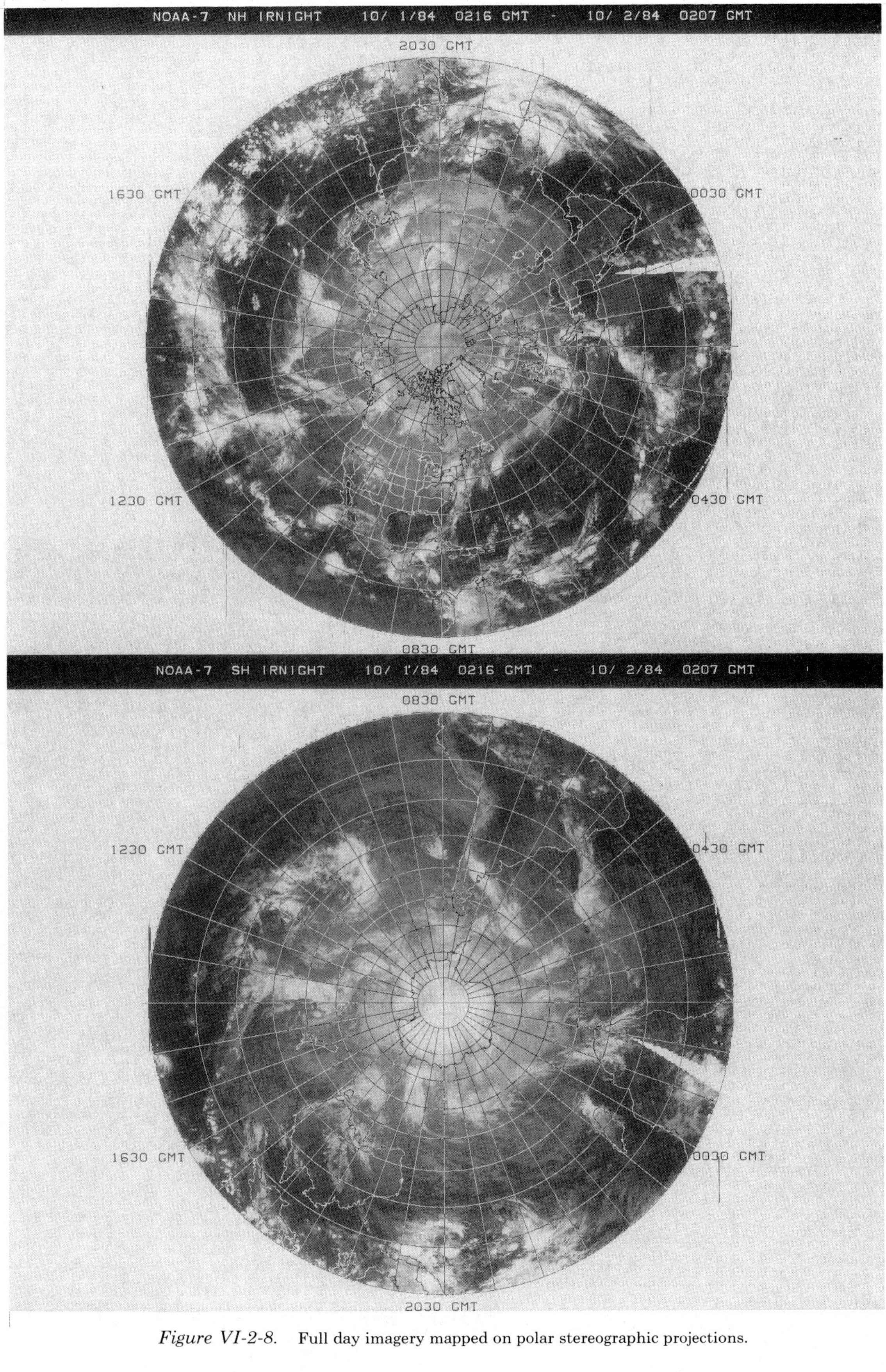

Figure VI-2-8. Full day imagery mapped on polar stereographic projections.

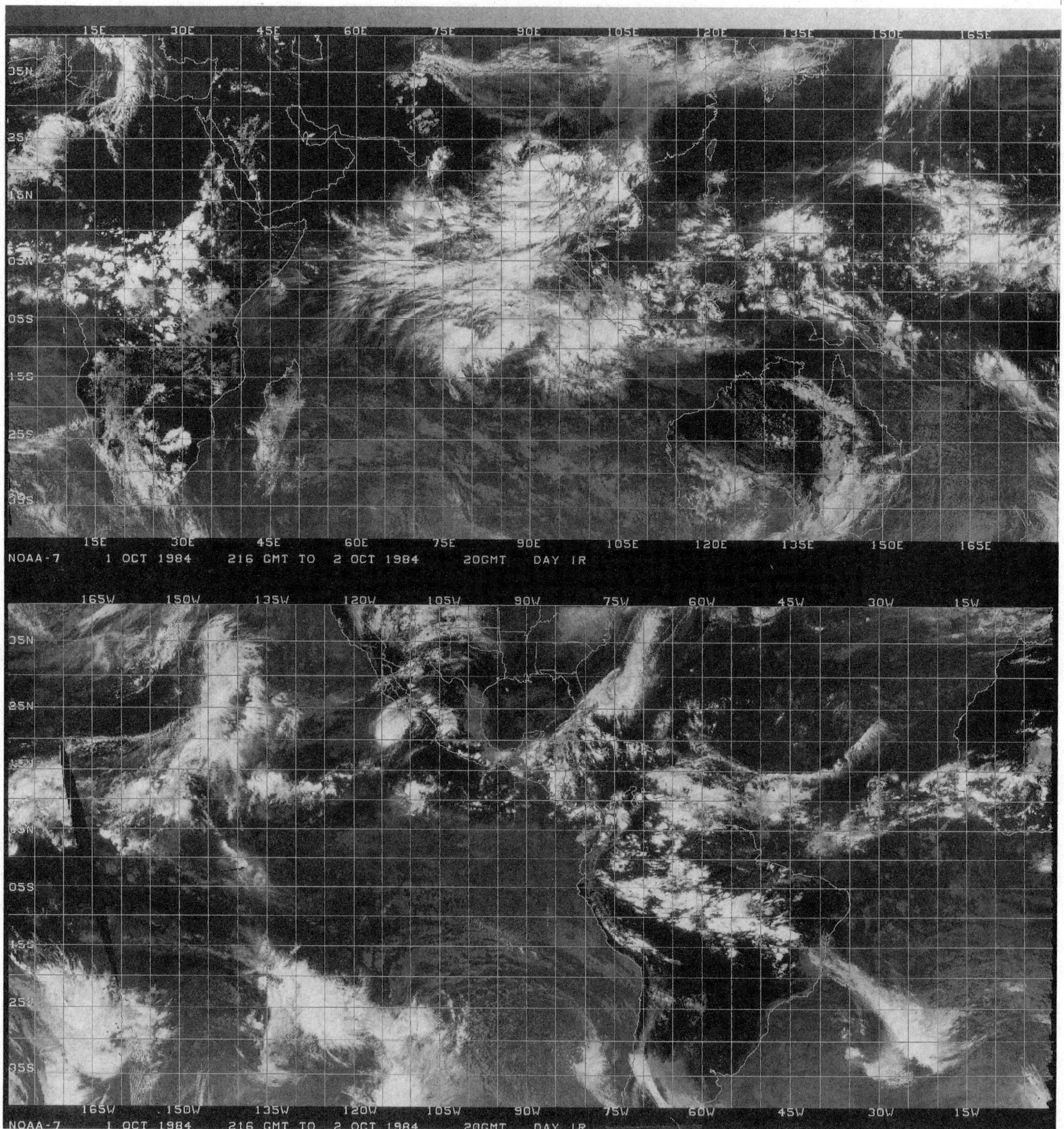

Figure VI-2-9. Full day imagery mapped on a Mercator projection.

tions of specified geographical areas with approximate resolutions of 1, 2, 4, or 8 km. A total of seven 1-km resolution sectors and/or a total of four 2-km resolution sectors cover the contiguous 48 states; two standard 4-km resolution sectors cover the U.S., Atlantic, and Pacific from 50°W to about 170°W. These sectors are available every 30 min. An additional 1-, 2-, or 4-km resolution sector is transmitted to the Weather Service Forecast Office in Puerto Rico every 30 min. Similar resolution sectors are also available for the

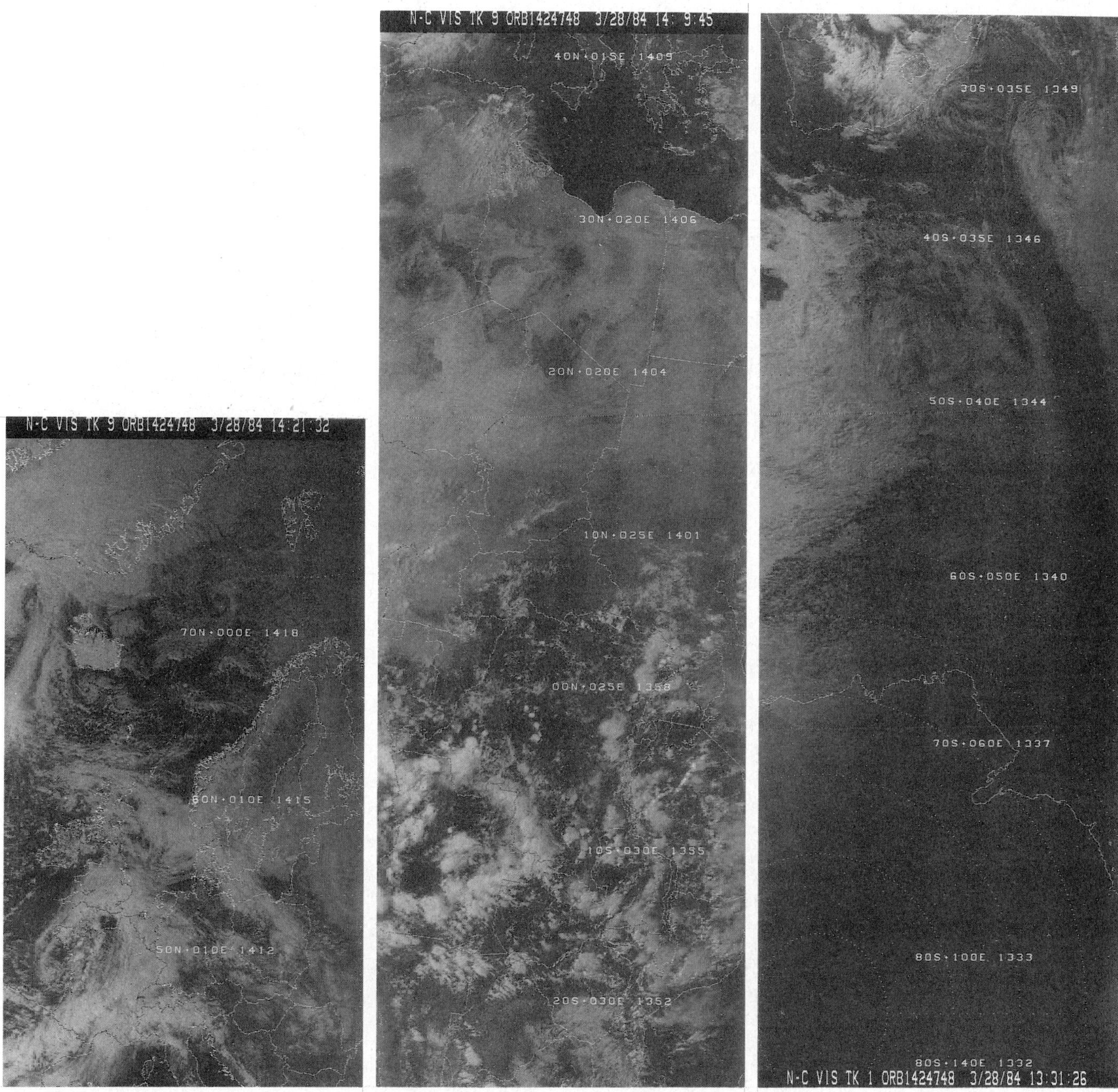

Figure VI-2-10. Stretched-gridded imagery.

Alaskan and Hawaiian areas. All standard sectors have automatically implanted grids.

The sectorizer operates in a serial fashion, i.e., it does not have the capability to receive and transmit data simultaneously. This causes data transmission from the CDDF to be on a near-real time basis. The sectorizer uses about 9 min to receive these data and about 17 min more for handling and transmission of selected sectors. This is completed well within the normally scheduled 30-min spacecraft imaging interval. Partial disk picture operation and limited scanning for short-interval viewing are used to monitor severe weather. This is called Rapid Interval Scan Operations (RISOP). Numerous sectorizers are located within the CDDF to produce sectors required by users. Sectorizers are controlled by a central console and automated scheduler. The CDDF has the capability of temperature enhancement for emphasis of specified features in the IR imagery. These enhanced images are available to all users through the normal distribution system.

Chapter VI-3

SATELLITE DATA DISTRIBUTION

VI-3.1 INTRODUCTION

After more than 25 years of operating meteorological satellites, the United States now maintains a ground data handling and distribution system that has incorporated many technologies developed during the period. Because this system represents a considerable investment, future upgrading and replacements must not only be carefully planned, but accomplished with minimum disruption to ongoing services.

VI-3.2 USE OF THE CONVENTIONAL TELEPHONE SYSTEM FOR DATA COMMUNICATIONS

The principal data distribution system currently in use is a network of voice-grade commercial telephone circuits for analogue facsimile transmission. At the time the network was started, after the launch of the first geosynchronous meteorological satellite in the early 1970s, system planners had to consider an immediate order-of-magnitude increase in data volume. Two major system problems in the early 1970s led to the data distribution system NOAA has in place today. These were: limited availability of wideband communications facilities and excessive costs of data-storage systems. Also, the market for meteorological satellite data was very uncertain at that time. This hampered efforts to justify a large-scale development program.

VI-3.2.1 Hierarchical Facsimile Broadcast

The system in place today is hierarchical. The data are acquired centrally and are partitioned into segments compatible with voice quality line capacity. The partitioned data, encoded into an analogue facsimile format, are distributed to users through six regional hubs. By use of this dissemination scheme, the NESDIS GOES-Tap Program distributes imagery to users linked to any of the regional hubs.

At the Central Data Distribution Facility (CDDF), the full Earth view (full disk) digital GOES data are processed for GOES-Tap distribution. The full disk is sectorized into geographical areas, image features are enhanced, and Earth location grids are added. Time, date, and image scale data are written onto each image caption, and each "sector" is then converted to an analogue facsimile signal. Combinations of these sectors are then distributed to the six regional hubs, or Satellite Field Distribution Facilities (SFDF). At the distribution facilities, the data are fed into touchtone-activated distribution equipment, to which users have access.

Private users can be connected to the touchtone-activated equipment of the GOES-Tap Program system for an initial fee (to cover equipment costs); an annual fee is also charged to cover equipment maintenance and circuit testing costs. Each user provides a private telephone line and touchtone transmitter. This transmitter sends control signals to the SFDF equipment, where specific sector data are transmitted in accordance with prearranged schedules (Clark, 1983). Users are also responsible for their own display devices.

A secondary user may, with the permission of a primary GOES-Tap user, have a receive-only extension. By using this method, the secondary user avoids the fees but is restricted to receiving only data selected by the primary GOES-Tap user. These programs have been operating since 1974. Currently there are more than 300 GOES-Tap customers and more than 500 secondary users.

VI-3.2.2 Facsimile Encoding

The facsimile encoding device used by NESDIS is controlled by a sectorizer computer. The sectorizer accumulates the real-time digital GOES data on digital tapes. The sector parameters (VIS or IR image channel, resolution, image scale, sector location, etc.) are predetermined. All data editing is done before being written onto tape. After the data for a complete sector are accumulated, the data are transmitted to the field distribution facilities.

All sectorizer signals are routed to a computer-driven communications switching and buffering system in the CDDF which controls all facsimile traffic to field distribution facilities.

VI-3.2.3 Data Dissemination

The voiceband private telephone lines used to distribute the imagery were chosen because of their widespread availability and relatively low cost (per cycle of bandwidth). All telephone line communication of satellite facsimile data is over a series of voiceband facility segments connected end-to-end. These private line segments are hard-wired together permanently, rather than temporarily switch-connected as an ordinary telephone line. While the different private line segments may have somewhat different qualities and characteristics, the overall end-to-end characteristics and qualities of each private hookup remain constant. This constancy permits normalization of frequency response and delay distortion by the use of conditioning units (hardware devices provided by the communications carriers that are installed at each end of the private line). By testing the private line from end-to-end when it is installed, characteristics can be determined and deficiencies compensated for by adjusting the conditioning units. The private lines that comprise the "trunk" network between the CDDF and the field-distribution stations are conditioned to more rigid specifications than most GOES-Tap private lines.

VI-3.2.4 Facsimile Receivers and Recorders

Most end users have facsimile receivers or recorders. The simplest types use an electrostatic stylus to charge a special paper, scan line by scan line. The paper is processed by passing it through a solution of suspended minute magnetic particles which become affixed to the surface; heat drying completes the process. Other devices use light emitting components to expose photographic paper or film.

These hard copy products can be interpreted directly or used as image input for more sophisticated devices. One such device currently in use scans the images with a TV camera and stores them on disk or solid-state memory. When the stored images are replayed in a time-lapse sequence, they produce animation of the features of the images. Images can also be computer-enhanced in color or in black and white to accentuate features of special interest. Some computer based display systems bypass the hard copy process and digitize the facsimile signal directly for image storage or processing.

VI-3.2.5 Summary

This distribution system has provided satellite imagery to a large number of users very effectively and at minimal cost for more than a decade. However, there are several shortcomings. Conversion of the data to analogue form and phone line impairments degrade the quality of the data to less than that of the original digital data.

VI-3.3 DIGITAL COMMUNICATIONS

From the beginning of the GOES-Tap network, all satellite products have been produced, transmitted, and processed as analogue facsimile products. Analogue transmissions were cost effective and the hard copy satellite images became vital to the meteorological community. Recent advances in telecommunication equipment, microprocessor based systems, and the quantitative use of satellite data, have made digital data affordable and more functional.

The enormous technological advances in microcomputer and mass storage devices have now made the storage and computational equipment to process digital satellite data affordable for users. Microcomputer (and minicomputer) systems, with mass digital storage, permit use of digital satellite imagery for all the same purposes as hard copy images. These uses include single image analysis, animation to determine cloud motion, and long-term archival. In addition, digital data can be processed quantitatively.

VI-3.3.1 Database Access

Digital meteorological data are of optimum value only if readily accessible. Near-real time access to remotely sensed data has traditionally been available only to users close to the mainframe computer that receives the data. In recent years, however, there has been increased remote user access via remote terminals.

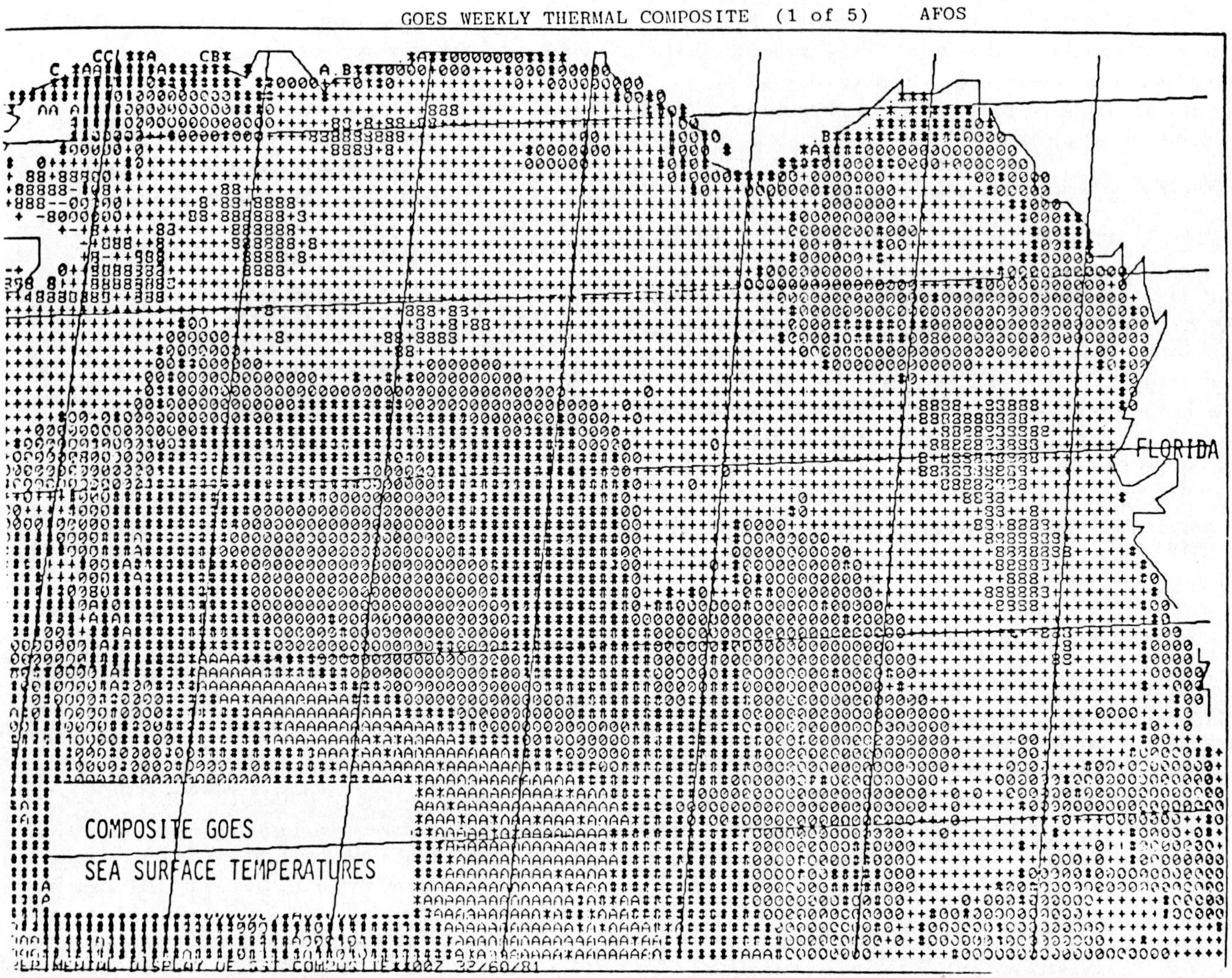

Figure VI-3-1. Character print map of composited satellite IR data as it would appear on a remote printer.

CHARACTER	VISSR IR COUNT VALUES	TEMPERATURE °C
C	61–62	26.3–25.8
·	63–64	25.3–24.8
B	65–66	24.3–23.8
*	67–68	23.3–22.8
A	69–70	22.3–21.8
#	71–72	21.3–20.8
0	73–74	20.3–19.8
+	75–76	19.3–18.8
8	77–78	18.3–17.8
–	79–80	17.3–16.8
6	81–82	16.3–15.8
=	83–84	15.3–14.8
4	85–86	14.3–13.8

Terminals can be connected to the mainframe computer via a remote terminal controller which is connected to the mainframe by modem and a dedicated telephone or microwave link (data transfer rates range from 48 to 56 kbps). Printout can be routed to a remote printer connected to the same terminal controller.

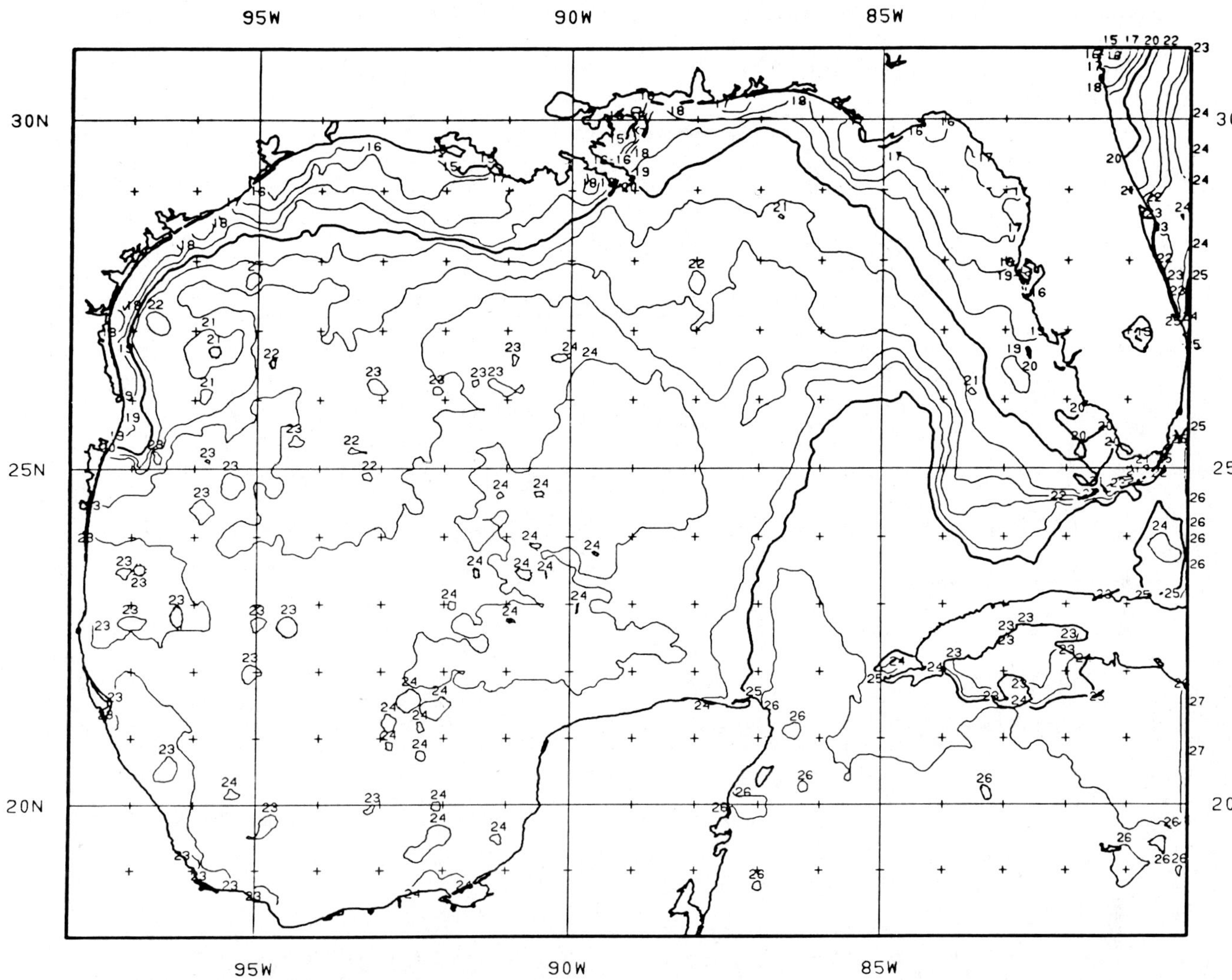

Figure VI-3-2. Sea surface temperature (SST) contour chart generated remotely by a graphics display device.

Terminals can also be connected to the mainframe computer via non-dedicated dial-up telephone lines (data transfer rates are generally 0.3 to 1.2 kbps).

In the NESDIS system, any operational dataset is accessible by remote terminals. Raw and processed satellite data, image composites, and grid point analyses are all sources for remote terminal display. Output forms include:

- Listings of observations within selected geographic regions or time intervals
- Plotted raw data values
- Character print maps or composites
- Plotter/photographic contour or gray scale charts

Figure VI-3-1 shows an example of a character print map of composited satellite IR data as it would appear on a remote printer. An example of graphic output that can be generated by a remote graphics display device is shown in Figure VI-3-2. Although most functions available to the local user can be duplicated remotely, remote terminals do have the following shortcomings:

- Since the communications rate is low and satellite data volume is high, remote site output is very limited.
- Tape handling, delivery of long printouts, and use of special output devices require special arrangements with the mainframe computer site.
- Diagnosing problems is difficult for remotely located personnel.

With an interactive Time Sharing Option (TSO) terminal, a user can initiate a program or procedure, interact with it to alter the selection of data or the type

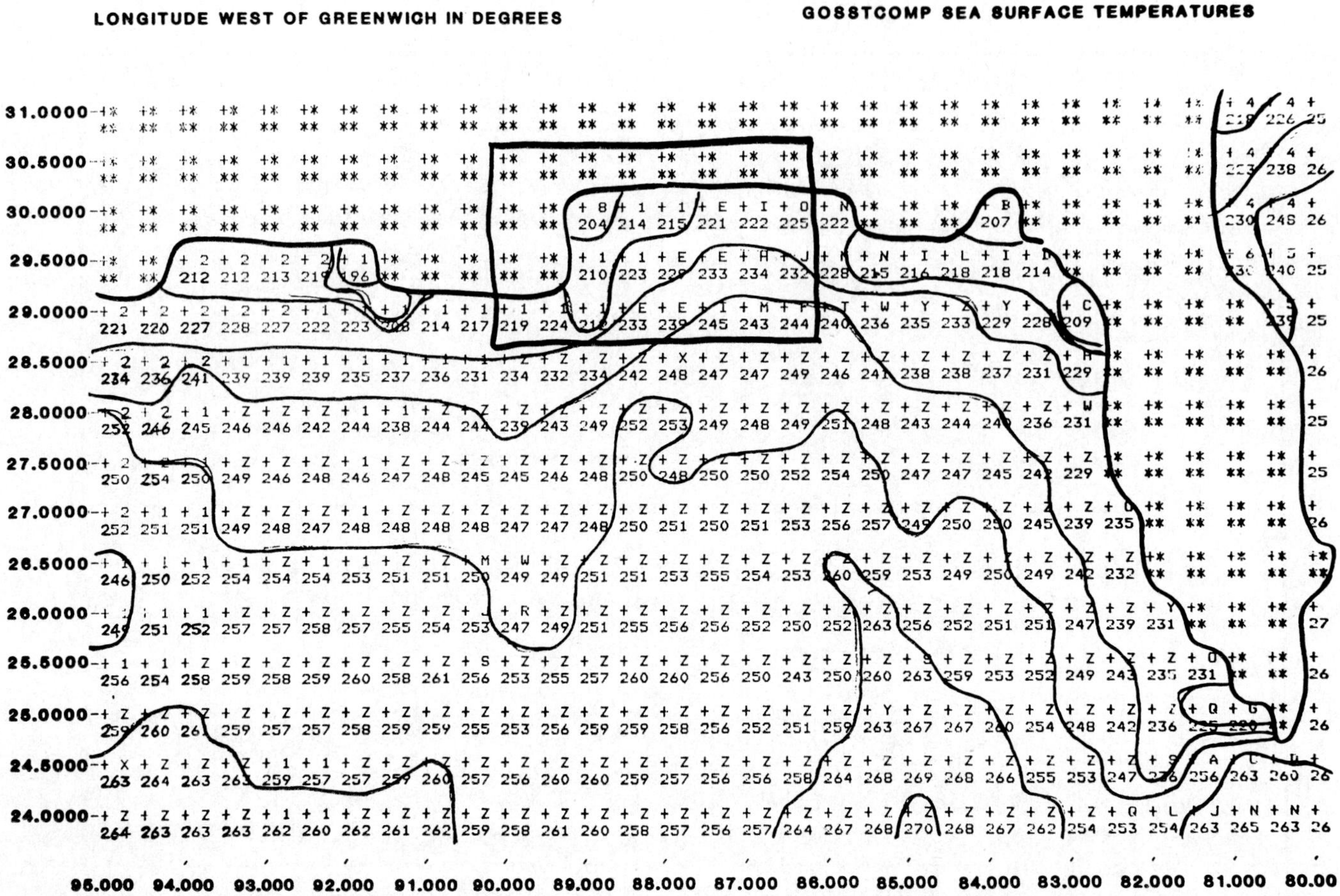

Figure VI-3-3. Sea surface temperature (SST) grid point map display produced by an interactive terminal and a remote printer.

of operations performed, and receive the final output within a few minutes.

With a TSO terminal, a user can be given a list of options to select such things as a particular geographic region, grid point resolution, and time period. The chosen data are then retrieved and displayed at the user's terminal. Time consuming jobs, such as those with large amounts of printout, can be submitted through the terminal for batch execution. Figure VI-3-3 shows an interactively processed grid point map display. Such a reduced resolution display can be used for an overview of a large region. Full resolution displays can later be used for detailed examination.

The number of TSO terminals that can be established depends on the amount of mainframe computer resources available. Their low cost and portability make TSO terminals attractive for use at remote sites. The latest innovation in interactive terminal on-line access is the availability of microcomputers, which can be used in a fashion similar to a TSO terminal or for more sophisticated graphics and image display.

VI-3.4 FUTURE COMMUNICATIONS

Current technology and telecommunications facilities are very different from those of the early 1970s. Common carrier networks have expanded greatly. An oversupply of communication satellite transponders is predicted and plans have been made to install wideband fiber-optic links between most major cities in the United States. New technology has made available affordable high performance computers, low cost mass memories, inexpensive satellite terminals, and small image processors for satellite data users. Meteorological techniques also have become more sophisticated and effective in recent years.

Satellite data no longer are a novelty. They are now a primary and, in many cases, an essential tool in weather forecasting and environmental analyses. Many users seek higher resolution, quantitative measurements, more timely delivery, many more kinds of data, and effective means of comparing and evaluating products from different sources.

Many efforts are underway to satisfy the newer needs of satellite data users. A configuration being considered to satisfy these needs is the adaptation of a Direct Broadcast Satellite (DBS). Such a satellite is now in use to provide audio to radio stations throughout the United States. A DBS system could offer satellite data users a complete ground terminal through an antenna interface, and a mix of analogue and digital products. It would cost less than $10,000 (versus the $75,000 cost for a single stretched VISSR terminal). Some 2000 DBS ground terminals are now in use for radio network program distribution. The system is digital, with a net transmission rate of 7.68 Mbps. Equipment for digital transmission of two GOES data streams and 24 3-kHz analogue channels could be developed.

Considering all the increases in amounts and kinds of satellite data planned for the next decade or two, a distribution system capable of providing timely delivery of this information will be paramount to the success of future meteorological satellite programs.

Chapter VI-4

SATELLITE DATA ARCHIVING: DATA MANAGEMENT AND USER SERVICES[1]

VI-4.1 INTRODUCTION

The functions of data archiving and management range from the establishment of policies governing data acquisition, retention, and access, to the functions of data reception, quality control, cataloging, preservation of data on archive media, and a variety of services provided to near-real time and retrospective users (as opposed to real time, operational users) of archived data and information. These functions are evolving to keep pace with:

- New user requirements for services, including improved access to data catalogs, growing quantities of digital data, and improved quality control of data.
- New storage, data processing, and communications technology.
- Implementation of new satellite systems and instruments, which leads to new data and products.

The paramount objective of any data archive program is to preserve data and to maximize their availability and usefulness to the broadest possible community of users. Meeting this objective requires that data be acquired, preserved, organized, and cataloged. Also, users must be provided with ready access to archived data, derived products, and documentation of data formats, processing algorithms, instrument histories, and other information.

[1] Primary contributor: Gregory Hunolt

The responsibility for archiving satellite data within NOAA resides with the Satellite Data Services Division (SDSD) of the NESDIS National Climatic Data Center (NCDC). In contrast to its parent organization, NCDC, and the other two NESDIS data centers (the National Oceanographic Data Center and the National Geophysical Data Center), the SDSD satellite data base is not oriented to any one discipline. SDSD serves users whose interests are almost evenly divided among meteorology, oceanography, and land sciences.

VI-4.2 USER REQUIREMENTS FOR SERVICES

The concept of the archive user as a person content to wait a year for a piece of paper to peruse is rapidly giving way to the concept of the computer-equipped user desiring rapid, even on-line access to digital data, both in near-real time and retrospectively. Nevertheless, the archive-user community remains a very heterogeneous group. The archive center must be responsive to the needs of the small user as well as the large, or "high-tech," user. The satellite-user community includes private citizens, attorneys, individual university researchers, large research organizations, small businesses, large businesses, and federal, state, local, and foreign government agencies. Table VI-4-1 categorizes several types of users of the NESDIS satellite database. Table VI-4-2 is a companion table illustrating the relative demand by users for the different types of data available from SDSD.

Table VI-4-1 *SDSD data requests by user type (FY1986)*

USER TYPE	DIGITAL DATA ORDERS	DIGITAL DATA PCT./TOTAL	DIGITAL DATA TAPES	HARD COPY IMAGERY ORDERS	HARD COPY IMAGERY PCT./TOTAL	HARD COPY IMAGERY PHOTO IMAGES
1. U.S. Government	668	67.2	2359	666	46.4	10045
2. University Research	107	10.8	307	194	13.5	2237
3. Private Research	24	2.4	85	28	1.9	283
4. Private Industry	61	6.1	140	275	19.2	1529
5. Miscellaneous U.S.	5	0.5	7	110	7.7	362
6. Foreign Government	22	2.2	63	20	1.4	735
7. Foreign Private	107	10.8	368	143	10.0	1136

The research-user community is also very diverse in the scientific disciplines represented. Although the archived data have been collected largely by meteorological satellites, only 40% of the applications of the data have been meteorological. Another 40% have been oceanographic and the remaining 20% distributed between land sciences, especially agricultural monitoring, and miscellaneous uses. Currently, land applications are growing most rapidly.

User requirements for products as measured by data and products provided by SDSD have evolved significantly over the past decade. Requirements for data in digital form have grown dramatically as the rapid evolution of data processing and display technology has reduced the cost to users of implementing systems capable of supporting sophisticated analyses of digital data. At the same time, the newest instruments (such as AVHRR and VAS) have been engineered to support quantitative applications. These instruments can be calibrated precisely, they possess very high signal-to-noise ratios (in the case of VAS they are operated in a special dwell sounding mode that ensures the quality of the data), and their data are quantified to sufficient precision (10 bits or more) for detailed analysis. Table VI-4-3 shows the changes in the amounts of photographic and digital products provided by SDSD between 1978 and 1987. The dramatic increase in demand for digital data is evident, as is the recent decline in demand for image products. This is due at least in part to the switchover from photographic negatives to UNIFAX prints for operational NOAA imagery.

VI-4.3 SATELLITE DATA MANAGEMENT OPERATIONS MODEL

This section presents a generic data management operations model as a frame of reference for subsequent description of the structure and content of a satellite data management plan. The scope of this discussion is limited to a model of the operations of a non-real-time, retrospective data management system such as SDSD. A comprehensive data management operations model would also include the real-time operational data management functions, such as those of the NESDIS OSDPD described in Chapters VI-1, VI-2, and VI-3.

VI-4.3.1 Elements of the Data Management Operations Model

The elements of the data management operations model trace the operational process from the sources of data to the products provided by the data management system to its users. In general, a source of data is normally a sensor or instrument measuring one or more parameters, or a combination of instruments

Table VI-4-2 *SDSD user requests by data type*

DIGITAL DATA TYPE	DIGITAL ITEMS REQUESTED	DIGITAL PCT. OF TOTAL	IMAGERY DATA TYPE	IMAGERY ITEMS REQUESTED	IMAGERY PCT. OF TOTAL
1. AVHRR	3194	80.5%	1. VISSR	1763	48.2%
2. TOVS	340	8.6%	2. AVHRR	1288	35.2%
3. CZCS	129	3.3%	3. CZCS	272	7.4%
4. SAR	68	1.7%	4. SAR	168	4.6%
5. VISSR	45	1.1%	5. VHRR	51	1.4%
All Other	183	4.6%	All Other	119	3.3%

(For orders handled October 1, 1985 through April 15, 1987.)

Table VI-4-3 *Trends in SDSD data requests—digital vs. image*

FISCAL YEAR	DIGITAL TAPES	PHOTO PRODUCTS
1978	654	31,638
1979	1,349	28,785
1980	1,624	22,154
1981	3,070	25,989
1982	3,163	26,758
1983	4,089	22,957
1984	5,010	21,730
1985	4,613	25,654
1986	6,270	13,368
1987	6,966	6,064

making measurements together, perhaps supplemented by human observations. In the case of satellite data, a source is an instrument making measurements of emitted or reflected radiances from which environmental parameters can be derived. The output from an individual source, such as an AVHRR instrument, generally consists of time and/or location tagged sets of data values accompanied by auxiliary information. Each such set of parameters is called an "observation." Observations may be received by the archive in digital or non-digital form. The flow of such observations from a source constitutes a "data stream." Data streams from multiple sources of the same type, i.e., members of the same source class, are aggregated into an overall data stream for the class of sources. A "platform" is what the source "rides" on, e.g., a satellite or ship. Platforms (such as the NOAA series of polar orbiting satellites) may bear multiple instruments, and hence may be associated with multiple data streams. Identification of the platform type and its characteristics is important, but each data stream is defined by its source. For example, AVHRR data are defined by the instrument, not by the particular satellite, such as NOAA-9. For fixed platforms, such as measurements from surface weather stations or seismic observatories, the platform may be the station or observatory.

The archive database built and maintained by the retrospective data management system (e.g., by SDSD for environmental satellite data) is referred to as the "primary" database. The primary database, built from incoming data streams, consists of a set of datasets, each containing a type of data produced from a data stream. Each dataset contains digital or non-digital data of the same type with common content and format, such as AVHRR Level 1B GAC, generated from data received over an interval of time. The data in each dataset consist of a series of granules or packets of digital or non-digital data stored as separately indexed elements. The database generally includes three levels of storage:

- On-line or active—data on media mounted for immediate access to take care of most rapid turnaround of on-line or off-line requests.
- Secondary on-line—data on media mounted in a device not ready for immediate access, but accessible without manual intervention in tens of seconds to a minute.
- Off-line—bulk storage of unmounted media, digital and non-digital.

Migration of data between levels of storage must take into account user requirements for access to the data. For example a subset of the overall database requirements for immediate on-line access by some users exists. An ideal system would offer this capability since the requirements for it will gradually grow over coming years as more and more users acquire data processing and communications capabilities.

The contents of the database are described in a catalog containing three levels of descriptive information. Level 1 directory information is a brief narrative description of each dataset. Level 2 catalog information is a detailed description of each dataset and products available from the dataset, including generic information about the granules making up the dataset. Level 3 inventory information describes each individual granule within each dataset, such as date, time, specific location or coverage.

An archival product is anything, digital or non-digital, produced from a granule or combination of granules for on-line or off-line delivery to a user. Examples of such products are:

- Full or selective copies of a digital granule delivered on magnetic tape, floppy disk, or WORM optical disk.
- Photographic prints produced from negatives or from digital data.
- Photocopies of paper charts.

The terminology used in this chapter conforms to the data systems lexicon being developed by NOAA, NASA, and USGS with international participation. The current definitions of dataset, data granule, and the three levels of catalog information are contained in the glossary.

VI-4.3.2 Data Management Operational Processes

There are three basic processes (Figure VI-4-1) that occur within a data management operation such as that of SDSD: ingest, data set maintenance, and archival product generation.

In the ingest process an incoming digital or non-digital data stream is converted into a series of granules to be added to one or more datasets resident in the primary database. The ingest process includes quality

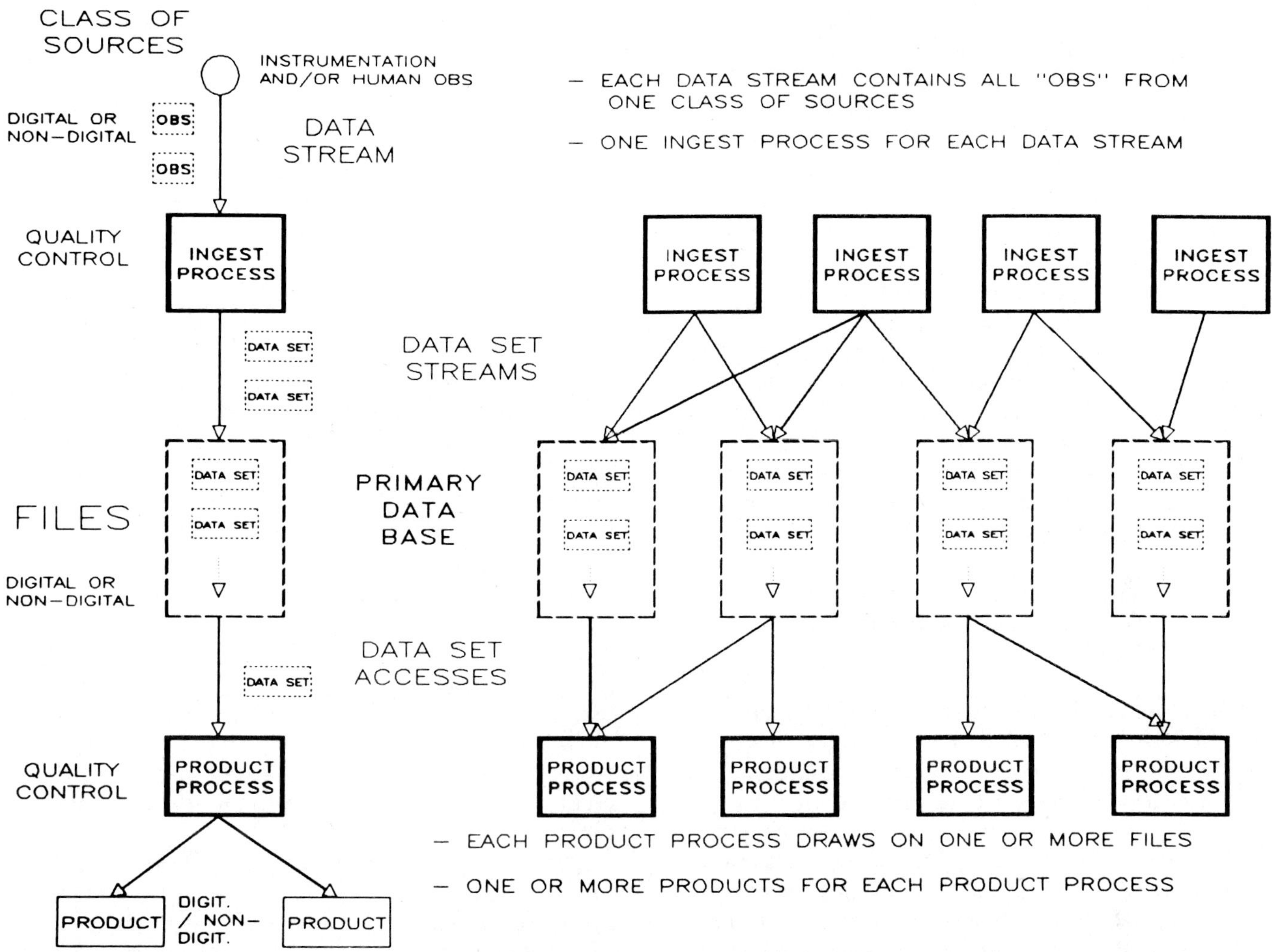

Figure VI-4-1. Data management operations model.

control, validation, reformatting, blocking the data into granules, and generating the corresponding inventory entry for each granule. There is an ingest process associated with each data stream or flow of observations from all members of a given class of sources. Each ingest process may produce granules to be added to more than one dataset. For example, three datasets might be built from a given data stream. One might be organized by geographic areas, another as time series, and the third might contain summary information such as averaged values and derived statistics. Each dataset receives its separate flow of granules from the ingest process. An ingest process may produce non-digital granules from a digital data stream. Non-digital products may include publications of tabulated data values or analyzed charts that are to be cataloged and held available for users in the primary database.

Dataset maintenance includes a variety of processes that manage the primary and active databases, including staging of granules from the primary to the active database, migration of granules between storage levels within the primary database, or planned purging of granules from the primary database. Some staging of granules to the active data base is done on a routine, scheduled basis according to provisions of the data management plan; other staging is done in response to ad hoc user requests.

In the product generation process, digital or nondigital granules from a given dataset are accessed and products are generated, quality controlled, and delivered to users. These archival products may be copies of granules, selective extracts from granules, or statistical summaries generated from groups of granules. A product generation process is associated with each dataset on the primary database. Any given product generation process may draw on granules from more than one dataset.

The left side of Figure VI-4-1 illustrates a simple, linear case; the right side of the diagram depicts the complexity that occurs in practice. Sources provide data in the form of observations, with the flow of observations from a given source or class of sources constituting a data stream. Data streams are handled by ingest processes, which feed one or more primary database datasets with new granules. The datasets are maintained and held accessible on the primary database and, as appropriate, the active database. In response to user requests, product processes access the database, read granules from one or more datasets, and generate the products to be delivered to the data user.

Figure VI-4-2 elaborates on this process to show the dataset maintenance process, handling staging of selected granules from selected datasets to the active database, and disposal of selected granules. Figure VI-4-2 also illustrates the three level catalog, receiving updates as data are ingested or as datasets are maintained. The data management plan, which governs the conduct of the processes shown, is also highlighted.

VI-4.4 SATELLITE DATA AND PRODUCTS FROM A DATA MANAGEMENT PERSPECTIVE

The data management functions of data ingest, dataset maintenance, and archival product generation must be considered in the context of the nature of satellite data and derived products that comprise the primary database.

The characteristics of satellite borne instruments, their data and products, and their applications are discussed extensively in other chapters. The description here is intended to present a data management perspective, focusing on those characteristics of satellite data and derived products that are critical for data management planning.

VI-4.4.1 Characteristics of Satellite Data and Derived Geophysical Products

In general, environmental satellite instruments are sensitive to electromagnetic radiation reflected or emitted from the Earth—either land, sea, clouds, or combinations thereof. The response of the instrument to the received energy is a time varying analogue signal that is converted to digital form as a series of arbitrarily scaled values. These values are the basic raw data, observations from which geophysical products are derived.

While data are defined in terms of their source instruments, geophysical products are defined in terms of their applications by users, e.g., sea surface temperature for oceanographers or vertical profiles of atmospheric temperature or humidity for meteorologists. Raw data from the source instruments are subjected to calibration and correction. These adjusted data are then used individually or in combination with data from other channels of the same or other instruments to produce geophysical parameters through use of physically and/or statistically based algorithms. These algorithms are subject to continuing research and refinement, especially in the case of newer and more complex instruments. This whole process is essentially irreversible. Thus, there is a data management issue of determining what combination of raw data, adjusted data, and final geophysical products must be preserved to meet user needs.

The processes described above should include quality control and validation. Quality control involves internal checks on data quality, such as calibration range checks or transmission error detection, while validation involves comparison of the data or derived parameters with an independent measurement. Quality control and validation may be performed operationally by data providers or by the archive during or after ingest.

The raw data, adjusted data, and derived geophysical products are all candidates to be added to the archive's primary database. From the point of view of the data management system, an incoming data stream might include raw data from the instrument more or less directly, adjusted data, and derived geophysical parameters produced by an operational data processing center.

The nature of satellite data is very different than, for example, the nature of Social Security or financial information. The latter involve a large number of discrete types of information which have unique and usually unrelated values grouped into a variety of types of records. Such data are extremely heterogeneous, and each item of information is of individual importance. Thus, the level at which such a database is managed is on the level of records and their constituent elements.

Satellite data, on the other hand, are relatively homogeneous and interdependent. For example, one scan of AVHRR HRPT data includes 2048 groupings of five channel data values. The data values are related in that neighboring values do not differ greatly as the instrument scans a gradually changing field of view. Also, the responses of multiple channels are not completely independent as they view identical areas. Records of satellite data usually include the data values for each scan line of the instrument's field of view. These records are aggregated into sets that may consist of all of the data from one complete orbit, one scan of the Earth disk, or some other arbitrary grouping. The level at which a satellite database is managed is thus at the level of the granule, a collection of internally homogeneous records rather than on the record and element level.

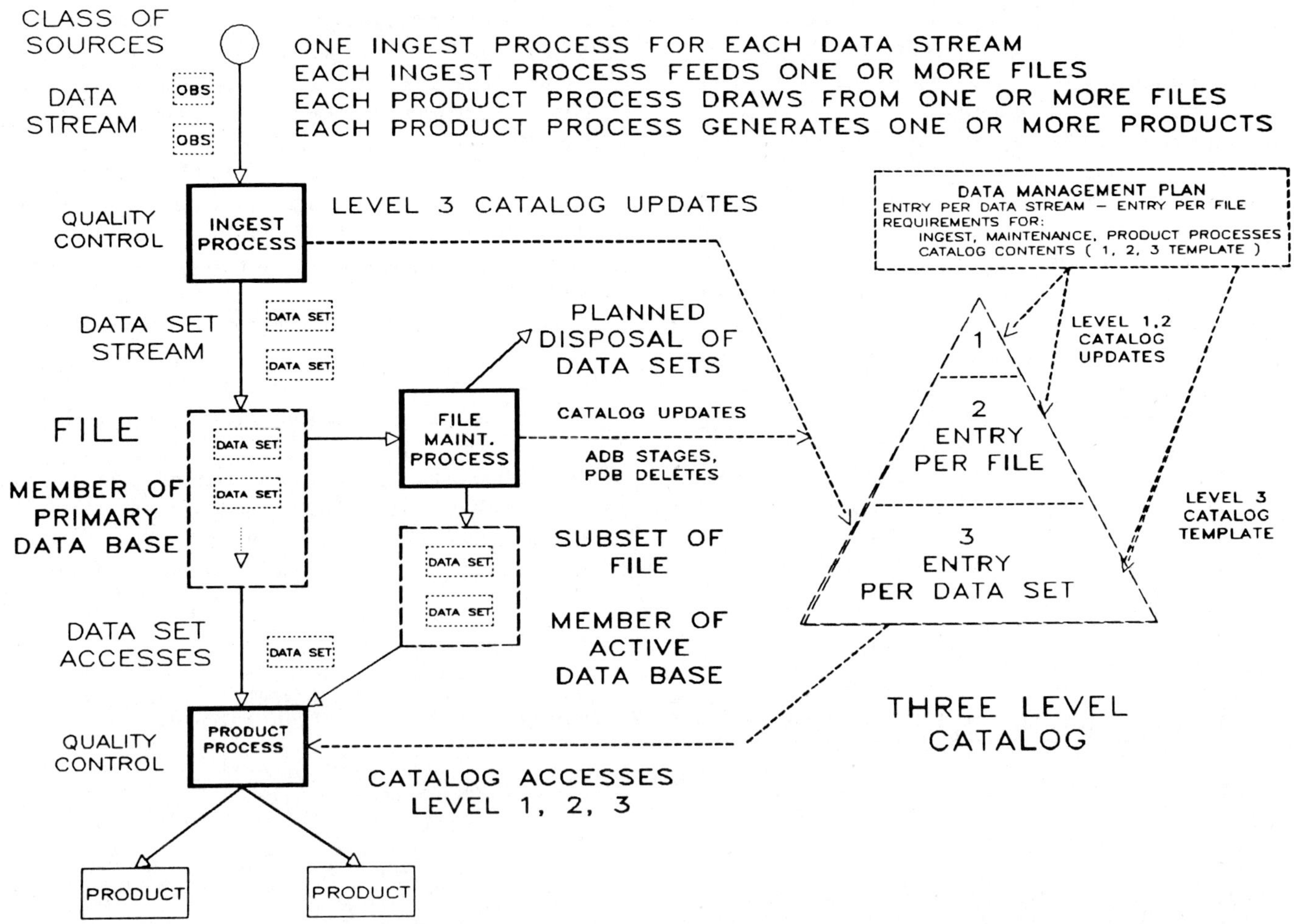

Figure VI-4-2. Data management operational processes.

The reliability with which data can be extracted from the archive media must be considered. The acceptable frequency of erroneous bits must be established, and potential archive systems measured against the standard set. After application of error detection and correction techniques, error rates of one bit error per 100 million bits read, or better, are typical. In the case of geophysical products derived from satellite data, the same standard must apply. Therefore, only high quality archive media have ever been used. In the case of raw or adjusted satellite data, however, the internal homogeneity of the data allows a lesser standard, error rates on the order of one bit error per million or one per 10 million, to be acceptable. This may make possible a considerable economy in designing and implementing an archive system, although the design must not overlook the requirement that it must be possible to extract data from the archive medium years later without degradation in the bit error rate.

As noted, quality control of the data entering the archive data management system is also of importance. Ideally, quality control measures should be built into the operational system that generates the data to be transferred to the archive. For example, SDSD, in conjunction with OSDPD, has built into the polar orbiter level 1B archive system a read-back quality check that allows bad archive tapes to be regenerated while the source data are resident on the operational, short-term, database.

VI-4.4.2 Volume of Satellite Data and Information

Data management planning must take into account the volume of raw data, adjusted data, and geophysical products that are candidates for inclusion in the archive database. The raw or adjusted data are aggregated into granules by the ground data handling system, and derived information is produced in the form of discrete geophysical products, often in a form which emulates conventionally observed data. The aggregation of the data and the sizes and frequencies of the granules thus become critical planning factors for the archive's primary database. The volume of derived products is typically at least two orders of magnitude less than the raw or preprocessed data volume. In the case of AVHRR data collected by the NOAA operational polar orbiters, the size of a GAC granule is 40 million bytes. This amounts to an average daily volume of 500 million bytes or an annual volume of 204 billion bytes. This contrasts with the sea surface temperature products produced from AVHRR data which amount to a volume of only a few million bytes per year.

Tables VI-4-4 and VI-4-5 list the environmental satellites and instrumentation from which SDSD has received and archived digital and non-digital data and derived geophysical products.

Table VI-4-6 lists the sizes and annual volumes of digital satellite granules and derived products that are currently resident in the SDSD satellite archive's primary database, or are to be added in the future.

Table VI-4-7 lists the current holdings and annual volume of satellite imagery received by SDSD. The film negatives of operational NOAA polar orbiter and GOES imagery were by-products of an operational NESDIS function. With the completion of a transition to direct facsimile production of hard copy prints, or to digitally generated video imagery, negatives are no longer produced. The SDSD archive of NOAA operational data has become all digital, with image products produced for users from digital data instead of by photographic reproduction from archive negatives. The primary digital archive is supplemented by a dataset of hard copy facsimile imagery for use as a browse data set.

VI-4.4.3 Summaries and Other Internal Products

The archive data management system must be capable of producing summaries and other special products. These may be specially sampled granules, such as the global scale AVHRR granule being produced by

Table VI-4-4 *Summary of polar orbiter data archived by SDSD*

SATELLITE NAME	LAUNCH DATE	DATES OF DATA ARCHIVED	FORM	INSTRUMENTS
TIROS-1	4/1/60	4/1/60–6/14/60	I	Vidicon
TIROS-2	11/23/60	11/23/60–9/27/61	I	Vidicon, IR Radiometer
TIROS-3	7/12/61	7/12/61–1/23/62	I	Vidicon
TIROS-4	2/8/62	2/8/62–6/18/62	I	Vidicon, IR Radiometer
TIROS-5	6/19/62	6/19/62–5/14/63	I	Vidicon
TIROS-6	9/18/62	9/18/62–10/21/63	I	Vidicon
TIROS-7	6/19/63	6/19/63–2/26/66	I	Vidicon, IR Radiometer
TIROS-8	12/21/63	12/21/63–2/12/66	I	Vidicon
TIROS-9	1/22/65	1/23/65–9/9/66	I	Vidicon
TIROS-10	7/2/65	7/2/65–4/2/66	I	Vidicon
ESSA-1	2/3/66	2/4/66–10/6/66	I	Advanced Vidicon Camera System (AVCS)
ESSA-3	10/2/66	10/4/66–6/1/67	B	AVCS, Low Resolution Radiometer (LRIR)
ESSA-5	4/20/67	6/1/67–12/3/68	B	AVCS, LRIR
ESSA-7	4/16/68	4/16/68–3/31/69	B	AVCS, LRIR
ESSA-9	2/26/69	4/1/69–11/15/72	B	AVCS, LRIR
ITOS-1	1/23/70	4/28/70–6/17/71	B	Scanning Radiometer (SR) (VIS and IR)
NOAA-1	12/11/70	4/26/71–6/20/71	B	SR
NOAA-2	10/15/72	11/16/72–3/19/74	B	SR, Very High Resol. Radiometer (VHRR), Vertical Temperature Profile Rad. (VTPR)
NOAA-3	11/6/73	3/26/74–12/17/74	B	SR, VHRR, VTPR
NOAA-4	11/15/74	12/17/74–9/15/76	B	SR, VHRR, VTPR
NOAA-5	7/29/76	9/15/76–3/16/78	B	SR, VHRR, VTPR
GEOS-3	4/9/75	4/14/75–12/1/78	D	Radar Altimeter
Seasat	6/27/78	7/7/78–10/9/78	B	SAR, SMMR, SCATT, ALT, VIRR
Nimbus-7	10/24/78	10/28/78–6/23/86	B	Coastal Zone Color Scanner (CZCS)
TIROS-N	10/13/78	10/30/78–11/1/80	B	Advanced VHRR (AVHRR), TOVS (HIRS, MSU, SSU)
NOAA-6	6/27/79	6/27/79–6/20/83	B	AVHRR, TOVS
		6/21/84–11/17/86	B	
NOAA-7	6/23/81	6/23/81–2/18/85	B	AVHRR, TOVS
NOAA-8	3/28/83	6/20/83–6/12/84	B	AVHRR, TOVS
NOAA-9	12/12/84	12/28/84–Present	B	AVHRR, TOVS
NOAA-10	9/17/86	10/21/86–Present	B	AVHRR, TOVS

Note: a. Form: I—Image, D—Digital, B—Both Image and Digital b. Does not include DMSP Film and Digital data.

Table VI-4-5 *Summary of geostationary data archived by SDSD*

SATELLITE NAME	LAUNCH DATE	DATES OF DATA ARCHIVED	FORM	INSTRUMENTS
ATS-1	12/6/66	1/1/67–10/16/72	I	Spin Scan Cloud Camera (SSCC)
ATS-3	11/5/67	3/2/68–9/2/74	I	SSCC & 2 color images
SMS-1	5/17/74	6/27/74–1/7/76	B	VIS—IR Spin Scan Radiometer (VISSR)
SMS-2	2/6/75	3/10/75–8/4/81	B	VISSR
GOES-1	10/6/75	1/8/76–3/15/80	B	VISSR
		8/ /84–2/3/85	B	VISSR VIS data only
GOES-2	6/16/77	8/15/77–9/15/80	B	VISSR
GOES-3	6/15/78	7/13/78–3/5/81	B	VISSR
GOES-4	9/9/80	3/5/81–6/1/83	B	VISSR Atmospheric Sounder (VAS) (& VISSR format data)
GOES-5	5/15/81	8/5/81–Present	B	VAS, VISSR format
GOES-6	4/28/83	6/1/83–7/30/84	B	VAS, VISSR format
GOES-7	2/26/87	3/25/87–Present	B	VAS, VISSR format
METEOSAT-2		7/1/83–Present	D	VISSR equivalent, ISCCP Data Set
GMS-1		1/21/84–6/30/84	D	VISSR Equivalent, ISCCP Data Set
GMS-2		7/1/83–1/30/84	D	Same as GMS-1
GMS-3		9/27/84–Present	D	Same as GMS-1

Note: a. Form: I—Image, D—Digital, B—Both Image and Digital b. Does not include DMSP Film and Digital data.

SDSD for the International Satellite Cloud Climatology Project (ISCCP), or statistical summaries, such as histograms generated from original data. The system must do this as cost effectively as possible, and the design of such internally generated products must take into account real or extrapolated user needs. These needs must be considered when the data stream sub-plan for each data type is updated; the sub-plan will specify any summary products to be generated as data are ingested by the archive system.

The archive system must also allow archive users to develop new products by refining current or developing new algorithms or products. This requires preservation of the raw data values themselves and possibly some of the adjusted data as well as the original geophysical products. It is critical that all information, such as calibration coefficients or tables, that allow raw data values to be regenerated from adjusted data be preserved. Thus SDSD has collected and preserved Level 1B granules of raw data values with Earth location and calibration information appended from the AVHRR and TOVS instruments flown on-board the NOAA operational polar orbiters, and adjusted full resolution GOES VISSR data, and the information needed to recover actual raw data from the adjusted data. In addition, the geophysical products obtained from these data are preserved.

VI-4.4.4 Primary Database Structure

The overall structure of the satellite primary database may be thought of as a simple hierarchy, with a very broad base. The base represents the very large volume of raw (Level 0) or adjusted (Level 1) satellite data. Intermediate levels represent summary or sampled products, or intermediate products such as direct conversions of data values to derived parameters in the original satellite geometry at original or sampled resolution (Level 2). A small apex represents derived geophysical parameters, perhaps mapped to a standard grid or map projection (Level 3). This is essentially the structure of the data stream sub-plans for all of the types of satellite data managed by SDSD.

In some cases intermediate products, such as the sampled GOES data provided to SDSD by OSDPD, are delivered to the archive center; in other cases the archive center creates data and products for the intermediate levels by summarizing or sampling the raw or preprocessed data, as done by SDSD to produce sampled AVHRR data for the ISCCP. In either case, the intermediate levels represent a particular combination of spatial and temporal resolution and information content intended to meet a user requirement.

The geophysical products at the top of the hierarchy may be those of value for climate research or other long-term applications, where the need is for a consistent, long-term information base but not necessarily for very fine spatial or temporal resolution. Retention of the top of the hierarchy must be permanent. The broad base of the hierarchy, raw or adjusted data, may not need to be retained permanently. A sampled raw or preprocessed granule may be sufficient as the lowest level of the hierarchy to be retained indefinitely. In the case of AVHRR data, NOAA plans to establish global 4-km resolution sampled data (GAC granules) as the base of the permanent archive, and to retain only se-

Table VI-4-6 *SDSD digital data base summary*

(CURRENT SIZE, ESTIMATED GROWTH, GB = 1 BILLION BYTES)

SOURCE/DATA STREAM	STATUS	CURRENT VOLUME, GB	GROWTH GB/YR	1998 PROJ. VOLUME, GB
POES AVHRR GAC Level 1B	A-92	1890.00	459.90	5109.30
POES AVHRR HRPT/LAC L-1B	A-92	900.00	547.50	4732.50
POES K-L-M AVHRR GAC Level 1B	93–95	0.00	459.90	1379.70
POES K-L-M AVHRR HRPT/LAC L-1B	93–95	0.00	547.50	1642.50
Polar Plat. AMRIR GAC L-1B	96–98	0.00	1214.14	3642.41
Polar Plat. AMRIR HRPT/LAC 1B	96–98	0.00	1445.40	4336.20
POES & ff AVHRR Mapped	A	13.33	4.00	65.33
POES & ff AVHRR Heat Budget	A	0.19	0.06	0.94
POES & ff AVHRR SST	A	2.19	0.53	9.06
POES & ff AVHRR Veg Index	A	1.82	6.50	86.32
POES HIRS, MSU, SSU Level 1B	A-92	314.84	50.00	664.84
POES K-L-M HIRS Level 1B	93–95	0.00	45.00	135.00
POES K-L-M AMSU Level 1B	93–95	0.00	81.00	243.00
Polar Plat. AMSU Level 1B	96–98	0.00	81.00	243.00
POES K-L-M SBUV Level 1B	93–95	0.00	5.00	15.00
POES TOVS Soundings	A	4.83	3.00	43.83
GOES VISSR Data (Mode A)	A-87	48760.80	1320.00	51400.80
GOES VAS Data (Mode AA)	A-87	8726.40	480.00	9686.40
GOES VAS Data (Mode AAA)	87–89	0.00	7008.00	21024.00
GOES-NEXT Imager/Sounder	90–98	0.00	14140.00	127260.00
GOES VISSR Synoptic Scale	A	204.12	30.24	597.24
GOES VAS Soundings	87–98	0.00	110.00	1320.00
GOES VISSR/VAS Winds	A	0.09	0.01	0.20
DMSP SSM/T Data	A	0.40	1.46	19.38
DMSP SSM/T Soundings	87–98	0.00	2.19	26.28
DMSP SSM/I Data	87–98	0.00	21.00	252.00
DMSP SSM/I Bright. Temp	87–98	0.00	40.00	480.00
DMSP SSM/I Antena Temps.	87–98	0.00	40.00	480.00
DMSP SSM/I Product	87–98	0.00	20.00	240.00
DMSP/FNOC MCSST	88–98	0.00	14.60	160.60
Nimbus-7 CZCS	A-87	655.53	114.00	883.53
Seasat Scatterometer	S	6.40	0.00	6.40
Seasat Altimeter	S	11.30	0.00	11.30
Seasat Syn. Apert. Radar	S	18.10	0.00	18.10
Seasat Micro. Radiometer	S	7.50	0.00	7.50
GEOS-3 Altimeter	S	0.80	0.00	0.80
Geosat Altimeter	86–88	0.00	1.30	3.90
ERS-1 Altimeter	90–93	0.00	1.30	5.20
ERS-1 Along Track Scan. Rad.	90–93	0.00	1.30	5.20
ERS-1 Microwave Sounder	90–93	0.00	1.30	5.20
ERS-1 AMI Scatterometer	90–93	0.00	8.00	32.00
ERS-1 AMI Syn. Apt. Radar	90–93	0.00	2737.50	10950.00
NROSS Scatterometer	91–94	0.00	20.10	80.39
NROSS Altimeter	91–94	0.00	48.63	194.50
NROSS SSM/I (data & products)	91–94	0.00	121.00	484.00
NROSS Low Frq. Micr. Rad.	91–94	0.00	55.11	220.43
NROSS Level 2 Products	91–94	0.00	8.90	35.59
SMS/GOES FGGE Data	S	664.00	0.00	664.00
SMS/GOES GATE Data	S	128.00	0.00	128.00
METEOSAT ISCCP B1	A	18.38	14.00	200.38
GMS ISCCP B1	A	18.38	14.00	200.38
RADARSAT Synthetic. Apt. Radar	93–98	0.00	2737.50	16425.00
Polar Plat. ERBI	96–98	0.00	45.00	135.00
Polar Plat. Along Track Scan. R.	96–98	0.00	1.30	3.90
Polar Plat. Scatterometer	96–98	0.00	45.00	135.00
Polar Plat. Global Ozone (GOMR)	96–98	0.00	105.00	315.00

Table VI-4-6 *(Continued)*

(CURRENT SIZE, ESTIMATED GROWTH, GB = 1 BILLION BYTES)

SOURCE/DATA STREAM	STATUS	CURRENT VOLUME, GB	GROWTH GB/YR	1998 PROJ. VOLUME, GB
Polar Plat. Altimeter	96–98	0.00	90.00	270.00
Polar Plat. SSM/I Data & Prod.	96–98	0.00	121.00	363.00
TOTAL DIGITAL DATA VOLUME:		62347.40		267079.54

Note: Data Volumes are approximate. In the Status Column, A indicates currently active data streams (data being received now), S indicates Static data streams (data no longer being received), and for future data streams the whole years when they are expected to be active are indicated.

POES (Polar Orbiting Environmental Satellites) is a generic term for NOAA operational polar orbiters (which include several series: TIROS, ESSA, NOAA).

YEAR BY YEAR PROJECTED GROWTH—SDSD DIGITAL DATA

YEAR	ESTIMATED GROWTH, GB	PROJECTED CUM. TOTAL, GB
Start (4/86)	—	62,347
End of 1986	3,046	65,394
End of 1987	10,288	75,682
End of 1988	8,388	84,070
End of 1989	8,387	92,457
End of 1990	18,268	110,725
End of 1991	18,522	129,247
End of 1992	18,522	147,769
End of 1993	21,341	169,110
End of 1994	18,591	187,701
End of 1995	18,337	206,039
End of 1996	18,695	226,386
End of 1997	18,695	246,733
End of 1998	18,695	267,080

lected subsets of the 1-km AVHRR data (HRPT/LAC granules) permanently.

VI-4.5 DATA MANAGEMENT SYSTEM TECHNOLOGY

Archive technology has evolved considerably over the years during which environmental satellites have been operating. Practical implementation of a dramatic new technology, optical storage of digital and analogue data, promises to transform data management practices in the near future. Innovations in magnetic tape technology and new magnetic tape alternatives continue to emerge. This section covers the archive technologies that have been used by SDSD—film and magnetic tape—and the potential of the new optical archive technology.

VI-4.5.1 Photographic Film

The original archive medium for satellite data and products was photographic film; indeed SDSD began as a small group working within the operational satellite photographic laboratory, collecting and cataloging film negatives produced as by-products of the laboratory's service of providing photographic prints to operational users. The basic archive item was a 25 × 25-cm film sheet, which provided very high resolution capability. Summary products in the form of reels of microfilm were produced from the large negatives. Film product archives date back to the TIROS-1 meteorological satellite and continue to the present time (Tables VI-4-4 and VI-4-5).

The advantages of the film archive are high image quality, in terms of resolution, and dynamic range and reliable long-term retention. The film negatives were filed in individual protective acetate sleeves and the earliest are still useful today.

The primary disadvantages are the inherent inflexibility of film and the impossibility of recovering complete digital data from film negatives. The inflexibility derives from the fact that the archive imagery is of scenes of interest to the original operational users of the data, enhanced to meet their needs. The retrospective users of these data have to accept prints made from the archive negatives, regardless of how different

Table VI-4-7 *SDSD non-digital data summary*

SOURCE/DESCRIPTION	MEDIA	CURRENT VOLUME	GROWTH UNITS/YR	EST. 1998 VOLUME
POES AVHRR/VHRR HRPT/LAC				
VIS, IR Imagery	Negatives	1,035,300	5,000	1,040,300
VIS, IR Imagery	UNIFAX Prints	20,000	29,000	397,000
POES AVHRR/SR				
Mapped VIS, IR Mosaics	Negatives	25,000	—	25,000
Mapped VIS, IR Mosaics	UNIFAX Prints	1,500	1,500	21,000
Mapped VIS, IR Mosaics	35mm Reels	150	5	215
Stretched/Gridded VIS/IR	Negatives	225,000	—	225,000
Stretched/Gridded VIS/IR	UNIFAX Prints	1,200	1,200	16,800
Stretched/Gridded VIS/IR	35mm Reels	2,000	—	2,000
SST Imagery	Negatives	3,500	500	10,000
SST Imagery	Paper Charts	10,000	500	16,500
Ocean Current Analysis	Paper Charts	2,000	260	5,380
POES AVHRR/VHRR				
Snow and Ice Cover	Paper Charts	2,000	260	5,380
POES AVHRR				
Vegetation Index Imagery	Negatives	600	416	6,008
GOES VISSR				
VIS, IR Imagery	Negatives	1,671,100	30,000	1,701,100
VIS, IR Imagery	35mm Reels	1,000	—	1,000
ATS Spin Scan Cloud Camera				
VIS, IR Imagery	Negatives	31,000	—	31,000
VIS, IR Imagery	35mm Reels	11	—	11
VIS, IR Imagery	16mm Reels	80	—	80
DMSP OLS				
VIS, IR Imagery	Transparencies	1,250,000	25,000	1,575,000
Nimbus-7 CZCS				
VIS, Chloro., Sediment Img.	Negatives	55,900	5,000	60,900
Seasat. Synth. Apt. Radar				
Surface Roughness Imagery	70mm Strips	478	—	478
Surface Roughness Imagery	Negatives	200	—	200

their applications are from the original operational use for which the products were designed.

Film has been almost completely phased out of operational use within NESDIS. Shortly it will no longer be used as an archive medium for new data. Currently, paper products produced by a facsimile device are being collected to provide a means of production of image products, pending implementation of a capability to produce imagery directly from digital data.

VI-4.5.2 Magnetic Tape

Magnetic tape has been the archive medium for preservation of digital data since the time of the first meteorological satellites in the early 1960s (Tables VI-4-4 and VI-4-5). Several magnetic tape systems have been used. They differ in physical characteristics, such as size of reels, hence length of tape per reel, and width as well as in recording format, and technique, and error detection and correction methods employed.

Magnetic tape has not been classically regarded as an "archive" medium in the same sense that photographic film has, but magnetic tapes of an age approaching 20 years have been successfully read. Reliable retention for at least ten years is commonplace. Magnetic tape does require some care. It is important that the environment in which the tapes are stored be stable in terms of temperature and relative humidity. Also tapes should be sealed from dust and other airborne matter.

A. Computer Compatible Tape. Archiving of digital data began in the mid-1960s using standard Computer Compatible Tape (CCT) reels containing ½-in wide × (typically) 2400-ft long magnetic tapes with data recorded in seven parallel tracks (six data tracks plus a parity) containing 6-bit binary or character values in a format designed to be used with industry standard tape drives. The tapes were initially written at densities of 200 and 556 bpi along the tape, then 800 bpi. The next advance was nine-track tape, which allowed parallel recording of eight bit bytes, compatible with eight bit byte oriented computers. Next followed the introduction of nine-track, 1600-bpi phase encoded data recording. In addition to doubling the amount of data that could be stored on one tape, the combination of the phase encoded recording technique and improved error detection and correction provided much more reliable performance than earlier systems.

The current state of the art in reel-to-reel CCT is the nine-track, 6250-bpi group encoded system. Again, improvements in both the recording technique and error detection and correction have made the current standard significantly better than 1600 bpi. The long-term read reliability of CCT, especially 6250-bpi CCT, is excellent for long-term high quality data storage, given proper care of the tapes.

SDSD currently retains approximately 100,000 CCT, mostly 6250 bpi and 1600 bpi; some 800 bpi and some seven-track, 800 bpi and 556 bpi are included. Conversion of the data on these tapes to cartridges will be essential if these data are to be preserved. Archiving of large quanitities of NOAA polar orbiter adjusted data (Level 1B) on 6250-bpi CCT began during 1985. This coincided with the phaseout of the special tape system, the Ampex Terabit Memory System, previously used to store these data.

B. Computer Compatible Cartridge Systems. The cartridge system was developed by IBM as the successor to the 1600-bpi or 6250-bpi CCT system as the basic magnetic tape computer peripheral. The system was announced in 1984 as the IBM 3480 cartridge system, fully supported by IBM system software. This system offers several important advantages over conventional CCT systems. The recording technique has been further refined, and a more sophisticated error detection and correction technique has been employed that promises to make the data recording reliability of the cartridge system better than the 6250-bpi CCT. The fact that the tape is enclosed in a plastic cartridge aids in protecting the tape from dust contamination in storage. The initial version offers a capacity per cartridge about 20% greater than a 6250-bpi CCT in a significantly smaller package ($1 \times 4 \times 5$ in for a cartridge as opposed to $1 \times 10 \times 10$ in for a CCT reel), an important consideration when the storage of tens of thousands of cartridges is planned. Also, cartridge drives are smaller and less expensive to maintain.

The 3480 system records data on 18 parallel tracks at a linear recording density of 38,000 bpi (compared to 6250 bpi for the most densely recorded CCT). The instantaneous transfer rate is as fast as 3 million bytes per second, comparable to magnetic disk rates and up to more than twice the rate possible with current CCT systems.

The NESDIS operational data processing system and the SDSD archive system switched over from 6250 bpi CCT to the cartridge system for archiving of NOAA polar orbiter Level 1B granules in 1985.

C. Ampex Terabit Memory System. The TBM system was one of two magnetic tape systems derived from video tape recording techniques that have been used for large volume data archiving. SDSD had a bad experience with the TBM system and this experience contains many lessons for archive system planners.

Perhaps foremost among the problems with TBM was the poor acceptance of TBM technology as compared with CCT technology. CCT became and remains an industry standard medium for data storage and exchange, supported by all equipment manufacturers for many years. An archive system must be assured of long-term support. One- or few-of-a-kind systems may offer technical elegance, but may not be commercially viable in the long term, and therefore may not be supported for long. Such possible unsuitability for data archive applications should be borne in mind when alternative magnetic tape systems or the new optical systems are considered for long-term archive use.

The TBM system was put into use as a dual-purpose system, as both an on-line buffer for the operational satellite data processing system and as an archive device. Level 1B granules collected by the NOAA polar orbiter AVHRR and TOVS instruments were written on TBM tapes operationally and accessed from the TBM by operational product processing systems. Depending on the data type, a TBM tape would remain on the system for a number of days before becoming full, at which point the tapes were unloaded and turned over to SDSD. SDSD filled user requests for Level 1B granules either while a TBM tape was still on the system (near real-time—within minutes to several days after acquisition) or reloaded the TBM tapes and recovered granules retrospectively. Level 1B granules were recorded on the TBM system from the launch of TIROS-N in October 1978 until early April 1985 (when the switch to 6250-bpi CCT was made).

A fundamental difficulty that plagued the TBM from the point of view of SDSD was its dual use; maintenance and operations procedures were tailored to the requirements of the higher priority "real time" operational function versus the archive function. This adversely impacted the services provided by SDSD to its users. But had it not been usable as a dual-purpose system, there would likely have been no comprehensive archive of the high volume Level 1B granules at all. Thus a major lesson from the TBM experience is that an archive system should be independently managed and operated to ensure that archive preservation and user services functions are not subordinated to other functions.

NESDIS transferred AVHRR GAC granules from the TBM system to 6250-bpi CCT for permanent retention. During this process, over 95% of the granules recorded on an average TBM tape were retrieved. Since a number of TBM tapes were not usable at all, however, the final overall successful retrieval rate was only 85%. The read reliability had been considered to be at least equal to CCT (error rates of one per 10 billion were cited), but the loss of whole tapes, let alone whole granules, for whatever reason, was not acceptable performance.

The last TBM system was operated by the National Center for Atmospheric Research (NCAR) in Boulder, Colorado. The most complete possible set of TOVS data and some AVHRR GAC data that SDSD was unable to recover from the TBM tapes were recovered by NCAR prior to their shutting down of their own TBM and their switch to IBM 3480 cartridges for data storage.

Through the combined efforts of NESDIS and NCAR, approximately 90% of the AVHRR GAC (global, 4-km resolution) and TOVS Level 1B granules and only about 5% of the AVHRR (specific areas, 1-km resolution) granules of the data written to TBM tapes from late 1978 until April 1985 have been preserved.

D. University of Wisconsin Videocassette System. The Space Science and Engineering Center (SSEC) of the University of Wisconsin developed a unique approach to recording the GOES digital data (stretched) with specially modified U-Matic (¾ in) Sony commercial videocasette recorders. The technique involves using special front-end electronics to reformat the GOES data and convert them to a modulated FM signal, and adjustment of the tape recorder operation to allow synchronization of the recorder with the GOES data. The recorder uses two video heads, one of which is in position to record or play back at any time. Data are recorded as a series of slant tracks as described previously for the TBM system. The data flow and tape head assembly motion are synchronized so that the transition between tracks occurs when a synch pattern, as opposed to actual data, is recorded; therefore no data are lost. The result is a very inexpensive, high density system, with the data written on one videocassette equal to about 50 6250-bpi CCT, or over 4 billion bytes of data. The system does not employ any read-after-write error detection and correction, which is impossible given the geometry of the recorder system and a major limitation from the archival point of view. Recorded data problems are discovered only when the cassettes are played back. The read reliability of the system has been tested and found to be suitable for data recording. The error rates are on the order of one bit in 10 million, an order of magnitude better than the estimated error rate inherent in the incoming data stream.

GOES data recording began in 1978 in preparation for the Global Weather Experiment (GWE) in 1979 and has continued under NESDIS funding from that time through the present. Years of experience in retrieving data from the videocassettes has demonstrated the reliability of the approach.

E. High Density Instrumentation Tape Systems. These systems generally use multitrack recording of Pulse Code Modulation (PCM) baseband signals on ½ to 2 in wide magnetic tape. They generally are not computer compatible and grew out of laboratory based systems used for recording of instrument responses or telemetry. An instrumentation tape system was in use for GOES data archiving for a period of time, and a more advanced 28-track instrumentation tape system is presently used for Landsat data recording. The most advanced systems offer sophisticated error detection and correction, comparable to CCT in quality, and the ability to record data at the very high rate of hundreds of millions of bits per second. Such systems are very expensive.

VI-4.6 IMPACT OF NEW PROCESSING AND COMMUNICATIONS TECHNOLOGY

There has been an explosion in the capacity of data processing technology available at relatively low cost to satellite data users. Personal computers with megabytes of internal memory capacity, disk storage of 60 million bytes or more, and very powerful image display and data processing capability are commonly available. As noted above, small optical disk systems offering hundreds of millions of bytes of storage are available as personal computer peripherals. In addition, sophisticated data display and analysis software packages are becoming widely available, often packaged with hardware as complete functional workstations. The growth of user requirements for rapid access to digital data follows these developments.

Rapid development of Earth and satellite based telecommunications technology, and especially the proliferation of commercially available telecommunications utilities, have made it possible for data management systems to be interconnected and to offer connections and on-line services to data users that would have been impossible or prohibitively expensive five or ten years ago.

It is now possible for data management functions such as data ingest, cataloguing, storage, and supporting user access to be performed at geographically distributed locations and for the data base itself to be geographically distributed. The data management system itself can function as a single, logically integrated system with physically distributed nodes. It is equally possible for the data management functions to be centralized at a single physical location even if they are managed from a number of geographically distributed sites. Either option can be supported by telecommunications—the design question of the degree of physical distribution of a future data management system will turn on other factors.

BIBLIOGRAPHY AND REFERENCES FOR SECTION VI

Abel, P. and A. Gruber, 1979: An improved model for calculation of longwave flux at 11 μm. *NOAA Technical Memorandum NESS 106*, Dept. of Commerce, Washington, DC, 24 pp.

Anding, D. and R. Kauth, 1970: Estimation of sea surface temperature from space. *Remote Sensing of Environment, l*, 217-220.

Barnes, J. C. and M. D. Smallwood, 1982: TIROS-N series direct readout mean analysis. *Oceanogrphic Monthly Summary, Vol IV, No. 8*, Dept. of Commerce, NOAA, NWS, Washington, DC, 12 pp.

Bell, A. and J. Clark, 1984: New global sea surface temperature monthly mean analysis. *Oceanogrphic Monthly Summary, Vol IV, No. 8*, Dept. of Commerce, NOAA, NWS, Washington, DC, 12 pp.

Bell System, 1973: Data communications using voiceband private line channels. *Bell System Technical Reference 41004*.

Brower, R. L., H. S. Gohrband, W. G. Pichel, T. L. Signore, and C. C. Walton, 1976: Satellite derived sea surface temperatures from NOAA spacecraft. *NOAA Technical Memorandum NESS 78*, Dept. of Commerce, Washington, DC, 74 pp.

Chahine, T., 1975: An analytical transformation of remote sensing of clear-column atmospheric temperature profiles. *J. Atmos. Sci., 32*, 1946-1952.

Clark, J. D. (ed.), 1983: The GOES User's Guide. Dept. of Commerce, NOAA, NESDIS, Washington, DC, 163 pp.

Conlan, E. F., 1973: Operations products from ITOS scanning radiometer data. *NOAA Techincal Memorandum NESS 52*, Dept. of Commerce, Washington DC, 57 pp.

Dave, J. V., 1964: Meaning of successive iteration of the auxiliary equation in the theory of radiative transfer. *J. of Astrophys.*, *140*, 1292-1303.

Deschamps, P. Y. and T. Phulpin, 1980: Atmospheric correction of infrared measurements of sea surface temperature using channels at 3.7, 11, and 12 μm. *Bound. Layer Meteor., 18*, 131-143.

Escobal, P. R., 1976: *Methods of Orbit Determination*. Robert E. Krieger Pub. Co., Huntington, NY, 479 pp.

Fischer, H., 1979: Determination of atmospheric temperature profiles and turbidity from satellite observations. *Proc. from the Technical Conf. on Use of Data from Meteorological Satellites, Lannion, France, 17-21 September*, European Space Agency, Paris, France, 145-153.

Fleming, H. E., M. D. Goldberg, and D. S. Crosby, 1988: Operational implementation of the minimum variance simultaneous retrieval method. *Proc. of the 3rd Conf. on Satellite Meteorology and Oceanography, Jan. 31-Feb. 5, Anaheim, CA*, Amer. Meteor. Soc., Boston, MA, 16-19.

Gonzales, R. C. and P. Wintz, 1977: *Digital Image Processing*. Addison-Wesley Pub. Co., Reading, MA, p. 431.

Green, R., G. Hughes, C. Novak, R. Schreitz, 1975: The automatic extraction of wind estimates from VISSR data. *NOAA Technical Memorandum NESS 64*, Dept. of Commerce, Washington, DC, 94-101.

Gruber, A., M. Varnadore, P. Arkin, and J. Winston, 1986: Monthly and seasonal mean outgoing longwave radiation and anomalies. *NOAA Technical Report NESDIS 26*, Dept. of Commerce, Washington, DC, 120 pp.

———, I. Ruff, and C. Earnest, 1983: Determination of the planetary radiation budget from TIROS-N satellites. *NOAA Technical Report NESDIS 3*, Dept. of Commerce, Washington, DC, 12 pp.

———, 1977: Determination of the Earth-atmosphere radiation budget from NOAA satellite data. *NOAA Technical Report NESS 76*, Dept. of Commerce, Washington, DC, 28 pp.

Hambrick, L. and D. Phillips, 1980: Earth locating image data of spin-stabilized geosynchronous satellites. *NOAA Technical Memorandum NESS 111*, Dept. of Commerce, Washington, DC, 98 pp.

Hayden, C. M., L. F. Hubert, E. P. McClain, and R. S. Seaman, 1979: Quantitative meteorological data from satellites. *Chapter 3, Technical Note No. 166, WMO No. 531*, World Meteor. Organ., Geneva, Switzerland, 60-86.

Hoke, J. E., J. L. Hayes, L. G. Renninger, 1981: *Map Projections and Grid Systems for Meteorological Applications*. Air Force Global Weather Central TN-79/003, Offutt AFB, NE, 86 pp.

Hughes Aircraft Company/Space and Communications Group, 1980: *Geostationary Operational Environmental Satellite (GOES) Spacecraft Operations Manual*, Natl. Aero. and Space Adm., GSFC, Greenbelt, MD, 187 pp.

Hughes, G., C. Novak, and R. Schreitz, 1978: Automated techniques for the detection and displacement measurement of selected cloud imagery observed in geostationary satellite data. *Proc. of the Symp. on Automatic Imagery Pattern Recognition: Emerging Patterns in AIPR, April 3-4, Gaithersburg, MD*, Assoc. of Electronic Industries and Natl. Bureau of Standards, 81-88.

Hunt, G. E., 1973: Radiative properties of terrestrial clouds at visible and infrared thermal window wavelengths. *Quart. J. Royal Meteor. Soc., 99*, 346-369.

Kellaway, G. P., 1964: *Map Projections.* Methuen and Co. Ltd, London, England, p. 127.

Klenk, K. F., V. G. Kaveeshwar, P. K. Bhartia, R. D. McPeters, P. M. Smith, and A. J. Fleig, 1982: Total-ozone determination from the Backscatter Ultraviolet (BUV) Experiment. *J. Appl. Meteor. 21*, 1672-1684.

Lauriston, L., G. J. Nelson, and F. W. Porto, 1979: Data extraction and calibration of TIROS-N/NOAA Radiometers. *NOAA Technical Memorandum NESS 107*, Dept. of Commerce, Washington, DC, 73 pp.

Mazur, W. E., June 1983: Implementation of the new VISSR/VAS data transmission system. (unpublished manuscript). Dept. of Commerce, NOAA, NESDIS, Washington, DC, 26 pp.

McClain, E. P., W. G. Pichel, C. C. Walton, Z. Ahmad, and J. Sutton, 1983: Multi-channel improvements to satellite-derived global sea surface temperatures. *Adv. Space Res., 2*, Pergamon Press, New York, NY, 43-47.

———, 1981: Multiple atmospheric window techniques for satellite-derived sea surface temperatures. *Oceanography from Space*, J. F. R. Gower (ed.), Plenum Press, New York, NY, 3-85.

———, 1979: Satellite derived Earth surface temperature, quantitative meteorological data from satellites. *WMO 531*, Geneva, Switzerland, 60-86.

——— and P. G. Abel, 1977: Remote sensing of ocean temperature. *Proc. of AIAA/AUG/AMS Conf. on Satellite Applications to Marine Operations, Nov. 15-17, New Orleans, LA*, 209-217.

McDonnell, P. W. Jr., 1979: *Introduction To Map Projections*. Marcel Dekker, Inc., New York, NY, 174 pp.

McMillin, L. M., 1978: An improved technique for obtaining clear radiances from cloud-contaminated radiances. *Mon. Wea. Rev., 106*, 1590-1597.

———, 1975: Estimation of sea surface temperatures from two infrared window measurements with different absorption. *J. Geophys. Res., 80*, 5113-5117.

Miller, D. E., 1979: Quantitative analysis of radiance data. *Proc. from the Technical Conf. on Use of Data from Meteorological Satellites, Lannion, France, September 17-21*, 77-83.

Nappi, A., A. Swaroop, N. Grody, and J. Weinman, 1986: Precipitation effects on microwave sounding unit data. *Proc. of the 2nd*

Conf. on Satellite Remote Sensing and Application, May 13-16, Williamsburg, VA, Amer. Meteor. Soc., Boston, MA, 114-117.

NOAA/NASA/Hughes Aircraft Company/Space and Communications Group, 1980: *GOES D, E, F Data Book*, Natl. Aero. and Space Adm., GSFC, Greenbelt, MD, 100 pp.

Novak, C. S. and M. T. Young, 1976: The operational processing of winds estimates from cloud motions. *COSPAR Symp. on Meteorological Observations From Space: Contributions to the First GARP Global Experiment, June 8-10, Philadelphia, PA*, NCAR, Boulder, CO, 214-221.

Orlin, Hyman, 1980: *Mathematics of Map Projections*. George Washington Univ., Washington, DC, p. 120.

Phillips, N., 1979: An evaluation of early operational temperature soundings from Tiros-N. *Bull. Amer. Meteor. Soc., 60*, 1188-1197.

Prabharkara, C., G. Dalu, and V. G. Kunde, 1974: Estimation of sea surface temperature from remote sensing the the 11 to 13 μm window region. *J. Geophys Res, 79*, 5039-5044.

Reale, A., D. Gray, and M. Chalfant, 1986: Higher resolution operational retrievals. *Proc. of the 2nd Conf. on Satellite Remote Sensing and Application, May 13-16, Williamsburg, VA*, Amer. Meteor. Soc., Boston, MA, 16-19.

Rodgers, C. D., 1976: Retrieval of atmospheric temperature and composition from remote measurements of thermal radiation. *Rev. Geophy. Space*, *14*, 609-624.

Schneider, J. R., 1976: Guide for designing, RF ground receiving stations for TIROS-N. *NOAA Technical Report NESS 75*, Dept. of Commerce, Washington, DC, 136 pp.

Schneider, W. H., P. K. Bhartia, K. F. Klenk, and C. L. Mateer, 1981: An optimum statistical technique for ozone profile retrieval from backscattered UV radiances. *Proc. of the 4th Conf. on Atmospheric Radiation, June 16-18, Toronto, Canada*, Amer. Meteor. Soc., Boston, MA, 33-37.

Schwalb, A., 1987, 1979, 1978: The TIROS-N/NOAA A-G satellite series (NOAA E-J) - Advanced TIROS-N (ATN). *NOAA Technical Memorandum NESS 95*, Dept. of Commerce, Washington, DC, 75 pp.

———, 1972: Modified version of the improved TIROS operational satellite (ITOS D-G). *NOAA Technical Memorandum NESS 35*, Dept. of Commerce, Washington, DC, 48 pp.

Smith, E. and D. Phillips, 1972: Measurements from satellite platforms. *Annual Scientific Report on NAS5-11542*, 1971-1972. Space Science and Engineering Center, Univ. of Wisconson, 1-53.

Smith, W. L. and H. M. Woolf, 1981: Vertical mass and moisture structure from Tiros-N. *Bull. Amer. Meteor Soc, 62*, 388-393.

———, H. M. Woolf, and C. M. Hayden, 1979: The Tiros-N operational vertical sounder. *Bull. Amer. Meteor. Soc., 60*, 1177-1187.

Strong, A. E. and J. A. Pritchard, 1980: Regular monthly mean temperatures of Earth's oceans from satellites. *Bull. Amer. Meteor. Soc., 61*, 553-559.

Taylor, S. L., P. K. Bhartia, V. G. Kaveeshwar, K. F. Klenk, A. J. Fleig, and C. L. Mateer, 1980: Role of multiple scattering in ozone profile retrieval from satellite measurements in the ultraviolet. *Remote Sensing of Atmospheres and Oceans*, A. Deepak (ed.), Academic Press, New York, NY, 219-231.

Taylor, V. R. and L. L. Stowe, 1984: Reflectance characteristics of uniform Earth and cloud surfaces derived from Nimbus-7 ERB. *Jour. Geophys. Res., 89*, 4987-4996.

Thekaekara, M. P., 1974: Extraterrestrial solar spectrum. *Appl. Opt., 13*, 518-522.

———, 1971: Solar electromagnetic radiation. NASA Space Vehicle Design Criteria (Environment), NASA SP-8005.

Walton, C., 1980: Deriving sea surface temperatures from TIROS-N data. *Remote Sensing of Atmospheres and Oceans*, Academic Press, New York, NY, 547-579.

Weinreb, M. P. and M. L. Hill, 1980: Calculations of atmospheric radiances and brightness temperatures in infrared window channels of satellite radiometers. *NOAA Technical Report NESS 80*, Dept. of Commerce, Washington, DC, 40 pp.

Werbowetzki, A. (ed.), 1981: Atmospheric sounding user's guide. *NOAA Technical Report NESS 83*, Dept. of Commerce, Washington, DC, 82 pp.

Williamson, L. E., 1977: Calibration technology for meteorological satellites. *Atmospheric Science Laboratory Monograph Series*, U. S. Army Electronics Command, White Sands Missile Range, New Mexico, 139 pp.

World Meteorological Organization, 1972: Comparative sea-surface temperature measurements, Marine Science Affairs. *Report No. 5, WMO No. 336*, Geneva, Switzerland, 27 pp.

Wrublewski, T., 1983, *Memorandum (September 23, 1983)*, Satellite Navigation and Calibration Branch, National Environmental Satellite, Data, and Information Service, NOAA, Washington, DC.

Wydick, J. E., P. Davis, and A. Gruber, 1987: Estimation of broadband planetary albedo from operational narrowband satellite measurements. *NOAA Technical Report NESDIS 27*, Dept. of Commerce, Washington, DC, 32 pp.

Contributors to Section VI

Anderson, Ralph K.
Banks, Barbara A.
Burdsall, Ellen L.
Callicott, William M.
Chalfant, Michael W.
Ellickson, James K.
Gray, Donald
*Hamilton, Gerald G.
*Hunolt, Gregory
Kahwajy, Faris T.
Koffler, Russell
Mack, Elizabeth G.
Mairs, Robert L.
Maturi, Eileen
McClain, E. Paul
Pichel, William G.
Planet, Walter G.
Schreitz, Richard
Shellman, Marcus B.
Silva, H. James
Staton, Carl P.
Tharps, Debbie T.
Watkins, Benjamin
Welch, Melvin E.
Wydick, James E.
* (Moved to a new position)

NOAA/NESDIS
Washington, DC 20233

Section VII

APPLICATIONS OF SATELLITE DATA IN METEOROLOGY

A good set of observations has always been the fundamental requirement for meteorological progress. The first TIROS satellite substantiated this hypothesis and ushered in a new era in the atmospheric sciences. The early images received from the weather satellite showed that they could be used immediately for meteorological analysis and qualitative interpretation. From the beginning, satellites provided global coverage that enabled meteorologists to observe synoptic scale features. Satellites have also come to play a major role in mesoscale meteorology, particularly due to the time and space resolution of satellite imagery and soundings. These previously unavailable observations provided new insights into mesoscale convective processes and led to improvements in short-range severe storm forecasting. In recent years, better instrumentation and increased quantitative data have allowed for various satellite products, many of which were discussed in Section VI.

This Section describes in detail various applications of satellite data in meteorology, such as derivation of meteorological parameters, synoptic- and convective-scale weather analysis and forecasting, and tropical, marine, and aviation applications.

Chapter VII-1

BASIC IMAGE ANALYSIS AND INTERPRETATION[1]

VII-1.1 INTRODUCTION

Satellite radiometric measurements can be converted into images of clouds, land, oceans, atmospheric water vapor, dust, and smoke. Much effort has been expended in learning to interpret these remotely sensed images, particularly with respect to identification and development of significant cloud types and patterns. This knowledge has given meteorologists a better understanding of synoptic-scale storm evolution and the atmospheric processes that control mesoscale weather.

The first TIROS meteorological satellite images were obtained by video (VIS band) cameras and reproduced as photographs or converted to nephanalyses. Procedures long used for the interpretation of aerial photography were immediately applied to satellite imagery. Digital processing was introduced in the mid-1960s (Bristor et al., 1966) and observations from contiguous polar satellite passes were computer composited to permit meaningful meteorological interpretation of large-scale cloud systems. The coming of age of operational IR imagery in the early 1970s made possible quantitative analyses of these new data in terms of cloud heights, sea surface temperatures, and storm pattern definition. The large amounts of new data also increased the need for computer handling. As more data channels were added to satellites, multispectral image analysis became important. Advances in computer technology now permit real-time interactive analysis of imagery and the TV screen is rapidly becoming the principal medium of image display.

To interpret satellite imagery an analyst must be familiar with: (1) all scales of meteorological processes; (2) basic elements of image interpretation; (3) radiative transfer processes in the earth-atmosphere system; (4) digital processing techniques and their use for increasing information content of imagery; and (5) capabilities of image display media.

VII-1.2 BASIC ELEMENTS OF INTERPRETATION

Image analysis requires the use of universally applicable basic concepts (Estes et al., 1983). In achromatic (gray tone) displays, each tone represents a level of the energy sensed by the satellite. The relative brightness of a feature is important for identification in both VIS and IR data. The relative brightness of a feature depends on the properties of the feature, the spectral interval of the sensor, and the display convention. Thus, in a VIS range (0.7 μm) satellite image, highly reflective areas appear white and areas of low reflectivity are dark. In an IR image (11.5 μm) the energy sensed is proportional to the temperature of the emitting surface, i.e., the higher the temperature, the brighter the response. The convention for displaying

[1] Primary Contributor: Ralph K. Anderson

IR data is to reverse the pattern, to show lower temperatures as white and higher temperatures as black.

The degree of contrast, or gray tone difference, between a cloud and its background is important. The greater the contrast, the easier it is to identify a cloud feature. In IR images, the gray tone of a low cloud layer is close to the gray tone of the Earth's surface because the temperature difference between the two emitting surfaces is small. As such small contrasts make these clouds difficult to detect, computer enhancement techniques are used to facilitate detection of features in the imagery.

The abruptness of tone contrast is another important concept of interpretation. Sensors with high spatial resolution produce images in which tone contrast is sharp; this allows for discrimination of fine detail. Resolution depends on the geometrical properties of the imaging system and is described in terms of Instantaneous Field Of View (IFOV) (Simonett, 1983), usually given as a distance in kilometers. The absolute size of a detectable element will vary since IFOV of a sensor depends on satellite altitude, detector size, and focal length of the optical system. Values stated for sensor resolution (IFOV) are at satellite subpoints. For oblique viewing angles, the resolution deteriorates (Figures VII-1-1 and VII-1-2).

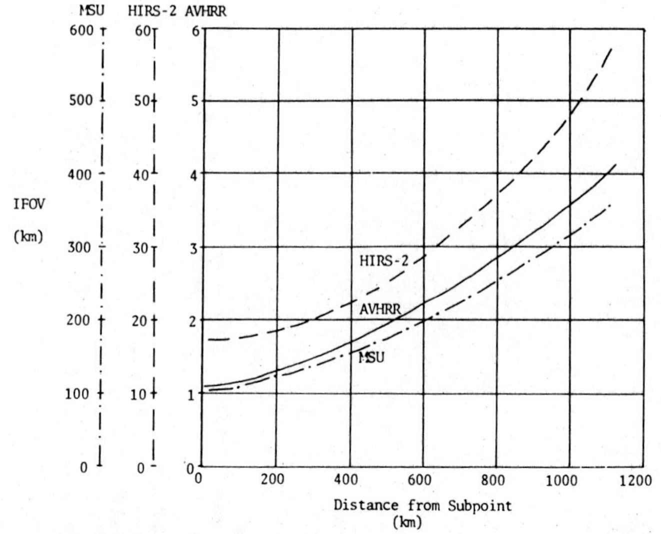

Figure VII-1-1. Effect of oblique viewing angles on the size of the instantaneous field of view for NOAA polar orbiting satellites (850-km altitude). Curves are shown for three instruments, each with a different IFOV. Scale for each curve is to the left of the diagram. The curves show the change in sensor resolution (increase in IFOV) for distance measured in kilometers along a scan line away from the satellite subpoint.

Simonett (1983) points out that the spatial resolution of an observing system must be within certain limits if one is to discern and analyze phenomena of interest. There are three steps to a given analysis task: (1) detection to determine the presence or absence of a cloud pattern or formation, e.g., a synoptic-scale storm can be seen with relatively low-resolution imagery; (2) identification when there is enough information to identify specific features; and (3) analysis, which requires that resolution be sharp enough to identify cloud types and mesoscale organization. The resolution needed for a given task is a function of feature scale and image processing constraints. To move from detection to identification, the resolution should improve about three times, by a factor of 10.

Radiometric resolution is determined by the number of discrete levels into which a signal can be divided. The dynamic range of a sensor is determined by the maximum radiance value that it can measure. This is currently 320 K for IR detectors on the NOAA and GOES satellites. The lowest radiance measured is that of space. The radiances that can be digitized on-board the satellite range over 167 K, however, with limits defined by the coldest cloud tops and the warmest land surfaces. The radiances are converted on-board to 256 values (8-bit digital data) for the GOES VAS and 1024 (10 bits) for the AVHRR. Since the sensor response is nonlinear, GOES radiance values equate to 0.5-K temperature intervals for temperatures above 254 K and to 1.0 K intervals for temperatures below 254 K. Radiometric resolution also is the degree to which a sensor can convert received radiation into the digitized electrical signals for transmission. For GOES, the sensor response does not match the sampling rate of the sensor signal. This means that the sensor will not respond to the minimum or maximum temperature of a scene unless that value remains constant over a distance equivalent to that of approximately 2 IFOVs. This effect causes thunderstorm cloud top features to appear slightly east of their actual positions in IR imagery and east of where they appear in VIS imagery of the same scene. This effect is small and can be neglected in operational applications.

Spectral resolution, determined by the bandwidth of the sensor channels, becomes important in multispectral analysis. Narrow bandwidths, which provide higher resolution, are important for detecting spectral signatures unique to the emitting surface. Differentiation between WV and cirrus cloud is an example (Fritz and Rao, 1967). Figure VII-1-3 shows the spectral response and the bandwidths of the five NOAA AVHRR channels. The lower part of Figure VII-1-4 shows the bandwidths of several GOES VAS channels.

Analysts must be aware that the relative brightness of a gray tone is a psychological concept or a sensation associated with light stimulus to the eye. Because the eye is very adaptable, absolute brightness cannot be

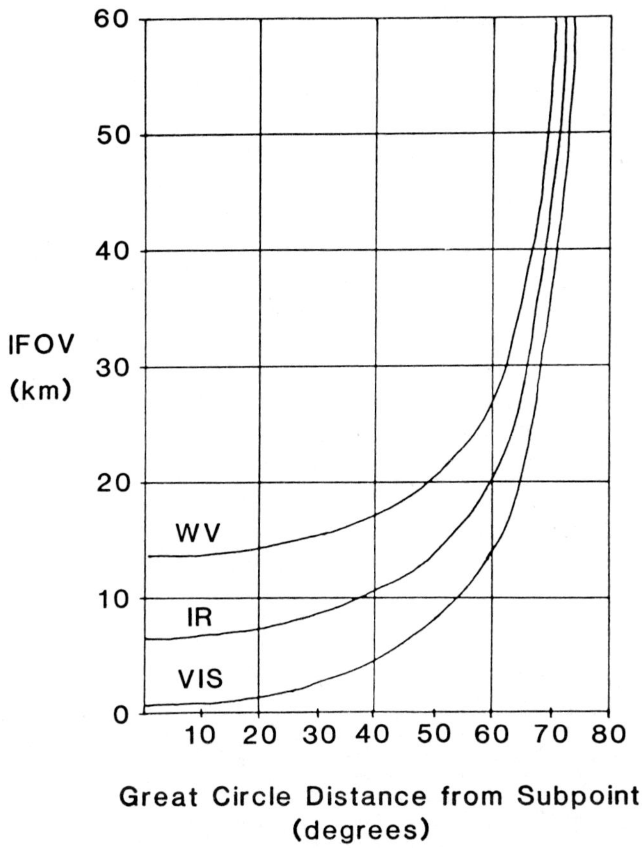

Figure VII-1-2. Effect of oblique viewing angles on the size of the instantaneous field of view for GOES satellites (38,800-km altitude). Curves are for the VIS 1 km, the IR 8, and 14 km (WV) detectors. The curves show the decrease in resolution (increase in IFOV) away from the satellite subpoint at distances measured in degrees along any great circle passing through the subpoint.

judged accurately. Two sources of equally intense light often do not appear equally bright to the eye, because the perceived grayness of an object is influenced by a contrasting gray tone in the surrounding area (Hall, 1979). For this reason, analysts should not rely on their perceptions alone when comparing the gray tones of clouds in different parts of an image.

Equal differences in gray tone intensity are more detectable by the eye in the medium tones than in the dark ranges (Graham, 1965). Thus, the differences in gray tones in areas where all tones are very dark, as in the displays of thermal IR values for a warm surface and low clouds, will be very hard to detect. Enhancement schemes can correct for this by displaying the imagery in lighter shades of gray so that features can be more easily identified. In general, identification of separate gray tones is better accomplished by the use of objective means, such as a densitometer or computer processing.

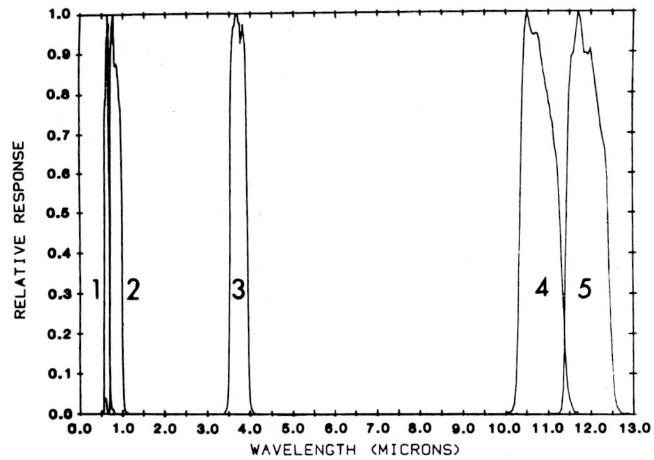

Figure VII-1-3. Spectral response of the TIROS-N/NOAA AVHRR channels. The spectral intervals of each channel are as follows:

1. 0.58 to 0.68 μm
2. 0.725 to 1.1 μm
3. 3.55 to 3.93 μm
4. 10.3 to 11.3 μm
5. 11.5 to 12.5 μm

The accurate Earth location of thunderstorm cloud tops in VIS and IR imagery is often needed for tracking severe convective storms with GOES imagery. With oblique satellite viewing angles, a location error is in-

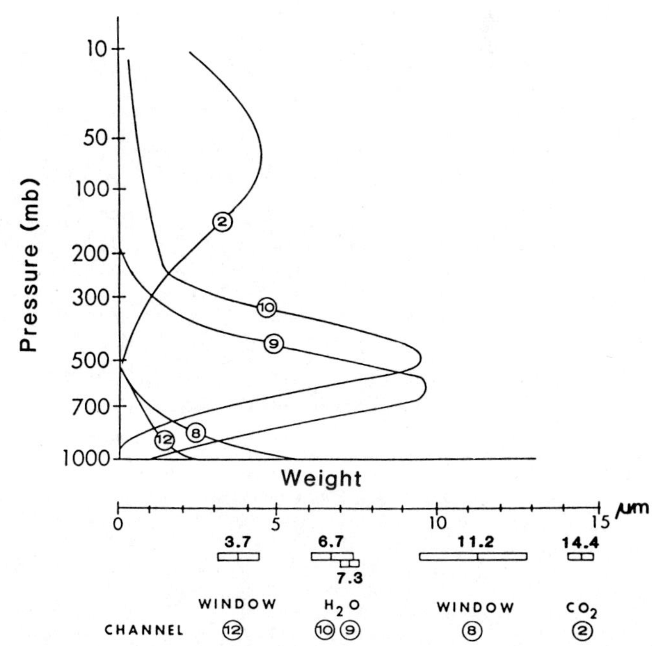

Figure VII-1-4. Weighting functions and bandwidths for some VAS channels. The bandwidths shown for each channel are defined by the spectral lines equal to 50% of peak energy at the central value of the channel.

troduced for tall clouds. The tall cloud appears to be farther away from the satellite subpoint than it actually is. Figure VII-1-5 shows the amount of displacement error for a 12-km (40,000 ft) cloud top in GOES oblique viewing angles. The arrows on the diagram point toward the satellite subpoint; corrections for the actual cloud position are made in that direction.

VII-1.3 CHARACTERISTICS OF VISIBLE IMAGERY

Reflected shortwave solar radiation from clouds and the Earth surface, and some scattering from the atmosphere, appear in VIS imagery. Figure VII-1-6 shows the energy distribution for solar radiation. The brightness of VIS imagery is an approximation of the Earth's albedo; light tones represent areas of high albedo and dark tones areas of low albedo. The brightness of a cloud as viewed from space is affected by the angular position of the cloud in relation to the sensor and to the Sun and by the reflectivity of the cloud itself (Conover, 1962). Reflectivity, in turn, is related to cloud thickness, particle size and distribution, and composition (ice or water).

Shadows produced by low sun angles are quite helpful in analyzing cloud imagery. Fujita (1963) and Oliver et al. (1964) showed that tall cumuliform cloud elements, or sharp cloud edges, produce shadows that are easily detectable. The apparent width of a shadow depends on both the Sun's angle and the viewing angle of the satellite relative to the Sun. A detectable shadow usually indicates the presence of a layer of middle or high clouds or a cumulus cloud of considerable vertical development. The most easily detectable shadows are those cast on a deck of lower clouds, on light colored

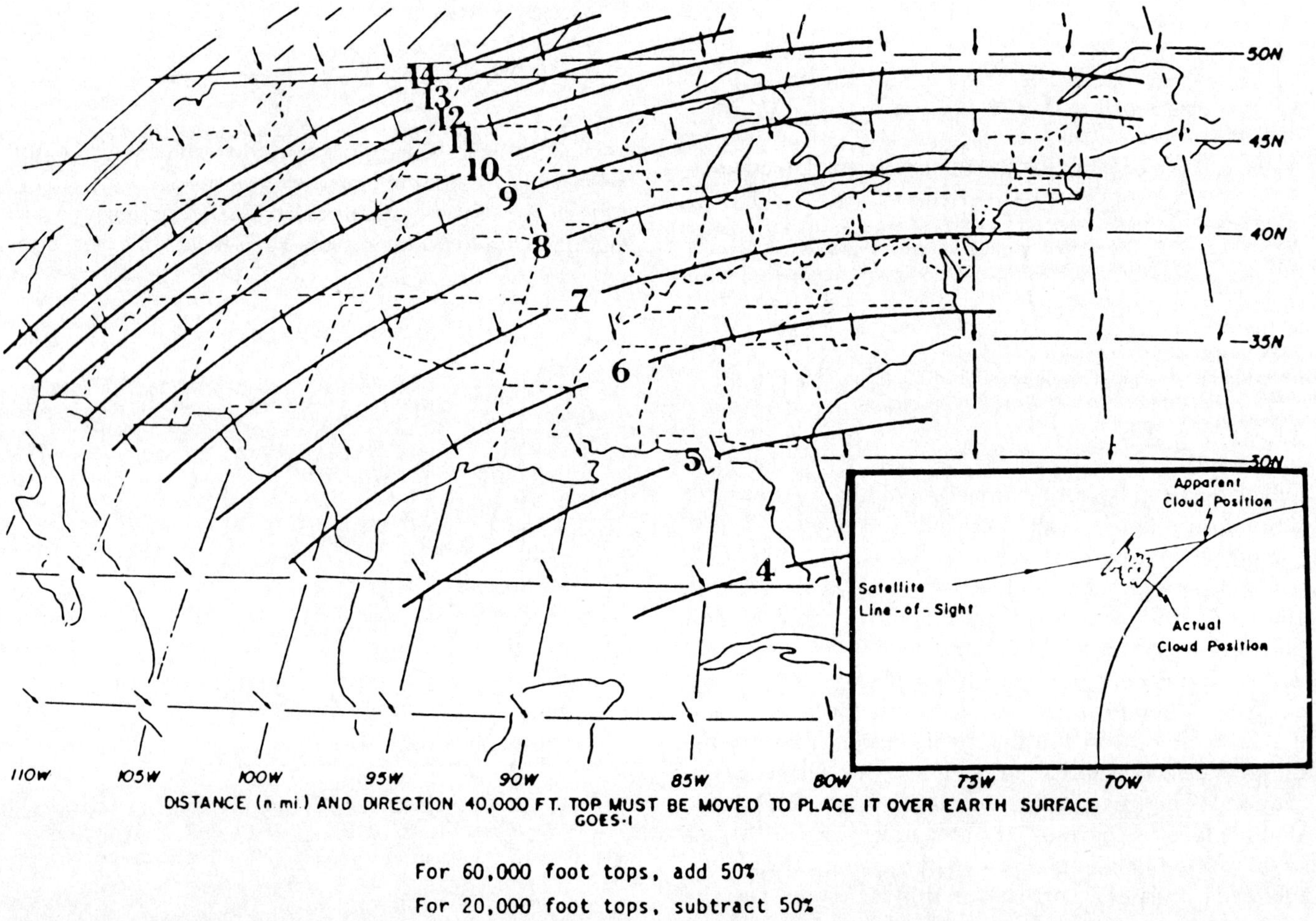

Figure VII-1-5. Earth location error of tall clouds in GOES imagery. The inset diagram, lower right, shows how the apparent cloud position can be significantly different from the actual one for tall clouds located toward the edges of GOES images. The curves are lines of equal displacement error for a satellite at 75°W and are labeled in nautical miles. The values are for a cloud top of 12 km (40,000 ft); for 6-km tops subtract 50%, for 18-km tops add 50%. To correct a cloud top position for a given height it should be displaced the appropriate distance toward the satellite subpoint in the direction of the small arrows (Weiss, 1978; courtesy of J. Purdom).

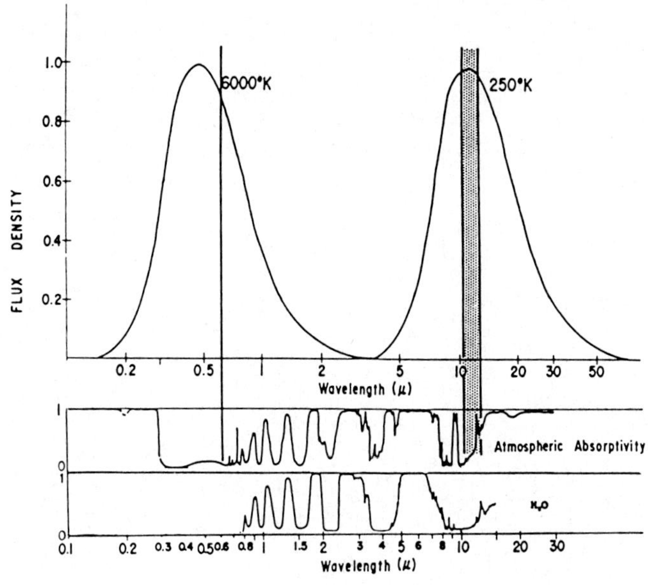

Figure VII-1-6. Emission and absorption spectra. The upper portion of this diagram shows normalized emission spectra for blackbody radiation at 6000 K, the temperature of the Sun, and for a body of 250 K corresponding roughly to the radiation emitted by the Earth and atmosphere. The radiation of the Earth and Sun covers two separate spectral ranges with little overlap. The middle part of the diagram shows the absorption spectrum of the atmosphere. The atmosphere is much more transparent to incoming solar radiation than it is to outgoing (IR) terrestrial radiation. The shaded band is the IR region (10.5 to 12.5 μm) commonly used for IR cloud imagery. The line at .63 μm is centered in the region where VIS measurements are made. The absorption spectrum for WV (H_2O) in the atmosphere appears at the bottom of the diagram (Fleagle and Businger, 1963).

terrain, on snow, or on ice covered oceans. Shadows are never seen on dark water surfaces. Figure VII-1-7a shows a cloud shadow where the Sun's angle is low.

There is more diffuse reflection from cloud surfaces near the Sun's specular point. Because clouds are quite variable, this effect is usually not obvious in the imagery. Variations in reflectivity also depend greatly on the physical characteristics of the clouds. Hewson (1943) has shown that the reflectivity of clouds with only small water droplets (5 μm), and a liquid water content of 1 gm^{-3}, increases rapidly until the clouds reach 200 m in thickness; above this thickness the increase is small. With less liquid water content (0.1 gm^{-3}) and the same droplet size, albedo increases less rapidly but continues with increasing cloud thickness. With larger droplets, the albedo is more sensitive to cloud thickness regardless of liquid water content. In general, cloud brightness viewed from space is associated with total cloud thickness. Frontal clouds, for example, appear quite light in tone. An exception is stratocumulus where small water droplets make the clouds highly reflective even though the cloud layer is not thick. Studies (Conover, 1962) show that striations of white on gray observed in large-scale cloud systems correspond to undulations in the cloud tops with brighter bands corresponding to thicker clouds. These patterns can also be seen when low cloud bands lie below a translucent overcast layer.

Water clouds always appear brighter than ice clouds of the same thickness under similar conditions of illumination. Table VII-1-1 (Conover, 1965) shows the average albedos of various cloud and terrestrial surfaces as measured from satellite picture data. These values, although from an early satellite, are useful for comparing relative reflectivities.

The relative brightness of land surfaces in VIS imagery depends on soil composition, type of vegetation, and snow cover. The most highly reflective land surfaces are alkali or salt flats and the sparsely vegetated soils of desert regions. Heavily vegetated areas appear dark, especially in the 0.58 to 0.68 μm spectral range of the AVHRR Channel 1 sensor, because of the low reflectivity of green plants at this wavelength. This is not the case in the AVHRR Channel 2 VIS imagery where land surfaces appear much brighter and there is greater contrast between land and water surfaces. Channel 2 is excellent for clearly identifying lakes and coast-line detail, incidentally facilitating accurate gridding of imagery.

The tone of snow in VIS imagery varies from bright white to light gray, depending on the nature of the terrain, vegetation, the age of the snow, and illumination (Figure VII-1-8a). In areas where terrain is relatively flat and without trees (steppe, tundra, fields of pack ice), snow surfaces appear uniformly white. In areas covered by extensive forests, a snow covered surface appears mottled with brighter patches in treeless areas and dark patches in forested areas. Knowledge of terrain features is important for cloud analysis over snow covered areas where clouds and snow often are equally reflective. Variations in terrain and vegetation produce uniquely recognizable surface patterns in the snow cover; these features disappear or are partially obscured when cloud covered. When clouds are stationary this is the only way of detecting clouds over snow in VIS range images.

Oceans, lakes, and rivers appear dark except when affected by Sun glint. Sun glint can range in appearance from a small bright spot to a large dull area with diffuse boundaries. Size and brightness of a Sun glint are dependent on the roughness of the water surface (McClain and Strong, 1969; Strong and DeRycke, 1973). The higher the waves the larger and more diffuse the Sun glint area. Little Sun glint can be detected in extremely rough seas; in calm seas, Sun glints are the brightest features in VIS satellite imagery (Strong and Ruff, 1970).

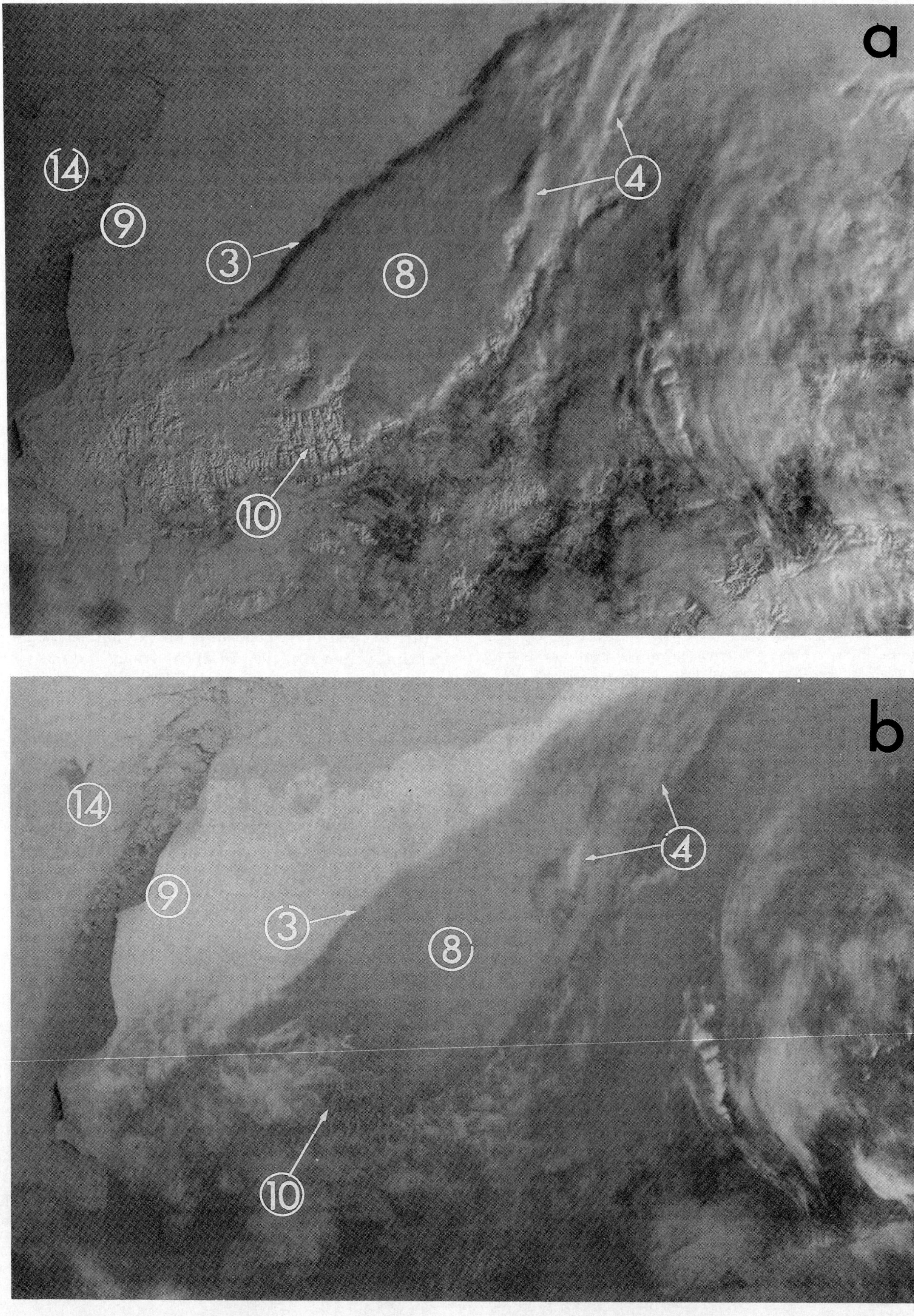
a
14
9
3
8
4
10
b
14
9
3
8
4
10

Table VII-1-1 *Average albedo in percent for various cloud and terrestrial surfaces as measured from satellite pictures*

FEATURE	ALBEDO
Cumulonimbus—large and thick	92
Cumulonimbus—small, top estimated 6 km	86
Cirrostratus—thick with lower clouds and precipitation	74
Cumulus and stratocumulus—MCO* over land	69
Stratocumulus—MCO* over land	68
Stratus—thick, approximately 0.5 km, over ocean	64
Sand—White Sands, New Mexico, U.S.A.	60
Stratocumulus masses within cloud sheet, over ocean	60
Snow, 3–7 days old mostly covering mountain above timber	59
Stratus—thin, over ocean	42
Cirrus—alone, over land	36
Cirrostratus—alone, over land	32
Cumulus of fair weather—MCO* over land	29
Sand—valleys, plains, and slopes	27
Sand and brushwood	17
Ocean—Gulf of Mexico—Sun glint	17
Coniferous forest	12
Lake—Great Salt Lake, U.S.A.	9
Ocean—Gulf of Mexico	9
Ocean—Pacific	7

* (MCO—cloud cover greater than 80% [Conover, 1965]).

A Sun glint always occurs between the satellite subpoint and the solar subpoint and is coincident with them when they are at the same point (Figure VII-1-9). For geostationary satellites, Sun glint moves between 23°N and 23°S with the seasons, and from east to west during daylight hours. The shape of these Sun glint areas is circular when the satellite and solar subpoints coincide, and is progressively oval as the distance between the points increases. The shape of Sun glint patterns is much different in polar orbiting satellite imagery because the Sun-satellite geometry is constantly changing rather than being relatively fixed as with a geostationary satellite (Figure VII-1-10). These glint patterns also move with the seasons. The glint appears on the eastern side of the polar orbiter subpoint track where it is local morning and on the western side where it is local afternoon.

VII-1.4 CHARACTERISTICS OF INFRARED IMAGERY

Infrared sensors measure the energy emitted by the Earth and its atmosphere. The amount emitted is dependent on the temperature of the radiation source. IR imagery prepared for meteorological use shows cold areas as light tones and warm areas as dark tones. This convention permits easy comparison between VIS and IR imagery. The Earth's and atmosphere radiate at an average temperature of 250 K with maximum energy flux at about 10 μm (Figure VII-1-6). At this wavelength, the Earth surface and thick clouds are assumed to be blackbody radiators so the radiation values from these surfaces can be converted to temperatures.

Measurements of radiation from clouds or the Earth's surface are made at wavelengths which fall in "atmospheric windows" where absorption by atmospheric gases is at a minimum. Figure VII-1-6 shows the absorption spectrum (H_2O) for the atmosphere as a whole. Atmospheric windows centered near 4 and 11 μm can be used to calculate the temperatures of cloud tops and the Earth's surface.

Measurements made at wavelengths where the atmosphere absorbs outgoing radiation provide imagery of the different gaseous constituents of the atmosphere, depending on the spectral interval selected. Figure VII-1-6 also shows that water vapor is responsible for a large part of atmospheric absorption.

Three factors control the effect of absorption on the energy reaching the satellite.

The first two factors, the degree of absorption at each wavelength being sensed, and the distribution of the absorbing gas within the atmosphere, are numerically described by a calculated term, the weighting function (see Chapters II-2 and VII-5). The majority of weighting functions, such as those in Figure VII-1-4, are calculated for a standard atmosphere. When distribution of the absorbing gas varies, as in the case of water vapor, this standard weighting function does not correctly represent relative contributions of individual atmospheric layers, and must be recalculated for each profile. The third factor, the temperature along the path length of the radiation, is expressed as the Planck function.

The lower part of Figure VII-1-4 shows the bandwidths of several VAS channels. Together, the two water vapor Channels (9 and 10) range from 6.0 to 7.6 μm. Water vapor has an absorption level of 100% at 6 μm but only about 50% at 7.3 μm. Thus more energy is sensed at the satellite from the lower atmosphere in Channel 9 (7.3 μm) than in Channel 10 (6.7 μm). For a standard atmosphere, the energy measured at 6.7 μm comes from a layer centered near 400 mb, and measurements at 7.3 μm from a layer centered near 600 mb. Thus, the imagery at 6.7 μm is more sensitive to middle and high tropospheric water vapor than that at

←

Figure VII-1-7. Cloud and surface features; northwest Alaska and Arctic Ocean, NOAA-5, 2033 UTC, February 26, 1977; (a) 1-km VIS image, (b) 1-km IR image, [3] shadow, [4] highlights, [8] altostratus, [9] snow on land, [10] snow covered mountains, [14] ice.

Figure VII-1-8a. Cloud and surface features; view of southwest Alaska and Bering Sea, NOAA-5, 2033 UTC, February 26, 1977, 1-km VIS image; [1] cloud line, [2] wave clouds, [3] shadow, [4] highlights, [5] striations, [6] cumulus congestus, [7] cumulonimbus, [10] snow on mountains, [11] thin altostratus, [12] stratus, [13] thin cirrus, [18] anvil cirrus.

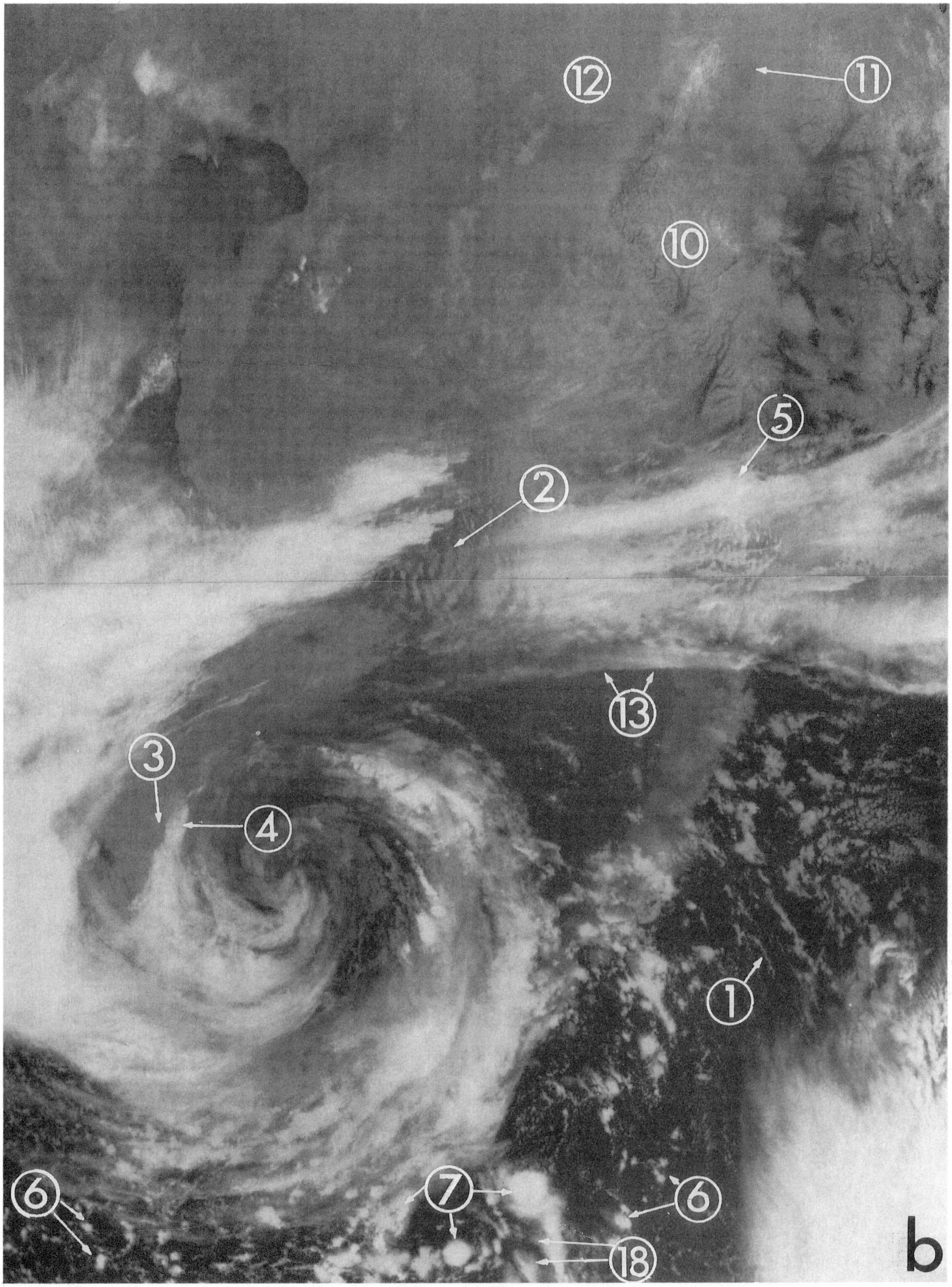

Figure VII-1-8b. Cloud and surface features; southwest Alaska and Bering Sea, NOAA-5, 2033 UTC, February 26, 1977, l-km IR image. See Figure VII-1-8a.

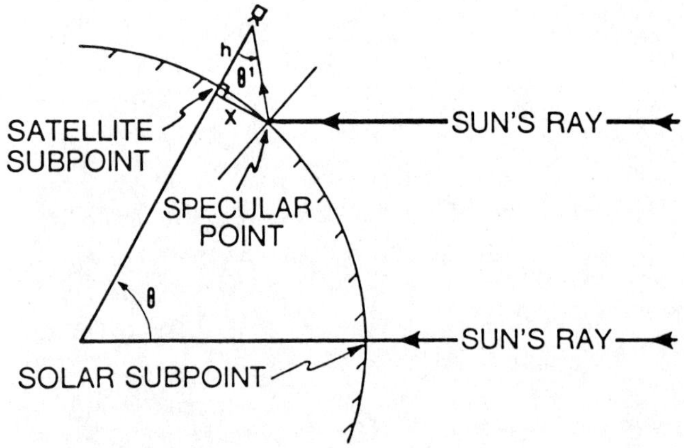

Figure VII-1-9. Relationship between satellite subpoint, solar subpoint, specular point, satellite sensor viewing angle ° of specular point, and angular distance between solar and satellite subpoints ° (Fett and Mitchell, 1977).

7.3 μm. Imagery at 7.3 μm is used to detect middle clouds and low clouds (when air is cold and dry) and shows low level moisture patterns, features not usually seen on the 6.7 μm channel imagery. Figure VII-1-11, parts (d) and (e) show WV imagery of the same scene from Channels 9 and 10 of VAS.

Two aspects of the moisture distribution within a satellite Field of View (FOV) affect the 6.7 μm radiance measured by the satellite: the total amount of water vapor in the atmospheric column and its vertical location. When radiant transfer is simulated, these effects are calculated as the weighting function for a given layer. The temperature of the water vapor, represented by the Planck function, is multiplied by the weighting function to calculate the radiance contributed by the layer. The satellite measures the net 6.7 μm radiance emitted through the atmospheric column. Dark pixels represent profiles where the air aloft is relatively dry and the sensor detects radiation from the warmer moisture at lower levels. Conversely, colder moisture concentrated at mid- and high levels produces lower radiances represented as light shades; here the sensor cannot detect the radiation emitted by the warmer moisture at lower levels.

Horizontal variability of these net radiances (from varying water vapor profiles) defines the patterns observed in the imagery. For instance, an extensive region of high-level moisture can produce light-toned patterns modulated by variations in low level moisture. Moreover, the moisture patterns at higher latitudes are most distinct in the summer when the temperature differences between levels are the greatest. Water vapor imagery (6.7 μm) has proved useful in locating jets; these are regions of much contrast in temperatures and in water vapor profiles.

Imagery from the 14.5 μm CO_2 Channel is sometimes used to observe stratospheric temperature patterns. Carbon dioxide is assumed to be uniformly distributed in the atmosphere so the return is representative of atmospheric temperature. Figure VII-1-12 shows TOVS Channel 3 radiances over the U.S. during the winter. The darkest grays represent the warmest stratospheric air; the lightest the coldest. The warm area is part of the circumpolar winter stratospheric warm belt found on 100-mb charts. The weighting function for this channel peaks near this level. Figure VII-1-4 shows the weighting function for VAS Channel 2 (14.4 μm) which is similar to that of TOVS Channel 3. The strong gradient on the Equatorward side of the warm belt in Figure VII-1-12 correlates with the mid-latitude 300-mb jet in the baroclinic zone above the polar front (Brodrick, 1978) and the tropopause topography in the same region (Anderson et al., 1982).

Middle and high clouds are more easily identified in IR imagery than in VIS imagery. Gray shades representing these clouds range from white to medium gray. Usually there is a large contrast between the bright grays of upper clouds and the darker grays of the much warmer Earth or lower clouds. During winter in polar regions the difference between the temperature of the Earth's surface and that of middle and high clouds is small; thus it is difficult to differentiate between clouds and the Earth's surface. To further complicate the situation, clouds above a polar snow surface can appear darker than the ground when a strong temperature inversion is present. For example, note the contrast between (8) and (9) in Figure VII-1-7b. If a strong surface inversion is present at night, the tops of fog and stratus are warmer than the surrounding terrain, and so appear darker than the surrounding area even in the absence of snow cover. In IR imagery, this is called "black fog" or "black stratus."

It is often difficult to differentiate between cirrus and middle clouds in the IR because of the transmission characteristics of ice crystals. Cirrus is translucent to IR radiation in the atmospheric window centered at 11.5 μm, and transmissivity depends on its vertical thickness and ice concentration within the cloud. Thin cirrus is much more translucent to IR radiation than thick cirrus. For example, a cirrus sheet of a given ice concentration and particle size transmits 80% of the radiation from below if it is 0.5 km thick but only 20% when it is 5.0 km thick (Jacobowitz, 1970). A satellite IR radiometer senses a single integrated value that is a combination of low radiation from cirrus tops and higher radiation coming up through the cirrus. Thus, some cirrus cloud top temperatures derived from IR data are too high and cloud top heights calculated from them, assuming blackbody emission, are too low (Figure VII-1-13).

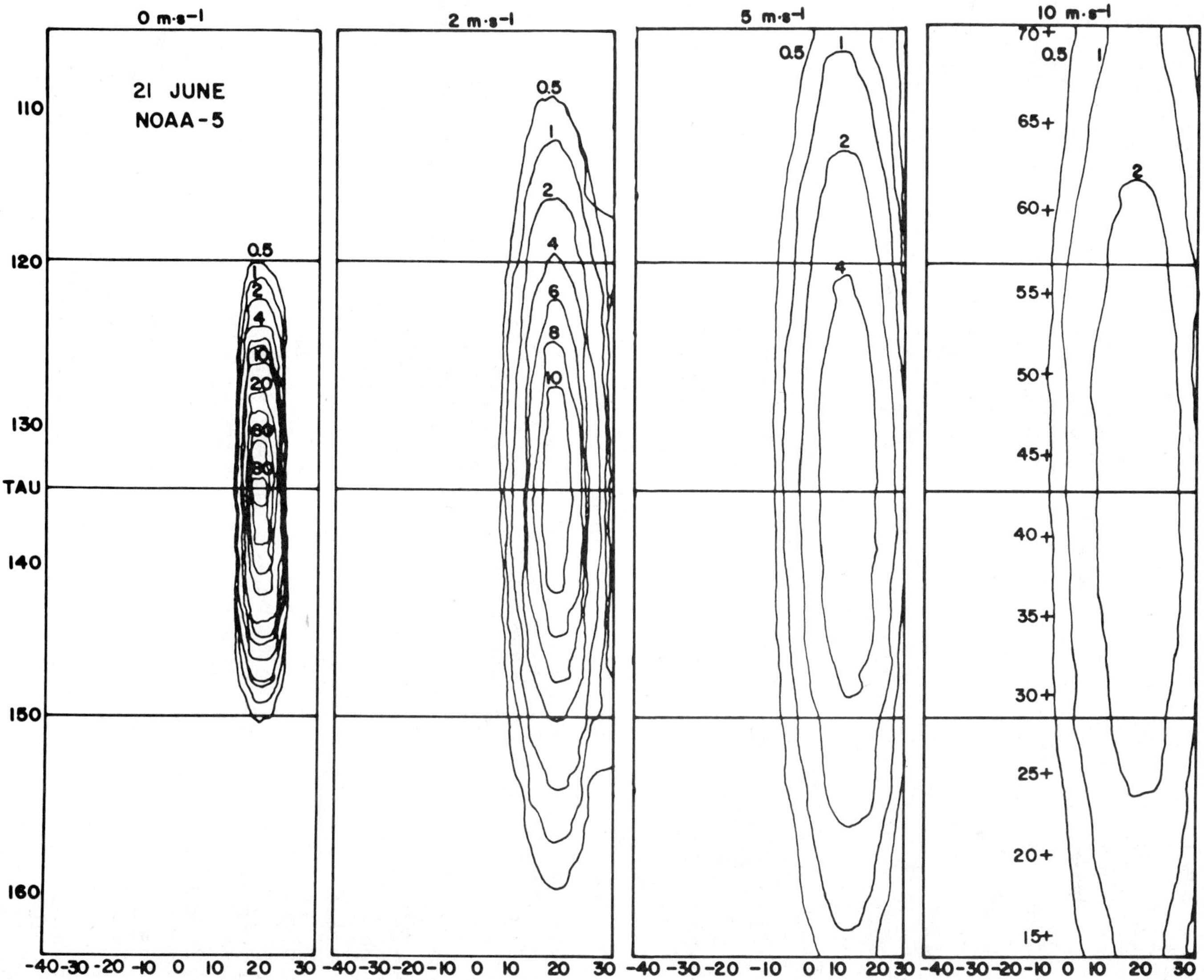

Figure VII-1-10. Computed patterns of reflected Sun glint for a polar orbiting satellite. The computation is for June 21 in the Northern Hemisphere. Diagrams are shown for 0, 2, 5 and 10 m s^{-1} wind speeds. The contours are in isophotes and represent the percentage of bidirectional reflectance from the water surface. Tau (left ordinate) shows the position in degrees latitude along the satellite subtrack measured from the ascending node at the equator. The equivalent latitude appears in the right hand diagram. Abscissa is degrees off nadir (courtesy of A. Strong).

A small amount of atmospheric absorption, mainly by CO_2 and H_2O vapor, takes place in the 11.5 μm window, causing a radiating cloud surface to appear colder in the satellite data than it actually is. Figure VII-1-14 shows this effect for polar satellites; it is most pronounced near the horizon, where the path of the radiation is quite long, and in the tropics where water vapor concentrations are high.

The appearance of imagery produced from the AVHRR Channel 3 (3.7 μm) and VAS Channel 12 (3.9 μm) is often difficult to interpret. The wavelengths of these channels are at the long wavelength end of the solar spectrum and at the short wavelength end of the blackbody emission spectrum for Earth temperatures (Figure VII-1-6). The satellite detects a combination of both reflected and emitted radiation during daylight at this wavelength. Note that the low clouds over the Great Lakes, the Dakotas, and the oceans in Figure VII-1-11 appear darker (warmer) in the 3.9 μm image than in the 11.2 μm image due to the contribution from reflected solar radiation at 3.9 μm. The gray tone of clouds depends on their absorption and emission characteristics at the different wavelengths.

Coast-lines, mountains, lakes, rivers, snow, ice, and ocean currents can all be seen in IR imagery. The larger the temperature difference across the boundary of a surface feature, the more apparent the feature is. Temperature differences vary with latitude, season, and

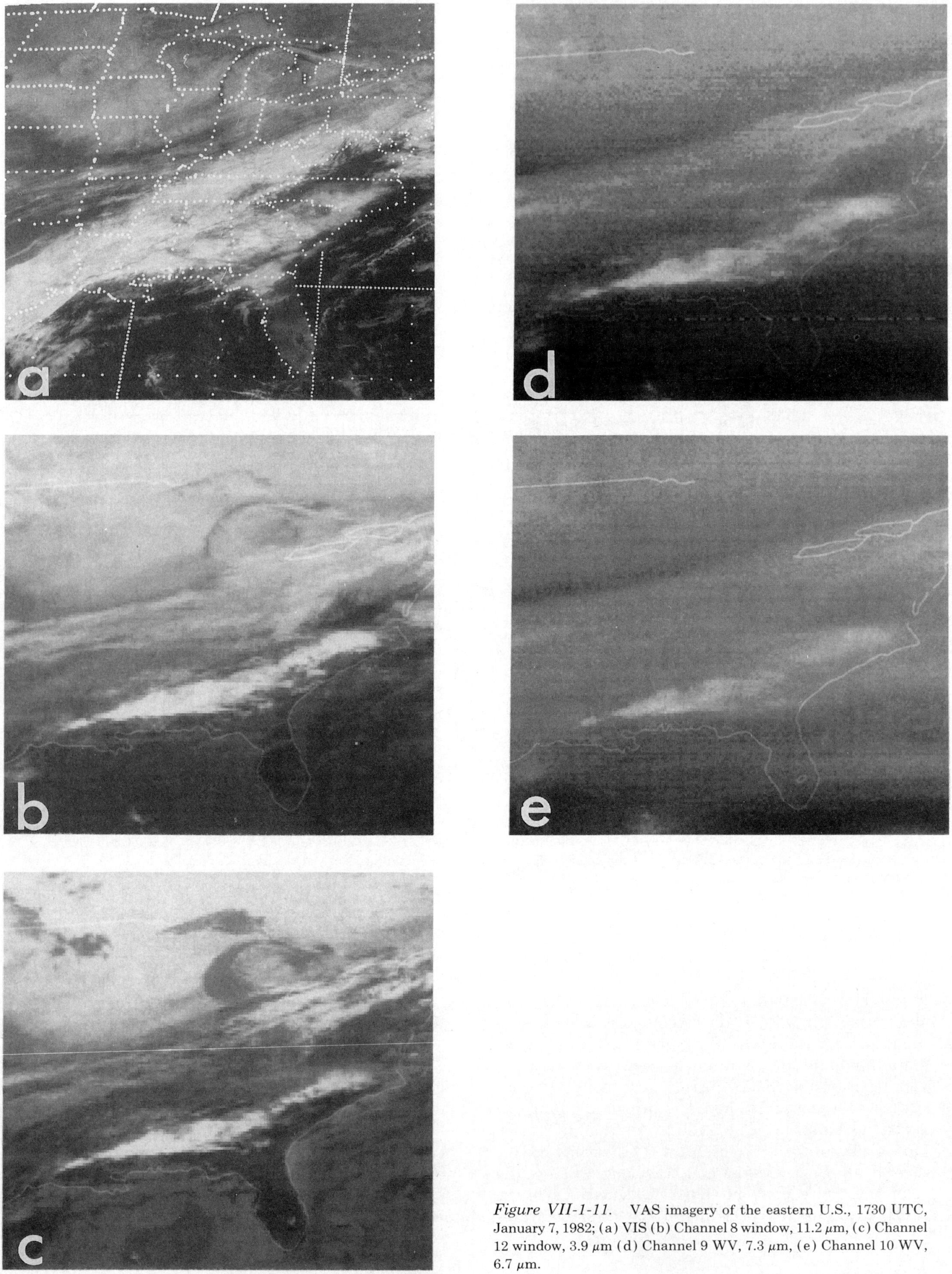

Figure VII-1-11. VAS imagery of the eastern U.S., 1730 UTC, January 7, 1982; (a) VIS (b) Channel 8 window, 11.2 μm, (c) Channel 12 window, 3.9 μm (d) Channel 9 WV, 7.3 μm, (e) Channel 10 WV, 6.7 μm.

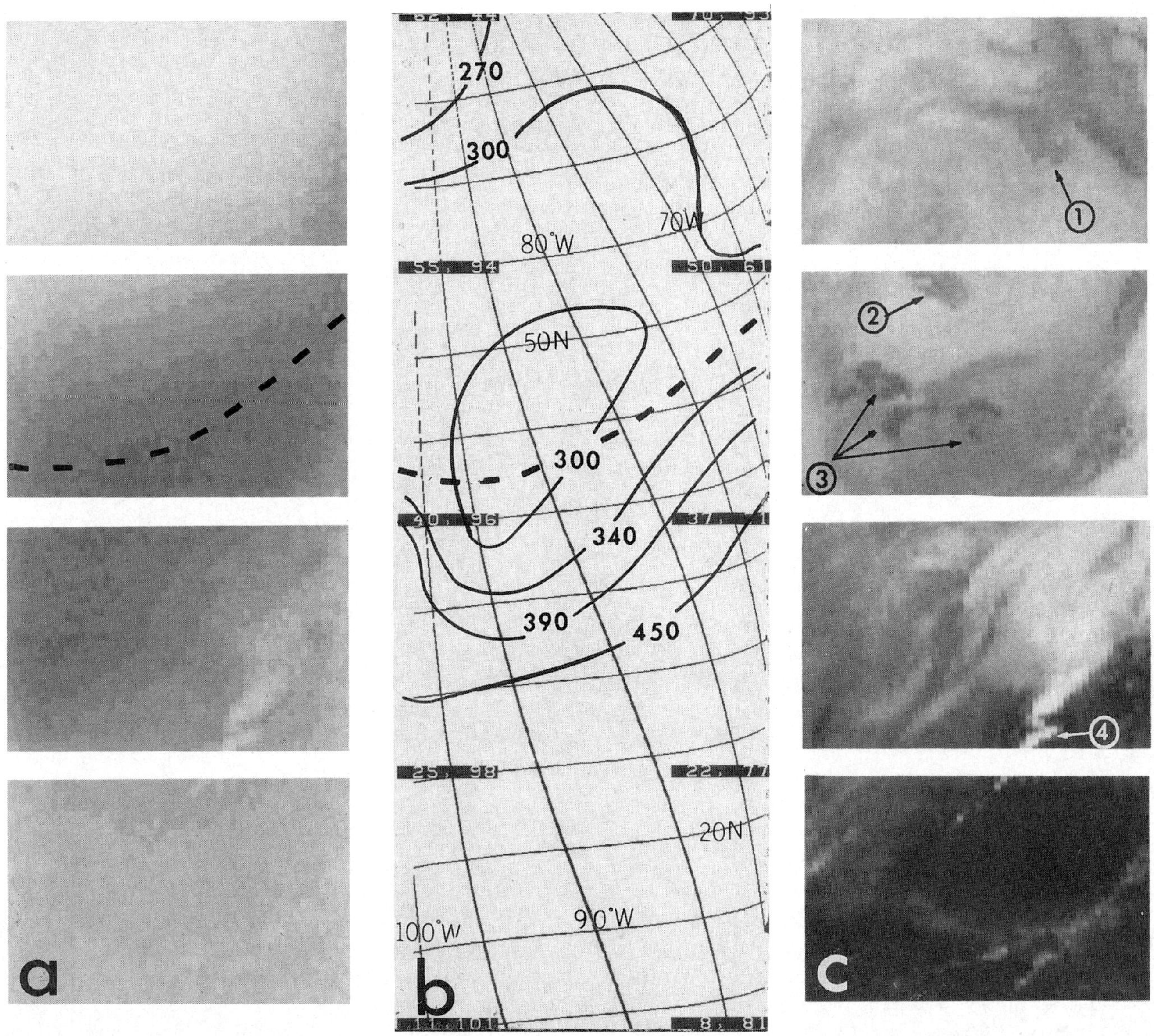

Figure VII-1-12. (a) Imagery from Channel 3 (14.5 μm) IR of the TOVS for 0831 UTC, January 14, 1982. The darkest gray shades are the warmest stratospheric air; the whitest, the coldest. The axis of the circumpolar winter stratospheric warm belt is shown by the dashed line. Cold cloud tops contaminate the thermal pattern between 25° and 35°N and 75° and 85°W. (b) NMC tropopause analysis for 1200 UTC, January 14, 1982. Contours are labeled in hundreds of feet. (c) Imagery from the TOVS window Channel 8 (11.1 μm) IR for 0831 UTC, January 14, 1982. Visible landmarks are [1] Ungava Bay, [2] Hudson Bay, and [3] the Great Lakes. A line of thunderstorms [4] appears in the Gulf of Mexico (Anderson et al., 1982).

time of day, e.g., coastlines are clearest in the summer daytime when land-sea temperature differences are at a maximum. The gray shade contrast at a land-water boundary often reverses at night. In daytime the land usually appears darker than adjacent water; at night the water, although remaining approximately the same gray shade, appears darker than the land which has cooled by radiation.

Some mountain ranges are easily seen in IR imagery. Those with a large temperature difference between the high elevations and lower adjacent land and those with steep slopes display a strong temperature gradient in the IR image. The Sierra Nevada in California, which rises from near sea level, is more obvious in IR imagery than the equally high and cold Rocky Mountains over the central U.S.

Snow covered terrain usually appears whiter in IR imagery than its surroundings. The IR brightness of a snow area depends, among other things, on the presence and type of vegetation and on how much of the

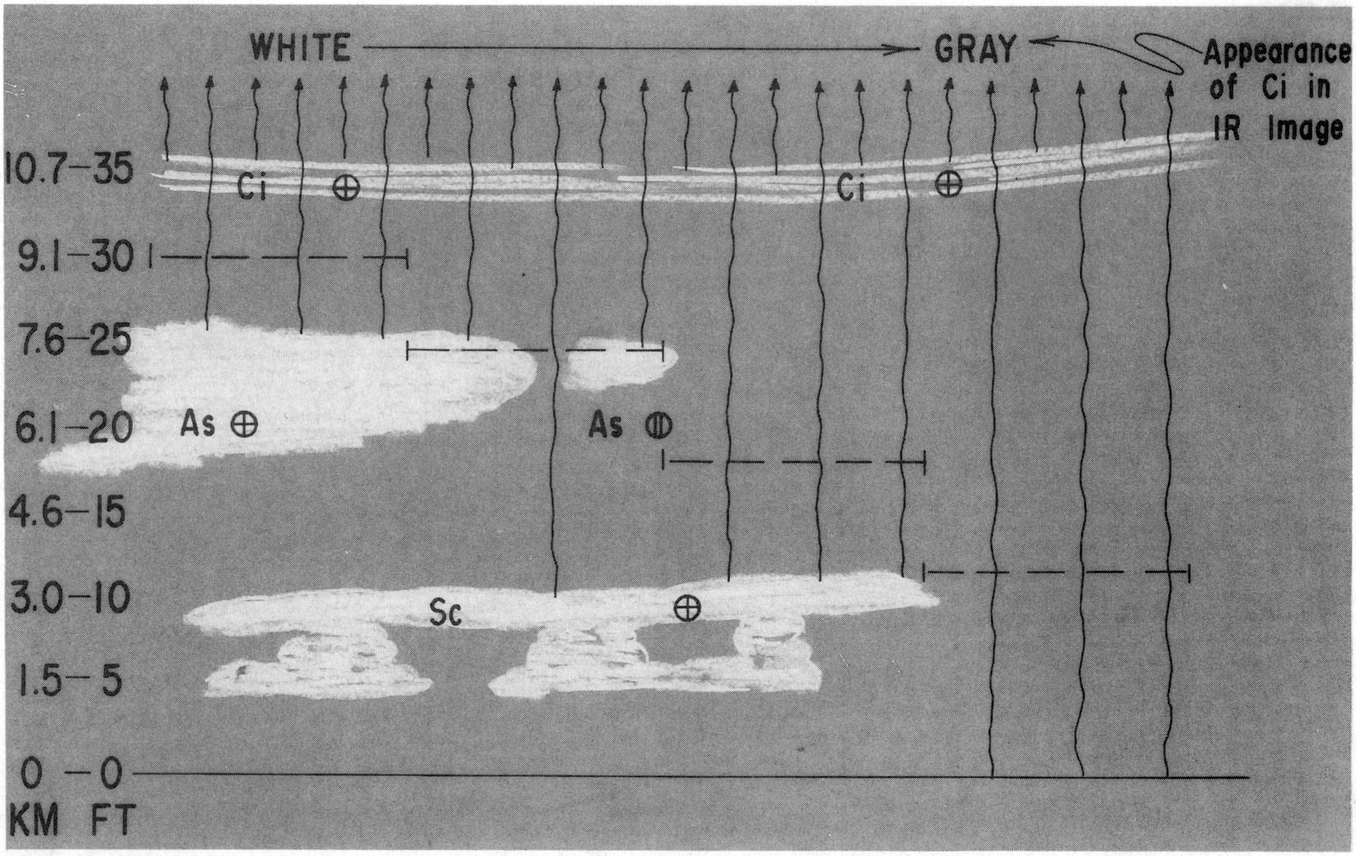

Figure VII-1-13. Schematic showing how the gray tone and apparent height of a uniform layer of cirrus varies depending on the temperature of the underlying cloud tops and the Earth's surface. The height of the clouds appears on the left. The dashed lines represent the apparent height of the cirrus (Anderson et al., 1974).

vegetation is snow covered. Light snow on the tops of high mountains is less detectable in IR than in VIS imagery because the temperature difference between cold land and snow is small, while the difference in albedo between snow and land is large.

VII-1.5 CLOUD IMAGE ANALYSIS TECHNIQUES

VII-1.5.1 Feature Enhancement by Digital Processing

This is a widely used technique that alters imagery to make interpretation easier. It is designed to improve the contrast, brightness, and sharpness of images. Enhancement is generally accomplished using digital data and computers, although it can be done with electronic analogue circuitry. Each digital image data point (pixel) has a number (0 to 255, for 8-bit data) to which can be assigned a gray tone or color corresponding to measured radiances. Enhancement is a gray level adjustment that accepts a digital image as input and produces an output image with each pixel gray level changed according to some rule. Look-up tables specify the relationships between input and output values; a graph, called an enhancement curve, shows this relationship.

Examples of IR enhancements that have proven useful are shown in Figure VII-1-15 for comparison. The IR data in Figure VII-1-15b are displayed as received from the satellite; the output values are the same as the input values so the enhancement curve is linear. The temperatures of the clouds and ocean in this scene range from about +25°C to −75°C; the extremes of the sensor range from +56°C to −110°C. Because these extremes were not recorded, there are no completely black or white tones in the image, only light to dark grays.

The ZA curve (Figure VII-1-15c) is used to stretch the contrast of an average GOES scene over the range of meteorologically significant data (+30°C to −75°C) by making use of the entire black-to-white dynamic range of display devices. To do this, all pixels higher than 30°C appear black, all those lower than −75°C appear white, and two steep slopes or ramps are intro-

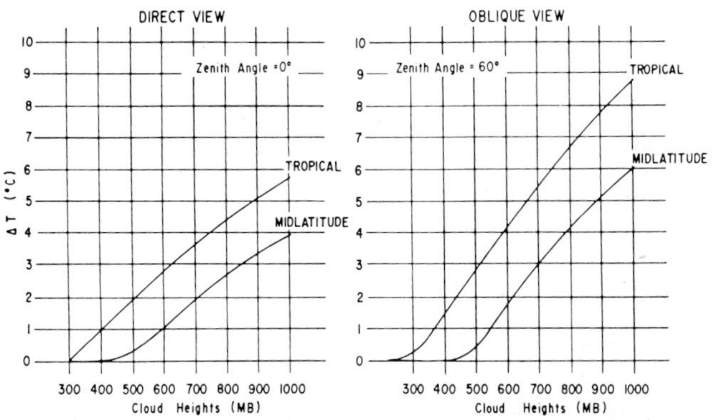

Figure VII-1-14. Error introduced in cloud top temperatures due to WV attenuation. These graphs apply to polar orbiting satellites and show errors directly under the satellite (zenith angle = 0°) and for a more oblique viewing angle (zenith angle = 60°). The curves represent typical air masses for the tropics and mid-latitudes. When the satellite sensor is pointing straight down, there is no error if the clouds are at or above 300 mb. For the same case, clouds at 1000 mb will appear to be 4°C colder than they actually are at mid-latitude and 6°C colder in the tropics. The errors become greater for oblique viewing angles. These curves are based on a radiative transfer model and are calculated for the 11.5 μm channel of ITOS-1. The different altitude of TIROS-N should not affect the results greatly (Anderson et al., 1974).

duced for high and low end enhancement. Data in the temperature range from 6°C to −55°C are not enhanced so the curve is linear for this interval.

The MB curve (Figure VII-1-15d) is the same as the ZA curve for temperatures higher than −32°C. Temperatures between −32°C and −59°C are divided into four segments with a single gray shade assigned to each. This contours the cold cloud tops in this temperature range. Gray shade reversal is used to increase the contrast between each segment of the curve. The last ramp of the curve covers a narrow range of temperature (16°C) and is designed to detect small temperature differences of overshooting tops in thunderstorm anvils. This effect is seen in Figure VII-1-15d, where small-scale detail is visible in the hurricane cloud canopy immediately south of the hurricane eye. The coldest top is on the west side of the eye wall. Chapter VII-6 covers uses of the MB curve for precipitation estimation.

The BB curve (Figure VII-1-15e) was designed for hurricane analysis. Here the objective is to show clearly the synoptic-scale banding in the storm; this is done by contouring the cold portion of the curve on the basis of normal hurricane cloud top temperatures. Gray tone reversal is again used to increase contrast between temperature segments. At the warm end of the curve there is a very steep ramp, from +28°C to +9°C, to permit analysis of the low level cumulus cloud organization around the storm; examples of clouds detectable with this enhancement are visible in the lower right-hand corner of Figure VII-1-15e. These lower clouds can be seen in the VIS image but not in the unenhanced, ZA, or MB displays. Applied uses of the hurricane analysis curves are discussed in Chapter VII-7.

The above enhancements are used primarily for gray tone hard copy displays. The paper or film used to record these data has a unique response to light, voltage, etc. It is often useful to tailor enhancement of imagery to match the linear portion of the film or paper response curve to produce a display with maximum contrast.

Color television displays are now frequently used for both operational and research analysis of satellite imagery. Color displays are superior to black and white displays because the eye can more easily differentiate between colors than tones of gray. This makes it possible to show larger temperature ranges in detail without repeating gray tones. Color has three main attributes: hue (red, orange, blue, etc.); saturation, which determines the mixture of the color with white (the more white the lower the saturation; pink is unsaturated red), gray, or black (high saturation); and brightness (luminosity), the degree to which colors appear to emit more or less light.

In developing color enhancement for cloud displays, it is best to select color combinations that do not detract from the continuity of the cloud forms. This is especially important when creating animated color displays.

VII-1.5.2 Edge Enhancement

There are standard algorithms (Hall, 1979) which either sharpen or smooth edges. Purdom and Vonder Haar (1983) have pointed out the need for smoothing 8-km IR imagery when used in conjunction with 1-km VIS imagery or mapped radar. When small areas are enlarged, the blockiness of 8-km IR pixels makes it difficult to follow the continuity of cloud top topography. Smoothing the IR greatly increases its usefulness when displayed and annotated in enlarged format (Figure VII-8-25a).

VII-1.5.3 Animated Imagery

This is produced from GOES VIS, IR (11.2 μm), and 6.7 μm WV imagery. The time interval of the imagery is dictated by the scale and imagery resolution of the features to be analyzed. One to two-hour intervals have proven satisfactory for animated 14-km, 6.7 μm WV imagery analysis; 3–5 min intervals are best for analyzing convective scale storms with 1-km GOES VIS imagery. The shorter the interval between images the more the cloud motions are slowed and the easier

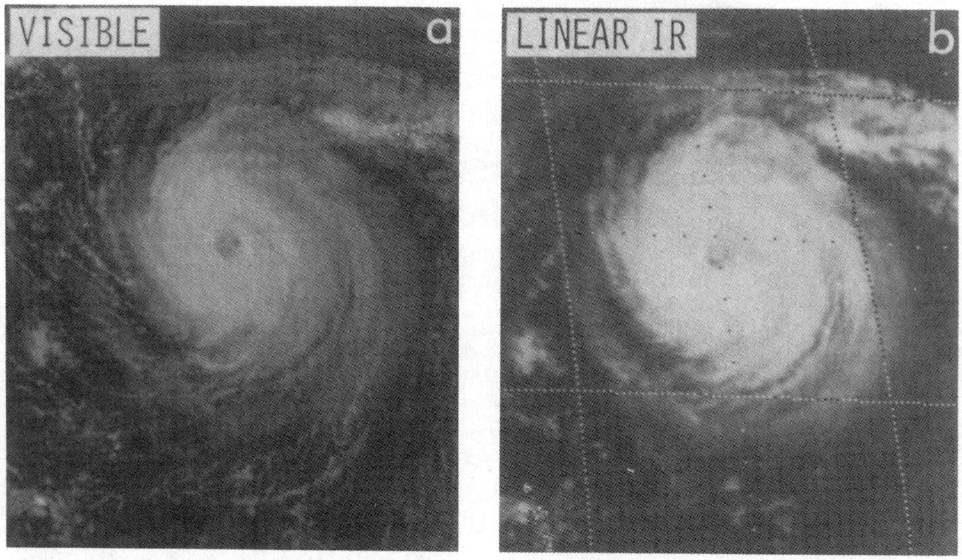

Figure VII-1-15. GOES imagery enhancement for an Atlantic hurricane; (a) VIS, (b) unenhanced IR (11.2 μm).

it is to follow the motion, growth, and decay of cloud features. When analyzing small, moving cloud elements, such as in fields of cumulus, it is important to remember that when the distance that elements move in the time lapse between images is similar to the distance between elements, the apparent motion of the elements may be reversed or otherwise incorrect.

Animated satellite imagery normally shows clouds moving with respect to the Earth. A different approach, an analysis technique called "follow motion," has been used to follow the relative motions of clouds. To use this technique, all images in an animated sequence are registered on a particular cloud feature that is moving relative to the Earth. This shows the motion of the clouds and atmosphere relative to the particular feature which appears to stand still—a LaGrangian frame of reference. The technique removes translational motion and permits relatively easy identification of rotation, convergence, and deformation. It has also been used to determine various scales of motion, and is most useful in mesoscale analyses of small areas. For example, squall lines are held still in the animated display so areas of maximum convergence into the squall line can be identified by following low cloud motions.

VII-1.5.4 Multispectral Composite Images

These are useful when it is desirable to view two types of imagery at the same time. In one composite, data of one kind are replaced by data of another kind for portions of the image. An example is the composite VIS image which shows VIS data everywhere except where temperatures are lower than −43°C in the IR data. In such areas the VIS data are replaced by enhanced IR imagery. This procedure is particularly useful for convective storm analysis. The IR portion provides unique information on cloud top topography and convective cell development of thunderstorm complexes and the VIS portion unique information about the low-level wind fields. Compositing of this kind can also be used for WV imagery. In this case, less noisy, better calibrated 11 μm IR data are used to replace the 6.7 μm data in areas of cold cloud tops. At high levels, the 6.7 μm sensor detects only clouds since insignificant amounts of water vapor aloft are present. Figures VII-10-5 and VII-10-11 show examples of composite images.

Another type of compositing can be done on video displays. The VIS image pixels of an image are colored to match the temperature of corresponding IR pixels. Such composites provide information over the entire image about cloud top temperatures as well as the detailed cloud structure shown in the high resolution VIS data.

VII-1.5.5 Time Composites of Imagery

In early photography methods, a series of registered images were photographed using the same negative to produce an image of average brightness. Miller and Feddes (1971) constructed monthly global composites of VIS digital data for 1400 LST. These showed the cloud climatology of large scale features such as the InterTropical Convergence Zone (ITCZ). More recently, Weaver and Kelly (1982) constructed multiday digital composites showing diurnal cloud changes over Colorado. The data are stratified according to synoptic

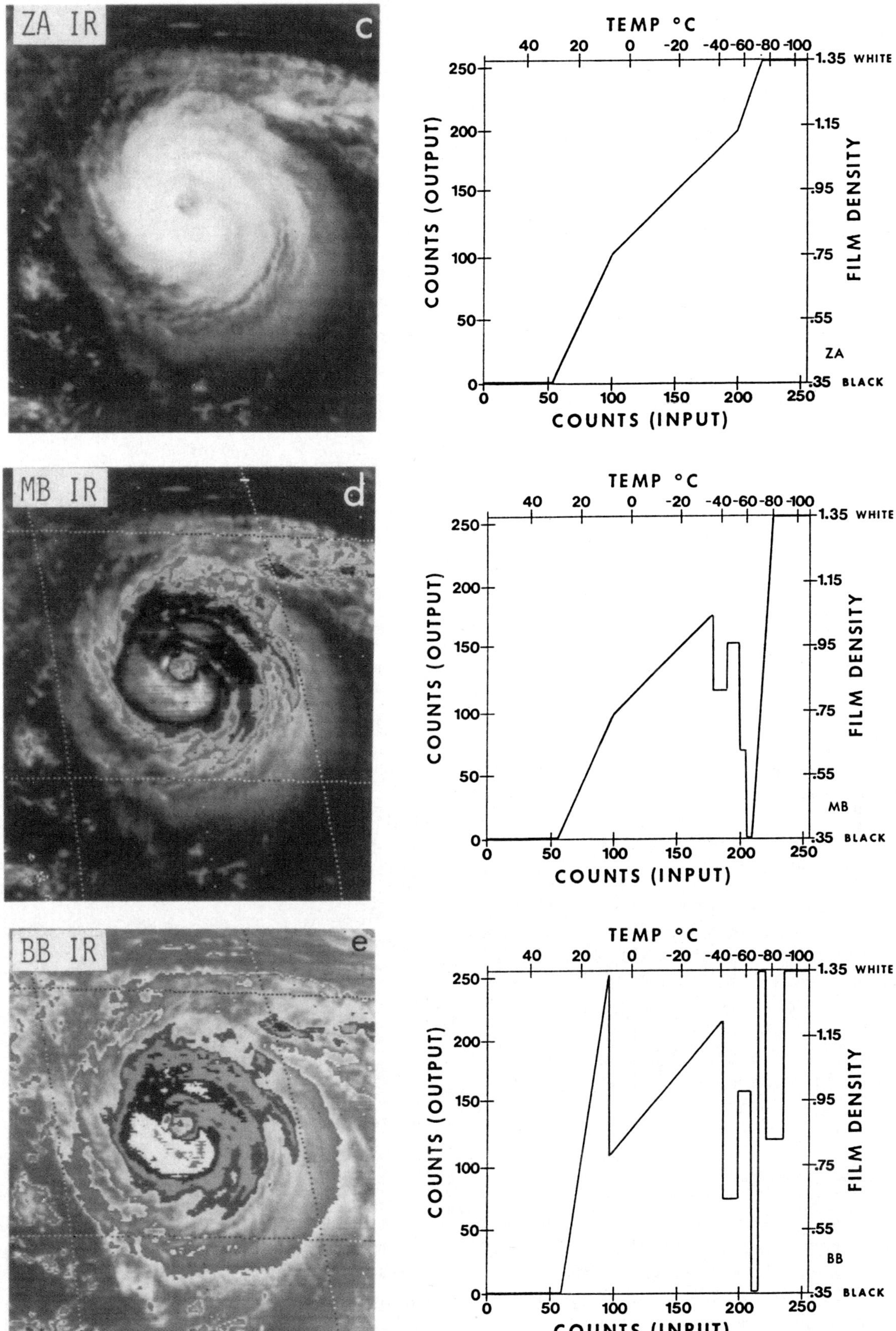

Figure VII-1-15. (c) general purpose meteorological IR enhancement—ZA, (d) cold cloud top enhancement—MB, (e) hurricane enhancement—BB.

Table VII-1-2 *Cumuliform cloud characteristics in satellite imagery*

CLOUD TYPE	CHARACTERISTIC SHAPE OR PATTERN	CHARACTERISTIC TONE		EXAMPLES/ FIGURE NUMBER
		VIS	IR (11.5 μm)	
Cumulus	Small elements or groups of elements. If cumulus elements are below sensor resolution, land or ocean areas appear mottled or slightly brighter than if cloud free.	Medium brightness.	Dark tone; usually difficult to detect.	VII-1-16 [I]
Cumulus-congestus	Irregularly shaped elements or group of elements. Frequently organized into lines or open cellular patterns in areas of cold air advection over oceans	Bright.	Relatively bright; depends on coldness of cloud top.	VII-1-8 [6] VII-1-16 [A]
Cumulo-nimbus	Globular or carrot shaped depending on vertical wind shear. With shear: upwind edge sharp; opposite downwind anvil edge indistinct.	Very bright; distinct shadows and highlights when sun angle low.	Bright white in center or adjacent to upwind boundary of anvil. Darker toward downwind edge.	VII-1-8 [7] VII-1-16 [H]
Strato cumulus	Closed cellular pattern (VIS imagery) over oceans in areas of cold air advection. Cell size varies depending on inversion height. Ship trails sometimes present. Gravity waves often present when viewed in motion. Mountain wave and mesoscale lee vortex patterns may be present in clouds.	Center of cells in pattern bright, become gray toward edges where cloud thins. If cells are below sensor resolution they appear smooth as if stratus.	Surface uniform dark gray; cellular structure not evident. Can be difficult to detect when cloud tops are low.	VII-1-16 [B]

situation and then used as a mesoscale forecast tool. Such cloud composites clearly show the effects of terrain on mesoscale cloud distribution for given flow and stability conditions.

VII-1.5.6 Multispectral Digital Cloud Data Analysis

This has been attempted using discriminant analysis techniques. The results show that high, middle, and low clouds can be separated reasonably well, but differentiating between the different cloud types within these broad categories is less successful. Cirrus clouds, in particular, are difficult to distinguish from no cloud situations. A more detailed description of computer cloud classification techniques is given in Chapter IX-2.

VII-1.6 CLOUD AND CLOUD SYSTEM IDENTIFICATION

Cloud and cloud system identification are an integral part of many meteorological applications of satellite data. The elements of basic imagery interpretation (Estes et al., 1983) are routinely used. These include: size and shape (the geometric arrangement of cloud forms); texture and pattern recognition, which relate to the spatial arrangement of cloud elements; indications of height, revealed directly in IR imagery and deducible in VIS imagery; and the locational elements of site and association that help identify clouds related to orographic effects or to particular areas of large synoptic-scale circulations.

Clouds viewed from early satellites were initially classified into three general categories based on appearance: cumuliform, stratiform, and cirriform. These categories are useful for converting low-resolution VIS imagery into nephanalysis for operational uses. Cumuliform clouds reflect varying degrees of atmospheric instability. They are identified by banded, cellular, or wave cloud patterns and by irregularly shaped cloud elements of varying size (Conover, 1962). Stratiform clouds, generally associated with a stable atmosphere, include thick fog, stratus, altostratus, and nimbostratus. They appear smooth and lack organized patterns. Cirriform clouds are recognizable by their distinct fibrous texture and sometimes striated appearance. The above cloud categories were developed for operational use and, when combined with estimates of cloud amount, provide a simple method for making large-scale nephanalyses (U.S. Weather Bureau, 1963).

Table VII-1-3 *Stratiform cloud characteristics in satellite imagery*

CLOUD TYPE	CHARACTERISTIC SHAPE OR PATTERN	CHARACTERISTIC TONE		EXAMPLES/ FIGURE NUMBER
		VIS	IR (11.5 μm)	
Fog-Stratus Combination	Smooth tops; boundaries often sharp and defined by topography.	Uniform medium gray tone if thick; mottled tone if thin.	Uniform dark gray tone; varies with season. Difficult to detect; can appear darker than surrounding land at night.	VII-1-8 [12]
Altostratus and Altocumulus	Smooth tops, boundaries can be ragged or smooth, often in layers. When in combination with deep convection, tops appear bumpy if low sun angle. Cellular structure of altocumulus too small for sensors to resolve so cannot be distinguished from altostratus. Gravity waves common when layers thin.	Light gray; appears mottled or striated depending on thickness or layered structure.	Uniform medium gray tone; depends on height. Can appear darker than surrounding snow covered land in polar regions.	VII-1-7 [8] VII-1-8 [11]

The advent of 1-km imagery in the 1970s made it possible to identify a number of specific cloud types that are described by the international cloud classification system (World Meteorological Organization, 1975). The characteristics of the individual cloud types (Anderson and Veltishchev, 1973) making up the cumuliform, stratiform, and cirriform categories are summarized in Tables VII-1-2, VII-1-3, and VII-1-4. Examples of each cloud type in the tables appear in Figures VII-1-7, VII-1-8, and VII-1-16. Additional examples of clouds and cloud systems appear in Brimacombe (1981).

Clouds are organized into distinctive and commonly observed formations related to atmospheric circulations on all scales. Cloud shapes are the result of dynamic processes. To interpret cloud imagery, an analyst should be familiar with the atmospheric processes responsible for various formations and be able to recognize the cloud forms associated with them.

VII-1.6.1 Mesoscale Patterns and Forms

A. Cloud Lines consist of convective cloud elements between 30 and several hundred kilometers in length (Figure VII-1-8).

Table VII-1-4 *Cirriform cloud characteristics in satellite imagery*

CLOUD TYPE	CHARACTERISTIC SHAPE OR PATTERN	CHARACTERISTIC TONE		EXAMPLES/ FIGURE NUMBER
		VIS	IR (11.5 μm)	
Cirrus Fibratus	Banded fibrous structure 50–100 km in width. Features of underlying terrain and cumulus clouds are sometimes detectable through clouds.	Dark gray to gray tone depending on underlying surface.	Light gray in tone. Fibrous structure not as evident as in VIS.	VII-1-16 [F] VII-1-8 [13]
Cirrus Spissatus	Cloud elements formed by thick dense convective cirrus, cloud elements often are globular or elongated and often have detectable shadows.	Light gray to white.	White tone.	
Anvil Cirrus	Smooth textured except where overshooting tops cast shadows. Sharp distinct upwind edge to plumes. Downwind edge filmy, fibrous, and indistinct.	Brightest over generating cell with decreasing brightness downwind.	Brightest over generating cell with decreasing brightness downwind.	VII-1-8 [18]
Cirro-stratus	Generally smooth and uniform tops but sometimes fibrous in appearance. May be in long bands or an extensive sheet.	Appears light gray when thin and whiter as thickness increases.	Varies from white to gray. Difficult to distinguish from middle clouds.	VII-1-16 [G]

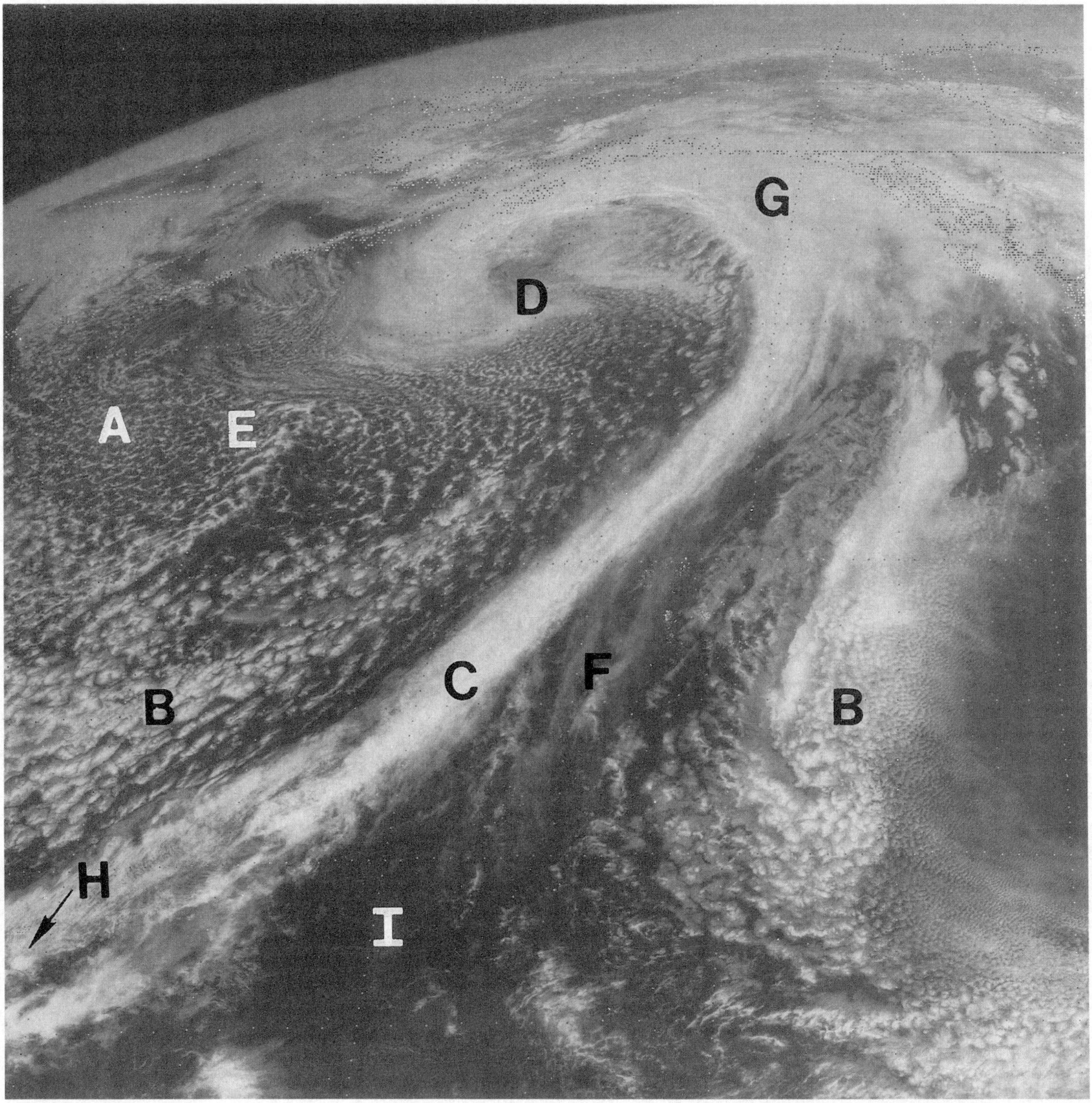

Figure VII-1-16a. Clouds and cloud formations, GOES-West, 2145 UTC, April 19, 1982, 2-km VIS image; [A] open cellular clouds, [B] closed cellular clouds, [C] cloud band, [D] cloud vortex, [E] vorticity center, [F] cirrus fibratus, [G] cirrostratus, [H] cumulonimbus, [I] cumulus.

B. Cellular Cloud patterns (open and closed) are the result of mesoscale convective mixing within the large-scale flow (Hubert, 1966). Open cell patterns are quasi-hexagonal in shape with a cloudless space inside and a ring of convective clouds at the edges. The cloud ring consists of several dozen cloud elements merging into one another or separated by cloud-free spaces. These open cells vary from 10 to 100 km in diameter. Transitional cellular cloud patterns occur between regular cells and cloud lines. In satellite imagery, an open cellular field has less than 50% cloud cover. Closed cell patterns are almost hexagonal in shape and are

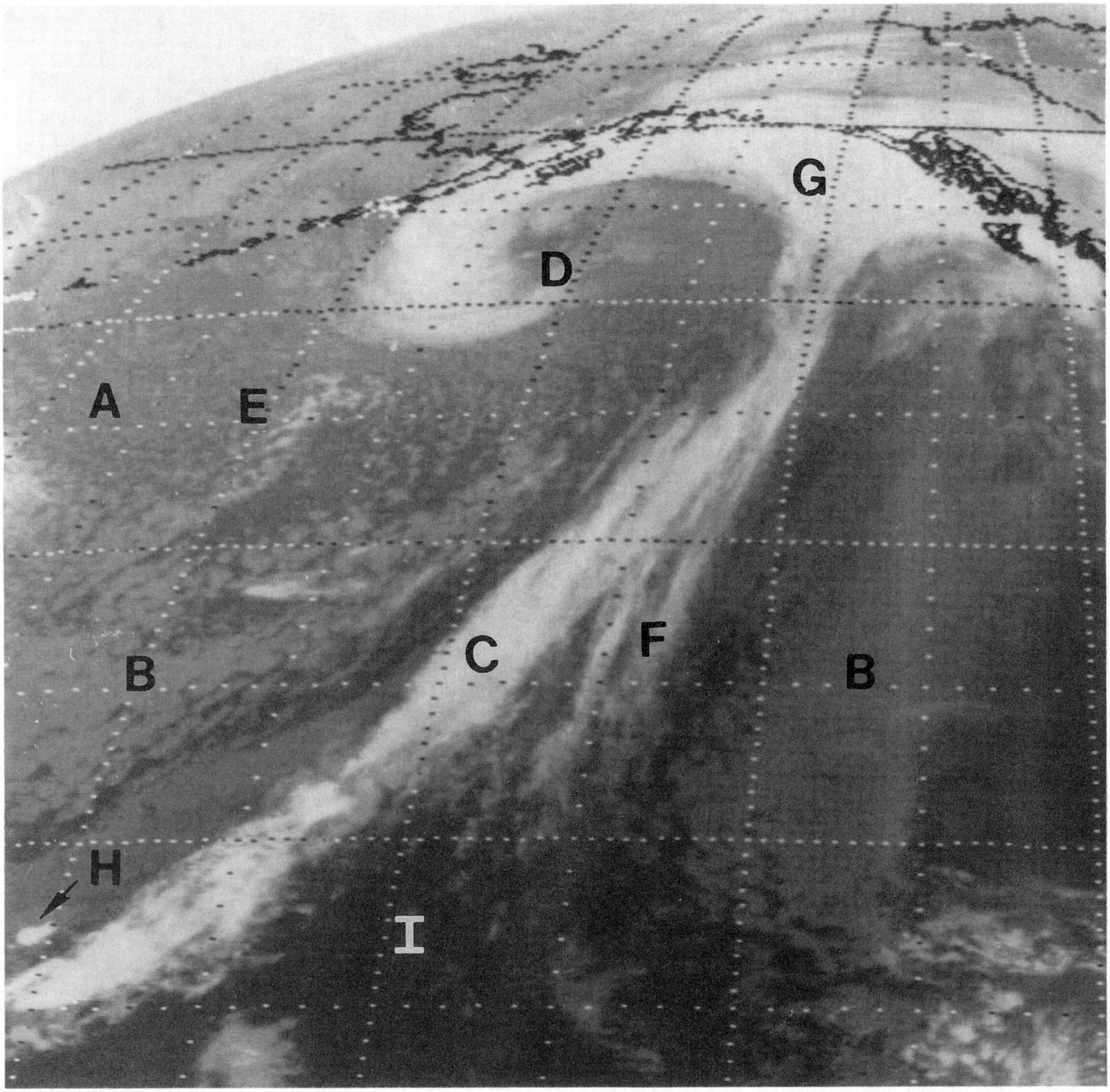

Figure VII-1-16b. Clouds and cloud formations, GOES-West, 2145 UTC, April 19, 1982, 8-km (11.2 μm) IR image; [A] open cellular clouds, [B] closed cellular clouds, [C] cloud band, [D] cloud vortex, [E] vorticity center, [F] cirrus fibratus, [G] cirrostratus, [H] cumulonimbus, [I] cumulus.

bounded by a cloudless space. Closed cellular fields form a near-overcast condition in which the cells vary from 10 to 100 km in diameter (Figure VII-1-16).

C. Wave Clouds (Conover, 1964), generally caused by terrain barriers (Figure VII-1-8), appear as a pattern of vermiculated bands that range in size from a few to tens of kilometers. Bands and clear spaces of the same size alternate. The appearance of wave clouds may vary, depending on the shape of the barrier and the speed, direction, and shear of the winds. Wave clouds also may occur below inversions; these are similar in appearance to orographic wave clouds.

D. Mesoscale Vortices (Hubert and Krueger, 1962) are cloud formations consisting of one or more spirals, each with a different amount of cloudiness. These formations range in size from 50 to 200 km. Mesoscale vortices can be caused by obstacles or by shear or thermal effects. Small chains of orographic mesoscale vortices may form to the lee of individual mountain peaks in fields of stratocumulus and stratus. A vortex chain may stretch as far as 1000 km. Orographic vortices also may occur in the wake of high capes or peninsulas, and can also be produced by horizontal wind shear along coast lines or on the cold air side of frontal cloud bands.

VII-1.6.2 Synoptic Scale Cloud Patterns

A. Cloud Bands (Figure VII-1-16) are defined as continuous, relatively broad cloud formations characterized by a distinct long axis with or without curvature. The ratio of length to width is at least 4:1 and the width is usually greater than 1° of latitude.

B. Cloud Vortices (Figure VII-1-16) consist of one or more spiral cloud bands which differ by amount and type of cloudiness and converge toward a common center (Widger, 1964; Anderson et al., 1974). Such patterns are associated with large-scale vortices. "Vortex" is used to describe both tropical and extratropical cloud systems. One of the cloud bands in well developed extratropical spiral patterns is often associated with a front. The horizontal extent of large-scale vortices range from 600 to more than 1500 km. Organized spiral or concentric lines of cumulus often are called vorticity centers (Anderson et al., 1966) because of the associated cyclonic relative vorticity. Secondary vorticity centers embedded in a large-scale cyclonic circulation often produce areas of large bright cloud elements. Comma shaped formations occur in areas of strong mid-tropospheric advection of positive relative vorticity.

Chapter VII-2

IMAGE DISPLAY AND ANALYSIS SYSTEMS

VII-2.1 INTRODUCTION

Conceptually, display and analysis systems consist of four major components: data acquisition, data processing, information products, and judgment. The systems concept of display and analysis is described in detail by Green (1983). Suomi (1977) pointed out that this technology is really only a tool and the human mind provides the vital element—reaction to the information presented. The fundamental concept of a display and analysis system is to convert data into useful information. The complexity and capabilities of each system are determined by technology, functional requirements, and financial resources. When all aspects and requirements are fully considered and met, the end result is an optimal system.

Display and analysis systems are usually designed for research or for specific operational applications. Research oriented systems encompass a wide range of technology and have broad and more costly functional requirements than operational systems. Operational systems with well-defined functional requirements generally do not allow for significant growth or expansion. Operational use of satellite data has shifted away from a system where hard copy satellite imagery is the primary display to systems using computers and electronic displays (Gird and Bradford, 1985).

Educational institutions have historically used innovative designs to fulfill their display and analysis needs. New approaches to displaying satellite data were the underlying theme when the Man-computer Interactive Data Access System (McIDAS) was developed (Suomi et al., 1983). McIDAS has two general purposes: to demonstrate technology for processing satellite data and to serve as a learning tool for university students. This trend has continued at other universities. Dutton (1983) described a new concept for universities that involved sharing and combining resources. In contrast, step by step procedures can be found on how to build a low-cost satellite display system using small computers (Summers and Gotwald, 1981). Through such efforts, a few high schools use satellite displays as an integral part of their curriculum; private citizens and organizations are also starting to develop these procedures for home and small office use (Ryan, 1983). Satellite applications software and a wide variety of direct broadcast receiver equipment are now available and affordable (Popham, 1984).

VII-2.2 HARD COPY DISPLAY

Satellite images today are very different from the photographic snapshots of the early TIROS satellites. An image is composed of a series of radiometer measurements formed into an array of data; the value of each data point is represented by a gray shade or color. Most hard copy devices process one line at a time. The paper or film negative is then advanced (usually by rotating a drum) to receive successive lines. Electronic signals form the image gray shades by varying the light intensity exposed to photographic film, varying contact with electro-sensitive paper, or by altering the electrostatic attraction of colored toner particles to the paper.

Several types of display devices are used to form hard copy images. The facsimile recorder was one of the first used to display weather satellite data and weather maps. This device uses electro-sensitive, chemically treated paper to form the image.

The first film recorder used in the meteorological satellite program was a Polaroid camera mounted in front of a Cathode Ray Tube (CRT). At the 1964 New York Worlds Fair, this device was used to give the public its first look at weather satellite imagery in real time.

A large variety of other film recorders have been used to record satellite imagery. There are now film recorders that transfer imagery at a very high rate from any type of video screen to hard copy. In general, the quality of most film recorders still exceeds the best electrostatically produced images.

By the early 1970s, laser technology was used in display devices. A Laser Facsimile (LASERFAX) system was used in the GOES Data Distribution System. The LASERFAX was connected directly to a commercial telephone line and was compatible with the 120 rpm facsimile signal routinely transmitted by this system.

Also in the early 1970s, a high-resolution photographic display was developed for the GOES satellite series. This display could be connected directly with the GOES stretched VISSR data link and could display the complete VISSR image at 1-km resolution. This system saw little operational use, however, because of its complexity and deficiencies.

VII-2.3 ANIMATION DEVICES

The geostationary satellite provided views of the Earth every 20 min. This raised the possibility of making films using sequential images that, in effect, could "animate" the passage of cloud patterns over the Earth. To make such films, each of the images was placed before a movie camera and several frames of the film were exposed. Registration of each image was required to compensate for variations in satellite orbit and attitude. After the entire sequence was photographed and processed, the movie film was spliced into a loop which could be continuously reviewed. Between 1973 and 1981, 16 mm film loops, covering the Western Hemisphere from near 60°N to 60°S, were prepared routinely by NESS. They were used for briefing meteorologists at the National Meteorological Center (NMC), the Washington, DC National Weather Service Forecast Office, and TV stations in the area and for deriving cloud motion winds (see Chapter VII-4). Loops covering enlarged smaller areas, such as those around tropical storms, were often prepared to obtain better display resolution and cloud definitions. Film loops have been replaced in recent years by more efficient electronic video animation.

VII-2.4 VIDEO DISPLAYS

The first operational video imaging system used by meteorologists was the Television Picture Terminal (TPT). Designed by the Space Science and Engineering Center at the University of Wisconsin, TPT recorded IR analog signals of SMS and GOES imagery. The data were written onto a video disk for storage and playback. The TPT stored up to eight half-hourly images, with the latest overwriting the oldest to form a continuously updated 4-hour loop. The TPT had limited application because only a single sector could be animated at one time.

The Electronic Animation System (EAS) was developed in 1981 to meet the need for loops of longer duration and broader coverage. The EAS uses a video camera and light table to record hard copy GOES images onto an analogue video disk storage device. Several animated sequences are constructed onto each of two separate channels. Playback of the sequences onto standard black-and-white video monitors can be remotely controlled; animation speed and direction can be changed by a meteorologist or operator at a data interpretation station. These features have made the EAS a flexible and very reliable system that offers good image quality. Unfortunately, the necessity for manual entry and use of hard copy introduces unwanted delays of 30 to 40 min when compared to automated systems.

Automated systems can be classified into two groups: those using disk storage and those using solid-state memory. Both types convert the GOES-Tap analogue input signal to a digital signal and sample or average the full resolution data stream to create a lower resolution image.

Disk-based systems record the facsimile format analogue signal from the GOES-Tap system serially onto a digital disk. Hundreds of satellite images can be stored; other features make multiple loop storage, loop speed variation, and false coloring possible. Although such devices are limited by a single playback restriction because they are mated with the GOES-Tap system, they can display data in nearer to real time than the EAS and so can access, record, and sequence any available GOES sector.

The second type of automated video display system also uses the analogue GOES-Tap signal, but stores the images in Random Access Memory (RAM). A microprocessor samples every third pixel of every third line of the full resolution data stream to compress the GOES sectors into a television format. The number of images that can be stored is limited only by the amount of RAM available. Again, the input is limited by the number of GOES sectors available from GOES-Tap analogue signals, but meteorologists can view imagery in near-real time.

One drawback of video display systems, as opposed to film recorders, is the resolution of the video image. Video image quality is limited by the horizontal and vertical density of the television output, and by its digital "depth" which determines the number of discrete steps in the image gray scale. Photographic images remain superior in quality especially when studying small-scale phenomena observable in the high-resolution GOES VIS data.

VII-2.5 OPTICAL METHODS FOR MERGING SATELLITE AND OTHER DATA

Satellite image geometry differs significantly from conventional meteorological map projections. Prior to the use of computer systems for rectifying and gridding satellite data, it was difficult to register satellite imagery with base maps. The Zoom Transfer Scope (ZTS), an optical system by Bausch and Lomb, was one of the first instruments that permitted the registration of unrectified satellite imagery. The ZTS enables an operator to view two scenes, such as a satellite image and a map, in superimposition. The analyst, looking through binocular eyepieces, can use controls to scale and stretch the satellite image so that landmarks on the image are in registration with the base map. Then the analyst can trace the desired satellite information onto the base map.

The ZTS has been used extensively to rectify satellite images with hydrologic river basin maps (see Chapter VIII-2). Once the snow covered areas are shaded in, an electronic density slicer is used to determine the percentage of snow cover in the basin. The Image Rectifying Video Analyzer (IRVA) consists of a ZTS, two video cameras, a light table, a video display device, and a videotape recorder. With the IRVA, the eyepiece has been modified for viewing by one of the video cameras rather than by an analyst. The other video camera is used to view a map on a separate light table. The video images of the satellite image and map are then merged and displayed together on the video display screen. The ZTS controls can still be used to scale and stretch the satellite image for registration by an analyst, and the videotape recorder can be used to store the display for later playback or for use with other stored scenes.

VII-2.6 INTERACTIVE COMPUTER ANALYSIS

Video display and animation systems have become extremely valuable to users of weather satellite data. The devices described above have made satellite imagery available for interpretative analyses more quickly than their film loop predecessors. The strength of automated video systems lies in the timeliness and quality of the satellite image display. The limitations of these systems, quickly apparent at the operational level, stimulated accelerated development of interactive video analysis display systems. The early versions of these devices were used to produce the same types of analyses that had been made for years with hard copy imagery. New techniques and analysis schemes under development required more than a visual examination of analogue data. To get more information from each image, interactive systems gave the analyst the ability to manipulate the digital data stream to extract quantitative analyses.

VII-2.6.1 Concept for Environmental Applications

The ability of a skilled analyst to develop empirical patterns and relations from satellite images has far surpassed the development of mathematical models and computer technology. Interpretation of satellite data, together with other environmental data, is recognized as a means to effectively reduce the enormous volume of data into low volume/high information formats. The degrees of judgment and skill required depend on the type of information used and its particular application.

Man-machine interactive computer devices give analysts the capability to assimilate and process enormous amounts of real-time data. By using such devices, analysts can enhance, animate, track, and manipulate data fields to determine the state of the environment. Powerful, complex applications software is used to derive and display products from a wide range of satellite and conventional databases (Smith et al., 1984). The most successful application of interactive interpretation processing techniques include satellite navigation and image gridding, satellite derived cloud motion winds, atmospheric soundings, and precipitation estimates.

VII-2.6.2 Navigation and Gridding

The VISSR Image Registration and Gridding System (VIRGS) provides GOES orbit, attitude, and gridding information. In terms of hardware and basic software, VIRGS, a derivative of the McIDAS, was designed and built by the Space Science and Engineering Center (SSEC) at the University of Wisconsin. One was installed at the World Weather Building in Camp Springs, Maryland in July 1978; operational use of the system began in May 1979.

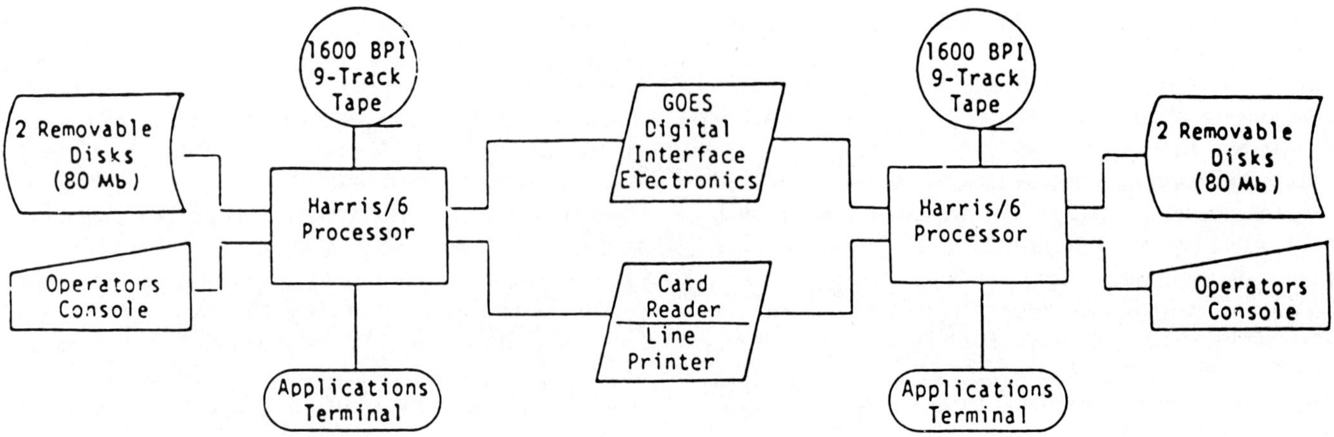

Figure VII-2-1. Schematic of the VIRGS and IFFA hardware configuration (Clark and Borneman, 1984).

The VIRGS consists of a small computer (Harris/6 with 64,000 words of memory), an on-line 40 megabyte disk, and an interactive display console for the analyst (Figure VII-2-1). The important characteristic of VIRGS is the elaborate set of applications software available to the analyst for processing and displaying the data. Earth location of a picture element is computed by using orbital parameters of the satellite and scanning geometry of the sensing instrument (Hambrick and Phillips, 1980). Accurate VISSR image location by use of McIDAS was first done in 1974 by the use of interactive commands to adjust the orbit attitude parameters iteratively. The VIRGS obtains the central Earth location for GOES daily, using four basic steps:

- Measurement of star and landmark positions in the VIS imagery.
- Determination of misalignment angles and attitude from star observations and of the orbit from landmark observations.
- Production and transfer of Earth location parameters for the GOES VISSR documentation block and the locations of standard satellite grid points.
- Monitoring and updating the accuracy of grid points throughout the day.

The human analyst plays a strong interactive role in this daily VIRGS operation by:

- Scheduling the ingest of image sectors containing stars; positioning the cursor controlled by a joystick to the bright spot of the star.
- Selecting cloud-free areas on the Earth that show favored landmarks; scheduling the ingest of the corresponding sectors; and positioning the cursor over the features.
- Judging an adequate set of star and landmark observations and entering commands for machine determinations of misalignment, attitude, and orbital parameters.
- Verifying that residual errors derived by computer programs are suitably small and navigation files have been updated.
- Entering execute commands to produce files, checking the quality of files, and transferring them to the Command and Data Acquisition station.
- Routinely checking the accuracy of orbital parameters and embedded grids currently on-line (i.e., in the VISSR data stream) and generating corrections as required.

An important feature of the VIRGS operation is the closed loop formed when the VISSR data, appended with the VIRGS-produced Earth location information (orbital parameters and embedded grids), are ingested and displayed. When ingesting VISSR data, VIRGS has the capability to monitor the performance of the system, end to end. The VIRGS operator was the first to become aware of degradations in the Earth location accuracy. In most cases, the operator corrected errors quickly through coordination with CDA station operators.

When the VIRGS became operational in May 1979, the orbit altitude determination software was not fully developed. Hence, its level of accuracy during the first several months of operation was variable, with errors ranging from 3 km on some days to 10 km on others. By August 1, 1979, the software had been revised to the point that altitude and misalignment parameters could be accurately determined from star positions alone, without using landmark measurements. Additional refinements enabled the VIRGS operation to satisfy all the geostationary navigation and gridding requirements, eliminating the need for ground ranging stations previously in use for determining orbit parameters.

VII-2.6.3 Wind Estimates

The extraction of wind vectors from satellite data started with the subjective analysis of cirrus anvil plumes in the tropics using single images from polar orbiting satellites. Animation of geostationary images by using film loops made it possible to estimate wind vectors by tracking the motion of clouds (Chapter VII-4). More precise information about cloud vector heights was obtained by using GOES IR cloud top temperatures. Neither manual nor completely automatic techniques can be used exclusively for accurate determination of the type of cloud, its characteristic emissivity, and its equivalent blackbody temperature. Hence, the Man-Machine Interactive Processing System (MMIPS), an image processing system, was developed to provide selectable computations of equivalent blackbody temperatures from the GOES digital data (Bristor, 1975).

Incoming GOES VIS and IR data were reduced in volume before being transferred into the MMIPS mass storage (Young et al., 1972). Once the data were transferred, they were automatically displayed as a full image on the CRT. The meteorologist selected a 1400 × 1400 sample set from the original image to be divided further into smaller sectors, by moving a cursor with a trackball device. A target cursor and a grid overlay, centered about the target, were superimposed over the Earth image; the grid defined sector boundaries.

Once sectors were selected, a stored gray scale, automatically provided within the initialization mode, converted the data to 4-bit intensity values (16 shades of gray). The final step in initialization was to display the desired sector on the CRT; six unique frames were available. The meteorologist could assign between one and six sectors to specific CRT frames. The selected sectors were then transferred for display.

During processing, MMIPS was used to display representative cloud top temperatures. Satellite data were converted from encoded values to Kelvin temperatures by means of a look-up table. The analyst decided whether the observed blackbody temperature of a cloud was the cloud top temperature by judging whether the cloud was sufficiently dense to block radiation from below. When cloud tracers were identified, the analyst would also determine the temperature of those tracers by using the processing mode of MMIPS. Centering the cloud tracer within a target cursor of 17 lines × 17 elements (126 × 126 km), the analyst could display an array of temperature values on the CRT.

VII-2.6.4 Soundings Edit

Interactive satellite soundings editing capability was developed on the VIRGS as part of a satellite Soundings Improvement Project (SIP). The purpose of SIP was to improve automated quality control information with evaluation by a human meteorologist. (The sounding derivation process is described in Chapters VI-2 and VII-5.)

VII-2.6.5 Precipitation Estimates

The application of the precipitation estimation technique of Scofield and Oliver (see Chapter VII-6) has benefited, beginning in August 1983, from the use of the Interactive Flash Flood Analyzer (IFFA). This operational system has significantly improved the timeliness and accuracy of precipitation estimates.

Like VIRGS, IFFA is a refinement of the University of Wisconsin McIDAS. Multiple image and graphic frames allow meteorologists to magnify and compare consecutive VIS and IR images with surface observations to locate the areas of heaviest precipitation precisely and to take storm motion into account.

GOES digital data are received on IFFA hardware in real time, processed with a Harris/6 computer and stored on two 80-megabyte disks. The IFFA operational configuration consists of two complete, redundant systems. In case of hardware failure, the disks can be transferred to the backup system and operational work can be resumed. The applications terminal (Figure VII-2-2) consists of an Intel 8080 command processor connected to a color monitor and an alphanumerical CRT

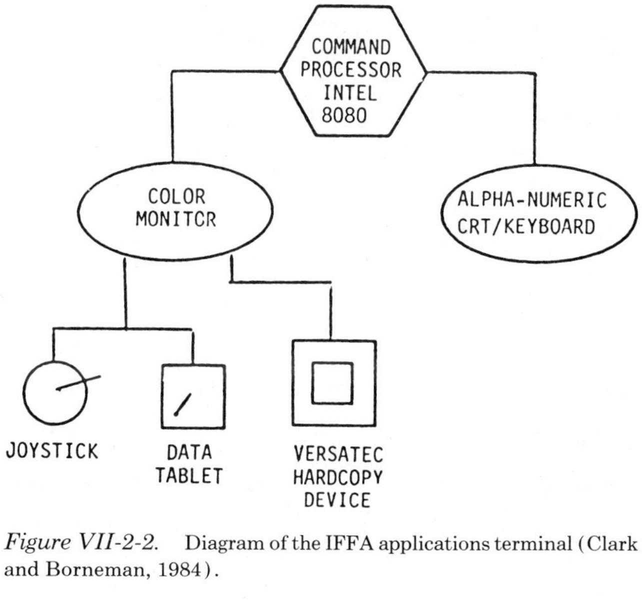

Figure VII-2-2. Diagram of the IFFA applications terminal (Clark and Borneman, 1984).

screen with a keyboard. A joystick and a data tablet are used to position the cursor on the screen to view enlarged areas and to draw isohyets. A graphic printer produces a permanent record of the estimates.

A typical IFFA schedule of procedures is shown in Figure VII-2-3. To begin the estimation procedure, the meteorologist displays the GOES data on the color video monitor and checks the GOES navigation to ensure that the Earth location accuracy is within 8 km. Next, the meteorologist selects an area in which rainfall estimates appear to be needed. The GOES images for the area are enlarged and sequences of both VIS and IR images are created. At this point, current surface and upper air observations may be plotted on the graphics frame in the GOES projection.

Comparing clouds in consecutive images on the CRT, the meteorologist determines rainfall amounts using the NESDIS operational precipitation estimation techniques (Chapter VII-6). An example of an operational IFFA product is shown in Figure VII-2-4.

VII-2.6.6 Outlook for Image Processing Systems of the Future

Racer and Gaffney (1984) have provided some initial concepts for the application of expert systems technology to interactive weather data analysis systems. They suggest a procedure that will formalize the capture of empirical knowledge and procedures used by meteorologists and that will develop a "weather information processing executive." This executive would orchestrate the interaction and provide guidance to the analysts. If these concepts are incorporated into a systematic approach for generic software design, the knowledge base that is developed will outlast the life cycle of the hardware system or the original contributing scientists.

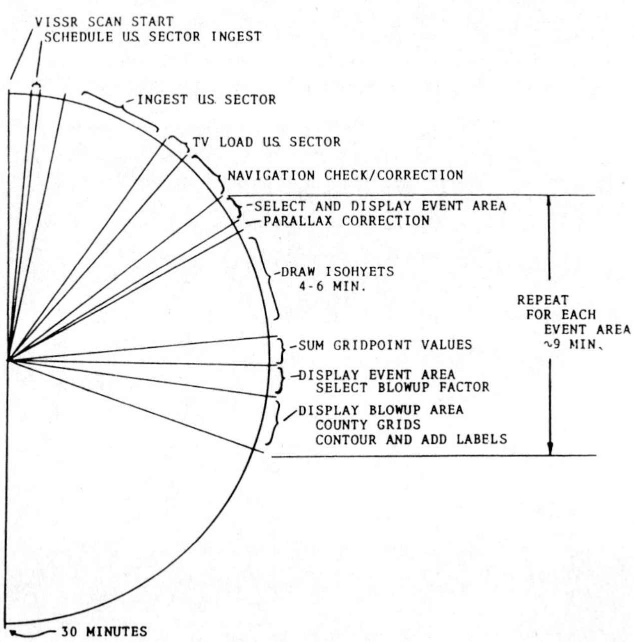

Figure VII-2-3. Typical half hour schedule for IFFA operations.

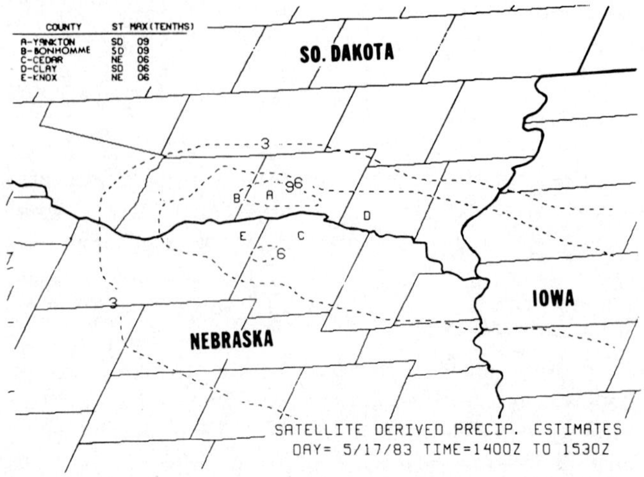

Figure VII-2-4. Example of a hard-copy product from IFFA (Clark and Borneman, 1984).

The Program for Regional Observing and Forecasting Services (PROFS) provides an example of generic software designs for environmental applications. A User Interface Management System has been developed by Kasik (1982) and applied to the PROFS interactive workstation (Leserman, 1984). The meteorologist is presented with a conceptually singular interface that can be used in any possible application. This approach is based on the principle that the entire user interface can be defined by a set of related tables which link the keyboard and cursor controls to the applications software and databases. Through a menu editor, the user can change the interface menu and procedures associated with a particular operation without reprogramming. In this manner, the system learns how the professional needs to interface with evolving databases and evolving applications.

The use of video analysis techniques for operational derivation of satellite products and for applied research into new satellite products has been very successful. The capability to combine many datasets rapidly and accurately on one interactive workstation has opened a new era in satellite analysis.

Chapter VII-3

STEREOSCOPIC MEASUREMENTS[1]

VII-3.1 INTRODUCTION

Stereoscopic observations of clouds from meteorological satellites are useful as an analytical tool. Stereoscopic observations from geostationary satellites cover a much wider area with better temporal resolution than stereoscopy from lower altitude satellites. Cloud height estimates using satellite stereoscopy generally can be made with vertical resolution of about 500 m and horizontal resolution of 1 km. Stereoscopically derived heights are made with much higher confidence than heights estimated from IR techniques (see VII-4.4).

VII-3.1.1 The History of Meteorological Stereoscopic Measurements

The earliest meteorological stereoscopy was done with ground-based cameras; the first reference to aerial stereoscopy in the literature (Schereschewsky, 1921) mentions the use of a side-looking camera. The forerunner of satellite stereoscopy was the use of a vertically pointing camera on a high-flying aircraft (Roach, 1967). Stereoscopic observations from satellites were first made using the early Vidicon camera (Ondrejka and Conover 1966). The only high quality meteorological stereoscopy from low orbit described in the literature was done from Apollo and Skylab (Shenk et al., 1975; Black, 1977). Conventional cameras on the space shuttle now make this kind of stereoscopy possible almost routinely.

The use of simultaneous images for stereoscopy from a pair of Geosynchronous Earth Orbiters (GEO) was proposed by W. E. Shenk (private communication to R. Wirth, December 12, 1974) and Bristor and Pichel (1974). Minzner et al. (1976) used these proposals to demonstrate the first stereoscopic, quantitative height analysis. Bryson (1978) and desJardins (1978) independently developed digital processing techniques to remap stereoscopic images. These techniques made possible precise height measurements and spectacular displays of stereograms (Hasler, 1981). Fujita (1982b) described an alternate method of stereoscopic analysis. Stereoscopic images from dissimilar cameras, e.g., those on the Japanese GMS and the U.S. GOES, have been described by Hasler et al. (1983) and Fujita and Dodge (1983).

The most recent development in meteorological stereoscopy as described by Hasler et al. (1983) uses a low Earth orbiter, such as the polar orbiters, and a geostationary orbiter to obtain stereoscopic images. Using red-green glasses to view stereoscopic pairs has led to the effective use of simultaneous VIS and IR image pairs in a synthetic stereo presentation (Hasler et al., 1981). Both true and synthetic stereoscopic data also can be presented using perspective display techniques developed by Hasler et al. (1985).

[1] Primary Contributor: Arthur F. Hasler

VII-3.1.2 The Importance of Stereoscopic Measurements

Accurate cloud heights are important for a host of applications (Hasler, 1981; Hasler and Morris, 1986; Mack et al., 1983a; Rodgers et al., 1983; see also Chapter VII-4). The stereoscopic measurement of cloud heights gives information independent of other measurements. The information is derived from direct geometric measurements of data and does not depend on the complications involved in the use of IR window and multichannel techniques. With the stereoscopic technique applied to VIS data, heights of clouds or cloud features with resolutions as small as 1 km can be measured. With the multichannel IR measurements, the horizontal resolution is no better than 5 km. The resolution of the IR sensor on current GOES satellites is only 8 km.

VII-3.1.3 Temporal and Spatial Coverage

The biggest problem with the stereo technique to date is the lack of routine observations. Stereoscopic measurements of cloud heights require simultaneous high-resolution images from two satellites. In principle this is possible during the daylight hours in the overlap regions of geostationary satellites. In practice, because of conflicting operational requirements, this has been possible only for special experimental periods and only over the GOES-East–GOES-West overlap region whenever one of the satellites has been in the 15-min scan mode. Other targets of opportunity occur whenever a polar orbiter passes under the viewing area of a geostationary orbiter. Small regions of simultaneous coverage are produced on many polar orbiter passes every day. The location of this stereo region can be adjusted by changing the image time of the geostationary orbiter. Stereoscopic measurements may also be made at night, but with reduced resolution.

VII-3.2 MEASUREMENT TECHNIQUES

VII-3.2.1 Geometry

Stereoscopic height measurements require angular measurements of the target from two satellites. A simplified two-dimensional diagram (Figure VII-3-1) shows how clouds at sea level (C_o), at a low elevation (C_h), and at a greater elevation (C_H) would be observed by two satellites. Only C_o appears at its true location on the Earth. The rays passing from GOES-East and GOES-West through C_h and C_H intersect the Earth at points that increase in separation (parallax) with increasing cloud height. With accurate knowledge of the positions of the satellites at the time of observation and the location of the points where the rays intersect the Earth, the height of the cloud can be calculated.

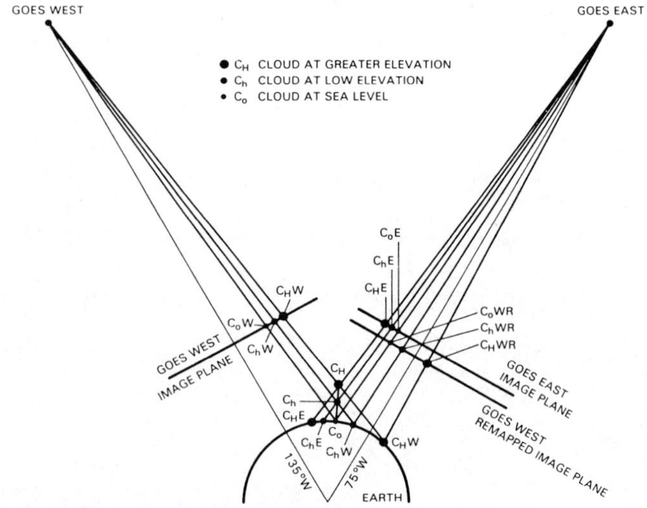

Figure VII-3-1. Simplified two-dimensional diagram illustrating how cloud height is observed by two geostationary satellites. Remapping of the image of one satellite into the coordinate system of the other satellite is also shown. When the GOES-East image is superimposed on the remapped GOES-West image, sea level features (C_o) should line up. The separation (parallax) between the two different images of the same cloud (C_hE-C_hWR, C_HE-C_HWR) is proportional to its height.

Figure VII-3-1 illustrates the ideal case in which a simultaneous image pair from two geostationary satellites is available. Since such image pairs are not always to be had, however, a geostationary orbiter and a polar orbiter can be used together to measure stereo heights over a small area such as that shown in Figure VII-3-2. The resulting stereo is more complex than that obtained using two geostationary orbiters, because the

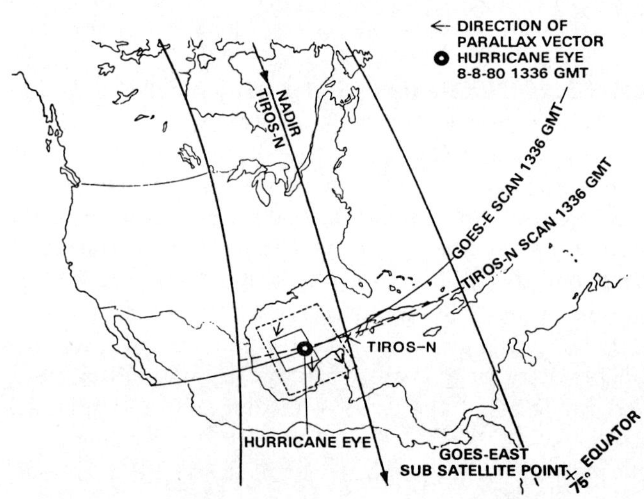

Figure VII-3-2. Diagram of the TIROS-N and GOES-East scan pattern over Hurricane Allen indicating how the parallax varies (see arrows).

parallax varies greatly over a typical 500 × 500-km area (solid box in Figure VII-3-2). Figure VII-3-2 also shows the subsatellite points and how the parallax varies over an area of a typical useful image pair (solid box) and an area four times larger (dashed box). The arrows indicate the direction of offset or parallax for features above sea level. The TIROS-N satellite scans at a rate of 360 lines per minute while GOES scans at 800 lines per minute; both obtain data with resolutions of about 1 km. Thus, with perfect synchronization at the center of the image, the synchronization error at the top and bottom of the solid box will not exceed what is tolerable for accurate stereo measurements (Hasler et al., 1981).

By adjusting the image start time of the geostationary satellites, good geostationary-polar orbiter stereo pairs can be obtained worldwide between 60°N and 60°S. There are, however, singular points near the Equator where the parallax is very small. Even Meteosat, which scans south to north, can be used with low level polar orbiters with daylight ascending nodes. The north/south scan rate of a geostationary satellite can also be matched to that of the lower level orbiter by repeating scans. This will permit synchronization within a few seconds over large areas.

VII-3.2.2 Remapping

The wide separation between the two satellites requires the remapping of the images from one satellite into the coordinate system of the other for viewing and measurement. With computer remapping of digital images, stereo-image time sequences can give great insight into atmospheric dynamics and can be used for precise measurements of cloud geometry. Generally a standard bilinear remapping is performed (Dalton et al., 1979) using the most accurate possible landmark (Chen, 1980). Figures VII-3-3a and VII-3-3b show a remapped Japanese GMS image and a GOES image of a large central Pacific cloud cluster in the ITCZ. This stereo pair can be viewed two-dimensionally with a mirror stereoscope.

Figure VII-3-1 also shows a simplified two-dimensional representation of how the remapping works. Sea level features (C_o) appear at the same location in original and remapped images (C_oE, C_oWR), while features above sea level (C_h, C_H) are displaced an amount proportional to the height of the two features (C_hE-C_hWr, C_HE-C_HWR). This parallax gives the stereoscopic impression when the images are displayed.

VII-3.2.3 Display

To prepare a stereo image pair for optimum viewing and measurement, several image cleanup and enhancement procedures are used. These include noise removal, brightness normalization, and gray scale adjustments. For maximum measurement accuracy the image pair is enlarged by a factor of 2 and smoothed. Many stereo display techniques are available (Hasler, 1981; Hasler, 1983; Warner, 1984; Dunne and Burgess, 1978). Viewing stereo pairs (displayed in red and green) with inexpensive, cardboard red/green glasses (Hasler et al., 1981) is most practical and effective.

VII-3.2.4 Height Measurement

Stereo height measurement is accomplished on an interactive image processing system such as the Atmospheric and Oceanographic Information Processing System (AOIPS). The parallax can be measured by viewing the pair with red/green glasses, shifting one image to coincide with the other, and measuring the amount of the shift. A cross-correlation algorithm can be used to measure the parallax of large, well-defined features. The most effective technique, however, is to use image alternation and then align the feature while keeping track of the amount of shift. This method permits measurement within 1-pixel even for poorly defined features.

Once the parallax has been determined, the result is used with a height calculation algorithm, an iterative technique (desJardins, 1978).

VII-3.2.5 Height Contouring

Contouring is done with 1-pixel accuracy on the AOIPS interactive image display (Mack et al., 1983a). The images are displayed and the feature for which the height is desired is identified. The system then computes the shift between images and one image is automatically shifted to the proper parallax. The two images are then alternated with a frequency between 2 and 8 s^{-1} and contours are drawn on the television monitor along lines of local null motion (constant parallax). The contour lines are then automatically smoothed and plotted on both the remapped and original images. Figure VII-3-4 shows an example of a stereo contour analysis superimposed on an original image.

VII-3.2.6 Cloud-Wind Height Assignment

Software routines on AOIPS are available to integrate the stereo height measurements into a cloud tracking package. Thus, cloud winds may be assigned accurate heights when simultaneous stereo images are available.

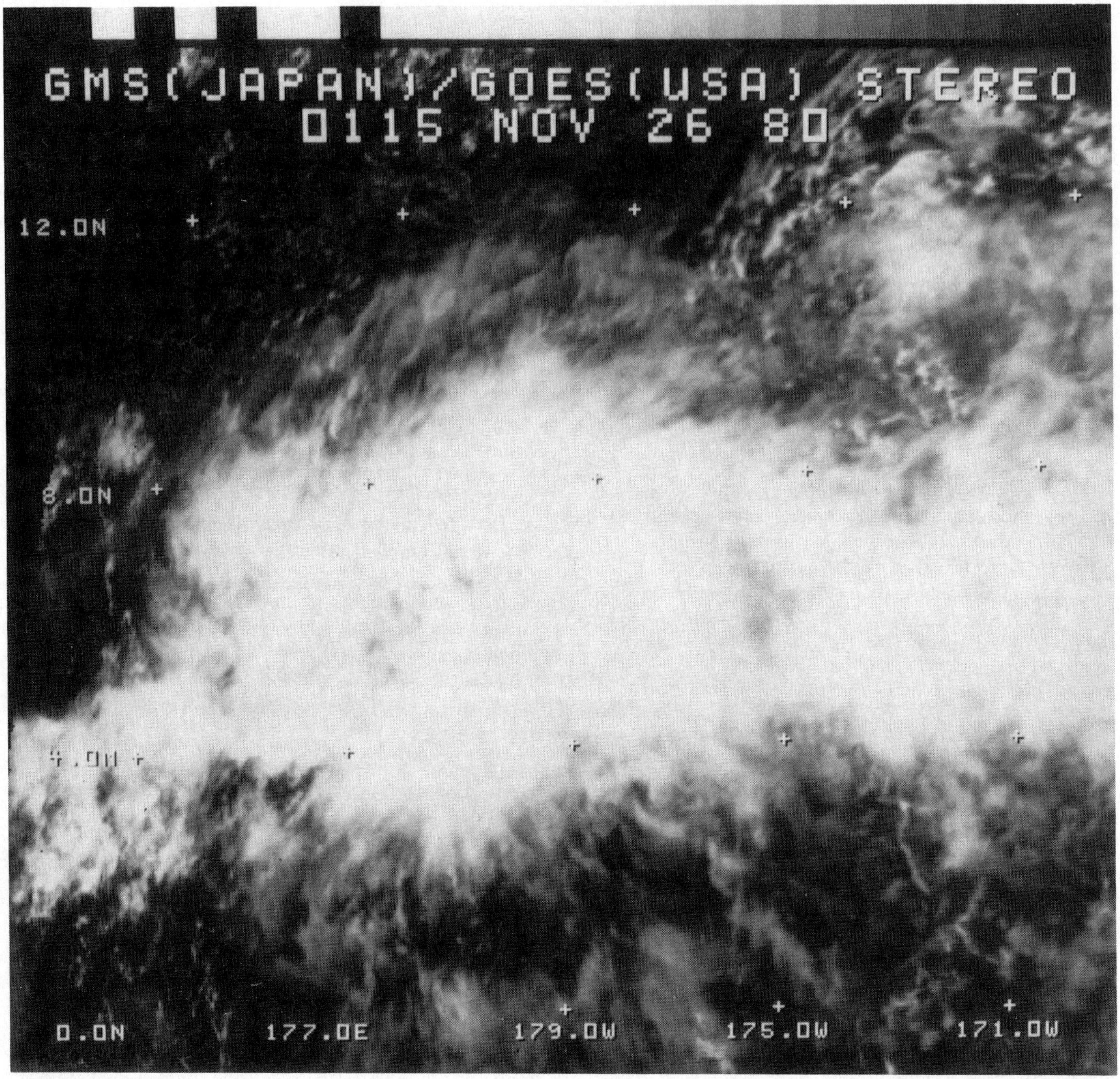

Figure VII-3-3a. Japanese GMS image remapped to a GOES-West image (see Figure VII-3-3b) of an ITCZ cloud cluster at 0115 UTC, November 26, 1980.

VII-3.3 ERROR ANALYSIS

Stereo height measurements have the following sources of error (Hasler, 1981; desJardins, 1978):

- refraction of light by the atmosphere
- irregularities in the images
- uncertain location of images
- nonlinearity of the images
- problems in remapping
- inaccuracies in parallax measurement because of differences in viewing angles and operator errors
- synchronization uncertainties

Synchronization of the images to within 20 sec prevents significant degradation of height measurements for typical maximum east-west winds of up to 25 m s^{-1}.

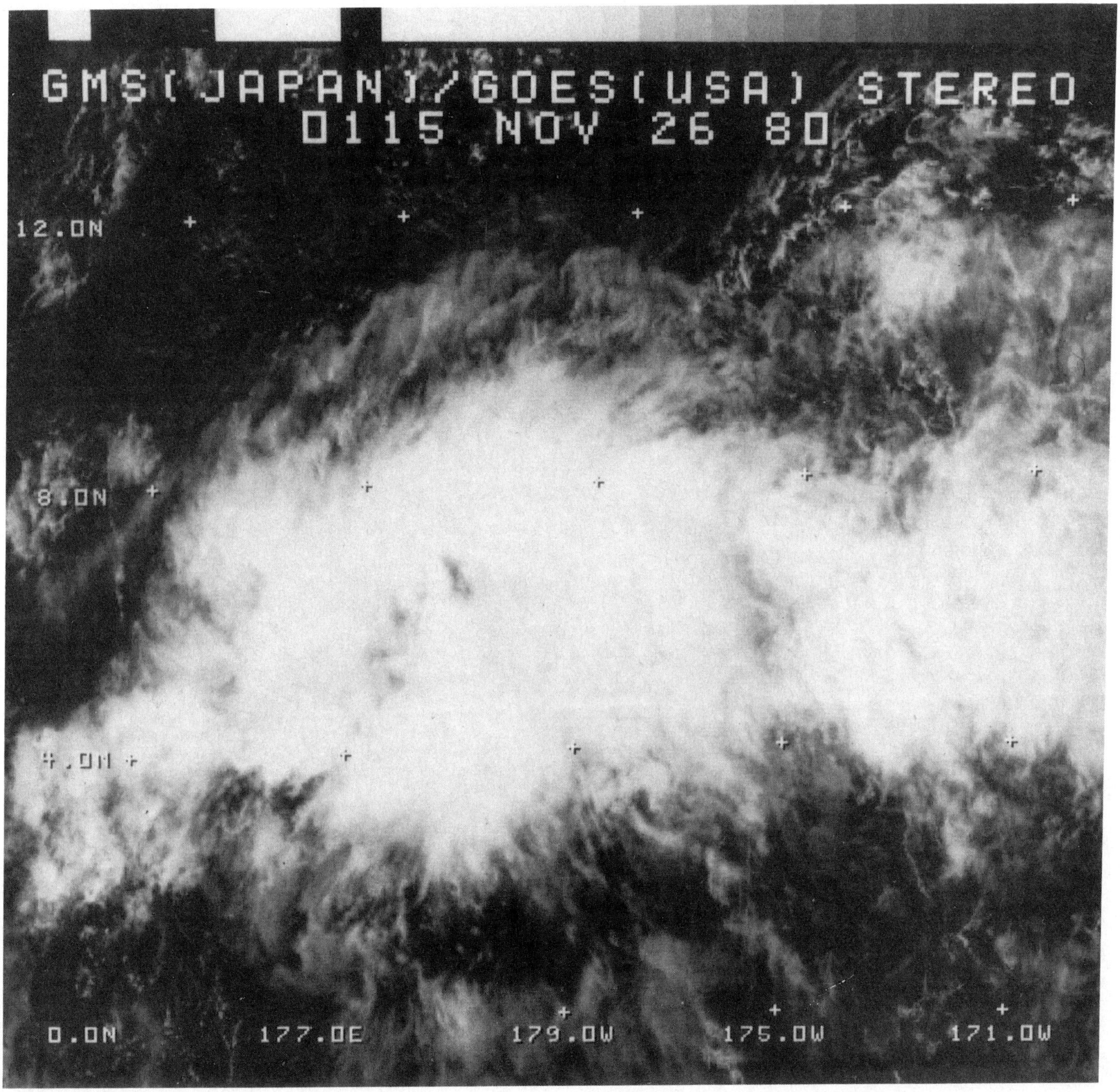

Figure VII-3-3b. GOES-West image of same ITCZ cloud cluster as in Figure VII-3-3a; together they make a stereo pair. This pair may be viewed in 3-D with a mirror stereoscope.

Verification of stereo heights using mountain lakes as a measuring standard have shown that an absolute accuracy of 500 m is easily obtainable if careful location procedures are used. Relative accuracy of about 150 m is possible under ideal conditions if sub-pixel measurements are made (Hasler, 1981). Rodgers et al. (1983) reported on a verification experiment using an aircraft-borne, vertically pointing lidar to check the height measurements of actual clouds. In flights over Hurricane Frederic (1980), a standard deviation of 500 m was found between the satellite stereo and the lidar cloud height measurements (Rodgers et al., 1983).

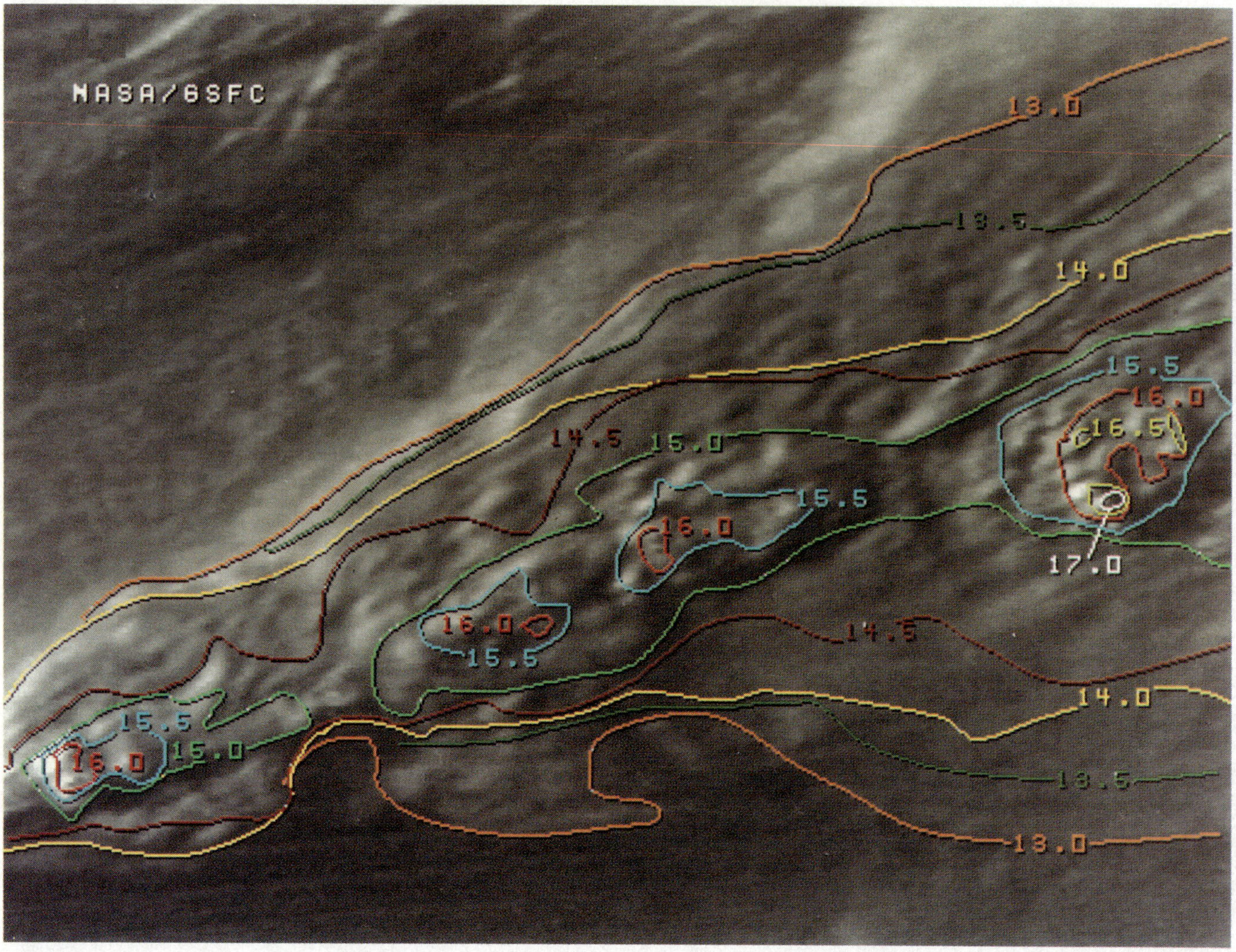

Figure VII-3-4. Stereo cloud top height contour analysis (km above sea level) superimposed on one of the two simultaneous GOES images from which it is derived. The analysis shows an Oklahoma tornadic thunderstorm complex at 0051 UTC, May 3, 1979, with a tornado on the ground under the 17-km tower on the east end of the squall line at the time of observation. The analysis was made using the on-screen interactive image alternation and shift technique on the Atmospheric and Oceanographic Information Processing System (AOIPS) at NASA/GSFC.

VII-3.4 APPLICATION TO METEOROLOGICAL PROBLEMS

VII-3.4.1 Three-Dimensional Cloud Geometry

Stereoscopic cloud height measurements from satellites have been used for numerous studies of intense convective clouds in severe local storms, tropical cyclones, and along the ITCZ.

A. Severe Local Storms. Satellite stereoscopic observations of severe local storms have been documented frequently. Figure VII-3-4 shows a detailed contour analysis of a tornadic thunderstorm on May 3, 1979, from GOES-East and GOES-West. This is similar to those done by Mack et al. (1983a) except that it was done by the on-screen contouring technique already described. Mack et al. (1983a) determined, by comparing stereo analyses with GOES IR data, that use of IR temperatures leads to gross underestimation of cloud top heights in severe storms, especially for immature ones, probably because of the low resolution of the IR data. Heymsfield et al. (1983a, 1983b, 1983c) used stereo cloud top heights to formulate new theories about the dynamics of severe thunderstorms and their interactions with the tropopause and stratosphere. Hasler and Adler (1980), Hasler (1981), Mack et al. (1983b), Hasler et al. (1983), Adler et al. (1983), and

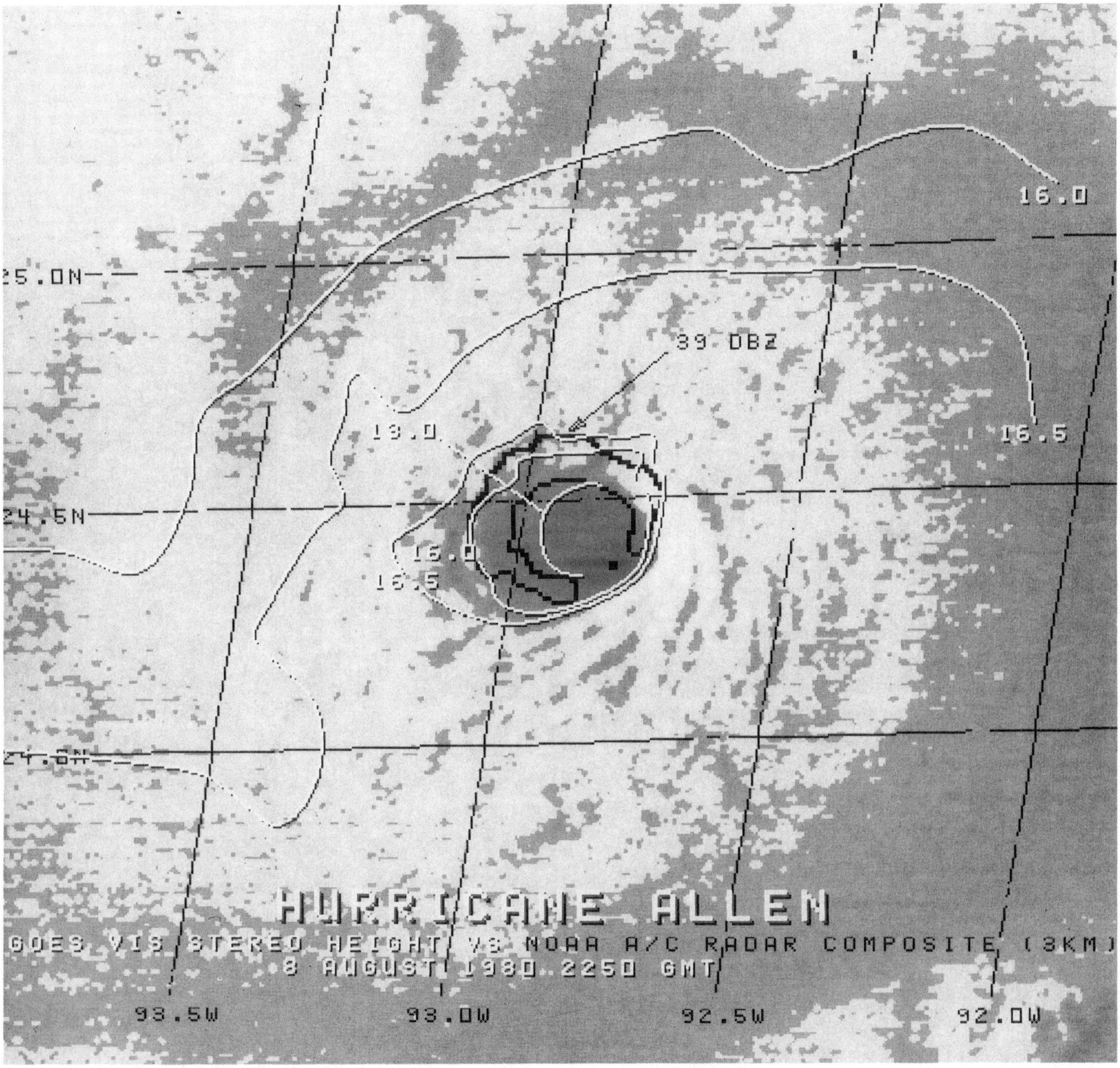

Figure VII-3-5. Stereo cloud top height contour analysis (in km above sea level) of the central dense overcast and the eye wall of Hurricane Allen superimposed on the radar reflectivity from NOAA aircraft and the GOES-East VIS image.

Adler and Mack (1984) also described stereo cloud top height analyses of severe local storms, comparisons with other kinds of data, such as radar, and inferences that can be drawn about storm structure.

B. Tropical Cyclones. In Figure VII-3-5, stereo cloud top height contours of Hurricane Allen (1980) are superimposed on the GOES-East VIS image and the 39-dB radar reflectivity contour from NOAA reconnaissance aircraft (Hasler and Morris, 1986). These data indicate that the most intense convection, detected by radar, lies under the outer edge of the sloping eye wall rather than under the highest clouds. This is shown more clearly in Figure VII-3-6. Rodgers et al. (1983) used both the stereo cloud top heights and IR temper-

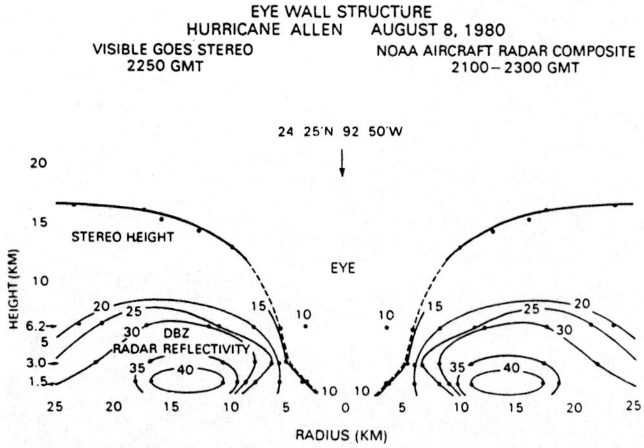

Figure VII-3-6. Cross section of Hurricane Allen (shown in plane view in Figure VII-3-5) depicting the slope of the eye wall. Above the 12-km level, the cloud top height is determined stereoscopically from a GOES-East and GOES-West image pair. The lower portion of the eyewall is defined by composited NHRL aircraft radar reflectivity at the 1.5-, 3.0-, and 6.2-km flight levels. The cross section is symmetrical and was derived by averaging the east-west and north-south cross sections.

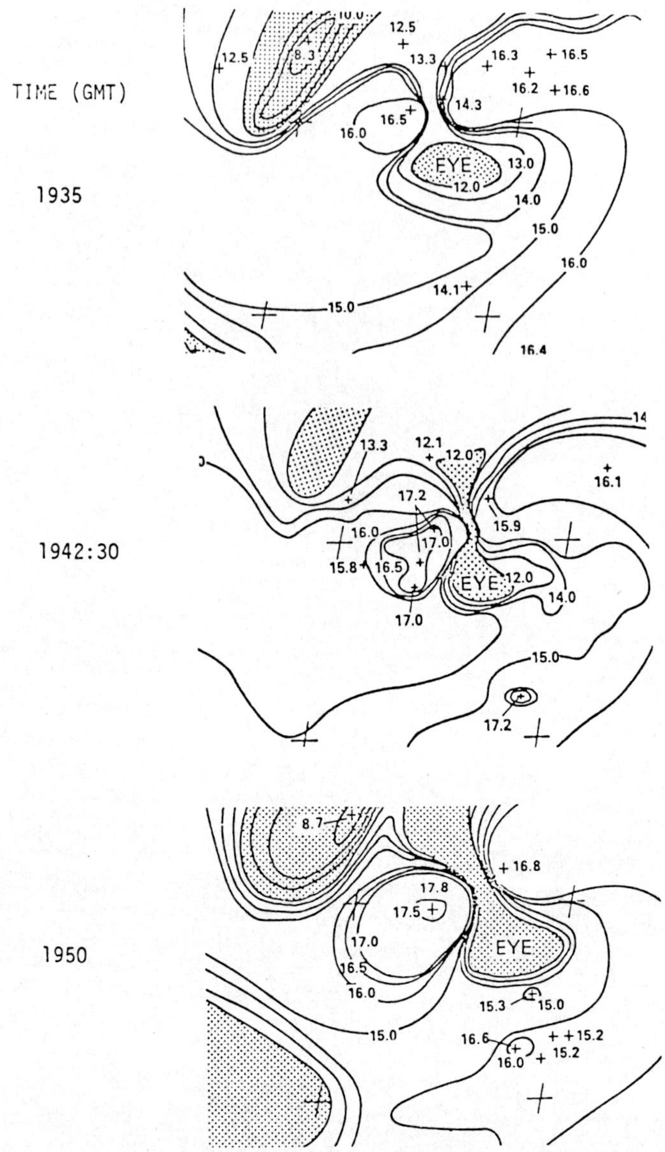

Figure VII-3-7. Time sequence of analyses of stereo cloud top heights (in km above sea level) for Hurricane Frederic on September 12, 1979. This sequence shows the growth of an intense convective cloud in the eyewall.

atures of the clouds to estimate tropical cyclone intensity. Further uses of stereo height measurements for tropical cyclones are described by Hasler (1981), Hasler et al. (1983), Mack et al. (1983b), and Black (1983).

C. Vertical Motion Estimates. Before stereo measurements were available the only relatively direct atmospheric motion measurements that could be made by using satellites were in the horizontal plane—the wind measurements inferred from cloud motions. With accurate vertical measurements from stereoscopic images it is now possible to measure convective-scale vertical motions. Direct cloud top vertical motion measurements have been used to estimate the intensity of thunderstorms. Figure VII-3-7 shows stereo contours of a rapidly growing cloud in a hurricane eye wall.

VII-3.4.2 Cloud-Wind Height Assignment

Stereo heights were first assigned to cloud-winds by Hasler (1981), but the first important use in research was by Hasler and Morris (1986) in their study of the wind field of Hurricane Frederic; Figure VII-3-8 shows the cloud-winds with stereo heights from this study. In the original study, the low-level cyclonic inflow winds (1–5 km) were plotted in red at heights starting at 1.4-km, middle level winds (5–10 km) in yellow, and upper-level winds (above 10 km) in blue. Much information can be obtained by this method. It reveals that the winds of the apparent anticyclonic outflow layer range from heights of 5.5 km west of the eye to 16.5 km south of the eye.

VII-3.4.3 Other Applications

In the absence of direct measurements, Hasler (1981) and Rodgers et al. (1983) have shown that a combination of stereo and IR images can be used to make atmospheric temperature estimates that are higher in resolution than IR soundings. Cloud emis-

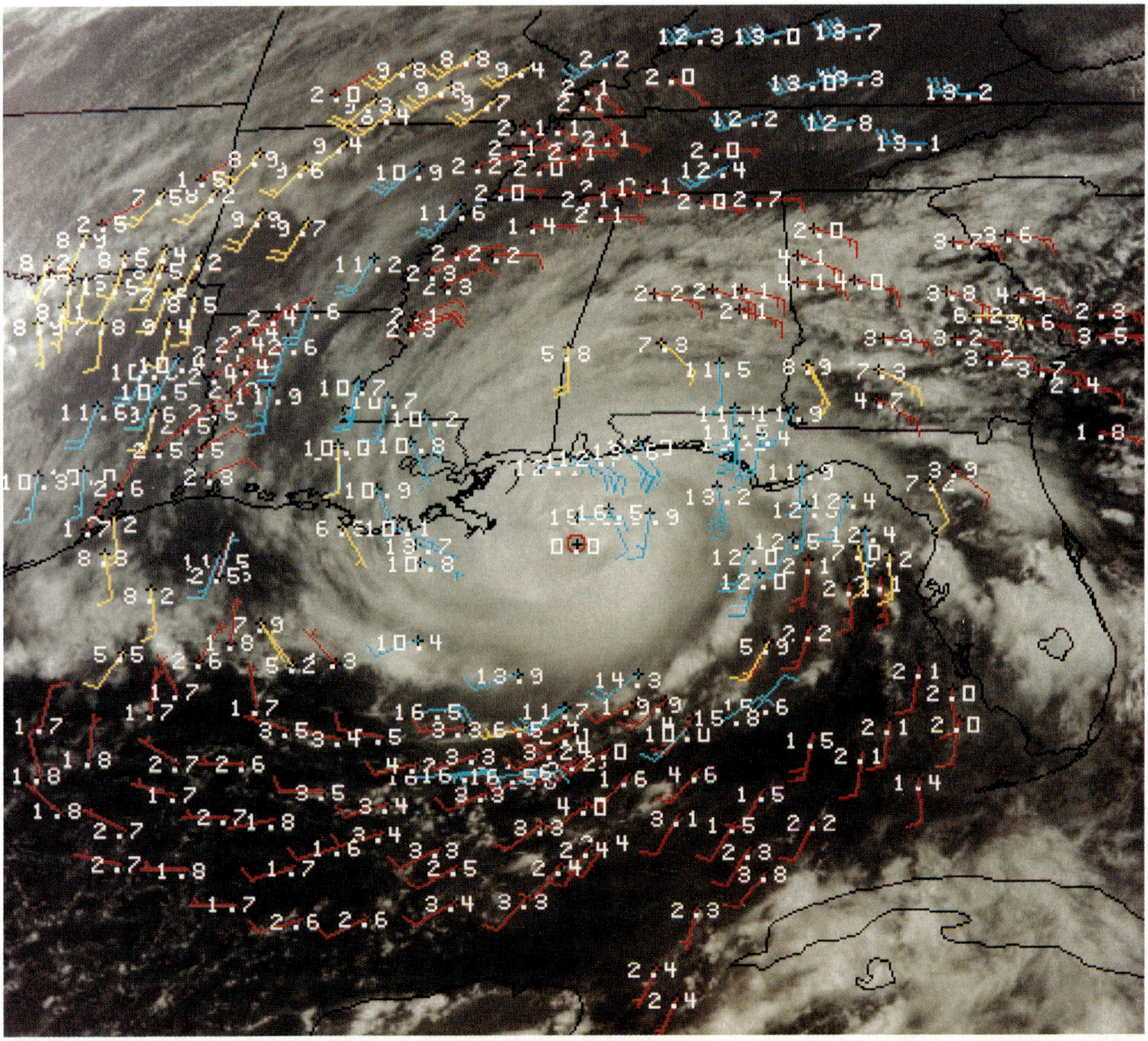

Figure VII-3-8. Cloud winds (m s^{-1}) and stereoscopically measured top heights (km) for Hurricane Frederic at 1922 UTC, September 12, 1979, superimposed on a 4-km resolution VIS image from GOES-East. Wind barbs for clouds with tops below 5 km are coded in red, 5–10 km in yellow, and above 10 km in blue.

sivity estimation is another use for the stereo/IR data combination (Szejwach et al., 1983). Stereo heights have been compared with IR temperatures and radar reflectivity in studies of severe local storms and hurricanes (Hasler, 1981; Hasler and Morris, 1984; Hasler and Morris, 1986).

Chapter VII-4

CLOUD DRIFT WINDS[1]

VII-4.1 INTRODUCTION

Geostationary satellites have a distinct advantage over polar orbiting satellites in that they furnish the meteorologist with frequent images of fixed areas. Following the launch of the NASA Applications Technology Satellite (ATS), it was quickly shown that animating these images permits ready observation of motions and changes in cloud and weather systems. Quantitative estimates of winds obtained by measuring cloud motions are now an important satellite product. While these estimates only supplement more conventional forms of wind measurement over many areas, they provide the only wind information over many other areas.

As geostationary satellites were added around the Equator, the potential for nearly complete coverage in the tropics and middle latitudes was realized. At present, five satellites provide this coverage continuously, except when satellite failures occur. Satellites in this system are the Geostationary Meteorological Satellite (GMS) of Japan (Himawari, or Sunflower); the two U.S. GOES (GOES-West and GOES-East); the European Space Agency's Meteosat; and Insat, the satellite operated by the Indian Space Research Organization and India Meteorological Department.

Practical cloud motion coverage provided by the five satellites extends only to about 50° of great circle arc from the subpoint before the oblique view degrades the resolution. Thus, there is no high latitude wind coverage from geostationary satellites. Nevertheless, this broad-scale coverage provides a valuable new source of winds for use as input to numerical models. In most cases, these winds are derived operationally at least twice a day from each satellite. An example of the density of coverage appears in Figure VII-4-1.

Researchers also use satellite winds for many specialized purposes. Usually they derive wind vectors to satisfy particular needs; for instance, they may wish to derive them more frequently than twice a day, in greater density, or over very short intervals to define mesoscale wind fields and variations. Many groups have devised and experimented with various techniques. Among those who have contributed much to this effort are NESDIS, the Space Science and Engineering Center (SSEC) at the University of Wisconsin, the Department of the Geophysical Sciences of the University of Chicago, NASA Goddard Space Flight Center (GSFC), the Laboratory of Dynamic Meteorology of the University of Paris, the Meteorological Satellite Center (MSC) of Japan, and the European Space Agency (ESA).

This chapter provides an overview of the derivation of winds from the sequential imagery of geostationary satellites. A more detailed and comprehensive discussion of the process has been given by Hubert (1979).

VII-4.2 NATURE AND PECULIARITIES OF CLOUD MOTION WINDS

The concept of deriving winds from cloud motion is simple. It involves tracking a cloud from one image to

[1] Primary Contributor: Linwood F. Whitney, Jr.

succeeding ones and calculating the distance and directional angle traversed in the time between them. Accurate knowledge of the geographical locations at the first and last picture is important, but accuracy of a wind vector depends more on relative registration of images than on absolute location. However, accurate absolute location is important for certain meteorological purposes (e.g., mesoscale analysis).

Earth location of an image element is computed from the knowledge of the orbital parameters of the satellite and the scanning geometry of its spin-scan camera (Smith and Phillips, 1972; Hambrick and Phillips, 1980). Since these satellites are never perfectly geostationary, the parameters are neither fixed nor always known as well as required. Thus, Earth locations must be refined by visual or objective means, such as using identifiable landmarks. For visual refinement, image landmarks are registered with those on computer-drawn grid overlays. For automatic refinement, previously observed digital patterns of well-defined landmarks are matched to those of the image by pattern correlation. In both cases, landmark adjustments correct Earth location over the entire image.

Tracer clouds are then selected. The best tracer is a cloud that is carried passively with the wind. Any development or decay of the cloud will produce error as will any cloud motion controlled by dynamics rather than advective transport. Thus, all but passive tracers should be avoided.

When observing an animated image sequence, skilled meteorologists integrate a wealth of background information (Hubert and Whitney, 1971) to assist them in deciding the legitimacy of cloud motion as a wind estimate. They are able to observe cloud growth or decay during motion, to identify cloud types by behavior and pattern, to distinguish between layers and to assess height qualitatively from relative motions and the gray shade of the IR cloud image, and to associate cloud appearances and behaviors with terrain effects or meteorological systems. The analyst can, for example, separate an individual cloud motion from the motion of a front or squall line, identify and separate anvil growth from the motion of jet stream cirrus, detect gravity wave clouds, or observe the motion of lower clouds through high, thin cirrus. Even though this subjective procedure can produce error, the advantage of human judgment in cloud tracking is enormously important.

In some automated techniques, however, the human factor is not involved. Automation is used to generate large numbers of cloud vectors rapidly. The argument is made that the automatic technique, by correlating arrays of digital values, is actually looking at a cloud field rather than an individual cloud patch, and thus growing and decaying patches or small scale perturbations within the field tend to be smoothed in the vector calculation. Although this argument may have some credibility, it is not always true. It does apply when cloud systems, such as cloud lines, fronts, or wave patterns, dominate a scene to the extent that the system, rather than the motion of the clouds comprising the scene, is tracked. As a result, other methods, such as editing, become substitutes for manual selection.

Resolution affects the scale of motion represented by a cloud vector. Since subpoint resolution varies from 1 km in VIS imagery to 8 km in the IR, the size of detectable cloud elements must lie in this dimensional range. Also, oblique views from the satellite exponentially increase the minimum detectable cloud size in the radial dimension; for example, it is doubled at a 52° geocentric angle.

Several pixels are usually required to give a cloud its definition, shape, and appearance. As a result, usually only large cloud masses are available for tracking. Automated techniques typically measure the mean motion of pixel arrays of from about 30 to 150 km across.

These characteristics make wind estimates based on cloud motion vectors somewhat different than conventionally measured winds, which are basically *in situ* measurements. Satellite cloud winds, which represent motions of very large volumes over substantial periods, are insensitive to microscale motions that affect conventional observations. As a consequence, cloud motions can and should be expected to deviate from conventional wind measurements.

Finally, the heights at which cloud motion vectors are estimated are crucial. No matter how accurately the motion is measured, the cloud altitude must be carefully chosen or serious errors can occur.

VII-4.3 METHODS OF MOTION MEASUREMENT

Cloud motions are measured manually or automatically using either VIS or IR data. VIS data afford higher resolution; IR data provide nighttime coverage and height information. Some operational systems use IR exclusively since it works equally well day or night.

In manual tracking, images are carefully registered by reference to identifiable Earth features and the motions are revealed by viewing the images in rapid succession. Cloud motions are selected for which suitable tracers exist. End points are marked from frame to frame, and the vectors are calculated (Young et al., 1972; Chang et al., 1973; Young, 1975; Chapters VI-2 and VII-2).

In automatic tracking, cloud patterns of the first and last images of a sequence are matched mathematically to provide the best fit possible. The displacement re-

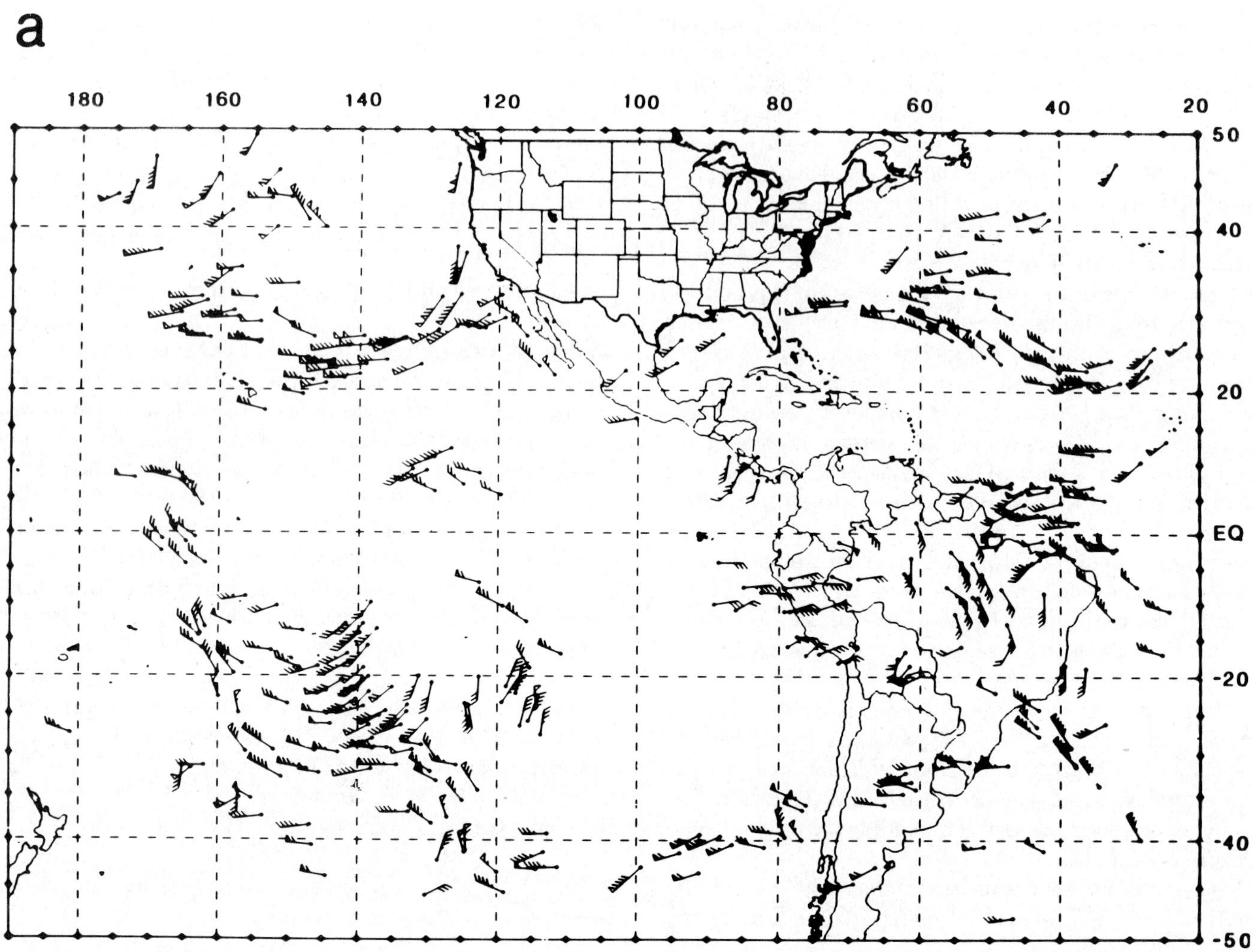

Figure VII-4-1a. High level cloud drift winds derived during one day from GOES-East and GOES-West. Multiple winds at a given location result from multiple derivation times. A full barb is 10 kt; a flag, 50 kt.

quired to achieve the best fit provides the motion. Although other automated procedures are available (Wolf et al., 1977; Endlich and Wolf, 1981; Haas, 1981), the matching of pixel arrays is the most widely used. Matching consists of selecting a digital pixel array of predetermined size from one image and searching within a larger array in the final image for the best fit. The best fit is the position of the first image relative to the last where the two are best matched by pattern correlation (Leese et al., 1971; Sitbon, 1972; Green et al., 1975; Chapter VI-2) or by the least mean or mean square difference (Suomi, 1975; Smith, 1975).

The searching process consists of stepping one pixel at a time through all rows and columns and calculating the mathematical match at each position. When the best match is found, the displacement between the centers of the two arrays provides the cloud motion (Figure VII-4-2). Computation time can become a problem when large numbers of vectors are needed, so many schemes have been devised to minimize calculation time. For example, in the United States (Green et al., 1975), a first guess wind obtained from the most recent analysis field provides a starting point for a reduced search area. Another strategy, introduced in Japan (Hamada and Watanabe, 1978), reduces the resolution of both arrays by averaging groups of pixels within the arrays and finds a rough best fit; then the scheme reverts to full resolution permitting "fine tuning" of the fit. Another strategy, employed by ESA (Sitbon, 1972), finds the rough fit with IR data and fine tunes with higher resolution VIS data.

The main advantage of automation is its ability to measure large numbers of vectors over extensive areas without the repetition inherent in manual measurement. The analyst, however, more ably judges which vectors are more reliable. A system combining the best

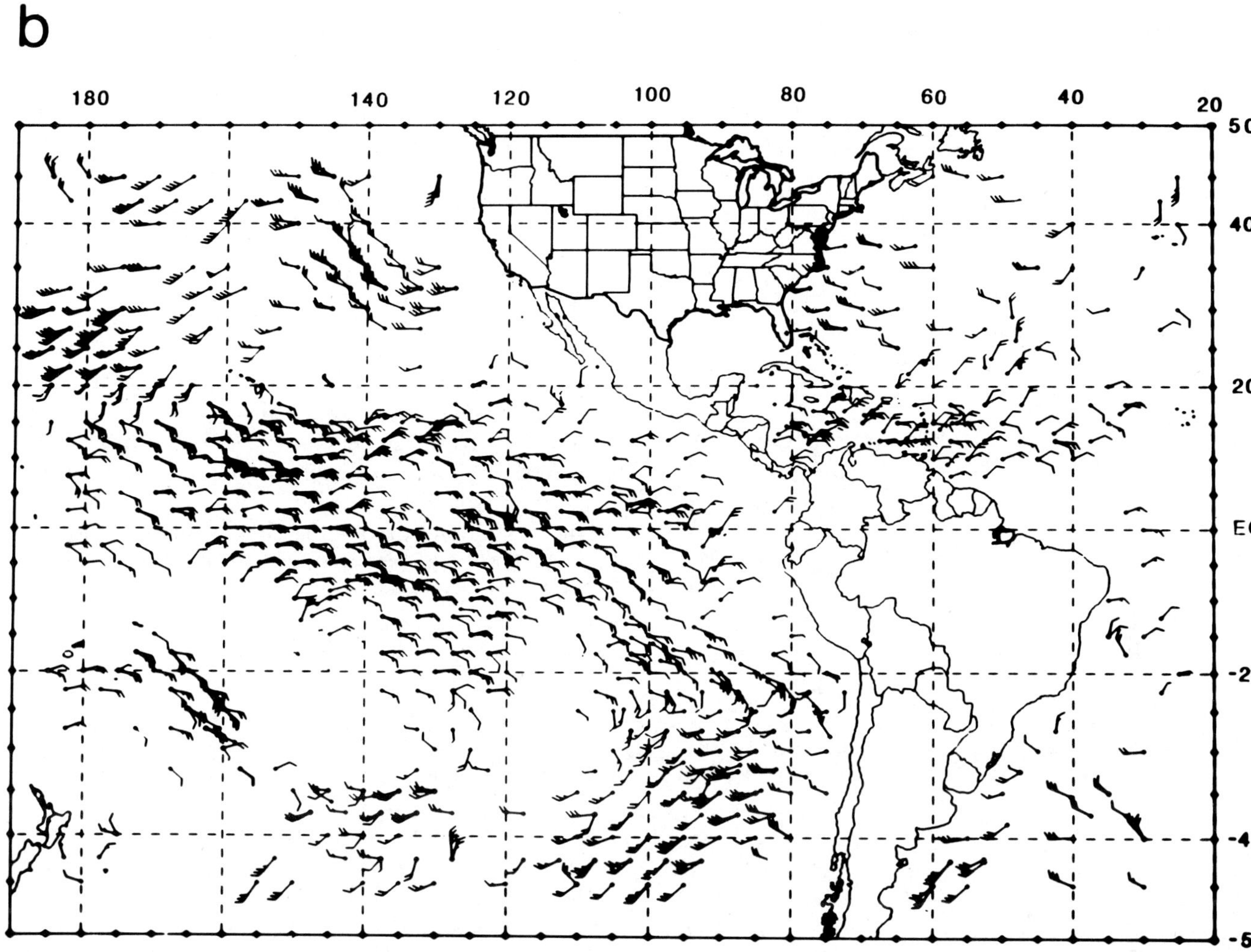

Figure VII-4-1b. Low level cloud drift winds derived during one day from GOES-East and GOES-West. Multiple winds at a given location result from multiple derivation times. A full barb is 10 kt; a flag, 50 kt.

attributes of both manual estimation and automation is the most desirable. Current interactive electronic display devices do permit this as an option (Chatters and Suomi, 1975). The analyst selects the clouds; their motions are measured and calculated by computer. Even so, the selection alone can be fatiguing and time consuming for the operational analyst who is confronted by the need to select clouds over vast areas two or more times daily.

Thus, operational agencies elect to measure most motions automatically. One approach, favored in the United States and Japan, uses automatic derivation at predetermined locations for motions at low levels, where clouds are numerous, and manual tracking at high levels, where there are generally fewer clouds and where human judgment is more necessary to identify the clouds. The ESA also uses fixed locations, but measures all Meteosat cloud motions by cross correlation (Bowen et al., 1979; Bridge, 1980).

VII-4.4 HEIGHT ASSIGNMENT

Several strategies exist for determining the heights of clouds. Methods are described below.

VII-4.4.1 Approximation

In this, the simplest method, height is approximated by establishing the level of best agreement between satellite winds and the most recent analysis. A variation of this procedure requires the operational analyst to maintain familiarity with current information on cloud height. This subjective approach is imprecise and often questionable, particularly over areas where sparse conventional data are not available as complementary information.

VII-4.4.2 Climatology

This method is based on a history of the level at which cloud motion best agreed with collocated sound-

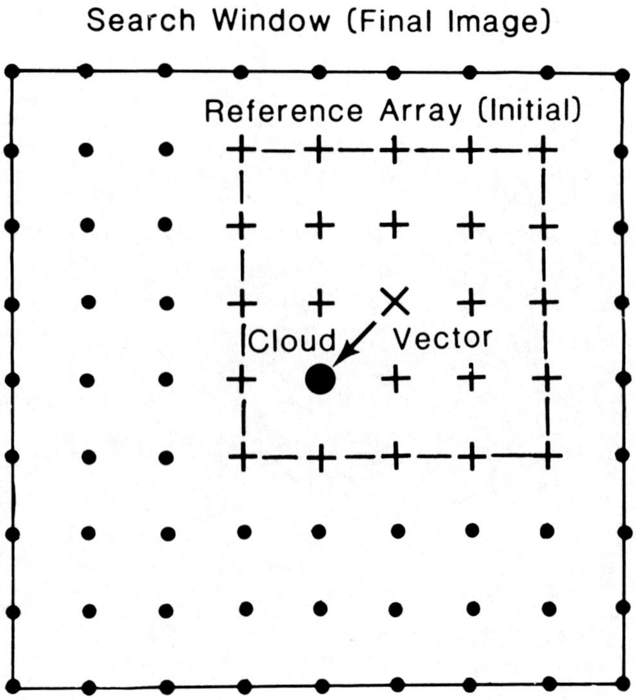

Figure VII-4-2. Schematic illustration of vector derivation of cloud drift by matching patterns of two images. The pattern arrays of both images are centered originally at the same geographic location. The position of best correlation defines the displacement between image centers and, thus, the cloud vector.

ings. Through analysis of the frequency distribution of heights, a preferred level for high, middle, or low clouds may emerge. Such preferred levels may be refined further as to season and geographical location (Hamada, 1982).

VII-4.4.3 Shadows

Use of cloud shadows was one of the earliest methods suggested for measuring cloud height (Fujita, 1963). Knowledge of the locations of the cloud, sun, and satellite, and of the length of the cloud shadow can be used to calculate the cloud height, but the limitations of this method are severe. Measurement and calculation are slow and its use is preferable at times of low solar elevation. Moreover, shadows are most readily seen only over light backgrounds, usually over lower clouds whose heights may be unknown. Use of shadows is most useful for specialized research, such as calculating the heights of convective towers.

VII-4.4.4 Stereoscopy

Overlapping coverage between satellites provides the opportunity to calculate cloud heights by stereoscopy. With knowledge of satellite positions and apparent positions of the cloud, parallax error can be converted into height (Hasler, 1981; Fujita, 1982b; Chapter VII-3). The operational use of stereoscopy has been prevented thus far by limitations in applying the technique to cloud motions. For one thing, stereoscopic capability is limited to the zone of overlapping coverage between adjacent geostationary satellites. This zone is greater for VIS imagery coverage of GOES type spacecraft, but considerably less for the lower resolution IR imagery. This is unfortunate since primary reliance is on IR data in operations.

Stereoscopic measurement depends on simultaneous scanning of a given area by both satellites. Without such viewing, any cloud movement during the time elapsed between overlapping pictures leads to a false height. Variations in scheduled picture times, in satellite attitude, or in operating design can limit or preclude synchronized coverage. Even when pictures are scheduled for the same start time, synchronization is not assured over the entire region of overlap because attitude differences can cause discrepancies in the time each scan reaches a given location. So start times must be closely controlled to permit simultaneous coverage. Stereoscopic coverage is further limited because Meteosat scans from south to north, GOES and GMS scan from north to south, and GMS scans at a different rate than GOES-West.

More basic limitations are lack of efficient real-time communications between agencies and lack of equipment (and time) to make stereoscopic measurements operationally.

Stereoscopic height measurement remains a valuable research tool, however. Studies of cloud heights in convective environments (Mack et al., 1983a; Fujita, 1982b) have been made over the United States using imagery from GOES-East and GOES-West. Techniques such as stereoscopy and shadows are appealing because they permit direct measurements of height using the image geometry, and so are unaffected by the problems of indirect measurements.

VII-4.4.5 Window Channel Temperatures

Infrared window data are most commonly used for determination of the heights of cloud tops. By assuming a cloud emissivity of 1, the equivalent blackbody temperature of a window channel observation of a cloud covered scene can be converted readily to pressure-altitude, or height, using the latest vertical temperature profile at the cloud location. This is generally quite reliable for all but cirriform clouds, which are often partly transparent to radiation from below. Subjective estimation of emissivity based on the apparent thickness of the cirrus (Young, 1975) can be made to derive a more accurate cloud height. This procedure also requires data on radiance from below the cloud to provide a good estimate of cloud top temperature and height. Because of the dubious accuracy of emissivity estimates of thin clouds, operational analysts generally avoid them by selecting thicker cirrus clouds estimated to have emissivities close to unity, or by estimating height by some altogether different procedure.

VII-4.4.6 Multiple Channels for High Clouds

In an extension of the IR window emissivity technique, ESA has devised a means to objectively derive the height of high, thin clouds by using both IR and WV channels of Meteosat (Bridge, 1980; Bowen et al., 1979; Szejwach, 1982). This approach permits calculation of cloud top temperature using a formulation for each channel based on the radiances of cloud and background measured by each channel. With the assumption that the emissivity of cirrus is the same in both channels, emissivity need not be specified and cloud top temperature and height are readily obtained. Since the WV channel can measure radiance only in the middle and upper troposphere, this technique is limited to higher clouds. Through use of this technique for heights of high, thin clouds, ESA has found an acceptable way to keep operational derivation of cloud motion winds entirely objective.

VII-4.4.7 Multiple CO_2 Channels

In a technique conceptually similar to the foregoing, Menzel et al. (1983) proposed using three CO_2 absorption bands from the GOES VAS instrument to measure cloud top height. A ratio of differences between cloud and nearby clear air radiances for a pair of the three channels is equated to a radiative transfer formulation. As in the preceding technique, all emissivity values cancel under the assumption of equal emissivities among the three CO_2 bands of similar frequency. Radiances of cloud and clear area measured by the CO_2 channels together with soundings, from either radiosondes or VAS, permit calculation of pressure height in increments of 50 mb. The apparent advantages of this technique over those discussed earlier lie in the abilities to measure height regardless of atmospheric pressure level and to obtain results more directly from theory. The technique also permits objective comparison among the heights obtained for the various channel pairs to select the best result or, in the event of failure, to provide a single window IR measurement as a backup procedure. Recently, the technique was put into use by NESDIS at the University of Wisconsin to obtain heights for the specialized winds used in analyses of severe convective storms and hurricanes.

VII-4.5 EDITING AND QUALITY CONTROL

Many schemes are used to minimize errors. The ones more commonly used are both objective and subjective. Both are incorporated in most automatic cloud tracking systems. When tracking is manual, subjective procedures are used.

VII-4.5.1 System Checks

These are objective checks that rule out automated vector measurement whenever certain system failures, such as missing scan lines or inaccurate image registration, occur.

VII-4.5.2 Geographical Selection

Terrain and diurnal heating can affect cloud motion and growth and so prevent passive cloud motion, especially at low levels. Because of this, some automatic tracking systems, such as those of the United States and Japan, limit cloud vectors to ocean areas. Although some land coverage is lost, vectors are obtained over oceans where meteorological data are typically sparse.

VII-4.5.3 Comparison of Cloud Vector Sets

Some automatic tracking systems measure cloud vectors over two consecutive sequences, thus providing a basis for comparing the two sets of cloud motions. When the difference between such a pair of vectors or their components depart from some selected threshold limit, they are eliminated. Each pair that falls within acceptable limits is averaged to provide a final vector. This procedure is used on the McIDAS system at the University of Wisconsin, is practiced operationally by ESA, and is available as an option for manually measured high level vectors on the VIRGS system at NESDIS.

VII-4.5.4 Neighbor Check

Here each vector is compared to its immediate neighbors, and those with large deviations are either removed or flagged for a second look.

VII-4.5.5 Comparison with Existing Analyses

In this comparison technique, cloud vectors exceeding some threshold of difference from a recent, or first guess, wind field are either removed or are flagged for manual scrutiny. Flagging is probably better because, after examination, any vector whose departure from the pre-existing wind field seems justified is used. The larger the justified departure, the greater the contribution toward improving the analysis.

VII-4.5.6 Wind Analysis by Conditional Relaxation

This editing procedure (Thomasell, 1979) is a sophisticated advance over that in VII-4. 5. 5. It is a completely objective, iterative procedure designed to delete vectors inconsistent with a series of successively refined analyses of all of the most recent data available at a given level. Each reanalysis results from a relaxation of smoothed vorticity and divergence fields obtained from the previous analysis of surviving vectors. This approach is especially well suited to regularly renewed vector sets of fairly uniform coverage such as might be derived by objective cloud tracking. For this reason, NESDIS uses this procedure for editing operational, low-level GOES vectors.

VII-4.5.7 Character of the Correlation Pattern

The matrix pattern of the variable (correlation coefficient, mean difference, or mean square difference) employed to judge the match between pairs of pictures is used to determine the uniqueness of the vector. If the position indicating the selected vector displacement does not exhibit a markedly better match than others, the vector becomes questionable. More than one distinct peak indicates vector ambiguity as does the lack of a peak or of a gradient around the peak. Large values everywhere may indicate great uniformity in the cloud field; small values everywhere may indicate a lack of cloud persistence. Vectors failing the test are either discarded or flagged for attention by an analyst.

VII-4.5.8 Subjective Editing

All cloud tracking systems are edited subjectively. Inspection by an analyst is the most easily used quality control technique. In automatic tracking, inspection is often used as the final screening after all objective editing or flagging is completed. The analyst critically examines flagged vectors for potential elimination but can delete any vector that departs unjustifiably from other indicators of the true situation. In manual tracking, cloud motions are, in effect, edited during manual selection but frequently also undergo final inspection to remove suspicious vectors.

VII-4.6 VALIDATION

After cloud motion winds became routinely available, many investigators tested their reliability. These tests contrasted cloud motions with various forms of "groundtruth" including rawinsondes, aircraft, conventional wind analyses, and other cloud motions. These results are always open to criticism because little satisfactory groundtruth exists. Each groundtruth source has either measurement problems or characteristics that conflict with other observations. Other differences are introduced by time and location separations tolerated between vector sets; such separations may be necessary to obtain enough data to make evaluations. Despite this, comparison of satellite wind data and more conventional wind measurement is one of the few available ways of learning about the reliability of satellite winds.

Comparison with rawinsondes is the most used approach because balloon measurements are readily available and have long represented the standard for conventionally measured wind, temperature, and humidity profiles. Mean absolute vector differences between cloud and balloon winds are generally estimated to range anywhere from 2 to 5 m s^{-1} at low levels and from 3 to 10, or more, m s^{-1} at high levels (Serebreny et al., 1970; Hubert and Whitney, 1971; Suchman and Martin, 1976; Whitney, 1984). These ranges in vector differences are probably attributable to variations in the severity of editing and selection procedures, in the approach of the measurement technique, and in the variations in climatic character of the wind.

Other studies using different forms of groundtruth are more encouraging. Hasler et al. (1979) tested cloud winds against simultaneous aircraft winds in the immediate environment of the same cloud. Mean absolute differences ranged from 0.9 to 1.7 m s^{-1} in trade cumulus, but up to 3.6 m s^{-1} in low frontal clouds. Tropical winds at high levels differed by a mean of only 1.7 m s^{-1}. Although these comparisons were made with what is probably the best groundtruth, sample sizes and geographical coverage were necessarily limited. The logistics of obtaining such carefully coordinated measurements preclude coverage over the broad range of possible wind regimes and climatic conditions. Nevertheless, these results suggest that satellite winds may be better in some cases than direct comparisons with conventional data indicate.

Another approach used at NOAA/NMC (and elsewhere) to evaluate the utility of cloud winds has been to compare their differences from a model 6-hour forecast with similar differences computed for rawinsondes. Such a comparison (for the standard deviation of the differences) is shown in Figure VII-4-3 for cloud winds produced by NESDIS, ESA (Meteosat), and JMA, and for rawinsondes for the zonal (on the left) and meridional (on the right) wind components. The differences in such a comparison also include the error in the forecast as well as in the wind measurement (rawinsonde) or estimate (cloud winds). In addition, for the results shown here, all available rawinsonde and cloud winds were used without regard to the reasonableness of the individual observations minus forecast differences.

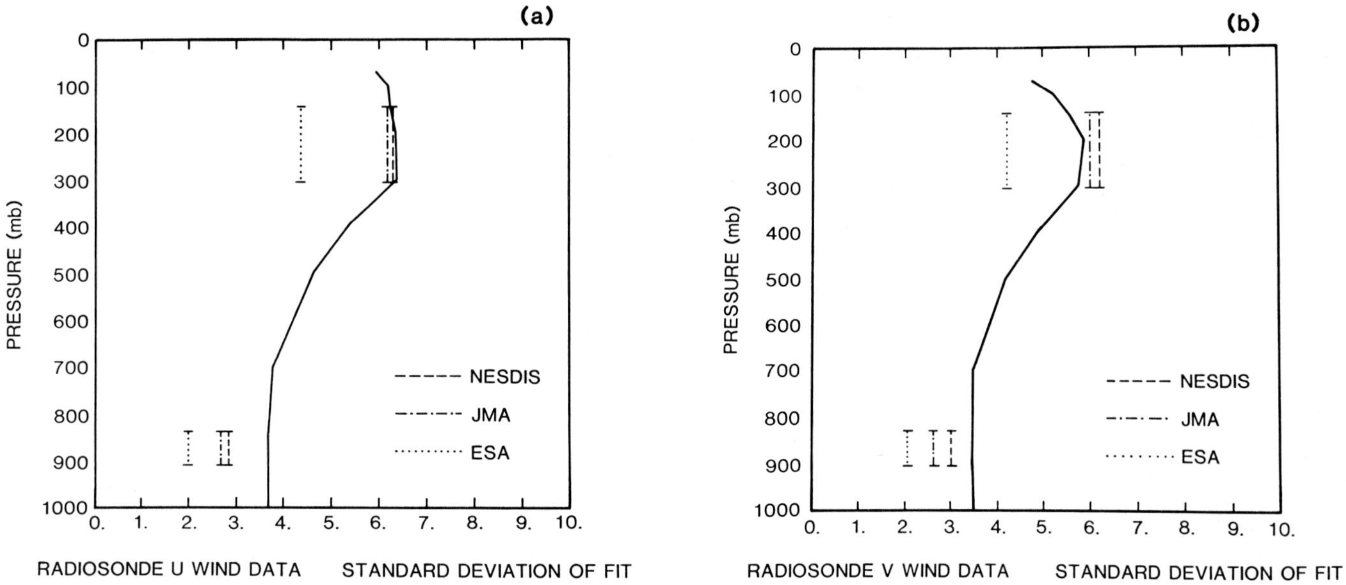

Figure VII-4-3. Standard deviation of the observation minus 6-hour forecast differences (NMC medium range forecast model used in the global data assimilation) computed for data collected between January 18 and February 15, 1989 (for data collected within ±3 hours of 00Z and 12Z) from 50°S to 50°N in m s^{-1}. a) *U* wind component, b) *V* wind component (courtesy of C. Dey and W. Baker).

Also, the mean differences, which are quite significant for high-level cloud winds, are not shown. With those caveats in mind, it may be seen in Figure VII-4-3 that the standard deviation of the differences between the cloud winds and the 6-hour forecast are in close agreement with those computed for rawinsondes for high-level (150 to 300 mb) NESDIS and JMA winds with the rawinsondes being slightly closer to the forecast for the meridional wind component. A significantly closer agreement ($\sim$1.8 to 2.0 m s^{-1}) is noted for the differences between the ESA winds and the forecast. For the low-level cloud winds (850 to 900 mb), data from all three producers exhibit better agreement with the forecast than do the rawinsondes, with the ESA winds showing the best agreement.

In a somewhat similar approach, Hubert and Whitney (1974) found that where rawinsonde and VIS cloud motions were compared to their combined analysis, differences were greatly reduced from those where datasets or their separate analyses were directly compared. This indicates that the two wind types are more compatible than direct comparison suggests.

The longest and probably the best record of satellite wind evaluation comes from the semiannual comparisons between rawinsondes and operational winds measured from GMS, GOES-East and West, and Meteosat. These comparisons have been made since 1979 by the Coordination for Geostationary Meteorological Satellites (CGMS), an international working group of geostationary satellite operating agencies created to promote information exchange. Comparisons are of two types: satellite-to-satellite and satellite-to-rawinsonde. Comparison of satellite winds can be made only in regions of overlapping view, while comparisons with rawinsonde data can be done anywhere in the useful field of view. In addition, comparisons are made among data pairs that are only approximately collocated, because data sites and times are uncoordinated between individual satellites and between satellites and rawinsondes.

Figures VII-4-4 and VII-4-5 show historical results of GOES and GMS wind comparisons given by Whitney (1983; 1984) and Hamada (1984). Among collocated satellite winds, root mean square (RMS) vector differences of high clouds rarely exceed 10 m s^{-1} and of low clouds, 6 m s^{-1} (Figure VII-4-4). Among collocated cloud and rawinsonde vectors, RMS vector differences are larger (Figure VII-4-5). Recently these differences have stabilized at around 10 to 15 m s^{-1} at high levels and 4 to 7 m s^{-1} at low levels. The much larger differences of the early days of GMS were decreased by refinements in procedures. Since then differences have remained similar for all satellites. Meteosat winds, though unavailable during much of the early historical period, also show similar results.

Further evidence that satellite winds (in this case GOES) may differ appreciably from rawinsonde winds appears in Table VII-4-1 which shows correlation coefficients varying from high to low values for both speeds and directions. Note that correlation coefficients at low

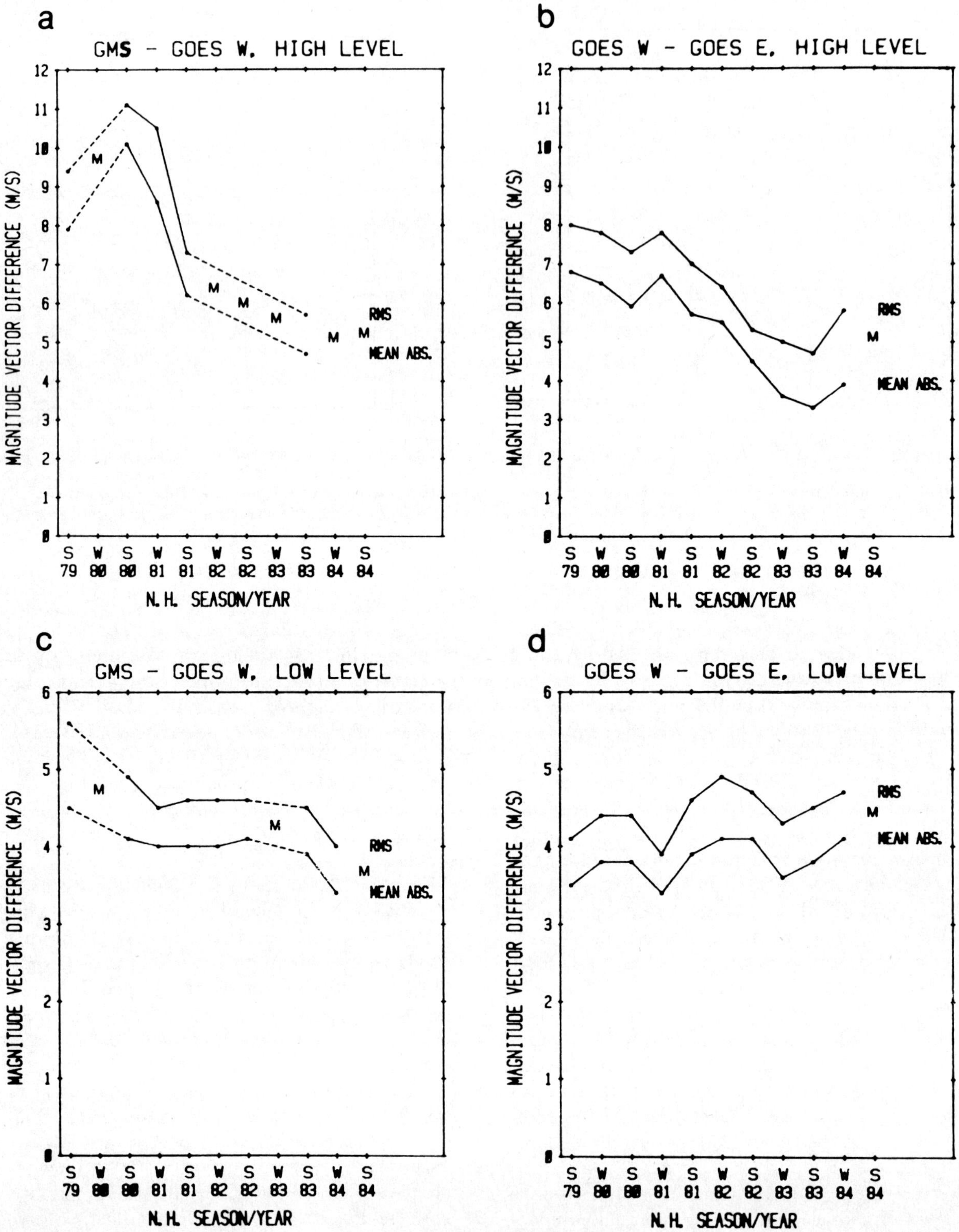

Figure VII-4-4. Historical record of vector differences from the intercomparison of cloud drift winds among satellites in overlap areas. Times of missing data are represented by the letter M. Comparisons of GMS and GOES-West winds at high and low levels appear in (a) and (c) comparisons of GOES-West and East winds in (b) and (d). Note scale differences.

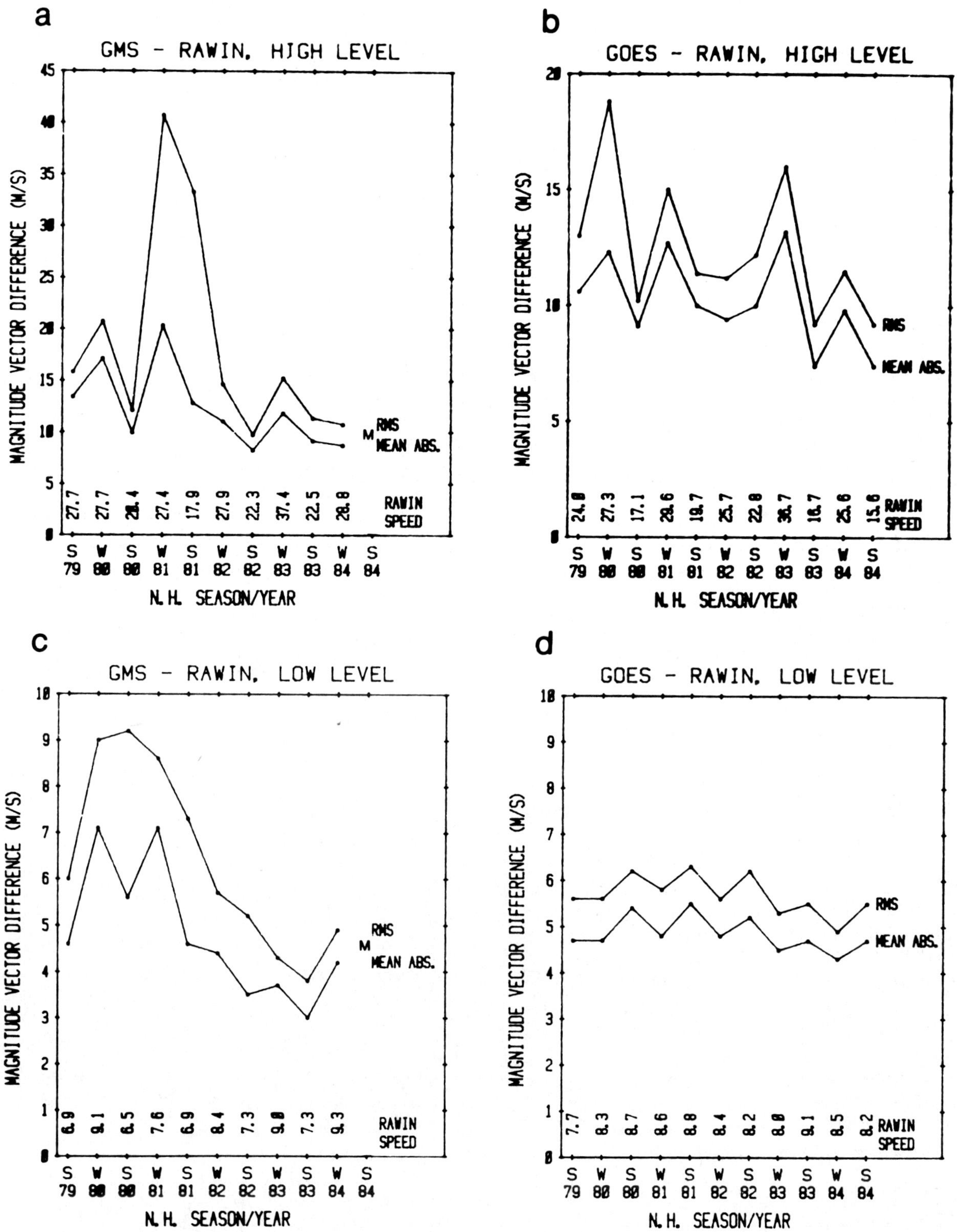

Figure VII-4-5. Historical record of vector differences from the comparisons of winds from satellites and rawinsondes. GMS and rawinsonde comparison at high and low levels appear in (a) and (c); GOES and rawinsonde comparisons in (b) and (d). Note scale differences.

levels, particularly for speeds, are worse than at high levels, which is consistent with the results shown in Figure VII-4-5 (c) and (d) where RMS vector differences approach the mean speed of the low level rawinsonde winds.

It is difficult to account for the contribution of each of the many sources of differences between cloud and rawinsonde winds. The evidence thus far suggests that the major sources of difference are improper height assignment to cloud vectors (Hubert and Whitney, 1971; Hubert 1979; Bengtsson et al., 1982a) and lack of exact space-time collocation with rawinsondes. These deficiencies become particularly serious in the presence of strong shear. Shear, both in the vertical and horizontal, is usually strongest and occurs most often in the upper atmosphere. Where there is strong vertical shear, incorrect height assignment to a satellite wind may result in a comparison with a rawinsonde wind of a very different speed or direction. Similarly, space or time separation between a cloud and a rawinsonde balloon can cause large differences when there is strong horizontal shear.

At low levels, the deviations of cloud winds from rawinsonde winds probably arise largely from the shear (height) problem. Efforts to assign cloud levels by using IR temperatures were often unsuccessful because whole cloud motion does not relate only to the wind at cloud top. Investigators have found that convective clouds tend to move with the wind at cloud base (Hubert and Timchalk, 1972; Hasler et al., 1979), at some sub-cloud level (Fujita et al., 1975), or at some level between top and base. For these reasons, NESDIS and MSC resorted to using a single level for low cloud winds (900 mb for GOES and 850 mb for GMS).

Table VII-4-1 *Correlation coefficients relating winds from GOES and rawinsondes during several comparison periods. W refers to winter (82 means 81–82, etc.) and S to summer.*

SEASON-YEAR	W82	S82	W83	S83	W84	S84
High level						
Number	60	94	92	166	90	169
Direction	.70	.92	.75	.92	.92	.93
Speed	.75	.76	.80	.82	.81	.79
Low Level						
Number	—	—	604	568	419	525
Direction	—	—	.88	.75	.80	.86
Speed	—	—	.42	.39	.44	.37

Table VII-4-2 *Vector differences between winds from GOES and rawinsondes at operationally assigned levels (AL) and level of best fit (LBF) during three comparison periods. Units are m s*$^{-1}$.

	SUMMER 83		WINTER 84		SUMMER 84	
	HIGH	LOW	HIGH	LOW	HIGH	LOW
Number	166	568	90	419	169	525
RMS at AL	9.1	5.5	11.5	4.9	9.2	5.5
RMS at LBF	5.2	2.8	6.5	2.9	4.8	3.2
Mean at AL	7.4	4.7	9.8	4.3	7.4	4.7
Mean at LBF	4.1	2.3	5.1	2.4	3.8	2.6

Examination of the Levels of Best Fit (LBF)[2] for recent GOES-to-rawinsonde comparisons shows an appreciable reduction in RMS and mean differences over those at operationally Assigned Levels (AL) (Table VII-4-2) for both high and low level clouds.

Although differences between satellite and rawinsonde winds are not true measures of motion accuracy, they are legitimate measures of the quality of the satellite wind. Without accurate heights, however, vectors are misplaced and so introduce errors in the data field and its analysis.

In further consideration of satellite-to-satellite comparisons (type 1), they have consistently been proven to be superior to those of satellite to rawinsonde (type 2). This is somewhat surprising because winds involved in the type 1 comparisons are at the very limit of their reliability in the area of overlap. On the other hand, the type 2 comparisons use winds generally more central to the satellite scene. Yet, the deficiency in the type 1 comparisons is not as important as other more favorable factors, among which one stands out. Collocated cloud pairs of type 1 are most likely to be found at or close to the same level because cloud elements at the same layer usually are viewed by both satellites. Thus, the height discrepancy problem that plagues type 2 comparisons is frequently eliminated in type 1 comparisons. Type 1 comparisons also benefit somewhat because the volume and intervals of measurements are far better matched than satellite-to-rawinsonde comparisons.

[2] Level of Best Fit (LBF) is a term used to describe the rawinsonde level at which the vector difference between cloud and balloon winds is a minimum (Hubert and Whitney, 1971).

Chapter VII-5

ATMOSPHERIC SOUNDINGS[1]

VII-5.1 INTRODUCTION

One of the most remarkable achievements of the weather satellite program is the ability to determine the vertical distribution of atmospheric temperature from remote observations of the Earth-atmosphere emission spectrum. This permits the use of only one instrument on a polar orbiting satellite to provide twice daily measurements of the three-dimensional temperature distribution over a large part of the globe. This chapter describes the methodology for deriving vertical soundings from the satellite radiance measurements, compares their accuracy with that of conventional radiosonde measurements, and describes their impact on weather forecasting models.

VII-5.2 METHODS FOR DERIVING REMOTE SOUNDINGS

VII-5.2.1 Basic Physical Principles

The atmosphere is partially or totally opaque for most of the spectrum of radiation emitted by the Earth surface. Water vapor, carbon dioxide, ozone, and other gases absorb the surface radiation and re-emit it at temperatures prevailing in the atmosphere. The amount of radiation absorbed depends on the absorbing characteristics and the amount of the gases in the atmosphere. At wavelengths of weak absorption, the surface radiation is absorbed in the atmospheric layers close to the surface and re-emitted to space from these low layers. At wavelengths of strong absorption, the surface radiation goes through a process of absorption and re-emission as it is transferred through the atmosphere and is essentially emitted to space from the higher atmospheric layers. At wavelengths of moderate absorption, the effective emission levels are at intermediate altitudes of the atmosphere. The spectrum of radiation measured by a satellite observing across such an absorption band would thus furnish information on thermal conditions at the respective effective levels of emission. Hence, the possibility exists of deducing vertical temperature profiles from remote spectral observations.

A simplified version of the integral equation representing this radiative transfer process can be written as

$$I(\nu) = B[\nu, T(x_0)]\tau(\nu, x_0) - \int_0^{x_0} B[\nu, T(x)] \times \frac{d\tau(\nu, x)dx}{dx}, \quad \text{(Equation VII-5-1)}$$

where $I(\nu)$ is the radiance or intensity of radiation emitted by the Earth-atmosphere system at wavenumber and observable by the satellite, B is the Planck function, T is temperature, x is an independent variable representative of the vertical dimension (e. g., altitude, pressure, ln pressure, etc.), and $\tau(\nu, x)$ is the transmittance of the atmosphere at wavenumber from height x to the top of the atmosphere (x_0 is at the Earth surface and 0 is at the top of the atmosphere). The

[1] Primary Contributor: George Ohring

transmittance depends upon the magnitude of the gas absorption coefficient and the amount of absorbing gas above the level x. The first term on the right represents the contribution of the surface to the observed radiance, the second term the contribution of the atmosphere. It is assumed that the surface emissivity is equal to 1.0. In the case of an atmospheric window, $\tau(\nu, x_0) = 1.0$ and the second term is negligible.

VII-5.2.2 Weighting Functions and Temperature Inferences

Equation VII-5-1 shows that the atmospheric component of the observed radiance consists of contributions from each atmospheric layer of the form of the Planck function times $d\tau(\nu, x)/dx$. Thus, $d\tau(\nu, x)/dx$ can be considered as a weight and its variation with x as a weighting function. Weighting functions in the 15 μm CO_2 absorption band for the various wavelengths (or wavenumbers) observed by the TIROS Operational Vertical Sounder (TOVS) are plotted in Figure VII-5-1. For any wavenumber, the weighting function indicates the portion of the atmosphere that contributes to the intensity observed.

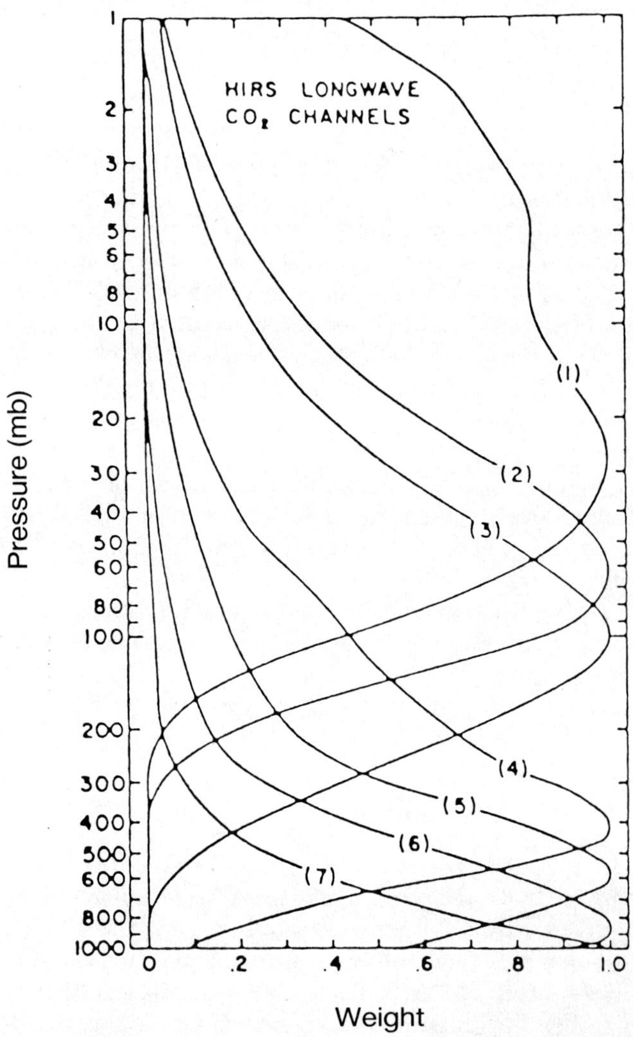

Figure VII-5-1. Weighting functions for various wavenumbers of the 15 μm carbon dioxide bands of the TIROS Operational Vertical Sounder (TOVS).

The characteristic shape of the weighting functions results from the following process. In regions of moderate absorption, radiation originating at low levels is absorbed by the overlying atmosphere and so contributes little to the emerging radiation. On the other hand, because of low densities, the contribution from high levels is also small. Thus, characteristically, there is a peak in the weighting function at some level within the atmosphere. For wavenumbers with weak absorption, the effective emission level is from the lower atmosphere, i.e., the peaks are at low atmospheric levels. For wavenumbers of strong absorption, the peaks are at high atmospheric levels. Thus, the radiance observed at a given wavenumber is associated with a specific atmospheric layer, or in other words, at each wavenumber the satellite "sees" to a different depth within the atmosphere. What the satellite sees is, from Equation VII-5-1, a function of the atmospheric temperature. Thus, if the weighting function of the absorption bands is known, the vertical profile of atmospheric temperature can be estimated from satellite observations. Carbon dioxide, uniformly mixed throughout the atmosphere, has absorption bands in the IR, and hence can be used for remote sensing.

VII-5.2.3 Effect of Clouds

One important atmospheric factor influencing radiance is the presence of clouds. Most low and middle (liquid droplet) clouds behave almost as blackbodies for IR radiation, i. e., their transmittances are close to zero. For cloud covered conditions, Equation VII-5-1, with x_o referring to the "surface" at the top of the cloud, would represent the appropriate radiative transfer equation. Remote sounding instruments in this case thus provide information about the atmosphere only at and above the cloud top. If the sky is partially covered with clouds, the contributions from the clear and cloudy portions of the atmosphere can be weighted, but the percentage of cloud cover must be estimated. High altitude (ice crystal) clouds not only do not behave as blackbodies but transmit IR with considerable variability—from 0.95 to 0.0 (Valovcin, 1968). Since the heights, amounts, and transmittances of ice crystal clouds are difficult to estimate and since they can obscure so much of the lower atmosphere from the satellite view, their presence severely hampers remote IR sensing schemes.

VII-5.2.4 Microwave Observations

Because the longer wavelengths of MW can penetrate most cloud systems, observations of thermal emission in the MW region can provide temperatures below clouds. Interpretation of MW observations is hampered by surface emissivity effects and sensitivity to precipitation. But these mechanisms also open possibilities for sensing surface properties (snow, sea ice, soil moisture) and precipitation.

VII-5.2.5 Retrieval of Other Meteorological Parameters

From the discussion, it is clear that remote sounding instruments can sense surface and cloud temperatures, as well as the vertical temperature profile. Also, once the temperature profile is determined, an integrated form of Equation VII-5-1 can be used to determine the distribution of absorbing gases with altitude. For example, if the temperature profile is obtained by observations of CO_2 absorption bands, the water vapor and ozone profiles can be inferred from observations in their respective IR bands.

Other important meteorological variables can be inferred through their connections with those parameters directly determined from the satellite observations. For example, the heights of constant pressure surfaces can be obtained from the temperature profile using an estimate of the pressure at some reference height, e.g., the surface. These heights, in turn, can be used to compute geostrophic winds. The implication is that any meteorological parameter correlated with temperature, water vapor, or ozone fields can be inferred from the satellite observations.

VII-5.2.6 Retrieval Methods

From a mathematical point of view, the temperature retrieval problem reduces to the solution of Equation VII-5-1. This type of equation, however, generally does not have a unique solution. Also, the presence of small observational errors can cause large errors in the temperature determinations. The mathematical nature of this instability has been described by Conrath (1967).

The physical nature of this instability can be seen by examining the graph of the weighting functions (Figure VII-5-1). The width of the weighting functions implies that the intensity of a particular wavenumber depends on the temperature at a number of levels on either side of the peak of the weighting function. Typically, the weighting functions have widths of the order of an atmospheric scale height (about 8 km). At each wavenumber the outgoing intensity is a weighted average of the temperature over an 8-km layer. In the first 30 km of the atmosphere only about four independent intensity observations can be obtained. To obtain the improved vertical resolutions needed for meteorological use, intensity observations at wavenumbers whose weighting functions overlap each other considerably must be used, although these observations contain much redundant information. Much of the information on the temperature profile is contained in the differences between the measured intensities. Small observational errors could lead to serious errors in the derived temperature profile.

To prevent incorrect, and possibly wildly incorrect, solutions for the temperature profile, some form of constraint must be used in arriving at solutions of the retrieval problem. Two general classes of retrieval techniques have been developed: statistical and physical.

Statistical techniques are based upon regression analyses that relate the temperature at each level to the observed radiance spectrum. The regression coefficients are generally determined from a historical set of concurrent observations of satellite radiances and radiosonde temperatures. The coefficients obtained are then applied to independent observations. In practice, the coefficients are based upon radiosonde-satellite matchups for five latitude bands and are updated weekly (see Chapter VI-2). The regression technique constrains the temperature retrieval to be in the "vicinity" of the mean temperature profile, thus insuring realistic solutions and avoiding instabilities. They have the additional advantage that the weighting functions need not be known. This scheme was used by NESDIS until it was replaced by a physical retrieval technique in October, 1988.

Physical retrieval techniques attempt to derive the temperature profile directly from radiance observations. In this method, a realistic (forecast or climatological) temperature structure is used as a first guess. It is then possible to derive a temperature profile that satisfies the observed radiance spectrum (Smith, 1970; Susskind et al., 1984). The method requires knowledge of the atmospheric transmittances, uncertainties which are overcome by estimating the transmittances using concurrent satellite radiance and radiosonde temperature observations.

These methods, if used with IR observations, assume clear-sky radiances. Although clouds are generally opaque to IR radiation, it is possible to make temperature soundings except under completely cloudy conditions. For partially cloudy conditions the satellite observed radiances are combinations of the cloudy and clear-sky radiances. By analyzing two nearby radiance observations of spots with different cloud amounts, and assuming that the vertical temperature profile is the same in both spots, one can derive clear-sky radiances for the region for use in the retrieval process (Smith, 1968; Susskind et al., 1984).

VII-5.3 APPLICATIONS OF SATELLITE SOUNDINGS

VII-5.3.1 Introduction

Perhaps the most important quantitative application of weather satellite observations is to numerical weather prediction. Predictions beyond a day or two require that the models be global in scale with temperature and humidity profiles among the key variables used in the prediction scheme. The conventional method for obtaining these variables is based largely on the radiosonde network. Figure VII-5-2a shows that the radiosonde station network leaves large gaps over oceans and unpopulated land areas. Satellite observations can fill these gaps (Figure VII-5-2b).

VII-5.3.2 Comparison of Satellite Soundings with Radiosondes

The accuracy of satellite soundings can be gauged by comparing them with those obtained by radiosondes. Table VII-5-1 shows the differences between layer temperatures obtained from the TOVS for the 30° to 60°N latitude zone over 1 year (Gruber and Watkins, 1982). The mean and RMS differences are tabulated as a function of pressure level.

These statistics are based upon collocated TOVS operational measurements and radiosonde observations. The spatial window for collocation was 1.5° × 1.5° for 30° to 40°N and 1° × 1° for 40° to 60°N. The 00 UTC and 12 UTC radiosonde observations were interpolated to the time of the satellite observation. The upper part of the table is for clear and partly cloudy retrievals, the lower half for totally cloudy retrievals based only upon the MSU of the TOVS system.

The mean or systematic errors are small. The satellite retrievals are generally best in the middle troposphere where the RMS errors for the clear/partly cloudy soundings vary between ~1° to 2°C. The retrievals deteriorate near the surface and tropopause where RMS differences grow to 2° to 3°C. For the cloudy cases the differences are greater by 1° to 2°C near the surface and by 0.5° to 1°C in the remainder of the troposphere.

Not all of the differences are due to satellite errors; the radiosonde measurement error approaches 1°C, and differences are introduced because of the small space and time offsets of the matchups. In an analysis that accounts for these contributions to the satellite-radiosonde differences, McMillin et al. (1983) concluded that the true errors in satellite temperature retrievals are between 1° to 1.5°C for those atmospheric layers where the satellite does well.

VII-5.3.3 Impact on Initial Conditions of Forecast Models

While comparisons of satellite and radiosonde temperatures furnish information on the accuracy of the satellite soundings, they do not indicate their value or usefulness in the forecast process. Currently, the major forecast centers, such as the National Meteorological Center (NMC) and the European Center for Medium-Range Weather Forecasts (ECMWF), use the satellite soundings only over the oceans and rely solely on the radiosonde network over land. It is difficult to determine if the added satellite soundings over the oceans improve the initial analysis. Without them, ocean analysis is based upon a first guess from a 6-hour forecast plus whatever conventional observations are available. The 6-hour forecast is generally quite good and, together with the available conventional data, provides for a reasonably accurate initial analysis.

Kalnay et al. (1985) evaluated the impact of satellite data on initial analyses by using the NASA Goddard Laboratory for Atmospheric Sciences global model. They calculated 6-hour forecast errors for two observing systems: the conventional system and the 1979 FGGE system, which included the conventional system plus satellite temperature soundings, satellite winds, and some non-conventional additional observations. Figure VII-5-3a shows the errors of 6-hour forecasts of 300-mb geopotential height for January 5-21, 1979, when only conventional data were used in the initial conditions. The "errors" over the oceans are small, but meaningless, since with few observations the 6-hour forecast itself defines the observed field. The major areas of interest are the west coasts of the continents; with weather systems propagating eastward and sparse data coverage over the oceans, the 6-hour forecasts have large errors there. Figure VII-5-3b is similar to Figure VII-5-3a except that the FGGE system is used to define the initial conditions. Comparison with Figure VII-5-3a shows significant reduction of errors on the west coasts of the continents. These reductions suggest that the satellite-enhanced FGGE data are more accurate for defining initial conditions over the oceans upstream from the coasts.

VII-5.3.4 Impact on Forecasts

The real test of the value of satellite soundings is their impact on forecast accuracies. Impact studies are generally conducted using two parallel sets of forecast runs: one using only conventional observations, the other using both conventional and satellite observa-

Figure VII-5-2a. Distribution of radiosonde stations around the globe (NOAA and NASA, 1987).

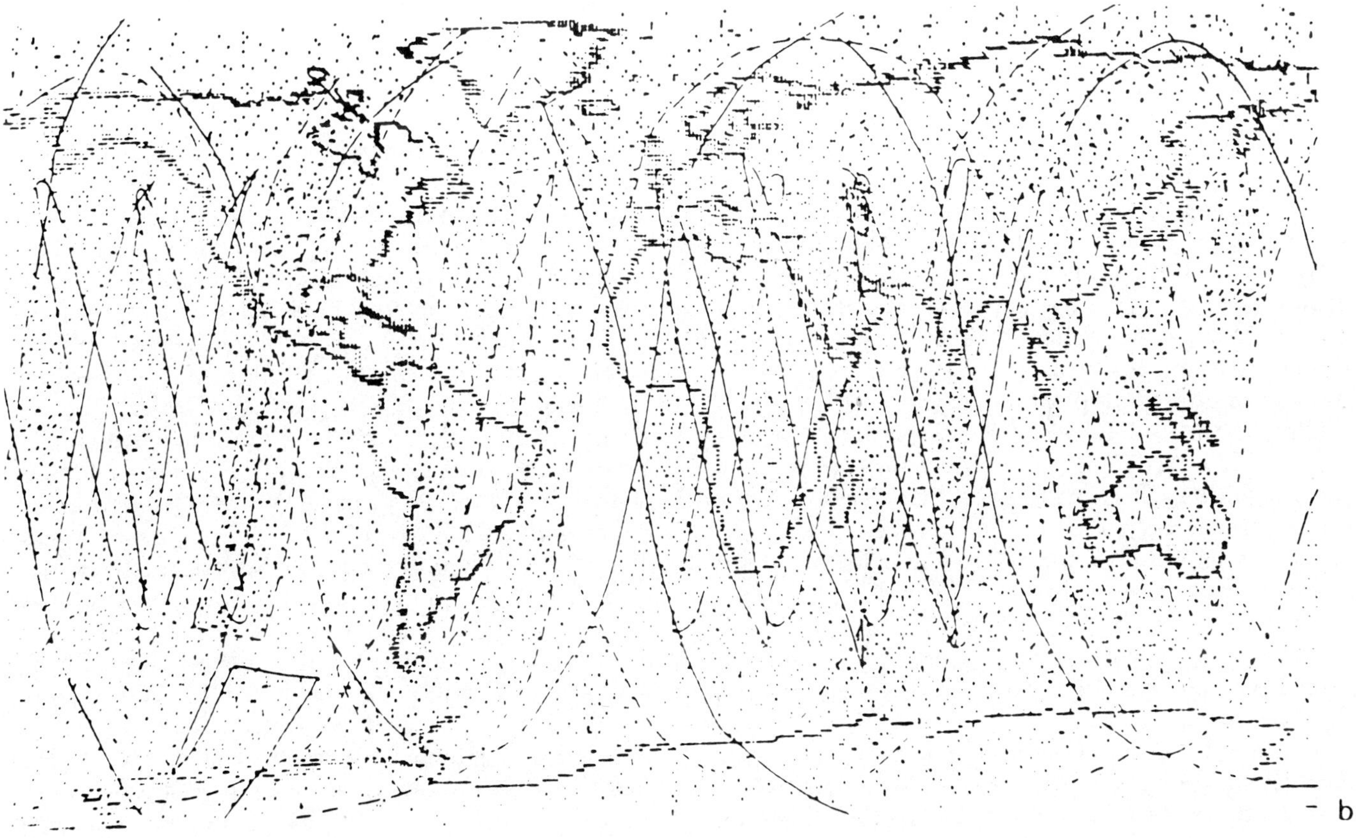

Figure VII-5-2b. Global coverage of the current two-satellite polar orbiting satellite system during a 24-hour period (NOAA and NASA, 1987).

Table VII-5-1 *Satellite-radiosonde temperature differences in °C. N is number of observations (Gruber and Watkins, 1982).*

CLEAR/PARTLY CLOUDY

SUMMER 79			FALL 79			Winter 79			SPRING 80			Layer
N	MEAN	RMS	N	MEAN	RMS	N	MEAN	RMS	N	MEAN	RMS	(cb)
334	−0.1	2.0	1390	0.7	2.7	917	−0.5	2.9	760	−0.1	2.5	100–85
991	0	1.3	3463	0.2	1.9	2417	−0.6	2.1	2041	−0.3	1.9	85–70
989	−0.3	1.0	3418	0.1	1.5	2390	−0.2	1.7	2020	−0.4	1.6	70–50
985	−0.5	1.4	3403	0.1	1.9	2367	0	1.9	2003	−0.6	2.0	50–40
986	−0.8	1.7	3381	0.1	1.9	2352	0.2	2.0	1998	−0.4	2.0	40–30
994	−0.9	1.8	3381	−0.1	2.1	2364	0.5	2.7	2002	−0.1	2.3	30–25
983	−0.4	1.9	3400	0.1	2.2	2393	0.6	2.5	2016	0.2	2.5	25–20
985	0.3	1.8	3422	0	2.2	2396	−0.3	2.2	2018	−0.6	2.4	20–15
981	0.4	2.0	3427	−0.2	1.9	2390	−0.4	1.9	2015	−0.4	1.8	15–10

CLOUDY

SUMMER 79			FALL 79			WINTER 79			SPRING 80			Layer
N	MEAN	RMS	N	MEAN	RMS	N	MEAN	RMS	N	MEAN	RMS	(cb)
137	2.8	3.9	535	−0.5	3.6	507	−1.4	4.0	527	0.1	3.8	100–85
293	−1.0	2.7	1488	−0.1	2.7	1333	−0.9	3.0	1388	−0.3	2.7	85–70
289	−0.9	1.7	1475	0.1	2.0	1322	−0.4	2.2	1374	−0.4	1.9	70–50
289	−1.3	2.3	1469	−0.2	2.5	1325	−0.2	2.4	1378	−0.8	2.6	50–40
291	−1.6	2.6	1457	−0.3	2.6	1311	0.2	2.5	1355	−0.9	2.8	40–30
293	−1.0	2.8	1457	−0.2	2.8	1305	0.8	3.1	1352	−0.6	3.2	30–25
291	0.7	2.6	1472	0.2	2.5	1311	1.0	2.9	1364	−0.1	3.0	25–20
287	0.5	2.2	1474	0.4	2.3	1314	0	2.3	1381	−0.3	2.5	20–15
282	−0.5	1.9	1473	0.1	1.9	1314	−0.3	1.9	1374	−0.3	1.8	15–10

tions. Both sets are then evaluated against verifying analyses.

Uppala et al. (1985) studied the impact of satellite soundings on the ECMWF forecasts. ECMWF uses an anomaly pattern correlation coefficient as a measure of forecast skill. This coefficient represents the correlation between two anomaly fields: one the departures from normal forecast values, and the other the departures from normal verification values.

The ECMWF analyses covered two periods: November 8-18, 1979 and February 22-March 7, 1979. During the first period, data from two satellites were available; during the second period, only one satellite was in operation. The ECMWF forecasts were made out to 10 days; however, the longer forecasts were not reliable because anomaly correlations generally dropped below 0.6, the limit at which forecasts are still considered useful. The 500-mb geopotential height was the verification parameter.

Figure VII-5-4 shows the mean anomaly correlation coefficients for the seven forecasts of the first period for both the Northern and Southern Hemispheres. The curve labeled *C* represents forecasts based upon the conventional observing system (surface observations, radiosonde observations, and aircraft wind observations); the curve *S* is for the conventional system plus satellite temperature soundings (but without aircraft winds). The satellite temperature soundings contribute to increased forecast accuracy. Using the 0.6 anomaly coefficient criterion, the limits of useful predictions were extended from 5.5 to 7 days in the Northern Hemisphere and from 3.5 to 5 days in the Southern Hemisphere.

For the second period, verifications indicated that satellite data had no significant impact on predictions in the Northern Hemisphere and only a modest (0.5 days) increase in predictability for the Southern Hemisphere. Uppala et al. (1985) attributed the lack of impact in the second period to two causes: (1) the availability of data from only a single satellite plus a gap in its coverage over a region in the eastern Pacific where considerable cyclonic activity was present, and (2) a general reduction in synoptic activity over the oceans. In explaining their results, Uppala et al. (1985) reasoned that satellites have the greatest potential for improving forecasts by providing observations over data sparse regions where cyclonic activity is developing.

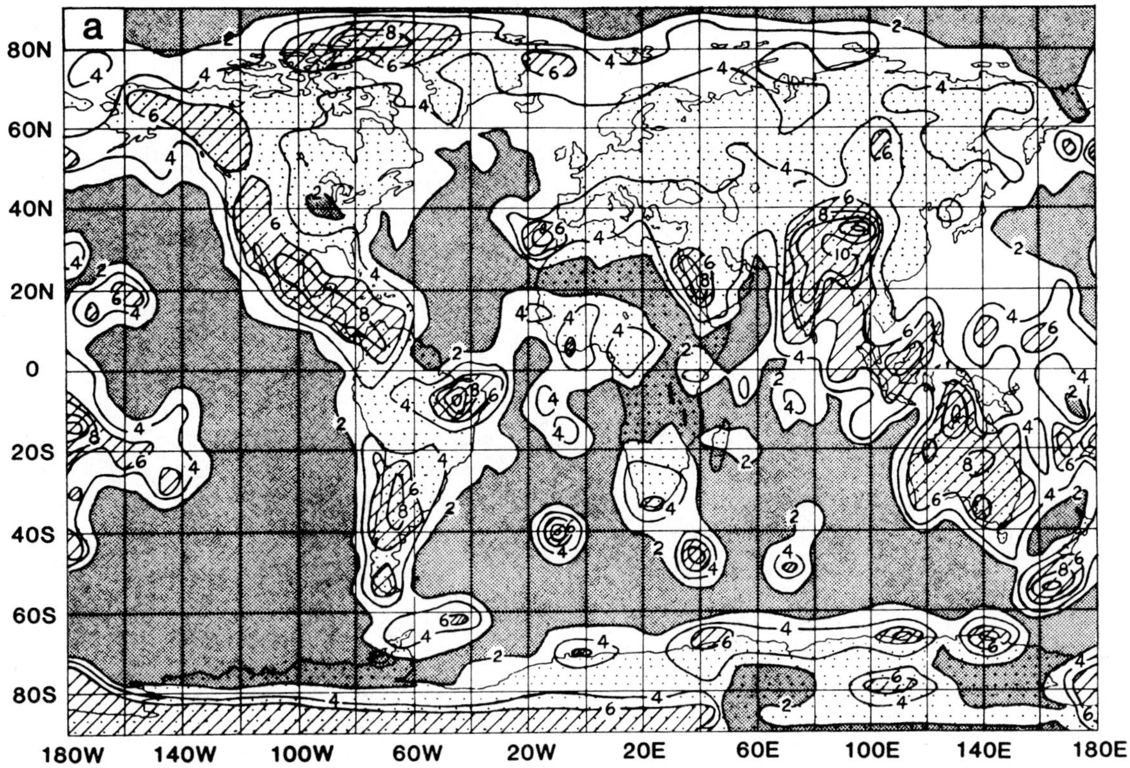

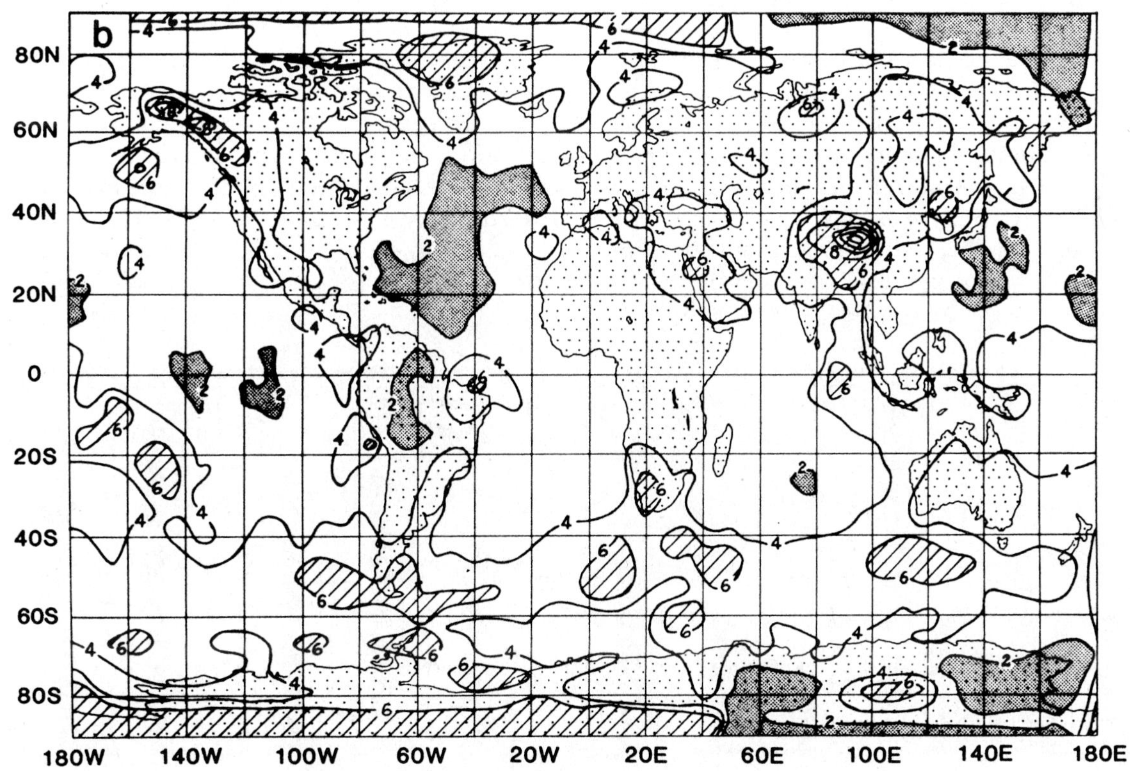

Figure VII-5-3. Root-mean-square errors of 6-hour forecasts of the geopotential height field. Errors less than 20 m are shaded; errors greater than 60 m are hatched. (a) RMS errors for forecasts based on observing system without satellite data (NoSAT), (b) RMS errors for forecasts based on FGGE observing system (Kalnay et al., 1985).

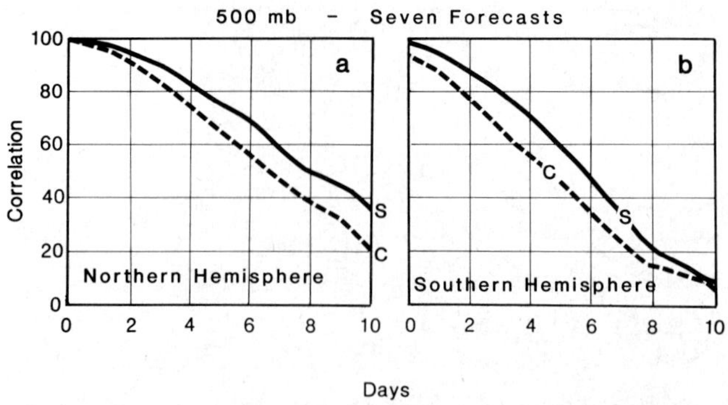

Figure VII-5-4. Anomaly correlations for forecasts of 500-mb geopotential height for two observing systems: conventional [C] and conventional plus satellite temperatures [S]; (a) Northern Hemisphere, (b) Southern Hemisphere (Uppala et al., 1985).

A similar series of forecast impact studies was conducted by Kalnay et al. (1985), whose evaluation of the impact of satellite soundings for defining initial conditions was discussed in VII-5.3.3. Their investigation covered two of the Special Observing Periods (SOP) of FGGE: SOP-1 (January 5-March 5, 1979) and SOP-2 (May-June 1979). During both SOP-1 and SOP-2, temperature soundings were available from only one satellite. The S1 skill score, a measure of the ability of the model to predict horizontal gradients, was used to evaluate the accuracy of the forecasts. Only those cases in which at least one of the forecasts retained useful skill were included in the analysis. The forecasts were verified over North America and Europe in the Northern Hemisphere and over South America and Australia in the Southern Hemisphere.

Figure VII-5-5 shows the number of times that one of the observing systems, FGGE or without satellite data (NoSAT), led to better forecasts than the other. Two forecasts were considered to be different if their S1 scores differed by more than two points. While the Southern Hemisphere results show that the FGGE observing system almost always produced a better forecast, the Northern Hemisphere data were not so impressive. They do show, however, that for most of the time the FGGE based forecasts were better than or just as good as the NoSAT forecasts. Similar results were obtained for SOP-2, except that effects in the Northern Hemisphere were somewhat smaller.

Larger effects in the Southern Hemisphere are to be expected since the satellite soundings, despite their errors, can make an important contribution to filling in the data gaps. Bourke et al. (1982) found that the introduction of satellite temperatures reduced errors in 48-hour sea level pressure forecasts over Australia from 7.5 to 5 mb.

In a study of the impact of satellite soundings on the numerical forecasts of the Israel Meteorological Service for a much smaller regional area, Thomasell et al. (1986) obtained results similar to those above. Although there were cases in which the impact was detrimental, in general it was beneficial. They found that positive impact depended on the distribution of satellite and conventional data with respect to specific synoptic features and on the ability of the forecast model to predict the development of those features accurately.

Comparison of the results of these three impact studies with a summary of earlier evaluations (Ohring, 1979) suggests that improvements in the satellite sounding instruments have led to more consistent positive impacts. However, there still remain instances, especially in the Northern Hemisphere, in which the satellite soundings lead to less accurate forecasts. There are two likely reasons for a degraded prediction: (1)

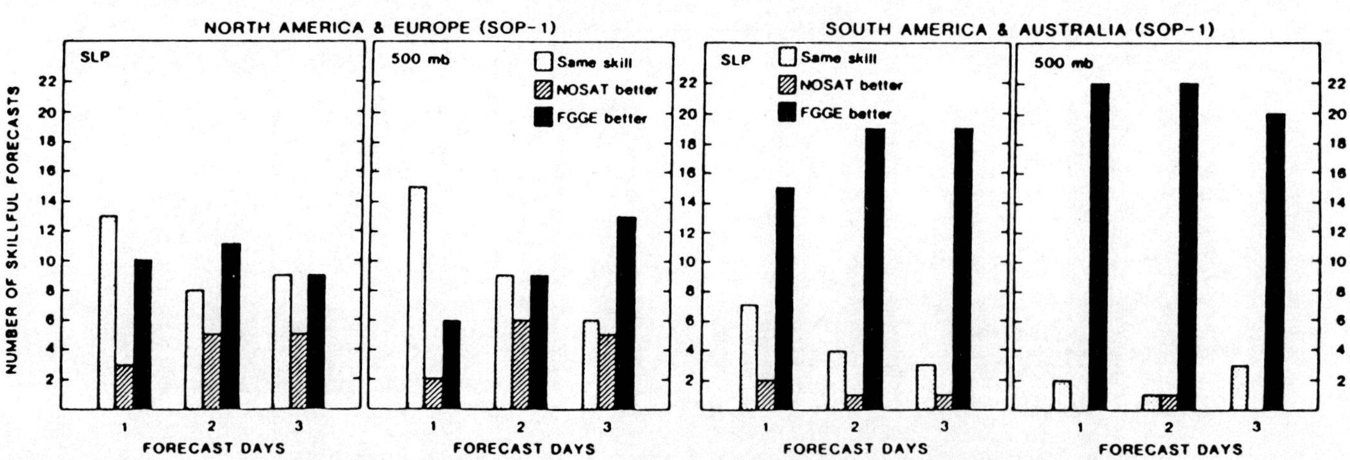

Figure VII-5-5. Number of cases of positive and negative forecast impact of the FGGE observing system during the period January 5 to March 5, 1979 (Kalnay et al., 1985).

the satellite soundings, because of their inherent errors, can worsen the depiction of the initial conditions, and (2) the information from the satellite soundings is not properly handled by the models.

Much remains to be done to improve the accuracies and the use of the satellite soundings. The Advanced Microwave Sounding Unit (AMSU), scheduled for flight in 1992, should improve temperature accuracies under cloudy conditions. Methods to improve the integration of satellite information into the prediction models must be worked upon. The asynopticity of satellite soundings is inconvenient for use in forecast models, which are basically designed for use with synoptic data. Four-dimensional data assimilation techniques are being developed to deal with this problem (Bengtsson et al., 1983). Another disadvantage of the satellite observations is their inherently poor vertical resolution. Yet the retrieval systems attempt to duplicate the capabilities of the radiosondes to measure the temperature at many levels of the atmosphere. Some recent work focuses on using only the basic information contained in the radiances (essentially mean layer temperatures) in the initialization process (Gal-Chen et al., 1986).

Chapter VII-6

PRECIPITATION

VII-6.1 INTRODUCTION

Information about precipitation is important for many disciplines and activities; among them are meteorology, climatology, hydrology, agriculture, energy, transportation, and recreation. Knowledge of extremes of precipitation are critical for the estimation and prediction of flooding, drought, and their effects on social and economic activities. The use of satellite VIS, IR, and MW measurements for estimating precipitation at synoptic and subsynoptic scales is described in this chapter. Larger scale precipitation estimates from satellites for climate diagnostics and modeling are treated in Chapter IX-4.

VII-6.2 PRECIPITATION ESTIMATION TECHNIQUES BASED ON INFRARED AND VISIBLE DATA

The geostationary satellites provide the most useful data for precipitation estimates; polar orbiter data are of somewhat more limited use. Satellite derived precipitation estimations can be used at resolutions ranging from mesoscale to climate scale (Barrett and Martin, 1981).

The two basic types of precipitation estimation techniques are "cloud indexing" and "cloud history." Cloud indexing, the principal method used where only polar orbiter data are available, involves characterizing a cloudy area by an index number based either on imagery or on quantitative radiances, and then empirically estimating its precipitation potential. The cloud history technique is used with geostationary satellite data, which permits following cloud fields throughout their life cycles.

Rainfall estimates can be based on the dominant synoptic or subsynoptic weather pattern, the amount of cloud cover, and the expected intensity of precipitation associated with the cloud types.

VII-6.2.1 Cloud Indexing Techniques

***A. The Follansbee Technique* (Follansbee, 1973)** is straightforward and simple. A 24-hour rainfall estimate for an area is based on a weighted sum of the percentages of the area covered by cumulonimbus, nimbostratus, and cumulus congestus. Empirical rainfall intensity weights are applied to these percentages: 1.0 for cumulonimbus, 0.25 for nimbostratus, and 0.02 for cumulus congestus. VIS and IR pictures are used to determine the cloud types and the percentages of the area occupied by each type. Local climatologies can be used to adapt the weighting factors to particular areas.

***B. The Earth Satellite Corporation (ESC) Technique* (Heitkemper et al., 1982)** operates on a 54-km grid and provides 6-hour rainfall estimates based on satellite observed IR cloud top temperatures and cloud areas, vertical motion from the NMC spectral model, and the time of day. The method requires four pictures a day from two polar satellites to compute complete estimates every 6 hours. When only two pictures are available, the 6-hourly intermediate cloud fields are interpolated

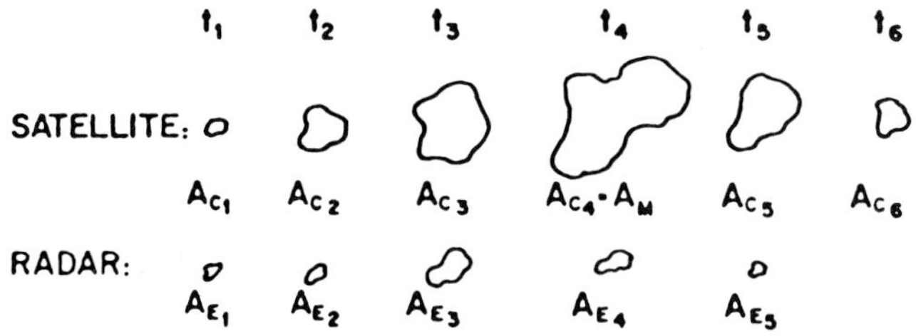

Figure VII-6-1. Idealized schematic history of a cloud or a satellite image and its associated radar echo (Griffith et al., 1978).

from 12-hourly satellite data and numerical prediction diagnostic fields.

C. The Bristol Technique **(Barrett et al., 1986)** uses polar orbiting or geostationary weather satellite VIS and IR imagery in either digital or pictorial form, and conventional synoptic data. Six-, 12-, and 24-hour rainfall totals are mapped over medium to large areas to produce better analyses than can be derived from satellite data or conventional data alone.

The central assumption of the Bristol Technique is that the total accumulation of rainfall is dependent on the cloud types within a specified area, the percentage of clouds within that area, a weighting factor based on rainfall data that indicates the rainfall potential of a cloud field, and a weight based on normalized mean monthly rainfall statistics for a given area.

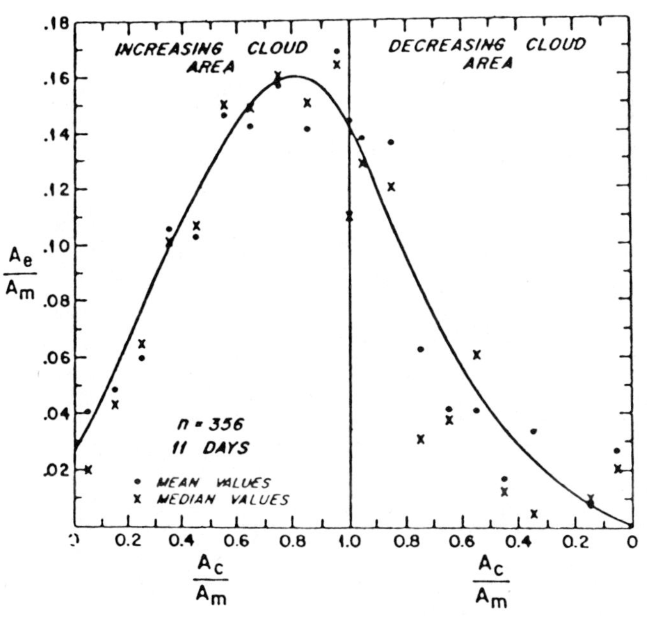

Figure VII-6-2. Cloud, echo area relationship for VIS data over Florida; A_c = cloud area, A_e = echo area, and A_m = maximum cloud area (Griffith et al., 1978).

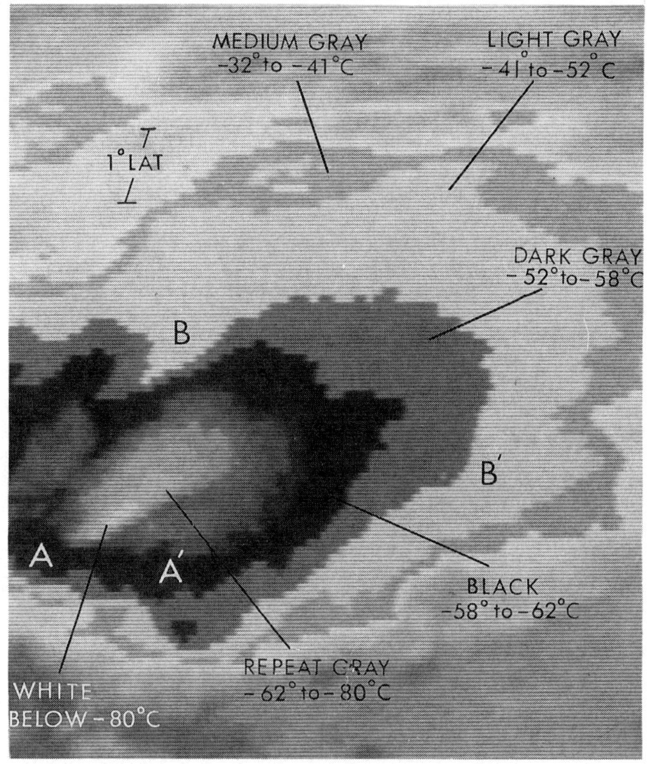

Figure VII-6-3. Enhanced IR imagery (MB curve), 0900 UTC, September 12, 1978.

VII-6.2.2 Cloud History

Cloud history techniques have been developed at NOAA/ERL by Woodley and Griffith and at NOAA/NESDIS by Scofield and Oliver. The former technique can be totally automated on a computer; the latter requires some interpretation by a meteorologist who can use an interactive computer system, such as McIDAS, for assistance.

A. The Woodley/Griffith Technique **(Griffith et al., 1978)** is a diagnostic method for estimating rainfall over large space (10^4 to 10^5 km^2) and time (12 to 24 hours) scales by the use of VIS and IR GOES imagery. Its algorithm is based on the knowledge that areas of active convection and rainfall, mainly in the tropics, are brighter (colder) than other areas in VIS (IR) pictures. The evolution of clouds seen on satellite images (and corresponding radar echoes) approximates the model of Figure VII-6-1. GOES images were calibrated from raingage adjusted 10-cm radar data over southern Florida and empirical relationships were established. These require time sequences of cloud areas measured from the satellite images at a specified threshold brightness (80 digital counts) or threshold temperature (253 K) to calculate rain volume over a given period.

CONVECTIVE STORM TECHNIQUE

STEP 1

RAINFALL IS COMPUTED ONLY FOR THE <u>ACTIVE</u> PORTION OF THE THUNDERSTORM SYSTEM:

THE FOLLOWING ARE CLUES FOR HELPING TO MAKE THIS DECISION.

- IR TEMPERATURE GRADIENT IS TIGHTEST AROUND STATION END OF ANVIL FOR A THUNDERSTORM SYSTEM WITH VERTICAL WIND SHEAR (IR).
- STATION IS LOCATED NEAR THE CENTER OF THE ANVIL WITH A TIGHT, UNIFORM IR TEMPERATURE GRADIENT AROUND ENTIRE ANVIL FOR A THUNDERSTORM SYSTEM WITH NO VERTICAL WIND SHEAR (IR).
- AN OVERSHOOTING TOP IS OVER THE STATION (VIS AND IR).
- ANVIL IS BRIGHTER AND/OR MORE TEXTURED (VIS).
- FROM COMPARING LAST TWO PICTURES: STATION IS UNDER HALF OF ANVIL BOUNDED BY EDGE WHICH MOVES LEAST (IR).
- STATION IS NEAR 300-MB UPWIND END OF ANVIL (IR, SKIP THIS CLUE IF NO UPPER AIR DATA AVAILABLE).
- STATION IS NEAR THE AREA OF LOW-LEVEL INFLOW (VIS).
- STATION IS LOCATED UNDER A RADAR ECHO.

STEP 2

HALF-HOURLY RAINFALL ESTIMATES IN INCHES ARE COMPUTED FROM THE FOLLOWING FACTORS:

FACTOR 1

CLOUD-TOP TEMPERATURE AND CLOUD GROWTH FACTOR [IR].
DETERMINE AMOUNT THAT THE COLDEST CLOUD TOPS INCREASED WITHIN HALF-HOUR.

	>2/3° LAT	>1/3° ≤2/3° LAT	≤1/3° LAT OR SAME	AREAL DECREASE OF SHADE OR WARMING FROM WHITE TO RPT GRAY OR WITHIN THE RPT GRAY	COLDEST TOPS 1 OR MORE SHADES WARMER
MED GRAY (-32 TO 41°)	0.25	0.15	0.10	0.05	T
LT GRAY (-41 TO -52°)	0.50	0.30	0.15	0.10	
DK GRAY (-52 TO -58°)	0.75	0.40	0.20	0.15	
BLACK (-58 TO -62°)	1.00	0.60	0.30	0.20	
RPT GRAY* (-62 TO -80°)	1-2.00	0.60-1.00	0.30-0.60	0.30	
WHITE (BELOW -80°)	2.00	1.00	0.60	0.40	0.10

*COLDER REPEAT GRAY SHADES SHOULD BE GIVEN HIGHER RAINFALL ESTIMATES.

OR

DIVERGENCE ALOFT FACTOR* [IR AND 200-MB ANALYSIS].

MED GRAY	LT GRAY	DK GRAY	BLACK	RPT GRAY	WHITE
0.15	0.30	0.40	0.60	0.60-1.00	1.00

*IR IMAGERY SHOWS EDGES OF THUNDERSTORM ANVIL ALONG THE UPWIND END FORMING A LARGE ANGLE OF BETWEEN 50-90 DEGREES POINTING INTO THE WIND; 200-MB ANALYSIS OFTEN SHOWS THESE STORMS JUST DOWNWIND FROM WHERE THE POLAR JET AND SUBTROPICAL JET SEPARATE.

GO TO FACTOR 2

Figure VII-6-4. Step 1 and Step 2 (Factor 1) of the Scofield/Oliver technique.

FROM FACTOR 1

FACTOR 2

OVERSHOOTING TOP FACTOR (VIS, IR). ADD TO THE OVERSHOOTING TOPS*:

MED GRAY	LT GRAY	DK GRAY	BLACK	RPT GRAY	WHITE
0.50	0.45	0.40	0.30	0.30	0.30

*HIGH-RESOLUTION VISIBLE IMAGERY IS THE BEST DATA FOR DETERMINING THIS FACTOR.

FACTOR 3

THUNDERSTORM OR CONVECTIVE CLOUD LINE MERGER FACTOR (IR, VIS). ADD 0.50 TO THE COLDER TOPS IN THE AREA OF THE MERGER.

FACTOR 4

SATURATED ENVIRONMENT FACTOR (IR, VIS). ADD TO THE COLDER TOPS STATIONARY FOR A GIVEN AMOUNT OF TIME:

	MED GRAY	LT GRAY	DK GRAY	BLACK	RPT GRAY	WHITE
≥1 HOUR BUT <2 HOURS	0.20	0.20	0.20	0.20	0.30	0.30
≥2 HOURS	0.40	0.40	0.40	0.40	0.50	0.50

FACTOR 5

MOISTURE CORRECTION FACTOR = PRECIPITABLE WATER (SFC-500MB) • RELATIVE HUMIDITY (SFC -500 MB)

STEP 3

FACTORS ARE SUMMED AND MULTIPLIED BY MOISTURE CORRECTION FACTOR:

TOTAL HALF-HOURLY CONVECTIVE RAINFALL ESTIMATES (IN INCHES) =

[(CLOUD-TOP TEMEPERATURE AND CLOUD GROWTH FACTOR OR DIVERGENCE ALOFT FACTOR)[1] + OVERSHOOTING TOP FACTOR[2] + MERGER FACTOR[3] + SATURATED ENVIRONMENT FACTOR[4]] • MOISTURE CORRECTION FACTOR[5]

END OF TECHNIQUE

Figure VII-6-5. Step 2 (Factors 2, 3, 4 and 5) and Step 3 of the Scofield/Oliver technique.

The algorithm uses hourly imagery. The estimated rain volume is the product of the cloud growth trend, the radar echo area, the time difference (usually 1 hour) between successive images, the fraction of the cloud system enclosed by each brightness or temperature contour, a weighting factor based on brightness or temperature, and a cloud model adjustment factor. The echo area is obtained from the curve in Figure VII-6-2; the cloud growth trend is computed from the slopes of the curve. Correction for the type of environment, ranging from very moist to very dry, is obtained from the cloud model adjustment factor described by

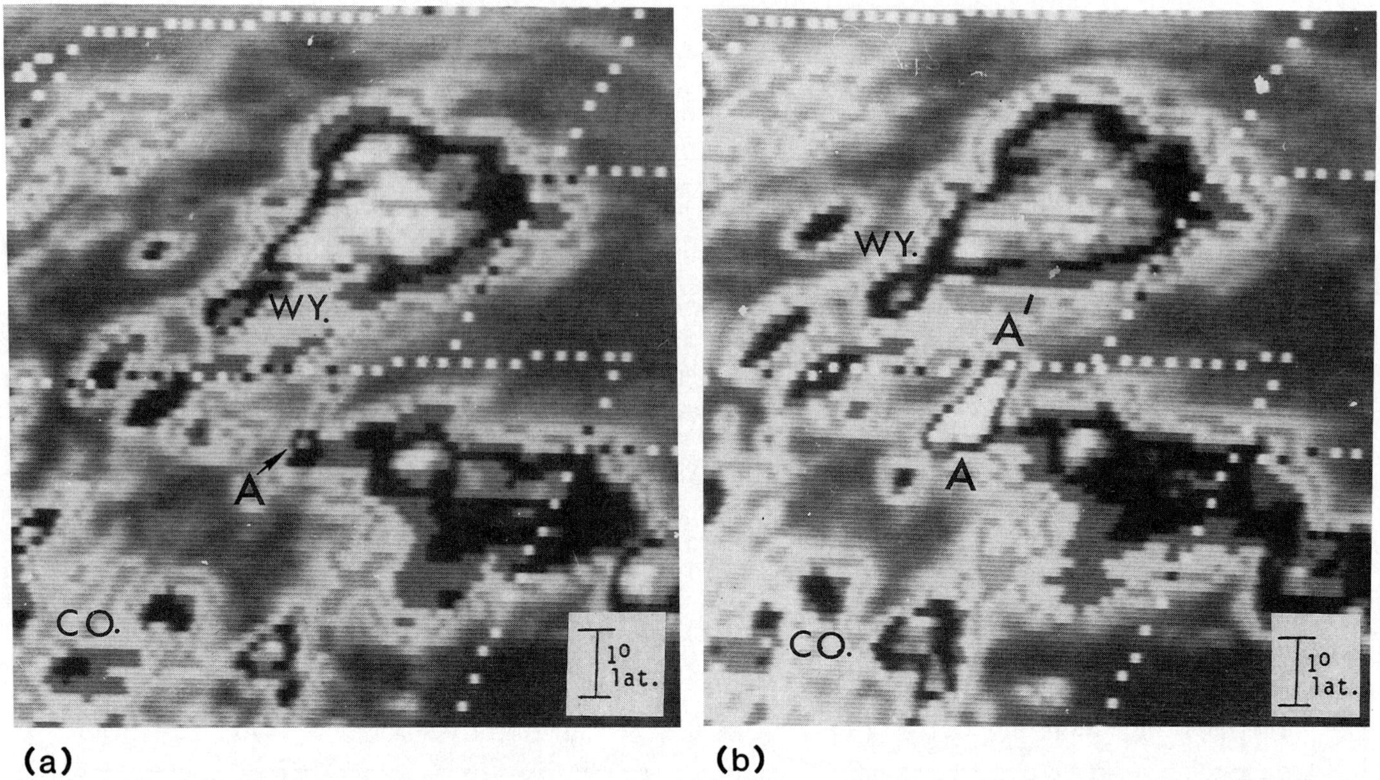

Figure VII-6-6. Enhanced IR imagery (CB curve), August 1, 1976, (a) 0030 UTC, (b) 0100 UTC.

Griffith et al. (1981). Simpson and Wiggert's (1969) one-dimensional cloud model is used to adjust the estimates.

B. The Scofield/Oliver Technique **(Scofield and Oliver, 1977; 1980; Scofield, 1987)** uses GOES imagery to make half-hourly or hourly rainfall estimates for deep convective systems of tropical air masses. It uses IR images (Figure VII-6-3) constructed by use of a digital enhancement (MB) curve to estimate convective storm intensity. The MB curve is explained by Clark (1983). Estimates of convective rainfall are computed by comparing the changes in cloud character that are observed between two consecutive images of enhanced IR and high-resolution VIS data. Procedures for using the technique are outlined in Figures VII-6-4 and VII-6-5. It is divided into three main steps:

- Identification of the active portion of the convective system where there are strong updrafts and heavy rain. For thunderstorm systems embedded in a moderate-to-strong vertical wind shear environment, the heaviest rainfall most often falls along the upwind edge of the anvil (between *A* and *A′* in Figure VII-6-3, while the inactive portion of the anvil is between *B* and *B′*). For thunderstorm systems embedded in a weak vertical wind shear environment, the heaviest rainfall most often falls near the center of the anvil.
- Computation of the half-hourly convective rainfall estimate for the active portion using five factors:

Factor 1: Cloud top temperature and cloud growth or divergence aloft (Figure VII-6-4). An example of a rapidly growing ($>\frac{2}{3}°$ latitude in $\frac{1}{2}$ hr) thunderstorm is shown in Figures VII-6-6a (at *A*) and VII-6-6b (between *A* and *A′*). A thunderstorm associated with strong divergence aloft is shown in Figures VII-6-7a and VII-6-7b.

Factor 2: Overshooting top (Figure VII-6-5). In VIS images, overshooting tops are quite bright and textured (Figure VII-6-8a at zero); in IR imagery, they are very small (pixel size area) and cold (usually lower than −62°C, see Figure VII-6-8b at zero).

Factor 3: Thunderstorm cluster or convective cloud line merger (Figure VII-6-5). The two IR satellite pictures in Figures VII-6-9a and VII-6-9b show a

Figure VII-6-7. (a) Enhanced IR imagery (MB curve), 0430 UTC, April 12, 1979. (b) Splitting jets are indicated in the 200-mb analysis, 1200 UTC.

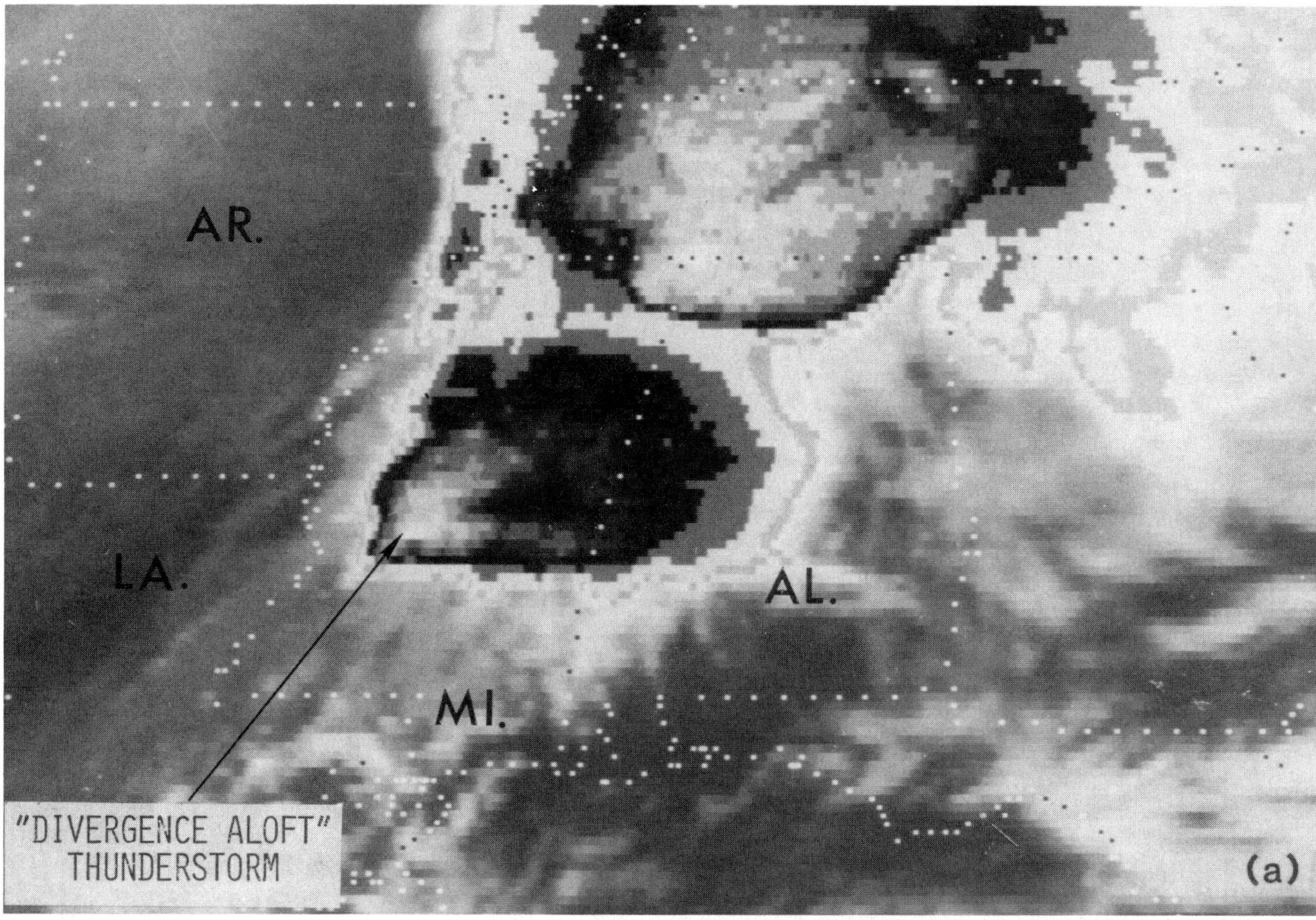
AR.
LA.
AL.
MI.
"DIVERGENCE ALOFT"
THUNDERSTORM
(a)

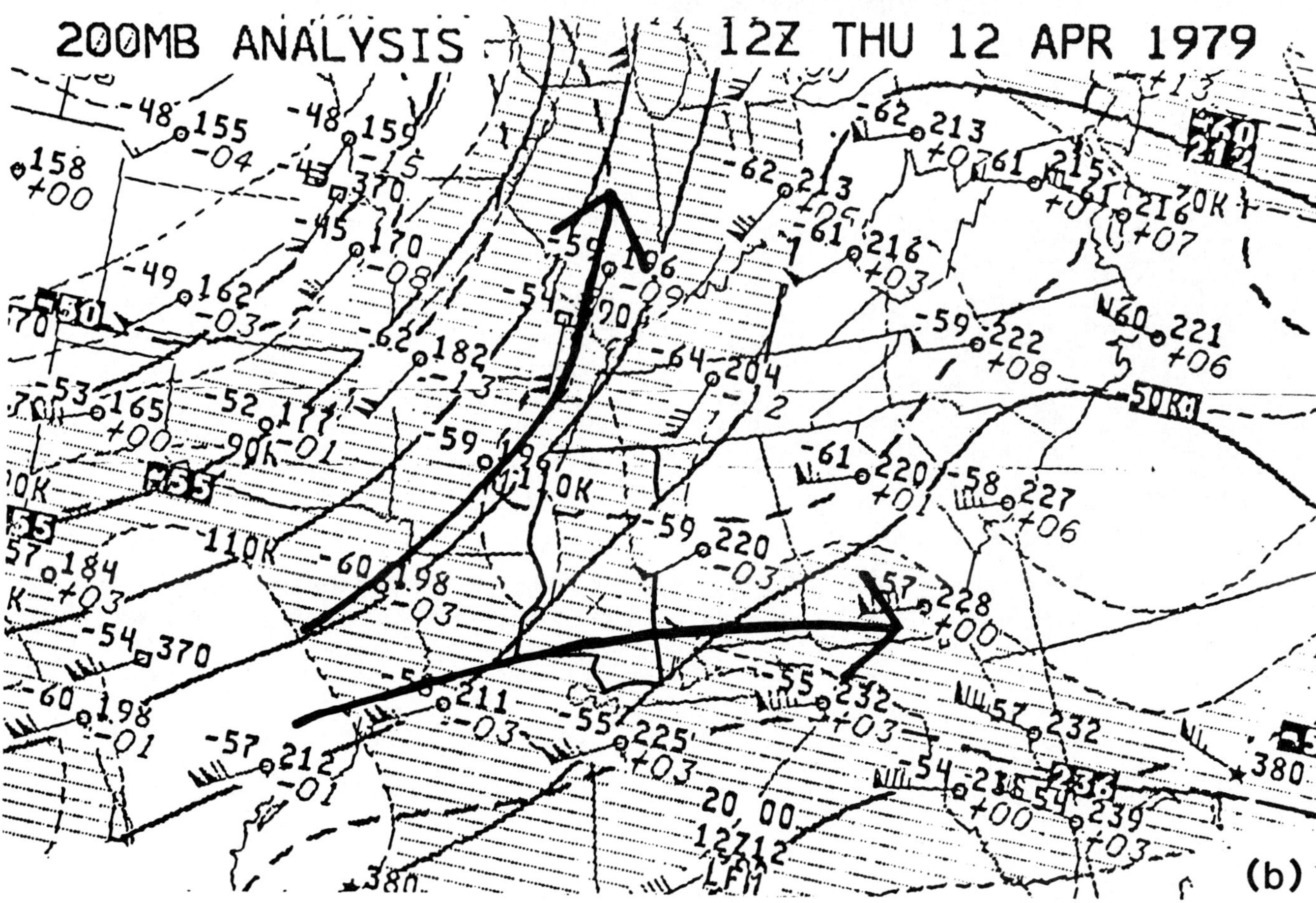
200MB ANALYSIS 12Z THU 12 APR 1979
(b)

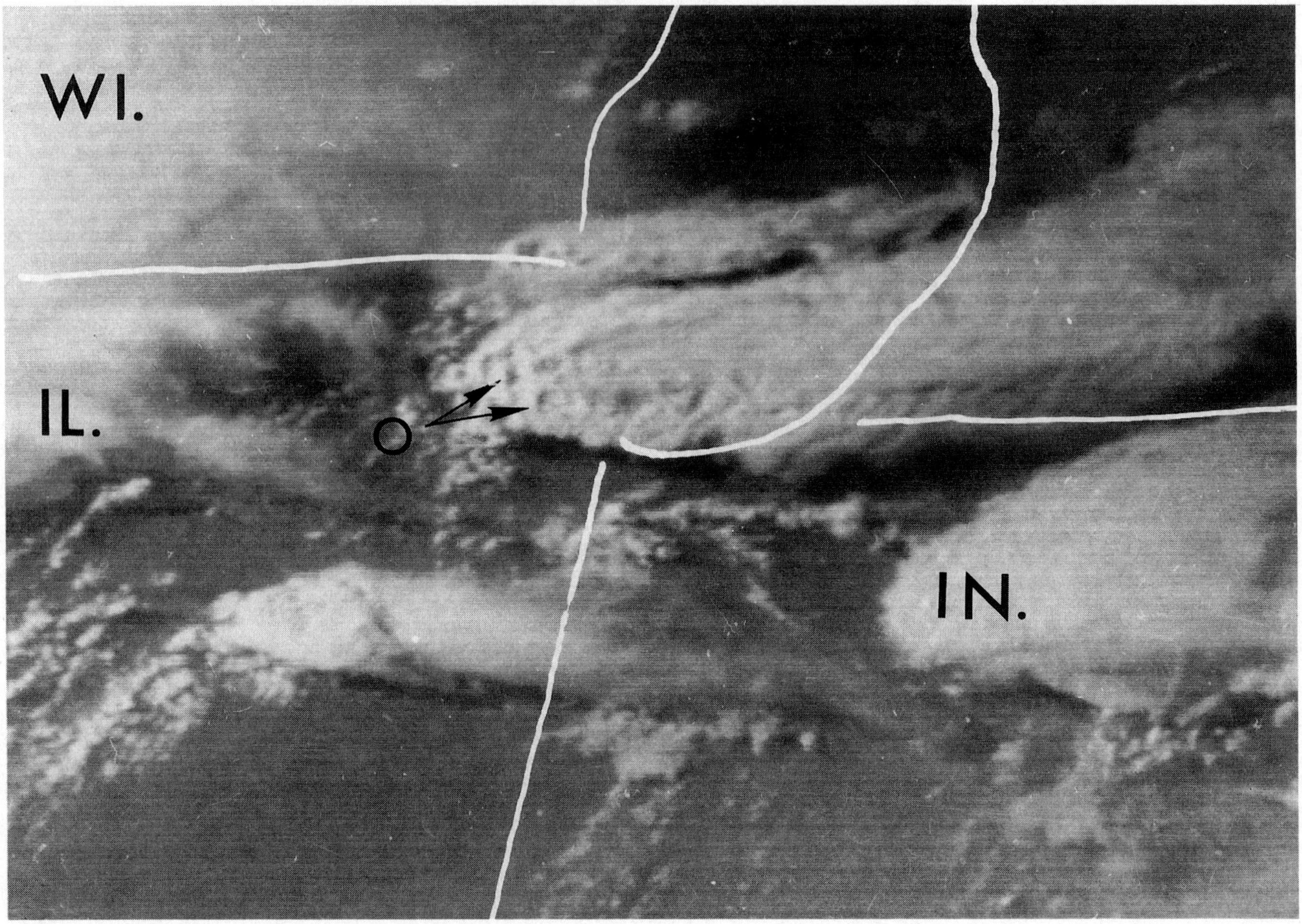

Figure VII-6-8a. Half-mile VIS imagery, 2300 UTC, June 13, 1976.

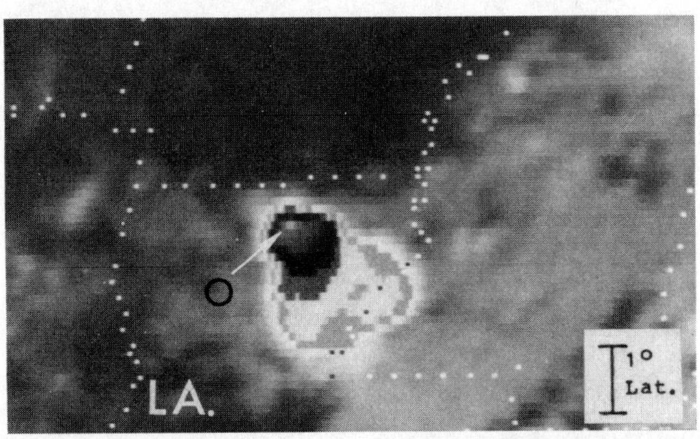

Figure VII-6-8b. Enhanced IR imagery (MB curve), 1200 UTC, September 15, 1978.

thunderstorm merger (*M*) taking place between clusters *A* and *B*.

Factor 4: Saturated environment (Figure VII-6-5). The saturated environment factor refers to those thunderstorms that are long-lived, cover large areas, and are stationary. These storms possess strong, relatively steady state updrafts and outflows, a large area saturated to great heights, and a lack of dry air entrainment.

Factor 5: Moisture correction (Figure VII-6-5). A moisture correction factor has been developed for adjusting estimates when the air is unusually dry or quite moist.

Figure VII-6-9. (a) Enhanced IR imagery (MB curve), 0930 UTC, September 9, 1980. (b) Enhanced IR imagery (MB curve), 1030 UTC, September 9, 1980.

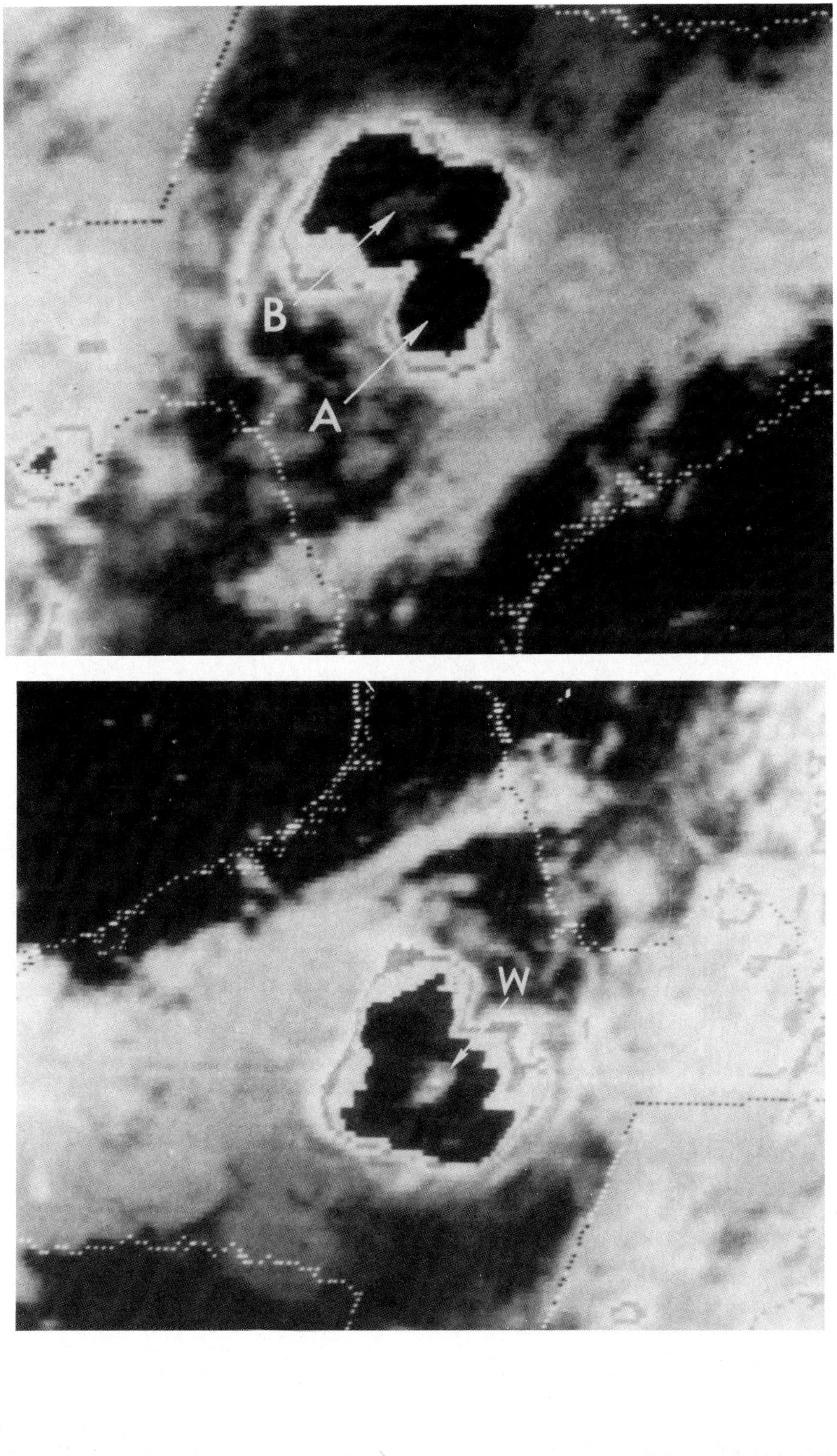
B
A
W

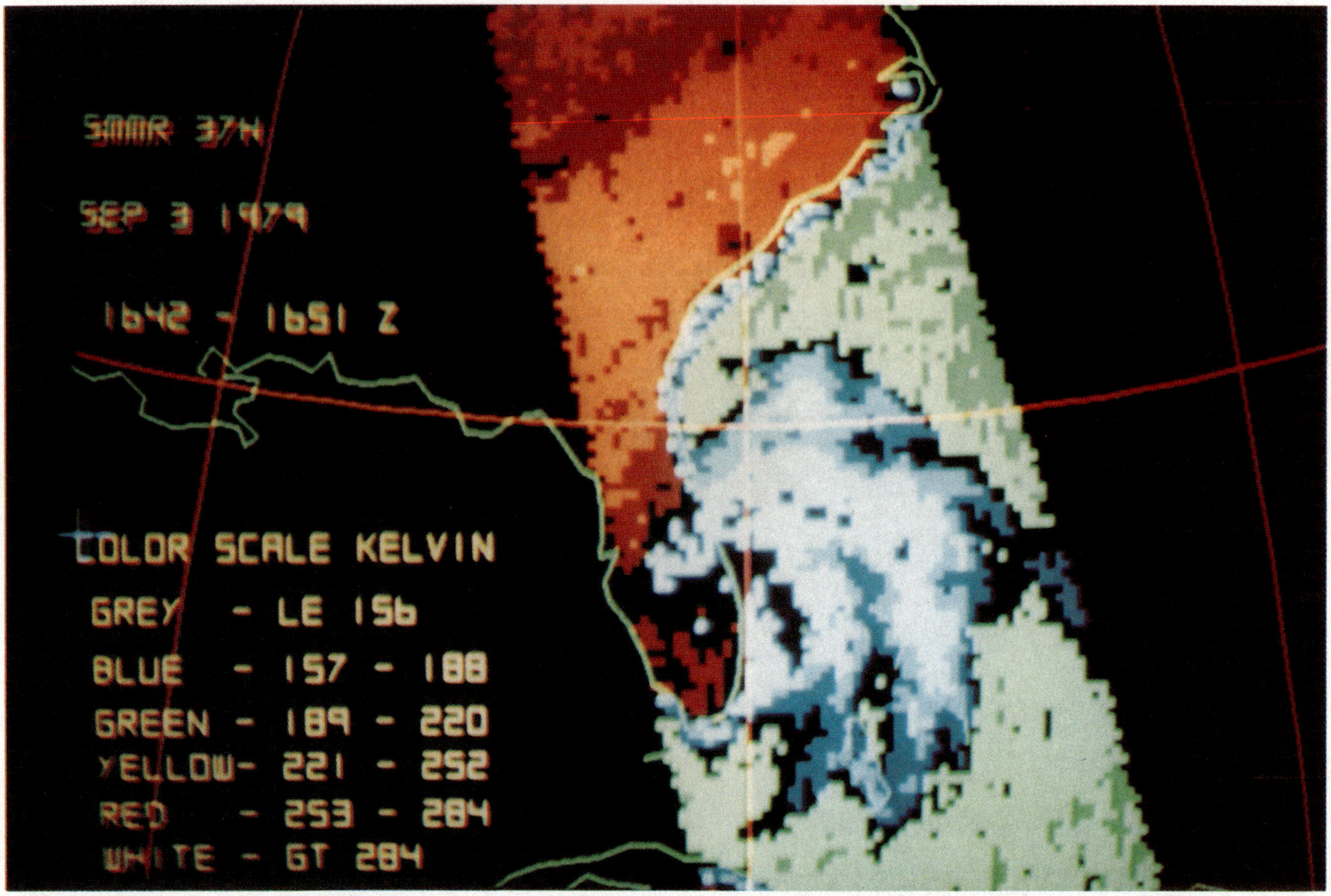

Figure VII-6-10. SMMR 37 GHz horizontally polarized brightness temperature image of Hurricane David, 1645 UTC, September 3, 1979.

- Summation of factors 1–4 and multiplication by the moisture correction factor (5).

A verification study (Field, 1985) of the operationally produced rainfall estimates (Borneman, 1988) from May-July 1984 was completed by comparing the total *maximum* rainfall estimates for a storm (usually a period of 1 to 6 hours) with the 24-hour cooperative rainfall reports available the next day. Events which included less than two or three rainfall reports per county and events where additional rain fell outside the period for which rainfall estimates were produced were eliminated from this verification.

The verification results show the average error of rainfall estimates for a precipitation event is about 30%. The absolute and percent error both increase as the magnitude of the event increases. For relatively small events (99 mm or less), there is a tendency to overestimate the rainfall; as the magnitude of the event increases there is a distinct trend toward underestimation of the rainfall. Operational users of this product should feel confident that if a rainfall estimate of 100 mm or more is made, a high probability exists that at least 100 mm will be observed. Verification results for 1985 and 1986 events are similar to those for 1984.

Precipitation estimates made by this method can be used in real time to make "nowcasting" estimates of flash flooding. The estimates are also extrapolated and adjusted to anticipate heavy rainfall and accompanying flash flooding for 3 hours after the latest imagery. Guidelines on doing this are given by Spayd and Scofield (1984a).

Other specialized precipitation estimation techniques of a similar nature have been developed for tropical and extratropical cyclones (Spayd and Scofield, 1984b; Scofield and Spayd, 1984).

VII-6.3 PRECIPITATION ESTIMATION BASED ON MICROWAVE MEASUREMENTS

Microwaves can "see" through clouds in the "window" frequencies below 50 GHz with rain being the major source of attenuation. Over sea surfaces, areas without rain have relatively low brightness temperatures in the MW measurements and they contrast strongly with the higher measurements from precipi-

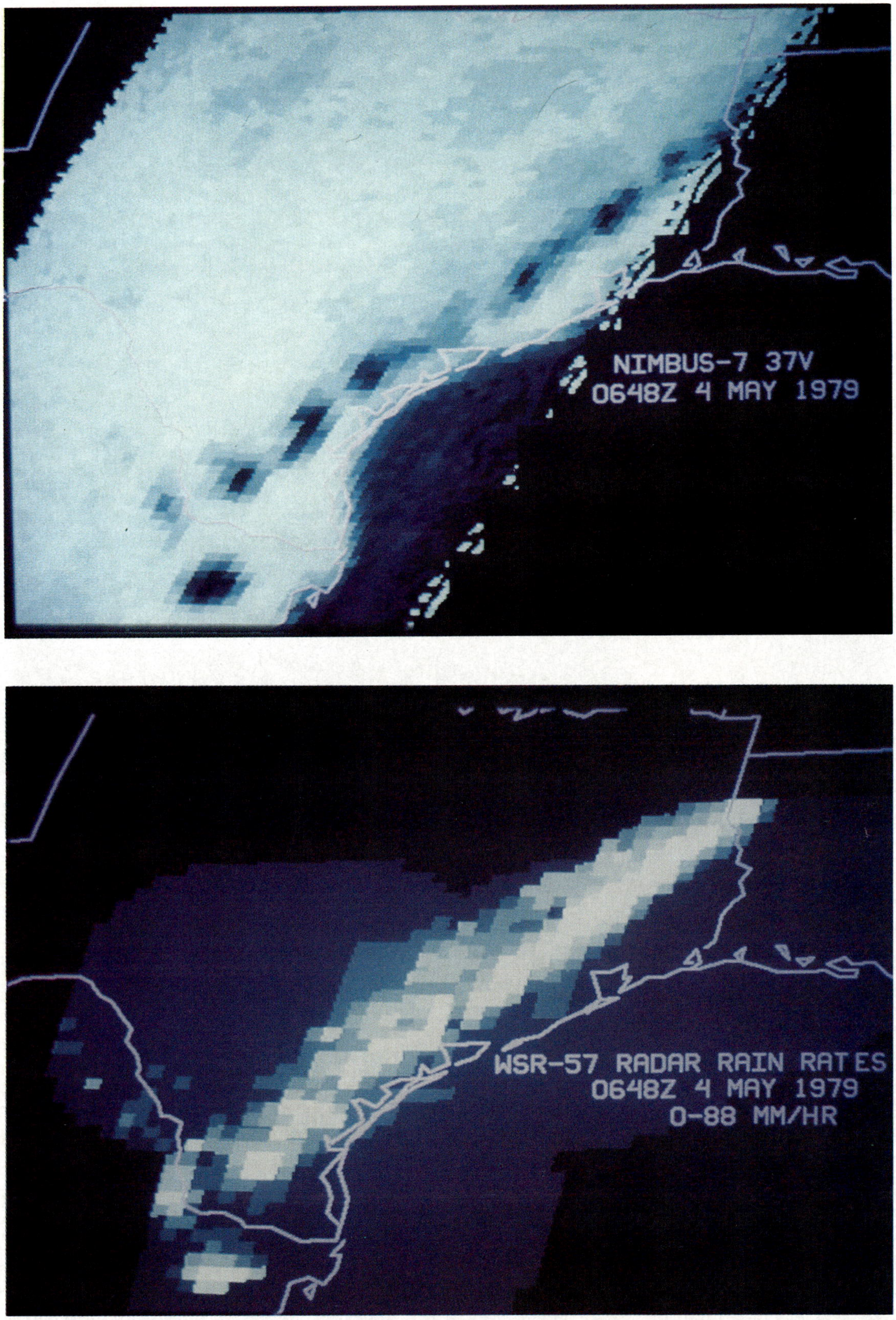

Figure VII-6-11. SMMR 37 GHz horizontally polarized brightness temperature image of a severe storm over Texas, 0645 UTC, May 4, 1979 (top). Digitized rain rates from radar data for the same time (botttom).

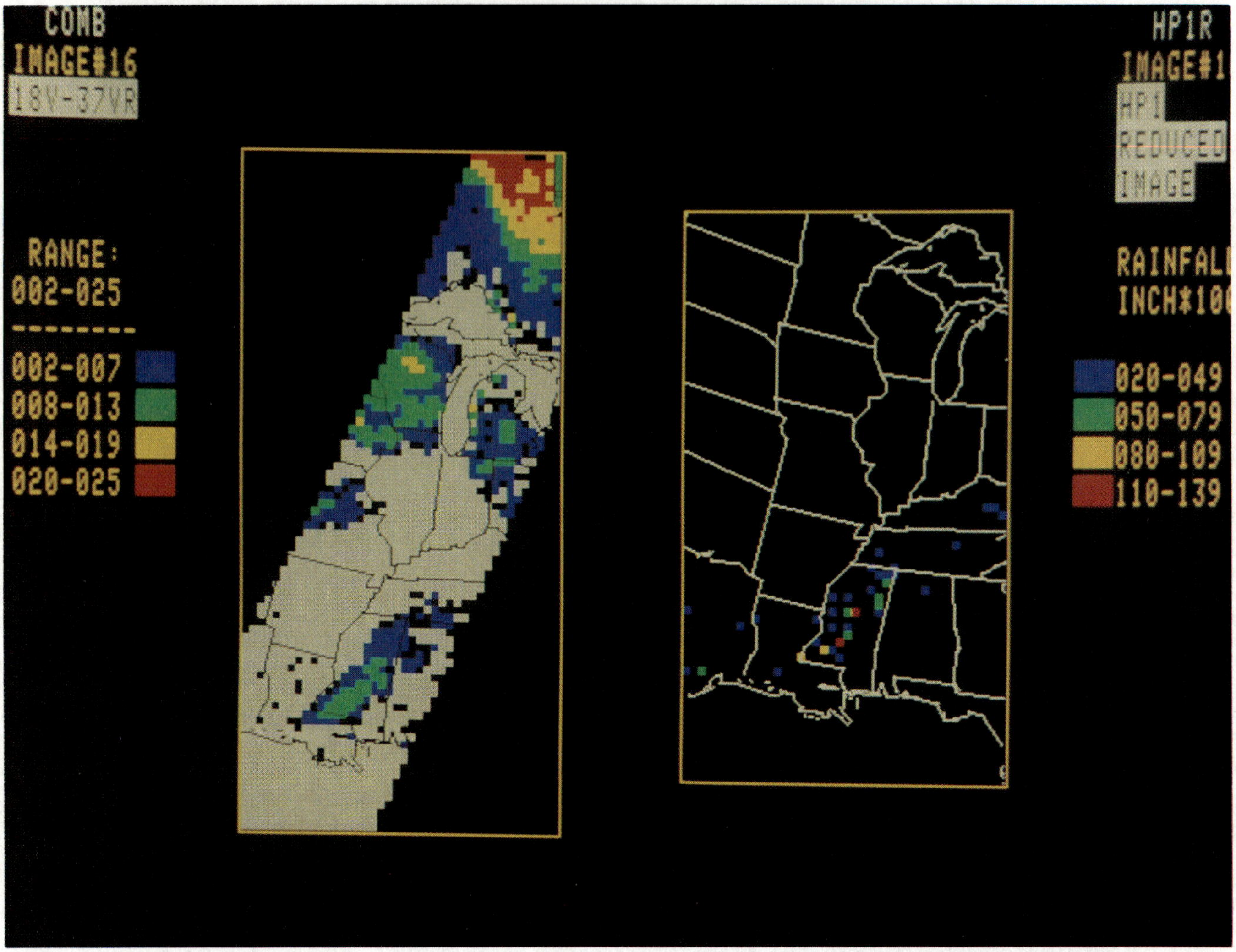

Figure VII-6-12. SMMR 18 GHz minus 37 GHz vertically polarized image of a storm over Arkansas, 1645 UTC, January 20, 1979 (left). Hourly raingage data for the same time (right).

tating regions. These large contrasts (>50 K) in brightness temperatures have stimulated much interest in using MW radiometry over oceans. Satellite and aircraft MW instruments have been used to obtain rain data for tropical storms (Allison et al., 1974; Adler and Rodgers, 1977; Jones et al., 1981), and measurement of the global distribution of rainfall over oceans has also been attempted from satellites (Rao et al., 1976). The most recent research indicates that a greater sensitivity to precipitation exists at frequencies above 50 GHz using the new SSMI instrument which can provide 85 GHz measurements. From data derived from the SSMI, the 85 GHz brightness temperatures have been found to be over 100 K lower when encountering convective rain cells over land or ocean. This strong sensitivity to precipitation is a result of the high spatial resolution of the 85 GHz channel (12 km) and the large scattering effect at 85 GHz by the ice particles at the tops of the rain cells.

VII-6.3.1 Single Frequency Technique

In most situations it is sufficient to relate rain intensity directly to brightness temperature over oceans. Effects on emissivity from varying sea surface roughness and the effects of cloud and water vapor absorption can usually be neglected. For example, Figure VII-6-10 shows the 37 GHz (horizontally polarized) brightness temperature (*Tb*) image of Hurricane David, 1645 UTC, September 3, 1979, from the Nimbus-7 SMMR. The spatial resolution is about 30 km. Heavy precipitation appears in the "white" spiral bands (*Tb* > 220 K) surrounding the hurricane, with less rain in

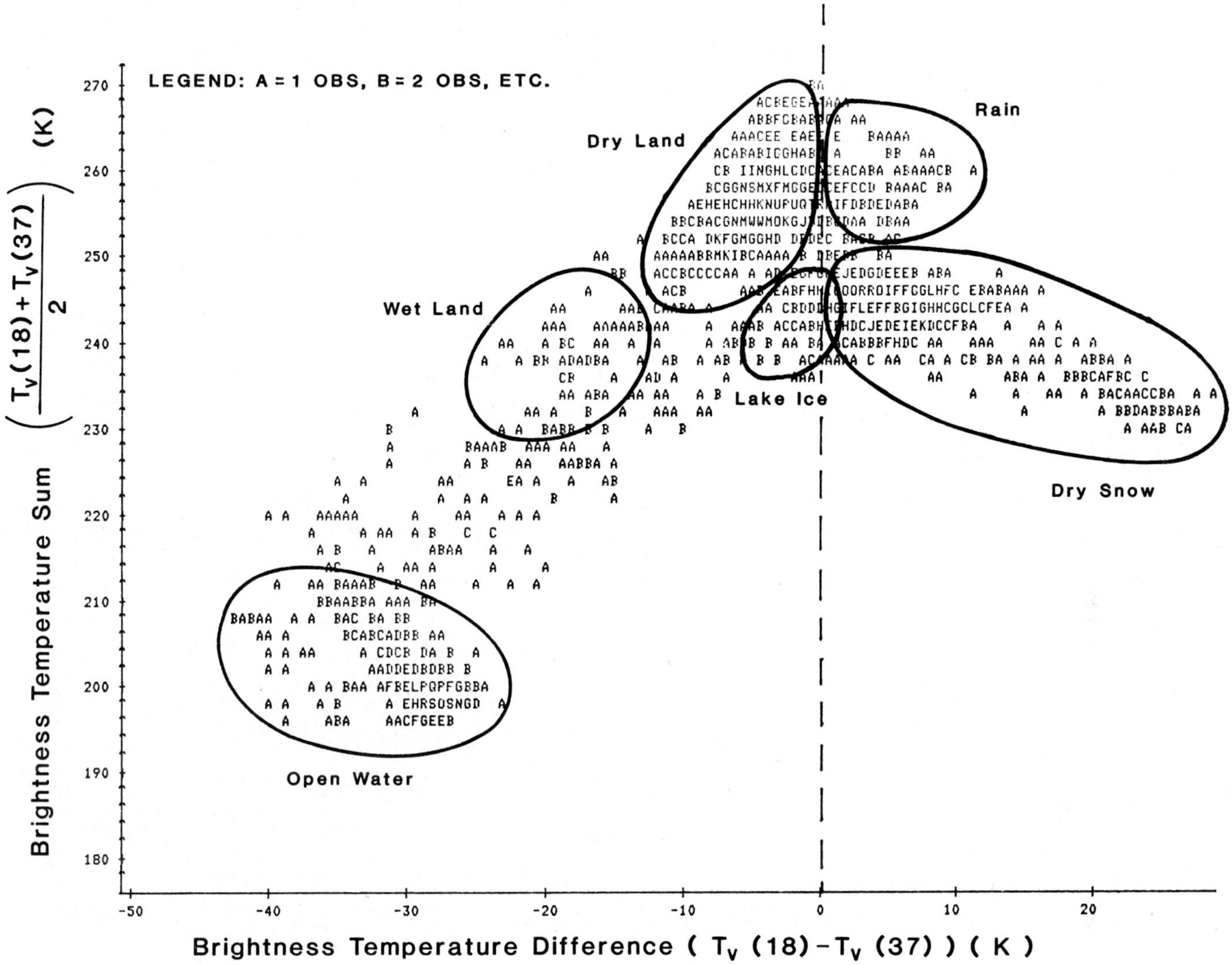

Figure VII-6-13. Cluster diagram based on SMMR 18 and 37 GHz vertically polarized brightness temperatures for January 20, 1979. Clusters corresponding to rain, snow, open water, dry and wet land, and lake ice are indicated.

the "blue" outer regions, while the "green" ocean (*Tb* < 180 K) is rain-free.

Over land, however, the high surface emission (*Tb* > 250 K) makes the contrast with precipitation ambiguous. Only in the case of heavy thunderstorms are MW measurements much lower than the surrounding brightness temperatures. The enhancement of intense storms is due mainly to the scattering of upwelling radiation by large ice particles at the tops of rain layers. This scattering effect increases at higher frequencies and was first noted when observing convective storms over the United States with SMMR measurements (Spencer et al., 1983).

Figure VII-6-11 shows an example of the 37 GHz (horizontal polarization) SMMR image for a severe storm over Texas on May 4, 1979, and its comparison with surface based radar data. Convective rain is readily identified by the low brightness temperature (<200 K; dark blue) at 37 GHz. In addition to the low brightness temperatures, a residual maximum polarization difference of about 15 K between the vertical and horizontal components was found and attributed to scattering by ice particles (Spencer et al., 1983).

VII-6.3.2 Frequency Screening

The emissivities of dry soil, vegetation, open water, melting snow, lake ice, and new sea ice either exhibit very little frequency dependence or increase with frequency. In contrast, rain cells produce brightness temperatures that decrease with increasing frequency because of scattering and thermal emission. Thus, it is

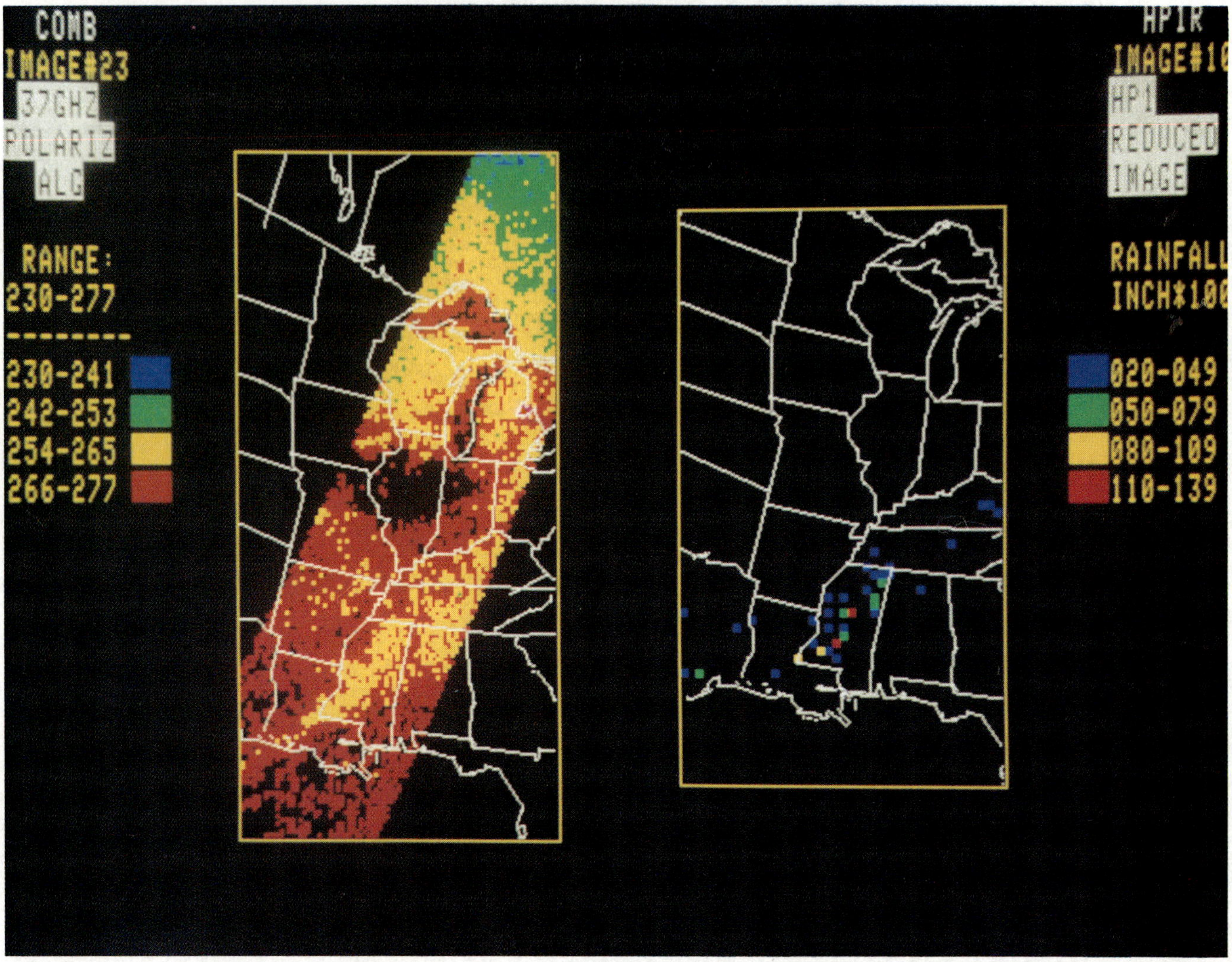

Figure VII-6-14. SMMR vertical and horizontal polarization differences displayed for January 20, 1979 (left) data. Hourly rain gage data for the same time (right).

possible to isolate precipitation by measuring the difference in brightness temperature at two frequencies. There are some difficulties, however; for example, the frequency response of dry snow is similar to that of rain, and so would not be screened out. Fortunately, this problem can be dealt with by examining sequential measurements and associating rapid changes with precipitation. Also, the sum of the brightness temperatures at the two frequencies can be used to make the final discrimination (e.g., rain from stratiform clouds produces higher brightness temperatures than dry snow).

The dual frequency approach is illustrated in Figure VII-6-12, which shows the difference between the 18 and 37 GHz (vertically polarized) brightness temperatures and the corresponding hourly rain gage data. The line of precipitation from Louisiana to Tennessee is identified in the MW image, with the larger difference values corresponding to the more intense rain. Also shown are large areas of positive difference surrounding the Great Lakes. This is a result of dry snow.

Separation between snow cover and rainfall is made by using the brightness temperature sum and difference fields as shown by the cluster diagram of Figure VII-6-13. The vertical and horizontal coordinate axes represent the sum and difference of the 18 and 37 GHz measurements, respectively. As in the imagery, the clusters corresponding to dry snow and rain lie in the upper right quadrant. Fortunately, however, these clusters are separated vertically by the higher brightness temperature of rain compared to dry snow. Where negative brightness temperature differences occur, the clusters are formed by dry soil, wet land along the Missississippi Delta, and open water in the Great Lakes and Gulf of Mexico. These regions are shown by the "white" areas in Figure VII-6-12 (left). The scattered points between the wet land and open water correspond to fields of view containing mixtures of different surfaces, e.g., land and water. Also, the small cluster denoted as lake ice arises from data over Hudson Bay, which is slightly north of the image shown in Figure VII-6-12.

VII-6.3.3 Polarization Algorithm

The frequency screening approach just described uses the correlation between emissivity at different frequencies to enhance precipitation signatures. Alternatively, vertical and horizontal polarization information can be used to isolate precipitation from surface features. A weighted difference between the vertical and horizontal brightness temperatures is useful for precipitation detection (Weinman and Guetter, 1977; Grody, 1984). This type of polarization algorithm used for the January 20, 1979, case produced the image of Figure VII-6-14. This image is based on 37-GHz measurements and therefore is higher in resolution than the image of Figure VII-6-12. Combined polarized brightness temperature <270 K (blue) corresponds to precipitation and snow cover.

Chapter VII-7

TROPICAL CYCLONE ANALYSIS AND FORECASTING[1]

VII-7.1 INTRODUCTION

The meteorological satellite is uniquely suited to the task of tropical cyclone surveillance. Since 1960 satellites have provided meteorologists with a wealth of information about the positions, intensities, and motions of tropical storms. Much of this knowledge is used for operational analysis and forecasting and plays a major role in hurricane warning systems around the world.

The use of satellite pictures for surveillance of tropical cyclones was recognized early in the satellite program (Sadler, 1962). It soon became apparent that the well organized spiral structures and eye characteristics visible in pictures of these intense storms were positively related to the associated high winds. Methods were devised for estimating their intensities by associating specific cloud features with estimates of storm strength based on post analysis and aircraft reconnaissance (Fett, 1964; Timchalk et al., 1965). The appearance of a storm's eye, its banding, and the size of its cloud pattern in a single VIS picture per day were used to estimate the intensity of the storm.

As visible imagery from polar satellites improved in the late 1960s and early 1970s, a technique was developed to improve the reliability and consistency of intensity estimates. This technique, which is still in use, considers the past history of a storm's cloud pattern to arrive at the current intensity analysis (Dvorak, 1972; 1975). The technique uses an empirical model that describes how a storm should appear on each successive day of development. The analyst applies rules by means of a decision tree to determine the storm intensity estimate.

In the mid-1970s, digital IR imagery from the GOES satellites became available and this was incorporated into the technique to make the operational determination of storm intensity more objective. The IR data allow cloud pattern evolution to be monitored both day and night and provide good estimates of cloud top heights for quantitative evaluation of storm intensity.

Several other techniques have been developed for forecasting tropical cyclone intensity using satellite data. These range from simple extrapolation of the observed trend in the storm development to more objective use of divergence, vorticity, and vertical shear calculated from cloud motions and to changes in IR cloud top temperatures. Recently, data from the MSU flown on NOAA polar orbiting satellites has been used to define the upper tropospheric warm core of the cyclone. MW monitoring of changes in the storm warm core aids in the detection of changes in the storm intensity.

Methods developed to predict the motion of storms based on satellite observations include simple extrapolation of the storm center from past movement in animated imagery, use of changes in cloud and moisture

[1] Primary Contributor: Vernon Dvorak

patterns in VIS, IR, and WV images, and use of satellite derived winds through a deep layer of the atmosphere.

VII-7.2 TROPICAL CYCLONE INTENSITY ANALYSIS

VII-7.2.1 Methodology Using Characteristics of Cloud Patterns of the Storm in Visible Imagery

Tropical cyclones develop as a spiral band of convective clouds becomes more and more curved and gradually forms a tight center, or eye. As the cyclone continues to intensify, the eye becomes more distinct in the increasingly dense overcast clouds that surround the storm center. The intensity of the cyclone is related to the cloud pattern, rather than the amount of cloudiness in the pattern. An investigation by Arnold (1977) found no significant differences in the amount of convective cloudiness between stages of the storm development. Cloud pattern characteristics of the storm center and the overcast around the center have long been used to estimate intensity. When these characteristic features of the cloud pattern are clear-cut and evolve systematically, they can be used for reliable intensity estimates. In many cases, however, cloud pattern changes are irregular and have no clear-cut evolution. There are also periods in the life cycle of a storm, such as the pre-storm stage and the weakening stage, when some cloud patterns are unrepresentative of intensity. These complexities are dealt with through the application of systematic procedures, cloud pattern descriptions, and empirical rules in the analysis technique of Dvorak (1984b), which is currently in use.

The Dvorak Technique outlines procedures and rules for locating the center of the storm (cloud system center), measuring cloud features which relate to intensity, classifying cloud systems as to stage of storm development, and applying a model for tropical cyclone development based on satellite experience. The most reliable satellite intensity estimates are made from quantitative measurements of clearly identifiable cloud features in the imagery. If cloud features are not clear-cut, storms are subjectively classified from conceptual models which relate intensity to cloud pattern evolution for the life cycle of different types of storms. If classification by cloud pattern is uncertain, the tropical cyclone development model is used and an intensity estimate is determined based on the previous history of the storm. The model is also used to set limits on estimates made from the first two procedures.

The technique is applied by first determining the cloud system center. The cloud pattern is then categorized as one of four basic types, each one of which has different cloud features that are measured. Measurements pertain to the major spiral cloud band of the storm and the relationship of the storm center to the area of dense overcast. The most common pattern, "curved band" type appears in Figure VII-7-1 which shows the evolution of this pattern according to the model for cyclone development.

There also are two alternative developmental patterns. One is the "central dense overcast" pattern, in which a dense overcast covers the central features of the tropical cyclone. For this pattern, the size of the overcast is directly related to the storm intensity. The other is the "shear" pattern, in which the vertical shear in the atmosphere inhibits the growth of the curved band of deep convection around the storm center. For this pattern the center is defined by organization in the low clouds and its proximity to the dense overcast is used to determine intensity. When the center becomes embedded in dense overcast clouds at the hurricane stage, the appearance of the center, its location in the central overcast, and the amount of banding around it are all used to estimate storm intensity. The more distinct the center, the more it is embedded, and the more banding around the central overcast, the greater is the storm intensity. For mature storms, there is an "eye" pattern; for this category, the distance of the eye from the central dense overcast is measured.

Descriptions of the cloud patterns and how they change with time are used to portray a model of tropical cyclone development as observed in satellite pictures. Figure VII-7-1 shows the curved band pattern changing at a typical daily rate of development. When the cloud system shows the same day by day increase in cloud band spiraling as that in Figure VII-7-1, the tropical cyclone is developing at the typical rate. When the

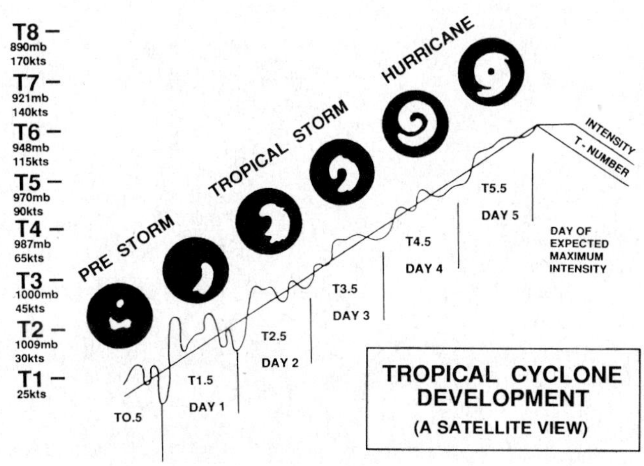

Figure VII-7-1. Model of tropical cyclone development used in intensity analysis. A schematic of curved band pattern evolution is shown.

curved band spiral develops faster or slower, a similar relative rate of growth is indicated. Storms of greater intensity than the peak shown in Figure VII-7-1 either develop rapidly or over a period of more than 5 days.

The diagonal line in Figure VII-7-1 illustrates the typical long-term growth rate of tropical cyclones, defined as one *T*ropical-Number (*T*-number) per day. Rapid or slow rates of growth usually occur at 1½ or ½ *T*-number per day, respectively. The empirical relationships between the *T*-numbers and the wind speed and central pressure are shown on the left in Figure VII-7-1. Examples of tropical cyclone cloud patterns classified at various stages of development (*T*-numbers) are shown in Figure VII-7-2.

The initial signs of tropical cyclone development first become recognizable about 36 hours before the storm is named. This stage (*T*1) is normally observed about 24 hours before significant surface pressure falls occur. The *T*1 classification is assigned when curved cloud lines or bands define a cloud system center near or within deep layer clouds for 12 hours. Once the *T*1 classification is assigned, the *T*2 classification should appear about 24 hours later. Faster or slower development of the cloud pattern warrants an upward or downward adjustment, respectively, of half of a *T*-number.

When the curved band spirals halfway around the center of the disturbance, the weak tropical storm stage (*T*2.5) has been reached. As the band completely encircles the center, the minimal hurricane stage (*T*4) has been attained. Once the "eye" is observed at stage *T*4.5, continued intensification is indicated by increases in: a) eye definition (warming), b) the embedment of

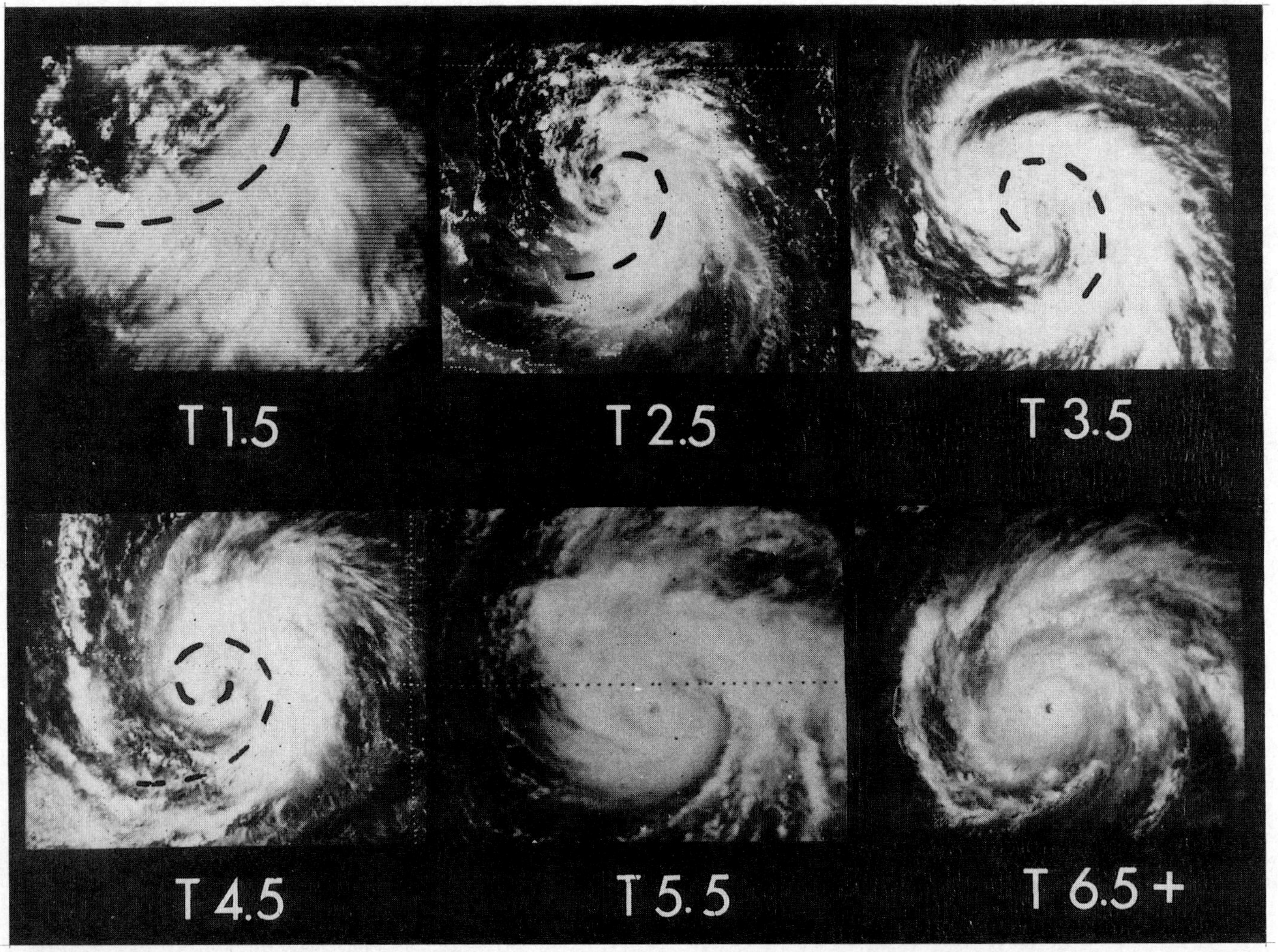

Figure VII-7-2. Examples of tropical cyclone cloud patterns in VIS imagery. The patterns show approximate daily changes from a prestorm (*T*1.5) pattern to a super hurricane (*T*6.5). The dashed line follows the curved band axis which spirals around the storm center as the cyclone intensifies.

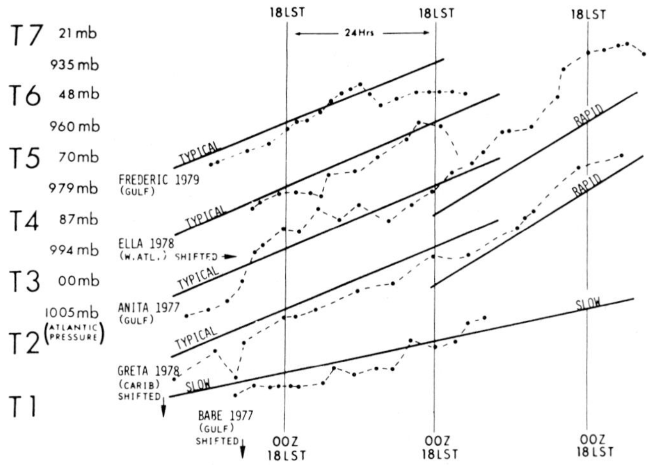

Figure VII-7-3. Plots of minimum central pressure for Atlantic hurricanes. Diagonal lines are rates of intensity change from the model of tropical cyclone development. The three rates of development are shown: slow, typical, and rapid.

the eye in the dense overcast, c) smoothness of the dense overcast, or d) the outer curved bands around the central overcast.

A tropical cyclone is expected to reach maximum intensity some 4 to 6 days after the *T*1 stage is seen, provided its environment remains favorable. Northward-moving storms usually "peak" in 4 days, northwestward moving storms in 5 days, and westward-moving storms in 6 days[2]. If the developmental cycle is interrupted at any stage by an unfavorable environment, the pattern will show dense (cold) clouds separating from the low level center, a non-diurnal lowering (warming) of the cloud tops, or a rapid expansion out from the center of dense (cold) clouds as the curved band dissipates. Indications of the weakening of a storm will be observed from 6 to 12 hours in advance of rises in the central pressure of the storm (Lushine, 1977). An analyst using this technique during a period of weakening usually maintains the intensity of the storm at the *T*-number already assigned, however, until subsequent imagery indicates that the storm is either continuing to weaken or that the weakening was only temporary. The tendency for the change in cloud pattern to precede the pressure change also shows up when a storm is developing and is most apparent when it is developing rapidly.

Frequently there are short-term changes—mainly diurnal—in the storm cloud pattern that may result in changes in estimates of intensity. Such changes are shown schematically by the wavy line in Figure VII-7-1. The amplitude of fluctuations, largest during the pre-storm stages of development, becomes less noticeable as the storm intensifies. The weakest patterns appear at night when high clouds are at a minimum (Oliver, 1974; Browner et al., 1977). The analysis technique deals with these fluctuations by allowing for little deviation from the typical rate of intensification, particularly during the prestorm stage of cyclone development.[3]

Figure VII-7-3 shows how changes in central pressure of several hurricanes compared with the rates of change for developing storms expected in the model used in the technique. Plots of the pressure data for all the hurricanes under frequent aircraft reconnaissance during the 1977, 1978, and 1979 seasons are shown. The average 24-hour pressure changes of most storms were well approximated by the model curve labeled as typical. The typical development rate of one *T*-number per day occurs about 70% of the time in the Atlantic region. The later stages of Anita and Greta follow approximately the rapid curve of 1.5 *T*-number increase per day. Babe shows a slow (0.5 *T*-number per day) rate of development. It is important to note that tropical storm growth as observed in changes in both central pressures and cloud patterns does not occur in a steady manner. The pressure falls often occur in surges followed by periods of little change or by pressure rises. The pressure plots also suggest a tendency toward a diurnal decrease in growth rate beginning near 1800 LST. In the sample shown in Figure VII-7-3, the average pressure fall during the 6 hours before 1800 LST was more than twice that after 1800 LST.

VII-7.2.2 Methodology Using Enhanced Infrared Imagery, Digital Infrared Data, and Microwave Data

Infrared pictures enable the analyst to monitor cyclone activity continuously over 24-hour periods. The IR image of a tropical cyclone appears similar to that seen in a VIS picture of the same storm. In fact, day-by-day changes in the VIS cloud patterns of Figure VII-7-1 can often be followed qualitatively in the IR pictures. There are two major differences between VIS and IR images insofar as intensity analysis is concerned. The first is that thin cirrus clouds, normally transparent in VIS images, are often opaque in the IR; when cirrus covers the boundaries of cloud features

[2] These directions refer to Northern Hemisphere storms. Substitute southward and southwestward for Southern Hemisphere storms.

[3] There is a type of "subtropical" cyclogenesis, however, for which the rule does not apply. It is a very rapidly developing storm in the strong vorticity environment of a cutoff cold low or a cold low in the process of cutting off. This pattern evolution is described in a classification technique for subtropical cyclones (Herbert and Poteat, 1975).

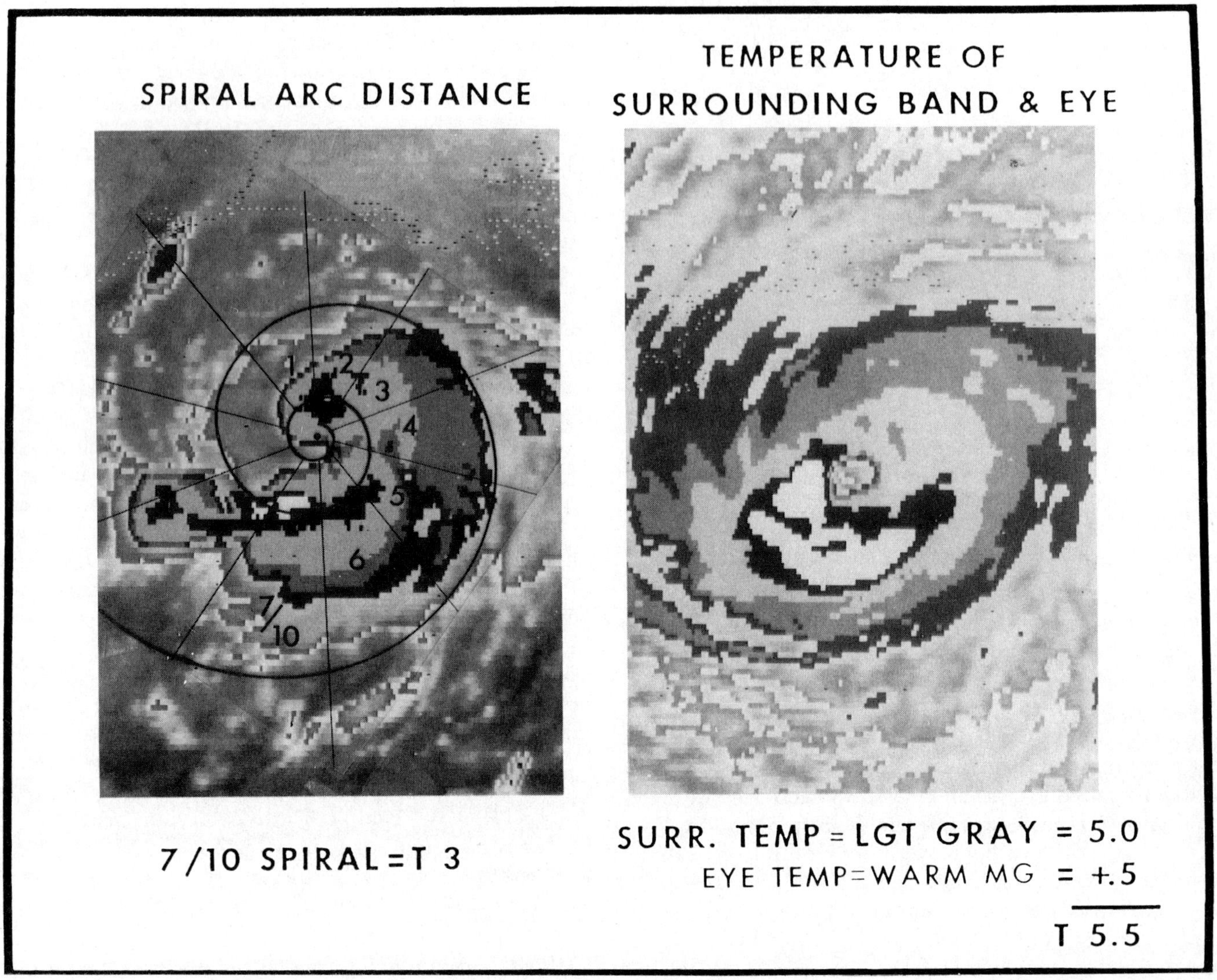

Figure VII-7-4. Enhanced IR pictures of a tropical storm (left) and a hurricane (right). In the left image, the center of the 10° spiral coincides with the cloud system center. The 10 radial lines are used to define the degree to which the cloud band wraps around the center. In this case, the degree of the spiral is 0.7 which equates to a storm intensity of *T*3. For the hurricane, the temperature of the surrounding band is light gray in tone and the eye temperature is medium gray. The *T*-number for these values is 5. 5.

used for measurement, the IR image furnishes either no measurement or an erroneous one. Thus, once a tropical cyclone has a central dense overcast, the unenhanced IR pictures cannot be used quantitatively in the intensity analysis. The second difference is that the low cloud lines, important for locating the storm center, may not be detectable in the IR picture. This problem is serious when unenhanced pictures are used, but it can usually be solved by the enhancement procedures described below.

A technique of using Enhanced Infrared (EIR) pictures for determining tropical cyclone intensity was added to the original technique (Dvorak and Wright, 1977). The method uses IR images that display cloud features in discrete gray shades representing known temperatures ranges. This enables the analyst to use the temperatures of the features as objective measures of intensity. The temperature of the eye and of the clouds surrounding the eye are of primary importance. These simple measurements replace several difficult subjective judgments required when VIS imagery alone is used to make intensity estimates.

During the early stages of cyclogenesis, the EIR method uses the curved band and shear pattern analyses of the VIS technique. The intensity estimate is determined by how much the curved band has spiraled around the system center, or by the appearance of the center near the overcast. An example of how the

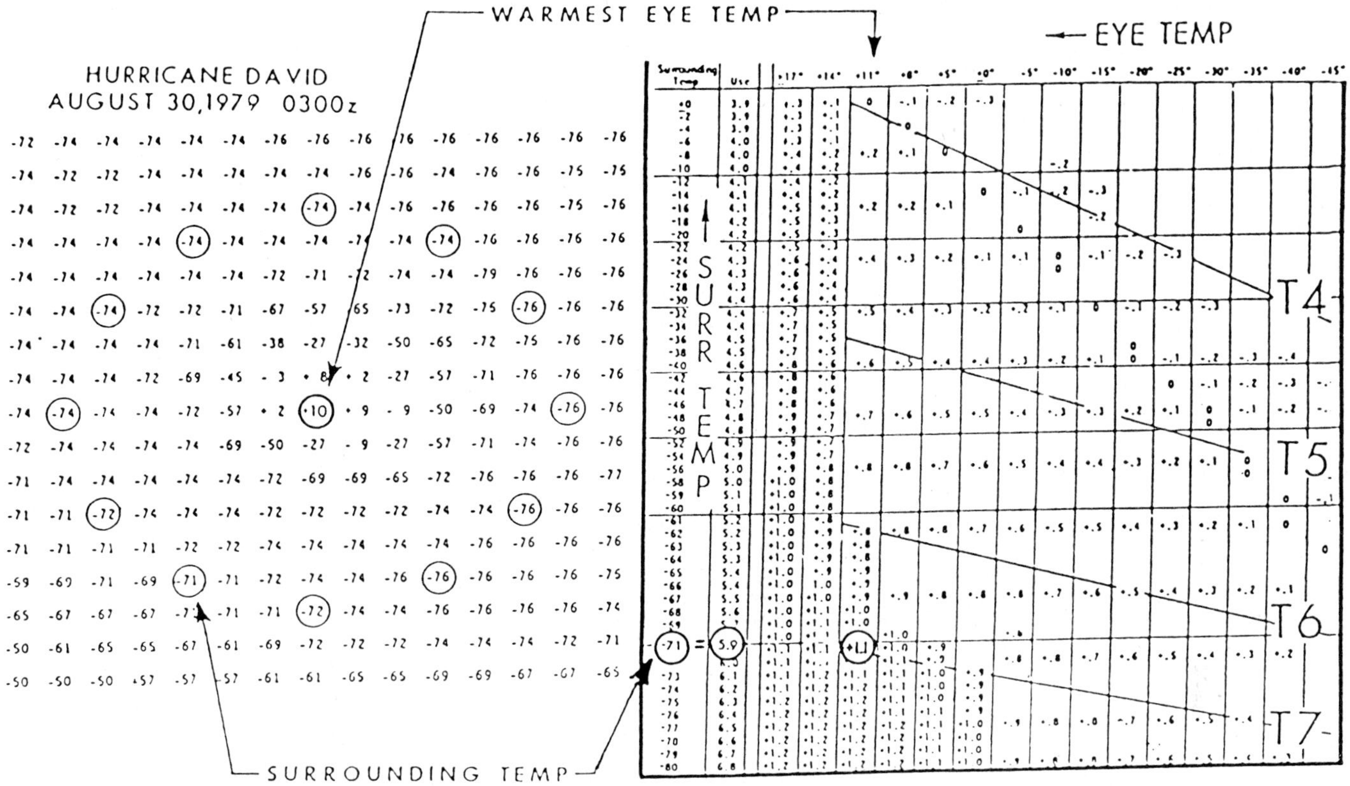

Figure VII-7-5. Method for determining hurricane intensity from digital IR data. The digital printout on the left provides two temperature values which when entered in the nomogram on the right result in an intensity estimate of *T*7. The surrounding temperature of −71°C contributes 5.9 while the relatively warm eye adds 1.1 to the *T*-number. Note that the value for the coldest surrounding band is determined by the highest temperature within that band.

amount of spiraling is measured in operational practice is shown in Figure VII-7-4 (left side). A 10° logarithmic spiral is laid over the curved band axis of the storm cloud pattern; then the number of tenths of the spiral covered by the cold portion of the band is counted. This spiral arc length is empirically related to the storm intensity. An arc length of 7/10 of a spiral represents a moderately strong *T*3 intensity.

When a curved band has encircled a storm center at the weak hurricane (*T*4) stage, the EIR technique becomes simpler and more objective than the VIS method. The analysis consists of making two simple measurements (Figure VII-7-4, right side). The first is the IR temperature of the coldest band (of a required minimum width) completely surrounding the eye. This surrounding temperature ("SURR TEMP" in the figure) shows as a light-gray shade. This temperature and the temperature of the eye contribute 5.0 and 0.5, respectively, to the intensity estimate (*T*5.5) of the storm. These values are obtained by use of the empirical relationships shown in the nomogram of Figure VII-7-5. A hurricane stronger than the one in Figure VII-7-4 (right side) would have either a lower surrounding temperature or a higher eye temperature. The technique also allows for a small increase in the intensity estimate when an outer band curves around the central cold overcast area.

When pictures are available every 6 hours or more frequently, a more precise analysis can be made by using several pictures instead of just one at analysis time. The average measurement of all the pictures taken during the 6-hour period ending at analysis time is used. For pictures with eyes, the average measurements can be used with the nomogram of Figure VII-7-5 to obtain a *T*-number. For weaker disturbances the average curved band arc length can be converted to a *T*-number by using relationships given by Dvorak (1984b).

The EIR technique has been simplified for use with storms of hurricane intensity. A digital printout of IR temperatures over the region of the storm is prepared and an overlay placed over it. The analyst then reads off two temperatures: that of the warmest point which is at the storm center and that of the point on a circle (about 55 km from the center) that defines the coldest surroundings. By using the nomogram of Figure VII-7-5, these two values yield the intensity. When intensities are calculated at short intervals and averaged over 3- to 6-hour periods, they are similar to those

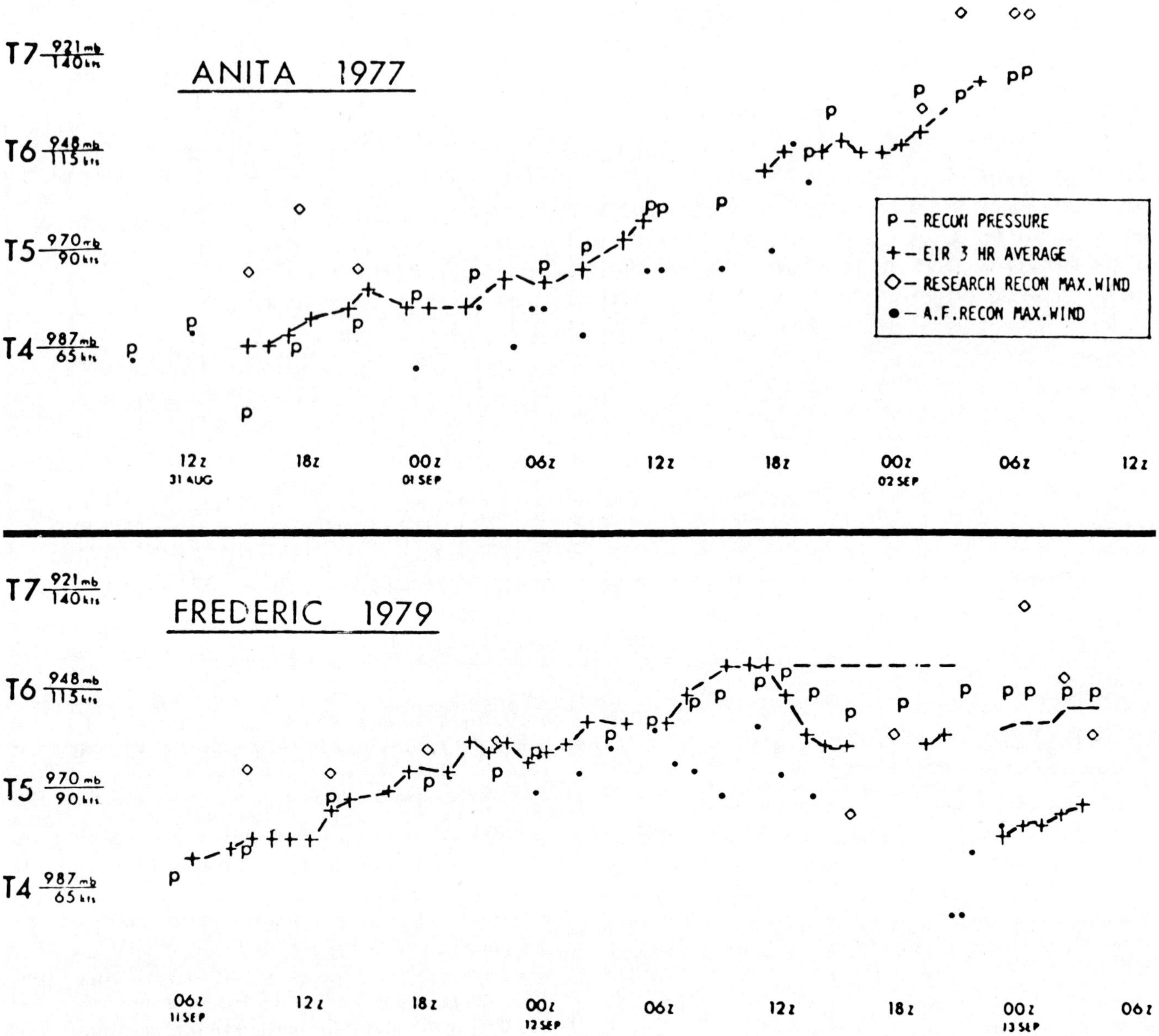

Figure VII-7-6. Comparison of digital IR intensity estimates (+) with those from aircraft reconnaissance central pressure (P), and maximum winds (◇ and·).

computed from reconnaissance dropsonde measurements. Figure VII-7-6 shows a plot of the intensity estimates from digital IR data and from reconnaissance measurements for hurricanes Anita and Frederic.

The digital technique was recently programmed on the McIDAS computer at the University of Wisconsin. With McIDAS, the analyst displays images at ½- to 3-hr intervals, places a cursor on the storm center, and the computer calculates the storm intensity. The system has been used since the 1984 hurricane season with good results.

The use of EIR pictures has resulted in the consideration of a new pattern type, the Central Cold Cover (CCC). The CCC is observed as a round, cold overcast near the storm center that develops rapidly as the curved band pattern dissipates. This can occur at any stage of storm development and may last for several hours or days. When it persists, it is related to an interruption in storm development; central pressure measurements show that no significant change in storm intensity takes place during CCC occurrence. Consequently, the appearance of the CCC is used to signal an interruption in storm development.

Another method for determining hurricane intensities, also programmed on the McIDAS, uses data from the MSU on the NOAA polar orbiting satellites (Vel-

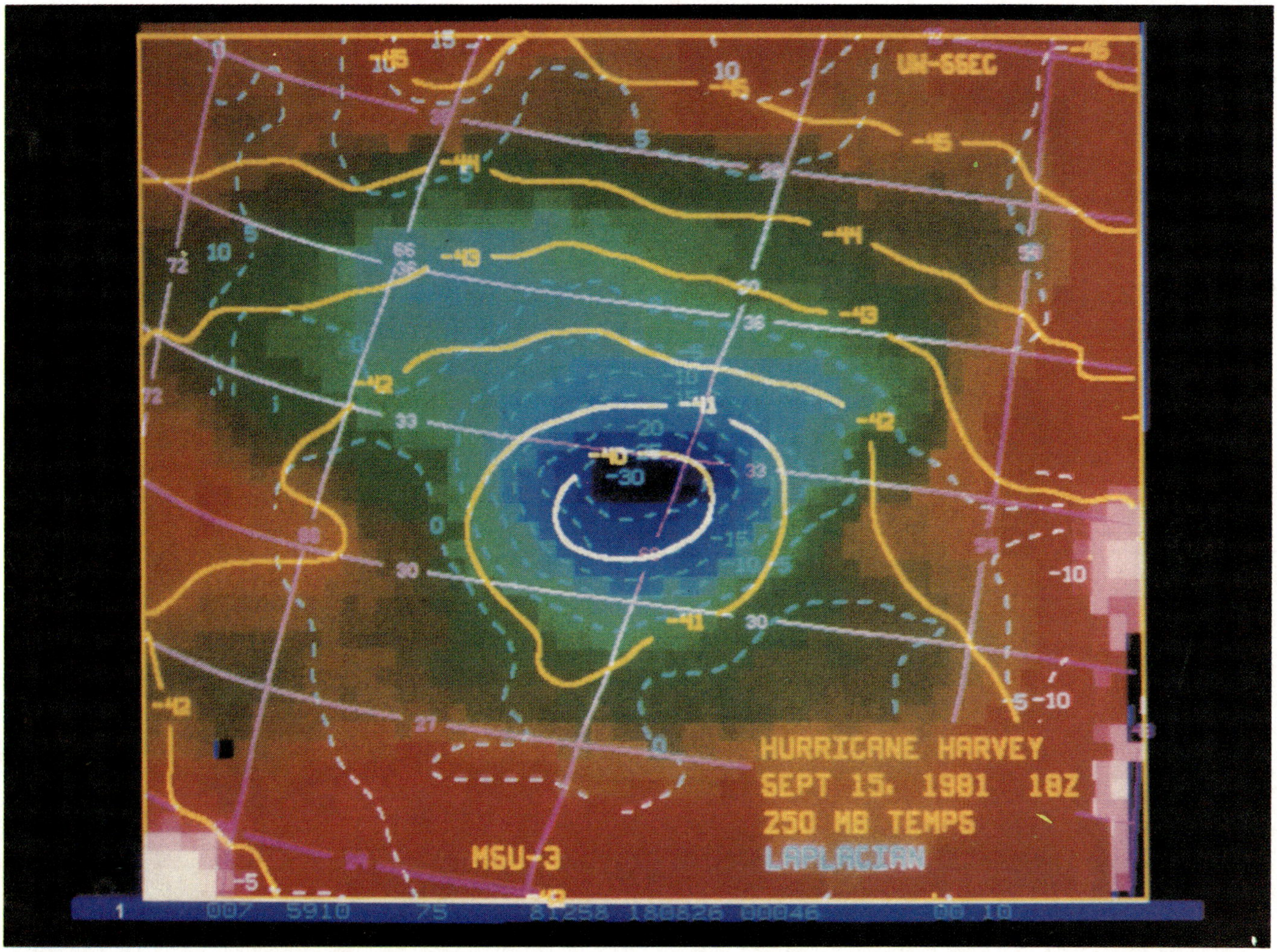

Figure VII-7-7. Hurricane Harvey, 1800 UTC, September 15, 1981. The solid lines show the 250-mb temperature field. The dashed lines are the Laplacian of the temperature field (Velden and Smith, 1983).

den and Smith, 1983). MW data are analyzed in real time to describe the 250-mb temperature field of the storm. Analyses of many tropical cyclones show good correlation between the temperature perturbation of the 250-mb warm core of the storm and its central surface pressure and maximum winds. For example, in the case of Hurricane Harvey (Figure VII-7-7), the high temperature in the warm core at the storm center indicates that Harvey has maximum sustained surface wind speeds of more than 100 kt. Also shown is the Laplacian of the temperature field, which also correlates with the maximum winds.

VII-7.2.3 Accuracy of Satellite Derived Intensity Estimates

The accuracy of tropical cyclone intensity estimates determined by satellite methods, or by any method, is difficult to evaluate. This is illustrated by the reconnaissance verification data in Figure VII-7-6. The regular U.S. Air Force reconnaissance maximum wind measurements (black dots) are often lower than the true wind speeds since the maximum wind area is not usually intersected by the aircraft flight path. On the other hand, research aircraft have recently been reporting unexpectedly high winds for short periods (see diamonds for Anita). The extremely variable wind data from reconnaissance aircraft have led to a heavy reliance on average maximum wind estimates inferred from reconnaissance pressure measurements. The post-season "best track" analyses of storms also rely heavily on the empirically derived central pressure/maximum wind relationship. The satellite intensity estimates relate best to the estimate from the central pressure measurements.

Studies by Erickson (1972) and Sheets and Grieman (1975) compared satellite estimates of storm intensities

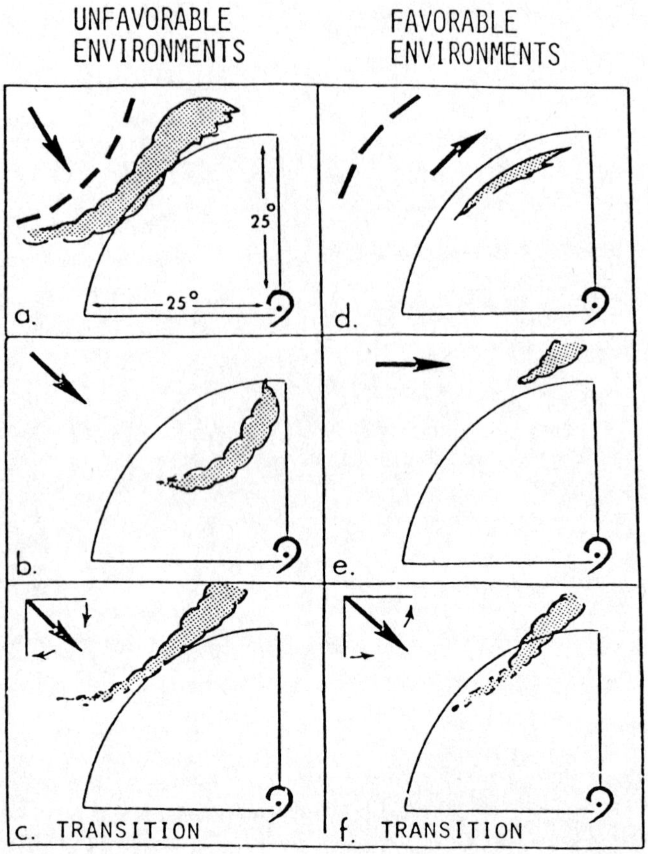

Figure VII-7-8. Illustrations of environments that are unfavorable (left) and favorable (right) for continued development of westward moving tropical disturbances. The storm is located at the lower right of each diagram. The arc is 25°lat. from the storm center.

with best track data. The estimates were made from once-a-day pictures of varying quality by a selected group of analysts experienced in satellite picture interpretation. The results showed an average difference of approximately 11 kt between satellite estimates and best track data. The two most recent studies of this kind (Shewchuck and Weir, 1980; Gaby et al., 1980), using better quality VIS and IR imagery and more than one observation a day, indicate an average difference between satellite estimates and the best track maximum winds of less than 8 kt in both the west Pacific and the Atlantic. This difference is quite small considering that the average error in estimating maximum winds from minimum central pressures is of the same magnitude (Atkinson and Holliday, 1977).

VII-7.3 TROPICAL CYCLONE INTENSITY FORECASTING

VII-7.3.1 Current Operational Methods

A 24-hour intensity forecast is made by first extrapolating along the modeled growth (or weakening) curve determined by the past history of the tropical cyclone. The forecast is modified when the cloud pattern is changing significantly or when the storm is entering or leaving an environment of strong vertical shear, stratocumulus (cold water), southward moving cirrus, or land.

Dvorak (1980) has also shown that cloud bands associated with the jet stream in the westerlies (as far as 25° lat. to the north and west) can be used to forecast development of westward moving tropical disturbances. Figure VII-7-8 shows upstream environments unfavorable for future development (left) and those favorable (right). Unfavorable environments have a strong (cold in IR pictures) cyclonically curved band entering the area (a), or forming within the area (b and c). Favorable environments are characterized by anticyclonically curved bands (d), no banding (e), or weakening cyclonically curved bands (f) within the area. The strong cyclonically curved bands (unfavorable environments) are related to a southward directed jet in the westerly stream north of the band (arrows indicate the axis of the maximum wind on the 250-mb surface). Weldon (1979) related such cloud band formation to frontogenesis and deformation in advance of the jet stream maximum. Anticyclonically curved bands, no banding, or weakening cyclonically curved bands appear to be related to a jet either directed away from the upstream environment of the disturbance or too far upstream to affect the environment. Parts (c) and (f) of Figure VII-7-8 show the transition states. In (c), the band forms southwestward as the jet turns more southward, and in (f) the band weakens to the north as the jet turns more northward.

VII-7.3.2 Objective Methods of Intensity Forecasting

Gentry et al. (1980) developed a technique to forecast 24-hour changes in maximum winds from IR observations of the equivalent blackbody temperatures of the storm cloud pattern. Use of regression equations, developed in their study, resulted in forecasts superior to those of persistence techniques.

Erickson (1974) found that numerical calculations of divergence, vorticity, and vertical shear using satellite observed cloud motions at high levels could serve as diagnostic supplements to other techniques. He applied polynominal fitting techniques to the fields of high cloud vectors obtained from ATS-3 satellite data. Relationships between some calculated parameters and cyclone intensity were found to be statistically significant, although the scatter of individual results was quite large. Similar studies, using better quality GOES wind fields, are continuing.

VII-7.4 TROPICAL CYCLONE MOTION ANALYSIS AND FORECASTING

Animated satellite picture sequences are excellent tools for tropical cyclone analysis. When such se-

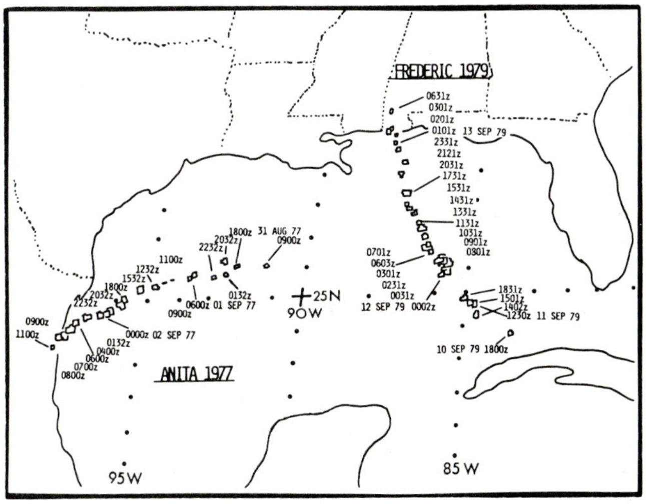

Figure VII-7-9. Tracks of Hurricanes Anita and Frederic traced from the inner temperature contours of the eyes in enhanced IR pictures.

quences are displayed automatically on a video screen, real-time analysis is possible. By watching the 6- to 12-hour past motion of the storm center on the video screen, forecasters can extrapolate its motion into the future. When shown to the public in warning areas along coast lines, such displays add credibility to the forecast of landfall issued by the warning center.

Sequences of individual enhanced IR pictures can be used to plot a detailed track of a storm, particularly for hurricanes with an eye showing. Figure VII-7-9 shows the tracks of hurricanes Anita and Frederic. The temperature contours around the eye were traced directly from enhanced pictures produced operationally at 1- to 3-hour intervals. In both storms the track looped as the eye first formed; then the track was clearly defined as the storms intensified and moved toward the coast. Note an oscillation in both tracks. The oscillation periods are not clear-cut, but appear to last about 10 to 12 hours.

Film loops can be used for quantitative estimates of winds connected with the storm and its environs. Wind vectors obtained this way are routinely used as input for analyses and for numerical forecasts of tropical cyclone motions. A recently developed technique that shows promise uses cloud displacement winds at 850, 400, and 200 mb; water vapor tracers; and satellite de-

Figure VII-7-10. A satellite derived combined wind set (upper level, approximately 200 mb) around Hurricane Debby, 0000 UTC, September 16, 1982. It consists of VAS gradient (blue), TOVS gradient (red), and cloud motion (yellow) winds in knots (Velden et al., 1984).

rived gradient winds to forecast tropical cyclone motion (Velden et al., 1984).

Figure VII-7-10 shows the combined vectors for mid-tropospheric levels around and over Hurricane Debby. The winds are used to derive a pressure-weighted deep-layer mean wind analysis. Once this analysis is complete, a simple trajectory model, taking a few seconds to run on McIDAS, is used to forecast the future storm position. The performance of this technique on four storms in 1984 and 1985 compared favorably with that of other operational models.

Several methods that use changes in the storm cloud pattern or cloud distribution have been developed. Lajoie and Nicholls (1974) found some relationship between the motion of storms near Australia and the distribution of convection in their cloud patterns. Rules were developed stating that cyclones move toward the most developed convective cluster near the downstream end of the outer cloud band and away from a sector devoid of cumulonimbus clusters.

Fett and Brand (1975) found that in western North Pacific storms the sense of rotation of major cloud features is related to a change in the track of the storm, i.e., when the storm features rotate clockwise, the storm will take a turn to the right. Although these techniques have been quite successful in individual cases, they are not widely used at tropical cyclone forecast centers.

Another recent experimental technique uses changes in the broadscale environment of a storm as well as changes in its moisture and cloud patterns to forecast a change in track. This method (Dvorak, 1984a) is based on changes in the upper level moisture patterns observed by the GOES VAS 6.7 μm channel. Eastern and central Pacific Ocean tropical cyclones of 1982 and 1983 were studied in developing this technique. Over 40 northwestward moving storms turned northward when a cyclonically curved moisture boundary approached the storm from the northwest while the storm pattern of moisture and cloudiness was building northward. Storms turned westward when a significant dissipation of the pattern of moisture and cloudiness occurred north of the storm center.

Chapter VII-8

CONVECTIVE SCALE WEATHER ANALYSIS AND FORECASTING[1]

VII-8.1 INTRODUCTION

Convective storms play an important role in meteorology. In general, the development of a convective storm array depends on the interaction of meteorological fields ranging from the synoptic-scale cyclone to the cloud condensation nucleus (Lilly, 1977). Little is known about why a particular storm forms and develops, or why a convective array behaves in a particular manner. Simpson and Dennis (1974) pointed out that more is known about synoptic-scale conditions and cloud microphysics favorable for convective development than is known about the mesoscale. The lack of information on the interaction between convective clouds and their mesoscale environment is mainly attributable to a gap in the meteorological observing capability that existed prior to high resolution geostationary satellite observations (Purdom, 1976). At the earlier time, meteorologists were forced to infer, rather than directly observe, most mesoscale processes. The cloud patterns in VIS and IR satellite images may be considered visualizations of mesoscale processes. When animated imagery is examined, the orientation, development, movement, and behavior of important mesoscale features can be seen.

For many mesoscale developments, the gradients of atmospheric parameters and their changes over short periods are important. The observational data from VAS, combined with other data, are helpful in detecting mesoscale gradients and their changes. GOES VAS sounding data are nearly instantaneous observations through a column in the atmosphere. They have moderate vertical resolution and high spatial resolution; and one uniformly calibrated sensor makes all the measurements. GOES VAS sounding data may be acquired over an area the size of the United States at hourly intervals in contrast to radiosonde observations taken once every 12 hours.

VII-8.2 THE INFLUENCE OF DIFFERENTIAL HEATING IN CONVECTIVE DEVELOPMENT

Differential heating appears to be a strong local trigger mechanism for convective development. Several phenomena associated with differential heating are well observed in satellite imagery: land-sea breezes, early morning cloud cover, and arc cloud lines. These represent the establishment of local convergence zones generated by differential heating. Other atmospheric properties help determine the effectiveness of local mechanisms in generating new deep convection. Among them are instability, large-scale dynamics, and low-level wind fields at and near the convergence zone.

VII-8.2.1 Sea, Lake, and River Breezes

Although convective cloud development attributable to terrain influences often is very complicated, many

[1] Primary Contributor: James F. W. Purdom

of the resulting cloud patterns are relatively easy to see in the 1-km resolution of GOES VIS imagery. Because the imagery is close in scale to that of cumulus clouds, and is available at short time intervals, convective development can be observed from its earliest stages through maturity.

The sea breeze is one of the most widely studied convective phenomena (Haurwitz, 1947; Estoque, 1962; Pielke, 1973, 1974). Several factors are involved in the development of the sea breeze:

(1) the shape of the coast line
(2) the direction and strength of the gradient wind
(3) friction
(4) the Coriolis effect
(5) air mass stability
(6) the land and water temperature difference

Items (4)-(6) are not discussed below, but one should realize the applicability of IR data to item (6) and the potential of VAS sounding data for item (5).

Different curvatures in a coast line cause differences in convergence along the sea breeze front, and so lead to a local strengthening or weakening of its cumulus activity. A small peninsula is generally an area of early strong convective development along the sea breeze front because the breezes from opposite shores merge near the peninsula center. Figures VII-8-1, VII-8-2, and VII-8-3 are examples of early to mid-afternoon sea breeze convection over Florida. Note the enhanced convection along the sea breeze front as well as the pronounced lack of clouds offshore. The effect of coast line curvature is most easily seen in the Gemini photograph (Figure VII-8-1); note the stronger convection in the Cape Canaveral peninsula region on the east coast. Immediately south of Cape Canaveral, where the

Figure VII-8-1. Gemini photograph of Florida, 1531 UTC, August 22, 1965 (south at top of photo).

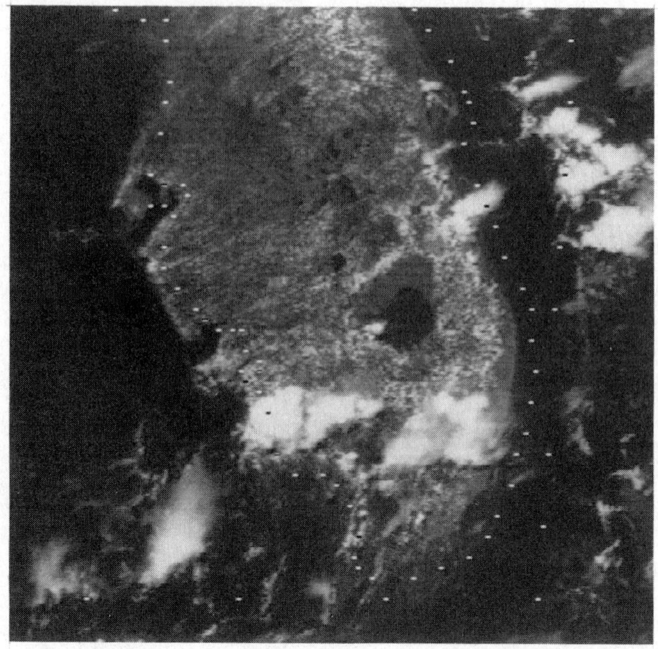

Figure VII-8-2. GOES-East 1-km VIS image, 1730 UTC, June 25, 1978.

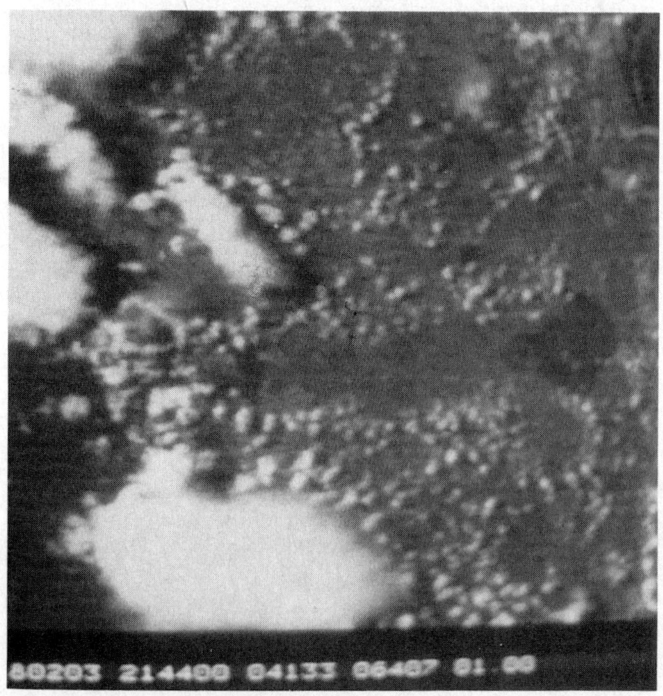

Figure VII-8-3. GOES-East 1-km VIS image, 2130 UTC, July 21, 1980.

Figure VII-8-4. GOES-East 1-km VIS image, 1701 UTC, 21 August, 1981.

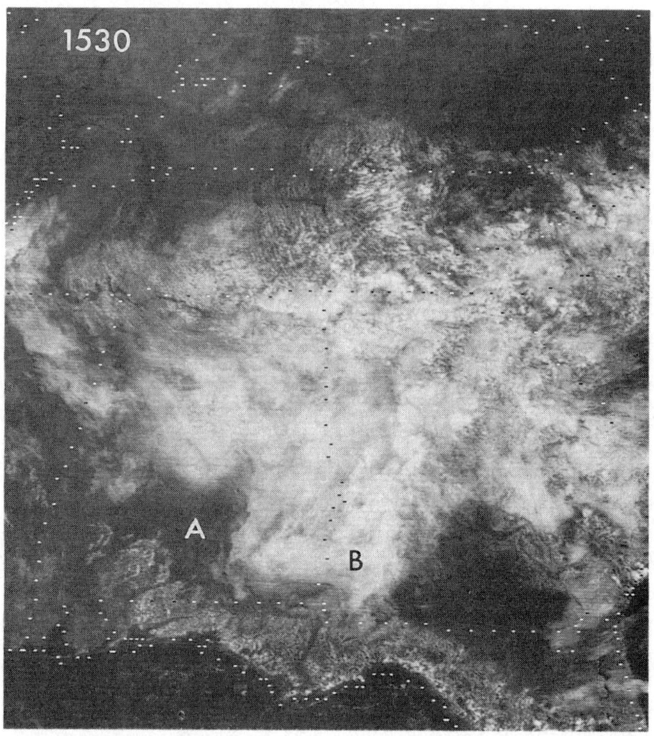

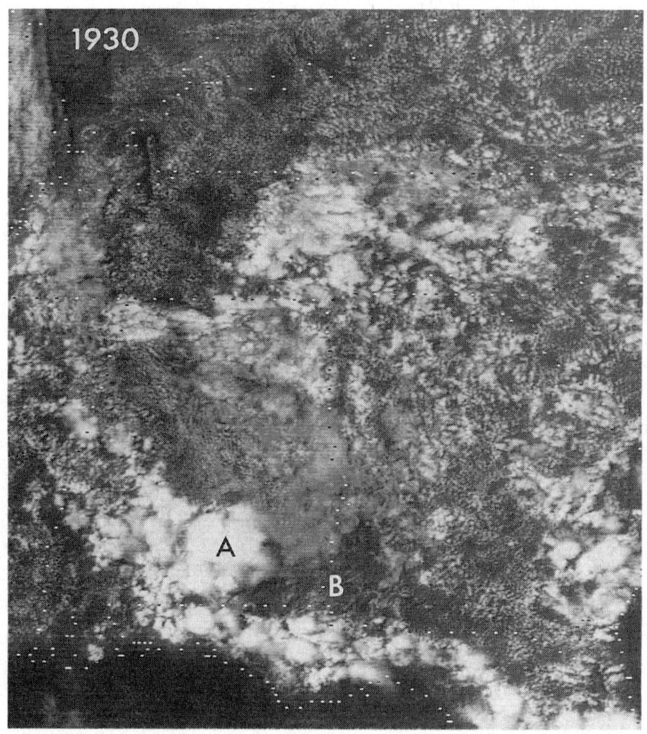

Figure VII-8-5. GOES-East 1-km VIS images, 1530 and 1930 UTC, May 27, 1977. Convective development over Alabama is due to early cloud cover. These images show the effect that early cloud cover can have on afternoon thunderstorm development. Note the early clear region (*A*) in southwest Alabama that becomes filled with strong convection during the day, while the early cloudy region (*B*) over the remainder of the state evolves into mostly clear skies. Also note how the strongest activity later in the day develops in the *notch* of the clear region in south central Alabama, as might be expected from merging cloud breeze fronts.

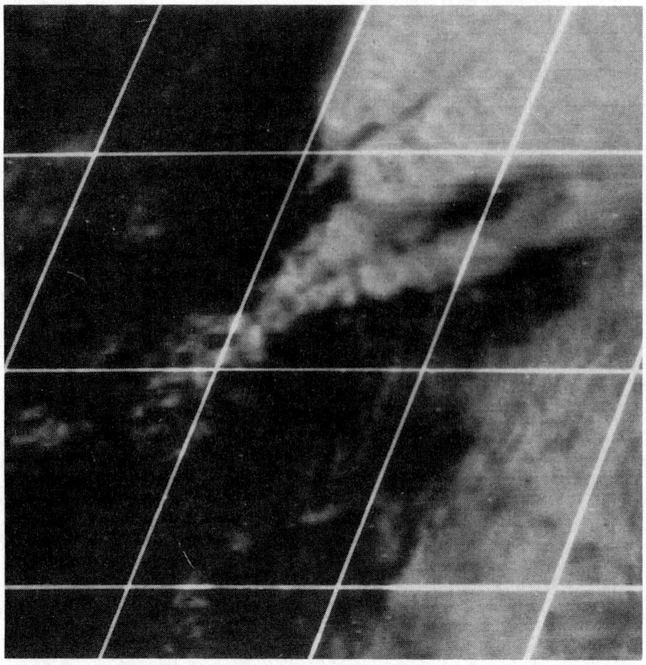

Figure VII-8-6a. GOES-East 1-km VIS image, 2126 UTC, April 10, 1979.

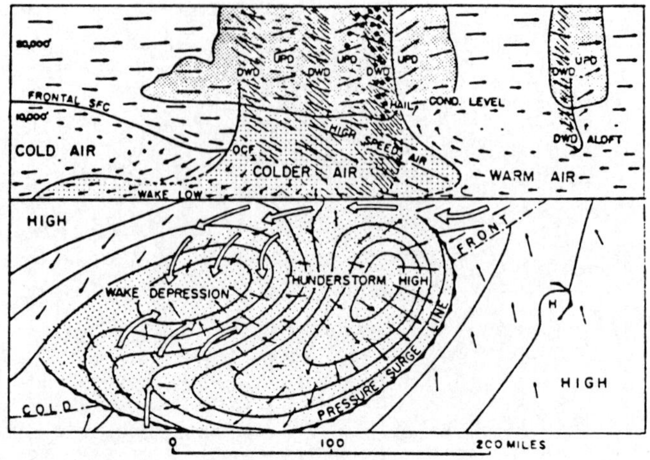

Figure VII-8-7. Illustration of the importance of rain cooled air in the production of the mesoscale high pressure system (Fujita, 1955).

coast line curvature is concave, sea breeze convection has penetrated several kilometers inland. Farther south, in the Miami area, convection is close to the ocean where coast line curvature is convex. Northward of Cape Canaveral, the convection is confined gradually closer to the coast despite its broad concave curvature. This occurs because the ridge of the Bermuda High on this day lies east to west across central Florida, causing weak westerly low-level winds that retard the mid-morning penetration of the sea breeze farther north.

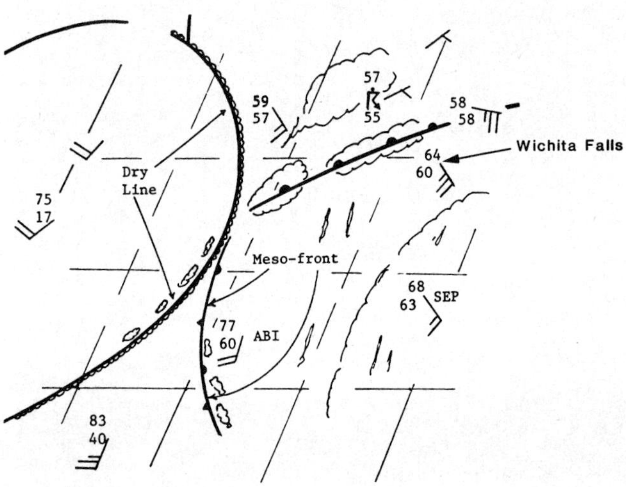

Figure VII-8-6b. Analysis of significant features and cloud patterns of Figure VII-8-6a. One of the storms that developed at the intersection of the dry line and the warm frontal boundary devastated Wichita Falls, Texas, later in the afternoon (see Figure VII-8-21a, b).

The same type of convective development also occurs regularly around lakes (Lyons, 1966). Figures VII-8-2, VII-8-3, and VII-8-4 are good examples of lake effects on summertime convective development. In Figure VII-8-2, light, low level, southerly flow helps generate a small area of clearing downwind from Lake Okeechobee. In Figure VII-8-3 a marked clearing extends westward of Lake Okeechobee indicating strong, low-level, southeasterly winds. Such clearing downwind of lakes, with strong convection bordering the clear region, is routinely observed in GOES imagery. Figure VII-8-4 shows the convective regime around the Great Lakes under northeasterly flow. Convection ends at the shoreline on the windward (east) side of the lake, but penetrates farther inland on the lee (west) side.

While the effects of oceans, lakes, and even cities (Changnon, 1976) on convective development have been studied extensively, relatively little work has been done on the effects of rivers and adjacent wetlands on convection. GOES imagery has revealed that such areas exert considerable influence on convective development during certain seasons. It is most noticeable when the low-level winds are either nearly calm or blowing parallel to a river. This river effect can be seen in Figure VII-8-1, in which the St. Johns River appears as a meandering line of clear skies running north-to-south in northeast Florida.

VII-8.2.2 Early Morning Cloud Cover

A. Weak Synoptic Forcing. The effects of early morning cloud cover on afternoon thunderstorm de-

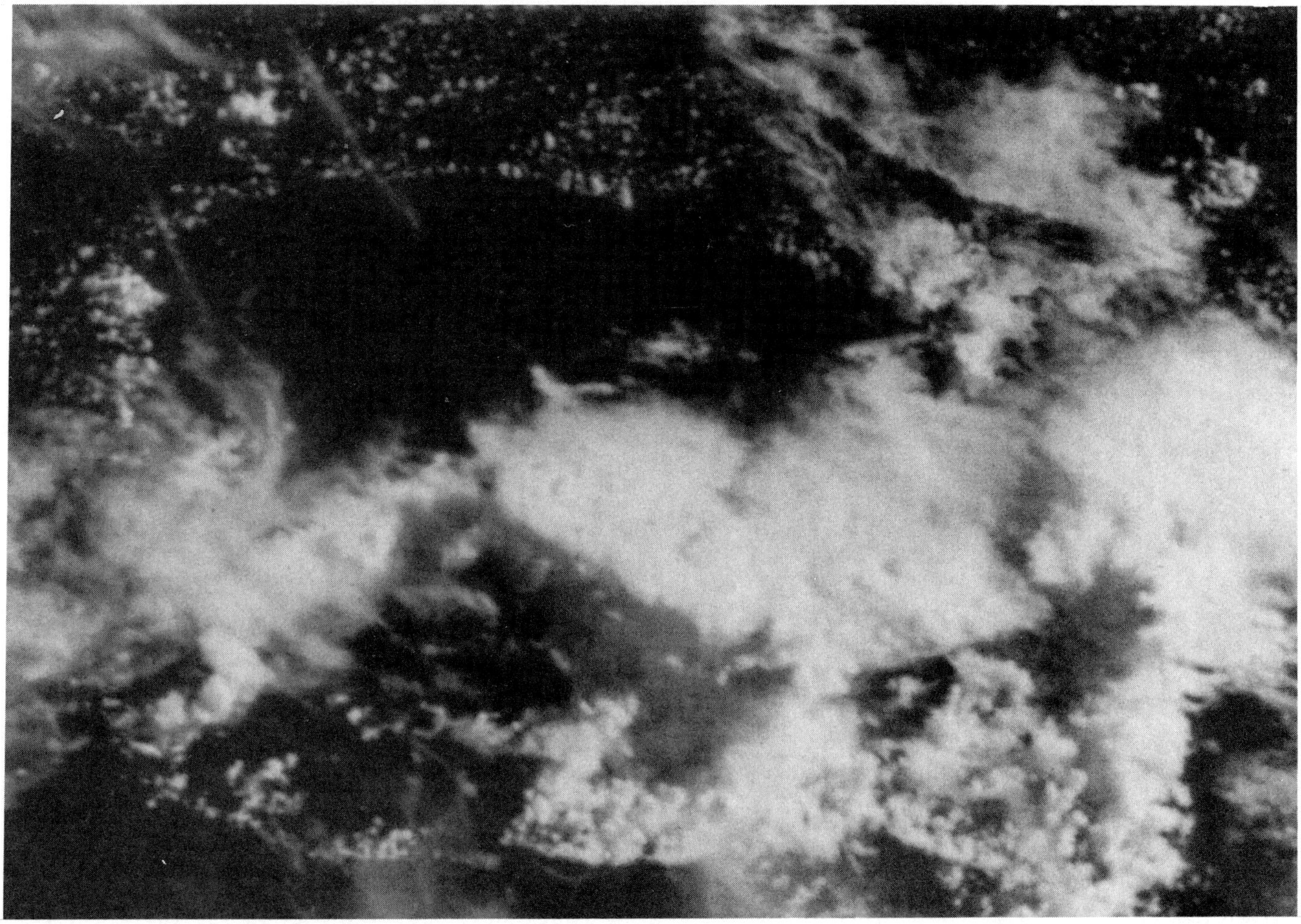

Figure VII-8-8. Photograph taken from a manned spacecraft of thunderstorm activity and resultant clearing within the cold dome.

velopment, under conditions of weak synoptic-scale forcing, are similar to those of the sea breeze. The first shower clouds form in the clear region near the boundary of the early morning cloud cover—a sort of cloud breeze front (Purdom and Gurka, 1974). Additionally, the slower heating rate in the early morning cloudy

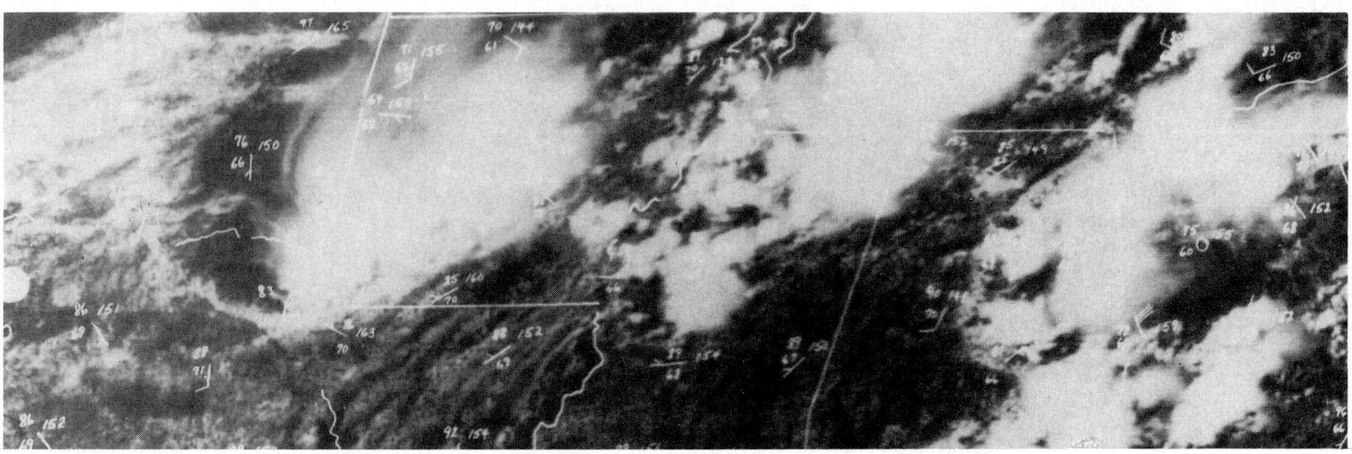

Figure VII-8-9. GOES-East 1-km VIS image, 2000 UTC, May 26, 1975.

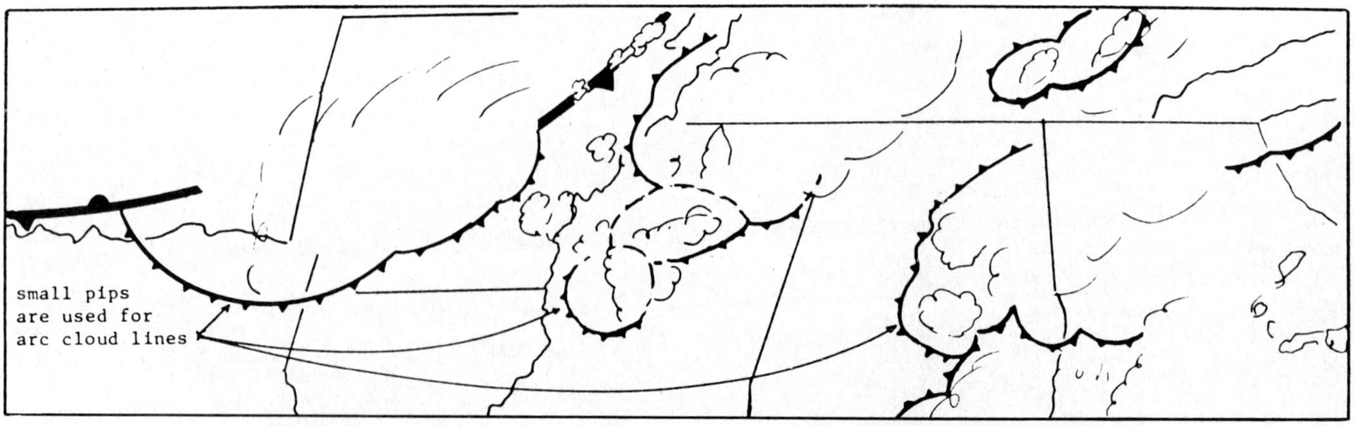

Figure VII-8-10. Analysis of arc cloud lines of Figure VII-8-9.

areas helps keep those regions free from convection for most of the day. Figure VII-8-5 is a good example of this phenomenon.

B. Strong Synoptic Forcing. Early morning cloud cover also can be important in providing the conditions for intense convection under strong synoptic-scale forcing. Purdom and Weaver (1982) showed the importance of mesoscale boundary interactions in focusing tornado activity over Texas on April 10, 1979; Figures VII-8-6a and VII-8-6b show the location of one of the associated mesoscale frontal boundaries. The most probable cause of the mesofront was differential heating associated with the cloudy (stratus) region to its east and the clear area to its west. The mechanism leading to the development of the mesofront and subsequent focusing of tornadic activity near it, was also important in the development of the warm sector instability. As the mesoscale front moved eastward, mostly clear skies developed in the warm sector. During that period Abilene (ABI) became clear with rapid surface heating, while Stephenville (SEP) retained its cloud cover and remained relatively cooler. Heated, more unstable, low-level air, similar in character to that near ABI, fed the development of intense thunderstorms and tornadoes.

Using a numerical boundary layer model, McNider et al. (1984) examined a case in which early morning cloud cover played an important role in the generation of a squall line in the presence of synoptic-scale forcing. They found that shading by cloud cover was fundamentally important to the formation of a local baroclinic zone and the development of low-level convergence.

The effect of early morning cloud cover can be complicated. It can act to set up or enhance a baroclinic zone through differential heating and it can set up local air mass instability. If an area remains cloudy, thunderstorms moving into the region may dissipate or

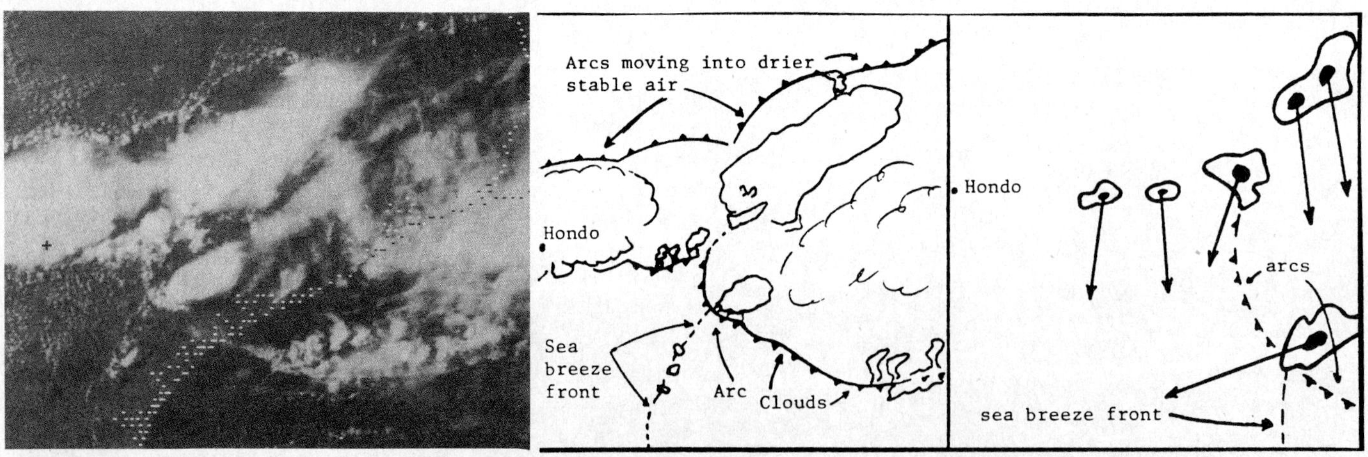

Figure VII-8-11. GOES-East VIS image for 2120 UTC, July 28, 1980 (left), analysis of the image (middle), and radar echo movements centered on 2200 UTC (right). The large thunderstorm generated at the sea breeze and arc cloud merger propagated (continuously regenerated) along the moving intersection of those two boundaries.

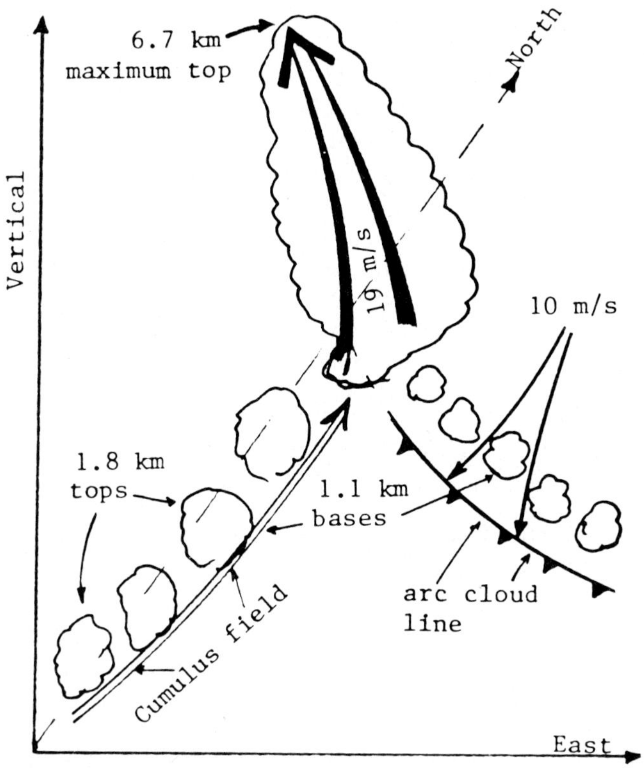

Figure VII-8-12. Adaptation by Purdom (1986) of numerical modeling results of Droegemeier and Wilhelmson (1983) to thunderstorm outflow collision with a convective cloud line.

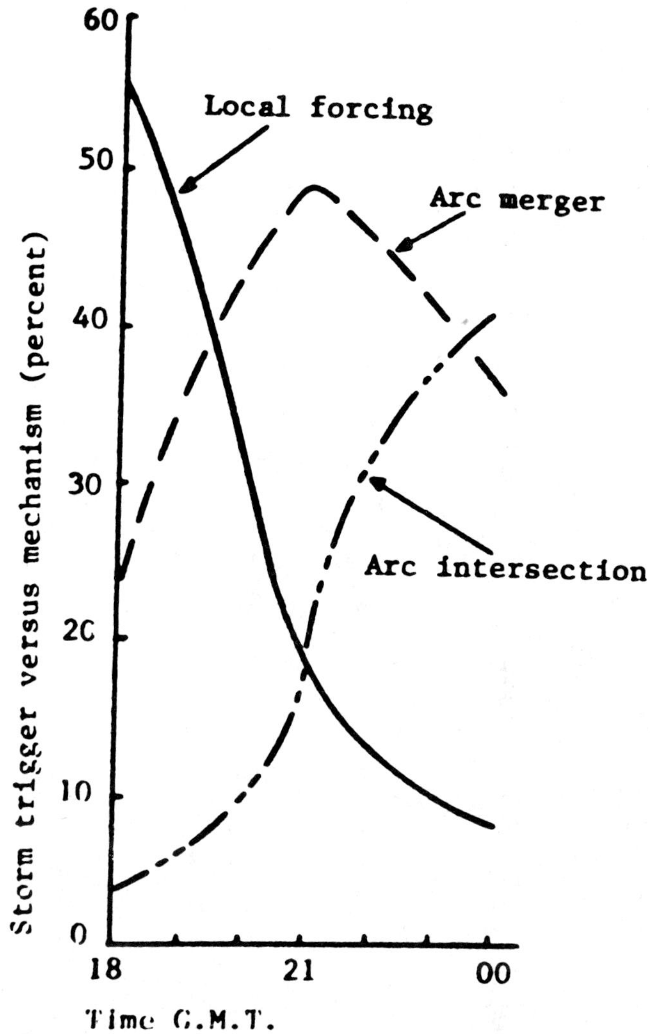

Figure VII-8-13. Diurnal variations of frequencies of convective generation mechanisms.

weaken considerably because of the negatively buoyant (cool) low level air. If the area clears an hour or two before thunderstorms move into it, there may have been sufficient heating and mixing at low levels to support explosive convection. VAS sounding data should be useful in assessing the convective potential of such mesoscale situations.

VII-8.3 ARC CLOUD LINES

Although the importance of downdrafts and cold outflows from thunderstorm cells and their possible influence on further storm development were recognized by Espy (1941) and Humphreys (1914), the first modern, indepth study of thunderstorm dynamics was done in Florida (1946) and Ohio (1947) by the Thunderstorm Project (Byers and Braham, 1949). They noted that the cold air outflow of a storm could cover an area larger than the storm, and that new cell development within a thunderstorm system could occur where the outflows of older cells collided. Subsequently, detailed analyses of mesoscale thunderstorm systems showed that a sharp pressure rise usually accompanied a system's arrival (Fujita, 1955; Fujita and Brown, 1958). Other phenomena usually occurring were a wind shift, an increase in wind speed, a temperature decrease, and rainfall. Fujita (1959) showed how evaporation of raindrops in the downdraft caused the development of a cold dome and mesoscale high pressure system beneath the thunderstorm. The leading edge of this "mesohigh" was the thunderstorm gust front. Figure VII-8-7 illustrates the development of the mesohigh and shows the pressure surge at its leading edge. In satellite imagery the leading edge of the mesohigh appears as an arc-shaped line of convective clouds moving out from a thunderstorm area (Purdom, 1973). These arc-shaped cloud lines of cumulus congestus and/or cumulonimbus play a key role in controlling the development and evolution of deep convection (Purdom, 1979).

Figure VII-8-8 is a photograph of an arc cloud line taken from a manned spacecraft. Note the large circular clear region caused by subsidence within the cold dome. Figure VII-8-9 shows the characteristic appearance of

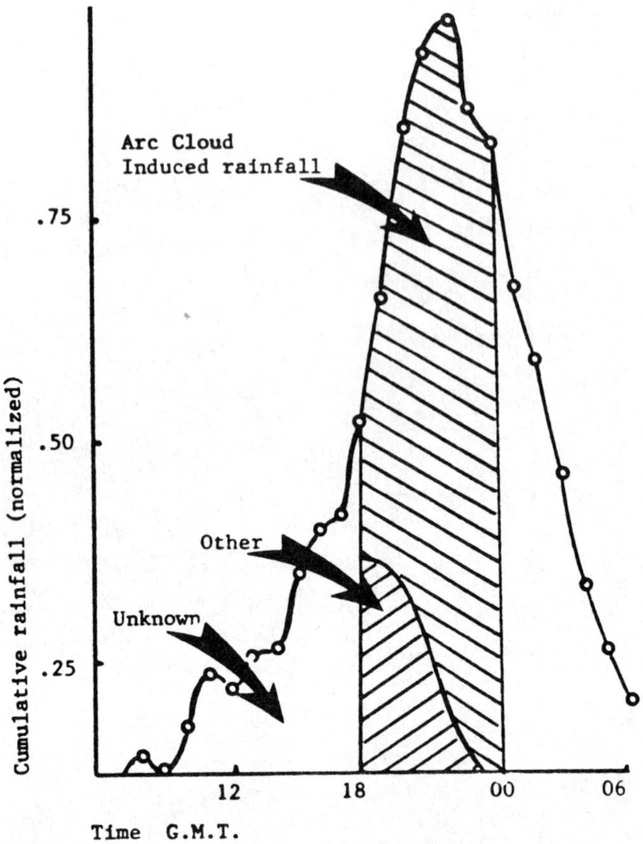

Figure VII-8-14. Diurnal variation of rainfall distribution by mechanism for same period as in Figure VII-8-13.

arc cloud lines in satellite imagery; Figure VII-8-10 is a mesoscale nephanalysis of this image. In this case, interaction between the arc cloud and the front in Oklahoma produced a large, severe thunderstorm (Purdom, 1983). Figure VII-8-11 shows another arc cloud line in GOES imagery; in this case, interaction of an arc cloud and a sea breeze front in Texas produced a large thunderstorm (Zehr, 1982). That thunderstorm moved to the the west, while other thunderstorms in the area moved to the south. The "anomalous" movement of the storm at the sea breeze / arc cloud junction can be explained as continuous regeneration along that moving intersection. An interesting point in this case is that the arc cloud lines were not detected by radar, so the satellite picture was needed to explain the anomalous movement of the storm.

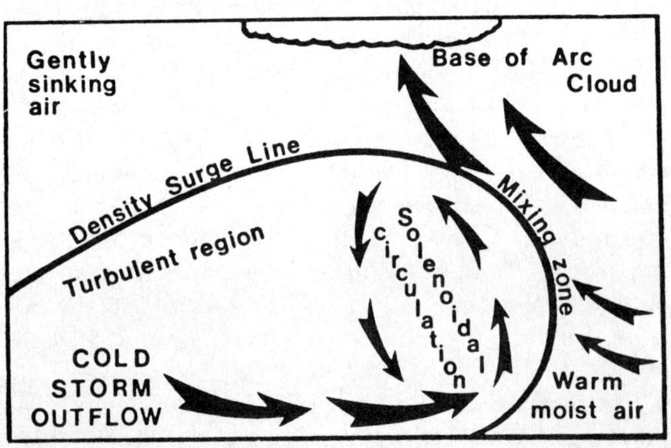

Figure VII-8-15. Schematic illustration of arc cloud line features as derived from aircraft penetrations.

Movies made with images taken at 3- to 5-min intervals[2] have provided much additional insight into the development of deep convection through interactions of arc clouds with each other or with other convective lines, areas, or boundaries (Purdom, 1979). Studies based on this rapid interval imagery have led to the following conclusions about these interactions:

- Thunderstorm produced outflow boundaries (arc clouds) are of primary importance in the formation and maintenance of strong convection.
- Outflow boundaries may maintain their identity as arc clouds for several hours after the generating convection has dissipated.
- Deep convective development along an arc cloud line is selective, i.e., it is highly favored where arcs intersect or where an arc moves into a convectively unstable region.
- In a weakly forced atmosphere, arc boundary interactions determine the location of most new late afternoon and early evening thunderstorms.

These observational findings have led to some numerical modeling studies of these interactions. Droegemier and Wilhelmson (1983), using a three-dimensional numerical cloud model, found that a deep convective cloud with maximum upward motions of about 19 m s^{-1} developed where the outflow interacted with the convective cloud line (Figure VII-8-12). They noted that this did not occur in a simulation where the outflow moved into a clear region.

Purdom and Marcus (1982) have classified the mechanisms of convective development, with weak synoptic scale forcing, over the southeastern United States on the basis of GOES imagery. They have found that convective scale interaction becomes the dominant mechanism following the initial convective developments in the early afternoon (Figure VII-8-13). A similar classification of convective rainfall for these same cases (Figure VII-8-14) shows the dominance of arc-generated rainfall.

[2] Since 1975 the GOES system has been operated in a special 3- to 5-min imaging mode on selected days during convective seasons. GOES imaging at 5-min intervals became a routine NESDIS operation during severe storm outbreaks in 1985.

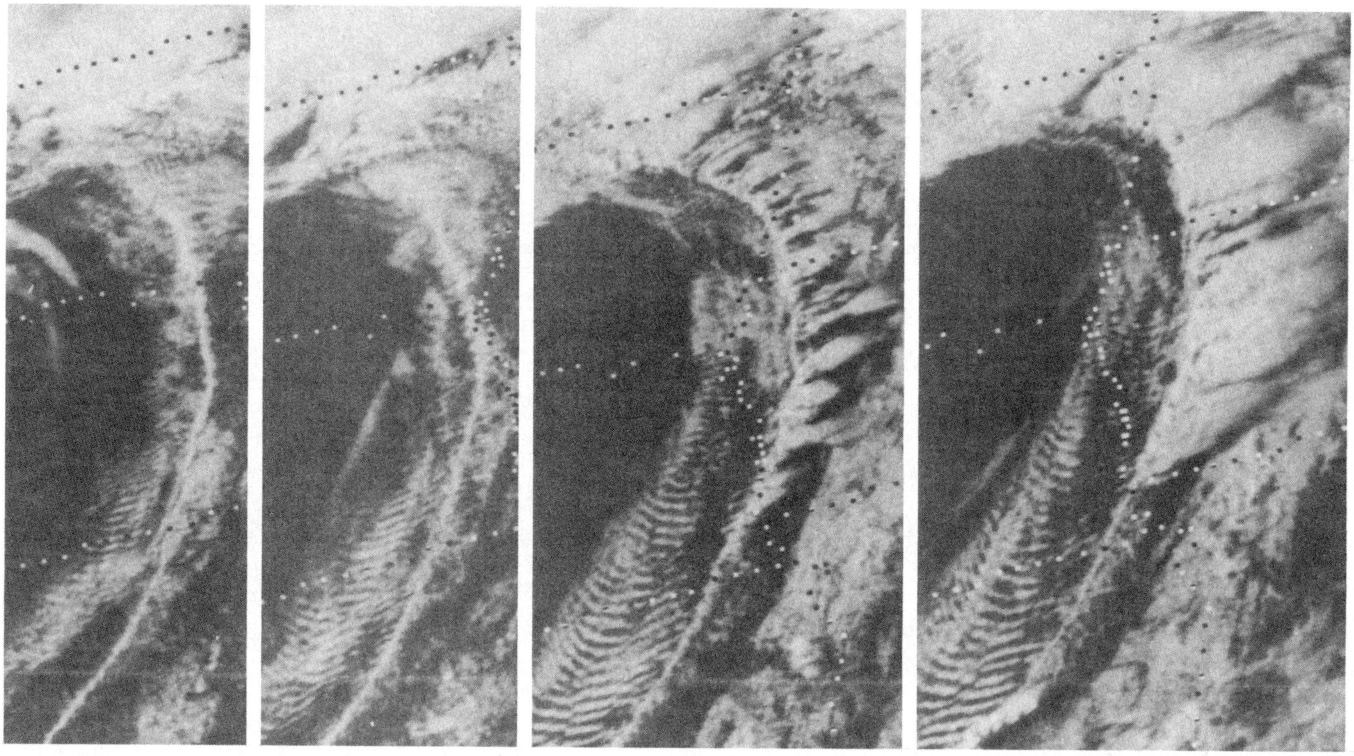

Figure VII-8-16. GOES-East 1-km VIS images, 1900, 2000, 2200, and 2300 UTC, June 14, 1976; a typical example of squall line development.

Figure VII-8-17. GOES-East enhanced IR image of a typical mesoscale convective complex (MCC) over the central United States, 1500 UTC, June 22, 1981.

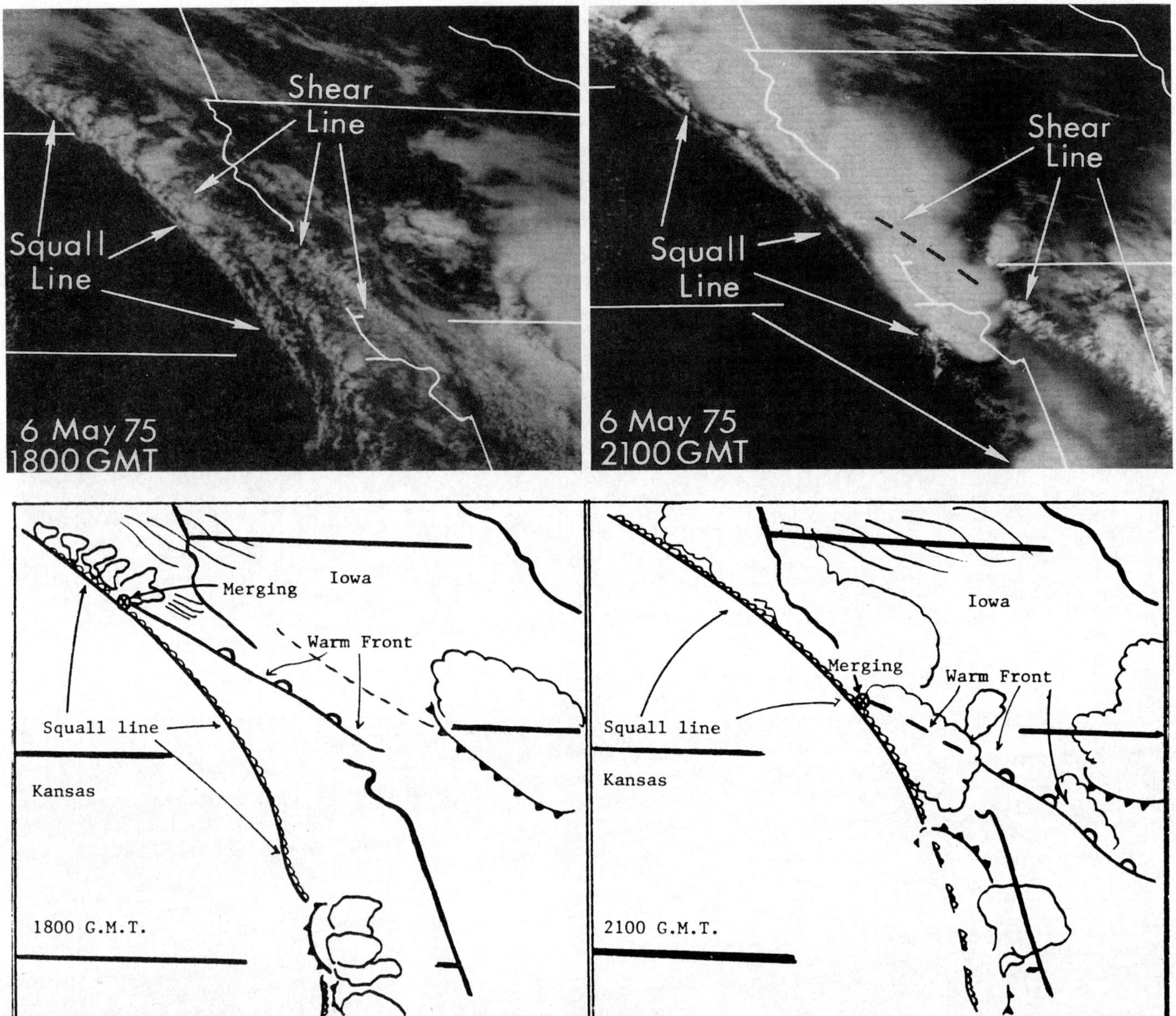

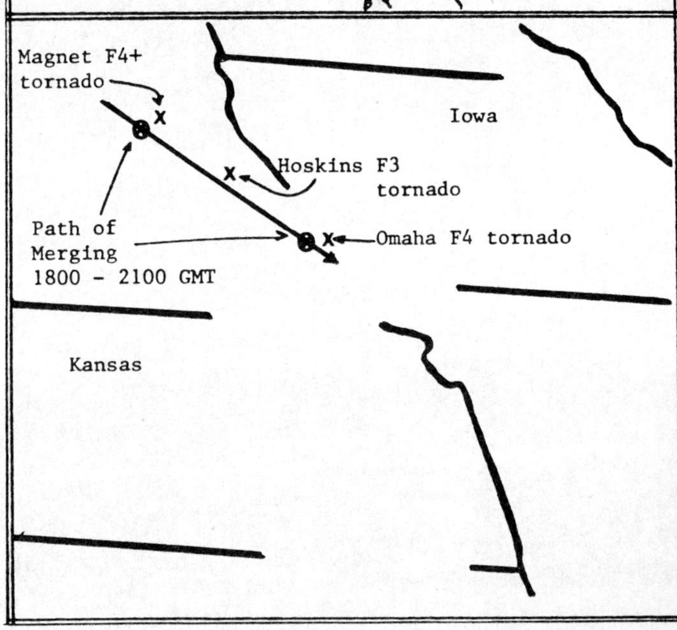

Figure VII-8-18. GOES images, 1800 UTC (upper left), 2100 UTC (upper right), May 6, 1975. Analyses of significant features derived by combining information from surface observations and GOES images are presented immediately below their respective images. Note the relationship, shown in the final panel (lower left), between tornado locations and the path of merging of the squall line and the warm frontal boundary.

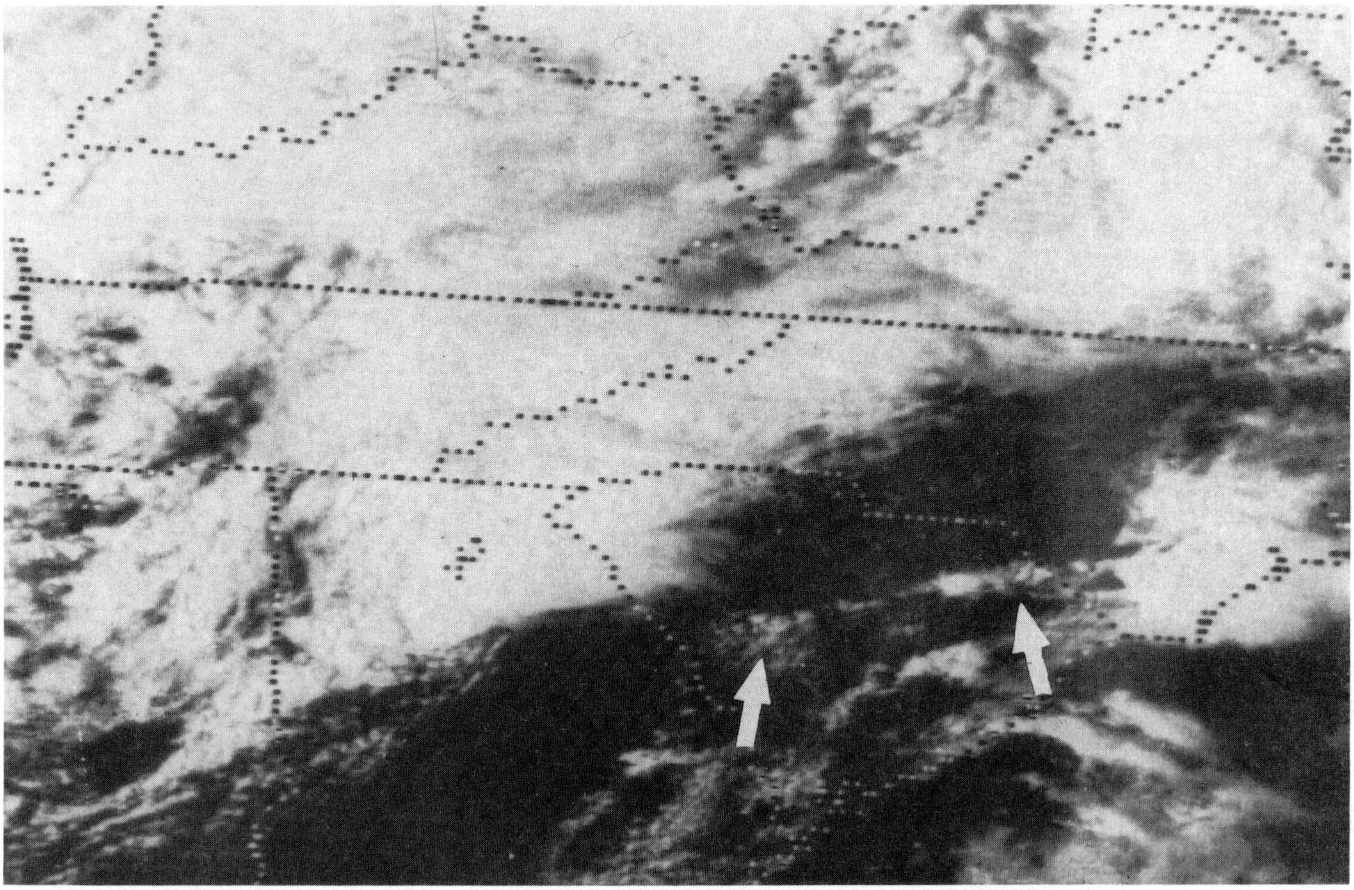

Figure VII-8-19a. GOES-East 1-km VIS image, 2030 UTC, March 28, 1984; arrows indicate outflow boundary.

The characteristics and interactions of arc cloud lines have been measured from aircraft by Sinclair and Purdom (1982 and 1984). Flights were made in the subcloud layer above the outflow frontal zone (density surge line) and beneath the base of the arc cloud (Sinclair and Purdom, 1982), and through the outflow frontal zone (Sinclair and Purdom, 1984). Analyses of these aircraft observations, shown schematically in Figure VII-8-15, indicate that arc cloud lines are important to the production of both vorticity and con-

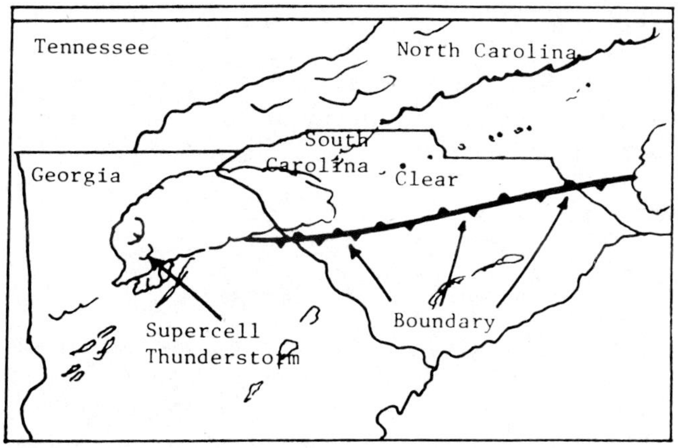

Figure VII-8-19b. Analysis of significant features and cloud patterns of Figure VII-8-19a.

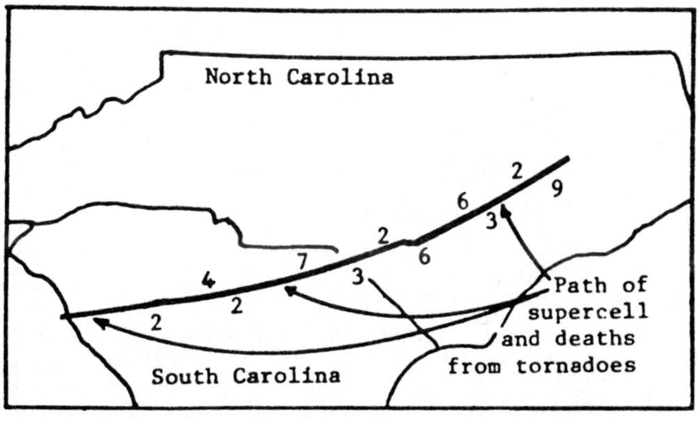

Figure VII-8-20. Path of super cell that produced significant tornadoes, and the number of deaths associated with those tornadoes, March 28, 1984.

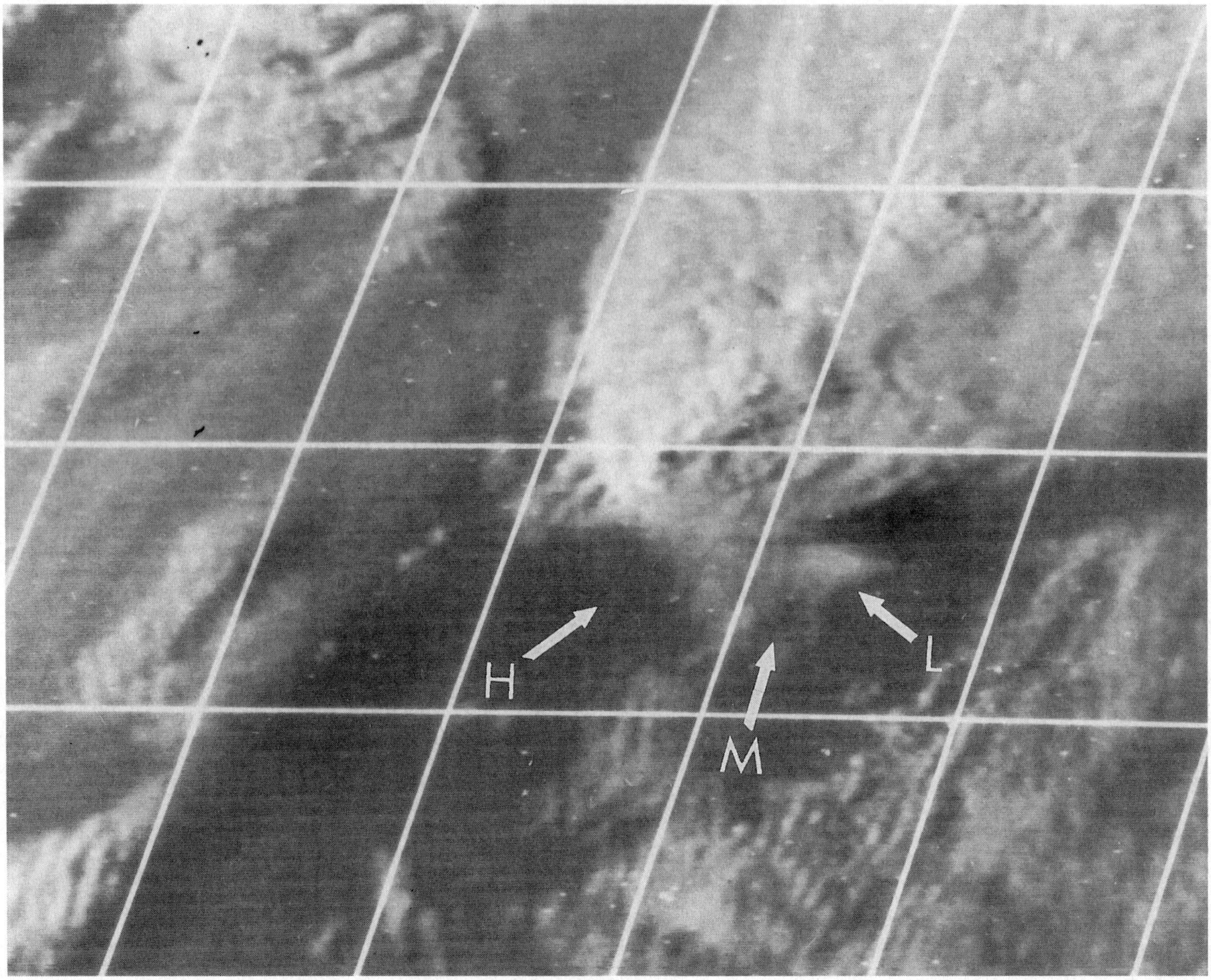

Figure VII-8-21a. GOES-East 1-km VIS image, 0015 UTC, April 11, 1979.

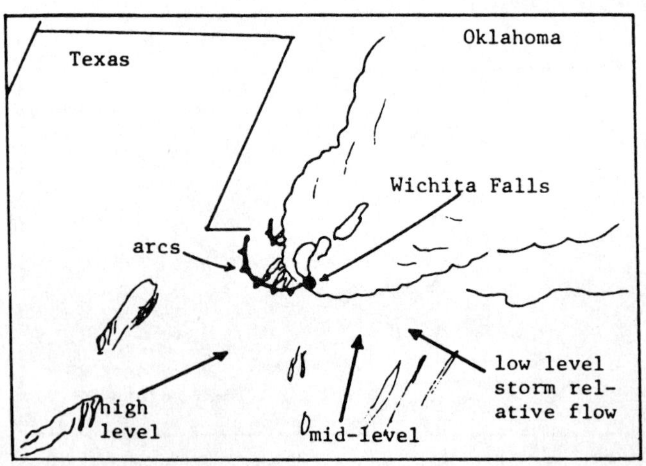

Figure VII-8-21b. Analysis of relative flow and significant features of the image in Figure VII-8-21a.

vergence. Also, the lateral extent of the vertical motion field along the arc (compared with that at the cloud scale) indicates that the main driving force for initial cloud development along the arc line is controlled by the interaction of the thunderstorm outflows with convectively unstable air.

VII-8.4 MESOSCALE CONVECTIVE SYSTEMS

VII-8.4.1 Squall Line Development

Generally, organized convergence lines that trigger strong convection are detectable in satellite imagery before deep convective activity develops along them (Purdom, 1976). An example is shown in Figure VII-8-16, where a Pacific front extending from eastern South Dakota and Nebraska into central Kansas de-

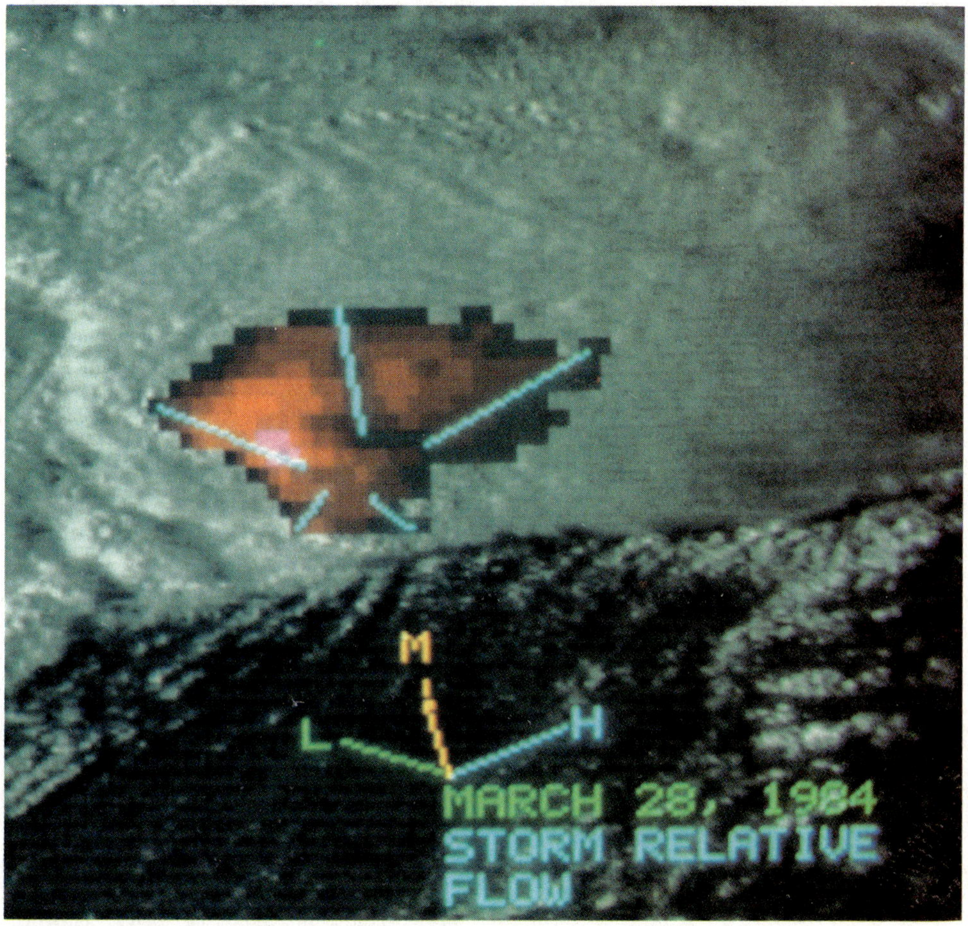

Figure VII-8-22. GOES-East 1-km VIS image, 2145 UTC, March 28, 1984 showing relative flow at low (*L*), middle (*M*), and high (*H*) levels, as well as −46°C isotherm divergence within anvil.

veloped into a line of severe thunderstorms with hail, severe winds, and tornadoes as it moved into Minnesota and Iowa.

VII-8.4.2 Mesoscale Convective Complexes (MCC)

A different type of highly organized mesoscale convection, the MCC, forms with weak synoptic-scale forcing (Maddox, 1980) and occurs most often over the central United States at night in spring and summer. They appear to have many characteristics similar to those of tropical convective clusters. MCC interact with and modify the larger scale environment in which they are embedded and affect downstream weather long after they dissipate. MCC were not recognized before GOES observations became available.

Figure VII-8-17 shows the characteristic MCC appearance in satellite imagery. MCC produce much of the beneficial rainfall over the U.S. Central Plains in summer and also are prolific producers of severe weather. Maddox (1980) has shown that of 43 MCC in the 1978 convective season there were 17 that caused flooding or heavy rain, 17 with tornadoes, 21 with severe wind, and 22 with hail; in only 8 of the 43 occurrences was there no report of severe weather.

VII-8.5 USE OF VISIBLE IMAGERY IN ANALYZING SEVERE CONVECTIVE STORM DEVELOPMENT

VII-8.5.1 Interactions Between Squall Lines and Other Boundaries

Under proper dynamic forcing, thunderstorms that form along organized convergence lines interact with other boundaries and severe storms develop (Miller, 1972). For example, on May 6, 1975, strong tornado activity that began in northeast Nebraska developed southeastward into the Omaha area along the intersection of a squall line and a warm front. Figure VII-8-18 shows how these features were tracked during the outbreak. The developing squall line appeared as a line of convective clouds along a surface windshift line between moist and dry air. The warm frontal boundary showed as another line of convection between warm moist air in southerly flow and slightly cooler and drier air in southeasterly to easterly flow. Other clues helpful for precisely locating the warm front were the transition from cumulus to stratus clouds across it and cloud track winds (not shown).

VII-8.5.2 Arc Clouds and Severe Storm-Tornado Development

The role of arc clouds from earlier convection in the development of tornado activity was vividly demonstrated by a storm system that moved through the southeastern United States on March 28, 1984 (Purdom, 1984). Thunderstorm activity during the early morning hours produced a well-defined outflow boundary from South Carolina across Georgia and into Alabama. This boundary moved slowly northeastward, and appeared in satellite imagery by midafternoon as a line of organized cumulus congestus extending across South Carolina into North Carolina (Figure VII-8-19). A large storm in northeast Georgia, at the junction of the convergence boundary and a mesolow, developed into a super cell that moved east-northeast along the boundary and produced most of the tornado activity (Figure VII-8-20).

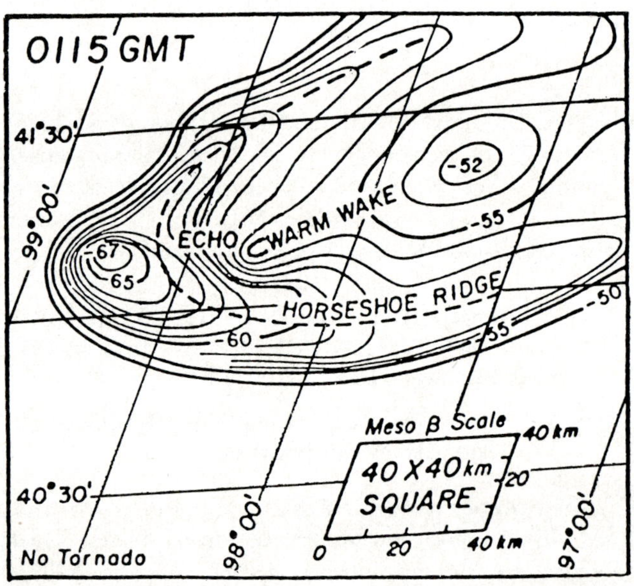

Figure VII-8-23. Isotherms of IR temperature of the Grand Island tornado cloud at 0115 UTC drawn at 1°C intervals. Horseshoe ridges (cold) are shown with dashed arcs; warm depression, with "warm wake" on the downwind side of radar echoes.

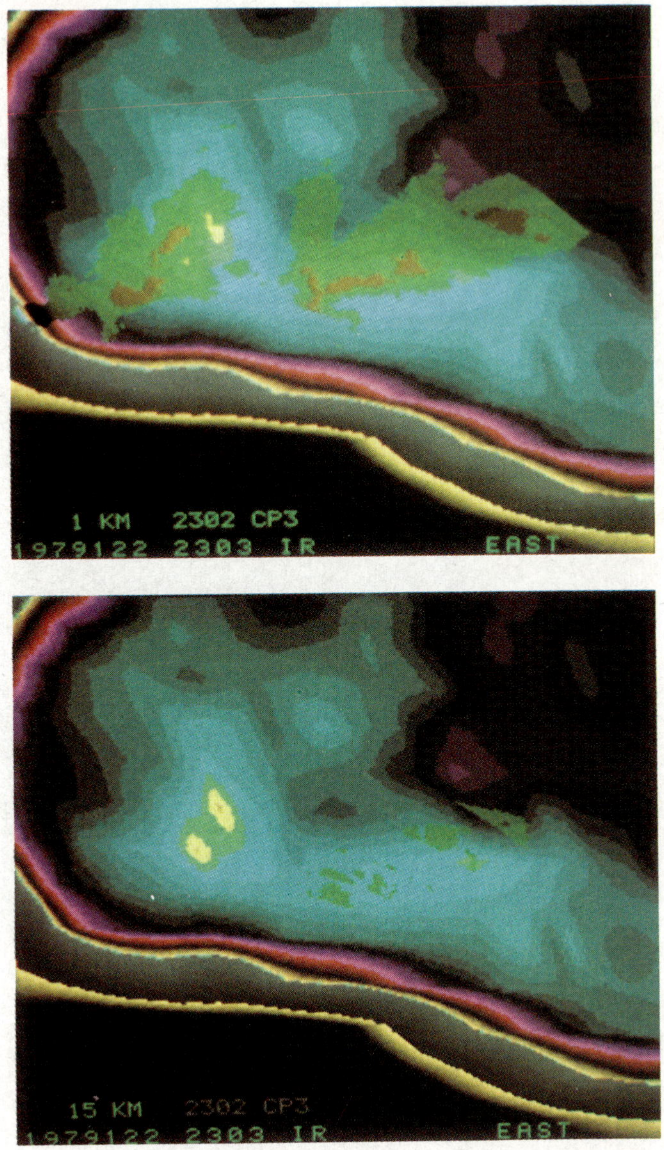

Figure VII-8-24. Satellite and radar composite for 2303 UTC, May 2, 1979 (1-km and 15-km CAPPI).

VII-8.5.3 Vertical Wind Shear

Vertical wind shear is important in determining the character of storms that evolve in a mesoscale environment (Newton, 1963) and in the development of rotation in growing thunderstorms (Klemp and Wilhelmson, 1978; Bleckman, 1981). Wind patterns in severe storm development undergo rapid changes. Rapid scan satellite imagery provides a means for observing these variations in vertical wind shear during severe storm development.

Purdom et al. (1984) used GOES-East images to analyze wind shear in the vicinity of an intense tornadic

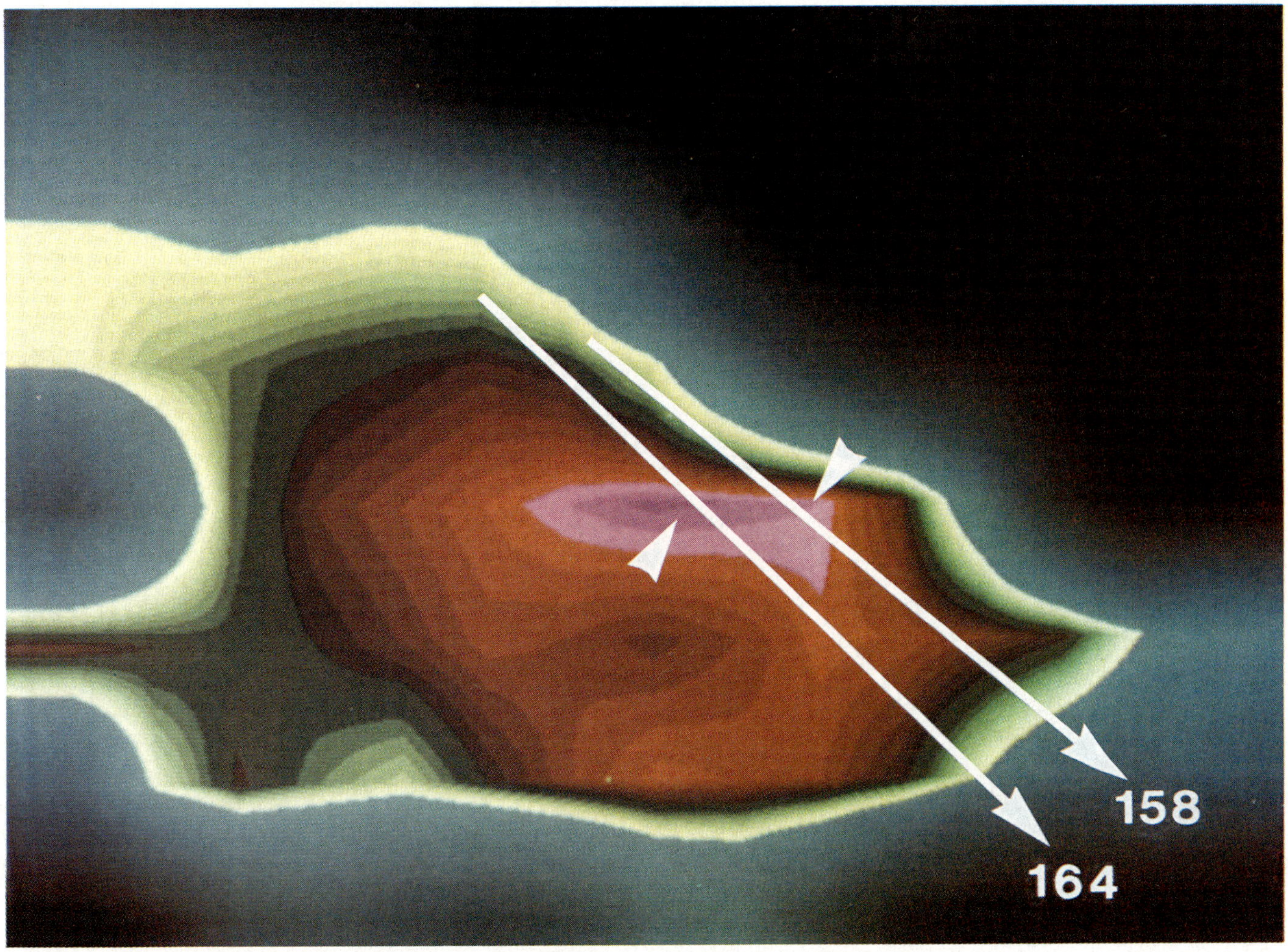

Figure VII-8-25a. Improved infrared product with 1°C isotherm contours for storm on May 19, 1982. Normal IR image was smoothed and "de-blocked" by a computer technique (Purdom and Vonder Haar, 1983).

thunderstorm. Figure VII-8-21 shows the relative flow derived from 3-min interval data at low, middle, and high levels near that storm. Figure VII-8-22 shows the relative flow for the March 28 super cell storm described in VII-8.5.2. These relative flows compare closely to those proposed by Browning (1964) for severe storms that travel to the right of the wind. The ability to diagnose relative flow is important for defining mesoscale regions favorable for the production of rotating thunderstorms and severe weather.

VII-8.5.4 Overshooting Thunderstorm Tops

Overshooting thunderstorm tops can be detected in VIS satellite imagery and related to severe weather. Early work was done by Pearl (1974) and Fujita et al. (1976) in which photographs of overshooting thunderstorm tops taken from Lear jets were compared with overshooting tops detected in geostationary satellite imagery. These overshooting tops were correlated with severe weather events. Similarly, Hasler (1981) correlated severe events with overshooting tops by combining GOES-East and GOES-West data to obtain stereoscopic measurements.

Overshooting thunderstorm tops are generally associated with updraft regions that penetrate well into the stratosphere. There is a serious limitation in using them as storm intensity indicators, however; they are detectable in VIS imagery only for the few morning or afternoon hours when the sun is low enough for the overshooting tops to cast shadows on the underlying anvil.

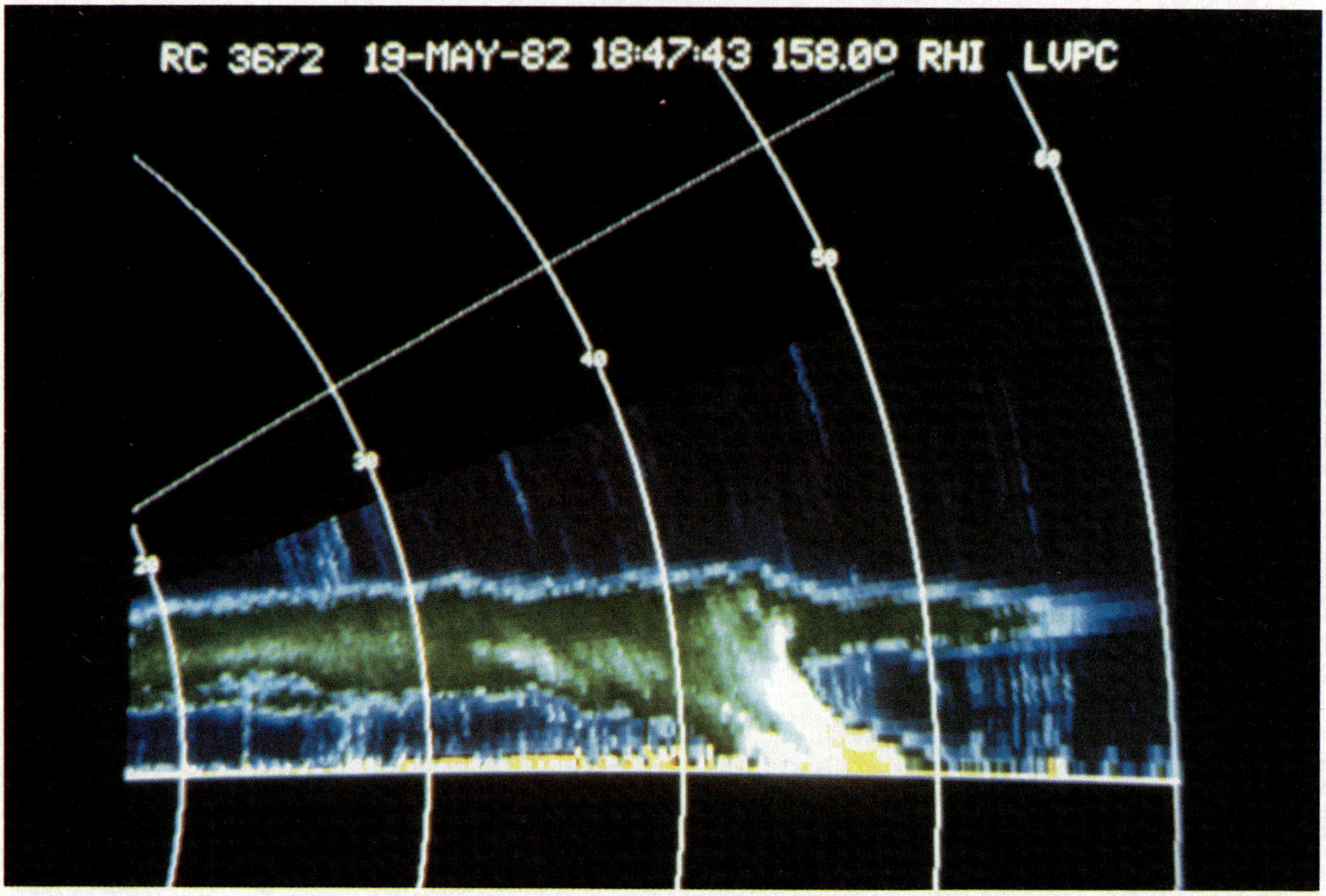

Figure VII-8-25b. RHI scan at 158° for storm in Figure VII-8-25a.

VII-8.6 USE OF INFRARED IMAGERY IN ANALYZING SEVERE CONVECTIVE STORM DEVELOPMENT

VII-8.6.1 Infrared Imagery Patterns

IR imagery provides more precise information than VIS imagery about increasing heights and areal coverage of cumulonimbus towers and anvils in severe storm developments. The areal expansion rate of isotherms less than tropopause temperature has been correlated with severe storm occurrence using GOES rapid scan IR data. Also, Pryor (1978) made a detailed case study using rapid scan GOES IR data to compare severe events with cloud top temperatures less than that of the surrounding tropopause. He found that only thunderstorms with IR cloud tops colder than the tropopause produced severe weather. Reynolds (1979) found a similar positive relationship between the occurrence of hail and IR tops colder than the tropopause. This relationship is definitely not universal, however. A number of such thunderstorms have *not* produced severe weather; and tornadoes and hail *have* occurred with thunderstorms that appeared rather ordinary in GOES imagery.

Fujita (1978) discussed anvil tops of severe storms with a particular pattern of IR temperatures that appears in enhanced GOES IR imagery. This pattern is illustrated in Figure VII-8-23 (Fujita, 1981a) which is a detailed analysis of GOES IR data for a storm that produced tornadoes at Grand Island, Nebraska. Note the cold "*V*" or "horseshoe ridge," the intense overshooting cold tops, and the downstream warm region ("wake"). Weaver and Purdom (1983) found a similar IR pattern for an intense tornado outbreak over Oklahoma. The same type of anvil signature has also been documented by Reynolds (1979) for severe hailstorms. With operational GOES data, McCann (1981) found a similar good correspondence between the *V* notch signature and severe weather. In general it has been found that the *V* or horseshoe ridge is a good indicator of severe storm activity, but it is not always present in tornadic storms.

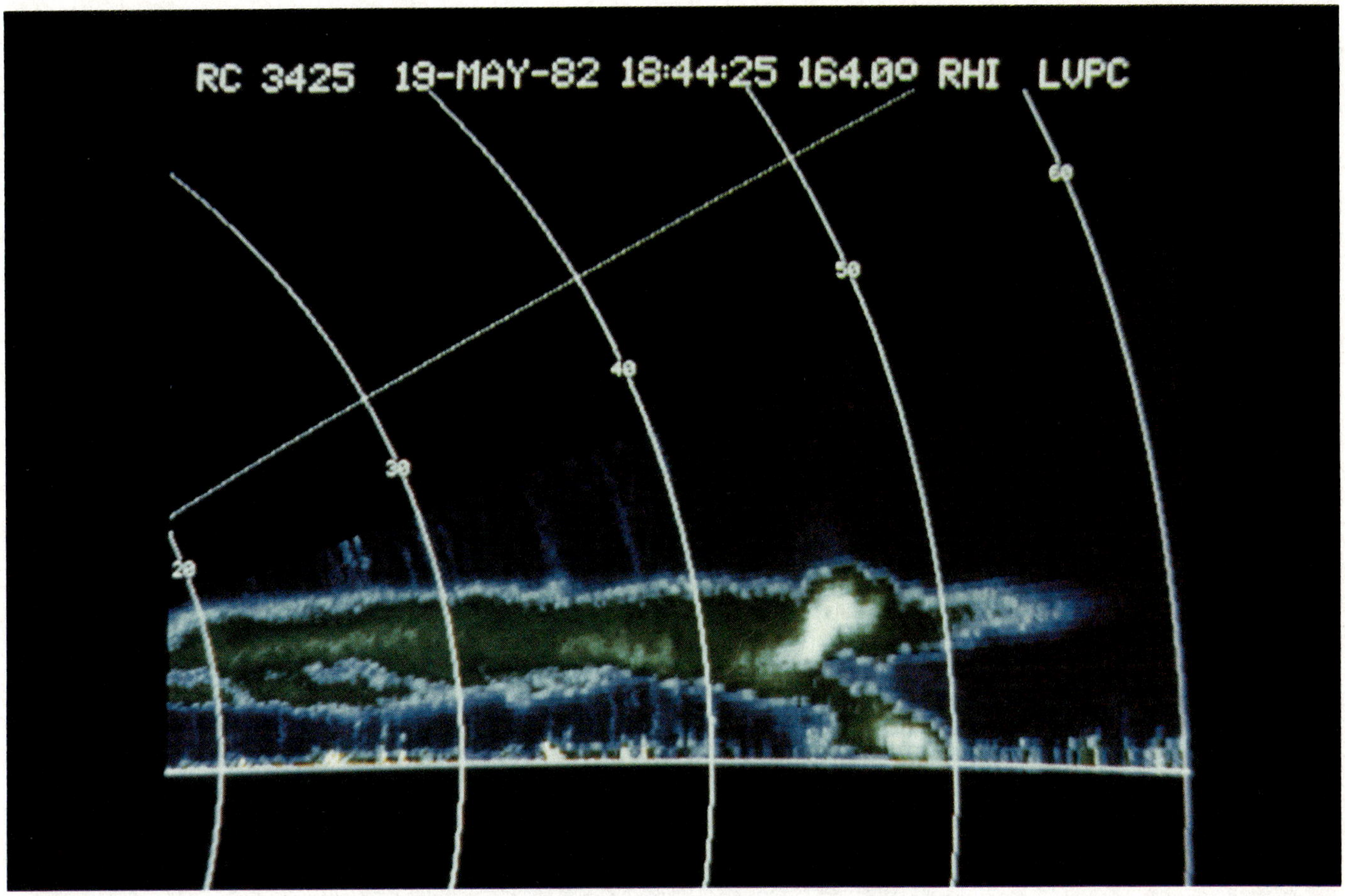

Figure VII-8-25c. RHI scan at 164° for storm in Figure VII-8-25a.

VII-8.6.2 Combining Infrared Imagery and Radar Data

The combined use of satellite IR and radar data is especially advantageous for monitoring severe storm development. Several investigations have been undertaken to diagnose thunderstorm intensity by using both types of data. Negri and Adler (1981) found that the location of satellite-detected thunderstorms coincided with radar echo locations and that radar reflectivity correlated with satellite-based estimates of intensity. There are similar findings on satellite defined thunderstorm tops and those defined by radar (Purdom et al., 1982; Green and Parker, 1983). Methods to combine satellite and radar data into a single meaningful product have been demonstrated by Reynolds and Smith (1979), who used a composite radar-satellite display to study severe storm development and convective rainfall. Later, Green and Parker (1983) used an interactive processing system to develop a composite satellite IR/Doppler radar image to relate radar echo changes to infrared cloud top behavior. Figure VII-8-24 shows 1-km and 15-km Constant Altitude PPI (CAPPI) radar scan echos mapped on a satellite infrared cloud top temperature field. This case has been studied using 3-min interval GOES IR data and 4-min interval CAPPI scan data at elevations from 1 to 17 km (Fujita, 1982a; Green and Parker, 1983). The findings point out some of the problems of combining satellite and radar data:

- When storms are forming in a line, the cloud top temperature downwind of the storms may appear higher than it actually is because of wake cirrus masking. Furthermore, satellite imagery at 3-min intervals is required to follow the evolution of these tops accurately and to observe the masking.
- The lowest satellite cloud top temperature in an anvil may not necessarily overlie the low level core of highest radar reflectivity.
- In measuring cloud top growth characteristics with satellite IR data, cirrus masking may make results less accurate for some storms. Radar CAPPI helps to match the correct IR and radar echo anvil areas.

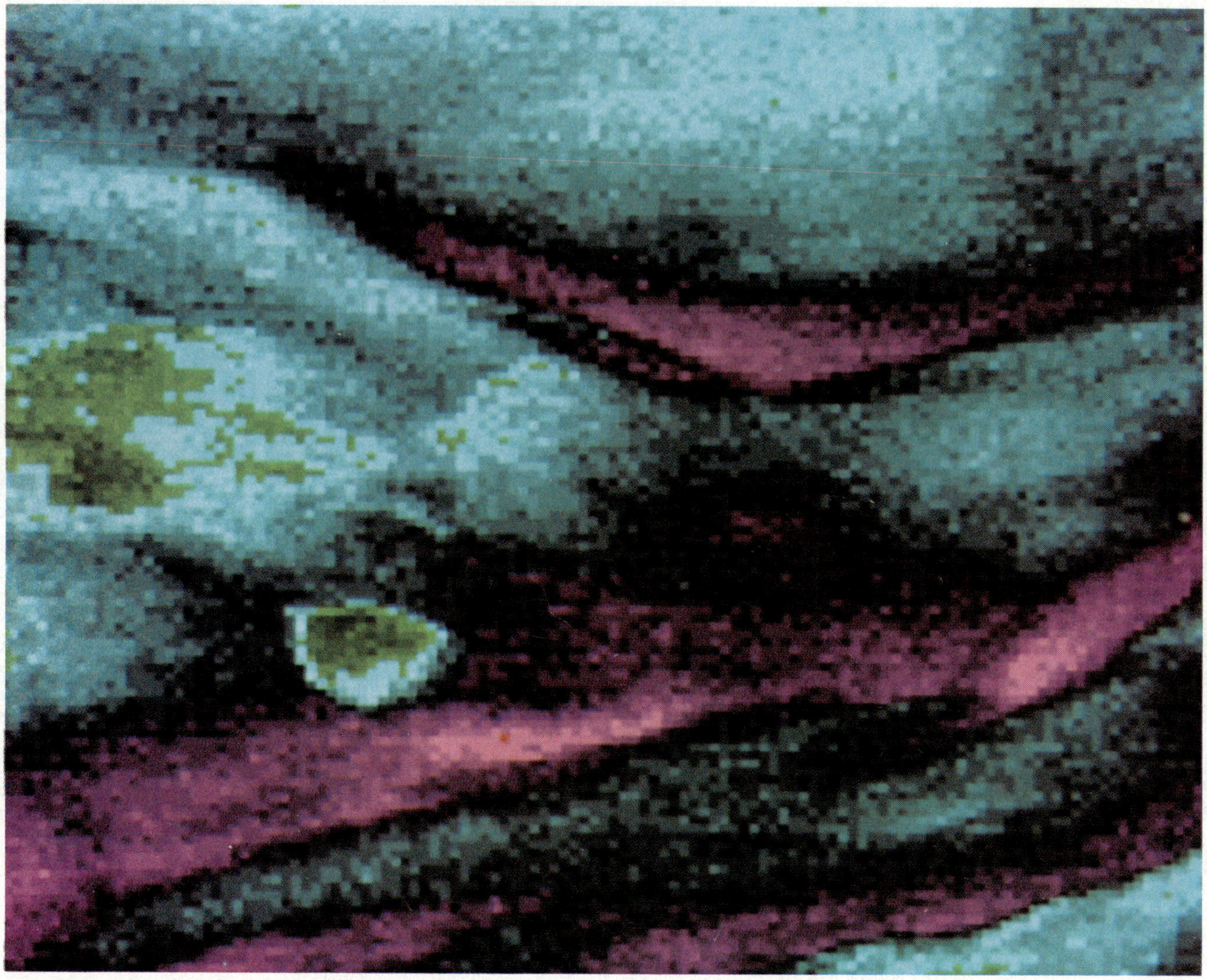

Figure VII-8-26. 6.7 μm WV image, 1730 UTC, July 22, 1981.

The problems of combining satellite and radar data become evident by inspection of Figure VII-8-25. Note that the strong echo return at upper levels for the 164° RHI scan matches the satellite IR cold cloud top area, while the maximum echo return at low levels (158° RHI) does not. This is because of vertical wind shear and the relationship of the radar echo at various altitudes to the dynamics of the storm. While the highest portion of a radar echo is correlated with the updraft, the lower portions of the echo correlate more closely with the downdraft and rainfall at the ground. Since vertical wind shear varies considerably from one type of thunderstorm regime to another, such information should be taken into account when satellite and radar data are combined.

VII-8.7 MULTISPECTRAL AND SATELLITE SOUNDING INFORMATION FOR MESOSCALE CONVECTIVE APPLICATIONS

VII-8.7.1 Moisture

The energy measured by each of the three GOES VAS WV channels (6.7, 7.3, and 12.7 μm) is influenced by the amount, distribution, and temperature of water vapor in the atmosphere, and by the surface and clouds. In the 6.7 μm channel data (Figure VII-8-26), which shows regions of middle level moisture and clouds, distinct patterns of moist areas (light tones) and dry areas (dark tones) are easily seen. These features, which are related to areas of synoptic and mesoscale advection

Figure VII-8-27. GOES-East 6.7 μm WV image and storm-relative flow, 2345 UTC, March 28, 1984.

and vertical motion, exhibit excellent spatial and temporal continuity in time-lapse films. Strong baroclinic regions (jet streams and vorticity maxima) often can be identified easily in cloud-free regions by sharp moisture gradients appearing in the 6.7 μm image (Anderson et al., 1982).

Figure VII-8-27 is a 6.7 μm WV image for the storm system of March 28, 1984. The dark region to the south of the storm system corresponds to the dry mid-level air in that region. The relative flow into the storm is from the SSE which indicates the likelihood of a sig-

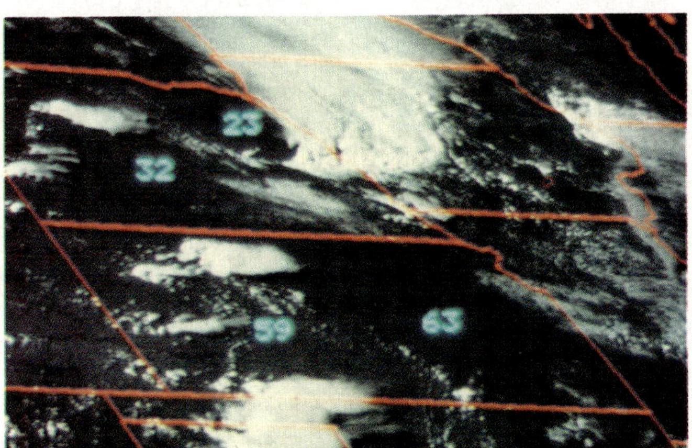

Figure VII-8-28. GOES-West 1-km VIS image, 2315 UTC, June 21, 1984. Numbers indicate locations of VAS soundings.

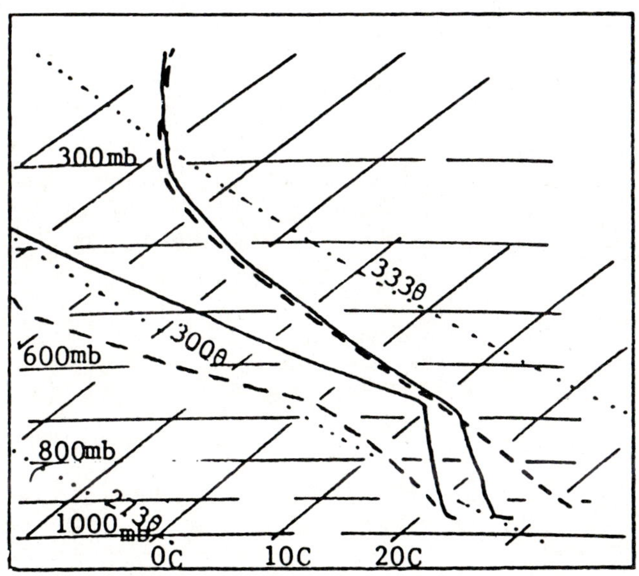

Figure VII-8-29. VAS soundings at locations 23 (T and Td solid lines) and 32 (T and Td dashed lines) in Figure VII-8-28.

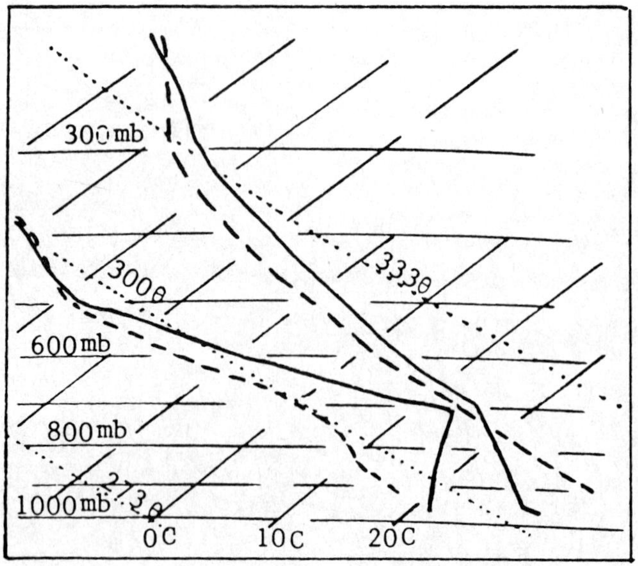

Figure VII-8-30. VAS soundings at locations 59 (*T* and *Td* dashed lines) and 63 (*T* and *Td* solid lines) in Figure VII-8-28.

nificant mid-level dry intrusion into the storm system. Such an intrusion could fuel the thunderstorm downdraft (outflow) and increase the possibility of tornado activity—a sort of forced dynamic-thermodynamic instability. The characteristic *V* notch at the thunderstorm top is detectable in this 6.7 μm image.

Except where there is significant low-level water vapor, the information in the 12.7 μm band (WV) is similar to the 11.2 μm window. Because of that factor, an interesting product may be derived from the difference in energy between the two bands. The product, called "split window" (Chesters et al., 1982), differentiates areas where water vapor extends over a deep layer and can support convection from those where water vapor is confined to a shallow layer and cannot support convection. This product is very useful over land during the afternoon when it is likely there will be a large difference in signals from the channels.

VII-8.7.2 Sounding Retrievals From VAS

GOES/VAS can detect small and significant temporal variations in atmospheric temperature and moisture (Smith et al., 1981). Initial research was concentrated on deriving vertical profiles of temperature and moisture (Smith, 1983). Significant results have also been achieved in deriving objective mesoscale convective forecast parameters from VAS data (Smith et al., 1984). VAS derived products provided in real time to the National Severe Storm Forecast Center in Kansas City include:

(1) upper and lower wind fields derived from VAS thermal field gradient winds and cloud drift winds
(2) analyses of lifted index
(3) 850- and 500-mb temperature fields
(4) total precipitable water
(5) a thermodynamic stability index
(6) statistical probability estimate of severe weather potential

Products (1), (2), (3), and (6) are prepared at 80-km resolution, products (4) and (5) at 7-km GOES/VAS resolution.

Although VAS derived products have proved useful in diagnosing the convective storm environment, there are still questions about the use of individual VAS soundings for such analyses. Figure VII-8-28 is a GOES image of the central United States on June 21, 1984: the numbers show the location of selected VAS soundings taken just prior to image time. Sounding 23 (within the cold dome) is representative of an air mass modified by thunderstorm outflow; sounding 32 is away from thunderstorm activity (Figure VII-8-29). Although the VAS information is presented for rather thick layers, the general characteristics expected are evident, i.e., there is a cooler and more stable low level airmass at location 23 than at 32. VAS sounding 63 is in cooler air behind a slow moving cold front moving westward across Kansas; sounding 59 is in the more unstable air west of the front (Figure VII-8-30). VAS soundings similar to these may be taken every hour over cloud free areas.

Chapter VII-9

USE OF SATELLITE IMAGERY IN SYNOPTIC-SCALE WEATHER ANALYSIS AND FORECASTING

VII-9.1 INTRODUCTION

The first observations from meteorological satellites were cloud pictures. These were put to use immediately and are still used daily for locating storms over ocean areas (see Chapters VII-7 and VII-10). To use cloud imagery successfully for analysis of synoptic-scale storms, two criteria must be met: (1) entire storms and their surroundings must be viewed approximately simultaneously, and (2) the frequency of observation must be no less than once every 12 hours and preferably once every 6 hours. The first criterion was met in 1966 with data from the Environmental Science Services Administration (ESSA) polar orbiters. In 1972, the NOAA series of polar orbiters provided images twice a day. In 1974 the GOES satellites increased the frequency of observations to every 30 min, permitting the production of time lapse animation of synoptic-scale weather systems.

Over the years much effort has been devoted to developing methods for converting imagery into quantitative values of height, thickness, pressure, moisture, and wind. This was the only way such information could be combined with surface, rawinsonde, and aircraft reports as input to numerical weather analyses. Manual techniques for obtaining quantitative information from cloud images were developed and used in the mid-1970s at the World Meteorological Centers in Washington and Moscow (Anderson and Veltishchev, 1973) and in Melbourne (Guymer, 1978). These manual efforts, in general, were discontinued as satellite soundings improved, cloud drift winds from geostationary satellites became operational, and objective analysis procedures of the numerical models became better able to ingest satellite soundings and wind data. Even so, IR, VIS, and, more recently, WV imagery continue to be powerful diagnostic tools for synoptic analysis, especially when animated.

VII-9.2 BASIS FOR APPLICATIONS OF SATELLITE IMAGERY

All applications of imagery to synoptic-scale analysis are based on the relationship between cloud organization and atmospheric parameters. Cloud pictures do not explicitly reveal the structure of the atmosphere; the translation of cloud features into conventional analysis parameters requires skilled interpretation by experienced analysts. Such interpretations are based on many studies that have identified qualitative and quantitative cloud-parameter relationships.

Early investigators systematically related cloud patterns to synoptic entities of the Norwegian frontal model, i.e., fronts, lows, jet streams, etc. These findings

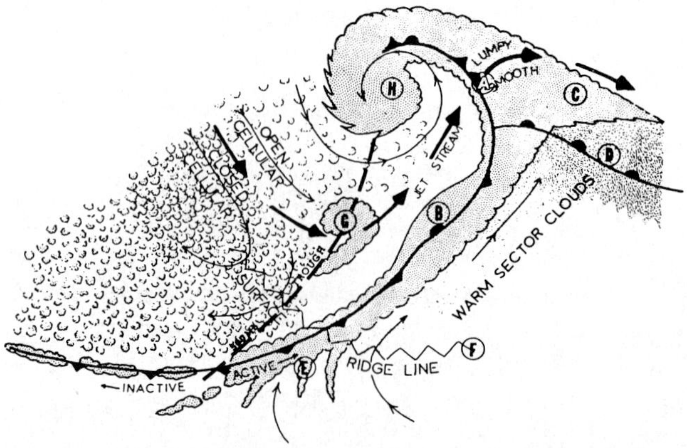

Figure VII-9-1. A schematic illustration relating surface and upper level flow, and surface frontal position to cloud formations observed in satellite visible pictures. [*A*] identifies the change in frontal cloud texture observed where the axis of the jet crosses a frontal band. [*B*] is a broadening of the frontal band indicative of a frontal wave which is forming in advance of a secondary vorticity center. [*C*] is an eastward extension of cirrus in the strong winds just north of the jet axis. [*D*] is the surface warm front position which can seldom be identified in cloud imagery. [*E*] identifies cloud bands merging into the front. These bands help define the position of the surface ridge line [*F*]. [*G*] is a comma-shaped cloud formation associated with a secondary positive vorticity center to the rear of the front. [*H*] identifies the primary cyclonic center (Anderson et al., 1966).

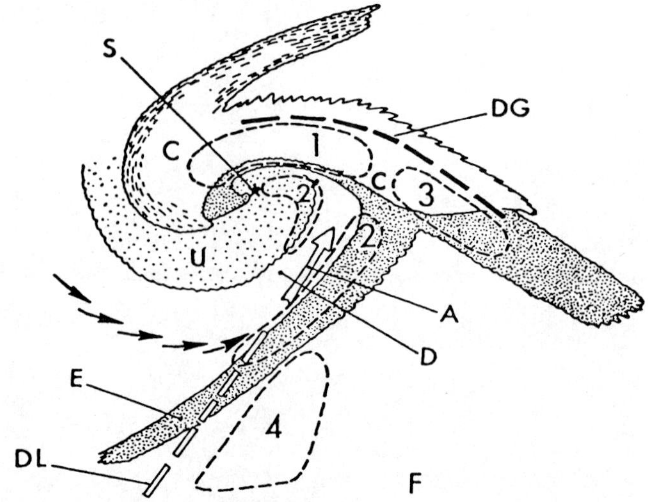

Figure VII-9-2. A developed storm cloud system over land. Arrow [*A*] is the upper-level maximum wind zone. Dashed line [*DG*] is a deformation zone in the basic upper air flow; it is transverse to the flow and immediately downstream from the wind maximum. The open dashed line [*DL*] is the the deformation zone in the basic flow which is upstream from the wind maximum and nearly parallel to the flow. Together these two features form a "*T*" pattern. The typical position of the surface low is at [*S*]. The area labeled [*C*] has high cold cirrus tops; the dark stippled area indicates lower middle clouds with cloud top heights decreasing southwestward along the band toward [*E*]. The [*U*] area is convective cloud. The numbers outlined by dashed lines locate areas where significant precipitation commonly occurs. Area 1 is diurnally persistent, stratiform type precipitation and is the most frequently observed location for significant precipitation. Areas 2 and 2′ are convective regions with diurnal maxima in the afternoon and evening. Area 3 is related to warm advection associated with the surface warm front. It usually consists of convection embedded in layered clouds and has a diurnal maximum in the late night and early morning hours. Finally, area 4 is a region of convection in the right rear quadrant of the jet. When low-level thermodynamic conditions are suitable, the convection is deep and diurnally persistent (Weldon, 1979).

were summarized by Anderson et al. (1966) and are shown in Figure VII-9-1. More recent analyses (Weldon, 1979; Carlson, 1980) have associated various features of the upper flow and precipitation regimes with different portions of the comma-shaped cloud system of the mature storm (Figure VII-9-2). The relationships between satellite observed clouds and features of the flow were developed empirically by comparing clouds with surface and upper-air observations over many years and have been used as aids in drawing surface and upper air charts over the oceans.

A next step was to develop models that related cloud patterns in VIS imagery to the stage of development of extratropical storms (Boucher and Newcomb, 1962; McClain and Brodrick, 1967; Anderson and Veltishchev, 1973). These models were modified for Southern Hemisphere use (Troup and Streten, 1972; Guymer, 1978). Troup and Streten also related surface pressure anomalies to the cloud pattern for each stage of storm development. Anomaly values were combined with mean sea level charts to obtain the central pressure of storms. Junker and Haller (1980) also developed an empirical relationship between central pressure and cloud-band curvatures. Figure VII-10-13 shows these pressure-cloud pattern relationships and a schematic for cloud pattern evolution derived from that of McClain and Brodrick (1967). Other models of storm development have also been studied. Anderson et al. (1974) described an occluded frontogenesis where the cloud system of a deepening low, poleward of the main polar front, merges with the frontal clouds to form the large scale spiral of a mature storm. Weldon (1975) modeled the structure and evolution of winter storms based on 30-min interval GOES IR imagery. These models relate storm cloud evolution to different upper-level flow and jet configuration. Weldon (1979) also identified and described a leaf-shaped formation of middle and high clouds that is often a precursor of surface cyclogenesis.

Nagle and Hayden (1971) developed a relationship between certain large scale cloud features and the zero relative vorticity contour at 500 mb. This relationship permitted the conversion of a cloud analysis into height information. The technique provided a systematic

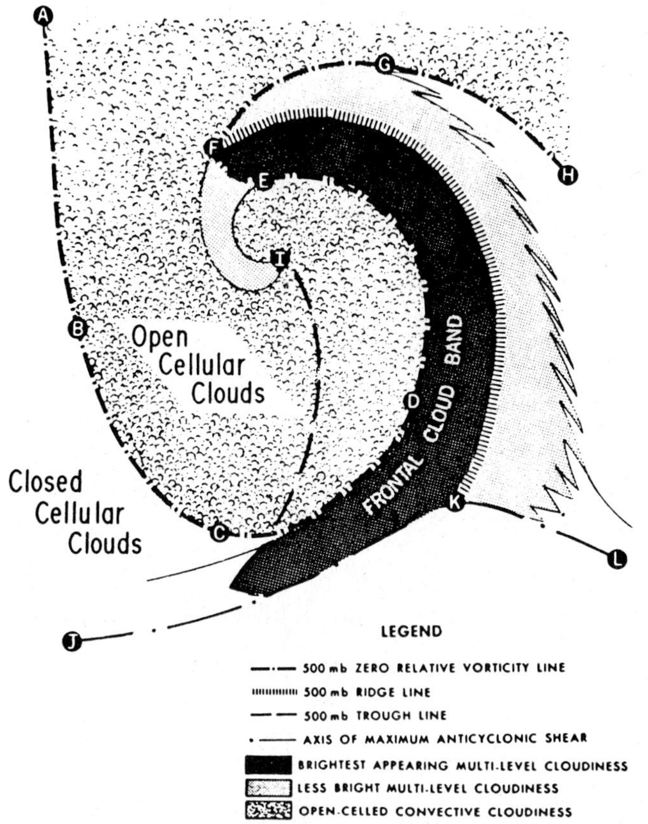

Figure VII-9-3. A schematic representation of the relationships between cloud patterns, 500-mb features and the 500-mb zero relative vorticity field obtained by a space smoothing analysis. The zero line lies along the boundary between open and closed cellular clouds in the cold air [*A* to *C*], along the trailing edge of the cold frontal cloud band [*C* to *E*], along the back edge of the brightest multilevel cloudiness [*E* to *F*], along the leading edge of the clouds [*F* to *G*], and then to [*H*] on the edge of the convective cellular cloud pattern. The center of the spiral cloud formation [*I*] corresponds to the positive vorticity maximum. The point marking the apex of the anticyclonic curvature along the leading edge of the frontal cloud band [*K*] corresponds to a maximum of negative vorticity. The leading edge of the frontal cloud band [points *J*, *K*, *L*] corresponds to the axis of anticyclonic shear (a ridge line in the relative vorticity field). The actual location of these features varies slightly from what is shown depending on the type of storm and stage of development (Nagle and Hayden, 1971).

means for introducing synoptic-scale disturbances seen in satellite images into the upper-air analyses. The procedure requires the separation of the 500-mb flow into its long- and shortwave components. The satellite derived zero relative vorticity line is used to modify the shortwave height field. The center of vorticity is defined by the cloud pattern and a height value is computed for this point. Figure VII-9-3 shows the cloud pattern and 500-mb vorticity relationships used in this technique.

A method for using imagery to deduce 1000- to 500-mb thermal patterns was described by Guymer (1978). Satellite cloud patterns are used to define a departure from the normal 1000- to 500-mb thickness field at 4-Dm intervals (Figure VII-9-4). The quantitative thickness anomaly field is combined with maps of climatological thickness values to provide numerical data for synoptic scale analysis. Guymer describes in detail the relationship between clouds and deep-layer thermal patterns.

Static images can be interpreted in terms of wind speed and direction by the appearance of cloud organizations and patterns. Cloud line structures, the effects of orography on clouds, and Sun-glint patterns all can be used to deduce surface flow. Jet stream cirrus formation and cirrus anvils are helpful in upper level wind analysis (Anderson and Veltishchev, 1973; Chapter VII-10).

The study of images in motion shows that deformation zones in the mid-tropospheric relative flow are important for explaining the production and organization of upper level cloud bands and boundaries. Cloud systems and their evolution have been found to be highly correlated with wind speed maxima and minima, the rotational field, and the deformation fields. An important concept developed by Weldon (1979) is that horizontal and vertical motions within the atmosphere are more important in defining cloud systems than the motion of the atmosphere with respect to the ground.

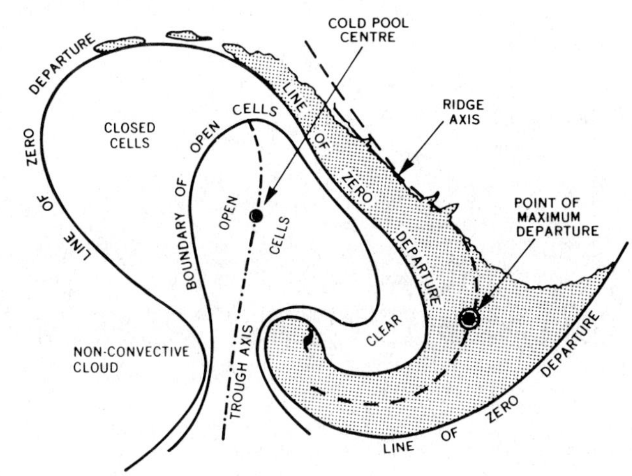

Figure VII-9-4. An analysis of a Southern Hemisphere storm in terms of cloud features which define four primary aspects of the 1000- to 500-mb thickness anomaly pattern. These are: 1) the line of zero thickness departure defined by the edge of the major spiral cloud band and by the trailing edge of the closed cellular cumulus; 2) the boundary of open cellular cumulus which outlines the deep cold air (−4 Dm thickness anomaly); 3) the cold pool center (−8 to −24 Dm anomaly depending on organization and cell size of cumulus); and 4) the thermal ridge axis and the point of maximum departure indicated by the major cloud spiral band (+4 to +16 Dm anomaly depending on cloud band characteristics). Courtesy of the Bureau of Meteorology, Australia; Zillman and Price, 1972.

Figure VII-9-5. GOES-West VIS image, 2145 UTC, September 21, 1982.

Since the 500-mb vorticity analysis represents the relative motion in the mid-troposphere, it is possible to use operationally produced charts to relate clouds to upper flow for diagnostic purposes and to forecast cloud pattern evolution by use of the predicted vorticity fields (Funk, 1982).

Criteria have been developed for estimating the mean relative humidity (RH) from the surface to 500 mb by means of manual analysis of cloud types, cloud combinations, and synoptic-scale cloud features (Smigielski and Mace, 1970). They defined 13 categories of cloud conditions that represent different profiles of vertical relative humidity. Vertical arrays of RH values can be estimated for use in numerical analysis by this process. Currently this technique is used to derive quantitative moisture data from satellite imagery for the NMC analyses.

VII-9.3 EXAMPLES OF APPLICATIONS OF IMAGERY

VII-9.3.1 Location and Identification of Weather Systems Using Imagery

With satellite imagery available, very few significant weather systems remain undetected, even over vast

Figure VII-9-6. GOES-West IR image, 2315 UTC, September 21, 1982.

data sparse regions. The images from GOES-West in Figures VII-9-5, VII-9-6, and VII-9-7 span a period of 2½ hours. They show the Pacific Ocean, with North America to the northeast and New Zealand near the southwest horizon. The figures illustrate the three most common types of imagery currently available for operational synoptic scale analysis. Numerous weather systems seen in these images may be identified by type, behavior, and relative strength through examination of the characteristics of their cloud and moisture patterns.

The long band of cloudiness north of Hawaii is associated with a frontal zone. In Figure VII-9-5 (VIS data), the band is wide and relatively bright, indicating thick clouds. This information, in combination with the cold cloud tops shown on the IR images, indicates that the frontal zone is vertically deep and is associated with an upper air trough and jet stream. Also present over the north Pacific are a "cutoff" low pressure system well west of California, a strong tropical cyclone west of Mexico, and a weak tropical weather circulation southwest of Hawaii. Identification of the cutoff cold

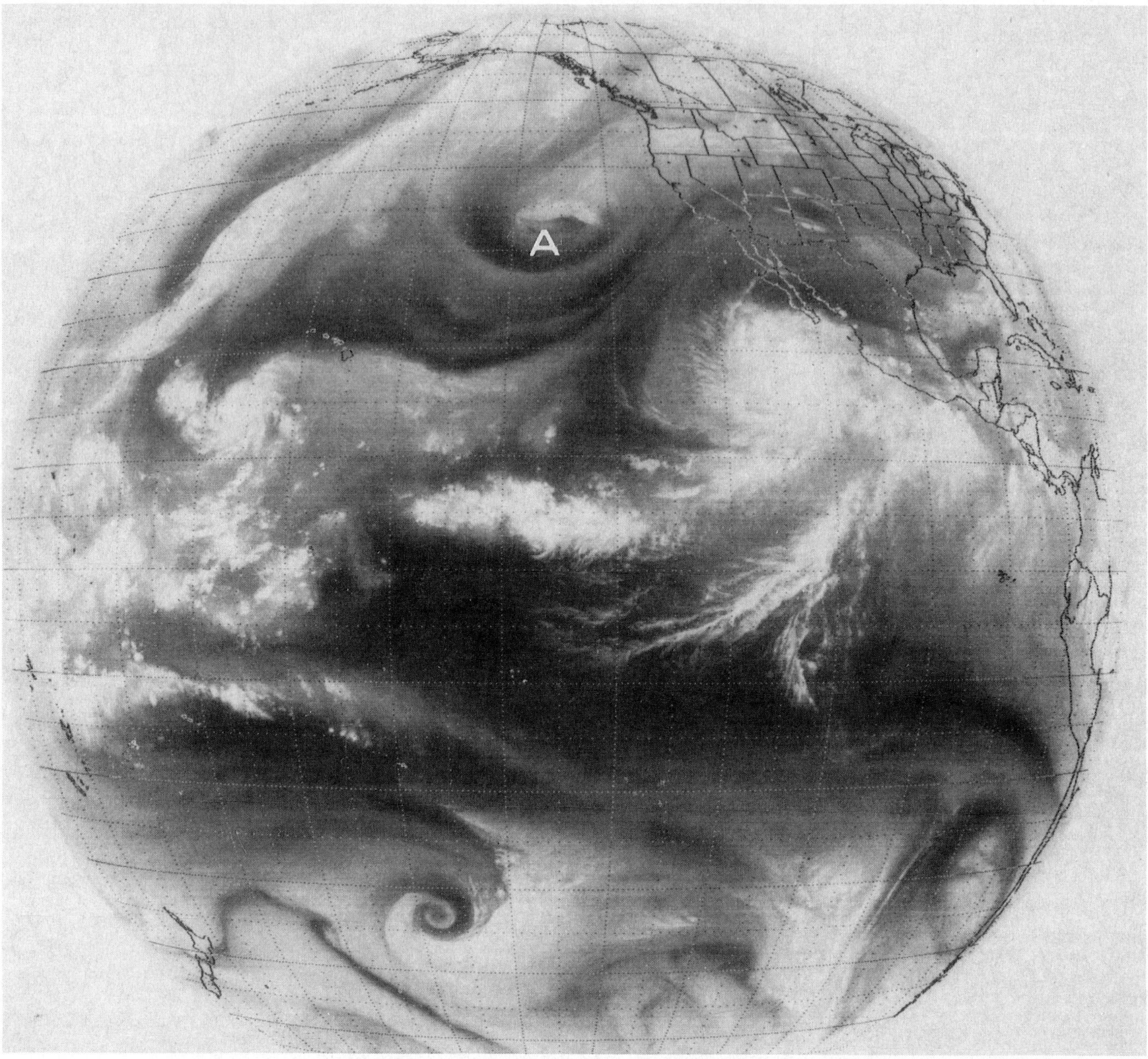

Figure VII-9-7. GOES-West 6.7 μm WV, 0015 UTC, September 22, 1982.

low (*A*) is easier when viewed on the 6.7 μm WV image of Figure VII-9-7. Since the 6.7 μm radiation is very easily absorbed by small quantities of moisture, the mid- and upper-tropospheric moisture patterns are revealed when no high cloudiness is present. The large region of high (cold) cloudiness and moisture associated with the tropical cyclone west of Mexico can be seen in both the IR and WV images. On the VIS image the bright clouds of this system show as a much smaller area. Significant precipitation is likely to be occurring in this area of thick cloudiness.

A well defined mid-latitude storm system is centered over the South Pacific near 32°S, 148°W. The spiral of alternating light and dark bands near the center, seen in the WV image, indicates that the system has a closed wind field in both the middle and upper troposphere. The cloud band curved around the poleward side of the storm is both bright (in the VIS image) and cold (in the IR image); this indicates a deepening storm in which the central pressure is likely to continue falling. Such a determination, although relatively reliable, is better estimated with a time sequence film loop.

Considerably more information is available from satellite imagery when it is observed in time-lapse motion. Observation of cloud elements and moisture parcels in motion permits easier identification of weather systems and their changes, and reveals low cloud circulations often difficult to detect in individual images.

VII-9.3.2 Detection of Clouds and Weather

Satellite images are useful for the detection of cloud conditions and other weather phenomena even when conventional observations are relatively dense. Imagery provides spatial continuity which improves one's ability to interpret conventional reports. Images also are available from geostationary satellites at more frequent intervals than routine weather reports, and are especially useful for observing cloud and weather conditions over mountains and other data sparse regions.

Figures VII-9-8 and VII-9-9 show cloudiness associated with frontal zones over the northeast Pacific Ocean. The cloud band of the frontal zone of February 17, 1982, (Figure VII-9-8) is very wide, and is vertically deep with a strong parallel jet stream and a deep cold upper-air trough on its west side. The cloudiness is bright on the VIS image, and has cold tops as indicated in the IR image. Such frontal zones produce relatively heavy precipitation for rather long periods as they pass over a particular location. Note in Figure VII-9-8a the long, thin, bright line of clouds on the southeast side of the cloud zone; this most likely marks the position of the surface front. The wide zone of overcast cloudiness and precipitation is on the cold side of the surface frontal position. By use of the satellite image, a surface ship about to encounter the frontal weather could be warned to expect cloudy weather and precipitation for some time after the frontal passage.

The front shown in Figure VII-9-9 is also well-defined and strong, but relatively shallow. The jet stream, with the deep-cold air on its left side, is not parallel or along the surface front as in the February 17, 1982, case. The IR image (Figure VII-9-9b) shows that the frontal zone clouds are relatively warm. A ship encountering the October 23, 1977, front would experience showers along and subsequent to frontal passage, but neither prolonged precipitation nor overcast conditions afterwards.

VII-9.3.3 Analysis of Upper-Air Circulation Systems Using Imagery

Satellite imagery may be used in combination with rawinsonde reports, aircraft reports, and other forms of satellite data to improve or correct upper-air analyses. Since geostationary satellite imagery is routinely available at 30-min intervals, it can be used to follow the movement of upper-air circulation features between rawinsonde reporting times and is particularly useful during situations when systems are changing rapidly.

Figures VII-9-10, VII-9-11, and VII-9-12 show 6.7 μm WV and IR images from GOES-West during 24 hours on December 4-5, 1981. Streamlines on the IR images are from 250-mb radiosonde and aircraft reports. Streamlines are not shown where wind reports are insufficient for reliable analysis.

In Figure VII-9-10a, a 6.7 μm WV image valid at 0015 UTC, December 4, 1981, (*J*), (*D*), and (*S*) are samples of three important types of boundaries commonly observed on WV imagery. As in the case of IR imagery, patterns on the 6.7 μm imagery do not represent any single level or surface. Upper-air circulation systems, however, especially those associated with the polar jet stream zone, extend through a relatively deep vertical layer. Therefore, the patterns on the WV imagery and their changes correlate well with the upper-air circulation systems and their changes. Boundary (*J*) on Figure VII-9-10a is located near the axis of a strong polar jet stream. Although data supported this relationship at the time, aircraft wind reports 12 hours later further confirmed the jet stream location. Boundary (*D*) lies along the "deformation zone," or stretching axis of the hyperbolic zone, formed by the flow turning northward around the closed anticyclone (*A*) and the flow turning southward along the closed cyclone (*C*) (Figure VII-9-10b). Such boundaries are commonly observed with upper-air anticyclogenesis, as the circulation along an upper-air ridge closes to form the anticyclone.

Boundary (*S*) extends across Hawaii northeastward to the base of the high amplitude upper-air trough. At the time of Figure VII-9-10a, aircraft enroute between Hawaii and the United States west coast encountered mainly southwesterly winds ranging to over 100 kt near the center of the route. An exception: near boundary (*S*) south of 25°N, the winds are light northeasterly.

During the 24 hours between Figures VII-9-10 and VII-9-12, the strong polar jet stream extended rapidly eastward. The anticyclone south of the jet stream elongated to the east-northeast, and boundary (*S*) extended northeastward along the "shear zone" on the south side of the anticyclonic flow. At the time of Figure VII-9-12a, winds, along the dark (warm) region along boundary (*S*) at 250 mb, range from 40 to 60 kt from the northeast or east-northeast. Winds just south of boundary (*S*) are westerly. Aircraft enroute from the Hawaiian Islands to the United States experienced headwinds during most of their flights.

These examples illustrate how imagery provides time and spatial continuity to the upper-air wind reports and additional understanding of the dynamics con-

Figure VII-9-8a. GOES-West VIS image, 2245 UTC, February 17, 1982.

Figure VII-9-8b. GOES-West IR image, 2345 UTC, February 17, 1982.

trolling changes in the upper-level flow. To an experienced observer, the imagery alone may be used for a qualitative upper-air analysis. When used in combination with aircraft or radiosonde reports, it becomes a means to follow features between upper-air observation times and to locate the features where conventional data are missing.

VII-9.3.4 Detection of Changes of Weather Systems

Although the movement of weather systems is important in weather forecasting, an equal or greater problem is variation in weather and circulation patterns as they move. Because satellite imagery provides excellent temporal and spatial continuity, rapid de-

Figure VII-9-9. (a) GOES-West VIS image, 2245 UTC, October 23, 1977. (b) GOES-West IR image, 2345 UTC, October 23, 1977.

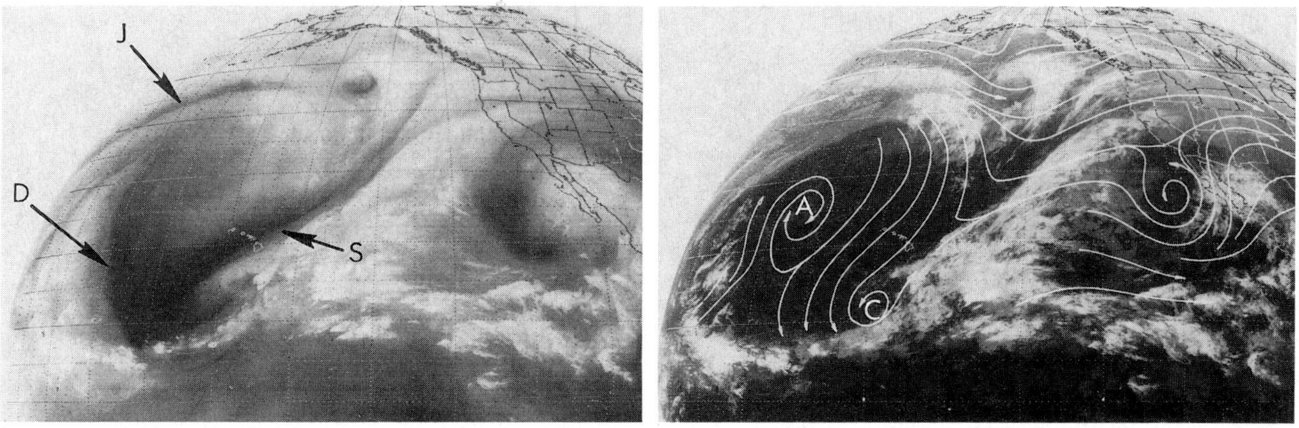

Figure VII-9-10. (a) GOES-West 6.7 μm WV image, 0015 UTC, December 4, 1981. (b) GOES-West IR image, 0045 UTC, December 4, 1981, streamlines of 250-mb winds.

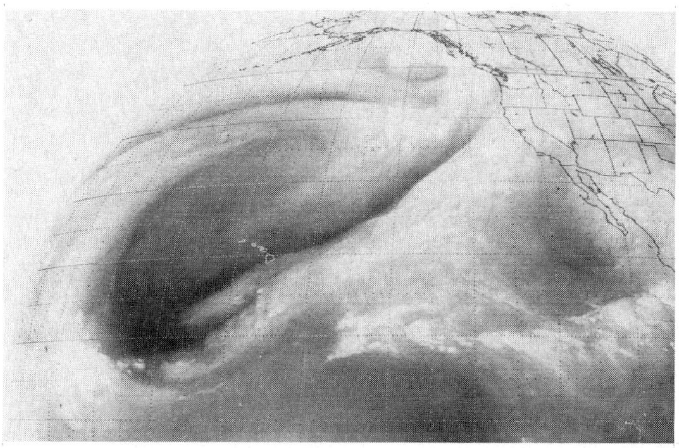

Figure VII-9-11a. GOES-West 6.7 μm WV image, 1215 UTC, December 4, 1981.

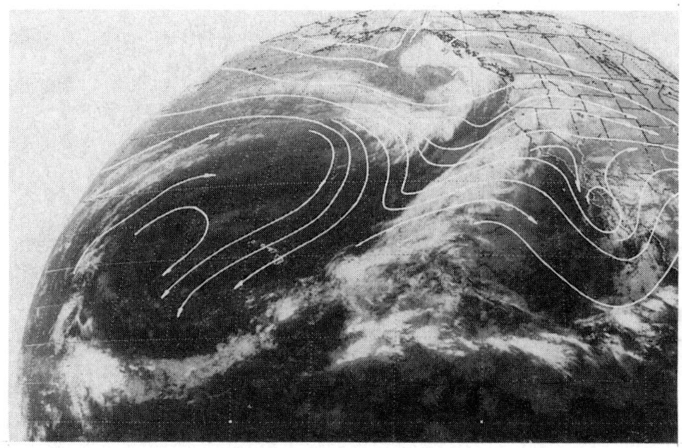

Figure VII-9-11b. GOES-West IR image, 1145 UTC, December 4, 1981, streamlines of 250-mb winds.

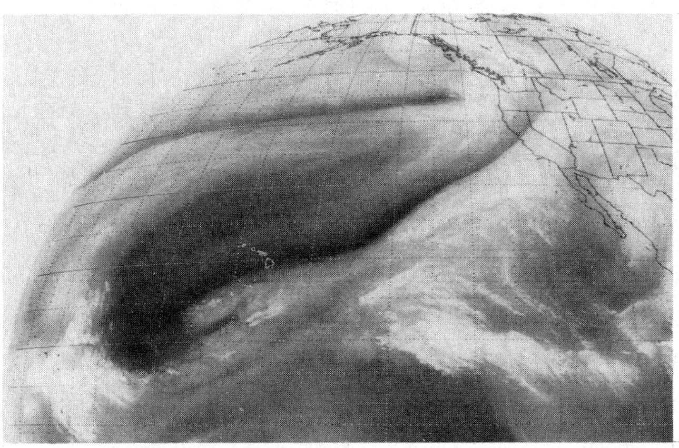

Figure VII-9-12a. GOES-West 6.7 μm WV image, 0015 UTC, December 5, 1981.

Figure VII-9-12b. GOES-West IR image, 0045 UTC, December 5, 1981, streamlines of 250-mb winds.

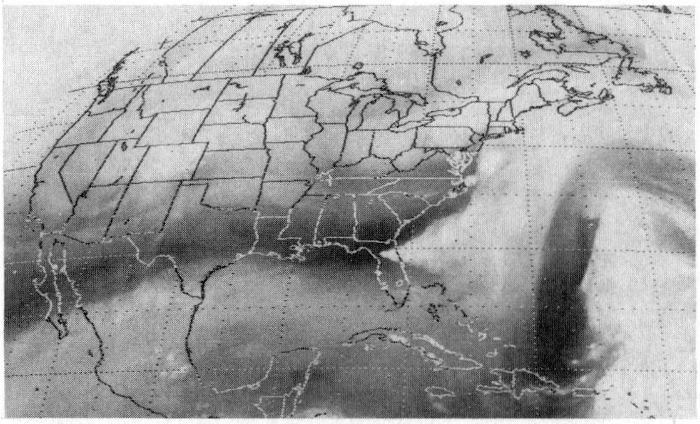

Figure VII-9-13a. GOES-East 6.7 μm WV image, 2300 UTC, May 16, 1983.

Figure VII-9-13b. GOES-East IR image, 0000 UTC, May 17, 1983.

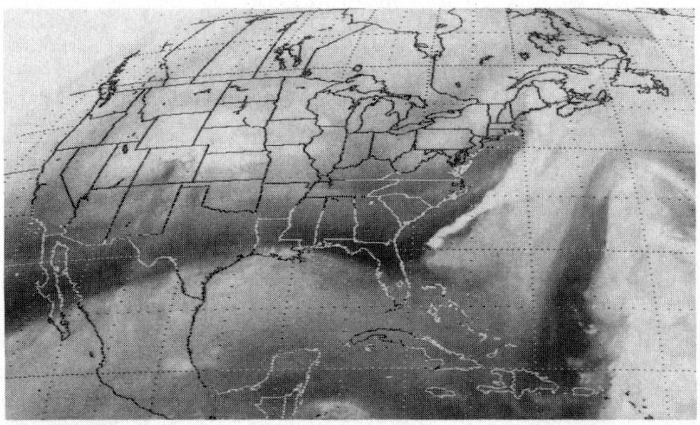

Figure VII-9-14a. GOES-East 6.7 μm WV data, 0500 UTC, May 17, 1983.

Figure VII-9-14b. GOES-East IR image, 0600 UTC, May 17, 1983.

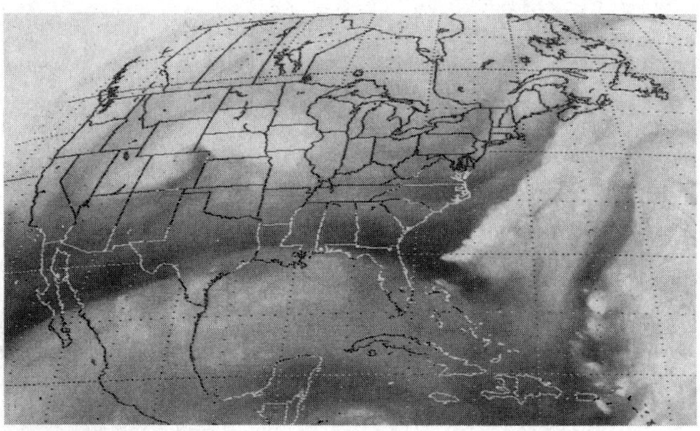

Figure VII-9-15a. GOES-East 6.7 μm WV image, 1100 UTC, May 17, 1983.

Figure VII-9-15b. GOES-East IR image, 1200 UTC, May 17, 1983.

velopments and changes can be observed in the intervals between radiosonde reports and computer forecasting runs.

One category of change that is a problem for computer model forecasting is the effect of intense convective weather on the upper-air wind and density

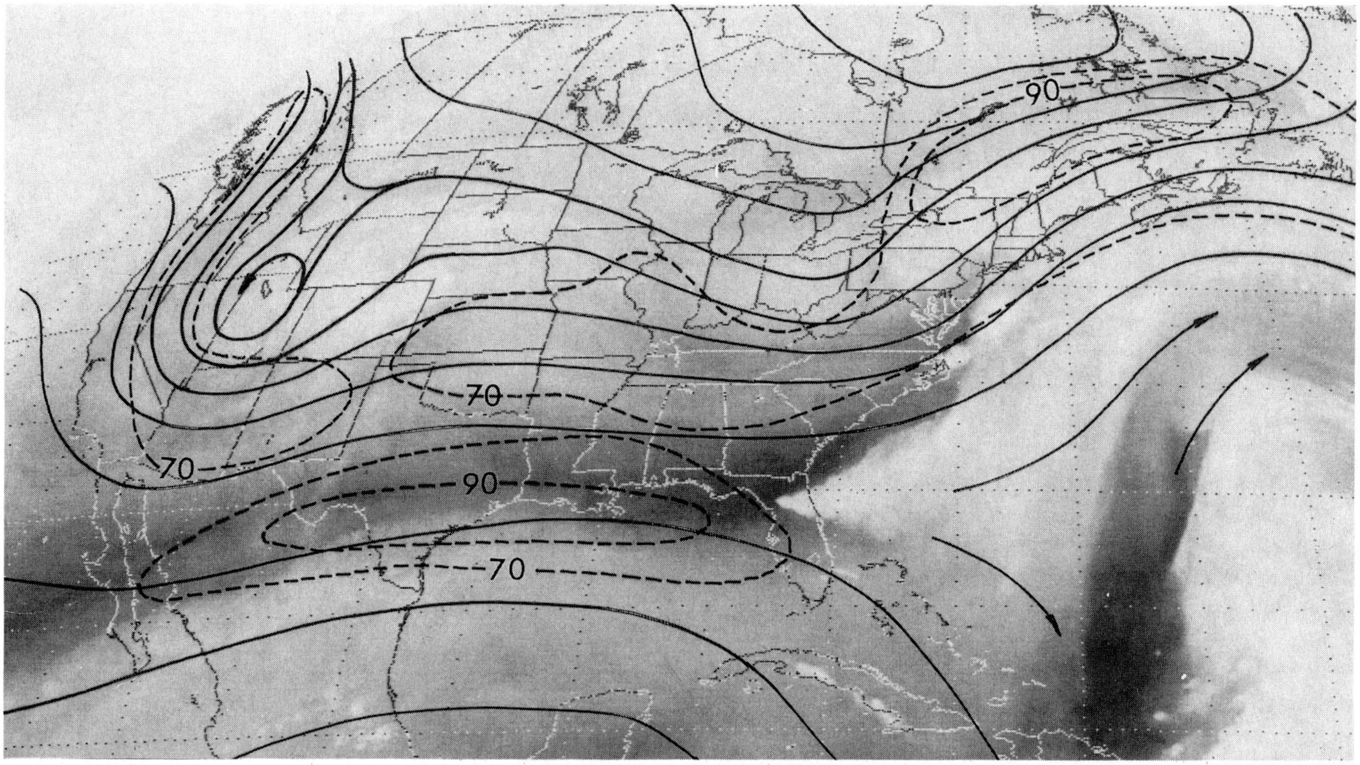

Figure VII-9-16. GOES-East 6.7 μm WV image, 2300 UTC, May 16, 1983. Overlay of 300-mb streamlines (solid lines) and isotachs (dashed line for 70 and 90 kt) from rawinsonde data at 0000 UTC, May 17, 1983. Same image as Figure VII-9-13a.

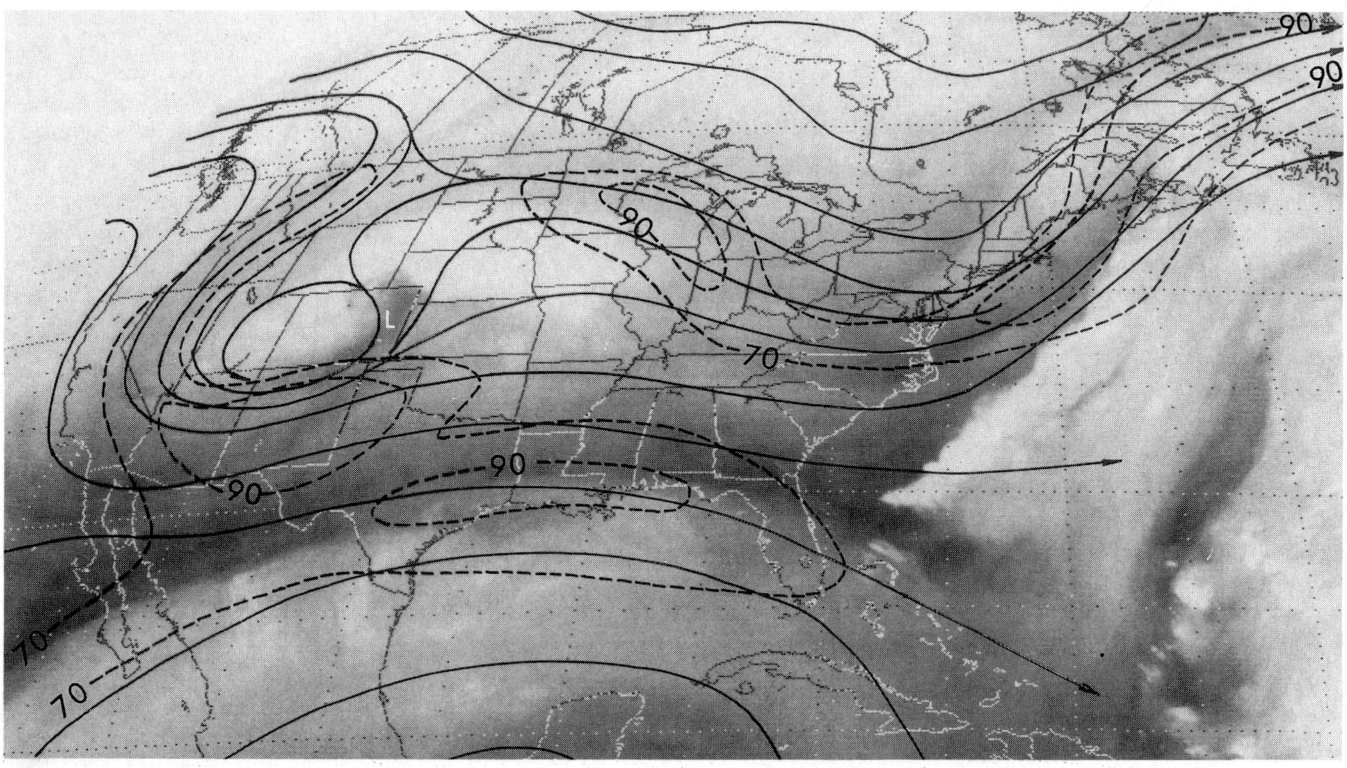

Figure VII-9-17. GOES-East 6.7 μm WV image, 1100 UTC, May 17, 1983. Overlay of 300-mb streamlines (solid lines) and isotachs (dashed lines for 70 and 90 kt) from rawinsonde data at 1200 UTC, May 17, 1983. Same image as Figure VII-9-15a.

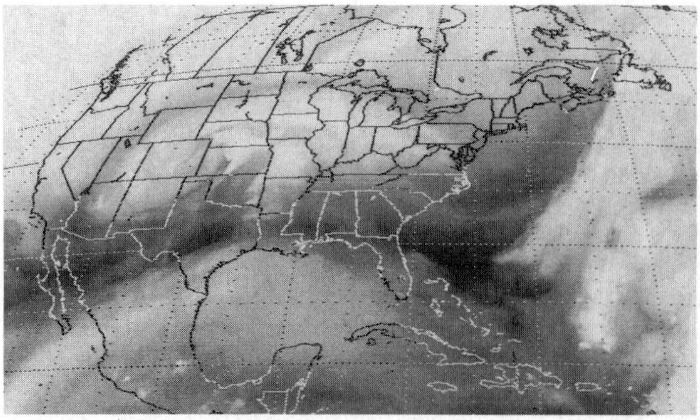

Figure VII-9-18a. GOES-East 6.7 μm WV image, 2300 UTC, May 17, 1983.

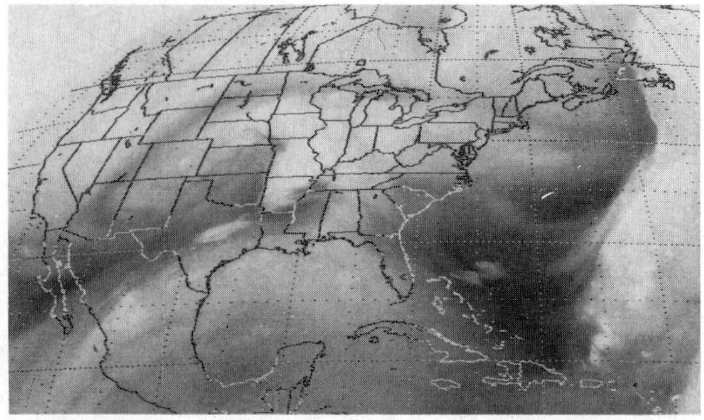

Figure VII-9-19a. GOES-East 6.7 μm WV image, 1100 UTC, May 18, 1983.

fields. This effect is particularly significant during the summer when temperature and height gradients are relatively weak, but it can also be significant during other seasons whenever convective weather is intense or rapidly developing. Figures VII-9-13 through VII-9-19 show an example of such a change.

Figures VII-9-13, VII-9-14, and VII-9-15 show 6.7 μm WV and IR images from GOES-East during 12 hours on May 16-17, 1983. The WV images for 2300 UTC, May 16, 1983, and for 12 hours later at 1100 UTC, May 17, 1983, are repeated in Figures VII-9-16 and VII-9-17 with 300-mb streamlines and isotachs superimposed. The wind data are those observed 1 hour later than the WV images.

Note the 300-mb wind field at 0000 UTC, May 17 (Figure VII-9-16). The jet stream is northerly over the northwestern United States, and southwesterly over the northeastern states, with a broad complex dual jet stream extending to the east from Arizona and northern Mexico. There is a distinct moisture boundary across Mexico and along the northern coast of the Gulf of Mexico near the axis of the southernmost of the two speed maxima across the southern United States. At 300-mb, the wind speeds along the southern maximum are greater than those farther north, but at higher altitudes—between 250- and 200-mb—the strongest winds are along the northern maximum.

During the 12 hours prior to Figure VII-9-16, a large region of convective weather formed on the east side of the western trough. The convection area produced a shortwave upper-air ridge in the region. Maximum wind speeds over the ridge are from 40 to 50 kt. During the 12 hours between 0000 UTC and 1200 UTC on May 17, the area of the convective cloud region increases slightly, the cloud tops become colder, and distinct borders form. By 1200 UTC, a "dry slot" forms at (*L*) (Figures VII-9-17 and VII-9-15b). The 300-mb wind speeds have increased significantly along the northern side of the convective cloud shield and downstream. During the 12 hours between the analyses, temperatures increased in the shortwave ridge warmed in the middle troposphere—primarily between 500- and

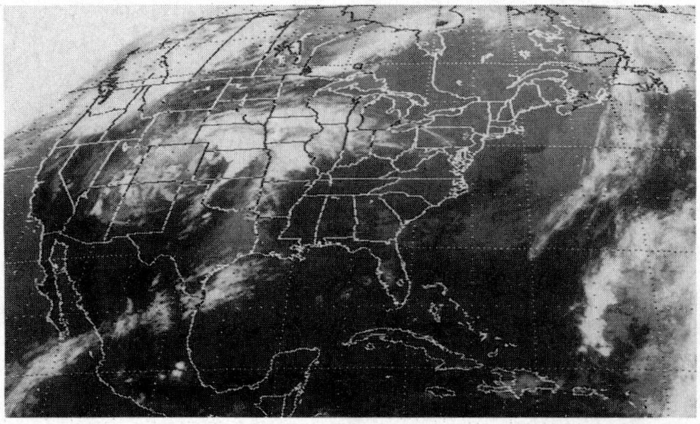

Figure VII-9-18b. GOES-East IR image, 0000 UTC, May 18, 1983.

Figure VII-9-19b. GOES-East IR image, 1200 UTC, May 18, 1983.

300-mb over eastern Nebraska and Iowa. Above, near 200-mb, significant cooling has occurred. The warming has produced larger temperature gradients and increased 300-mb wind speeds to its north, with smaller temperature gradients and decreased 300-mb wind speeds to its south where the northern branch of the dual jet stream had been 12 hours previously. Strong winds have been maintained above 200 mb, south and east of the cooling at 200 mb.

Imagery for 12 hours later, at 0000 UTC, May 18, 1983, is shown on Figure VII-9-18. Note that the northern border of the convective region has become even more strongly defined and extends eastward across the northeastern United States. At this time, the southernmost jet stream along the Gulf of Mexico coast and Florida has been maintained, but the strongest wind speeds are now along the newly formed northern jet stream. At 0000 UTC on the 17th, the largest 500-mb temperature gradients were over the southeastern states. At 0000 UTC, May 18, 1983, the maximum 500-mb temperature gradient is associated with the newly formed jet stream over the northeastern states (Figure VII-9-18). By 1200 UTC on the 18th (Figure VII-9-19), the effect of the new jet stream can be seen on the frontal zone cloudiness over the Atlantic Ocean.

Forecasters observing the satellite imagery between 0000 and 1200 UTC on the 17th (Figures VII-9-13b and VII-9-15b) would have been alerted to the impending change in the upper-air wind field as the intense convective region formed.

Chapter VII-10

AVIATION AND MARINE APPLICATIONS[1]

VII-10.1 INTRODUCTION

Meteorological satellites provide valuable information for specialized aviation and marine forecast applications. Satellite derived winds and temperature soundings are used in numerical analysis and prediction. This satellite information, therefore, reaches aviation and marine interests through the standard numerical guidance that is used in the preparation of aviation and marine forecasts and advisories. VIS and IR imagery, particularly from geostationary satellites, are used directly in providing the latest information on the location and development of enroute and terminal cloudiness, fog, convective storms, and atmospheric aerosols. In addition, for aircraft operations, the imagery can provide some useful inferences concerning the jet stream and turbulence. For marine interests, in particular, the imagery provides details of the motion and development of oceanic, synoptic, and mesoscale systems and their cloudiness that are unavailable from any other data sources.

VII-10.2 FOG AND STRATUS

Visible imagery is most often used to identify and monitor the dissipation of fog and stratus. IR imagery is used at night to observe fog and stratus formation. Animation of the satellite imagery aids in following changes in fog and stratus and in distinguishing fog from stratus (Gurka, 1975).

[1] Primary Contributor: Gary P. Ellrod

VII-10.2.1 Identifying Fog and Stratus Over Land

Fog in mountainous terrain often has a unique dendritic signature, easily identified in VIS images. If the fog is not confined to the valleys, however, the appearance is so like other cloud cover that it cannot be easily identified. Careful observation of some details can enable the analyst to distinguish fog and stratus from other clouds. For example, the tops of fog and stratus are smooth so that early morning VIS imagery shows neither the shadows nor highlights common among high or convective clouds. Fog and stratus are most easily identified in animated geostationary imagery because they move more slowly and often in a different direction than other clouds in the imagery. The lack of daytime development also is a clue to distinguishing fog and stratus from other clouds.

Figure VII-10-1a, an early morning VIS image over the eastern United States, shows the dendritic signature of the fog-filled Connecticut River Valley bordering New Hampshire and Vermont (*A*). Along the Appalachian Mountains in Pennsylvania (*B*), the valley fog appears elongated and conforms to the orientation of the valleys. Farther to the south (*C*), the river valleys do not confine the fog and stratus, so their appearance is similar to that of widespread cloud cover.

Enhanced VIS images are used to identify fog in the early morning hours (Gurka, 1978a; 1978b). For example, in Figure VII-10-2a, a standard VIS image, no fog is evident at *A*; in Figure VII-10-2b, however, an enhanced version of this image, the valley fog becomes evident.

Although IR images are also useful for detecting fog and stratus, they have limitations. In many cases, the temperature differences between the land or water surface and the fog or stratus tops are minimal and cannot be seen in standard IR imagery. Gurka (1980) developed an enhancement curve in the IR to improve discrimination in the higher temperature ranges. As long as there is a measurable temperature difference, a noticeable discontinuity can appear at the edges of the fog or stratus. Such discontinuities are most easily identified with animated geostationary imagery. Diurnal variations in surface temperatures and low level inversions may result in the cloud top temperature being higher, lower, or the same as the surface temperature. When there is a strong inversion at lower levels in the atmosphere, the fog and stratus tops are warmer and appear darker than the surrounding surface (referred to as "black stratus" [see VII-1.4]). Fog and stratus decks appear as a smooth, uniform gray shade since there is little cloud top temperature variation. A cloudfree surface, on the other hand, appears textured because of the effects of terrain features and boundaries between land and water. Multiple IR channels are also useful in detecting fog and low stratus. Eyre et al. (1984) developed a technique that employs color enhanced images based on differences between AVHRR Channels 3 and 4.

VII-10.2.2 Fog and Stratus Formation

Enhanced IR imagery can be used to identify locations of probable nighttime fog formation along coastal areas (Gurka, 1980). Soon after sunset, clear, moist areas often appear relatively darker (warmer) than surrounding clear, dry areas in the enhanced IR. Coast lines and reservoirs are generally discernible within the moist region. If the moist air is moving, its leading dark edge may be tracked inland in successive imagery. Changes in dewpoint temperatures are observed at surface stations with passage of the boundary. Gurka (1980) observed such dark regions extending from along the Gulf coast to far inland over the Central Plains.

As fog and stratus form, surface features become obscured. When these clouds form under an inversion, the dark areas often become more pronounced and the boundaries become better defined. The gray shade contrast at the edge of fog and stratus increases as the temperature difference between the clear, moist area and the cloudy area increases.

VII-10.2.3 Fog and Stratus Dissipation

Examples of the dissipation of fog and stratus are shown in Figure VII-10-1. Rapid dissipation of fog in valleys occurs at *A* and *B*. Where fog and stratus cover a large region such as *C*, less bright areas dissipate first. Gurka (1978a, 1978b) developed a technique to estimate fog dissipation time using enhanced VIS brightness differences within the cloud. This empirical scheme permits estimation of time of dissipation for any location within the fog area.

Gurka (1978c) made other studies of regional fog dissipation using animated geostationary imagery. He noted that an area of even brightness dissipates from the outer edges inward with calm or light winds; nonuniform dissipation occurs with increasing wind speeds. In the case of moving stratus, the upwind boundary appears to move faster than the downwind boundary because of the translation and dissipation processes taking place. Fog and stratus dissipation are also affected by the presence of water bodies (Weiss and Gurka, 1975), cloud cover aloft (Gurka and Oliver, 1974), and snow cover (Johnston, 1978). Gustafson and Wasserman (1976) observed that late dissipation of fog retards afternoon convection over the United States (*E* in Figure VII-10-1c). It should also be emphasized that afternoon convection which forms on ridges should not be confused with valley fog. Note the texture, shadows, and highlights associated with the convective cells at *F* in Figure VII-10-1c.

VII-10.2.4 Sea Fog and Coastal Stratus

Visible and enhanced IR images are both important for detecting and forecasting sea fog and coastal stratus. In VIS imagery, fog and stratus have a wide range of brightness even within the same sea fog deck (*D* in Figure VII-10-1). No definite shadows or highlights give texture to the fog top or edges. The boundaries have varying amounts of irregularity (compare *D* in Figures VII-10-1a through VII-10-1c). The shape of an area of sea fog is also known to change rapidly, depending upon coast line, sea surface temperatures, and wind conditions.

In IR imagery, sea fog exhibits relatively high temperatures. In areas of large sea surface temperature gradients, such as off the New England coast, the temperature of the sea fog may be higher, lower, or the same as that of the sea surface. Gurka (1982) has shown that deviations from the normal patterns of sea surface temperature along the New England coast generally indicate the existence of sea fog. Movement of boundaries seen in animated IR is also useful in identifying the extent of sea fog.

Gurka used 30-min interval geostationary satellite images to study movement of sea fog and found that extrapolation was useless. He noticed that acceleration

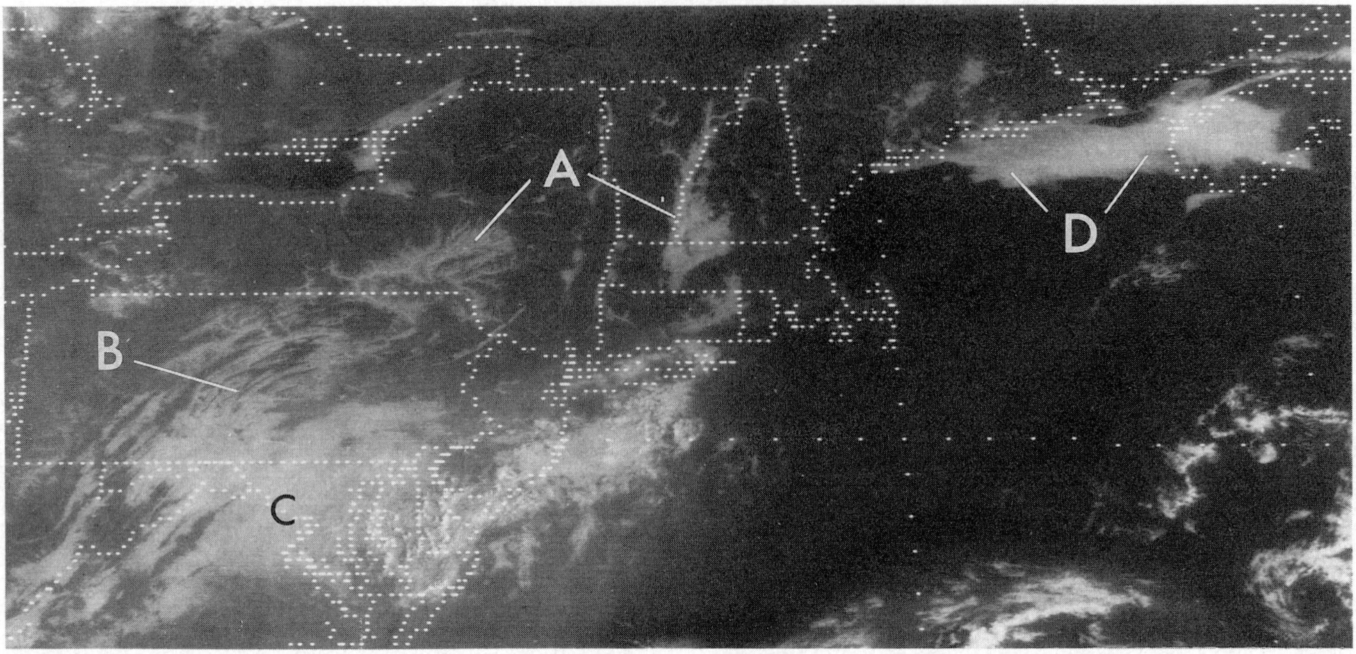

Figure VII-10-1a. GOES-East VIS image, 1301 UTC, October 6, 1982. Fog lingers in New York and New Hampshire-Vermont valleys [*A*] and Pennsylvania Appalachians [*B*]. Fog and stratus combine to form a deck at [*C*]. Sea fog hovers off Maine coast [*D*].

occurs as sea fog moves toward cooler waters, suggesting that the observed motion is the result of advection and growth or dissipation.

Coastal stratus also has a strong diurnal variability. Lee (1979), using satellite images along the California coast, demonstrated how heavy morning overcasts, often extending onshore, dissipated over both sea and land by afternoon.

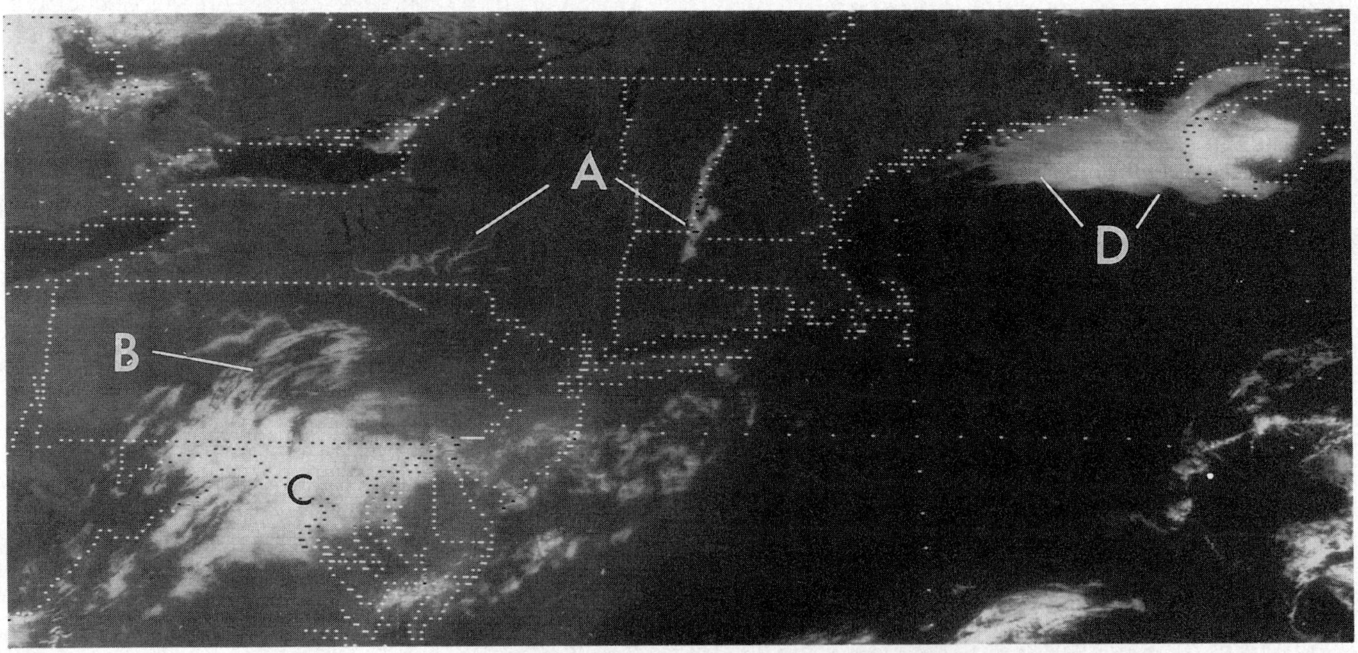

Figure VII-10-1b. GOES-East VIS image, 1501 UTC, October 6, 1982. Much of the valley fog has dissipated at [*A*] and [*B*]. The deck [*C*] is reduced in area; less bright regions dissipate first. Sea fog [*D*] demonstrates more brightness variation than does the land fog and stratus.

Figure VII-10-1c. GOES-East VIS image, 1801 UTC, October 6, 1982. Valley fog has dissipated in Pennsylvania Appalachians [*B*]. The fog and stratus deck [*C*] continues to dissipate inward from the outer edges. Afternoon convection is retarded in the area of early morning fog and stratus cover [*E*]. Afternoon convection forms on ridges [*F*] (not be confused with valley fog).

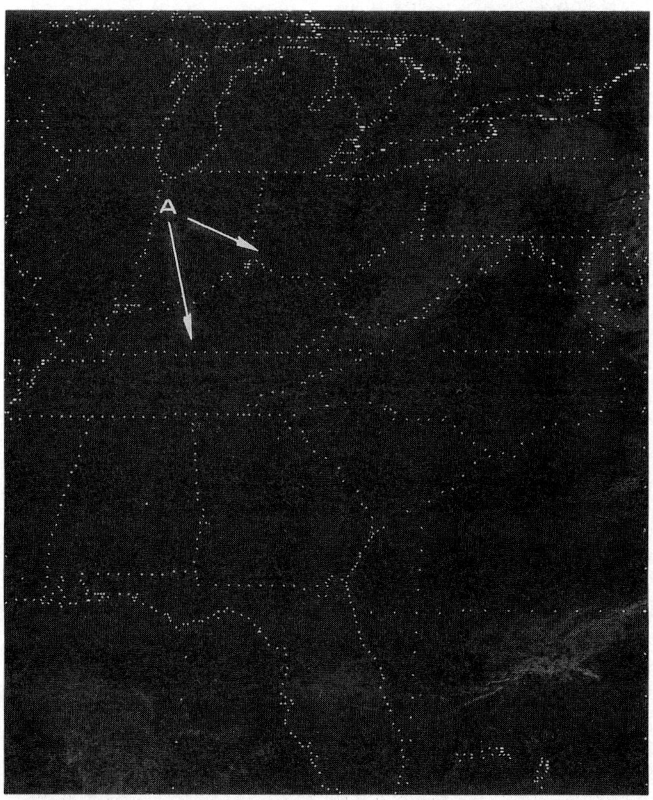

Figure VII-10-2a. GOES-East VIS image, 1230 UTC, October 20, 1977. Early morning valley fog is not evident at [*A*] in standard operational VIS imagery (Gurka, 1978b).

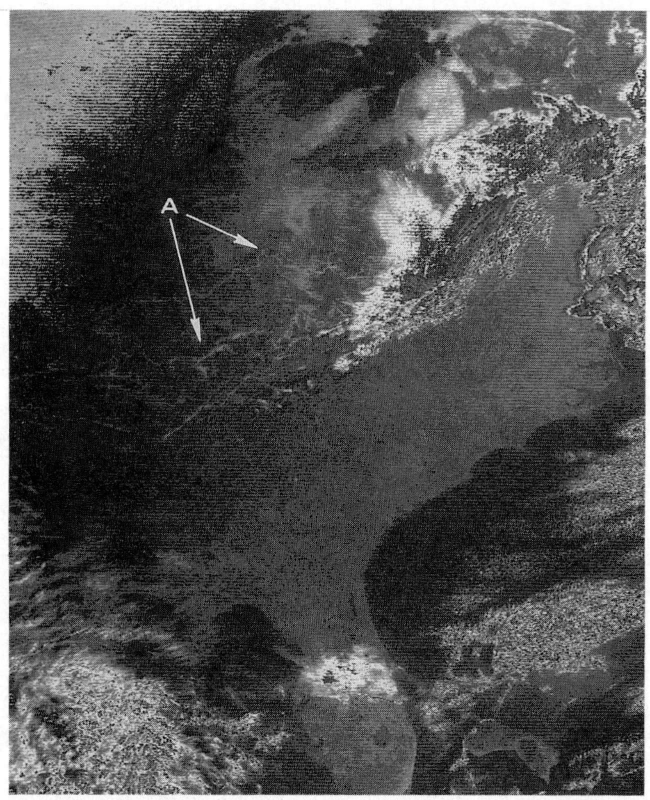

Figure VII-10-2b. GOES-East enhanced VIS image, 1230 UTC, October 20, 1977, of the same scene as in Figure VII-10-2a, provides much better definition of low-level features (Gurka, 1978b).

VII-10.3 ATMOSPHERIC AEROSOLS

The extent and thickness of atmospheric pollutants can be monitored with satellite VIS imagery (Parmenter, 1977; Gurka, 1977). Aerosol pollutants, principally haze, smoke, and blowing dust, are most easily observed in satellite imagery when the solar elevation angle is low, thus maximizing scattering and reflection of light. Figure VII-10-3 is an extraordinary early morning view of the eastern United States taken from GOES-6. The smooth, milky gray areas (*A*) indicate the presence of an extensive haze area where surface visibilities are generally about 7 km or less. These contrast with the dark areas (*B*) where visibilities are more unrestricted. The aerosol layers are not as bright as surrounding clouds but are approximately 5% brighter than clear air (Parmenter and Anderson, 1977).

When the concentrations of the aerosols are large enough to block outgoing longwave radiation from lower levels, they appear in IR images as gray regions similar to thin cirrus or middle clouds (see VII-1.4). Over land this is most apparent during daylight hours when the surface has warmed, and clear areas adjacent to aerosol blocking appear darker.

Dust or sand storms are common in arid or drought stricken regions. An example of how blowing dust can be monitored with both VIS and IR images is shown in Figure VII-10-4. Trajectory models can be used to predict the future areas that will be affected by restricted visibility aloft associated with the movement of these dust clouds (Gurka, 1977).

VII-10.4 THUNDERSTORM GUST FRONTS AND DOWNBURSTS

The gust front, the leading edge of strong, gusty downdraft thunderstorm winds, can often be detected in satellite images as a line of clouds, or arc line (Purdom, 1973; Chapter VII-8).

Gust front types associated with strong surface winds have been identified and classified (Gurka, 1976). The highest wind speeds are found in the part of the arc

Figure VII-10.3. Early morning view of haze and pollution [*A*] over the central and eastern United States; GOES-6 VIS image, August 15, 1984. Areas of unrestricted visibility [*B*] are also evident.

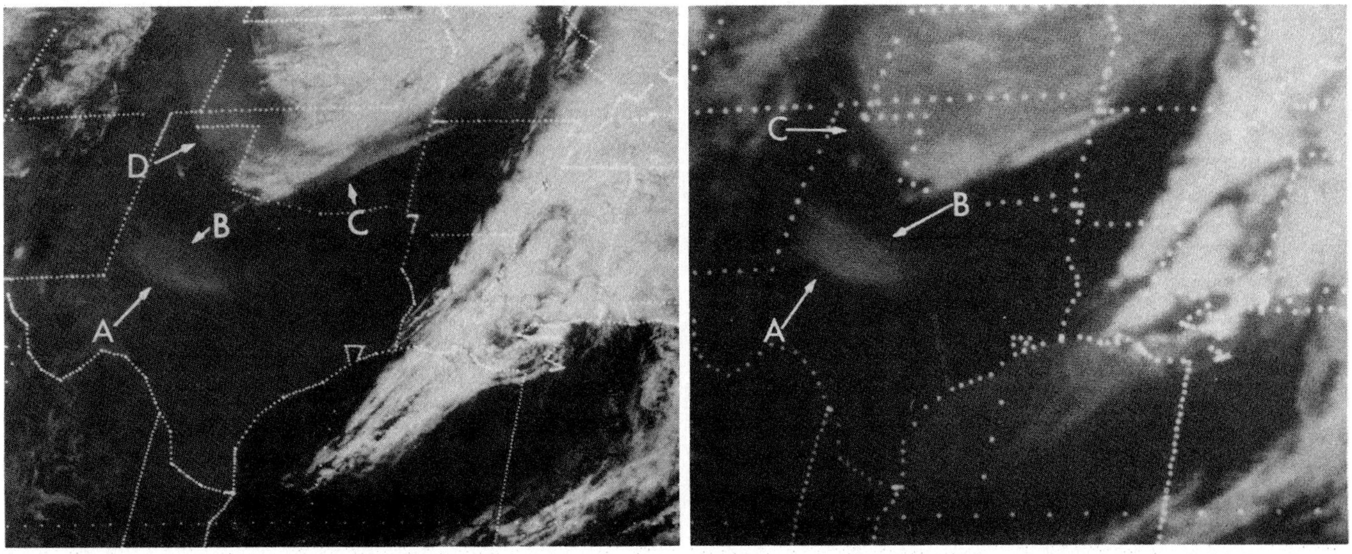

Figure VII-10.4. VIS (left) and IR (right) images of a dust storm over western portions of Texas, Oklahoma, and Kansas and over southeastern Colorado on February 23, 1977. Letters refer to the edges of the dust cloud.

closest to the most vigorous convection, i.e., near the region of strong cloud top temperature gradients in IR images. Also, fast-moving arc lines usually have strong low-level winds and wind shears. A gust front that satisfies most of these criteria is shown in the composite VIS/IR image of Figure VII-10-5.

Thunderstorm downdraft winds can cause considerable destruction. In addition, hazardous wind shears just above the surface can cause rapid changes in headwind components during aircraft landing and takeoff. A number of airline crashes (Caracena et al., 1983) have been attributed to these extreme downdrafts which have been called "downbursts" by Fujita (1978). Very small scale (5 km or less) diverging winds embedded within downbursts are called "microbursts." Figure VII-10-6 shows a schematic example of a downburst.

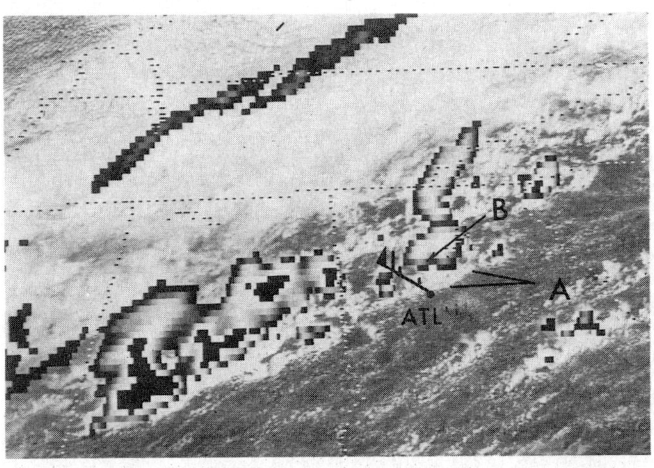

Figure VII-10.5. Winds gusted to 34 m s^{-1} (67 kt) at Atlanta, Georgia (ATL) shortly after the time of this composite VIS/IR image on July 27, 1984. The gust front [*A*] has extensive cloudiness behind it and is located near the south edge of the lowest thunderstorm top temperatures [*B*].

Although downbursts or microbursts cannot be seen in satellite data, one of the effects is sometimes observed: the rapid warming of the anvil top (Fujita, 1978). Temperature in the anvil may increase by 10°C in half an hour. It is not detected with all downbursts, primarily because of poor gray-scale temperature resolution or insufficiently frequent sampling by satellites. This warming usually occurs upwind of the coldest tops and the surface location of the downburst. Figure VII-10-7 shows a circular warm area developing in the anvil top of a huge thunderstorm system over Arkansas. Just prior to 0200 UTC, severe wind damage occurred at a location marked by the circle (also later just to the northeast of the warming).

Prior warning of the initial downburst in a strong thunderstorm is usually not possible by use of satellite data alone because of rapid development of the event. Since downbursts tend to be recurrent with long-lasting storms, the satellite signature may serve as a warning of future downbursts over periods of up to several hours.

VII-10.5 ENROUTE THUNDERSTORM AVOIDANCE

Aircraft in flight may experience extreme turbulence, heavy icing, hail, and lightning strikes in or near thunderstorms. Airborne weather radar is relied upon heavily to avoid the strongest thunderstorm cells in

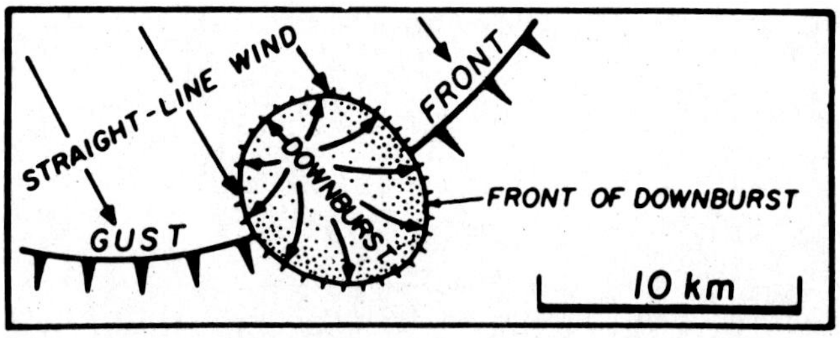

Figure VII-10-6. Schematic view of surface winds with a downburst near a thunderstorm gust front (Fujita, 1981b).

the path of an aircraft. Radar may be limited in range and difficult to interpret at times, however, because of signal attenuation by adjacent cells. Satellite imagery can also be used by air controllers in determining the safest path for current and future flights along air routes.

Thunderstorms (or cumulonimbus clouds) have distinctive signatures in satellite imagery and can usually be identified easily (see VII-1.6). Convection embedded in layered clouds is more difficult to detect but often displays the same characteristics.

Sometimes the organization of thunderstorm activity can be discerned more easily from satellite imagery than from radar because of signal attenuation and data gaps in the latter. Another advantage is that satellite imagery can detect developing convection (i.e., towering cumulus) before the initial radar echoes, thus providing a few more minutes of lead time for the issuance of forecasts or advisories. This is especially important in detecting the development of broken to solid lines or clusters of thunderstorms. When such activity lies across major airways, large flight deviations become necessary. Such deviations force controllers to require greater horizontal separation between aircraft or to completely reroute traffic.

Satellite derived cloud top temperatures can be used to estimate the heights of thunderstorms to determine if aircraft can fly above the cloud. Thunderstorm top estimates may be made either by simple techniques involving available conventional data (e.g., Ellrod, 1979) or more sophisticated methods using one-dimensional updraft models (Adler et al., 1982). Because of the slow response of the GOES IR sensor (see VII-1.2) and its relatively large (8 to 10 km) IFOV, the cloud top temperatures of intense thunderstorms may be overestimated by about 15°C on the average (Adler et al., 1982) with a resulting large (1 to 2 km) underestimate of the height of the storm tops.

VII-10.6 AIRCRAFT ICING

Ice accumulating on aircraft surfaces can cause rapid loss of lift and is especially dangerous for small fixed-wing aircraft flying at relatively low altitudes. Among the meteorological factors that determine the icing rate are cloud top temperatures, the horizontal and vertical extent of the clouds, and the liquid water in the air. All of these can be estimated by use of satellite measurements.

The cloud top temperature must be below 0°C for icing to occur. It has also been shown that as the top temperature in stratiform clouds falls below −15°C, the threat of severe icing greatly diminishes (Jones,

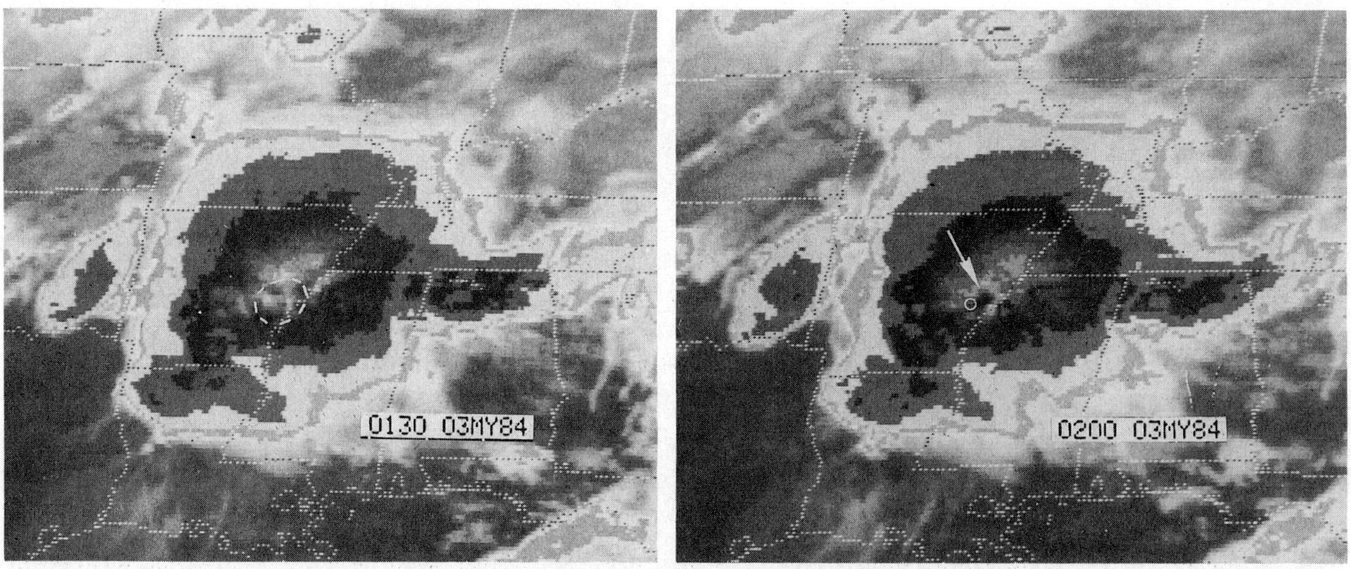

Figure VII-10.7. Enhanced IR images taken 30 min apart showing the signature of a downburst in a severe thunderstorm over the southern United States. A round, warm spot in the anvil (arrow) is seen shortly after the occurrence of wind damage (small circle).

1961). Infrared satellite images can be enhanced to show the 0°C to −15°C cloud top temperature range in which icing is possible. The meteorologist must determine from radiosonde data or animated satellite imagery that the observed clouds are thick stratiform rather than thin cirrus in which liquid water is not available for icing. Figure VII-10-8 is an example of an enhanced IR image with aircraft icing reports superimposed.

Embedded convection can produce strong updrafts that can cause moderate or severe icing at temperatures lower than −15°C. Such areas of convection can be found with VIS and IR imagery and radar reports.

The estimation of liquid water content, still developmental, involves the use of MW instruments, such as the TOVS, on the polar orbiting satellites. Grody and Shen (1982) developed one such method.

VII-10.7 JET STREAM IDENTIFICATION

Several empirical rules have been developed for locating the jet stream in satellite images (Oliver et al., 1964; Oliver et al., 1968; Anderson et al., 1974). The primary rule is that the jet stream is often located just on the poleward edge of a persistent shield of cirrus. The jet often marks the border between dry subsiding air on the poleward side and moist rising air on the Equatorward side. In VIS images, where the jet crosses a front, shadows are often cast on the lower frontal clouds poleward of the jet by the higher jet stream cirrus. In IR images, cloud tops appear brighter (colder) south of the jet, and a sharp line of demarcation is seen.

Common locations for a jet stream cirrus border are (1) over a low-amplitude upper-level ridge and (2) on

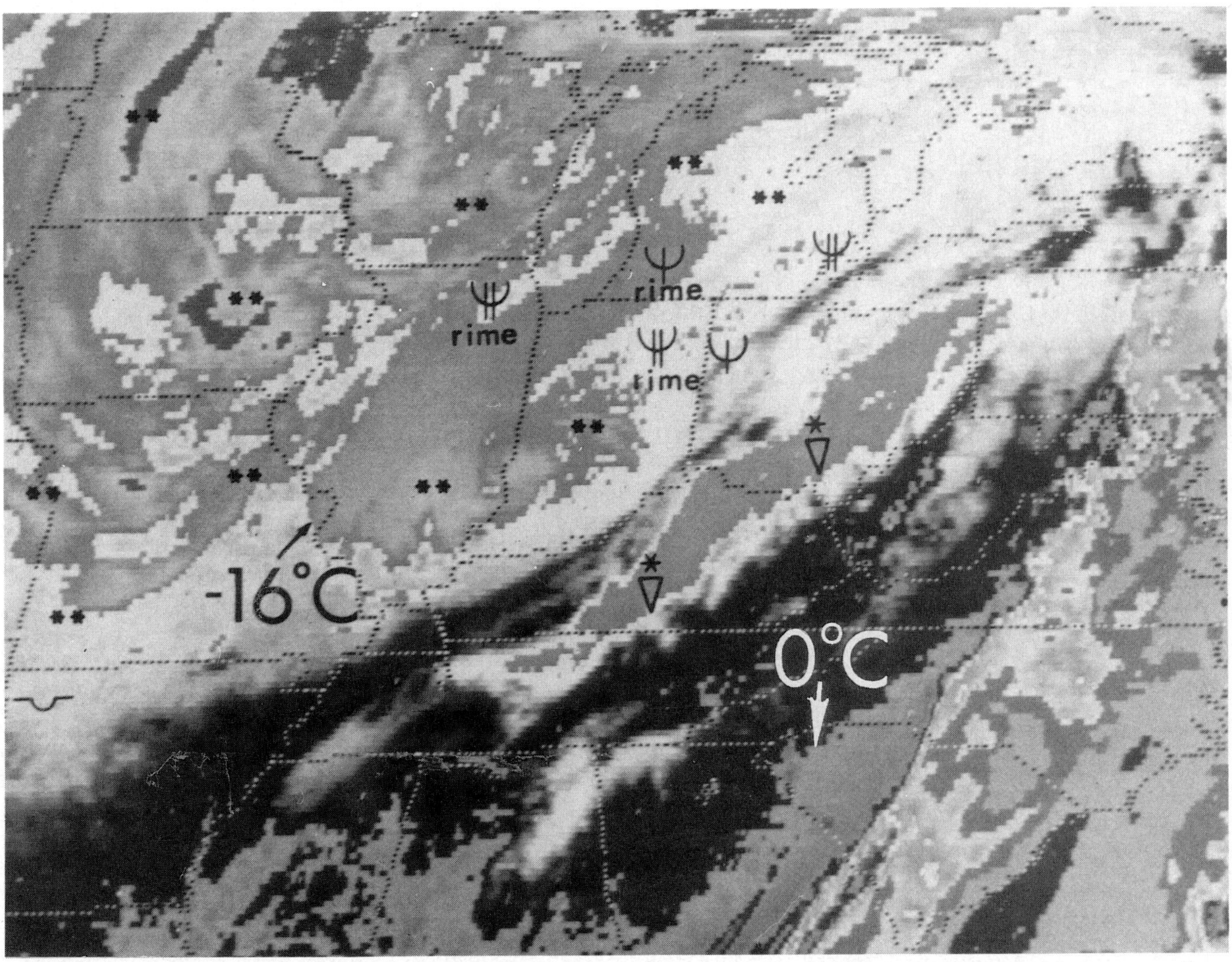

Figure VII-10.8. Enhanced IR image showing the 0°C to −16°C cloud top temperatures. Symbol refers to moderate or severe icing, refers to light icing as reported by aircraft. Areas of precipitation (medium gray areas) are less likely to have significant icing.

the east side of a high-amplitude trough (Weldon, 1979). Jet stream cirrus appear less frequently at the base and west side of a trough because there is little upper level moisture there. In these latter cases, a boundary may be identifiable in WV (6.7 μm) imagery.

In the absence of high clouds, the jet axis is often marked by a change in character of the low clouds from the poleward to the equatorward side. This is quite common over the ocean, in particular, and over land areas where there is abundant low-level moisture. The preferred location for this demarcation of low cloud types is along the east side of an upper ridge to the downstream trough position. The cold, unstable low-level air on the poleward side of the jet produces open cell convection patterns; on the equatorward side of the jet there is often a low-level inversion that caps convection and produces closed-cell cloud patterns. Over land, there often are no low clouds on the Equatorward side of the jet.

Finally, when middle or high level clouds extend across the jet stream, the cloud edges will move faster at the location of maximum winds, thus forming a U- or V-shaped cloud pattern that points downstream. This rule can usually be used with small-amplitude synoptic systems in zonal flow or on the east side of a longwave ridge.

It is estimated that the jet stream can be located by use of these three rules approximately 75 to 85% of the time (Weldon, 1979) when there are no conventional upper-wind data. Figure VII-10-9 shows a case where all three rules were used to locate two jet streams over the eastern North Pacific Ocean.

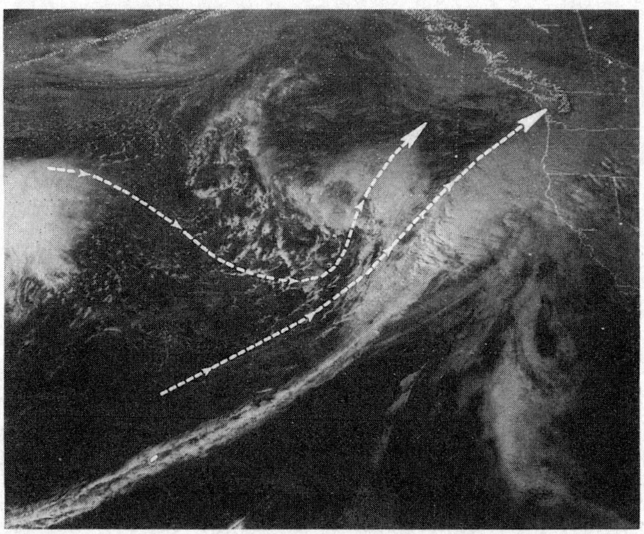

Figure VII-10.9. Locations of jet streams estimated using empirical rules described in text.

VII-10.8 LOW-LEVEL JET STREAMS

Bands in middle- or low-level clouds, often observed in VIS images, correspond to the occurrence of low-level jets (Beckman, 1982). The bands form during the morning and may persist well into the afternoon. Wavelengths are typically 8 to 20 km and the bands are nearly perpendicular to the cloud layer shear vector. Wind shears may range from 10 to 25 $\times 10^{-3}$ s^{-1}. Strong thunderstorms may develop on the north or west edges of these banded regions but not within them because of high stability. Figure VII-10-10 is an example of low-level jet stream cloud bands over the central United States, with strong, 30 to 50-kt low-level winds over most of the region.

Since these low-level jets generally form during the night and can produce hazardous vertical wind shears, it is important to detect them before early morning rawinsonde data or pilot reports are available. To observe them in IR data, it is necessary to use animated images and computed cloud drift winds. Parmenter (1976) described an example of the use of the cloud drift wind technique to identify a low-level jet in the western Gulf of Mexico.

VII-10.9 TURBULENCE

Non-convective atmospheric turbulence, with its small scale and short duration, is one of the most difficult hazards to aviation to detect and forecast. Early studies linked this type of turbulence at high levels primarily to curving jet stream segments, strong vertical and horizontal wind shear, and high static stability in the vicinity of upper atmospheric fronts (Endlich, 1964; Sorenson and Beckwith, 1975). Turbulence at levels below 500 mb is more likely to occur with mountain waves or with turbulent mixing caused by strong winds blowing over rough terrain.

Satellite imagery complements conventional weather data in turbulence analysis. Geostationary imagery enables features associated with turbulence to be followed and predicted with fair accuracy for a period of a few hours. Detection in satellite data is dependent on either (1) the presence of high-level cloud or moisture boundaries in VIS, IR, or WV images, or (2) the presence of smaller scale wave clouds oriented nearly perpendicular to the wind flow (usually observed only in high resolution VIS data). Cloud boundaries usually persist for 6 hours or more; wave clouds are more variable in nature.

In the first category, high-level turbulence occurs frequently in deformation zones in the upper flow pattern (Anderson et al., 1982); these zones often appear as cloud or moisture boundaries in the imagery (Wel-

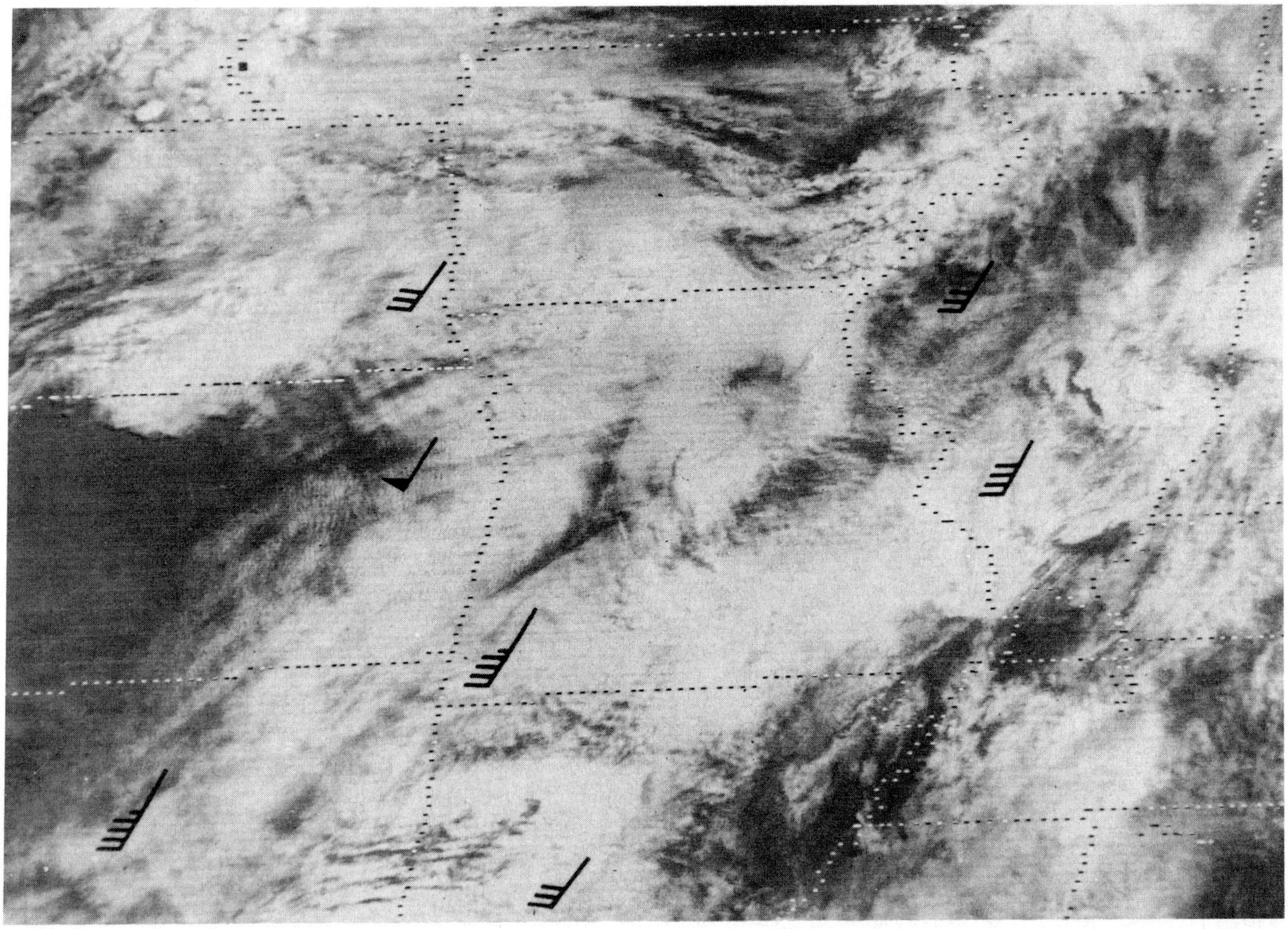

Figure VII-10-10. Bands in low clouds over the central United States in a VIS image, 1430 UTC, June 7, 1984. Winds shown are for 850 mb at 1200 UTC indicating a low level jet stream over eastern Oklahoma, eastern Kansas, and western Missouri. Thunderstorms developed later in the day along the western edge of the clouds. Cloud bands over Arkansas and eastern Oklahoma are probably mountain waves.

don, 1979). Deformation zones may be found along dense cirrus edges associated with upper lows, sharp troughs, or other vorticity maxima. Although the jet stream may occur along portions of these boundaries, that is not always the case. The turbulence typically occurs within 200 to 300 km of either side of the cloud edge.

Sharp, anticyclonically curved bulges along the edge of an otherwise straight segment of jet stream cirrus often produce turbulence (Ellrod, 1985). These rapidly moving, short wavelength features are probably related to geostrophic motion near jet streaks (wind maxima embedded in the jet stream current). Turbulence normally extends downwind from the crest of the cirrus bulge for several hundred kilometers.

Mountain waves appear often in VIS and sometimes in IR imagery as evenly spaced, parallel cloud lines perpendicular to the winds (Conover, 1964; Ernst, 1976). The wavelength of the clouds is directly proportional to the windspeed and instability (Fritz, 1965). Wavelengths have been seen to increase through the day, as diurnal heating proceeds and mean wind speeds increase because of turbulent mixing (Weiss, 1976). Long wavelengths are more likely than short wavelengths to be accompanied by significant low-level turbulence (Ellrod, 1985).

When a standing lee wave cloud consists of cirrus, the downward motion of the waves may not be sufficient to dissipate the ice crystals; the result is a long cirrus plume that may extend hundreds of kilometers downwind. There usually is no turbulence in these extensive plumes unless a narrow, stationary clear zone in the cirrus occurs in IR or WV images over the lee slopes of the mountain ridges (Ellrod, 1987). Mountain

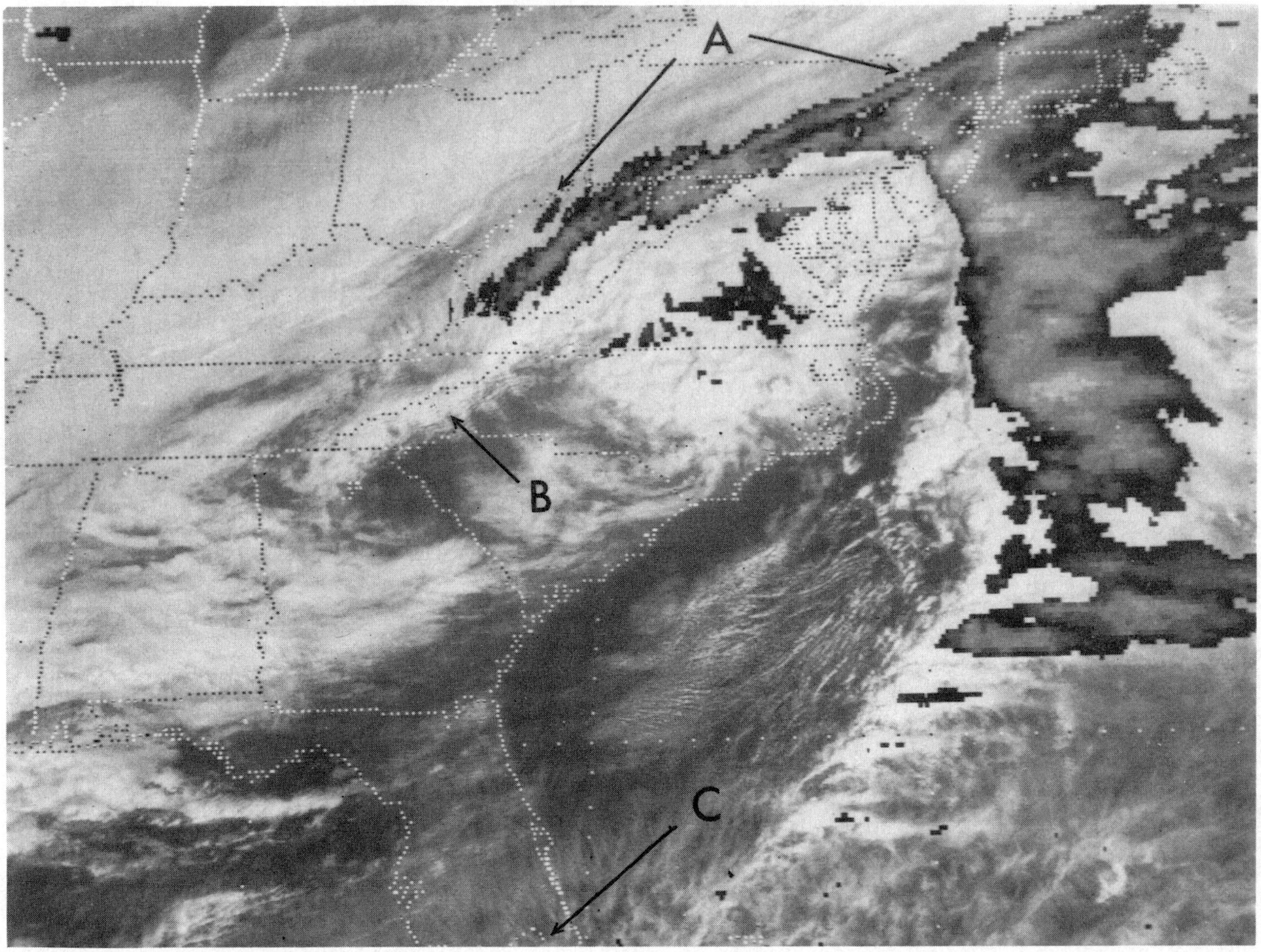

Figure VII-10-11. Indicators of possible turbulence in a composite VIS/IR image, 1500 UTC, February 23, 1983. A deformation zone [*A*], wave clouds in the lee of the Appalachian Mountains [*B*], and transverse bands in jet stream cirrus [*C*] are shown.

wave turbulence will sometimes propagate as high as the tropopause; thus, the heights of the tops of wave clouds are not always indicative of the altitudes affected by turbulence.

Short-lived waves, or billows, embedded in cirrus near the jet stream are good indicators of moderate to severe turbulence (Beckman, 1981). Often non-orographic, such billows are observable only in high-resolution VIS images. Their occurrence has been related to strong vertical wind shear in the presence of high static stability (Ludlam, 1967), resulting in a breaking wave phenomenon known as Kelvin-Helmholtz instability.

Wave-like clouds that are larger and more irregular than mountain wave clouds or billows are sometimes observed in cirrus on the anticyclonic side of the jet. They are often seen as jagged spikes or hooked filaments on the eastern or southern edge of jet cirrus. Their wavelengths are sufficiently long to be observed in both IR and VIS imagery. These clouds, known as transverse bands or cloud trails (Anderson et al., 1974), contain strong vertical and/or horizontal wind shears and high static stability (Viezee et al., 1967). The inference is that turbulence is likely in the presence of these transverse clouds.

Several features indicative of the likelihood of turbulence are pointed out in the composite VIS/IR image of Figure VII-10-11.

VII-10.10 ESTIMATING SURFACE WIND SPEED AND DIRECTION OVER OCEANS

In addition to cloud drift wind estimates (Chapter VII-4), satellite imagery offers several means of esti-

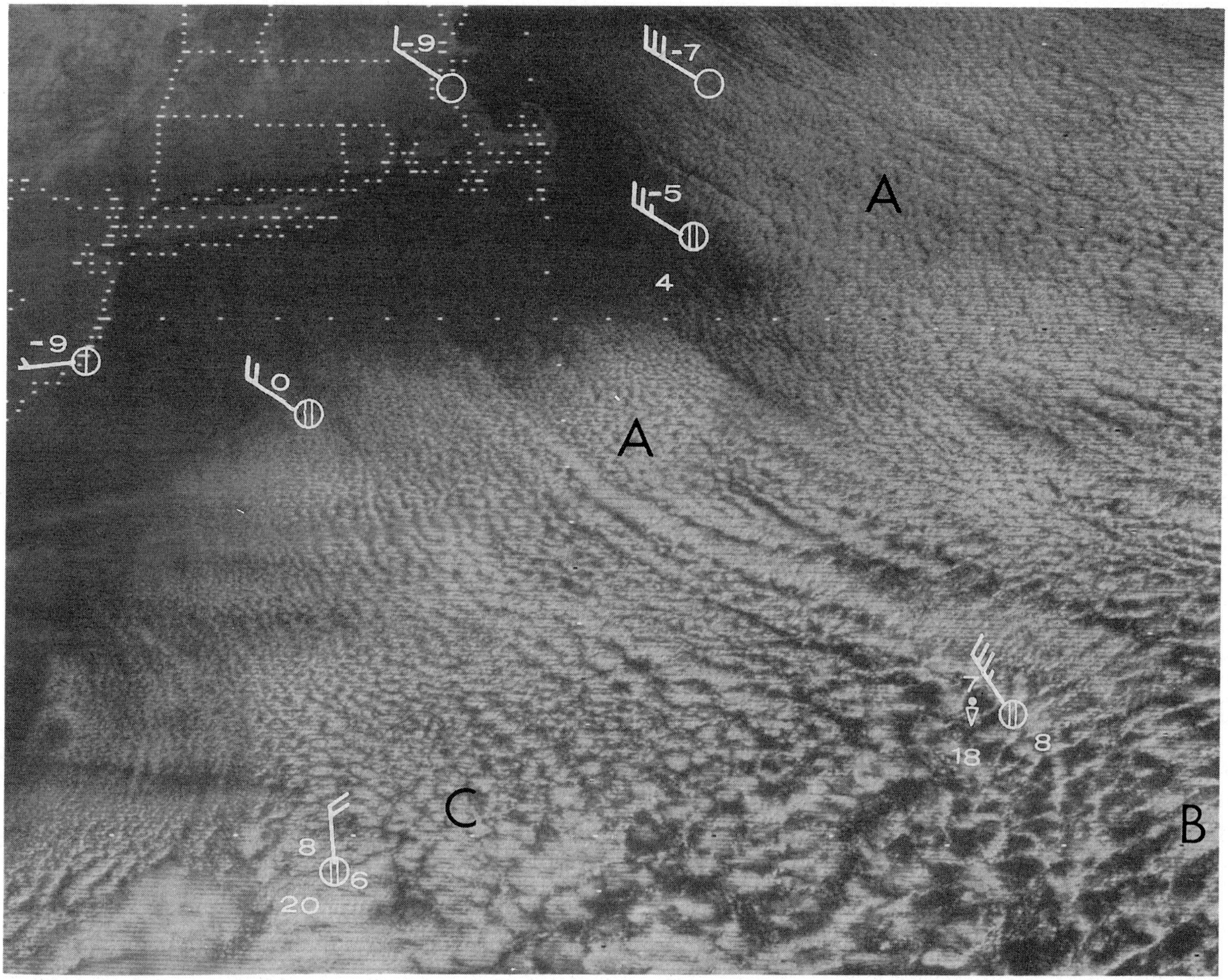

Figure VII-10-12. VIS image, 1400 UTC, February 26, 1982. A cold northwest flow is modified by warm ocean water resulting in cloud lines parallel to the wind [*A*] and open cell convection [*B*]. A region of more anticyclonic flow is dominated by closed cell convection [*C*]. Wind speeds (kt) increase offshore due to turbulent mixing. Temperatures (°C) of the air and sea surface are shown, along with wave heights (½ m).

mating wind and related sea state conditions over ocean areas. Specifically, these techniques involve analysis of: (1) low cloud patterns, (2) Sun glint, and (3) anomalous gray shades (Anderson et al., 1974; Fett and Mitchell, 1977; Fett et al., 1977).

When cold air flows off continents over relatively warm waters, the air is heated from below and turbulent mixing occurs. This brings air with higher momentum down near the surface, resulting in stronger surface winds than when the air was over land. Under these conditions, narrow cloud lines form parallel to the cloud layer wind shear. If there is little directional shear within the subcloud layer, the cloud lines approximate the direction of the surface wind. This assumption is usually good except in the vicinity of frontal zones and rapidly developing cyclones. Wind speeds associated with these cloud lines are usually 20 kt or more. However, near the center of a deep cyclone, winds of 50 kt or more are possible.

As the cloud lines move over progressively warmer waters offshore, the inversion rises and open cell convective patterns develop. Large diameter open cells occur mostly in cyclonic, unstable flow to the rear of polar fronts or cyclones. Open cell patterns also indicate wind speed of 20 kt or more. The larger the diameter of the cells, the lighter the winds. A rapid decrease in cell diameter within a short distance upstream indicates a rapid upstream increase in surface wind speed. This

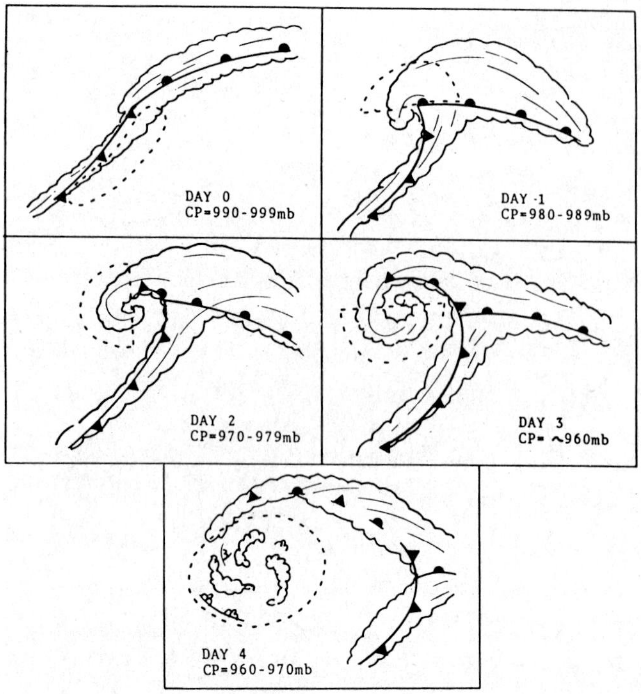

Figure VII-10-13. Schematic of satellite observed cloud pattern evolution for an intense oceanic cyclone developing at a normal rate. The region of strongest winds at each stage is enclosed by the dashed line. Minimum central pressures (CP) are also indicated.

transition, often observed in satellite images, is called a surge line.

Small diameter cells, open or closed, are seen in anticyclonic flow associated with subtropical high pressure areas; wind speeds are light, less than 20 kt. Figure VII-10-12 shows a typical variety of cellular patterns in an offshore flow of cold air; observed wind speeds are indicated.

Patches of sunlight reflected by the sea surface, or Sun glint (VII-1.3), can be used to infer information about surface winds and sea state. Small areas of brilliant Sun glint indicate nearly calm winds and a smooth sea surface acting like a mirror. Large areas of diffuse Sun glint are related to light winds (less than 20 kt). An elongated dark swath or indentation on either side of the brightest Sun glint also indicates calm wind conditions and usually marks the location of the surface high pressure ridge or equatorial trough. Little or no Sun glint is observed with extremely rough seas and strong winds. These reflectivity effects associated with capillary waves are also measurable using active MW measurements.

Strong winds and rough seas produce regions of mist and haze that cannot be seen in VIS or IR images, but do appear as light to occasionally dark-toned areas in the near IR channel (0.8 μm) from polar orbiting satellites. These are called anomalous gray shades, and are caused by strong scattering by the aerosols in the near-IR. Anomalous gray shades also can delineate moist areas associated with dissipating or developing clouds, regions of haze, blowing dust or smoke, and mesoscale wind patterns in the lee of small islands.

VII-10.11 INTENSE OCEANIC CYCLONES

The rapid, unexpected development of extratropical cyclones over oceans is especially important to mariners; ships are often unable to return to shore or to divert to a more favorable route before encountering dangerously strong winds and high seas. Anderson et al. (1974) described satellite observed cloud patterns usually associated with an extratropical cyclone developing at the normal rate. The main features are:

- an increase in the width of a cloud band associated with the surface warm front
- an increase in the cloud thickness in the same region (revealed by IR or VIS brightness)
- an anticyclonic bulge along the poleward boundary of a frontal band

Cloud pattern schematics of the evolution of the cloud band into the classic comma pattern are shown in Figure VII-10-13.

The characteristics of rapidly deepening cyclones, described in some detail by Sanders and Gyakum (1980), are associated with rapidly falling pressures (at least 1 mb $\cdot$ hr^{-1} for 24 hr) and hurricane-like wind and pressure fields. Most intense lows (960 mb or less) undergo explosive deepening. They often develop in regions of strong sea surface temperature gradients, such as along the Gulf Stream. In satellite images, these storms sometimes show spiraling convective bands and eye-like centers similar to those of tropical storms (Chapter VII-7). The evolutionary pattern of rapidly deepening storms is similar to that described by Anderson et al. (1974) except that it proceeds at a much faster rate.

The central pressure of oceanic extratropical storms can be estimated by using an empirical technique developed by Junker and Haller (1980). The central pressure is determined from the degree of curvature of the main baroclinic cloud band as it wraps around the surface center (Figure VII-10-13). This permits the reanalysis of surface pressures in data sparse areas to determine more realistic maximum surface winds.

The distribution of strong surface winds around an oceanic cyclone depends on the stage of development (Smigielski, 1973). During the initial stage, the strongest winds occur in the southeast semicircle of the low. They then rotate to the north, the west, and the south semicircles during the deepening and mature stages (Figure VII-10-13). As the low dissipates, the strongest winds are nearly evenly distributed around the center.

BIBLIOGRAPHY AND REFERENCES FOR SECTION VII

Achtemeier, G. L., 1983: The relationship between the surface wind field and convective precipitation over the St. Louis area. *J. Clim. Appl. Meteor., 22*, 982-999.

Adler, R. F. and R. A. Mack, 1984: Thunderstorm cloud top dynamics as inferred from satellite observations and models. *Proc. of the Conf. on Satellite/Remote Sensing and Applications, June 25-29, Clearwater Beach, FL,* Amer. Meteor. Soc., Boston, MA, 1945-1960.

———, M. J. Markus, D. D. Fenn, G. Szejwach, and W. E. Shenk, 1983: Thunderstorm top structure observed by aircraft overflights with an infrared radiometer. *J. Clim. & Appl. Meteor., 22*, 579-593.

———, M. J. Markus, D. D. Fenn, and W. E. Shenk, 1982: Thunderstorm top structure observed by aircraft overflights with an infrared radiometer. *Proc. of the 12th Conf. on Severe Local Storms, January 11-15,* Amer. Meteor. Soc., Boston, MA, 160-163.

——— and D. D. Fenn, 1979: Thunderstorm intensity as determined from satellite data. *J. Appl. Meteor., 18*, 502-517.

——— and D. D. Fenn, 1977: Satellite based thunderstorm intensity parameters. *Proc. of the 10th Conf. on Severe Local Storms, October 18-21, Omaha, NE*, Amer. Meteor. Soc., Boston, MA, 8-15.

——— and E. B. Rodgers, 1977: Satellite-observed latent heat release in a tropical cyclone. *Mon. Wea. Rev., 105*, 956-963.

Allison, L. J., E. B. Rodgers, T. T. Wilheit, and R. W. Fett, 1974: Tropical cyclone rainfall as measured by the Nimbus 5 electrically scanning microwave radiometer. *Bull. Amer. Meteor. Soc., 55*, 1074-1089.

Anderson, R. K., J. J. Gurka, and S. J. Steinmetz, 1982: Application of VAS multispectral imagery to aviation forecasting. *Proc. of the 9th Conf. on Weather Forecasting and Analysis, June 28-July 1, Seattle, WA*, Amer. Meteor. Soc., Boston, MA, 227-234.

———, et al., 1974: Application of meteorological satellite data in analysis and forecasting. *ESSA Technical Report NESC 51*, Dept. of Commerce, Washington, DC, 3-B-8 to 3-B-10; 3-D-1 to 3-D-11.

——— and N. F. Veltishchev (eds.), 1973: The use of satellite pictures in weather analysis and forecasting. *Technical Note 124,* WMO No. 333, World Meteor. Organ., Geneva, Switzerland, 275 pp.

——— and A. H. Smith, 1971: Application of meteorological satellite data in analysis and forecasting. *Supplement to ESSA Technical Report NESC 51*, Dept. of Commerce, Washington, DC, 6-A-1 to 6-C-18.

———, E. W. Ferguson, and V. J. Oliver, 1966: The use of satellite pictures in weather analysis and forecasting. *Technical Note 75,* No. 190, TP. 96, World Meteor. Organ., Geneva, Switzerland, 6-7, 18-30, 105.

Aoki, T., 1982: An improved method to retrieve the clear radiance from partially cloudy spots of radiometer on board satellite. *J. Meteor. Soc. Japan, 60*, 758-764.

Arnold, C. P., 1977: Tropical cyclone cloud intensity relationships. *Atmospheric Science Paper No. 277*, Dept. of Atmospheric Science, Colorado State Univ., Fort Collins, CO.

Atkinson, G. D. and C. R. Holliday, 1977: Tropical cyclone minimum sea level pressure/maximum sustained wind relationship for western North Pacific. *Mon. Wea. Rev., 105,* 421-427.

Baker, W. E., 1983: Objective analysis and assimilation of observational data from FGGE. *Mon. Wea. Rev., 111*, 328-342.

Barrett, E. C., G. D'Souza, and C. H. Power, 1986: Bristol techniques for the use of satellite data in rain cloud and rainfall monitoring. *J. Brit. Interplan. Soc., 39*, 517-526.

———, 1984: Development and testing of BIAS-3F: The Bristol/NOAA interactive scheme for satellite-improved rainfall monitoring. AgRISTARS Stage 4, Contr. Rpt. NA-83-DDC-00034, Univ. of Bristol, U.K., 90 pp.

——— and D. W. Martin, 1981: *The Use of Satellite Data in Rainfall Monitoring*. Academic Press, New York, NY, 340 pp.

Beckman, S. K., 1983: A technique for predicting backdoor thunderstorm outflow boundaries. *Satellite Applications Information Note 83-1*, Dept. of Commerce, NOAA, Washington, DC, 6 pp.

———, 1982: Relationship between cloud bands in satellite imagery and severe weather. *Proc. of the 12th Conf. on Severe Local Storms, San Antonio, TX,* Amer. Meteor. Soc., Boston, MA, 483-486.

———, 1981: Wave clouds and severe turbulence. *Natl. Wea. Dig., 6*, 30-37.

Bengtsson, L., M. Ghil, and E. Kallen (Eds.), 1983: *Dynamic Meteorology: Data Assimilation Methods*. Springer Verlag, New York, NY, 330 pp.

———, M. Kanamitsu, P. Kallberg, and S. Uppala, 1982a: FGGE 4- dimensional data assimilation at ECMWF. *Bull. Amer. Meteor. Soc., 63*, 29-43.

———, M. Kanamitsu, P. Kallberg, and S. Uppala, 1982b: FGGE research activities at ECMWF. *Bull. Amer. Meteor. Soc., 63,* 277-303.

Berroir, A., P. de Felice, L. Pontier, and P. Sitbon, 1983: Janus: an aircraft prototype of a low earth orbit split beam stereoscopic observing system. *Adv. Space Res., 2*, Pergamon Press, New York, NY, 125-131.

Black, P. G., 1983: Tropical storm structure revealed by stereoscopic photo- graphs from skylab. *Adv. Space Res., 2,* Pergamon Press, New York, NY, 115-124.

———, 1977: Some aspects of tropical storm structure revealed by hand held camera photographs from space. *Skylab Explores the Earth*, Natl. Aero. and Space Adm., Washington, DC, 417-461.

Bleckman, J. B., 1981: Vortex generation in a numerical thunderstorm mode. *Mon. Wea. Rev., 109*, 1061-1071.

Bohan, W. A., 1981: The importance of thunderstorm outflow boundaries in the development of deep convection. *WAB 456, 16mm Color Sound Movie*, The Walter A. Bohan Co., Park Ridge, IL, 29 minutes.

Borneman, R., 1988: Satellite rainfall estimating programs of the NOAA/NESDIS Synoptic Analysis Branch. *Natl. Wea. Dig., 13*, 3-11.

Boucher, R. J. and R. J. Newcomb, 1962: Synoptic interpretation of some TIROS vortex patterns: A preliminary model. *J. Appl. Meteor., 1*, 127-136.

Bourke, W., K. Puri, and R. Seaman, 1982: Numerical weather prediction studies from the FGGE Southern Hemisphere database. *Mon. Wea. Rev., 110*, 1787-1800.

———, K. Puri, R. Seaman, B. M. McAverey, and J. LeMarshall, 1982: ANMRC data assimilation for the Southern Hemisphere. *Mon. Wea. Rev., 110*, 1749-1771.

Bowen, R. A., L. Fusco, J. Morgan, and K. O. Roska, 1979: Operational production of cloud motion vectors (satellite winds) from Meteosat image data: Use of data from meteorological satellites. *ESA No. 143*, European Space Agency, Paris, France, 65-75.

Brandes, E. A., 1982: *Nowcasting*. Academic Press, Inc., New York, NY, 256 pp.

———, 1978: Mesocyclone evolution and tornadogenesis: Some observations. *Mon. Wea. Rev., 106*, 995-1101.

Brandli, H. W. and F. A. Lombardo, 1974: High level wave clouds not associated with terrain. *Bull. Amer. Meteo. Soc., 55*, 1472-1474.

Bridge, G. C. (ed.), 1980: *Meteosat system guide-Volume 5; Meteorological Products*. Meteosat Data Management Department, ESOC, Darmstadt, West Germany, 90 pp.

Brimacombe, E., 1981: Atlas of Meteosat Imagery. *ESA SP-1030*, European Space Agency, Paris, France, 494 pp.

Bristor, C. L., 1975: Central processing and analysis of geostationary satellite data. *NOAA Technical Memorandum NESS 64*, Dept. of Commerce, Washington, DC, 155 pp.

——— and W. Pichel, 1974: 3-D cloud viewing using overlapped pictures from two geostationary satellites. *Bull. Amer. Meteor. Soc., 55*, 1353-1355.

———, W. M. Callicott, and R. E. Bradford, 1966: Operational processing of satellite pictures by computer. *Bull. Amer. Meteor. Soc., 94*, 515-527.

Brodrick, H. J., 1978: The relationship between vertical sounder radiances and mid-latitude 300-mb flow patterns. *J. Appl. Meteor, 17*, 477-481.

Browner, S. P., W. L. Woodley, and C. G. Griffith, 1977: Diurnal oscillation of the area of cloudiness associated with tropical storms. *Mon. Wea. Rev., 105*, 856-864.

Browning, K. A., 1964: Airflow and precipitation trajectories within severe local storms which travel to the right of the winds. *J. Atmos. Sci., 21*, 634-639.

Bryson, W. (Univ. of Wisconsin, Madison, WI), 1978: Cloud height determination from geosynchronous satellite images. Ph. D. Dissertation, 91 pp.

Butler, D., 1984: Science and mission requirements working group, "earth observing system." *NASA Technical Memorandum 86129*, Natl. Aero. and Space Adm., GSFC, Greenbelt, MD, 51 pp.

Byers, H. R. and R. Braham, 1949: *The Thunderstorm*. Govt. Printing Office, Washington, DC, 287 pp.

Caracena, F., R. Maddox, J. Purdom, J. Weaver, and R. Green, 1983: Multiscale analyses of meteorological conditions affecting Pan American World Airways Flight 759. *NOAA Technical Memorandum ERL ESG-2*, Dept. of Commerce, Boulder, CO, 53 pp.

Carlson, T. N., 1980: Airflow through mid-latitude cyclones and the comma cloud pattern. *Mon. Wea. Rev., 108*, 1498-1509.

Chang, Y. M., J. J. Tecsun, and T. T. Fujita, 1973: METRACOM system of cloud- velocity determination from geostationary satellite pictures. *SMRP, 110*, Univ. of Chicago, Chicago, IL, 29 pp.

Changnon, S. A., 1976: Effects of urban areas and echo merging on radar echo behavior. *J. Appl. Meteor., 15*, 561-570.

Chatters, G. C. and V. E. Suomi, 1975: The applications of McIDAS. *IEEE Trans. Geosci. Elecs., GE-13*, 137-146.

Chen, L., 1980: *METPAK/Navigation System User's Guide and Study Report*. Natl. Aero. and Space Adm., GSFC, Greenbelt, MD, Contract NAS5-25548, General Software Corporation, Silver Spring, MD, 43 pp.

Chesters, D., L. W. Uccellini, and A. Mostek, 1982: VISSR Atmospheric Sounder (VAS) simulation experiment for a severe storm environment. *Mon. Wea. Rev., 110*, 198-216.

Clark, J. D. and R. Borneman, 1984: Satellite precipitation estimates program of the Synoptic Analysis Branch. *Proc. of the 10th Conf. on Weather Forecasting and Analysis, June 25-29, Clearwater Beach, FL*, Amer. Meteor. Soc., Boston, MA, 392-399.

———, 1983: *GOES User's Guide*. Dept. of Commerce, NOAA, Washington, DC, 79 pp.

Conover, J. H., 1965: Cloud and terrestrial albedo determinations from TIROS satellite pictures. *J. Appl. Meteor., 4*, 492-503.

———, 1964: The identification and significance of orographically induced clouds observed by TIROS satellites. *J. Appl. Meteor., 3*, 226-234.

———, 1962: Cloud interpretation from satellite altitudes. *Research Note 81*, USAF Cambridge Research Lab., Bedford, MA, 77 pp.

Conrath, B., 1967: Inverse problems in radiative transfer: A review. *NASA TMS-55857, Proc. of the 18th Intl. Astronautical Federation Congress, Sept. 24-30, 1967, Belgrade, Yugoslavia*, 339-360.

Dalton, J. T., M. L. desJardins, A. F. Hasler, and R. A. Minzner, 1979: Digital cloud stereography from geostationary orbit. *Proc. of the 13th Intl. Symps. on Remote Sensing of the Environment*, Environ. Res. Inst. of MI, Ann Arbor, MI, 1479-1488.

Darkow, G. L., 1968: The total energy environment of severe storms. *J. Appl. Meteor., 7*, 199-205.

desJardins, M., 1978: Three-dimensional observations of cloud motions using interactive stereo displays. (unpublished manuscript). Natl. Aero. and Space Adm., GSFC PIP presentation, Greenbelt, MD, July 10, 1978.

Doswell, C. A., III and L. R. Lemon, 1979: An operational evaluation of certain kinematic and thermodynamic parameters associated with severe thunderstorm environments. *Proc. of the 11th Conf. on Severe Local Storms, Oct. 2-5, Kansas City, MO*, Amer. Meteor. Soc., Boston, MA, 397-402.

Droegemier, K. and R. B. Wilhelmson, 1983: Three-dimensional numerical simulation of the interaction between a shallow cumulus field and thunderstorm outflow boundary. *Proc. of the 13th Conf. on Severe Local Storms, Oct. 17-20, Tulsa, OK*, Amer. Meteor. Soc., Boston, MA, 245-248.

Dunne, J. A. and E. Burgess, 1978: The voyage of Mariner 10. *NASA Report SP-424*, JPL, Pasedena, CA, 224 pp.

Dutton, J. A., 1983: *The UNIDATA System for University Weather Data Analysis and Modeling*. Natl. Center for Atmospheric Research, Boulder, CO, 26 pp.

Dvorak, V. F., 1984a: Satellite observed upper level moisture patterns associated with tropical cyclone movement. *Proc. of the 15th Conf. on Hurricanes and Tropical Meteorology, Jan 9-13, Miami, FL*, Amer. Meteor. Soc., Boston, MA, 227-234.

———, 1984b: Tropical cylone intensity analysis using satellite data. *NOAA Technical Report NESDIS II*, Dept. of Commerce, Washington, DC, 47 pp.

———, 1980: Tropical cyclone development as related to changes in the westerlies. *Proc. of the WMO Symp. on Typhoons, Oct. 6-11,* World Meteor. Organ., Geneva, Switzerland, 64-72.

———, 1979: Tropical cyclogenesis in the western Atlantic as observed in satellite infrared imagery. *Proc. of the 12th Technical Conf. on Hurricanes and Tropical Meteorology, April 25-27, New Orleans, LA*, Amer. Met. Soc., Boston, MA.

——— and S. Wright, 1977: Tropical cyclone intensity analysis using enhanced infrared satellite data. *Proc. of the 11th Technical Conf. on Hurricanes and Tropical Meteorology, December 13-16, Miami Beach, FL*, Amer. Meteor. Soc., Boston, MA, 268-273.

———, 1975: Tropical cyclone intensity analysis and forecasting from satellite imagery. *Mon. Wea. Rev., 103*, 420-430.

———, 1972: A technique for the analysis and forecasting of tropical cyclone intensities from satellite pictures. *NOAA Technical Memorandum NESS 36*, Dept. of Commerce, Washington, DC, 15 pp.

Eigenwillig, N. and H. Fisher, 1982: Determination of midtropospheric wind vectors by tracking pure water vapor structures in Meteosat water vapor image sequences. *Bull. Amer. Meteor. Soc., 63*, 44-58.

Ellrod, G. P., 1987: Identifying high altitude mountain wave turbulence and strong chinook wind events with satellite imagery. *Proc. of the AIAA 25th Aerospace Sciences Meeting, January 12-15*, Amer. Inst. for Aeronautics and Astronautics, Washington, DC, 7 pp.

———, 1985: Indicators of high altitude, non-convective turbulence observed in satellite images. *Proc. of the 2nd Int. Conf. on the Aviation Weather System*, Amer. Meteor. Soc., Boston, MA, 277-284.

———, 1979: Estimation of thunderstorm top heights using enhanced IR imagery. *Satellite Applications Information Note 70/5*, Dept. of Commerce, NOAA, NESS, Washington, DC, 6 pp.

Endlich, R. M. and D. E. Wolf, 1981: Automatic cloud tracking applied to GOES and Meteosat observations. *J. Appl. Meteor., 10*, 309-319.

———, 1964: The mesoscale structure of some regions of clear air turbulence. *J. Appl. Meteor., 3*, 261-276.

Erickson, C. O., 1974: Use of geostationary-satellite cloud vectors to estimate tropical cyclone intensity. *NOAA Technical Memorandum NESS 59*, Dept. of Commerce, Washington, DC, 37 pp.

———, 1972: Evolution of a techinique for the analysis and forecast of tropical cyclone intensities from satellite pictures. *NOAA Technical Memorandum NESS 42*, Dept. of Commerce, Washington, DC, 28 pp.

Ernst, J., 1976: SMS-1 nighttime infrared imagery of low level mountain waves. *Mon. Wea. Rev., 104*, 207-209.

Espy, J. P., 1941: *The Philosophy of Storms*. Charles C. Little and James Brown Co., Boston, MA, 347 pp.

Estes, J. E., E. J. Hajic, and L. R. Tinney (eds.), 1983: Fundamentals of image analysis of visible and thermal infrared data. *Manual of Remote Sensing*, Amer. Soc. of Photogram., Falls Church, VA, 987-1124.

Estoque, M. A., 1962: The sea breeze as a function of the prevailing synoptic situation. *J. Atmos. Sci., 19*, 244-250.

Eyre, J. R., J. L. Brownscombe, and R. J. Allam, 1984: Detection of fog at night using advanced very high resolution radiometer imagery. *The Meteorological Magazine, 113,* 264-271.

Fankhauser, J., 1976: Structure of an evolving hail storm, Part II: Thermodynamic structure and airflow in the near environment. *Mon. Wea. Rev., 104*, 576-587.

Fett, R. W. and W. F. Mitchell, 1977: Naval Tactical Applications Guide. Volume I, Techniques and Applications of Image Analysis. *NEPRF Technical Report 77-03*, Walter A. Bohan Co., Park Ridge, IL, 110 pp.

———, P. E. LaViolette, M. Nestor, J. W. Nickerson, and K. Rabe, 1977: Naval Tactical Applications Guide. Volume II, Environmental Phenomena and Effects. *NEPRF Technical Report 77-04*, Walter A. Bohan Co., Park Ridge, IL, 161 pp.

——— and S. Brand, 1975: Tropical cyclone movement forecasts based on observations from satellites. *J. Appl. Meteor. 14*, 452-456.

———, 1964: Some characteristics of the formative stage of typhoon development: A satellite study. *Natl. Conf. on Physics and Dynamics of Clouds, Chicago, IL, March 24-26*, Dept. of Commerce, U. S. Weather Bureau, Washington, DC, 10 pp.

Field, G., 1985: Verification of SABs satellite precipitation estimates for the 1984 convective season. *Natl. Wea. Dig., 10*, 39-44.

Fleagle, R. G. and J. A. Businger, 1963: *An Introduction to Atmospheric Physics*, Academic Press, New York, 336 pp.

Fleming, H. E. and M. L. Hill, 1983: An objective procedure for detecting and correcting errors in geophysical data, Part 2: Multidimensional applications. *J. Geophys. Res., 88*, 3859-3867.

Follansbee, W. A., 1973: Estimation of average daily rainfall from satellite cloud photographs. *NOAA Technical Memorandum NESS 44*, Dept. of Commerce, Washington, DC, 39 pp.

Fritz, S. and P. K. Rao, 1967: On the infrared transmission through cirrus clouds and the estimation of relative humidity from satellites. *J. Appl. Meteor., 6*, 1088-1096.

———, 1965: The significance of mountain lee waves as seen from satellite pictures. *J. Appl. Meteor., 4*, 31-37.

———, 1964: Pictures from meteorological satellites and their interpretation. *Space Science Reviews 3*, D. Reidel Pub. Co., Dordrecht, Holland, 541-580.

Fujita, T. T. and J. C. Dodge, 1983: Applications of stereoscopic height computations from dual geosynchronous satellite data/Joint NASA–Japan Stereo Project. *Adv. Space Res., 2*, Pergamon Press, New York, NY, 153-160.

———, 1982a: Infrared, stereo-height, cloud-motion, and radar-echo analysis of SESAME-day thunderstorms. *Proc. of the 12th Conf. on Severe Local Storms, January 11-15, San Antonio, TX*, Amer. Meteor. Soc., Boston, MA, 213-216.

———, 1982b: Principle of stereoscopic height computations and their applications to stratospheric cirrus over severe thunderstorms. *J. Meteor. Soc. Japan, 60*, 355-368.

———, 1981a: Mesoscale aspects of convective storms. *Proc. of the IAMAP Symp.: Nowcasting: Mesoscale Observations and Short-Range Prediction, Hamburg, Germany*, European Space Agency, Paris, France, 3-10.

———, 1981b: Tornadoes and downbursts in the context of generalized planetary scales. *J. Atmos. Sci., 38*, 1511-1534.

———, 1978: Manual of downburst identification for project Nimrod. *SMRP 156*, Univ. of Chicago, Chicago, IL, 104 pp.

———, G. S. Forbes, and T. A. Umenhofer, 1976: Close-up of 20 March 1975 tornadoes: Sinking cloud tops to suction vortices. *Weatherwise, 29*, 115-132.

———, E. W. Pearl, and W. E. Shenk, 1975: Satellite-tracked cumulus velocities. *J. Appl. Meteor., 14*, 407-413.

———, D. L. Bradbury, C. Merrino, and L. Hull, 1968: A study of mesoscale cloud motions computed from ATS-1 and terrestrial photographs. *Satellite and Mesometeorology Project Research Paper No. 71*, Dept. of Geophys. Sci., Univ. of Chicago, Chicago, IL, 25 pp.

———, 1963: Use of TIROS pictures for studies of the internal structure of tropical storms. *Mesometeorology Project Research Paper 25*, Univ. of Chicago, Chicago, IL, 21 pp.

———, 1959: Precipitation and cold air production in mesoscale thunderstorm systems. *J. Meteorol., 16*, 454-466.

——— and H. A. Brown, 1958: A study of mesosystems and their radar echos. *Bull Amer. Meteor. Soc., 39*, 538-554.

———, 1955: Results of detailed synoptic studies of squall lines. *Tellus, 7*, 405-436.

Funk, L., 1982: Operational use of the deformation zone concept in analysis and forecasting cloud boundaries. *Proc. of the 9th Conf. Weather Forecasting and Analysis, June 28-July 1, Seattle, WA*, Amer. Meteor. Soc., Boston, MA, 336-341.

Gaby, D. C., J. B. Lushine, B. M. Mayfield, S. C. Pearce, and F. E. Torres, 1980: Satellite classifications of Atlantic tropical and subtropical cyclones: A review of 8 years of classification at Miami. *Mon. Wea. Rev., 108*, 587-595.

Gal-Chen, T., B. Schmidt, and L. Ucellini, 1986: Simulation experiments for testing the assimilation of geostationary satellite temperature retrievals into a prediction model. *Mon. Wea. Rev., ll4*, 1213-1230.

Gentry, R. C., E. Rogers, J. Seranka, and W. E. Shenk, 1980: Predicting tropical cyclone intensity using satellite measured equivalent blackbody temperatures of cloud tops. *Mon. Wea. Rev., 108*, 445-455.

Gird, R. S. and R. E. Bradford, 1985: The evolution of interactive operations of the National Environmental Satellite Data and Information Service. *Proc. of the 1st Inter. Conf. on Interactive Information and Processing Systems for Meteorology, Oceanography and Hydrology, January 7-11, Los Angeles, CA*, Amer. Meteor. Soc., Boston, MA, 223-228.

Goody, R. M., 1964: *Atmospheric Radiation* (Vol. 1, Theoretical Basis). Oxford University Press, New York, NY, 436 pp.

Graham, C. H. (ed.), 1965: Visual space perception. *Vision and Visual Perception*, John Wiley and Sons, Inc., New York, NY, 504-547.

Green, R. N. and H. A. Parker, 1983: Application of satellite and radar data to severe thunderstorm analysis. *Proc. of the 13th Conf. on Severe Local Storms, October 17-20, Tulsa, OK*, Amer. Meteor. Soc., Boston, MA, 190-193.

———, G. Hughes, C. Novak, and R. Schreitz, 1975: The automatic extraction of wind estimates from VISSR data, Chapter 12, Central Processing and Analysis of Geostationary Satellite Data. C. L. Bristor (ed.), *NOAA Technical Memorandum NESS 64*, Dept. of Commerce, Washington, DC, 94-110.

Green, W. B., 1983: *Digital Image Processing*. Van Nostrand Reinhold Inc., New York, NY, 191 pp.

Griffith, C. G., J. A. Augustine and W. L. Woodley, 1981: Satellite rain estimation in the U.S. high plains. *J. Appl. Meteor., 20*, 53-66.

———, W. L. Woodley, P. G. Grube, D. W. Martin, J. Stout, and D. N. Sikdar, 1978: Rain estimates from geosynchronous satellite imagery: Visible and infrared studies. *Mon. Wea. Rev., 106*, 1153-1171.

Grody, N. C., 1988: Determination of surface and atmospheric parameters using the Advanced Microwave Sounding Unit. *Proc. of the 3rd Conf. on Satellite Meteor. and Ocean., January 31-February 5, Anaheim, CA,* Amer. Meteor. Soc., Boston, MA, 100-104.

———, 1984: Precipitation monitoring over land from satellites by microwave radiometery. *Proc. of the Inter. Geosci. & Remote Sensing Symp. (IGARSS,84), Strasbourg, France, 27-30 August*, (ESA SP-215), 417-423.

——— and W. Shen, 1982: Observations of Hurricane David (1979) using the microwave sounding unit. *NOAA Technical Report NESS 88*, Dept. of Commerce, Washington, DC, 52 pp.

Gruber, A. and C. Watkins, 1982: Statistical assessment of the quality of TIROS-N and NOAA-6 satellite soundings. *Mon. Wea. Rev., ll0*, 867-876.

Gurka, J. J., 1983: Training notes on CAT detection using satellite data. (unpublished manuscript). Dept. of Commerce, NOAA, NESDIS, SAL, Washington, DC, 12 pp.

———, 1982: The use of geostationary satellite imagery for observing and forecasting movement of New England sea fog. *Proc. of the 9th Conf. on Weather Forecasting and Analysis, June 28-July 1*, Amer. Meteor. Soc., Boston, MA, 143-151.

———, 1980: Observations of advection-radiation fog formation from enhanced IR satellite imagery. *Proc. of the 8th Conf. on Weather Forecasting and Analysis, June 10-13,* Amer. Meteor. Soc., Boston, MA, 108-114.

———, 1978a: The use of enhanced VIS imagery for fog detection and prediction. *Satellite Applications Information Note 78/4*, Dept. of Commerce, NOAA, NESS, Washington, DC, 6 pp.

———,1978b: The use of enhanced VIS imagery for predicting time of fog dissipation. *Proc. of the Conf. of Weather Forecasting and Analysis and Aviation Meteorology, October 16-19*, Amer. Meteor. Soc., Boston, MA, 343-346.

———, 1978c: The role of inward mixing in the dissipation of fog and stratus. *Mon. Wea. Rev., 106*, 1633-1635.

———, 1977: The February 23-24 dust storm as viewed from GOES-1. *Satellite Applications Information Note 77-6*, Dept. of Commerce, NOAA, NESS, Washington, DC, 5 pp.

———, 1976: Satellite and surface observations of strong wind zones accompanying thunderstorms. *Mon. Wea. Rev., 104*, 1484-1493.

———, 1975: Distinguishing fog from stratus on satellite pictures. *Satellite Applications Information Note 10/75-3*, Dept. of Commerce, NOAA, NESS, Washington, DC, 2 pp.

——— and V. J. Oliver, 1974: Fog persistence under a cirrus band. *Mon. Wea. Rev., 102*, 869-870.

Gustafson, A. V. and S. E. Wasserman, 1976: Use of satellite information in observing and forecasting fog dissipation and cloud formation. *Mon. Wea. Rev., 104*, 323-24.

Guymer, L. B., 1978: Operational application of satellite imagery to synoptic analysis in the Southern Hemisphere. *Technical Report 29*, Dept. of Science, Australian Bureau of Meteorology, Melbourne, Australia, 83 pp.

Haas, U. L., 1981: Cloud tracking from satellite pictures. *Scientific Report No. 2.*, Colorado State Univ., Ft. Collins, CO, Contract No. F19628-78-C-0207, 119 pp.

Halem, M., E. Kalnay, W. E. Baker, and R. Atlas, 1982: An assessment of the FGGE satellite observing during SOP-1. *Bull. Amer. Meteor. Soc., 63*, 407-426.

Hall, E. L., 1979: *Computer Image Processing and Recognition*. Academic Press, New York, NY, 26-30.

Hamada, T., 1984: The cloud wind estimation system (CWES) from the Geostationary Meteorological Satellite (GMS) at the Meteorological Satellite Center. *WMO Meeting on Calibration, Validation and Intercomparison of Satellite Products*, Geneva, Switzerland, 18 pp.

———, 1982: New procedure of height assignment to GMS satellite winds. *Japanese Meteorological Satellite Center Technical Note No. 5*, Meteorological Satellite Center, Tokyo, Japan, 91-95.

——— and Watanabe, K., 1978: Determination of winds from geostationary satellite data - present techniques. *WMO/UN Regional Training Seminar on Interpretation, Analysis and Use of Meteorological Satellite Data*, Tokyo, Japan, 20 pp.

Hambrick, L. N. and D. R. Phillips, 1980: Earth locating data of spin- stabilized geosynchronous satellites. *NOAA Technical Memorandum NESS 111*, Dept. of Commerce, Washington, DC, 106 pp.

Hasler, A. F. and K. R. Morris, 1986: Hurricane structure and dynamics from stereoscopic and infrared satellite observations and radar data. *J. Clim. Appl. Meteor., 25*, 709-727.

———, H. Pierce, and J. Dodge, 1985: Meteorological data fields in perspective. *Bull. Amer. Meteor. Soc., 66*, 795-801.

——— and K. R. Morris, 1984: Stereoscopic satellite observations of hurricanes: An update. *Proc. of the 15th Technical Conf. on Hurricanes and Tropical Meteorology, January 9-13, Miami Beach, FL*, Amer. Meteor. Soc., Boston, MA, 64 pp.

———, 1983: Advances in systems for interactive processing and display of meteorological data. *Proc. of the 9th Conf. on Aerospace and Aeronautical Meteorology, June 6-9, Omaha, NE*, Amer. Meteor. Soc., Boston, MA, 60-62.

———, R. Mack, and A. Negri, 1983: Stereoscopic observations from meteorological satellites. *Adv. Space Res., 2*, Pergamon Press, New York, NY, 105-113.

———, 1981: Stereographic observations from geosynchronous satellites: An important new tool for the atmospheric sciences. *Bull. Amer. Meteor. Soc., 62*, 194-212.

———, M. desJardins, and A. J. Negri, 1981: Artificial stereo presentation of meteorological data fields. *Bull. Amer. Meteor. Soc., 62*, 970-973.

——— and R. F. Adler, 1980: Cloud top structure of a tornadic thunderstorm from 3 min interval stereo satellite images compared with radar and other observations. *Proc. of the 19th Conf. on Radar Meteorology, Miami Beach, FL*, Amer. Meteor. Soc., Boston, MA, 405-412.

———, W. C. Skillman, W. E. Shenk, and J. Steranka, 1979: *In Situ* aircraft verification of the quality of satellite cloud winds over oceanic regions. *J. Appl. Meteor., 18*, 1481-1489.

Haurwitz, B., 1947: Comments on the sea breeze circulation. *J. Meteor., 4*, 1-8.

Hayden, C. M., L. F. Hubert, E. P. McClain, and R. S. Seaman, 1979: Quantitative meteorological data from satellites. *Technical Note No. 166*, WMO No. 531, World Meteor. Organ., Geneva, Switzerland, 102 pp.

———, 1971: Nimbus-3 "SIRS" pressure height profiles as compared to radiosondes. *Mon. Wea. Rev., 99*, 659-664.

Heitkemper, L., J. N. Cooper, E. S. Merritt, and D. Masonis, 1982: An interactive meteorological satellite rainfall diagnostic system designed for global agricultural applications. *AgRISTARS Contr. Rep. NA-81-SAC-00714*, Earth Satellite Corp., Chevy Chase, MD, 86 pp.

Herbert, P. J. and K. O. Poteat, 1975: A satellite classification technique for subtropical cyclones. *NOAA Technical Memorandum NWS SR-83*, Dept. of Commerce, Washington, DC, 25 pp.

Hewson,E. L., 1943: The reflection, absorption, and transmission of solar radiation by cloud. *Quart. J. Roy. Meteor. Soc., 69*, 47-62.

Heymsfield, G. M., R. H. Blackmer, and S. Schotz, 1983a: Upper-level structure of Oklahoma tornadic storms on May 2, 1979, I: Radar and satellite observations. *J. Atmos. Sci., 40*, 1739-1755.

———, G. Szejwach, S. Schotz, and R. H. Blackmer, 1983b: Upper-level structure of Oklahoma tornadic storms on May 2, 1979, II: Proposed explanation of "V" pattern and internal warm region in infrared observations. *J. Atmos. Sci., 40*, 1739-1755.

———, R. H. Blackmer, and S. Schotz, 1983c: Severe thunderstorm anvil top structure observed from satellite and radar. *Proc. of the 13th Conf. on Severe Local Storms, October 17-20, Tulsa, OK*, Amer. Meteor. Soc., Boston, MA, 182-185.

Hillger, D. W. and T. H. Vonder Haar, 1981: Retrieval and use of high-resolution moisture and stability fields from Nimbus-6 HIRS radiances in preconvective situations. *Mon. Wea. Rev., 108*, 1010-1028.

——— and T. H. Vonder Haar, 1979: Analysis of satellite infrared soundings of the mesoscale using statistical structure and correlation functions. *J. Atmos. Sci., 36*, 287-305.

Holle, R. L. and M. Maier, 1980: Tornado formation from downdraft interaction in the FACE mesonetwork. *Mon. Wea. Rev., 42*, 348-380.

Hubert, L. F., 1979: Wind derivation from geostationary satellites, Chapter 2: Quantitative meteorological data from satellites, *WMO Technical Note 166*, J. S. Winston (ed.), Geneva, Switzerland, 33-59.

——— and L. F. Whitney, Jr., 1974: Compatibility of low-cloud vectors and rawinsondes for synoptic scale analysis. *NOAA Technical Report NESS 70*, Dept. of Commerce, Washington, DC, 26 pp.

——— and A. Timchalk, 1972: Convective clouds are tracers of air motion. *NOAA Technical Memorandum NESS 40*, Dept. of Commerce, Washington, DC, 12 pp.

——— and L. F. Whitney, Jr., 1971: Wind estimation from geostationary- satellite pictures. *Mon. Wea. Rev., 99*, 665-672.

———, 1966: Mesoscale cellular convection. *Meteorological Satellite Laboratory Report No. 37*, Dept. of Commerce, Washington, DC, 68 pp.

——— and A. F. Krueger, 1962: Satellite pictures of mesoscale eddies. *Mon. Wea. Rev., 90*, 457-463.

Humphreys, W. J., 1914: The thunderstorm and its phenomena. *Mon. Wea. Rev., 42*, 348-380.

Jacobowitz, H. (Massachusetts Institute of Technology, Cambridge, MA), 1970: *Emission, scattering and absorption in cirrus cloud layers*. Ph. D. Dissertation, 181 pp.

Jarvinen, B. R. and C. G. Griffith, 1984: Forecasting rainfall in tropical cyclones using digitized infrared satellite data. *ERL Technical Paper*, Dept. of Commerce, Natl. Hurricane Center, Miami, FL, 12 pp.

Johnston, E. C., 1978: Effect of snow cover on dissipation on fog and stratus. *Satellite Applications Information Note, 78/5*, Dept. of Commerce, NOAA, NESS, Washington, DC, 4 pp.

Jones, R. F., 1961: Ice formation on aircraft. *WMO Technical Note 39*, World Meteor. Organ., Geneva, Switzerland, 35 pp.

Jones, W., P. G. Black, V. E. Delnore, and C. T. Swift, 1981: Airborne microwave remote-sensing measurements of Hurricane Allen. *Science, 214*, 274-280.

Junker, W. N. and D. J. Haller, 1980: Estimation of surface pressure by satellite cloud pattern recognition. *Proc. of the 8th Conf. on Weather Forecasting and Analysis, June 10-13, Denver, CO*, Amer. Meteor. Soc., Boston, MA, 119-122.

Kallberg, P., S. Uppala, N. Gustafsson, and J. Pailleux, 1982: The impact of cloud track wind data on global analysis and medium range forecasts. *Tech. Rpt. 34*, European Centre for Medium Range Weather Forecasts, Reading, England, 60 pp.

Kalnay, E., R. Atlas, W. Baker, and J. Susskind, 1985: GLAS experiments on the impact of FGGE satellite data on numerical weather prediction. *Proc. of the 1st National Workshop on the Global Weather Experiment*, National Academy Press, Washington, DC, 85-117.

Kasik, D. J., 1982: A user interface management system. *Proc. of the Association for Computer Machinery/Special Interest Group for Computer Graphics 82, Oct. 25-27, Dallas, TX.*

Kellogg, W. W., 1982: Early satellite program developments. The conception, growth, accomplishments and future of meteorological satellites. *NASA Conf. Pub. 2257*, Nat'l. Aero. and Space Admin., Washington, DC, 4 pp.

Klemp, J. B. and R. Rotunno, 1983: A study of the tornadic region within a supercell thunderstorm. *J. Atmos. Sci., 40*, 359-377.

——— and R. B. Wilhelmson, 1978: The simulation of three-dimensional convective storm dynamics. *J. Atmos. Sci., 35*, 1070-1096.

Koehler, T. L., J. C. Derber, B. D. Schmidt, and L. H. Horn, 1983: An evaluation of soundings, analyses, and model forecasts from TIROS-N and NOAA-6 satellites. *Mon. Wea. Rev., 111*, 562-571.

Lajoie, F. A. and Nicholls, 1974: A relationship between the direction of movement of tropical cyclones and the structure of their cloud systems. *Technical Report No. 11*, Australian Bureau of Meteorology, Melbourne, Australia, 22 pp.

Laurs, R. M., 1971: Fishery-advisory information available to tropical Pacific tuna fleet via radio facsimile broadcast. *Comm. Fish. Rev., 33*, 40-42.

Lee, T. F., 1979: GOES images show diurnal motion of California coastal stratus. *Report No. PMTC-TP-80-02*, Pacific Missile Test Center, Point Mugu, CA, 60 pp.

Leese, J. A., C. S. Novak, and B. B. Clark, 1971: An automatic technique for obtaining cloud motion from geostationary satellite data using cross correlation. *J. Appl. Meteor., 10*, 118-132.

———, C. S. Novak, and V. R. Taylor, 1970: The determination of cloud pattern motions from geosynchronous satellite image data. *Pattern Recognition*, Pergamon Press, London, 279-292.

Leserman, D., 1984: A new editor/composer and user interface management system, motivation and design. *Proc. of the 3rd Conf. on Interactive Meteorological Processing, Oct. 24, 1984*, Natl. Aero. and Space Adm., GSFC, Greenbelt, MD, 44.

Lilly, D. K. (ed.), 1977: *Project SESAME: Planning documentation volume*. Dept. of Commerce, NOAA, ERL, Boulder, CO, 308 pp.

Lorenz, D., 1983: Stereoscopic imaging from polar orbit and synthetic stereo imaging. *Adv. Space Res., 2*, Pergamon Press, New York, NY, 133-139.

———, A. F. Hasler, and R. F. Adler, 1983: Thunderstorm cloud top observations using satellite stereoscopy. *Mon. Wea. Rev., 111*, 1949-1964.

———, A. F. Hasler, and E. B. Rodgers, 1983: Stereoscopic observations of hurricanes and and tornadic thunderstorms from geosynchronous satellites. *Adv. Space Res., 2*, Pergamon Press, New York, NY, 143-151.

Ludlam, F. H., 1967: Characteristics of billow clouds and their relation to clear air turbulence. *Quart. J. Roy. Meteo. Soc., 93*, 419-435.

Lumb, F. E., 1964: The influence of cloud on hourly amounts of total solar radiation at the sea surface. *Quart. J. Roy. Meteo. Soc., 90*, 49.

Lushine, J. B., 1977: A relationship between weakening of tropical cyclone cloud patterns and lessening of wind speed. *NOAA Technical Memorandum NESS 85*, Dept. of Commerce, Washington, DC, 12 pp.

Lyons, W. A., 1966: Some effects of Lake Michigan upon squall lines and summertime convection. *SMRP 57*, Dept. of Geophysical Science, Univer. of Chicago, Chicago, IL, 22 pp.

Mack, R. A., A. F. Hasler, and R. F. Adler, 1983a: Thunderstorm cloud-top observations using satellite stereoscopy. *Mon. Wea. Rev., 111*, 1949-1964.

———, A. F. Hasler, and E. B. Rodgers, 1983b: Stereoscopic observations of hurricanes and tornadic thunderstorms from geosynchronous satellites. *Adv. Space Res., 2*, Pergamon Press, New York, NY, 143-151,

Maddox, R. A., 1980: Mesoscale convective complexes. *Bull. Amer. Meteor. Soc., 61*, 1374-1387.

———, 1976: An evaluation of tornado proximity wind and stability data. *Mon. Wea. Rev., 104*, 133-142.

Mancuso, R. L. and R. M. Endlich, 1966: Clear air turbulence as a function of wind shear and deformation. *Mon. Wea. Rev., 94*, 581-585.

McCann, D. W., 1981: The enhanced-V, a satellite observable severe storm signature. *NOAA Technical Memorandum, NWS NSSFC-4*, Dept. of Commerce, Washington, DC, Natl. Severe Storms Forecast Center, Kansas City, MO, 31 pp.

McClain, E. P. and A. E. Strong, 1969: On anomalous dark patches in satellite-viewed sunglint areas. *Monthly Weather Review, 97*, 875-884.

——— and H. Brodrick, 1967: Recent research on the application of meteorological satellite data to numerical weather analysis. *Air Weather Service Tech. Report 196*, Scott AFB, IL, 45-50.

McElroy, J. H. and S. R. Schneider, 1984: Utilization of the polar platform of NASA's Space Station Program for Operational Earth Observations. *NOAA Technical Report NESDIS 12*, Dept. of Commerce, Washington, DC, 67 pp.

McMillin, L. M., 1984: The operational TOVS retrieval method. *Proc. of the 1st Intl. TOVS Study Conference, Aug. 29 - Sept. 21, 1983, Igls, Australia*, Cooperative Institute for Meteorological Satellite Studies, Madison, WI, 142-147.

———, D. G. Gray, H. F. Drahos, M. W. Chalfant, and C. S. Novak, 1983: Improvements in the accuracy of operational soundings. *J. Clim and Appl. Meteor., 22*, 1948-1955.

———, 1978: An improved technique for obtaining clear radiances from cloud contaminated radiances. *Mon. Wea. Rev., 106*, 1580-1597.

McNider, R. T., G. Jedlovec, and G. Wilson, 1984: Data analysis and model evaluation of the initiation of convection on 24 April 1982. *Proc. of the 10th Conf. on Weather Forecasting and Analysis, June 25-29, Clearwater Beach, FL*, Amer. Meteor. Soc., Boston, MA, 543-549.

McPherson, R. D., 1975: Progress, problems and prospects in meteorological data assimilation. *Bull. Amer. Meteor. Soc., 56*, 1154-1166.

Menzel, W. P., W. L. Smith, and T. R. Stewart, 1983: Improved cloud motion wind vector and altitude assignment using VAS. *J. Clim. Appl. Meteor., 22*, 377-384.

Miller, D. B. and R. G. Feddes, 1971: *Global Atlas of Relative Cloud Cover, 1967-1970*. United States Air Force/Dept. of Commerce, Washington, DC, 237 pp.

Miller, R. C., 1972: Notes on analysis and severe-storm forecasting procedures of the Air Force Global Weather Central. *AWS Technical Report 200* (rev), Air Weather Service, Offutt Air Force Base, NE, 190 pp.

Minzner, R. A., W. E. Shenk, J. Steranka, and R. D. Teagle, 1976: Cloud heights determined stereographically from imagery recorded simultaneously by two synchronous meteorological satellites, SMS-1 and SMS-2. *EOS, 57*, 593.

Moses, J. F., 1985: Interactive global precipitation estimates. *Proc. of the Conf. on Interactive Information & Processing Systems for Meteorology, Oceanography, and Hydrology, Jan. 7-11, Los Angeles, CA*, Amer. Meteor. Soc., Boston, MA, 206-212.

———, 1984: Applications of automatic cloud detectors to interactive meteorological operations. *Proc. from the Conf. on Satellite/Remote Sensing and Applications, June 25-29, Clearwater Beach, FL*, Amer. Meteor. Soc., Boston, MA, 178-184.

Mosher, F. R. and J. T. Young, 1983: Real-time stereoscopic applications using the GOES operational satellites. *Adv. Space Res., 2*, Pergamon Press, New York, NY, 173-177.

——— and T. Stewart, 1981: Characteristics of water vapor tracked winds. *Final Report, Contract N00014-80-C-0632*, SSEC at Univ. of Wisconsin, Madison, WI, 50 pp.

Nagle, R. and C. M. Hayden, 1971: The use of satellite-observed cloud patterns in northern hemisphere 500-mb numerical analysis. *NOAA Technical Report NESS 55* , Dept. of Commerce, Washington, DC, 55 pp.

National Oceanic and Atmospheric Administration and National Aeronautics and Space Administration, 1987: A report to the Congress, *Space-Based Remote Sensing of the Earth*, U. S. Govt. Printing Office, Washington, DC, p. 27.

Negri, A. and R. Adler, 1981: Relation of satellite-based thunderstorm intensity to radar-estimated rainfall. *J. Appl. Meteor., 20*, 66-78.

Newton, C. W., 1963: Dynamics of severe convective storms. *Meteor. Monogr., 5*, No. 27, 33-58.

Normand, C. W. B., 1938: On instability from water vapor. *Quart. J. Roy. Meteor. Soc., 64*, 47-69.

Ohring, G., 1979: Impact of satellite sounding data on weather forecasts. *Bull. Amer. Meteor. Soc., 60*, 1141-1147.

O'Lenic, E., 1984: Results of experimental numerical forecasts using temperature profiles from geostationary satellites. *Proc. of the Recent Advances in Civil Space Remote Sensing, May 3-4, Arlington, VA*, SPIE, The Intl. Soc. for Optical Engineering, 116-121.

Oliver, V. J., 1974: *Hurricane Fifi: A film documentation*. NWS Hurricane Conference, Coral Gables, FL, Dept. of Commerce, NOAA, NESS, Washington, DC.

———, R. K. Anderson, and E. W. Ferguson, 1968: Cellular cumulus over the Pacific Ocean, picture of the month. *Mon. Wea. Rev., 96*, 470-471.

———, R. K. Anderson, and E. W. Ferguson, 1964: Some examples of the detection of jet streams from TIROS photographs. *Mon. Wea. Rev., 92*, 441-445.

Ondrejka, R. J. and J. H. Conover, 1966: Note on the stereo interpretation of Nimbus-2 APT photography. *Mon. Wea. Rev., 94*, 611-614.

Palmen, E. and W. E. Newton, 1969: *Atmospheric Circulation Systems*. Academic Press, Washington, DC, 238-241.

Parker, W. T. and R. D. Hickey, 1980: The Cheyenne tornado of 16 July 1979. *Natl. Wea. Dig., 5*, 45-62.

Parmenter, F. C., 1977: Monitoring air quality from satellites. *Mon. Wea. Rev., 105*, 789-792.

——— and R. K. Anderson, 1977: A satellite overview of inadvertant weather modification. *Proc. of the 6th Conf. on Inadvertant and Planned Weather Modification, Oct. 10-13, Champaign-Urband, IL*, Amer. Meteor. Soc., Boston, MA, 83-86.

———, 1976: Evidence of a low level jet stream in the western Gulf of Mexico. *Proc. of the 7th Conf. on Aerospace and Aeronautical Meteorology and Symp. on Remote Sensing from Satellites, November 16-19, Melbourne, FL*, Amer. Meteor. Soc., Boston, MA, 312-317.

Pearl, E. W., 1974: Characteristics of anvil-top associated with the Poplar Bluff tornado of May 7, 1973. *SMRP 119*, Univ. of Chicago, Chicago, IL, 12 pp.

Petersen, R. A. and A. Mostek, 1982: The use of VAS moisture channels in delineating regions of potential convective instability. *Proc. of the 12th Conf. on Severe Local Storms, San Antonio, TX*, Amer. Meteor. Soc., Boston, MA, 168-171.

Phillips, N., 1980: Two examples of satellite temperature retrievals in the North Pacific. *Bull. Amer. Meteor. Soc., 61*, 712-717.

Pielke, R., 1974: A three-dimensional numerical model of the sea breezes over south Florida. *Mon. Wea. Rev., 102*, 115-139.

———, 1973: A three-dimensional numerical model of the sea breezes over south Florida. *NOAA Technical Memoramdum ERL WMPO-2*, Dept. of Commerce, Boulder, CO, 136 pp.

Popham, R. W., 1984: A personal eye in space. *Weatherwise, 37*, 76-82.

Pruppacher, H. R. and J. D. Klett, 1978: *Microphysics of Clouds and Precipitation*. Reidel Pub. Co., Dordrecht, Holland, 714 pp.

Pryor, S. P. (Dept. of Atmos. Sciences, Colorado State Univ., Ft. Collins, CO), 1978: Measurement of thunderstorm cloud-top parameters using high-frequency satellite imagery. M. S. Dissertation, 101 pp.

Purdom, J. F. W., 1986: Satellite contributions to convective scale weather analysis and forecasting. *Proc. of the 11th Conf. on Weather Forecasting and Analysis, Kansas City, MO*, Amer. Meteor. Soc., Boston, MA, 295-314

———, 1984: Use of satellite soundings and imagery for nowcasting and very-short range forecasting. *Proc. of the Nowcasting II Symp., Norrkoping, Sweden*, European Space Agency, Paris, France, 99-111.

———, T. H. Vonder Haar, J. K. Stewart, and N. E. Leary, 1984: Diagnosing the severe thunderstorm environment by mesoscale cloud tracking—a new approach and new information. *Proc. of the Conf. on Satellite/ Remote Sensing and Applications, June 25-29, Clearwater Beach, FL*, Amer. Meteor. Soc., Boston, MA, 100-105.

———, 1983: Diagnosing the mesoscale state of the atmosphere using satellite data. *Proc. of the 13th Conf. on Severe Local Storms, October 17-20, Tulsa, OK*, Amer. Meteor Soc., Boston, MA, 167-170.

——— and T. Vonder Haar, 1983: *Enhancing GOES infrared imagery for Nowcast applications, Product Users' Guide*, Cooperative Institute for Research in the Atmosphere, Colorado State Univ., Ft. Collins, CO, 21 pp.

———, R. N. Green, and H. A. Parker, 1982: Integration of satellite and radar data for short range forecasting and storm diagnostic studies. *Proc. of the 9th Conf. on Weather Forecasting and Analysis, June 28-July 1, Seattle, WA*, Amer. Meteor. Soc., Boston, MA, 51-55.

——— and K. Marcus, 1982: Thunderstorm trigger mechanisms over the southeast United States. *Proc. of the 12th Conf. on Severe Local Storms, January 11-15, San Antonio, TX*, Amer. Meteor. Soc., Boston, MA, 487-488.

——— and J. F. Weaver, 1982: Nowcasting during the 10 April 1979 tornado outbreak: A satellite perspective. *Proc. of the 12th Conf. on Severe Local Storms, January 11-15, San Antonio, TX*, Amer. Meteor. Soc., Boston, MA, 467-470.

———, 1979: The development and evolution of deep convection. *Proc. of the 11th Conf. on Severe Local Storms, October 2-5, Kansas City, MO*, Amer. Meteor. Soc., Boston, MA, 143-150.

———, 1976: Some uses of high resolution GOES imagery in the mesoscale forecasting of convection and its behavior. *Mon. Wea. Rev., 104*, 1474-1483.

——— and J. G. Gurka, 1974: The effect of early morning cloud cover on afternoon thunderstorm development. *Proc. of the 5th Conf. on Weather Forecasting and Analysis, March 4-7, St. Louis, MO*, Amer. Meteor. Soc., Boston, MA, 58-60.

———, 1973: Mesohighs and satellite imagery. *Mon. Wea. Rev., 101*, 180-181.

———, 1971: Satellite imagery and severe weather warnings. *Proc. of the 7th Conf. on Severe Local Storms, Kansas City, MO*, Amer. Meteor. Soc., Boston, MA, 120-127.

Racer, R. and J. Gaffney, Jr., 1984: Interpretive processing/expert system: An initiative in weather data analysis and forecasting. *Natl. Wea. Dig., 9*, 31-45.

Rao, M. S. V., W. V. Abbott III, and J. S. Theon, 1976: Satellite-derived global oceanic rainfall atlas (1973 and 1974). *NASA SP-410*, Natl. Aero. and Space Admin., Washington, DC, 186 pp.

Reynolds, D., 1979: Observations and detection of damaging hailstorms from geosynchronous satellite digital data. *Proc. of the 11th Conf. on Severe Local Storms, October 2-5, Kansas City, MO*, Amer. Meteor. Soc., Boston, MA, 181-188.

——— and E. Smith, 1979: Detailed analysis of composited digital radar and satellite data. *Bull. Amer. Meteor. Soc., 60*, 1024-1037.

Roach, W. T., 1967: On the nature of the summit areas of severe storms in Oklahoma. *Quart. J. Roy. Meteor. Soc., 93*, 318-327.

Rodgers, C. D., 1976: Retrieval of atmospheric temperature and composition from remote measurements of thermal radiation. *Rev. Geophys. and Space Phys., 14*, 609-624.

Rodgers, E. B., R. Mack, and A. F. Hasler, 1983: A satellite stereoscopic technique to estimate tropical cyclone intensity. *Mon. Wea. Rev., 111*, 1599-1610.

Ryan, J. A., 1983: Using the home computer as a facsimile receiver. *Weatherwise, 35*, 308-310.

Saastamoinen, J., 1972: Refraction, photogrammetric engineering. *Photogrammetric Engineering, 38*, 799-810.

Sadler, J. C., 1962: The first hurricane track determined by meteorological satellite. *Indian J. Meteor. & Geographics, 13*, 29-44.

Sanders, F. and J. R. Gyakum, 1980: Synoptic-dynamic climatology of the "Bomba." *Mon. Wea. Rev., 108*, 1589-1606.

Sckereschewsky, P., 1945: Clouds and states of the sky. *Handbook of Meteorology*, Berry, Bollay and Beers (eds.), McGraw-Hill, New York, NY, 1068 pp.

———, 1921: L'Evolution de la Meteorologie. La Nature, 171-175.

Scofield, R. A., 1987: The NESDIS operational convective precipitation technique. *Mon. Wea. Rev., 115*, 1773-1792.

———, 1984: The NESDIS operational convective precipitation estimation technique. *Proc. of the 10th Conf. on Wea. Forecast, and Anal., June 25-29, Clearwater Beach, FL*, Amer. Meteor. Soc., Boston, MA.

——— and L. E. Spayd, Jr., 1984: A technique that uses satellite, radar and conventional data for analyzing precipitation from extratropical cyclones. *NOAA Technical Memorandum NESDIS 8*, Dept. of Commerce, Washington, DC, 51 pp.

——— and V. J. Oliver, 1980: Some improvements to the Scofield/ Oliver technique. *Proc. of the 2nd Conf. on Flash Floods, March 18-20, Atlanta, GA*, Amer. Meteor. Soc., Boston, MA, 115-122.

———, V. J. Oliver, and L. E. Spayd, 1980: Estimating rainfall from thunderstorms with warm tops in the infrared imagery. *Proc. of the 8th Conf. on Weather Analysis and Forecasting, June 10-13, Denver, CO*, Amer. Meteor. Soc., Boston, MA, 85-92.

——— and V. J. Oliver, 1977: A scheme for estimating convective rainfall from satellite imagery. *NOAA Technical Memorandum NESS 86*, Dept. of Commerce, Washington, DC, 47.

Serebreny, S. M., E. J. Wiegman, and R. E. Hadfield, 1970: Further comparison of cloud motion vectors with rawinsonde observations. *Final Report, ESSA Contract No. E210-69(N)*, Stanford Research Institute, Menlo Park, CA, 60 pp.

Sheets, R. C. and P. G. Grieman, 1975: An evaluation of the accuracy of tropical cyclone intensities and locations determined from satellite pictures. *NOAA Technical Memorandum ERL WMPO-20*, Dept. of Commerce, ERL, Boulder, CO, 36 pp.

Shenk, W. E., R. J. Holub, and R. A. Neff, 1975: Stereographic cloud analysis from Apollo 6 photography. *Bull. Amer. Meteor. Soc., 56*, 4-15.

Shewchuck, J. D. and R. C. Weir, 1980: An evaluation of the Dvorak technique for estimating tropical cyclone intensity from satellite imagery. *NOCC/JTWC 80-2,* Naval Oceanography Command Center Joint Typhoon Warning Center, Comnavmarinas, Guam, FPO, 25 pp.

Sikdar, D. N., V. E. Suomi, and C. E. Anderson, 1970: Convective transport of mass and energy in severe storms over the United States — An estimate from geostationary altitude. *Tellus, 22*, 521-532.

Simonett, D. S. (ed.), 1983: The development and principles of remote sensing. *Manual of Remote Sensing*, Amer. Soc. of Photogram., Falls Church, VA, 20-25.

Simpson, J. and Dennis, A. S., 1974: Cumulus clouds and their modification. *Weather Modification*, W. N. Hess (ed.), John Wiley and Sons, Inc., New York, NY, 229-280.

——— and V. Wiggert, 1969: Models of precipitating cumulus towers. *Mon. Wea. Rev., 97*, 471-489.

Sinclair, P. C. and J. F. W. Purdom, 1984: Aircraft penetrations of arc cloud lines. *Proc. of the Conf. on Satellite/Remote Sensing and Appplications, June 25-29, Clearwater, FL*, Amer. Meteor. Soc., Boston, MA.

——— and J. F. W. Purdom, 1982: Integration of research aircraft data and 3 minute interval GOES data to study the genesis and development of deep convective storms. *Proc. of the 12th Conf. on Severe Local Storms, January 11-15, San Antonio, TX*, Amer. Meteor. Soc., Boston, MA, 269-271.

Sitbon, P. (ed.), 1972: *Determination des champe de vents a partir de photographies de satellites geostationnair*. Laboratoire de Meteorologie Dynamique, Univer. of Paris, France, 57 pp.

Smigielski, F. J., 1973: Idealized development patterns of surface cyclogenesis. (unpublished report). Dept. of Commerce, NOAA, NESS, Washington, DC, 5 pp.

——— and L. M. Mace, 1970: Estimating mean relative humidity from the surface to 500 mb. *ESSA Technical Memorandum, NESCTM 3*, Dept. of Commerce, NOAA, Washington, DC, 12 pp.

Smith, E. A., 1975: The McIDAS system. *IEEE Transactions on Geoscience, GE-13*, 123-128.

——— and D. R. Phillips, 1972: Automated cloud tracking using precisely aligned digital ATS pictures. *IEEE Transactions on Computers, C-21*, 715-729.

Smith, W. L., G. S. Wade, W. P. Menzel, V. E. Suomi, F. J. Fox, C. S. Velden, and J. S. LeMarshall, 1984: Nowcasting—Advances with McIDAS III. *Proc. of the Nowcasting II Symp., Norrkoping, Sweden,* European Space Agency, Keplerlaan, The Netherlands, 433-438.

——— and J. F. W. Purdom, 1984: Aircraft penetration of arc cloud lines. *Proc. of the Conf. on Satellite/Remote Sensing and Applications, June 25-29, Clearwater Beach, FL*, Amer. Meteor. Soc., Boston, MA, 160-163.

———, 1983: The retrieval of atmospheric profiles from VAS geostationary radiance observations. *J. Atmos. Sci., 40*, 2025-2035.

———, V. E. Suomi, W. P. Menzel, H. M. Woolf, L. A. Stromousky, H. E. Revercomb, C. M. Hayden, D. N. Erickson, and F. R. Mosher, 1981: First sounding results from VAS-D. *Bull. Amer. Meteor. Soc., 62*, 232-236.

——— and H. M. Woolf, 1976: The use of eigenvectors of statistical covariance matrices for interpreting satellite sounding radiometer observations. *J. Atmos. Sci., 33*, 1127-1140.

———, 1970: A regression method for obtaining real time temperature and geopotential height profiles from satellite spectrometer measurements and its application to Nimbus-3 "SIRS" observations. *Mon. Wea. Rev., 98*, 582-603.

———, 1968: An improved method for calculating tropospheric temperature and moisture from satellite radiometer measurements. *Mon. Wea. Rev., 96*, 387-396.

Sorenson, J. E. and W. B. Beckwith, 1975: Clear air turbulence forecasting as practiced in airline operations. *Proc. of the FAA Symp. on Clear Air Turbulence Forecasting, August 12, Washington, DC*, 22 pp.

Spayd, L. E., Jr. and R. A. Scofield, 1984a: An experimental satellite derived heavy convective rainfall short-range forecasting technique. *Proc. of the 10th Conf. on Weather Forecasting & Analysis, June 25-29, Clearwater Beach, FL*, Amer. Meteor. Soc., Boston, MA, 400-408.

——— and R. A. Scofield, 1984b: A tropical cyclone precipitation estimation technique using geostationary satellite data. *NOAA Technical Memorandum NESDIS 5*, Dept. of Commerce, Washington, DC, 36 pp.

Spencer, R. W., W. S. Olson, W. U. Rongzhang, D. Martin, J. A. Weinman, and D. A. Santek, 1983: Heavy thunderstorms observed over land by the NIMBUS-7 scanning multichannel microwave radiometer. *J. Clim. &. Appl. Meteor., 22*, 1041-1046.

———, D. W. Martin, B. B. Hinton, and J. A. Weinman, 1982: Heavy thunderstorms observed over land by the NIMBUS 7 scanning multichannel microwave radiometer. *J. Clim. Appl. Meteor., 22*, 1041-1046.

Staelin, D. H., A. H. Barrett, P. W. Rosenkranz, F. T. Barath, E. J. Johnston, J. W. Waters, A. Wouters, and W. B. Lenoir, 1975: The scanning microwave spectrometer (SCAMS) experiment. *NIMBUS-6 User's Guide*, Contract NAS-5-20694, Goddard Space Flight Center, Greenbelt, MD, 227 pp.

Strong, A. E. and R. J. DeRycke, 1973: Ocean current monitoring using a new satellite sensing technique. *Science, 182*, 482-484.

——— and I. S. Ruff, 1970: Utilizing satellite-observed solar reflections from the sea as an indicator of surface wind speeds. *Remote Sens. Environ., 1*, 181-185.

Suchman, D. and D. W. Martin, 1976: Wind sets from SMS images: An assessment of quality for GATE. *J. Appl. Meteor., 15*, 1265-1278.

Summers, J. R. and T. Gotwald, 1981: *Building and Operating Weather Satellite Ground Stations*. Natl. Aero. and Space Adm., GSFC, Greenbelt, MD, 83 pp.

Suomi, V. E., R. Fox, S. S. Limaye, and W. L. Smith, 1983: McIDAS III: A modern interactive data access and analysis system. *J. Climate Appl. Meteor., 22*, 766-778.

———, 1977: An Introduction to McIDAS, interactive video displays for atmospheric studies. *Proc. of Workshop, Univer. of Wisconsin-Madison, June 14-16*, J. Hilyard (ed.), 322 pp.

———, (Principal Investigator), 1975: Man-computer interactive data access system (McIDAS). *Final Report, Contract NAS5-23296*, Univ. of Wisconsin, Madison, WI, 255 pp.

Susskind, J., J. Rosenfield, D. Reuter, and M. T. Chahine, 1984: Remote sensing of weather and climate parameters from HIRS2/MSU on TIROS-N. *J. Geophys. Res.*, 89, 4677-4698.

Szejwach, G., T. N. Sletten, and A. F. Hasler, 1983: The use of stereoscopic satellite observation in the determination of the emissivity of cirrus. *Adv. Space Res., 2*, Pergamon Press, New York, NY, 161-164.

———, 1982: Determination of semi-transparant cirrus cloud temperature from infrared radiances: Application to METEOSAT. *J. Appl. Meteor., 21*, 384-393.

Tepper, M., 1963: The status of the program in meteorological satellites. *Rocket and Satellite Meteorology*, John Wiley and Sons, New York, NY, 437-440.

Thomasell, A., Jr., A. Gruber, H. Brodrick, N. Wolfson, and Z. Alperson, 1986: The impact of satellite soundings on the Israel Meteorological Service. *Monthly Weather Review, 114*, 1251-1262.

———, 1979: Wind analysis by conditional relaxation. *NOAA Technical Report NESS 77*, Dept. of Commerce, Washington, DC, 23 pp.

Timchalk, A., L. Hubert, and S. Fritz, 1965: Wind speeds from TIROS pictures of storms in the tropics. *MSL Report No. 33*, U.S. Weather Bureau, Dept. of Commerce, Washington, DC, 33 pp.

Todd, D. K. (ed.), 1970: Surface water. *The Water Encyclopedia*. Water Information Center, Port Washington, NY, 169-171.

Tracton, S. M. and R. D. McPherson, 1977: On the impact of radiometric sounding data upon operational numerical weather prediction at NMC. *Bull. Amer. Meteor. Soc., 58*, 1201-1209.

Troup, A. J. and N. A. Streten, 1972: Satellite-observed southern hemisphere cloud vortices in relation to conventional observations. *J. Appl. Meteor., 11*, 909-917.

Uppala, S., P. Kallberg, A. Hollingsworth, and S. Tibaldi, 1985: Results from two recent observing system experiments. *Proc. of the 1st Natl. Workshop on the Global Weather Experiment*, National Academy Press, Washington, DC, 85-117.

U.S. Army Corps of Engineers, 1979: *Red River of the North Post-Flood Report 1978*. United States Army Engineer District, St. Paul, MN.

U.S. Geological Survey, 1978: Water resources review for April 1978.

U.S. Weather Bureau, 1963: Catalog of meteorological satellite data—TIROS IV television cloud photography. *Key to Meteorological Records Documentation 5.34*, Dept. of Commerce, Washington, DC, 1-6.

Valovcin, F. R., 1968: Infrared measurements of jet stream cirrus. *J. Appl. Meteor., 7*, 817.

Vaughn, W., 1982: Meteorological satellites - Past, present, and future. *NASA Conference Publication 2227*, Natl. Aero. and Space Adm., Washington, DC, 2 pp.

Velden, C. S., W. L. Smith, and M. Mayfield, 1984: Applications of VAS and TOVS to tropical cyclones. *Bull. Amer. Meteor. Soc., 65*, 1059-1067.

——— and W. L. Smith, 1983: Monitoring tropical cyclone evolution with NOAA satellite microwave observations. *J. Climate Appl. Meteor., 22*, 714-724.

Viezee, W., R. M. Endlich, and S. M. Serenbreny, 1967: Satellite-viewed jet stream clouds in relation to observed wind field. *J. Appl. Meteor., 6*, 929-935.

Wark, D. Q. and H. E. Fleming, 1966: Indirect measurements of atmospheric temperature profiles from satellites: I. Introduction. *Mon. Wea. Rev., 94*, 351-362.

Warner, C., 1984: Stereo-pair photographs of monsoon clouds. *Bull. Amer. Meteor. Soc., 65*, 344-347.

Weaver, J. F. and J. F. W. Purdom, 1983: Some unusual aspects of thunderstorm cloud top behavior on May 11, 1982. *Proc. of the 13th Conf. on Severe Local Storms, October 17-20, Tulsa, OK*, Amer. Meteor. Soc., Boston, MA, 154-157.

——— and F. Kelly, 1982: A mesoscale climatologically-based forecast technique for Colorado. *Proc. of the 9th Conf. on Weather Forecasting and Analysis, June 28-July 1, Seattle, WA*, Amer. Meteor. Soc., Boston, MA, 277-280.

Weinman, J. A. and P. J. Guetter, 1977: Determination of rainfall distributions from microwave radiation measured by the NIMBUS-6 ESMR. *J. Appl. Meteor., 16*, 437-442.

Weiss, C. E., 1978: Cloud location corrections near the horizon of an SMS image. *Satellite Applications Information Note 78-8*, Dept. of Commerce, Washington, DC, 8 pp.

———, 1976: Diurnal variation in the wavelength of lee standing waves. *Satellite Applications Information Note 76-14*, Dept. of Commerce, NOAA, NESS, Washington, DC, 7 pp.

——— and J. J. Gurka, 1975: Fog dissipation in the vicinity of the Chesapeake Bay. *Satellite Applications Information Note, 12/75-1*, Dept. of Commerce, NOAA, NESS, Washington, DC, 4 pp.

Weldon, R. W. and S. Holmes, 1989: Characteristics of water vapor imagery. *NESDIS Technical Memorandum* (to be published).

———, 1979: Cloud patterns and the upper wind field, Part IV. *Satellite Training Course Notes*, Dept. of Commerce, NOAA, NESS, Washington, DC, 80 pp.

———, 1975: The structure and evoluton of winter storms, Part II. *Satellite Training Course Notes*, Dept. of Commerce, NOAA, NESS, Washington, DC, 49 pp.

Wexler, H., 1954: Observing the weather from a satellite vehicle. *British Interplantery Soc., 13*, 269-276.

Whitney, L. F., 1984: Satellite-derived products, winds, NESDIS. *WMO Meeting on Calibration, Validation and Intercomparison of Satellite Products,* Geneva, Switzerland, 14 pp.

———, 1983: International comparison of satellite winds—An update. *Adv. Space Res., 2*, Pergamon Press, New York, NY, 73-77.

Widger, W. K., 1964: Synthesis of interpretation of extratropical vortex patterns as seen by TIROS. *Mon. Wea. Rev., 192*, 263-282.

———, 1961: Satellite meteorology—fancy and fact. *Weather, 16*, 47-55.

Wiesnet, D. R., D. F. McGinnis, and J. A. Pritchard, 1974: Mapping of the 1973 Mississippi River floods by the NOAA-2 satellite. *Water Resources Bulletin, 10*, 1040-1049.

Wilson, J. and R. Carbone, 1984: Nowcasting with Doppler radar: Forecaster-computer relationship. *Proc. of the Nowcasting II Symp., Norrkoping, Sweden*, European Space Agency, Keplerlaan, The Netherlands, 177-186.

Wolf, D. E., D. J. Hall, and R. M. Endlich, 1977: Experiments in automatic cloud tracking using SMS-GOES data. *J. Appl. Meteor., 16*, 1219-1230.

World Meteorological Organization, 1975: Manual on the observations of clouds and other meteors. *International Cloud Atlas, 1, WMO No. 407*, World Meteor. Organ., Geneva, Switzerland, 155.

Young, M. T., 1975: The GOES wind operation, Chap. 13. Central Processing and Analysis of Geostationary Satellite Data. C. L. Bristor (Ed.), *NOAA Technical Memorandum NESS 64*, Dept. of Commerce, Washington, DC, 111-121.

———, R. C. Doolittle, and L. M. Mace, 1972: Operational procedures for estimating wind vectors from geostationary satellite data. *NOAA Technical Memorandum NESS 39*, Dept. of Commerce, Washington, DC, 19 pp.

Zehr, R. M. and R. N. Green, 1984: Mesoscale applications of VAS imagery. *Proc. of the Conf. on Satellite/Remote Sensing and Applications, June 25-29, Clearwater Beach, FL*, Amer. Meteor. Soc., Boston, MA, 94-98.

———, 1982: Thunderstorm motion analyses. *Proc. of the 9th Conf. on Weather Forecasting and Analysis, Seattle, WA*, Amer. Meteor. Soc., Boston, MA, 455-458.

Zillman, and P. G. Price, 1972: On the thermal structure of mature southern ocean cyclones. *Aust. Meteor. Mag., 20*, 34-48.

Zipser, E. J., 1984: *The National STORM program; STORM central phase preliminary program design.* National Center for Atmospheric Research, Boulder, CO, 147 pp.

Contributors to Section VII

Chang, Alfred T. — NASA, Goddard Space Flight Center, Greenbelt, MD 20771

Hasler, Arthur F. — NASA, Goddard Space Flight Center, Greenbelt, MD 20771

Hayden, Christopher — NOAA/NESDIS/SAL, University of Wisconsin, 1225 West Dayton Street, Madison, WI 53706

Purdom, James F. W. — NOAA/NESDIS, Department of Atmospheric Sciences, Colorado State University, Fort Collins, CO 80523

NOAA/NESDIS, Washington, DC 20233:

Anderson, Ralph K.
Clark, J. Dane
Coldren, Raymond L.
*Dvorak, Vernon F.
Ellrod, Gary P.
*Gird, Ronald S.
Grody, Norman C.
Hawkins, Jamison S.
Holmes, Susan J.
*Jager, Gilbert
Matson, Michael
*McClure, C. Hume
*Moses, John F.
Ohring, George
Perkins, Marlin O.
Rao, P. Krishna
*Reinhardt, George
Scofield, Roderick A.
Wark, David Q.
Weaks, Marcia L.
*Whitney, Linwood F., Jr.

* (Moved to a new position.)

Section VIII

APPLICATIONS OF SATELLITE DATA TO LAND AND OCEAN SCIENCES

Although the first weather satellites were intended to be used primarily for meteorological purposes, it soon became apparent that the images could be useful for hydrologic purposes, especially for mapping snow cover and lake and sea ice. When, in the 1970s, the image resolution improved and quantitative forms of the data became available, derivations of sea surface temperature, land-surface temperature, and precipitation were developed. With the coming of multichannel digital measurements late in the 1970s, more and better applications to hydrology, oceanography[1], and the land sciences became possible. These include improved sea- and lake-surface temperatures, detection of fires, volcanic activity, and urban heat islands. Details of the various applications are described in the following chapters.

[1] Many ocean applications were developed in addition to SST, however, they are beyond the scope of this text and will be presented in a separate book.

Chapter VIII-1

THERMAL ANOMALIES[1]

VIII-1.1 INTRODUCTION

Since 1979, several papers have been published on the use of satellite data to detect volcanic eruptions and their ash clouds. Eruptions documented include: Augustine, Alaska, 1976 (Kienle and Shaw, 1979); Soufrière, St. Vincent, 1979 (Krueger, 1982); Alaid, U. S. S.R., and Pagan, Marianna Islands, both in 1981 (Sawada, 1983); El Chichón, Mexico, 1982 (Matson, 1984); and Galunggung, Indonesia, 1982 (Hanstrum and Watson, 1983). The El Chichón event is used here to show how eruptions can be monitored with meteorological satellite data.

Because of increased concern about the climatic and economic impact of forest fires and seasonal burning, monitoring is needed. Aircraft surveys are not economical. Environmental satellites with large swath widths, 24-hour coverage, and fire detecting sensors offer a means for efficient and economical detection and monitoring of fires in large areas.

The increased concern in recent years about human effects on the environment have been summarized by Landsberg (1970) and Kellogg (1977). Increased urbanization and industrialization in many cities have increased the intensity and extent of the associated positive thermal anomalies known as "urban heat islands." During the last decade or so, satellite thermal IR radiation data have been used in studies of urban heat islands. Satellite determination of land surface temperature for agricultural applications is discussed in Chapter X-2.

[1] Primary Contributor: Michael Matson

VIII-1.2 VOLCANOES

El Chichón (17.33°N, 93.20°W) first erupted on March 28, 1982. Three subsequent eruptions followed on April 3 and 4, 1982 (Matson, 1984). Figure VIII-1-1, a GOES IR image, shows the first of two eruptions on April 3, approximately 10 min after the eruption began. The whiteness of the top of the eruption cloud equates to equivalent blackbody temperatures lower than −69.7°C. GOES digital data an hour later gave a minimum temperature of −71.2°C in the cloud. Comparison with radiosonde data at Veracruz, Mexico (19.15°N, 96.12°W), for 1800 LST on April 2 showed that the satellite measured temperature of the eruption cloud corresponded to an altitude of either 15.2 or 18.8 km. The tropopause was at 17 km. Six hours after the first eruption of April 3, the El Chichón ash cloud had spread both to the northeast and the southwest (Figure VIII-1-2) and covered an area of 86,500 km^2.

The Veracruz upper air data at 1800 LST on April 3 showed southwest winds between 16.3 and 18.5 km. Northeast winds were present below 3 km and between 21–24 km. Figure VIII-1-2 shows that the edge of the ash cloud moving to the northeast had traveled approximately three times as far from the volcano as the edge moving to the southwest. The radiosonde data showed the southwest winds were approximately 2–3 times faster than the northeast winds. This indicates

Figure VIII-1-1. GOES IR image of the El Chichón eruption on, 0300 LST, April 3, 1982.

Figure VIII-1-3. GOES IR image of the El Chichón eruption, 2000 LST, April 3, 1982.

that the part of the ash cloud moving northeastward was near the tropopause (17 km), while the part moving southwestward was most likely near 20 km; the radiosonde data also indicates that the ash cloud could also be near the surface ($\sim$3 km).

El Chichón again erupted late on April 3 and early on April 4. The first eruption began at about 2000 LST (Figure VIII-1-3); at 2200 LST the lowest satellite temperature recorded was $-78.2°$C, corresponding to the tropopause temperature. The ash cloud spread over southern Mexico, Belize, and northern Guatemala (Figure VIII-1-4) covering an area of 192,000 km^2. From the Veracruz upper-wind data and the measured displacements of the cloud edges to the northeast and

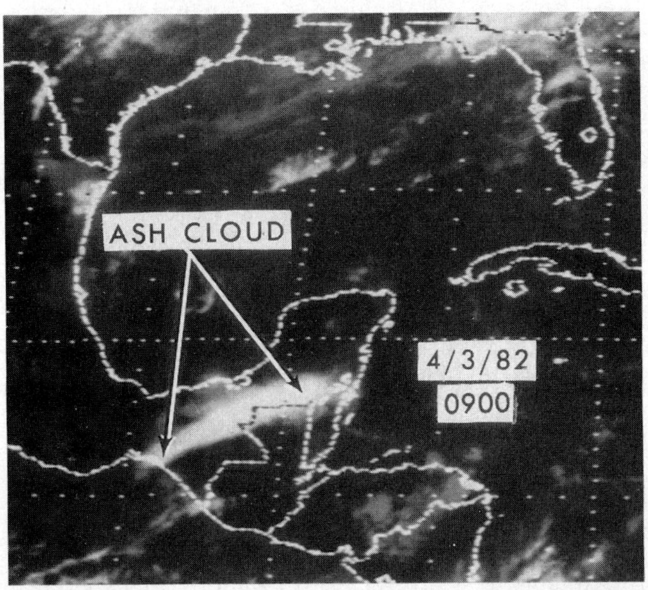

Figure VIII-1-2. GOES IR image of the El Chichón ash cloud, 0900 LST, April 3, 1982.

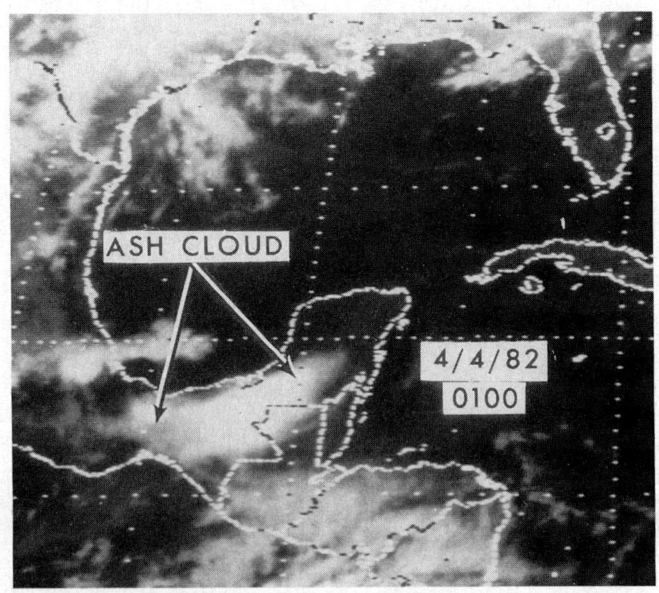

Figure VIII-1-4. GOES IR image of the El Chichón eruption, 0100 LST, April 4, 1982.

Figure VIII-1-5. GOES IR image of the El Chichón eruption, 0600 LST, April 4, 1982.

southwest, the northeastward moving cloud top was at about 17 km while the southwestward moving cloud top was in the stratosphere at about 23 km.

The largest eruption of all began at 0522 LST on April 4 (Figure VIII-1-5). The new eruption cloud merged with the ash cloud from the preceding eruption. The IR data again gave lowest satellite temperatures of −78.2°C (that of the tropopause). Figure VIII-1-6 shows the dispersal of the April 4 ash cloud, which covered 1,200,000 km^2. Once again the northeastward moving ash cloud was primarily tropospheric and the

Figure VIII-1-6. GOES IR image of the El Chichón eruption, 1730 LST, April 4, 1982.

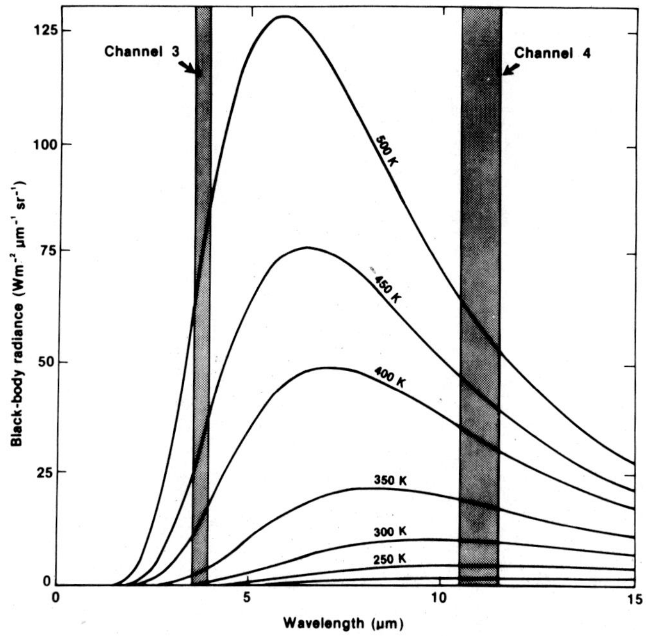

Figure VIII-1-7. Planck radiances for temperatures from 200 to 500 K. For a given increase in temperature the increase in area under the Channel 3 (3.7 μm) segments of the curves is much greater than under the Channel 4 (11 μm) segments.

southwestward moving cloud was stratospheric. The height of the stratospheric cloud derived for these periods immediately following the eruptions corresponds closely with satellite and lidar measurements of the El Chichón ash and aerosol cloud that subsequently circled the Earth (Robock and Matson, 1983; Thomas, 1983; DeLuisi, 1983; McCormick, 1983).

VIII-1.3 FIRES

Large areas of the world (e.g., Alaska, the western United States, western Canada, and Siberia) have a heightened potential for forest fires during summer. Limited rainfall and summer heat reduce soil moisture and cause systematic dehydration of underbrush and low vegetation from root through canopy. In such areas, fires can start easily, spread rapidly, and destroy thousands of hectares of forest. Although fire tower and aircraft surveillance have served to detect forest fires for many years, satellites can provide widespread surveillance over even the most remote regions.

Other high temperature targets (i.e., industrial, commercial, and residential) are also detectable from satellites. Although these may be at scales smaller than the resolution of satellite instruments, their intense heat allows for their detection.

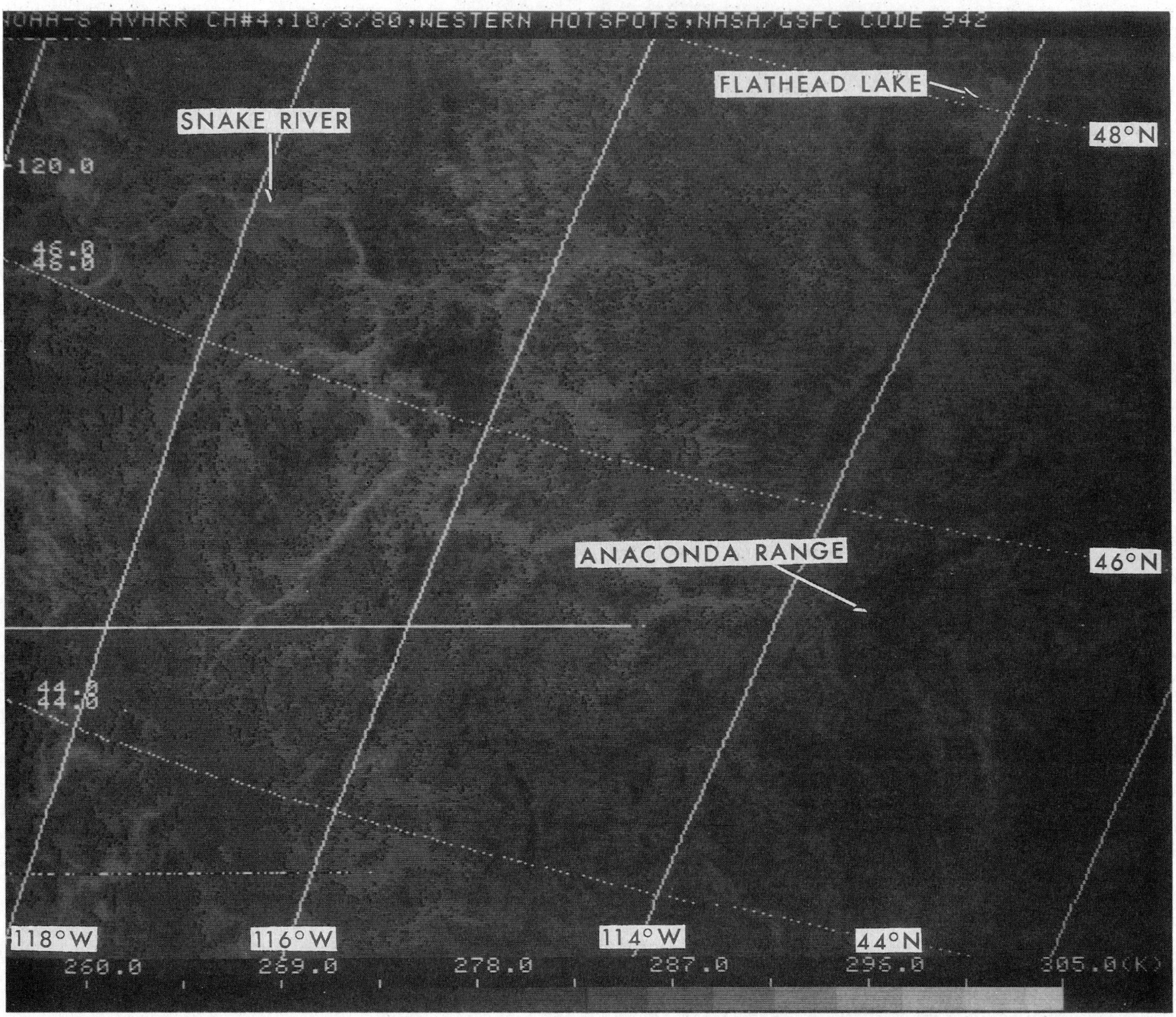

Figure VIII-1-8. NOAA-6 11 μm image of the western United States, October 3, 1980.

Matson and Dozier (1981) showed that simultaneous use of the 3.7 and 11 μm IR channels of NOAA polar orbiting satellites is capable of detecting very high-temperature areas, such as steel plants and waste gas flares. The 3.7 μm channel is particularly sensitive to high-temperature targets (Figure VIII-1-7). For high-temperature targets, the maximum blackbody radiance response shifts toward the 3.7 μm channel. Figures VIII-1-8 and VIII-1-9 show the images from a NOAA-6 pass over the western United States on October 3, 1980 (Matson and Schneider, 1984). Although the gray scale for both images is the same, the 3.7 μm image shows several hot spots (white) not apparent on the 11 μm image. Figure VIII-1-10 is a brightness temperature plot over two of these high temperature areas (targets 1 and 2) where the 3.7 μm temperatures are 16.2°C and 33.9°C, respectively, higher than the corresponding 11 μm temperatures.

Using algorithms developed by Matson and Dozier (1981), it is possible to use the temperature difference in the two channels to estimate the area and temperature of a hot target. For the two targets in Figure VIII-1-10, the areas and temperatures were 0.28 ha and 430° K for target 1, and 1.7 ha and 483° K for target

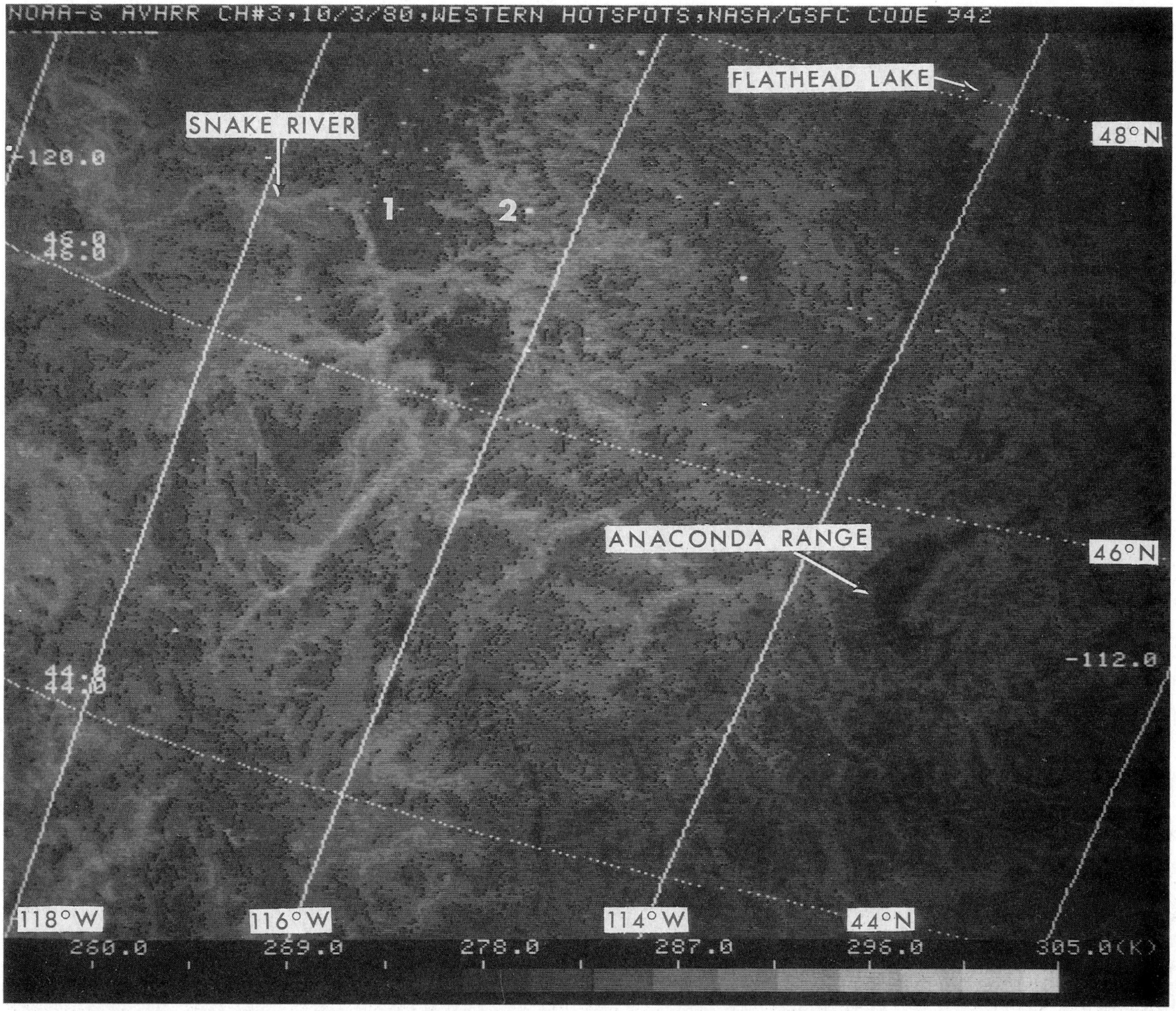

Figure VIII-1-9. NOAA-6 3.7 μm image of the western United States, October 3, 1980. Note the "hot" (black) targets 1 and 2.

2. As evidenced by the calculated target sizes, a target smaller than 1.1 km causes a response in a 1.1 $\times$ 1.1 km, 3.7 μm measurement. Most of the high-temperature targets in Figure VIII-1-9 were subresolution fires. Information from the U.S. Forest Service showed that the high temperature areas north of the Snake River were the result of private agricultural burning in the state of Washington (4.5 to 81 ha). The hot spots southwest of Flathead Lake were mostly the result of small U.S. Forest Service prescription burns in Idaho.

Not all hot spots are fires. Target 2 is the phosphorous plant operated by the FMC Corporation in Pocatello, Idaho. In using the 3.7 μm channel, it is therefore necessary to screen the imagery for industrial targets. This is relatively easy since the industrial targets appear on each daily image. Fires generally appear for a few days and change from day to day.

The utility of the NOAA 3.7 μm data for fire detection is demonstrated in two other cases. The first is the Dune Lake, Alaska, fire of late May 1981. Figures VIII-1-11 through VIII-1-13 show the Dune Lake fire on 3 successive days. The gray scale in these images ranges from black for hot to white for cold. It is easy to follow the progress of the fire (black areas) as it moves toward the north from May 22 to 24.

The second case is over Brazil near Manaus on Oc-

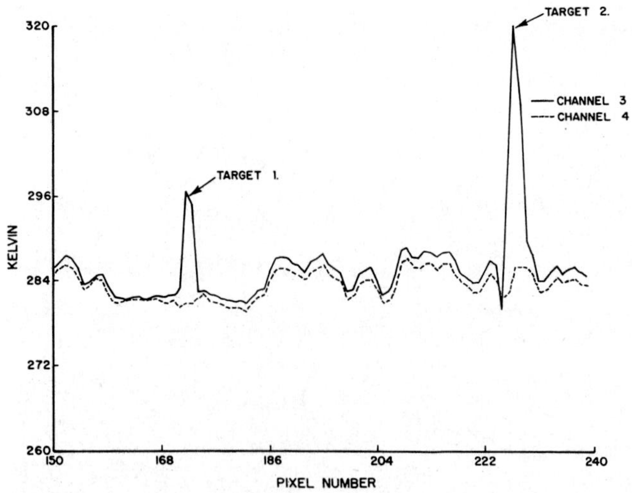

Figure VIII-1-10. 3.7 μm (Channel 3) and 11 μm (Channel 4) brightness temperature plot of hot targets 1 and 2 in Figure VIII-1-9.

Figure VIII-1-12. 3.7 μm IR image of the Dune Lake, Alaska, fire, May 23, 1981.

tober 17, 1981, (Figures VIII-1-14 and VIII-1-15). In the reflective IR (0.9 μm) channel (Figure VIII-1-14) a large smoke plume is apparent; the other white areas are cumulus clouds. The 3.7 μm channel (Figure VIII-1-15) clearly shows the fire generating the smoke plume, plus many other fires near the Solimoes River and Manaus (Matson and Schneider, 1984). These fires are a result of extensive slash and burn agriculture, timber clearing, and colonization in the area.

VIII-1.4 URBAN HEAT ISLANDS

Bornstein (1968) discussed the energy balances that lead to heat islands. The large heat capacity and high

Figure VIII-1-11. 3.7 μm IR image of the Dune Lake, Alaska, fire, May 22, 1981.

Figure VIII-1-13. 3.7 μm IR image of the Dune Lake, Alaska, fire, May 24, 1981.

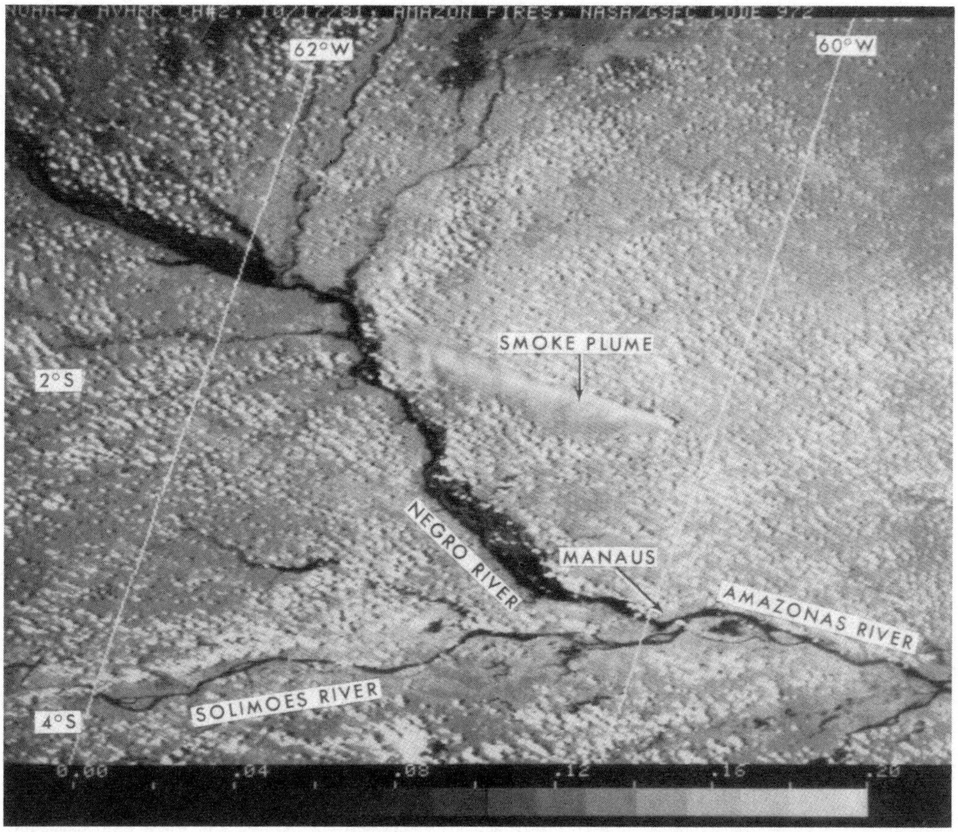

Figure VIII-1-14. NOAA-7 0.9 μm image of a fire plume near Manaus, Brazil, October 17, 1981.

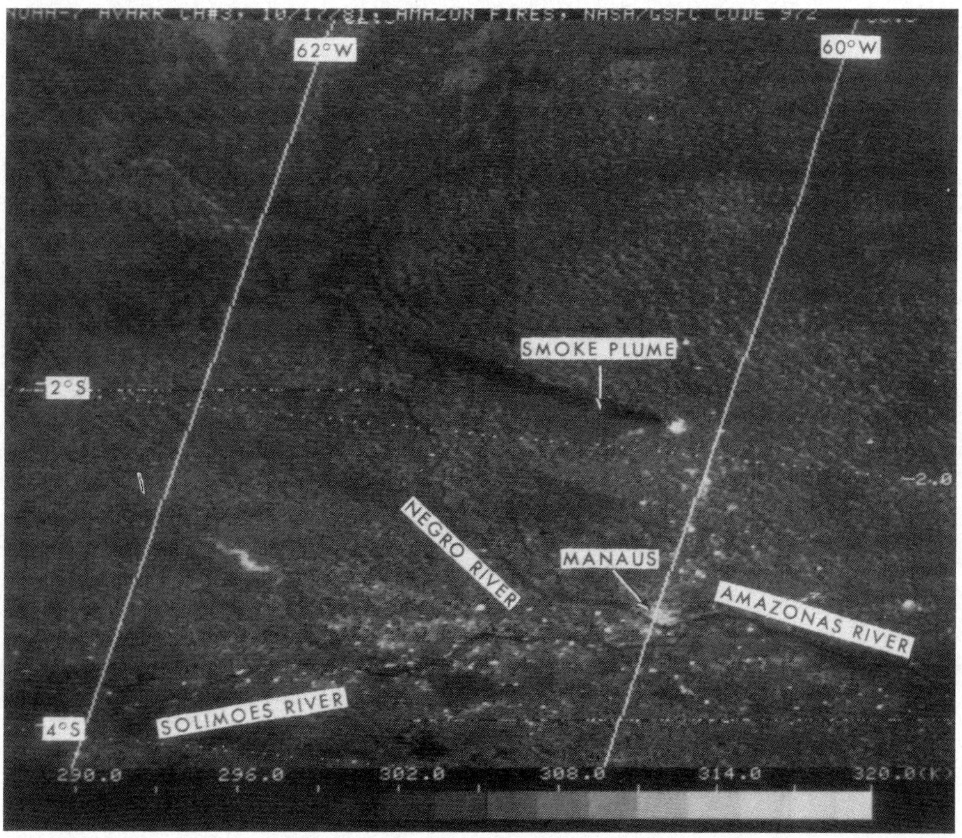

Figure VIII-1-15. NOAA-7 3.7 μm image of fire areas (black) near Manaus, Brazil, October 17, 1981. The Manaus urban heat island also appears white.

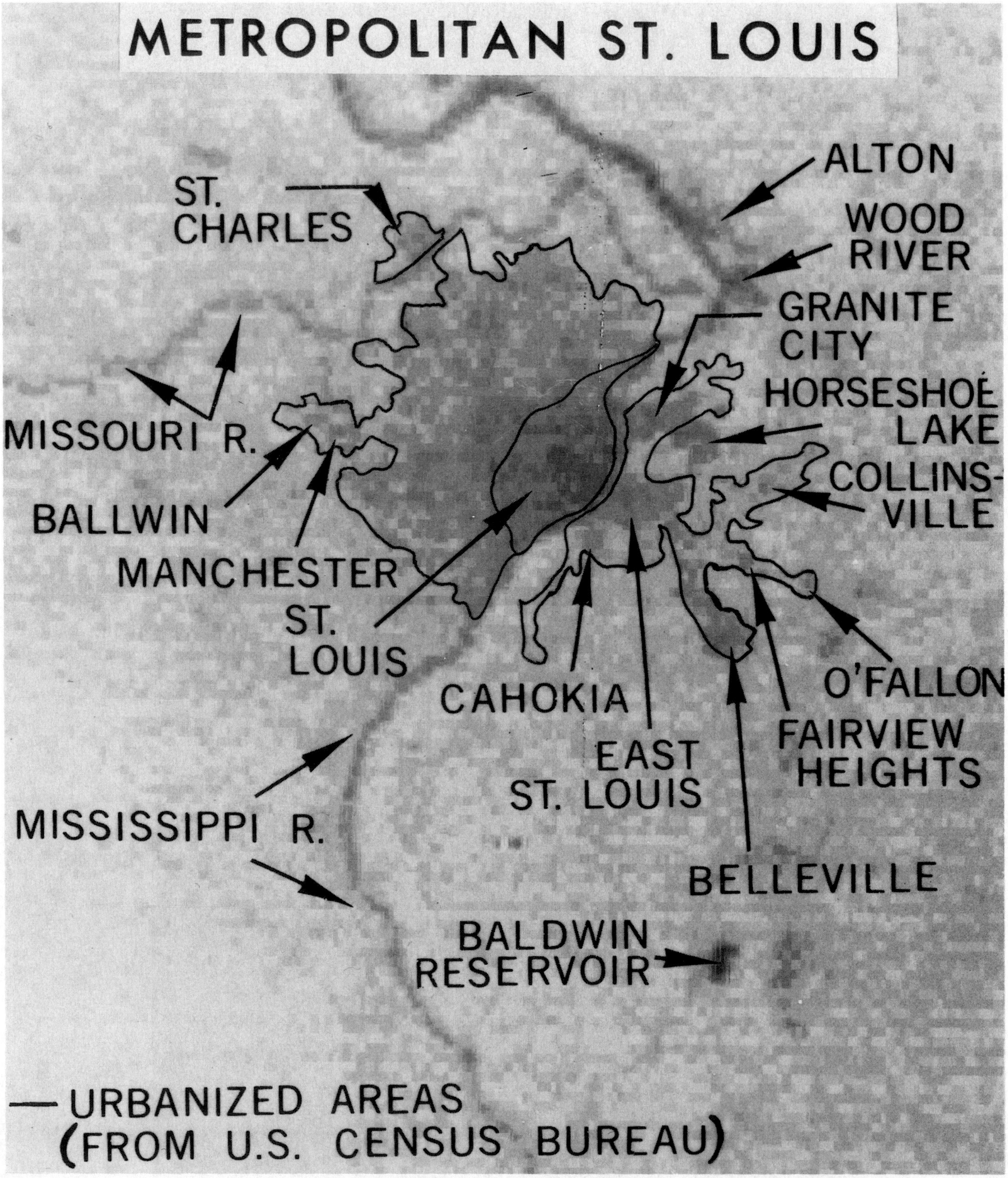

Figure VIII-1-16. Computer enhanced VHRR IR image of the St. Louis area, 0200 UTC, July 28, 1977. The gray scale has 22 steps from black to white as follows: 32, 31, 30, . . . 17, 16, 14, 12, 8, 4, 0°C. Urban boundaries are based on the 1970 Census Map of Urban Areas. The urban boundaries are comprised of political limits of incorporated areas as well as those of locally recognized communities. Population densities within these boundaries are generally >390 persons km^{-2}.

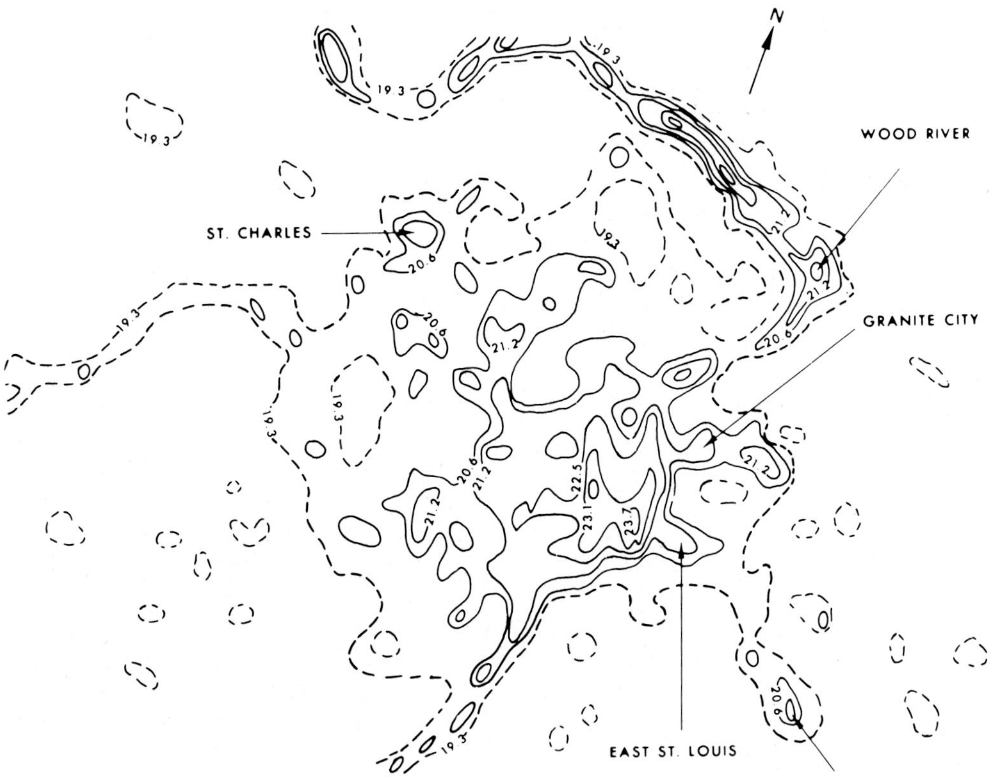

Figure VIII-1-17. Isotherm analysis of the digitized printout of satellite- derived temperatures (°C) for St. Louis and vicinity, 0200 UTC, July 28, 1977. Brightness temperatures have been nominally corrected for atmospheric attenuation.

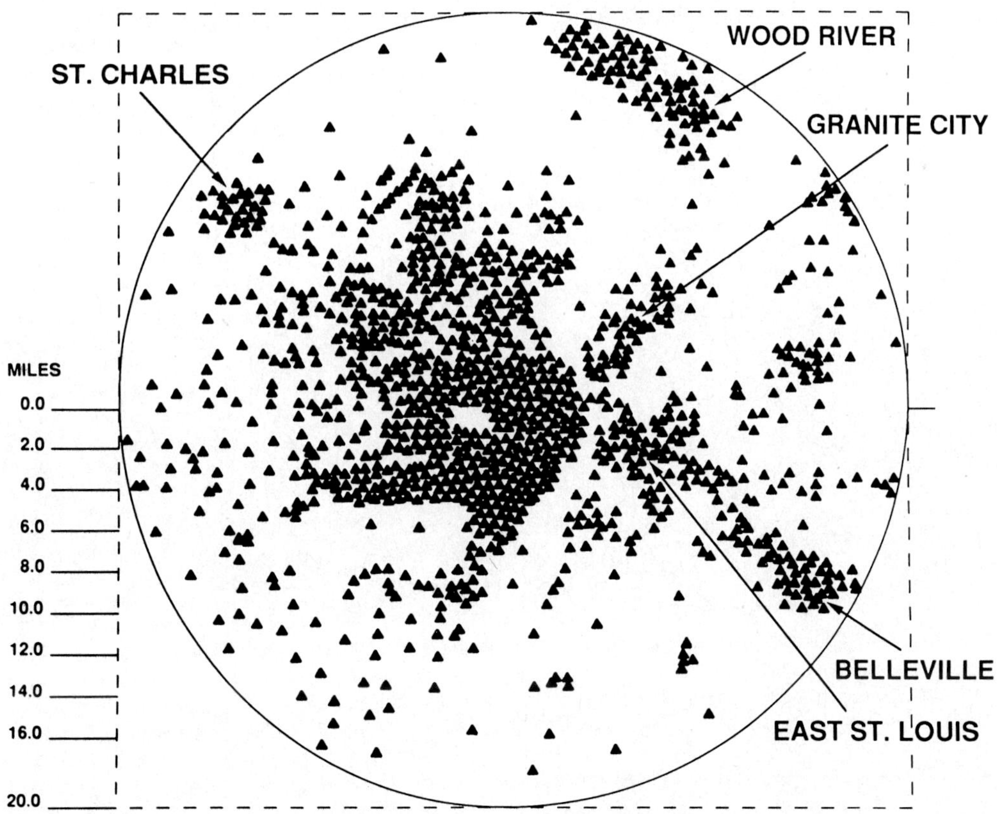

Figure VIII-1-18. St. Louis population density within a 32.2-km radius of the approximate center of St. Louis (38.65°N, 90.25°W). Each triangle represents 350 dwelling units or about 1000 persons. This map was obtained from the U.S. Census Bureau and is based on the 1970 census.

Table VIII-1-1 *Maximum heat island effect as determined from VHRR IR data, 0146 UTC, July 28, 1977. Population figures are for 1974 based on U.S. Bureau of the Census Standard Metropolitan Statistical Areas (SMSA) as defined in the 1970 census.*

CITY	MAXIMUM URBAN-RURAL TEMPERATURE DIFFERENCE (°C)	POPULATION ($\times 10^6$)
Louisville, KY	6.5	0.89
Baltimore, MD	5.2	2.14
Washington, DC	5.2	3.02
Cincinnati, OH	5.1	1.38
Indianapolis, IN	4.5	1.14
Dayton, OH	4.5	0.84
St. Louis, MO	4.4	2.37
Richmond, VA	3.8	0.57
Columbus, OH	3.3	1.07
Kansas City, MO	3.2	1.30
Petersburg, VA	2.6	0.04

conductivity of urban buildings prevent rapid cooling after sunset. A variety of other factors enter into heat island formation; examples are rapid precipitation runoff and reduction of evapotranspiration in urban areas, and waste heat from buildings. Landsberg (1970) also mentioned the increased "roughness" of cities, which causes a 25% reduction in wind speeds, and a lower albedo.

In 1971, the 5-year METROMEX project was started in the St. Louis area to study the effects of an urban industrial complex on the environment. Results of the project (Changnon, 1978; Huff and Vogel, 1978) showed that such an area acts as a catalyst for increasing summer thunderstorms and precipitation. Landsberg (1976) also summarized other effects of urban heat islands.

Since the early 1970s, satellite IR data have been used in studies of urban heat islands. Rao (1972) used 7.4-km resolution SR thermal measurements from ITOS-1 to show that the New York City-Philadelphia-Washington urban corridor could be roughly delineated as a high temperature island (Matson et al., 1978). Carlson et al. (1977) made morning and evening surface temperature analyses for the Los Angeles area using 1-km resolution VHHR IR data from NOAA-3. These analyses showed that the highest morning temperatures were over the industrial zone; in the evening, the highest temperatures were in the central business district and high density residential areas. Matson et al. (1978) and Matson and Legeckis (1980) developed methods for using satellite IR imagery to show urban heat islands in considerable detail.

Figures VIII-1-16, VIII-1-17, and VIII-1-18 show mappings of the heat island in the St. Louis area on July 28, 1977, and the associated population density of the area for comparison purposes. Table VIII-1-1 lists the differences between the highest satellite determined temperatures in the urban area and the average temperature of the surrounding rural areas for St. Louis and ten other cities on July 28, 1977.

Chapter VIII-2

SNOW COVER

VIII-2.1 INTRODUCTION

Snow cover plays an important role in climatic processes and weather forecasting. It affects surface and air temperature, atmospheric circulation patterns, storm tracks, cloudiness, surface albedo, evaporation, precipitation, water storage, and soil moisture (Kukla, 1981; Walsh, 1984), as well as energy requirements, transportation systems, food supplies, and winter recreation.

Attempts to determine snow cover and amount by use of images and quantitative data from satellites have been underway since TIROS-1 was launched. According to Popham (1968), the first operational test to use satellite imagery for snow surveillance was conducted in April 1963. In November 1966, the first satellite derived Northern Hemisphere Weekly Snow and Ice Cover Chart was produced by ESSA. Since then, continental snow cover has been monitored weekly. Barnes and Bowley (1974) made several evaluations of the use of TIROS and ESSA satellite sensors as tools for snow studies, and produced a handbook of techniques for such uses.

In 1972 and 1973, NESS experimentally mapped the American River basin above Fair Oaks, California, and transmitted snow-extent maps to the River Forecast Center (RFC) at Sacramento (Wiesnet and McGinnis, 1973). Shortly thereafter, several other RFC asked for snow mapping of river basins. Estimates of percent of basin covered by snow were derived from NOAA-2 VHRR imagery and were sent to the RFC within 30 hours after the satellite pass. By the late 1970s NESS operationally produced over 600 snow maps annually for 30 river basins. The program was transferred to the National Weather Service (NWS) in 1982.

Snow analyses have been made using data from NOAA polar orbiting satellites, Landsat, and GOES. Highly detailed imagery can be obtained in four spectral bands from the Landsat MultiSpectral Scanner (MSS). Snow maps of the American River basin from VHRR and MSS VIS data were shown to compare favorably (Wiesnet and McGinnis, 1973). The high-resolution Landsat data have limited value because of the satellite's 9- to 18-day coverage cycle in contrast to the daily coverage of NOAA polar orbiters. Limited geographic area coverage by Landsat also makes snow mapping difficult in larger basins (Wiesnet and McGinnis, 1973).

GOES data are valuable for snow mapping because of the high frequency of coverage. Because of the satellite's positioning, the resolution of GOES VISSR imagery decreases at higher latitudes, particularly at the edges of the image area. However, accurate snow maps can be made up to 52°N.

VIII-2.2 NORTHERN HEMISPHERE SNOW COVER MAPS

Northern Hemisphere Weekly Snow and Ice Cover Charts (Figures VIII-2-1 and VIII-2-2) are prepared using VIS sensors listed in Table VIII-2-1. These charts show the areal extent and brightness, but not the depth,

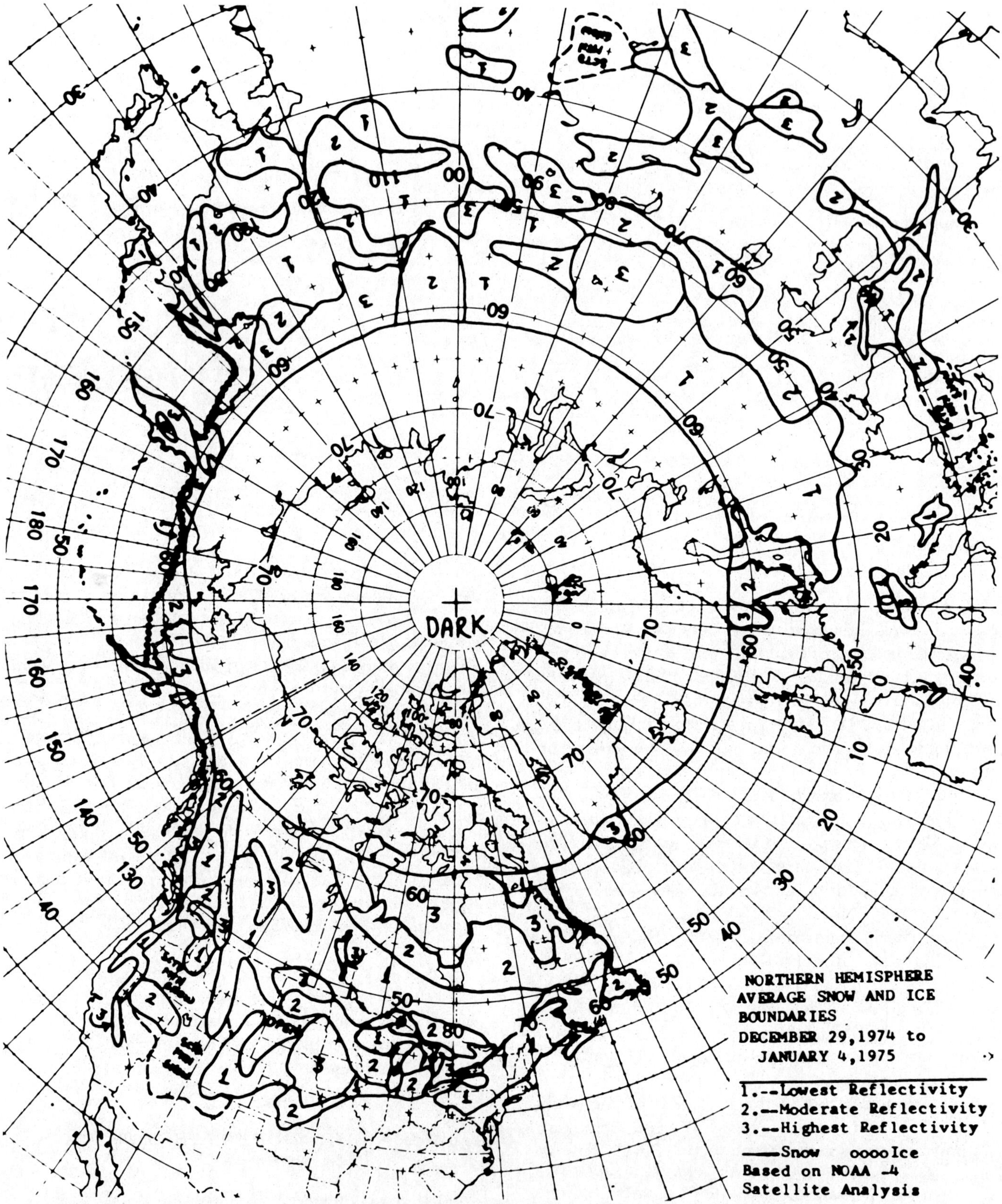

Figure VIII-2-1. Example of a Northern Hemisphere Weekly Snow and Ice Cover Chart generated prior to May 1982.

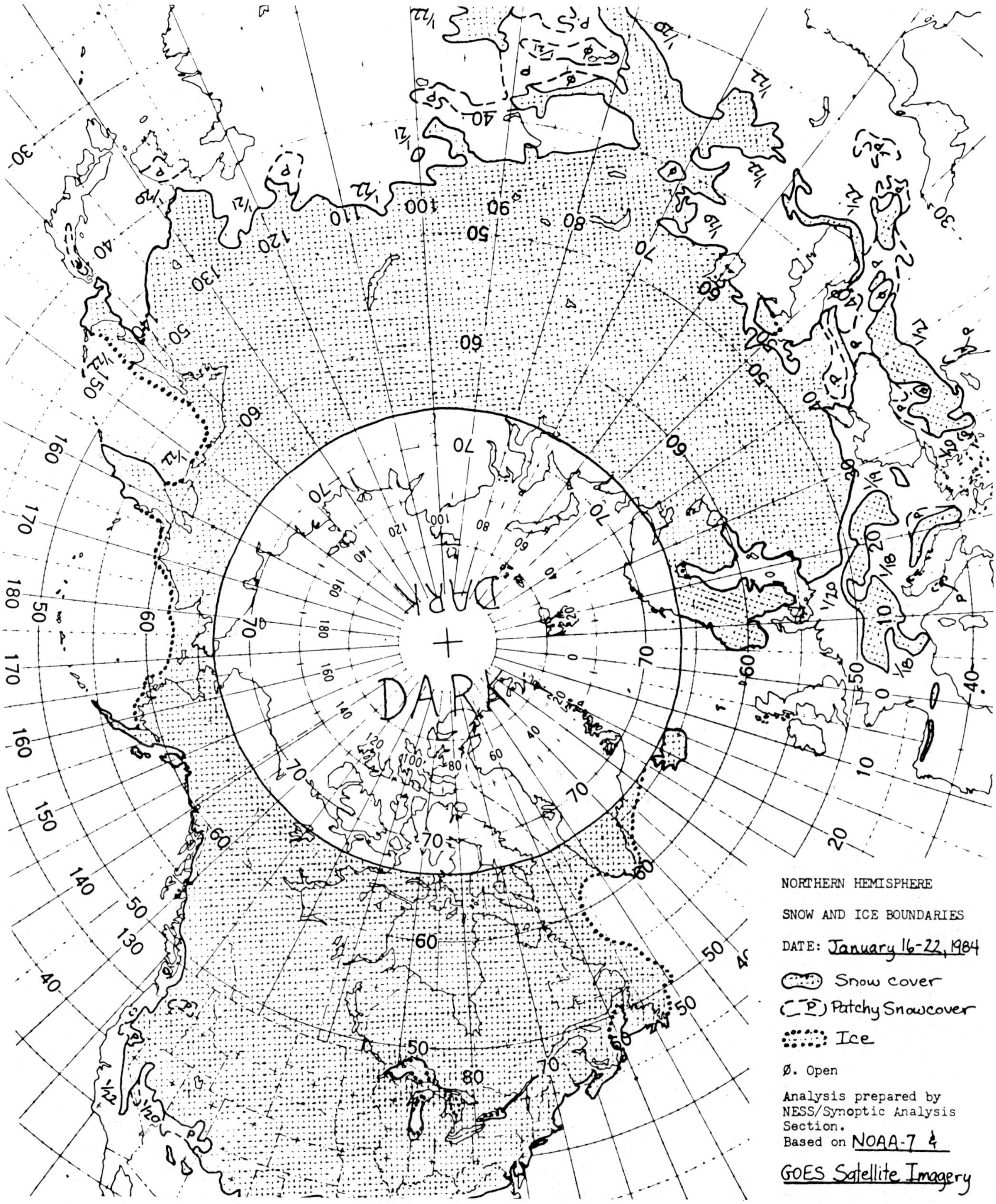

Figure VIII-2-2. Example of a Northern Hemisphere Weekly Snow and Ice Cover Chart produced after May 1982.

Table VIII-2-1 *ESSA and NOAA satellites and sensors used in mapping Northern Hemisphere snow and ice cover*

SATELLITE	SENSOR[1]	SPECTRAL BAND (μm)	SUBPOINT RESOLUTION (km)	PERIOD OF OPERATIONS
ESSA-3	AVCS	0.5–0.75	3.7	October 2, 1966–October 9, 1968
ESSA-4	APT	0.5–0.75	3.7	January 26, 1967–December 6, 1967
ESSA-7	AVCS	0.5–0.75	3.7	August 15, 1968–July 19, 1969
ESSA-8	APT	0.5–0.75	3.7	December 15, 1968–March 12, 1976
ESSA-9	AVCS	0.5–0.75	3.7	February 26, 1969–December 15, 1973
ITOS-1	AVCS	0.5–0.75	3.7	January 23, 1970–June 17, 1971
	APT	0.5–0.75	3.7	
	SR	0.52–0.73	3.7	
		10.5–12.5	7.4	
NOAA-2	VHRR	0.6–0.75	1.0	October 15, 1972–January 30, 1975
		10.5–12.5	1.0	
	SR	0.52–0.73	3.7	
		0.50–0.94	3.7	
		10.5–12.5	7.4	
NOAA-3	VHRR	0.6–0.75	1.0	October 6, 1973–December 17, 1974
		10.5–12.5	1.0	
	SR	0.52–0.73	3.7	
		10.5–12.5	7.4	
NOAA-4	VHRR	Same as NOAA-3		November 15, 1974–September 15, 1976
	SR	Same as NOAA-3		
NOAA-5	VHRR	0.6–0.75	1.0	July 29, 1976–March 1, 1979
		10.5–12.5	1.0	
	SR	0.52–0.73	3.7	
		0.50–0.94	3.7	
		10.5–12.5	7.4	
TIROS-N	AVHRR	0.55–0.90	1.1 or 4.4[2]	November 6, 1978–January 20, 1980
		0.725–1.10	1.1 or 4.4	
		3.55–3.93	1.1 or 4.4	
		10.5–11.5	1.1 or 4.4	
NOAA Series	AVHRR	0.55–0.68	1.1 or 4.4	July 16, 1979–Present
		0.725–1.10	1.1 or 4.4	
		3.55–3.93	1.1 or 4.4	
		10.5–11.5[3]	1.1 or 4.4	
		10.5–11.3[4]	1.1 or 4.4	
		11.5–12.5[4]	1.1 or 4.4	
GOES Series	VISSR	0.55–0.70	1.0 or 7.4	May 17, 1975–Present
		10.5–12.5	7.4	

[1] Cameras and sensors: AVCS, Advanced Vidicon Camera System; APT, Automatic Picture Transmission; SR, Scanning Radiometer; VHRR, Very High Resolution Radiometer; AVHRR, Advanced Very High Resolution Radiometer; VISSR, Visible and Infrared Spin Scan Radiometer.

[2] 1.1-km resolution for direct readout (HRPT) and Local Area Coverage (LAC). 4.4-km resolution for Global Area Coverage (GAC).

[3] Four-channel AVHRR.

[4] Five-channel AVHRR.

of continental snow cover. The analyst collects all data received during the current week, compares snow and ice cover with that on the previous week's chart, and makes changes. If an area is cloud covered for the entire week, the previous week's analysis of that area is entered on the new chart.

Prior to May 1982 (Figure VIII-2-1), solid lines on the map indicated boundaries between areas of substantially differing brightnesses of the snow field. The chart was divided into three categories of reflectivity: (1) lowest, (2) moderate, and (3) highest. These categories were subjective, but there were several guidelines for assigning them. First, the designated reflectivity of a given snow area was the average within that boundary. For example, an assigned category 2 could represent either the average of small areas of category 3 and small areas of category 1, or the category 2 appearance of the entire area. Second, the reflectivity assigned to an area was relative to that of surrounding areas at the time of observation. Third, when snow covered mountain tops were seen in large open areas, averaging less than category 1, the area was labeled "scattered mountain snow." Because of such subjectivity, this practice was discontinued in May 1982.

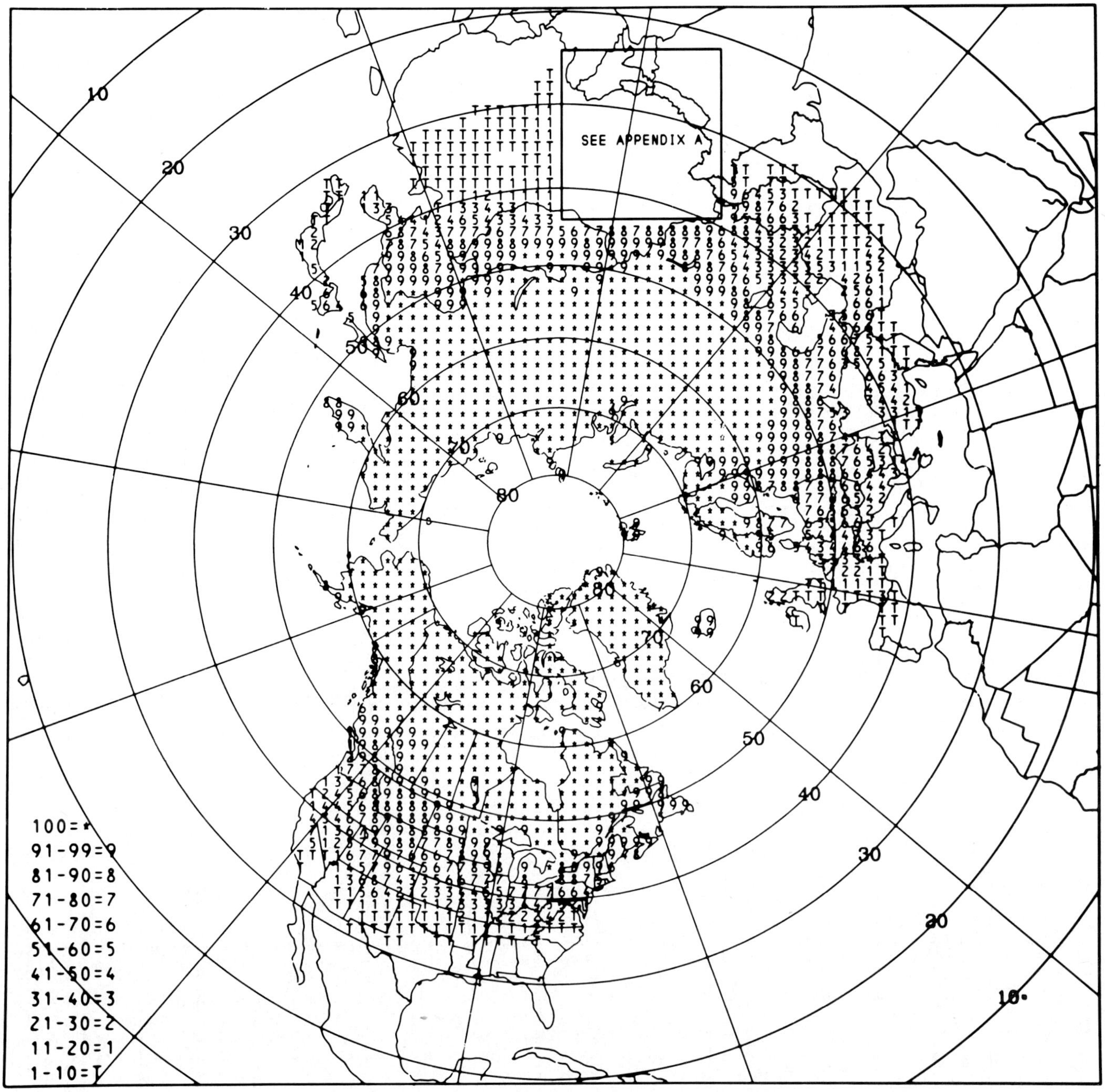

Figure VIII-2-3. Example of a 15-year Snow Cover Frequency Map (January).

Since then the charts have had only two categories: snow cover and patchy snow cover (Figure VIII-2-2). Dates of the latest cloudfree snow cover analysis during the week are also shown by figures representing the month and the day of the month (e.g., 1/18 is January 18, 1984).

In winter, snow is not evident on VIS imagery at high latitudes much of the time because of low solar illumination. At the winter solstice, the snow and ice chart is labeled "dark" north of about 55°N. In general, the land areas within the "dark" zone are assumed to be completely snow covered.

Since the analysis of snow cover is basically photo-interpretation, there have been some systematic changes between the earliest and latest charts. Part of these differences can be attributed to improved satellites and sensors and part to increased experience and more attention to detail, especially outside North

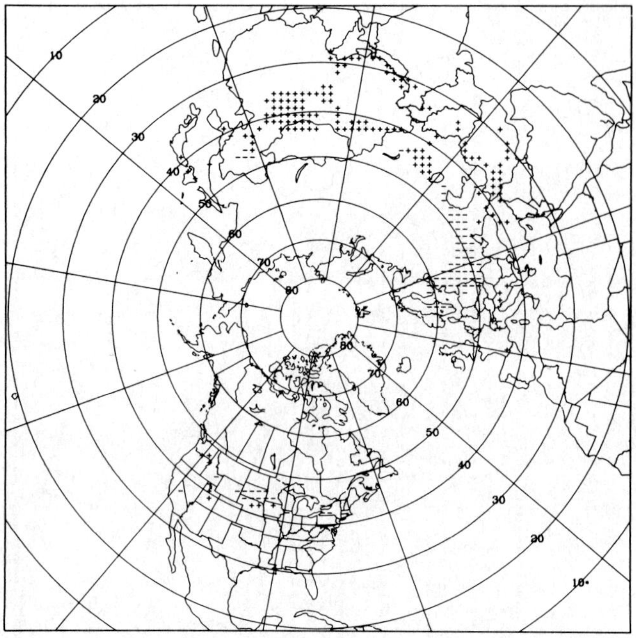

Figure VIII-2-4. Example of a Snow Cover Anomaly Chart, December 1986. Plus signs indicate above normal snow cover, minus signs below normal.

America. Early in the record, the Himalayan region was particularly susceptible to inconsistencies (Matson and Wiesnet, 1981).

Although there has been no comprehensive validation of the entire map series, several portions of the record have been examined. Kukla and Robinson (1979) analyzed the accuracy of weekly charts based on satellite data by comparing them with an independent daily analysis of snow cover based on satellite data plus ground station reports for the United States. Kukla and Robinson (1981) conducted similar comparisons for both North America and Asia. They showed that snow fields under persistent cloud systems were not shown on the weekly charts. This leads to an underrepresentation of snow extent, particularly in autumn. Dickson (1984) noted the same problem. Discrepancies in snow cover areas were also noted in the autumn NOAA charts between 1967 and 1972. In these cases, the relatively low quality of satellite imagery and lack of experience with snow recognition in poorly illuminated areas contributed to omission of snow fields over parts of north Asia and sometimes North America (Kukla and Robinson, 1981). Kukla and Gavin (1983) found that snow continues to be underestimated in poorly illuminated areas and is not always recognized in heavily forested areas.

VIII-2.3 DIGITIZED NORTHERN HEMISPHERE SNOW COVER DATA

Digitization of the Northern Hemisphere Weekly Snow and Ice Cover charts was accomplished by Dewey

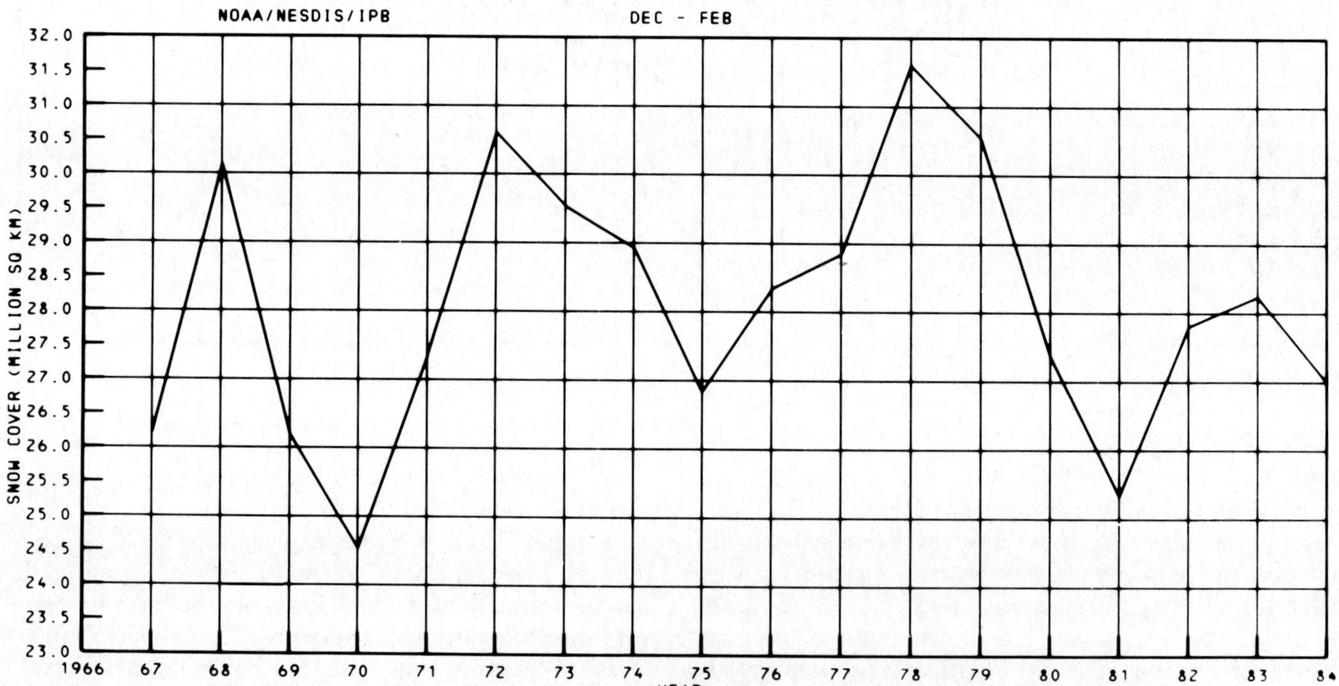

Figure VIII-2-5. Variation of Eurasian winter (December-February) average snow cover, 1966-1984.

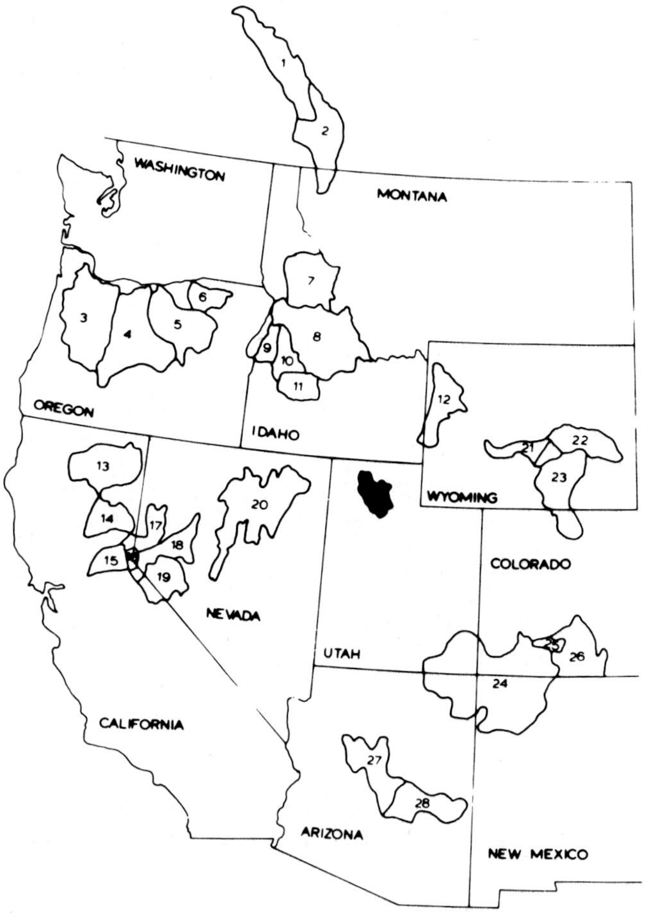

Figure VIII-2-6. Map showing location of operational snow cover basins.

and Heim (1981; 1982). Grid resolution ranges from 16,000 to 42,000 km^2; grid boxes are a nominal 2.5° lat. × 2.5° long. In the digitized analysis, a grid box 50% or more snow covered is considered to be completely covered; any grid box with less than 50% is classified as "no snow." All land grid boxes in the "dark" zone are assumed to be completely snow covered. Areas of sea ice were also digitized in the same way through 1980. The weekly digitized data are archived on magnetic tape and updated annually. Monthly values are derived from the weekly data. A grid box for the monthly values is considered snow covered if there is snow cover for at least half of the weeks comprising the month.

Several products can be created from the continental digitized snow cover database: snow cover frequency maps, snow cover anomaly maps, and snow covered area. Frequency maps can be made for a period of any length. For example, Figure VIII-2-3 shows the frequencies over 15 Januaries. The calculated frequencies are based on the weekly snow cover data, e.g., a grid box coded as 71 to 80% means that the grid box was snow covered 71 to 80% of all the weeks of the period.

Figure VIII-2-4 shows a snow cover anomaly map for December 1986. The snow covered areas are compared with the 15-year December climatology created from the satellite database.

The gridded, digitized snow cover data can be used conveniently to determine areal snow cover over any part of the Northern Hemisphere, weekly, monthly, and seasonally. Three such areas, North America, Eurasia, and the whole Northern Hemisphere have had such calculations made regularly. For example, a plot of the record of the area of snow cover over Eurasia for winter is shown in Figure VIII-2-5. See Chapter IX-4 for climatic analysis of longer term data records.

VIII-2.4 RIVER BASIN SNOW COVER

Snow cover data or snow maps for river basins are provided to water resource managers in numerous Federal, State and local agencies (Figure VIII-2-6). Area snow cover percentages are sent over the Radar Warning and Coordination (RAWARC) teletype circuit to NWS RFCS in Sacramento, Fort Worth, Salt Lake City, Kansas City, and Portland; and River District Offices in Great Falls, Phoenix, and Albuquerque. Snow maps are sent by various means to other agencies including the U.S. Geological Survey, Bureau of Reclamation, Corps of Engineers, Soil Conservation Service, and the U.S. Forest Service. Basin snow maps are made about once a week starting November 1 and continuing until the snow pack disappears. Over 600 snow cover measurements are made at NOAA/NESDIS each snow season, and data are provided to users within 30 hours of a satellite overpass for use in watershed runoff forecast models. A more detailed program description is given by Schneider (1980).

Most basin snow cover maps have been prepared by subjective interpretation of VIS (and sometimes IR) imagery. Experiments have been carried out, however, to map basin snow cover with the Man-computer Interactive Data Access System (McIDAS) (Gird, 1980). Also, an all digital snow mapping method using 4-km resolution GOES data was developed by Tarpley et al. (1980). A sample digital snow map for the Sierra Nevada is shown in Figure VIII-2-7.

Quality control techniques used in operational river basin snow mapping are described by Schneider et al. (1976). One method involves checking the snow maps produced from low resolution GOES and NOAA imagery against those produced from high resolution Landsat data. Figure VIII-2-8 shows a comparison of snow cover measurements made for the American River Basin in California from MSS, VHRR, and VISSR im-

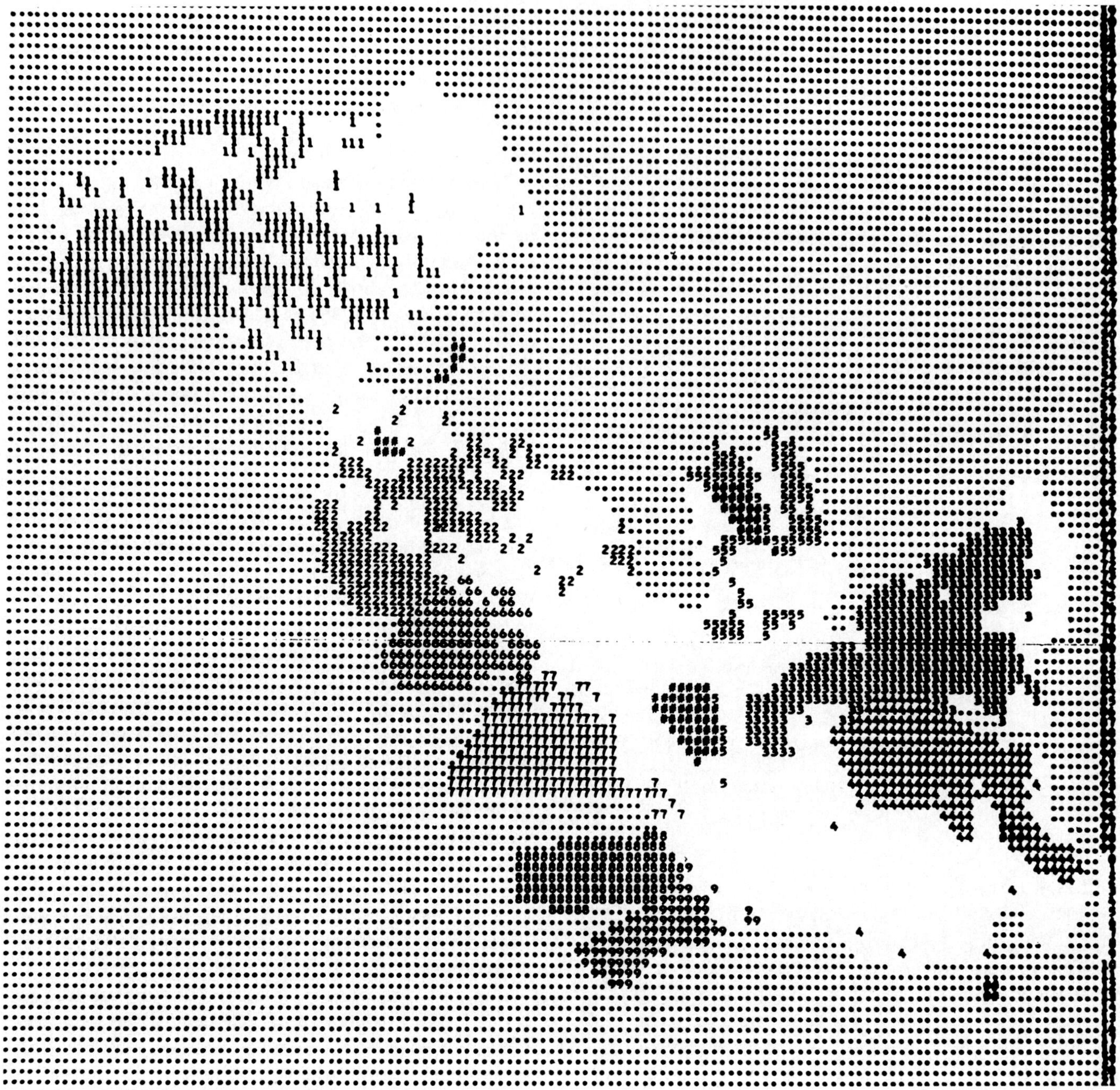

Figure VIII-2-7. Printout of digital snow maps for nine contiguous basins in the Sierra Nevada range. Basin snow cover percentages are as follows:

1 Sacramento	57.7
2 Feather	57.0
3 Carson	42.7
4 Walker	61.6
5 Truckee	64.2
6 Yuba	34.8
7 American	33.3
8 Mokelumne	22.4
9 Stanislaus	47.4

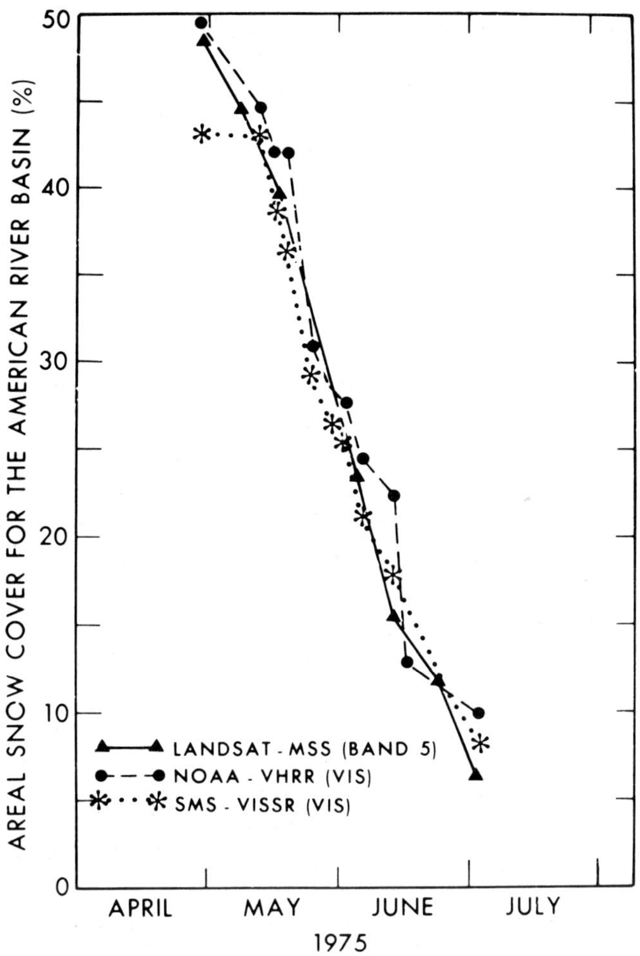

Figure VIII-2-8. Comparison of areal snow cover derived from Landsat MSS, NOAA polar orbiting VHRR, and geostationary SMS VISSR for the American River Basin in California. Results are generally close for all three sensors.

agery. Other quality control methods include use of computer enhanced imagery, ground based snow pack measurements, and aerial survey maps.

During the first decade (1974-1984) of operational snow mapping, many analyses were used to document cases of flooding or drought. The most memorable of the satellite snow cover maps were those of the spring of 1983, which showed snow packs over the western United States from 200 to 300% above normal. The melting of these snow packs resulted in flooding throughout the central and southern Rocky Mountain region. The spill gates of Hoover Dam were opened for the first time in 40 years, and residents along the lower Colorado River had to move to high ground. Only the early release of water from reservoirs of the high Sierra prevented severe flooding in California and Nevada, although saturated crop fields and mud slides were prevalent. Figure VIII-2-9a (May 7, 1983) compares this unusually heavy snow pack with that from a normal year (Figure VIII-2-9b).

The use of snow cover data for stream-flow prediction has been reported by many investigators. Figure VIII-2-10 shows the relationship between satellite snow cover estimates and measured runoff over a 6-year period for the Indus River, India. The Indus was chosen as an initial test site because of its large natural watershed. Correlations for the Indus were quite favorable (Rango, 1975). More recent investigations (Dillard, 1980; Shafer and Leaf, 1980; Martinec and Rango, 1980) confirm the utility of satellite snow cover data in seasonal runoff models.

VIII-2.5 SNOW DEPTH

The brightness of snow, indicated by the reflected radiation received by satellite VIS sensors, has been considered a possible indicator of snow depth (Barnes and Bowley, 1968; McClain, 1973). This hypothesis was tested on February 11, 1973, immediately after a heavy snowstorm struck the southeastern part of the United States (McGinnis et al., 1975). The storm produced record snow depths from Alabama northeastward through the Carolinas. Hodge (1973) described in detail the meteorological conditions and effects of the storm.

Figure VIII-2-11 shows the analysis of total snowfall in Alabama, Georgia, and South Carolina compiled by individual State climatologists from scattered point observations taken mainly by cooperative observers. Despite inconsistencies, values in excess of 30 cm were reported over a large area. Storm totals, because of the warm ground and settling of the snow, were somewhat greater than observations taken at 1200 UTC the morning after the storm. Rapid clearing followed the storm; at 1545 UTC, February 11, NOAA-2 viewed the cloud free, snow covered landscape (Figure VIII-2-12). The snow covered area seen by the VHRR shows a distinct western limit to the snow area with a highly reflective strip in its center (corresponding generally to the deepest snow shown in Figure VIII-2-11). Some land features, such as dark swampy or heavily forested areas adjacent to rivers, contrast sharply with the snow covered land. The more heavily forested Piedmont Plateau in the west is mottled in appearance and less bright than elsewhere. The 1-km resolution of the VHRR permits easy identification of many small rivers and also enables the analyst to determine the location of the western snow edge to within 8-km by using drainage features as local reference points.

Analysis of the brightnesses takes into account system limitations. Each pixel of the VHRR image covers an area of approximately 1 km^2 (at nadir) on the ground. The brightness value of each pixel represents

Figure VIII-2-9a. Landsat MSS scene of Uinta Mountains, Utah, May 7, 1983.

an average response that may include dark, forested regions and highly reflective open fields. The differentiation of land features by the differing reflectance of the snow tends to mask the relation of snow brightness to snow depth. Also, the aging of a snow pack is manifest in the reduction of surface brightness.

Digital computer printouts analogous to pictorial displays were produced from the VHRR analog tapes. The analog tapes were converted into nine track digital tapes containing every scan line generated by VHRR, and the information averaged or combined in various ways to meet the needs of the investigator. In digital printout, 36 different alphanumeric characters are used to represent the 256 possible gray scale steps of the VHRR image. (Note: The current AVHRR has 1024 possible gray scale steps.) For the February 11 image, only 63 gray levels were used to represent brightness variations ranging from dark swamps to bright, snow covered, cleared fields.

Using a 1-km resolution option, a grid of 201 VHRR (32×32 km) squares was fitted to Figure VIII-2-11; the greatest snow depth for each square was matched

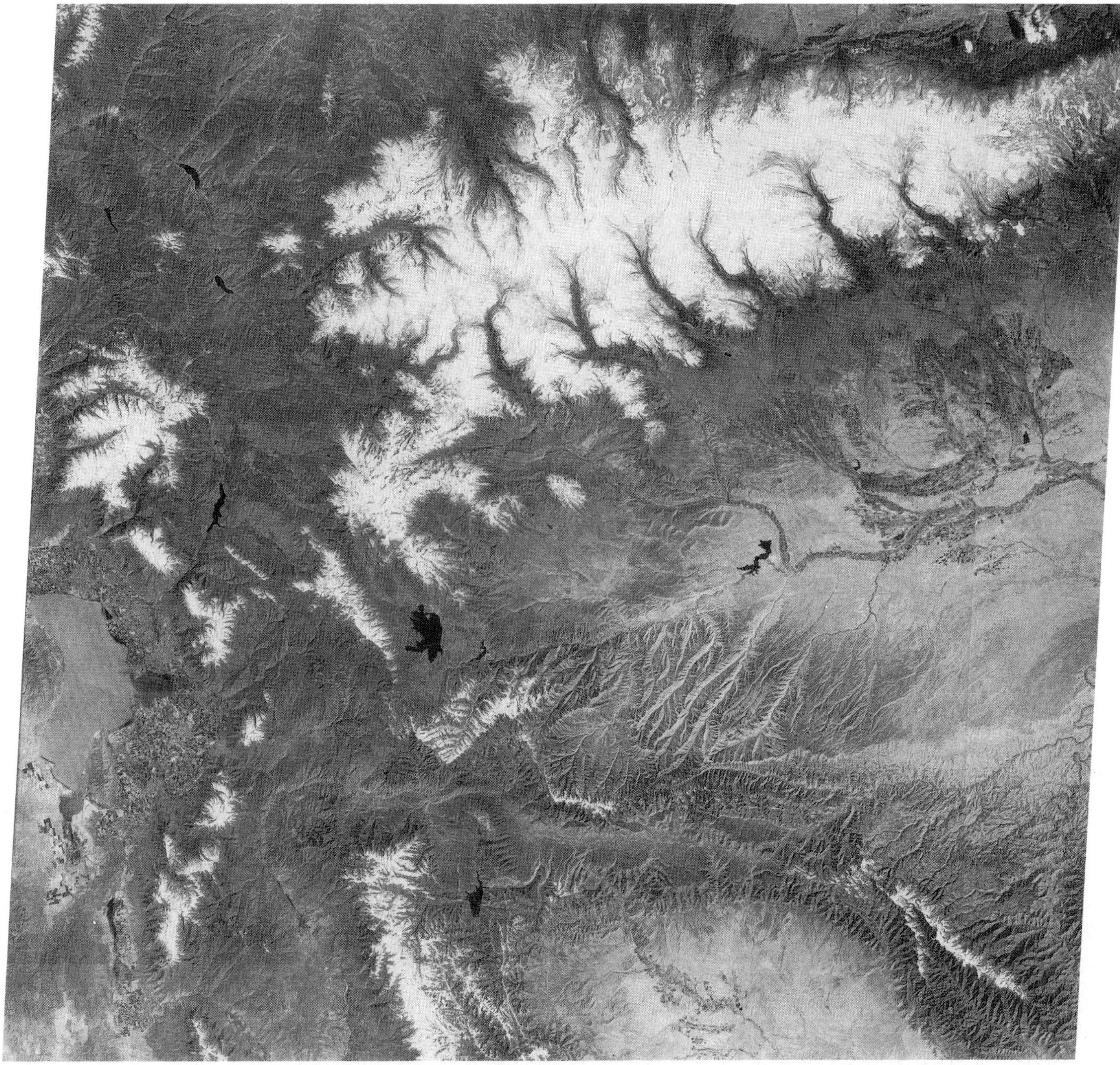

Figure VIII-2-9b. Landsat MSS scene of Uinta Mountains, Utah, May 14, 1976.

with the greatest brightness value. Figure VIII-2-13 shows the 201 data points and the curve of best fit. The correlation coefficient for this parabolic curve is 0.84. Rather small increases in depth when the snow is only several centimeters deep produce large increases in brightness. In this case, the snow covers more objects on the ground as it deepens: first blades of grass, then small shrubs, then larger objects. Once snow accumulates to about 25 cm, most small plants are covered; only the larger plants (corn stalks to trees) remain visible, and in most cases will remain uncovered except where the snow is extremely heavy.

Not all the radiation penetrating a snow pack returns to the surface and contributes to the albedo. Giddings and LaChapelle (1961) showed that absorption from an underlying black surface greatly reduced the albedo

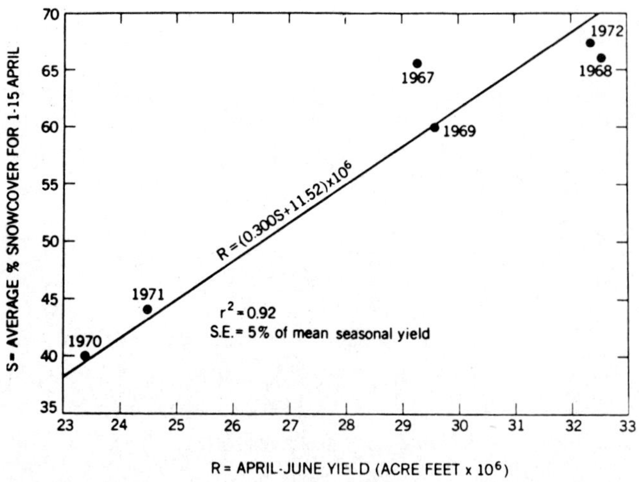

Figure VIII-2-10. Satellite derived snow cover estimates versus measured runoff for the Indus River, 1967-1972 (Rango, 1975).

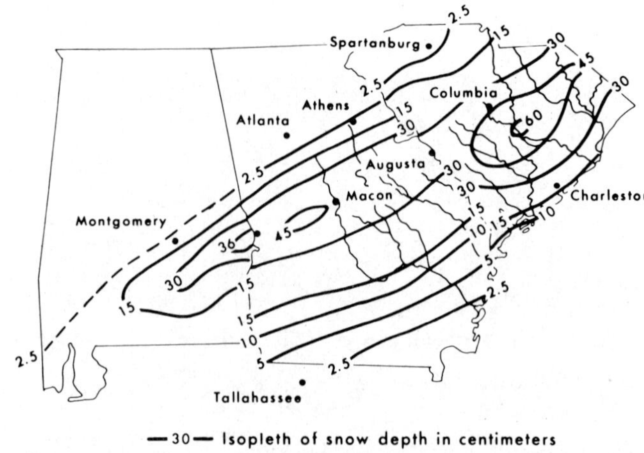

Figure VIII-2-11. Total surface observed snowfall for storm of February 9-10, 1973. The dashed line represents extrapolated data (McGinnis et al., 1975).

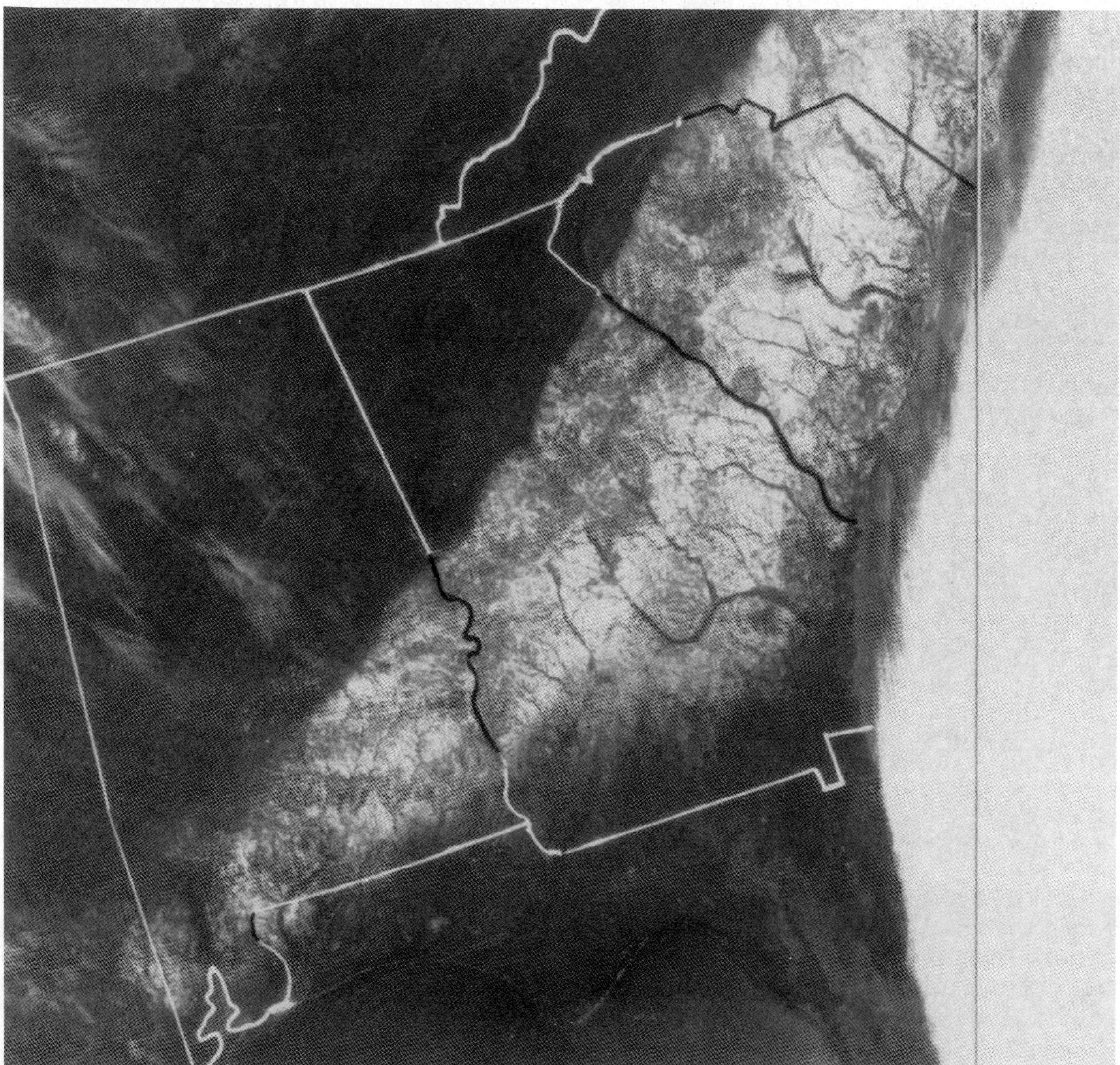

Figure VIII-2-12. Portion of the NOAA-2 VHRR VIS (0.5-0.6 μm) image, 1545 UTC, February 11, 1973 (McGinnis et al., 1975).

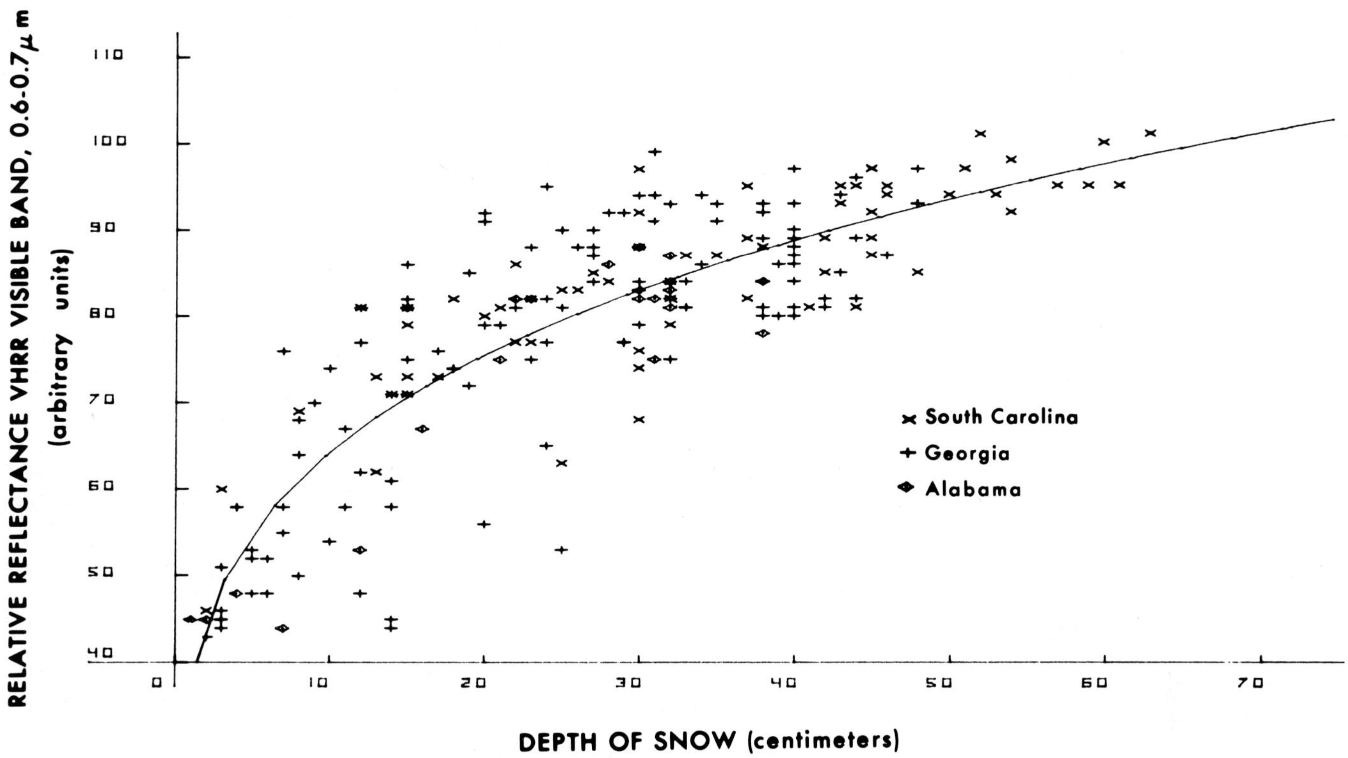

Figure VIII-2-13. Relation of relative VHRR VIS reflectance to depth of snow (McGinnis et al., 1975).

of aged snow with depths less than 20 cm (Figure II-2-2). Thus, variations in depth below 20 cm may be detected by changes in satellite brightness, although the differences become quite small between 10 and 20 cm.

For snow-free winter conditions, all types of land surfaces in the southeastern United States appear almost uniformly bright. When there is a deep snow cover, however, the varying land surfaces reflect very differently. Under the same depths of snow, open farmland appears very bright, wet swampy areas near major rivers are dark, and forested areas appear in differing intermediate tones depending upon the number and kinds of trees. Using the greatest brightness within a particular region helps to determine snow depth (i.e., selecting clearings and open areas, free of trees and other objects).

Additional case studies in the U.S. have not produced similar high (0.84) correlations between snow depth and snow brightness. There are several possible reasons for this: the ground cover in the Southeast may be uniquely different from that in the rest of the United States, the effect of changes in the snow pack with time with a resulting decrease in albedo may limit the rate at which brightness changes as the snow depth changes, and the variation of Sun angle with time of day (and seasonally) may limit illumination of the snow's surface.

VIII-2.6 MELT DETECTION

Strong et al. (1971) first suggested the use of multispectral satellite imagery to detect melting snow and ice packs by observing changes in the spectral brightness of these fields in the VIS and IR portions of the spectrum. Laboratory studies by O'Brien and Munis (1975) of various natural snow samples under different aging conditions resulted in curves similar to those shown in Figure II-2-1. For curves A, B, and C, when the snow surface temperature is below the melting point, there is little difference in the reflectance over the VIS and IR range. Once the snow surface temperature is raised above the melting point (curves D and E), however, the reflectance drops dramatically at wavelengths greater than 0.75 μm. Even refreezing of the melting snow surface (curve F) does not restore the reflectance to its pre-melting levels. Such reflectance differences provide evidence on the condition of the snow pack that could be used by water-resource planners in their estimation of expected runoff.

Chapter VIII-3

LAKES

VIII-3.1 INTRODUCTION

The use of data from NOAA polar orbiters in observing lake phenomena has been restricted mainly to the Great Lakes. Although the ice and surface water temperature analyses described here use the Great Lakes as examples, the analysis techniques can be used worldwide. Satellites have also been used for detecting heavy concentrations of floating phytoplankton (Great Lakes Res. Div., 1967) and algal blooms (Bressette and Lear, 1973; Strong, 1974). Satellites also permit the monitoring of changes in land-water boundaries as well as sedimentation and biological activity in water bodies. For example, NOAA and Landsat data have detected a wind-induced seiche in the Great Salt Lake and a dramatic change in the area of Lake Chad after a prolonged drought.

Although particular applications of satellite data and techniques to lakes are stressed here, it should be noted that most of the applications are the same as those for the oceans, such as the surface temperature discussed in Chapter VIII-4.

VIII-3.2 GREAT LAKES ICE ANALYSIS

The ice cover of the Great Lakes is related to the physical properties of each lake. For example, Lake Erie, a shallow lake, normally has a more extensive ice cover than Lake Ontario, a very deep lake.

Detailed ice analysis over the Great Lakes became feasible late in 1972 when high-resolution VIS and IR imagery from the VHRR (and later the AVHRR) became available. With two polar orbiting satellites in operation, four views of the Great Lakes could be acquired daily. In addition, GOES-East views the area every 30 min, but IR resolution is only 7.5 km.

NESS (later NESDIS) started operational ice analyses of the Great Lakes during the 1973-74 season. Analyses were produced twice weekly and distributed by facsimile and mail. Currently, the ice analysis of the Great Lakes is done by the Navy/NOAA Joint Ice Center at Suitland, Maryland, and distributed by facsimile. Although satellite data are the main sources of information for the analyses, reports from lakeside land stations, ship reports, and aerial reconnaissance are also used.

The Great Lakes ice analyses are done manually using photo-interpretation techniques. Ice concentration is expressed in percent of the lake surface covered. Figure VIII-3-1, an AVHRR VIS image from February 17, 1979, shows an extreme amount of ice in the Great Lakes. In fact, this is probably the greatest amount of ice ever observed by satellite in the lakes. Figure VIII-3-2 is the corresponding ice analysis. Lake Superior appears very bright; the analysis indicates approximately 100% concentration. The analyzed concentration in Lake Ontario is low, from 0 to 30% in the southwest part.

Figures VIII-3-3 and VIII-3-4 illustrate the other extreme of wintertime ice cover over the Great Lakes. On this day, February 21, 1975, there was a minimal

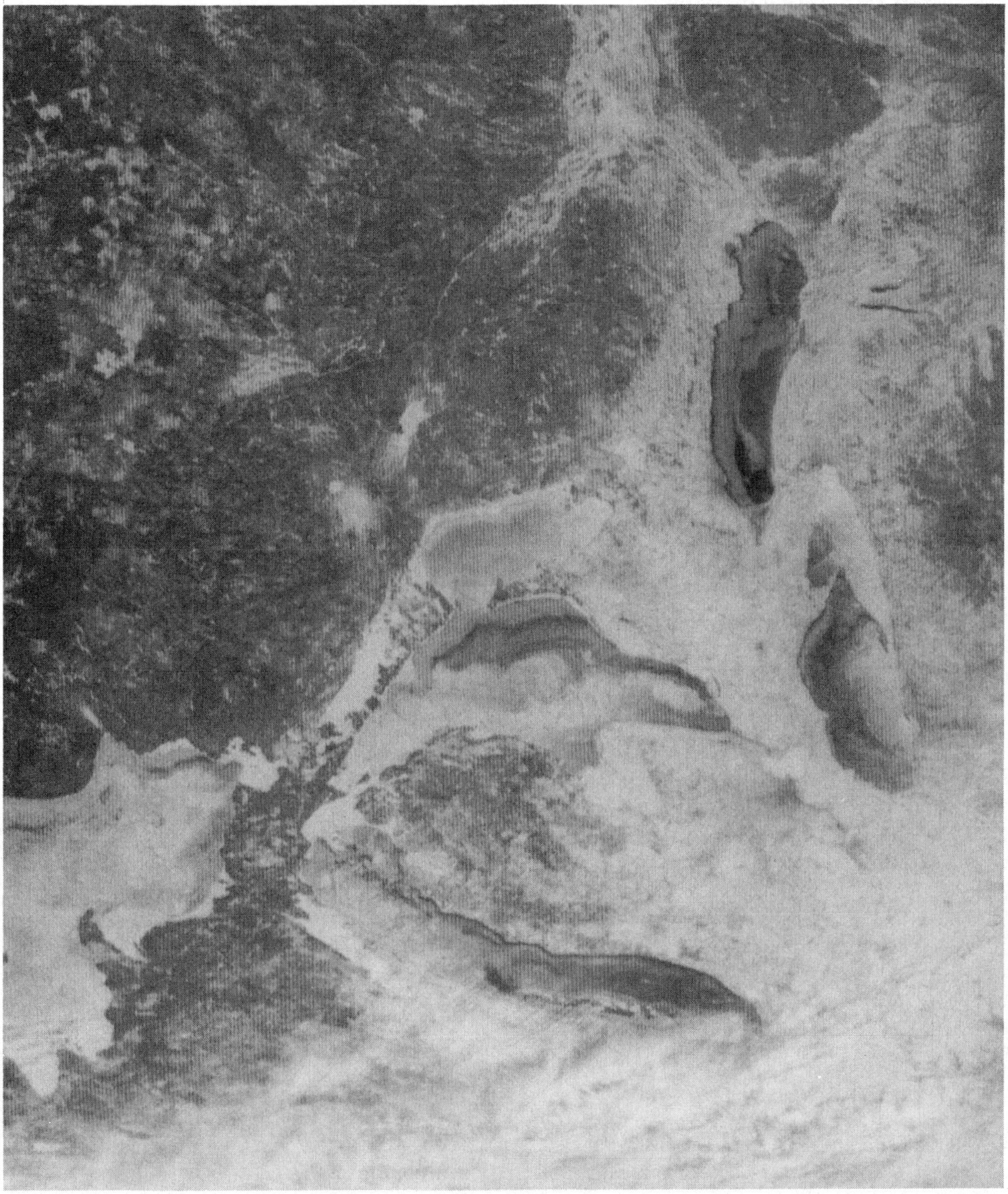

Figure VIII-3-1. AVHRR VIS image illustrating maximum ice cover in the Great Lakes, February 17, 1979.

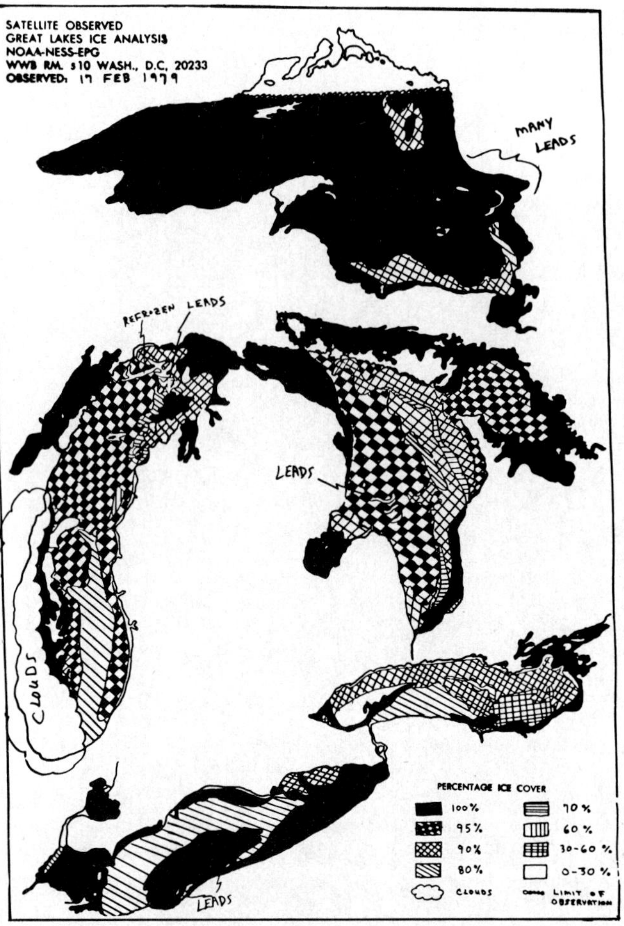

Figure VIII-3-2. Ice analysis corresponding to the AVHRR image in Figure VIII-3-1.

amount of ice. Only Lake Erie had an appreciable amount of ice. The dark areas are either ice free or very thin ice. The contrast in the ice scenes of Figures VIII-3-1 and VIII-3-3 is striking.

Figures VIII-3-1 and VIII-3-3 are mostly cloud-free; the only clouds visible are over western Lake Michigan (Figure VIII-3-1) and over eastern Lake Ontario (Figure VIII-3-3). Many times during the winter, clouds obscure the Lakes making it very difficult to obtain information for an ice analysis with a single image. During such times, images obtained over the course of a week or more are used to produce a composite ice analysis.

The NWS Central Region, the largest user of the Great Lakes ice analyses, is responsible for forecasting ice conditions in each of the Great Lakes. Satellite imagery received at Weather Service Forecast Offices gives the ice forecasters a firsthand look at some of the data used to prepare the ice analyses. NESDIS also provides a special enhancement of the AVHRR IR imagery as an aid to ice detection in the Lakes.

Ice analysts and forecasters use the ice analyses and satellite imagery for ship routing in the Lakes. During an 8-year Winter Navigation Demonstration Program conducted by the Corps of Engineers from 1972-79, these ice analyses helped prove the feasibility and economic advantage of maintaining commercial shipping on the Great Lakes during as much of the winter as weather and ice conditions permit. Researchers have also used satellite imagery and ice analyses gathered over the first 10 years of the operational analyses to aid in producing an ice climatology atlas for the Great Lakes (Assel et al., 1983).

VIII-3.3 GREAT LAKES SURFACE WATER TEMPERATURE ANALYSIS

Since August 1975, surface water temperature analyses over the Great Lakes have been made operationally at NESS/NESDIS during the ice-free season, usually April-December. This analysis is derived from digital AVHRR IR data.

The images of the Lakes are screened daily by Ocean Products Center personnel to select cloud-free areas. Figure VIII-3-5 is a typical AVHRR IR image used for producing a surface water temperature analysis. For the clear areas, the calibrated brightness temperatures are processed by computer and surface temperature values are mapped at an average resolution of 5 km. A multiwindow technique (McClain et al., 1983; and Chapter VIII-4) is used to correct for atmospheric attenuation and to convert brightnesses to surface water temperatures.

Figure VIII-3-6, the surface water temperature analysis for June 8, 1982, corresponds to the image of Figure VIII-3-5. Note the cold water (white in the image) in Lakes Superior, Michigan, Huron, and Ontario and the warm water (dark gray in the image) in Lake Erie, Green Bay, and Saginaw Bay. The numbers inside the little boxes in the analysis in Figure VIII-3-6 are buoy measurements of lake temperatures at a depth of 1 m. Generally, the satellite temperatures are within ±0.6°C RMS of the buoy temperatures. When storms and clouds prevail over the Great Lakes for extended periods, the analysis is incomplete or nonexistent over some Lakes. At such times, the few scattered buoys are the only sources of temperature. The Great Lakes surface water temperature analysis is produced twice weekly during the ice-free season and is distributed by facsimile and mail.

The Great Lakes surface water temperature analyses were originally made to test the accuracy of AVHRR IR sensors. At present, the analyses are used by the NWS Central Region to warn of storm conditions on the Lakes. A knowledge of Great Lakes water temperatures is an important forecast factor, especially be-

Figure VIII-3-3. AVHRR VIS image illustrating minimum ice cover in the Great Lakes, February 21, 1975.

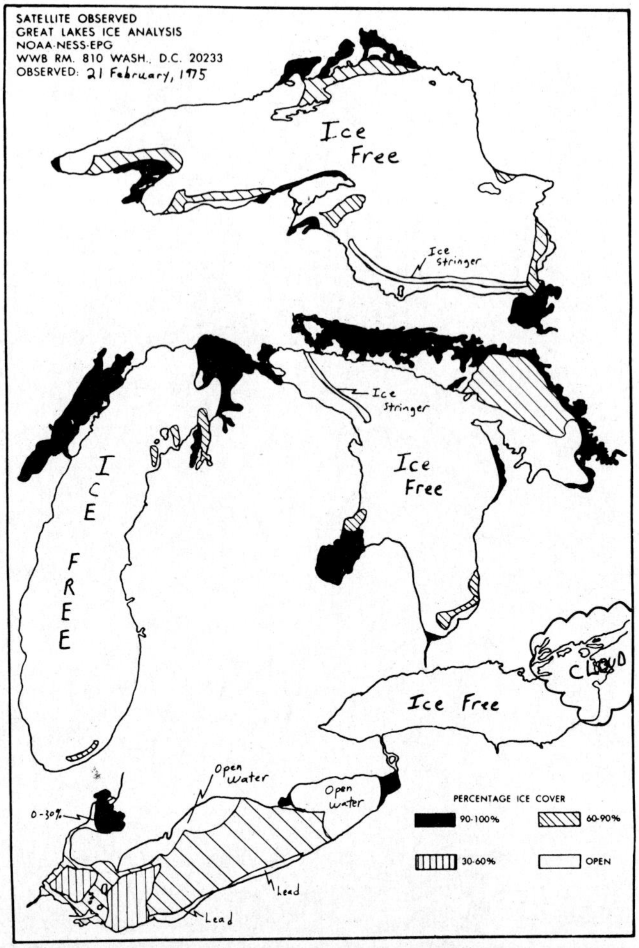

Figure VIII-3-4. Ice analysis corresponding to the AVHRR image in Figure VIII-3-3.

Figure VIII-3-5. AVHRR IR image showing variations in surface water temperature in the Great Lakes, June 2, 1982.

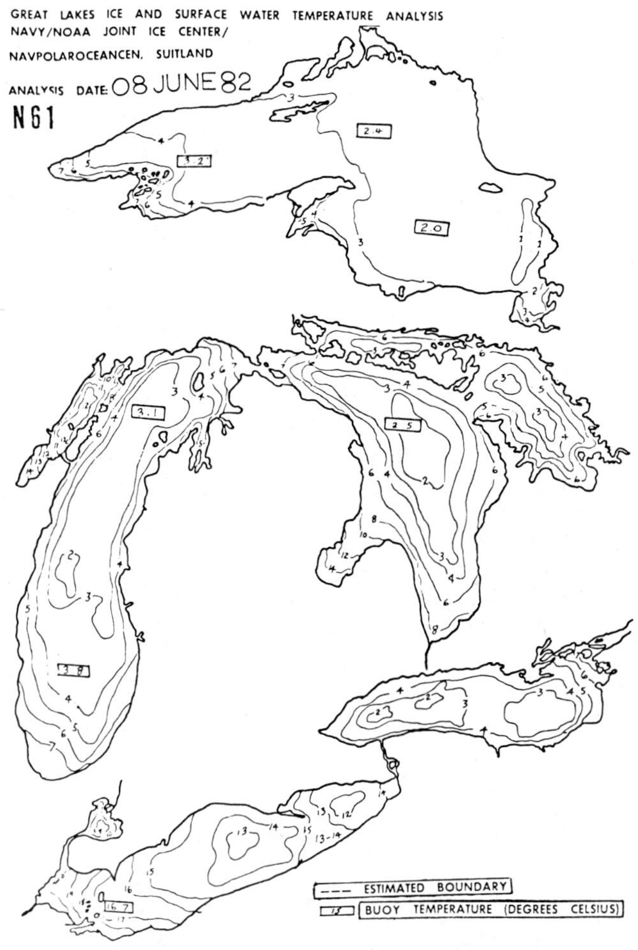

Figure VIII-3-6. Surface water temperature analysis of the Great Lakes six days later than the AVHRR image in Figure VIII-3-5, June 8, 1982.

tween late September and December when cold air outbreaks are influenced by the relatively warmer water. Accurate surface water temperatures also are a major factor in the prediction of ice formation, growth, and dissipation during the early winter and spring transition periods. Other applications include search and rescue operations (when hypothermia of accident victims is a major concern), research on circulation in the Lakes, heat budget analyses, and the effect of the Lakes on thunderstorms.

VIII-3.4 ALGAL BLOOMS AND WHITINGS

VIII-3.4.1 Fresh Water Observations of Algal Blooms

Reflective IR sensors on the Landsat MSS made possible detection of chlorophyll in floating algal material even under a thin layer of water. Typically, however, algal blooms are dry on the parts projecting above the water's surface. Extensive algal blooms are easily detected by the MSS.

A Landsat observation of an algal bloom over Utah Lake was obtained on September 12, 1972 (Figure VIII-3-7). It shows an anticyclonic gyre in this shallow ($\sim$3 m) lake. There is a reversal in contrast for the gyre at 0.7 μm. The gyre itself covers a large part of the surface area (about 375 km^2) of this lake, which is known for its high turbidity and summer algal blooms. Within one day of this satellite observation, a limnological cruise on the lake revealed a massive, highly variable algal bloom (see White and Rushforth, 1973).

VIII-3.4.2 Observations of Chemical Whitings

Chemical precipitations of calcium carbonate ($CaC0_3$), or "whitings," are a fine-grained chalky, white material that gradually sinks. Whitings in the Great Lakes have been described by Ayers et al. (1967) and Strong et al. (1974). The latter study was initiated because of an intense August upwelling along the eastern shore of Lake Michigan, which was noted in the IR data of the NOAA-2 satellite (Figure VIII-3-8). The extensive narrow white strip of temperatures lower than 12°C generally coincided with features seen in VIS imagery of Landsat 1 one day earlier (Figure VIII-3-9) when weather conditions were nearly identical. Data from the green channel (MSS/4, left) generally show considerably more turbidity than data from the red channel (MSS/5, right). This means that the turbidity resides not at the surface, but several meters below since radiant energy in the red channel comes from only the upper few meters of water, whereas the green channel is effective to depths capable of reaching down to 10 m.

Although coincident measurements of calcium carbonate concentrations in Lake Michigan were not available for these specific satellite observations, data from the NOAA Great Lakes Environmental Research Laboratory from 1965-75 show extensive supersaturation of calcium carbonate in Lakes Michigan, Ontario, and Erie in summer (Strong and Eadie, 1978). Other evidence that the satellite whiting observations are related to calcium carbonate has been obtained by Schelske and Roth (1973).

Another case of even more striking satellite observations of whitings is that of Pyramid Lake, Nevada, during the summer of 1978. The increasing coverage of the lake by calcium carbonate precipitate was monitored with data from Landsat and GOES. The satellite data were used to map the extent of lake surface covered by precipitate and to determine whether a relationship existed between reflectance and precipitate concentration in the lake (Berg et al., 1981a).

Whitings in Pyramid Lake were first observed on GOES satellite imagery on July 30, 1978. By August

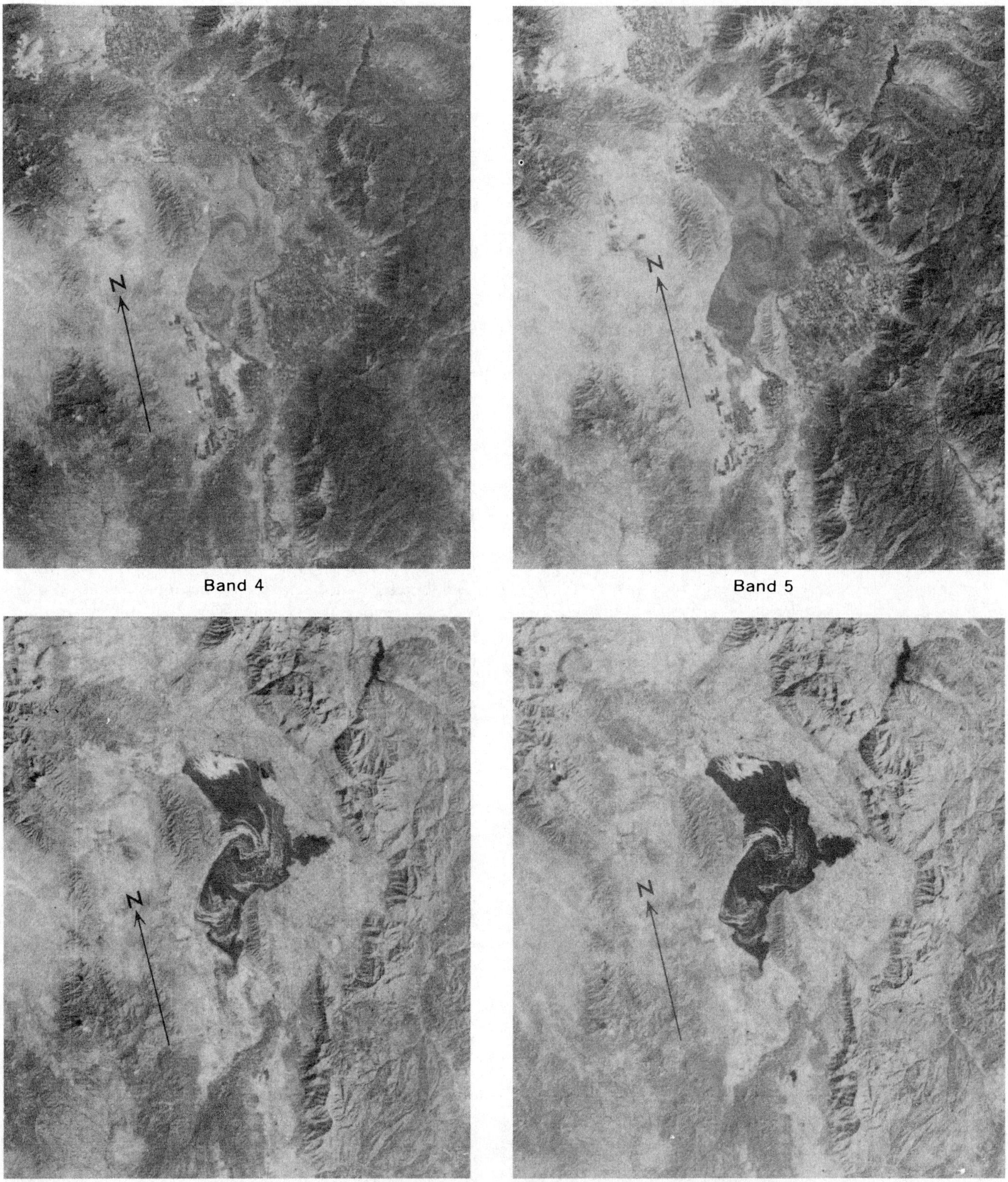

Figure VIII-3-7. Landsat-1 MSS images of Utah Lake (40°10'N, 111°45'W) where algal bloom covers large portion of lake, September 12, 1972. MSS images are 46 km on a side and centered on Utah Lake (Strong, 1974).

Figure VIII-3-8. NOAA-2 IR image, August 22, 1973, depicting upwelling (lighter tones) along eastern shoreline (right) of Lake Michigan.

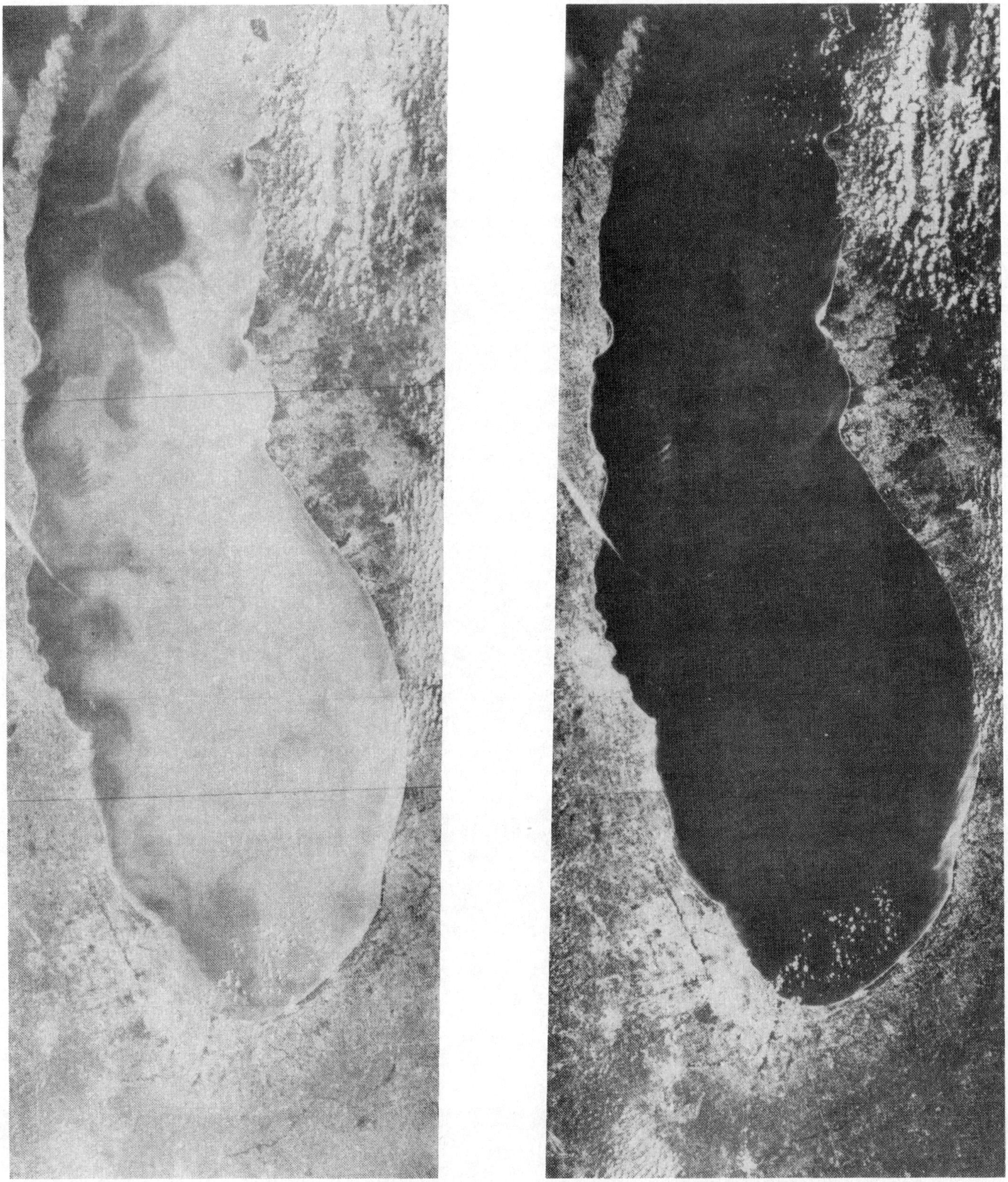

Figure VIII-3-9. Landsat-1 MSS/4 (0.5 to 0.6 μm) and MSS/5 (0.6 to 0.7 μm) image (left and right, respectively), August 21, 1973, 1500 UTC.

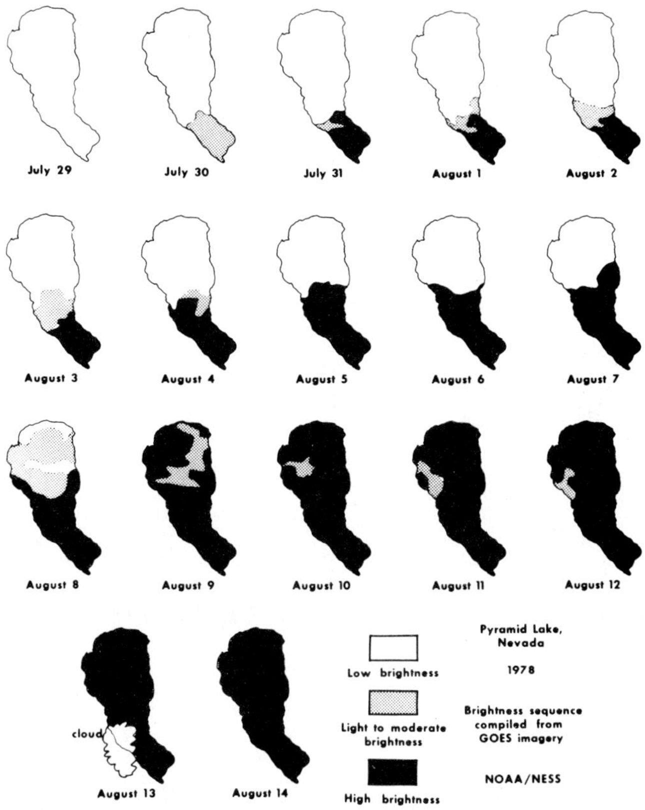

Figure VIII-3-10. The chronology of the 1978 whitings in Pyramid Lake compiled from GOES imagery (Berg et al., 1981a).

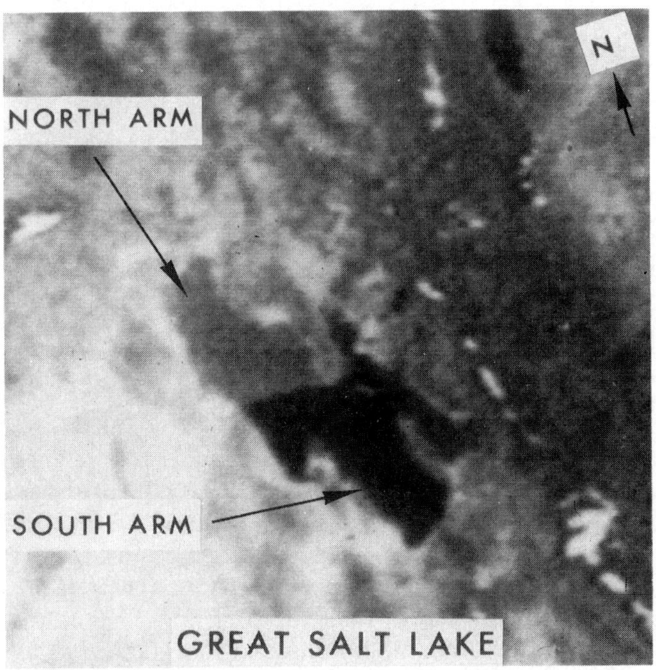

Figure VIII-3-11. NOAA-5 VHRR VIS (0.6 to 0.7 μm) image of Great Salt Lake, June 1, 1977, showing the lake as it typically appears. Note the two different near uniform brightnesses of the lake, i.e., the brighter north arm and the darker south arm. This image is not computer enhanced (Matson and Berg, 1981).

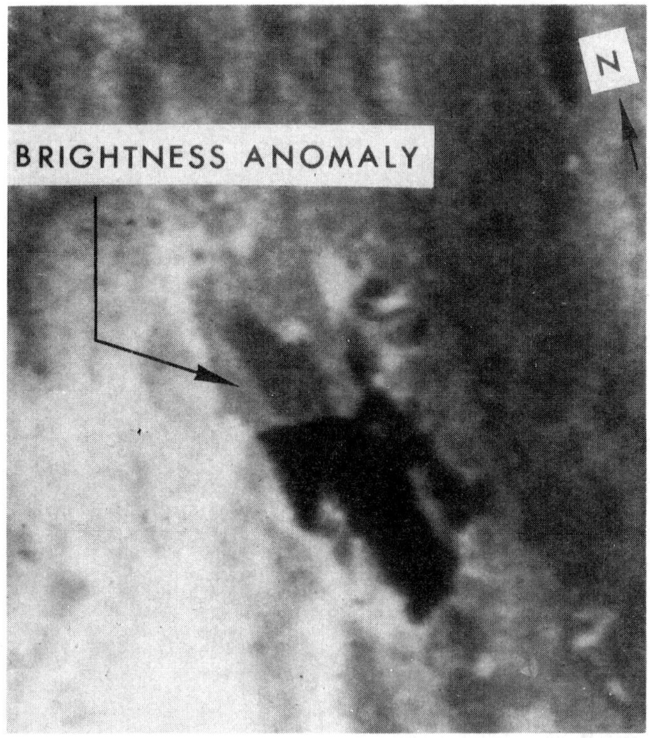

Figure VIII-3-12. The north arm positive brightness anomaly case of June 15, 1977, as detected by the NOAA 5 VHRR VIS sensor. This image is not computer enhanced (Matson and Berg, 1981).

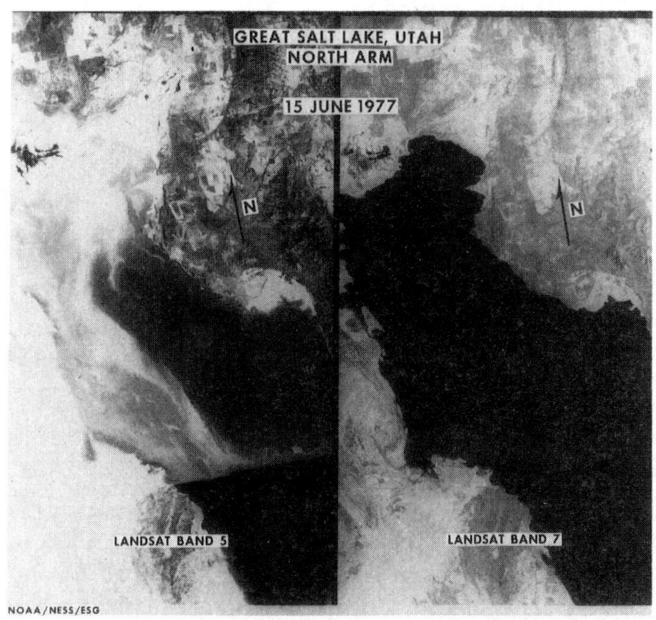

Figure VIII-3-13. Landsat-2 band 5 (0.6 to 0.7 μm) and band 7 (0.8 to 1.1 μm) computer-enhanced imagery of the North Arm anomaly case of June 15, 1977 (Matson and Berg, 1981).

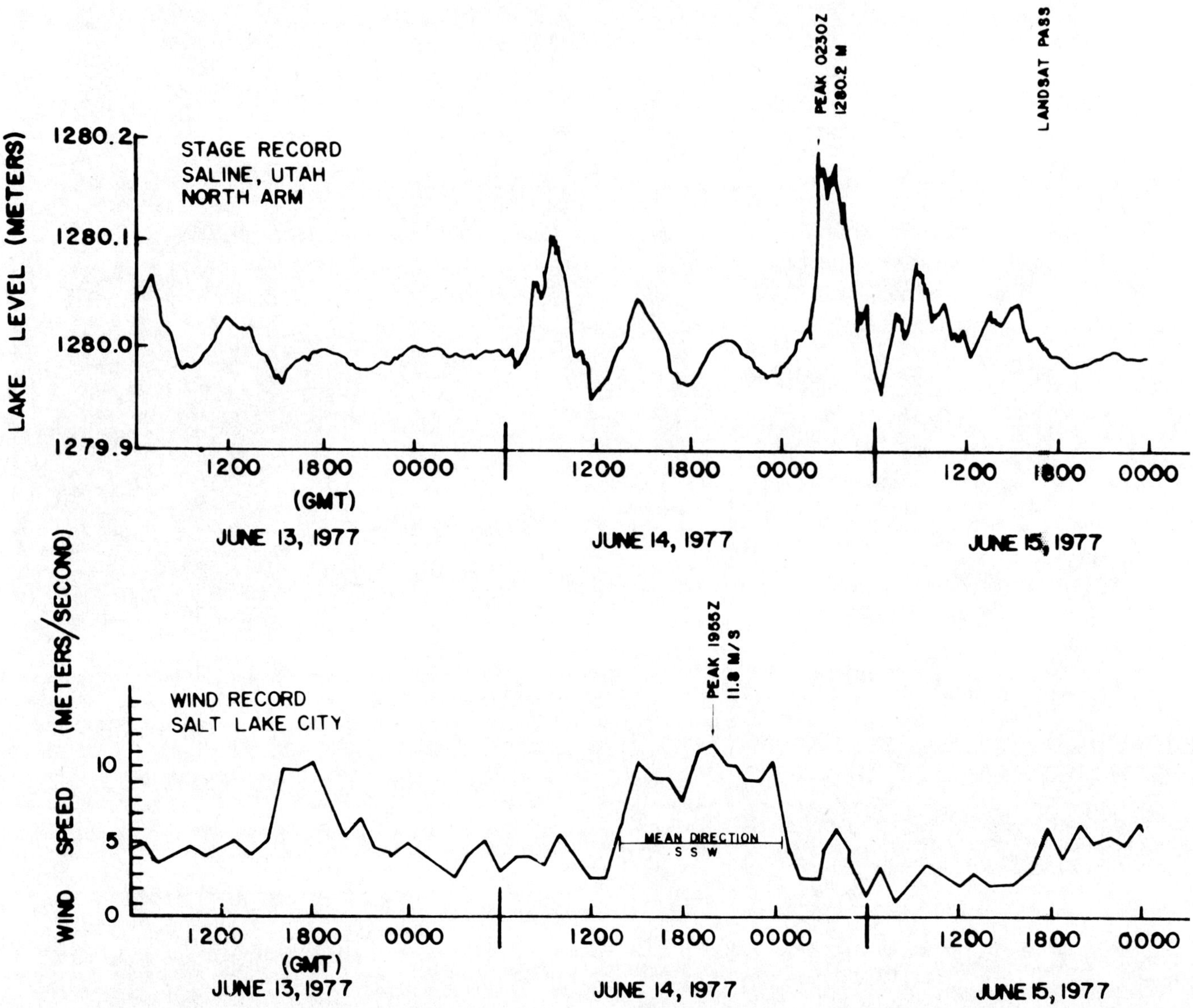

Figure VIII-3-14. Lake levels and wind velocities on Great Salt Lake (North Arm) for June 13-15, 1977 (Matson and Berg, 1981).

9, the entire lake had some evidence of whiting. GOES images taken from mid-morning to mid-afternoon were used to develop the whiting chronology shown in Figure VIII-3-10. The light-to-moderate and high brightness classifications are subjective, with low brightness representing a lack of observed whitings.

Observations of patterns of the whitings can serve as natural markers that can be used for charting near-surface current patterns in the lakes (Strong and Stumpf, 1975). Such estimates have been found to be in good agreement with previous drogue measurements under similar wind conditions (Ayers et al., 1958). Composite current charts would be restricted exclusively to nearshore areas (river plumes, etc.) were it not for the whiting phenomenon, which provides a tracer in mid-lake.

VIII-3.5 SEICHES AND WATER LEVELS

VIII-3.5.1 Seiches

Turbulence and mixing associated with lake seiches are important to the ecology of a lake because they affect dispersal and transport of heat, dissolved oxygen, and nutrients (Wilson, 1972). Wind stress or, rarely, an earthquake, can cause a sudden displacement of the water level from its equilibrium position. This is followed by a free damping oscillatory motion known as a seiche.

Figure VIII-3-11 is a typical NOAA-5 VIS (0.6-0.7 μm) image of Great Salt Lake on June 1, 1977 (Matson and Berg, 1981). The most noticeable feature is the difference in the surface brightness over the lake; the north arm is uniformly brighter by about five brightness

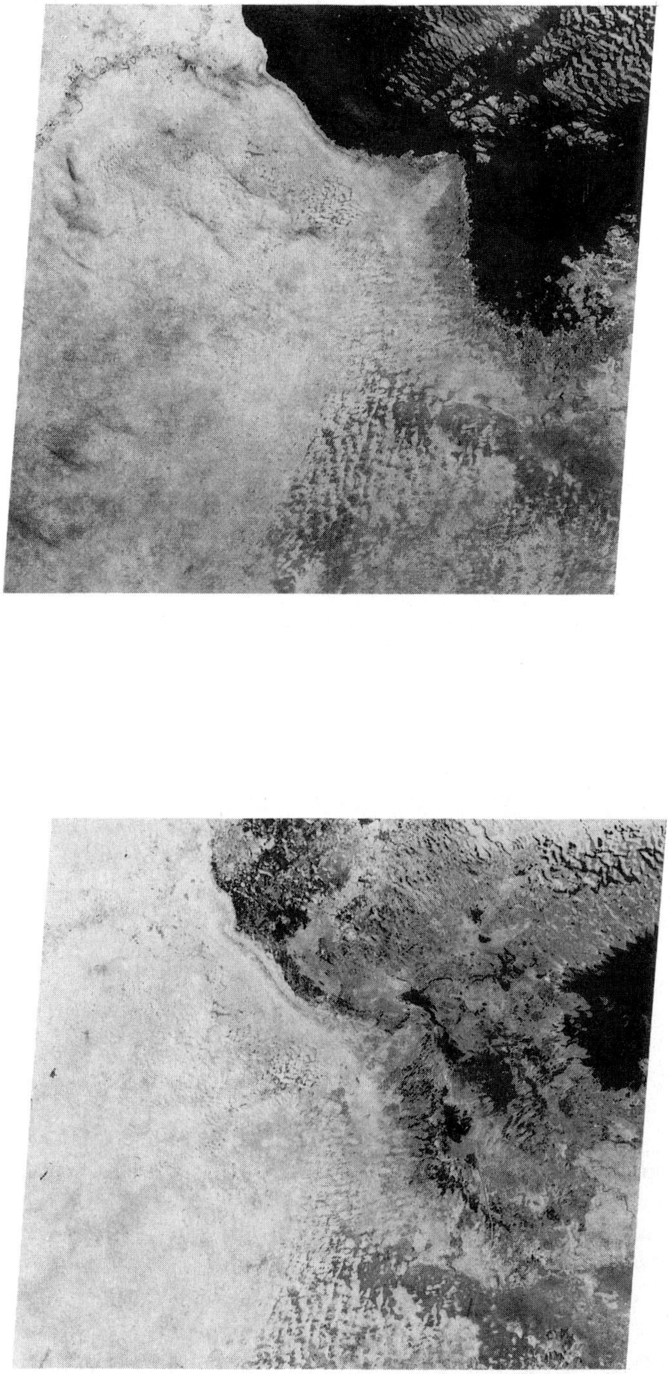

Figure VIII-3-15. Landsat MSS/7 (top) (0.8 to 1.1 μm) image of southwest portionof Lake Chad, December 8, 1972. Landsat MSS/7 (bottom) image of essentially same area (slightly shifted), March 13, 1979.

"counts" than the south arm. Schneider and McGinnis (1977) suggest that the brightness difference may be due to more turbid water in the north arm. Another possibility is the presence of red bacteria suspended throughout the water column of the north arm.

On June 15, 1977, an unusual positive brightness anomaly along the western edge of the north arm of Great Salt Lake was detected in NOAA-5 imagery (Figure VIII-3-12). This anomalous pattern was observed for 3 days after initial detection. Imagery from Landsat MSS for the same day (Figure VIII-3-13) also showed a sharply delineated brightness in band 5 over the north arm of the lake. Quantitative analysis showed the western part of the north arm was as much as 30 "counts" brighter than the eastern part. This bright area was virtually absent in band 7, implying that nearly all of the anomaly was due to a subsurface phenomenon.

Inspection of NOAA-3, 4, and 5 imagery for 1974-78 revealed 21 cases of such brightness anomalies. Lake level records for the north arm gage at Saline and meteorological data from Salt Lake City International Airport were examined for all 21 cases. Lake level and wind data for the June 15 anomaly are shown in Figure VIII-3-14. High winds persisted for 11 hours, beginning at 1500 UTC, June 14. The maximum sustained winds preceded by 6.5 hours a lake level peak at Saline of 20 cm above the base level of 1280.0 m. The Landsat image (band 5) of Figure VIII-3-13, showing the positive brightness anomaly, was obtained 15 hours later.

All 21 documented positive brightness anomaly cases in Great Salt Lake were associated with meteorological disturbances and all were preceded by a sequence of events similar to the one shown in Figure VIII-3-14. Apparently the brightness anomaly was in each case the result of a wind-induced seiche in the lake. The sustained northwesterly or southwesterly winds displaced the north arm water mass toward the eastern shore, resulting in lower water depth, turbulence, and mixing in the western third of the lake. The subsequent oscillatory motion of the water mass probably also caused further entrainment of the fine-grained sediments from the shallow western lake bottom. The resuspended sediment, perhaps combined with bottom reflectance, apparently produced the increased brightness observed by the satellite.

VIII-3.5.2 Water Levels

Lake levels respond to fluctuations in rainfall and runoff, tidal cycles, and human demand. Most changes in level are relatively small (i.e., from centimeters to several meters) and generally result in similarly small horizontal changes in surface area. Landsat, with its 30- to 80-m resolution, can detect small areal changes. Its near-IR bands, in particular, provide the best means of locating land-water boundaries. An example of a dramatic change in water extent and water level of Lake Chad is shown in Figure VIII-3-15. The large reduction in water area (dark areas) was caused by a significant decrease in rainfall and the resulting reduction of runoff into the lake (Schneider et al., 1985).

CHAPTER VIII-4

SEA SURFACE TEMPERATURE

VIII-4.1 BACKGROUND

VIII-4.1.1 Thermal Infrared Measurements

Satellite Sea Surface Temperature (SST) determinations from ocean surface emission in the thermal IR portion of the spectrum have been made for a number of years (McClain, 1979; Alishouse and McClain, 1985). A number of research and operational environmental satellites have carried scanning radiometer systems that measure emitted terrestrial and atmospheric IR radiation. The various sensors and their characteristics have been described in Section IV and by McClain (1980a,b). In most cases the atmospheric window from 10.5 to 12.5 μm has been used, but its use for determining absolute sea surface temperature has often been secondary to its use for imaging of clouds and thermal fronts associated with ocean currents and upwelling (Legeckis, 1978). Radiometers of improved quality, such as the AVHRR, with two and three atmospheric window channels, have permitted development of methods to retrieve absolute SST of far better accuracy than was possible with single window systems (Strong and McClain, 1984).

Immediately before AVHRR IR measurements were available, McClain (1979) reviewed the methods of deriving surface temperatures from single window measurements and gave examples of operational and research products. Correction for atmospheric attenuation, principally by water vapor, can amount to 8 to 10°C for warm, moist atmospheres in the 10.5 to 12.5 μm data (Maul, 1984). Empirical corrections have been generally unsatisfactory; even those based on moisture information from atmospheric sounders on the same spacecraft have been inadequate. Accurate corrections are possible with coincident *in situ* radiosonde measurements, a situation that is relatively rare over the open ocean (See also Chapter VI-2).

VIII-4.1.2 Microwave Measurements

Satellite-based SST dependent upon IR measurements suffer from the limitation that adequate sampling of cloudfree fields of view is required. Although they are limited by antenna to considerably larger footprints on the ocean surface than IR radiometers, MW radiometers can collect data useful for SST determination in all but convective rainfall conditions. Wilheit and Chang (1980) treat the physical background of SST retrievals from Scanning Multichannel Microwave Radiometer (SMMR) data, and recent results of such retrievals are discussed by Milman and Wilheit (1985). Radiative transfer and specific means for determining SST by use of the SMMR data are summarized by Alishouse and McClain (1985). Also, the Microwave Sounding Unit (MSU) has been used for SST determination in conjunction with HIRS on NOAA operational satellites (Susskind and Reuter, 1985).

VIII-4.2 SEA SURFACE TEMPERATURE DERIVATIONS AND THEIR APPLICATIONS

VIII-4.2.1 SST Derived from AVHRR Using Multiple Channel Infrared Techniques

The multichannel SST (MCSST) techniques make use of various combinations of the four or five channels of AVHRR in the initial and critical cloud filtering stages of the procedure (McClain et al., 1985). In the absence of clouds, ice, or relatively intense Sun glint, the oceans are quite dark compared with clouds, so appropriate thresholds have been established for the bidirectional reflectances in Channel 1 (0.58 to 0.68 μm) or Channel 2 (0.725 to 1.1 μm). The principal cloud tests, however, take advantage of the small-scale spatial uniformity of fields of ocean surface temperature and VIS reflectance, air mass moisture, and the low level of noise in the AVHRR data. In the daytime, the cloud filter rejects those small scanspot arrays (2 $\times$ 2 or 3 $\times$ 3) with variabilities of more than about 0.3% in reflectance. Nighttime cloud tests are based on temperature thresholds (many clouds are colder than the coldest ocean surface) and, particularly for clouds that do not fill the radiometer field of view, on temperature uniformity in the spatial sense. The nighttime cloud filter rejects scanspot arrays whose IR temperature varies by more than 0.2°C. Interchannel IR temperature comparisons are also useful for cloud detection, especially at night; and are critical in those cases that fail spatial uniformity tests (e.g., an extremely uniform cloud that completely fills the radiometer field of view). The reflectivity of water droplet clouds is greater at 3.7 μm than it is at 11 or 12 μm, which has the effect

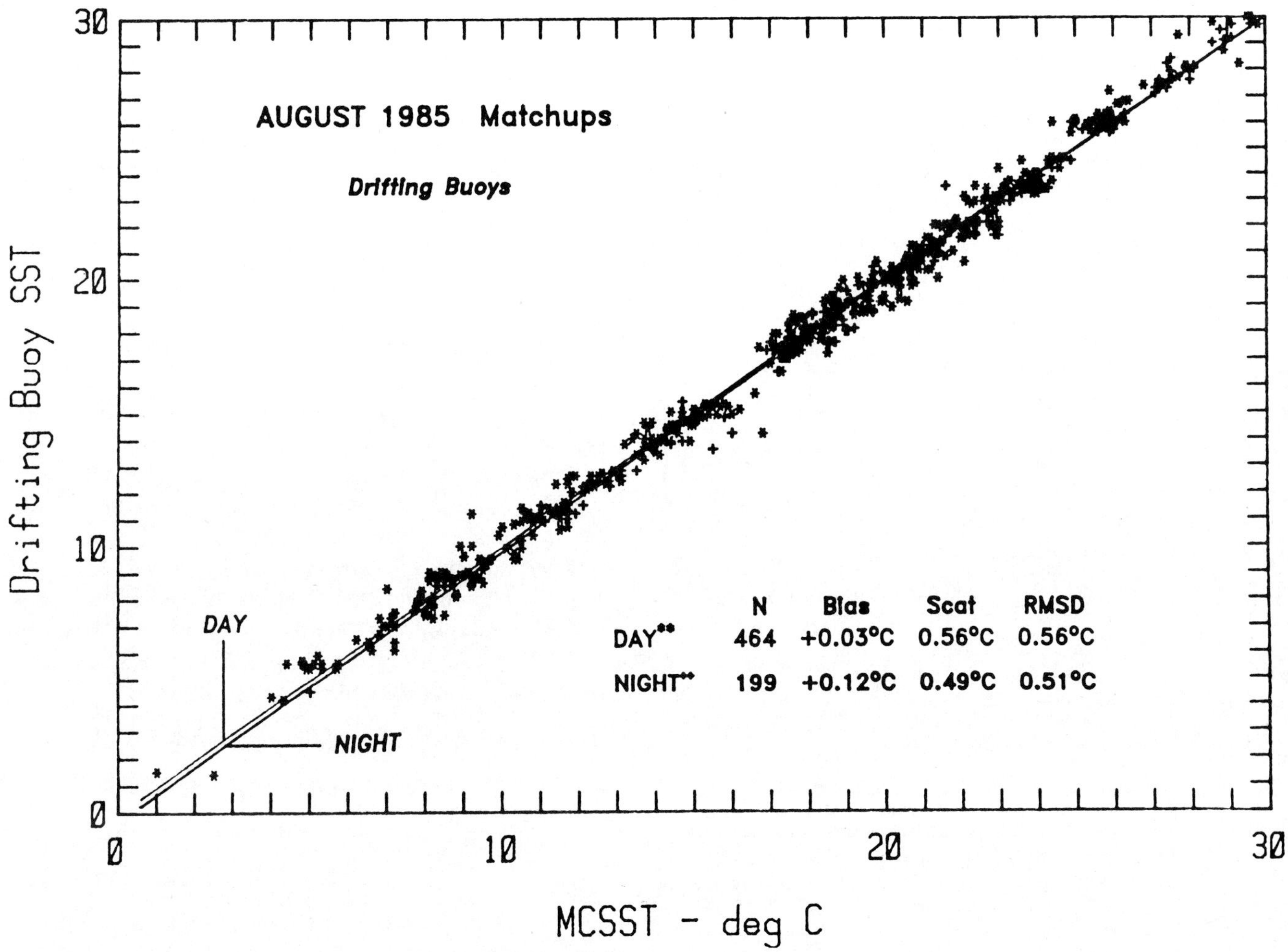

Figure VIII-4-1. Global MCSST matchup statistics with drifting buoys, August 1985.

of significantly lowering the 3.7 μm temperature relative to either of the other window temperatures and thus providing a means of nighttime cloud detection.

The fundamental basis for making atmospheric attenuation corrections to satellite-based IR temperatures is differential absorption in the various atmospheric window regions of the spectrum (McMillin and Crosby, 1984). The differing IR temperatures measured in the various channels as the radiometer views the same spot on the ocean surface lead to a highly linear relation between any pair of IR temperatures and the correction needed. Simulation experiments using radiosonde measurements from marine atmospheres and atmospheric transmittance models enabled development of the first practical MCSST estimates, but initial validation efforts indicated a small but significant temperature-dependent bias (McClain, 1981). Further tests using substantial numbers of satellite and fixed-buoy measurements that matched to within 25 km and 24 hours enabled the derivation of operational temperature dependent bias corrections for all the simulation equations. Operational production of MCSST began in November 1981. Typical of the operational MCSST equations are the following for AVHRR on NOAA-9 (IR temperatures in K yield MCSST in °C):

Daytime (using the split-window channels)

$$T_{11/12} = 3.64T_{11} - 2.66T_{12} - 267.96 \qquad \text{(VIII-4-1)}$$

Nighttime (using all three window channels)

$$T_{3.7/11/12} = 0.99T_{3.7} + 1.01T_{11} - 0.99T_{12} - 273.28 \qquad \text{(VIII-4-2)}$$

Figure VIII-4-1 shows daytime and nighttime drifting buoy matchups with MCSST calculated from NOAA 9 measurements for August 1985. Bernstein (1982b)

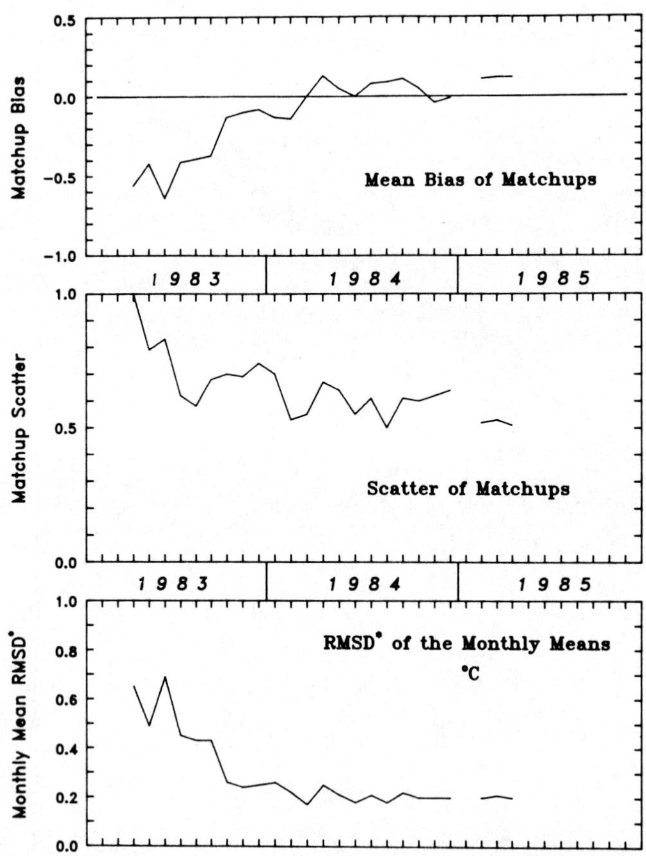

Figure VIII-4-2. Monthly mean biases and scatter (standard deviation of the difference) for MCSST compared with drifting buoy SST for 1983, 1984, and part of 1985. Daytime and nighttime data have been combined. The root mean square difference (RMSD*) is calculated for the monthly mean MCSST product assuming the number of independent observations (N) at each grid point is at least 10 (see text).

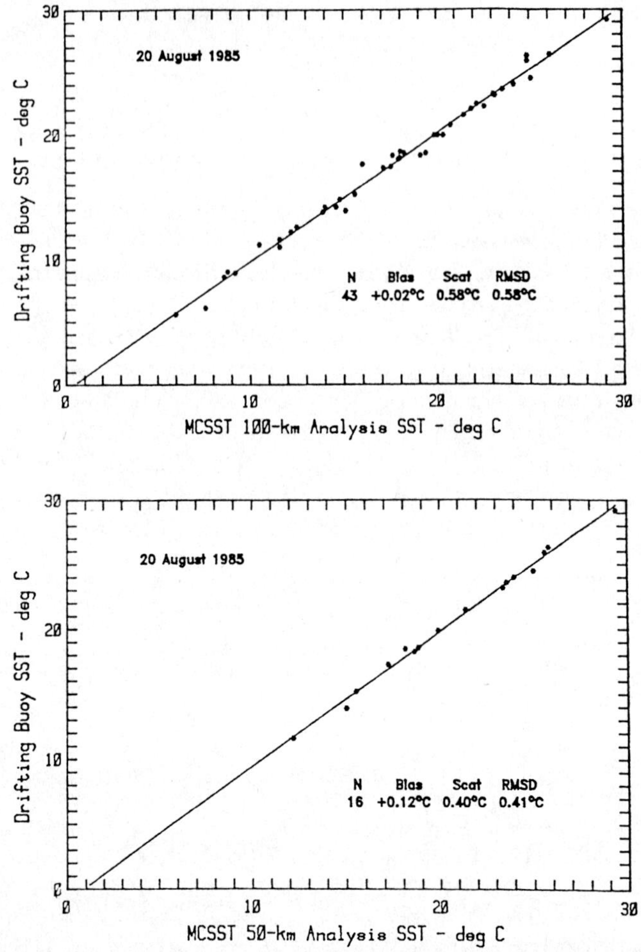

Figure VIII-4-3. Validations of MCSST 100-km and 50-km daily field analyses using drifting buoy SST within 3 days of August 20, 1985 (analysis date).

obtained comparable results using 3.7 and 11 μm data from NOAA-6 AVHRR in 1979-80. The "scatter" (standard deviation of the buoy-MCSST differences during a month) in 1983-84 does not exceed 0.7°C after the first three months of statistics (Figure VIII-4-2).

Measurements from MCSST 100-km and 50-km daily field analyses validated with drifting buoy data show biases within ±0.1°C and accuracies from 0.4 to 0.6°C, similar to the individual matchup statistics (Figure VIII-4-3). Matchup statistics (Table VIII-4-1) indicate no significant or systematic differences among the various regions or from the global values.

Monthly mean maps of MCSST and their anomalies are produced by NESDIS using the SST climatology of Robinson and Bauer (1976), and Robinson et al. (1979). Monthly means for December 1982 and December 1983 (Figure VIII-4-4) show an El Niño in December 1982 along the Equator west of South America.

Ocean currents can be followed in the MCSST charts. Equatorial ocean currents with westward propagating waves about 1000 km long have been detected on a thermal front in the eastern equatorial Pacific and Atlantic Oceans. Operational MCSST charts (e.g., Figure VIII-4-5) have permitted regular monitoring of these long waves (Legeckis et al., 1983).

Satellite derived SST have also been used in studies of the Gulf Stream (Legeckis and Bane, 1983), the Brazil/Falkland current (Legeckis and Gordon, 1982), the Leeuwin current off Australia (Legeckis and Cresswell, 1981), and the Kuroshio/Oyashio current (Vastano and Bernstein, 1984). MCSST patterns and ocean color imagery were used to investigate warm core rings along the Gulf Stream (Schink et al., 1982).

A study near the North Atlantic Current front showed that mesoscale eddies visible in the MCSST field matched the kinematic pattern of the surface layer under conditions of moderate stratification, but not in the presence of a strong seasonal thermocline (Hardtke and Meincke, 1984).

VIII-4.2.2 SST Derived from GOES VAS Data

The procedure used to derive SST from the GOES VAS multispectral data is similar to that described by McClain et al. (1985) for the AVHRR data. Empirical regression equations are generated by matching buoy SST reports with 3 × 3 arrays of 16-km clear field of view radiance measurements. The JPL Workshop results showed that VAS derived SST have a standard deviation of 0.8 to 1.0°C with a slight warm bias (Bates and Smith, 1985). A second and larger independent set of empirical matches and further improvements in cloud filtering are currently used to eliminate the bias. Table VIII-4-2 shows the validation of the VAS algo-

Table VIII-4-1 *MCSST matchup statistics with drifting buoy SST (°C) grouped by region and moisture, February–August 1985*

REGIONAL	DAY				NIGHT			
	NO.	BIAS	SCAT.	RMSD	NO.	BIAS	SCAT.	RMSD
WORLD	1625	+0.01	0.55	0.55	1042	+0.12	0.47	0.48
W. Trop. Pac.	189	−0.02	0.47	0.47	103	+0.22	0.33	0.40
E. Trop. Pac.	433	−0.15	0.53	0.55	290	+0.08	0.41	0.41
Trop. Atl.	424	+0.08	0.50	0.51	277	+0.12	0.49	0.51
Trop. Ind.	277	−0.03	0.50	0.50	163	+0.11	0.45	0.46
N. Atl. & Pac.	208	+0.18	0.57	0.60	115	+0.18	0.52	0.54
N. Pac.	112	+0.25	0.60	0.65	74	+0.16	0.57	0.59
N. Atl.	96	+0.10	0.52	0.53	41	+0.23	0.40	0.46
Southern Ocean	139	+0.08	0.62	0.63	113	+0.11	0.43	0.44
MOISTURE*								
Trop. Oceans								
0° <T_4–T_5 <1°	500	−0.08	0.46	0.47	218	+0.16	0.44	0.47
1° <T_4–T_5 <2°	694	+0.00	0.56	0.56	553	+0.09	0.45	0.46
2° <T_4–T_5 <3°	69	−0.01	0.62	0.61	38	−0.01	0.53	0.52

* Note: the greater the T_4–T_5 value, the greater the atmospheric column water vapor amount.

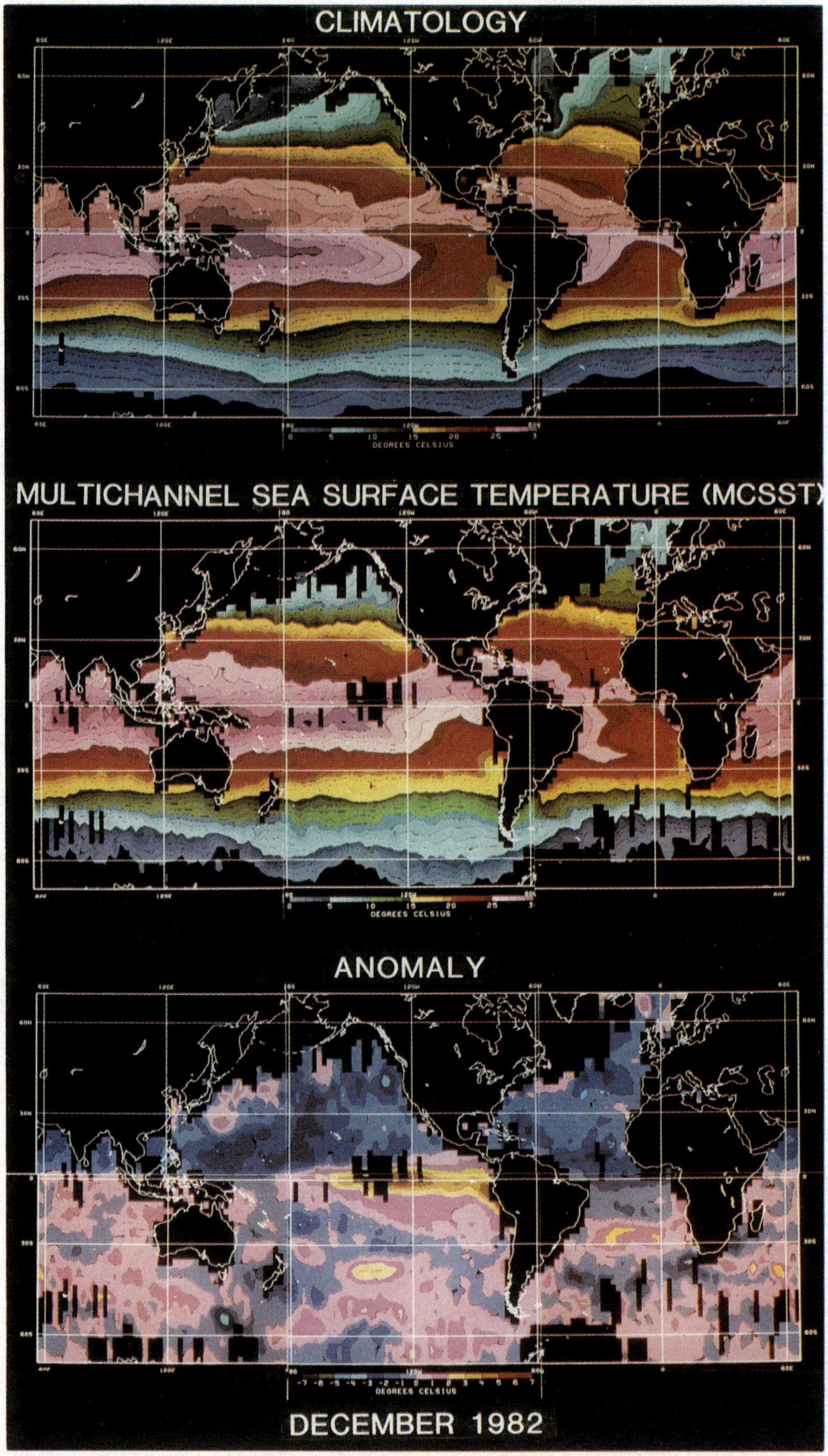

Figure VIII-4-4a. Climatology (top), monthly mean MCSST fields (middle), and anomaly fields (bottom) for December 1982. (Blue, 0 to 5°C; cyan, 5 to 10°C; green, 10 to 15°C; yellow, 15 to 20°C; red, 20 to 25°C; and pink, 25 to 30°C) Black areas are either land areas or are data voids at grid points.

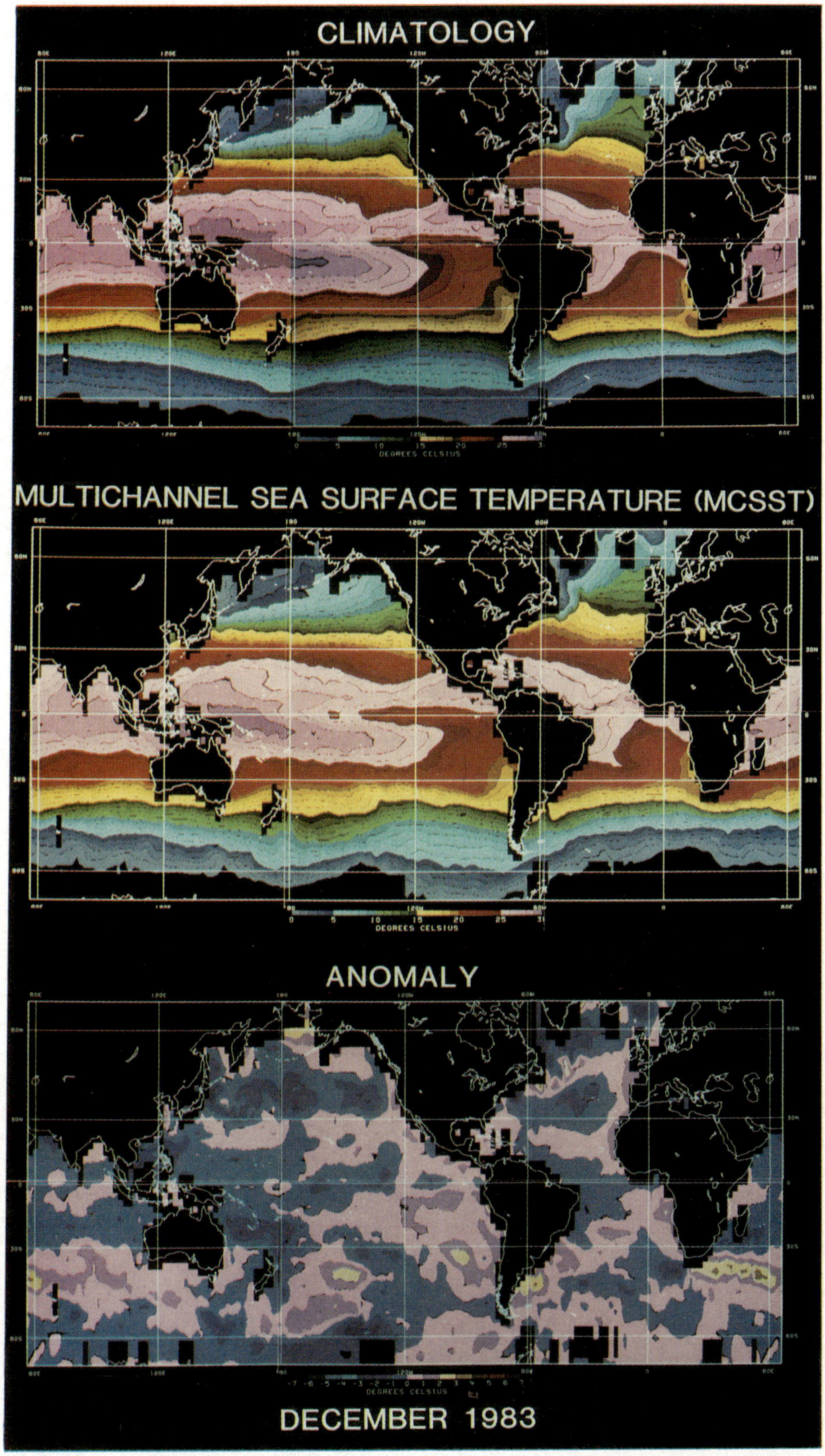

Figure VIII-4-4b. Climatology (top), monthly mean MCSST fields (middle), and anomaly fields (bottom) for December 1983. (Blue, 0 to 5°C; cyan, 5 to 10°C; green, 10 to 15°C; yellow, 15 to 20°C; red, 20 to 25°C; and pink, 25 to 30°C) Black areas are either land areas or are data voids at grid points.

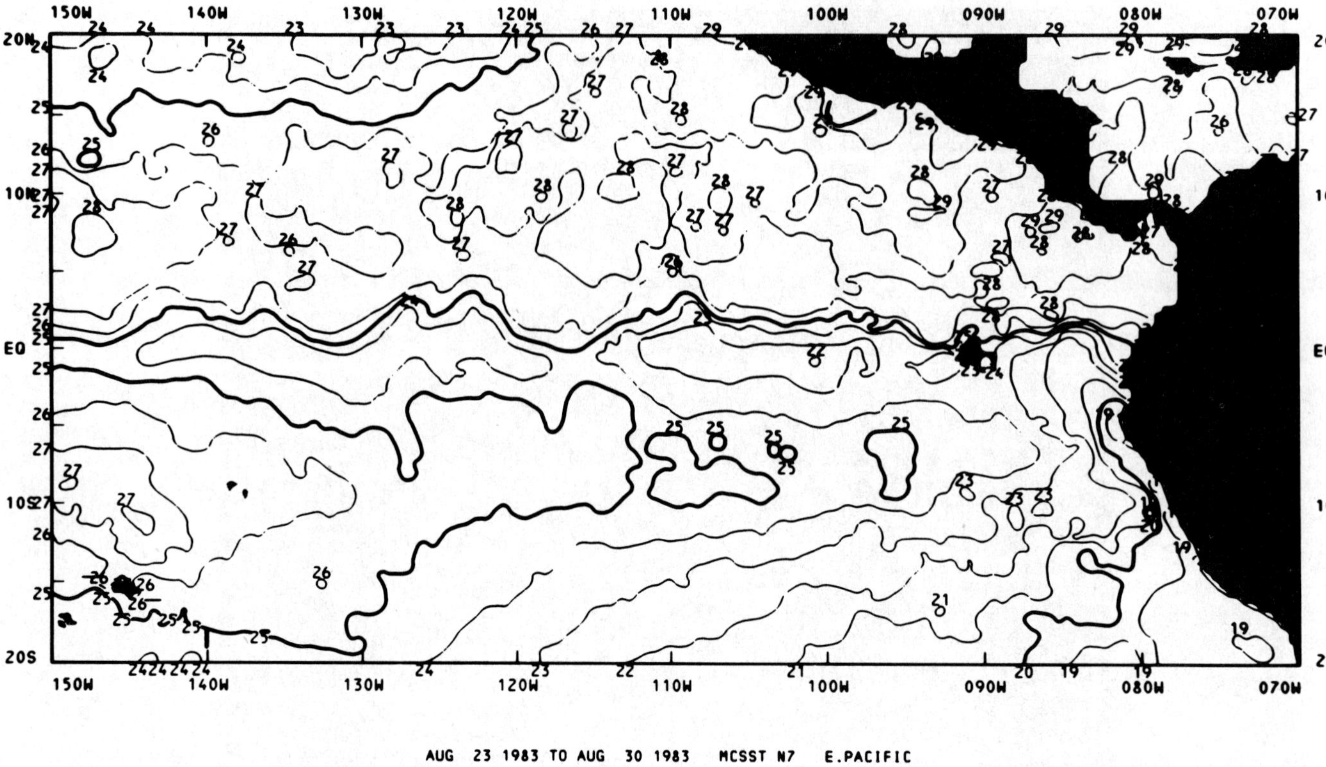

Figure VIII-4-5. 50-km grid MCSST chart for eastern equatorial Pacific, August 20-30, 1983. Note the long waves in the front forming the northern boundary of the tongue of cold water extending westward from South America.

rithms in comparison with NOAA fixed buoys in the Gulf of Mexico, the northwest Atlantic, and the Great Lakes.

VIII-4.2.3 SST Derived from Scanning Multichannel Microwave Radiometer (SMMR) Data

Satellite borne SMMR experiments to specify sea surface temperature are described by Gloersen and Barath (1977) and by Njoku et al. (1980). Two distinct methods, one a linear regression, were used to convert the Seasat SMMR data to SST. Initial development of the linear regression method is described by Wilheit and Chang (1980). The other algorithm was a non-linear, least-squares, iterative estimation technique (Bierman et al., 1978). Differences between observed and computed temperatures are described by Wentz (1983). Both methods yield SST of comparable accuracy; however, only the regression method was used in the final data processing and archiving. SST were derived from Nimbus-7 SMMR data through a regression method developed by Wilheit and Chang (1980), with revisions by Milman and Wilheit (1985).

The first SST results from Seasat SMMR (Lipes et al., 1979) were from the Gulf of Alaska Seasat Experiment, a special activity in September 1978 to collect high quality surface data for verifying SMMR SST estimates and other Seasat measurements. For a limited data set, comparisons showed a standard deviation of 1.5°C between SMMR estimates and surface observations. SMMR SST also had a cold bias of about 4°C. Subsequent work concentrated on determining and eliminating sources of the bias in the SMMR data (e.g., polarization mixing, Sun glint, rain, interference from ground transmissions, and proximity to land). When this was done, results improved dramatically

Table VIII-4-2 *VAS SST regional relations derived from 1982 matches with NOAA fixed environmental buoy SST (Bates and Smith, 1985)*

	CORRELATION COEFFICIENT	EXPLAINED VARIANCE °C	STANDARD ERROR °C
Two Channel	0.83	0.68	1.12
Three Channel	0.87	0.76	0.98

SST range 5°–30°C; number of matches = 205

SST (two channel) = $T_{11} + 1.2956(T_{11}-T_{12}) - 0.28$

SST (three channel) = $T_{11} + 0.798(T_{3.9}-T_{12}) + 0.63$

(Lipes, 1982); the standard deviation was reduced to 0.9°C and the bias was reduced to 0.1°C (warm).

Bernstein (1982a) compared temperature fields from nighttime SMMR data and from surface observations in the northwest Pacific. Figure VIII-4-6 shows SMMR SST versus ship SST; the agreement is good over a wide range. Figure VIII-4-7 compares mapped SST derived from Seasat SMMR data with mapped surface observations. Subsequently, Bernstein and Morris (1983) analyzed Pacific SST for the entire Seasat mission; it was found that corrections for cross-scan biases and monthly adjustments for Sun angle were required (Table VIII-4-3).

VIII-4.2.4 SST Derived from HIRS/MSU Data

Susskind et al. (1984) have used HIRS2/MSU sounding data to produce global sets of SST. Although IR and MW sounding channels have much lower resolution than AVHRR data, the multiplicity of simultaneous information available from the various sounding channels can be used to make consistent corrections for cloudiness and moisture. Thus useful SST derivations can be made from the HIRS IR window channels. Susskind and Reuter (1985) compared the HIRS2/MSU SST retrievals with those from AVHRR, VAS, SMMR, and ships, and found that they were of at least comparable quality. Errors were generally random in both space and time, although individual values were rather noisy.

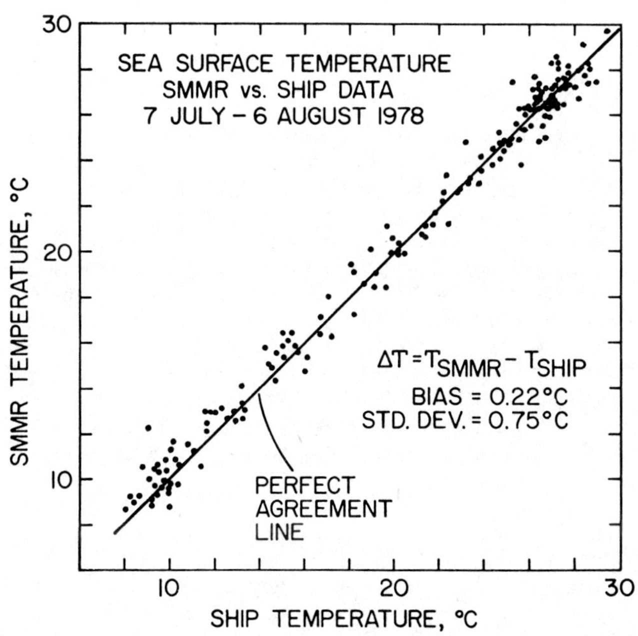

Figure VIII-4-6. SMMR derived SST (ordinate) vs ship derived SST (abscissa). Mean difference and standard deviation are shown (Bernstein, 1982a).

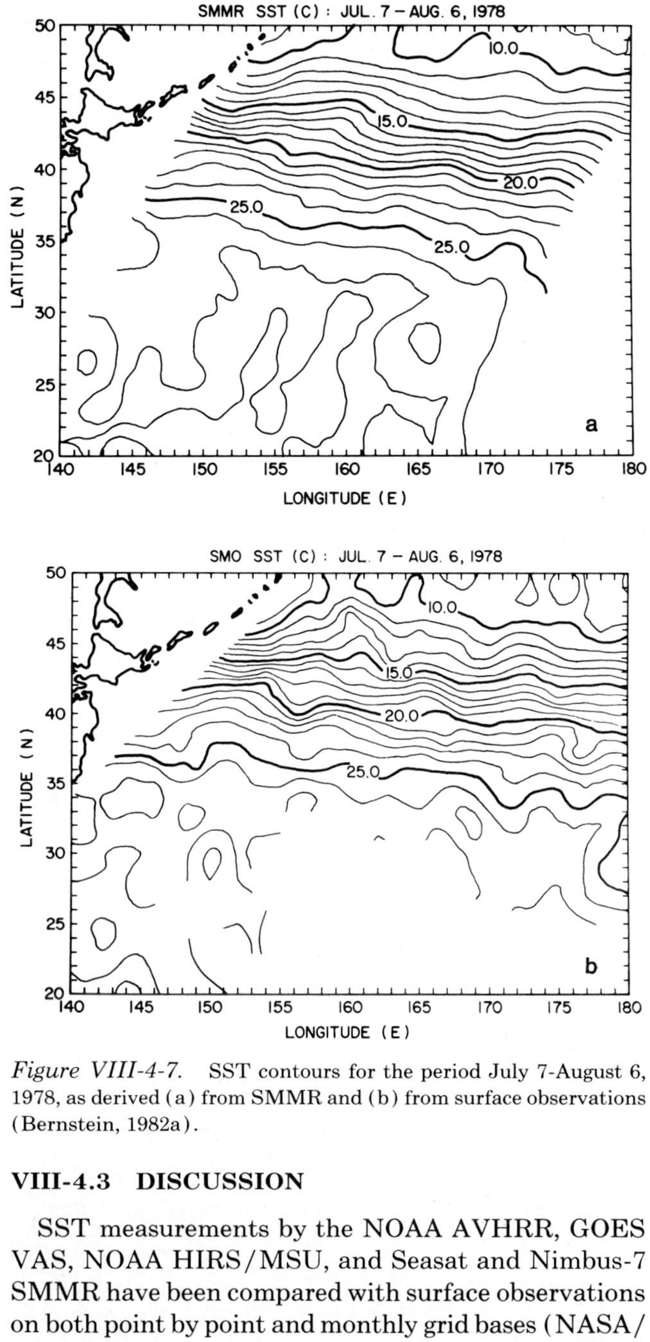

Figure VIII-4-7. SST contours for the period July 7-August 6, 1978, as derived (a) from SMMR and (b) from surface observations (Bernstein, 1982a).

VIII-4.3 DISCUSSION

SST measurements by the NOAA AVHRR, GOES VAS, NOAA HIRS/MSU, and Seasat and Nimbus-7 SMMR have been compared with surface observations on both point by point and monthly grid bases (NASA/

Table VIII-4-3 *Monthly mean SMMR and surface comparisons (Bernstein and Morris, 1983)*

PERIOD (1978)	BIAS (°C)	STANDARD DEVIATION (°C)
July 6–August 4	−0.48	0.52
August 5–September 3	1.04	0.46
September 4–October 3	0.09	0.61

Comparisons are gridded at 2° lat.-long. resolution; corrections for cross-scan biases and Sun angle have been made.

JPL, 1985). The agreement between averaged and gridded values is significantly better than the point by point comparisons. Possibly because of much smaller fields of view, point by point comparisons of SST derived from VAS and AVHRR with surface observations are better than for HIRS/MSU and SMMR data. All sets of SST determinations use multiple channels to correct for atmospheric effects. The SMMR and MSU data require correcting for surface emissivity as well.

In these comparisons it was assumed that surface *in situ* observations are correct, but comparisons of satellite derived SST with observations from nearly coincident "ships of opportunity" have suffered from two main problems: (1) the satellite obtains an areal average of the "skin" temperature of the sea, while the ship reports a point value of "intake temperature" (i.e., water drawn in to cool the engines) from a depth of 10 m or more; and (2) the ship data are notoriously variable and of uncertain quality. Temperatures measured in the first meter below the surface by buoys have proved to be considerably better for comparison purposes (Strong and McClain, 1984).

The most serious limitation of the MCSST technique is that essentially cloud-free Field of Views (FOV) are required, although multi-day compositing permits satisfactory mapping of the temperature field in all but the most persistently cloud covered regions. Anomalously high concentrations of atmospheric aerosol, such as those associated with severe volcanic eruptions or intense dust storms, occasionally degrade IR-based SST, but some progress has already been made in dealing with this problem (Walton, 1985).

Both the MW and IR results described here are state-of-the-art; each method has its strengths and weaknesses. The IR radiometer can achieve a spatial resolution that antenna size and detector sensitivity deny the MW radiometer. The MW radiometer easily penetrates cloud and fog that can block the view of an IR radiometer. User requirements for SST vary in terms of spatial resolution and frequency of coverage, or accuracy of temperature determinations versus an ability to detect gradients in SST. Thus, each technology can make significant contributions to meeting SST user requirements.

BIBLIOGRAPHY AND REFERENCES FOR SECTION VIII

Alishouse, J. C. and E. P. McClain, 1985: Sea surface temperature determinations. *Advances in Geophysics, 27*, Academic Press, New York, NY, 279-296.

Assel, R. A., F. H. Quinn, G. A. Leshkevich, and S. J. Bolsenga, 1983: *Great Lakes Ice Atlas*. NOAA, Great Lakes Environ. Res. Lab., Ann Arbor, MI, 115 pp.

Ayers, J. C., E. F. Stoermer, and P. McWilliam, 1967: Recently noted changes in the Biology-Chemistry of Lake Michigan. *Studies on Environ. and Eutro-phication of Lake Mich.*, Spec. Rpt. No. 30, Great Lakes Res. Div., Univ. of Michigan, Ann Arbor, MI, 95-111.

———, D. C. Chandler, G. H. Lauff, C. F. Powers, and E. B. Henson, 1958: *Currents and Water Masses of Lake Michigan*. Publ. No. 3., Great Lakes Res. Div., Univ. of Michigan, Ann Arbor, MI, 169 pp.

Barnes, J. C. and C. J. Bowley, 1974: *Handbook of Techniques for Satellite Snow Mapping*. ERT Doc. No. 0407-A, Environ. Research & Technology, Inc., Concord, MA, 95 pp.

——— and C. J. Bowley, 1968: *Operational Guide for Mapping Snow Cover from Satellite Photography*. Allied Research Associates, Inc., Concord, MA, 116 pp.

Bates, J. J. and W. L. Smith, 1985: Sea surface temperature: Observations from geostationary satellites. *J. Geophys. Res., 90*, 11609-11618.

Bathurst, R. G. C., 1971: Carbonate sediments and their diagenesis. *Developments in Sedimentology, 12*, Elsevier Pub. Co., New York, NY, 136-292.

Berg, C. P., S. R. Schneider, and D. L. Galat, 1981a: Calcium carbonate precipitation in Pyramid Lake, Nevada, as monitored by satellite: 1978 and 1980. *Proc. of the 15th Intl. Symp. on Remote Sensing of Environ.*, Environ. Res. Inst. of MI, Ann Arbor, MI, 721-731.

———, D. R. Wiesnet, and M. Matson, 1981b: Assessing the Red River of the North 1978 flooding from NOAA satellite data. *Satellite Hydrology*, M. Deutsch, D. R. Wiesnet, and A. Rango (eds.), American Water Resources Association, Minneapolis, MN, 309-315.

Berner, R. A., 1975: The role of magnesium in the crystal growth of calcite and aragonite in sea water. *Geochim. Cosmochim. Acta., 39*, 489-504.

Bernstein, R. L. and J. H. Morris, 1983: Tropical and mid-latitudeNorth Pacific sea surface temperature variability from the Seasat SMMR. *J. Geophys. Res, 88*, 1877-1891.

———, 1982a: Sea surface temperature mapping with the Seasat microwave radiometer. *J. Geophys. Res., 87*, 7865-7872.

———, 1982b: Sea surface temperature estimation using the NOAA-6 satellite Advanced Very High Resolution Radiometer. *J. Geophy. Res, 87*, 9455-9465.

Bierman, G. J., F. J. Wentz, III, and R. G. Lipes, 1978: Modern estimation techniques applied to microwave sensing of the marine boundary layer. *Proc. of the Asilomar Conf. Circuits, Syst. Computer, 12th IEEE Computer Sci., Pacific Grove, CA*, Inst. of Electrical and Electronics Engineers, New York, NY, 101-106.

Binenko, V. I., L. N. D'Yachenko, K.Ya. Kondrat'ev, and A. P. Chernenko, 1974: Detection of small incipient fires by infrared surveying. *Atmos. Rad. Studies*, K.Ya. Kondrat'ev (ed.), Keter Publ. House, Jerusalem, Israel. 203-212.

Bornstein, R. D., 1968: Observations of the urban heat island effect in New York City. *J. Appl. Meteor., 7*, 575-582.

Bressette, W. E. and D. E. Lear, 1973: The use of near-infrared photography for biodegradable pollution monitoring of tidal waters, Part II. *Proc. of the Symp. on Remote Sensing, Oct. 2-5, Orlando, FL,* Amer. Soc. of Photogram., Falls Church, VA, 906-937.

Brown, O. B., R. H. Evans, J. W. Brown, H. R. Gordon, R. C. Smith, and K. S. Baker, 1985: Phytoplankton blooming off the U.S. East Coast: A satellite description. *Science, 229*, 163-167.

Carlson, T. N., J. K. Dodd, S. G. Benjamin, and J. N. Cooper, 1981: Satellite estimation of the surface energy balance, moisture availability, and thermal inertia. *J. Appl. Meteor., 20*, 67-87.

———, J. N. Augustine, and R. E. Boland, 1977: Potential application of satellite temperature measurements in the analysis of land use over urban areas. *Bull. Amer. Meteor. Soc., 58*, 1301-1303.

Casper, V. L., 1965: A phytoplankton bloom intern Lake Erie. *Proc. of the 8th Conf. on Great Lakes Res*., Great Lakes Res. Div., Univ. Mich., Ann Arbor, MI, 29-35.

Chang, A. T. C., 1986: Nimbus-7 SMMR Snow Cover Data. *Glaciological Data Report GD-18*, Snow Watch '85, G. Kukla, R. Barry, A. Hecht, and D. Wiesnet (eds.), World Data Center for Glaciology, Boulder, CO, 181-187.

Chang, A. T. C., J. L. Foster, D. K. Hall, and A. Rango, 1980: Monitoring snow pack properties by passive microwave sensors on board aircraft and satellites. *NASA Conf. Publ. CP 2153*, NASA/GSFC, Greenbelt, MD, 235-248.

Changnon, S. A., 1978: Urban effects on severe local storms at St. Louis. *J. Appl. Meteor., 17*, 578-586.

Chave, K. E., 1970: Carbonate-organic interactions in sea water. *Organic Matter in Natural Waters*, D. W. Hood (ed.), Inst. Mar. Sci. (Alaska), Occasional Publ. 1, 373-385.

Chouret, A., 1974: Les effets de la srecheresse actuelle en Afrique sur le niveau du lac Tchad. *Cuh O. R. S.T. O. M., ser. Hydrol., XI*, No. 1, 35-46.

Chrystal, G., 1908-09: Investigation of the seiches of Loch Earn by the Scottish Loch Survey, Part III. *Trans. Roy. Soc. Edinburgh, 45 (2)*, 361-396.

Clarke, J. F. and J. T. Peterson, 1972: The effect of regional climate and land use on the nocturnal heat island. *Proc. of the Conf. Urban Envir. and 2nd Conf. Biometeor., Philadelphia, PA*, Amer. Meteor. Soc., Boston, MA, 147-152.

Crane, R. G., 1979: Inventory of snow cover and sea ice data. *Glaciol. Data Report GD-7*, World Data Center A for Glaciology, Univ. of Colorado, Boulder, CO, 170 pp.

DeLuisi, J., 1983: Measurements of the El Chichon dust cloud from Mauna Loa observatory. *Proc. of the 7th Annual Climate Diagnostics Workshop, October 17-22, 1982*, Natl. Cntr. for Atmos. Res., Boulder, CO, 383-385.

Dewey, K. F. and R. Heim, Jr., 1982: A digital archive of Northern Hemisphere snow cover, November 1966 through December 1980. *Bull. Amer. Meteor. Soc., 63*, 1132-1141.

——— and R. Heim, Jr., 1981: Satellite observations of variations in Northern Hemisphere seasonal snow cover. *NOAA Technical Report NESS 87*, Dept. of Commerce, Washington, DC, 83 pp.

Dickson, R. R., 1984: Eurasian snow cover versus Indian monsoon rainfall an extension of the Hahn-Shukla results. *J. Climate Appl. Meteor., 23*, 171-173.

——— and J. Posey, 1967: Maps of snow-cover probability for the Northern Hemisphere. *Mon. Wea. Rev., 95*, 347-353.

Dillard, J. P., 1980: Use of satellite data in runoff forecasting in the heavily-forested, cloud-covered Pacific Northwest. *NASA Conf. Publ. 2116*, Natl. Aero. and Space Admin., Washington, DC, 127-150.

Edgerton, A. T., F. Ruskey, D. Williams, A. Stogryn, G. Poe, D. Meeks, and O. Russell, 1973: Microwave emission characteristics of natural materials and the environment (A summary of six years research). *Final Tech. Rep. 9016 R-8*, Aerojet-General Corp., El Monte, CA.

Environmental Data Service, 1973: *Climatological Data-Alabama, 79 (2)*, 19 pp.; *Climatological Data-Georgia, 77 (2)*, 22 pp.; *Climatological Data-South Carolina, 76 (2)*, 17 pp., Dept. of Commerce, National Climatic Data Center, Asheville, NC.

Environmental Protection Agency, 1974: *Methods of Chemical Analysis of Water and Wastes*. EPA, Cincinnati, OH, 298 pp.

Fiedler, P., 1984: Satellite observations of the 1982-83 El Niño along the U.S. Pacific Coast. *Science, 224*, 1251-1254.

Follansbee, W. A., 1973: Estimation of average daily rainfall from satellite cloud photographs. *NOAA Technical Memorandum NESS 44*, Dept. of Commerce, Washington, DC, 39 pp.

Foster, J. L., D. K. Hall, and A. T. C. Chang, 1984: An overview of passive microwave snow research and results. *Rev. Geophys. Space Phys., 22*, 195-207.

Fung, A. K. and H. J. Eom, 1981: An approximate model for backscattering and emission from land and sea. Remote Sensing Lab., Univ. of Kansas, Lawrence, KS, AgRISTARS Rep. SM-K1-04049.

Giddings, J. C. and E. LaChapelle, 1961: Diffusion theory radiant energy distribution and albedo of snow. *J. Geophy. Res., 66*, 181-189.

Gird, R., 1980: Snow extent measurements from geostationary satellites using an interactive computer system. *NASA Conf. Publ. 2116*, Natl. Aero. and Space Admin., Washington, DC, 21-39.

Gloersen, P. and F. J. Barath, 1977: A scanning multichannel microwave radiometer for Nimbus-G and Seasat-A. *IEEE J. Oceanic Eng., 0E-2*, 172-178.

Great Lakes Res. Div., 1967: Lake Michigan NAVOCEANO flight. *Special Report No. 31*, Great Lakes Res. Div., Univ. of Michigan, Ann Arbor, MI, 52 pp.

Hamilton, G., 1984: Appendix B revision, NOAA-F/9. *NOAA Technical Memorandum NESDIS 107,* Dept. of Commerce, Washington, DC, 11 pp.

Hanstrum, B. N. and A. S. Watson, 1983: A case study of two eruptions of Mount Galunggung and an investigation of volcanic eruption cloud characteristics using remote sensing techniques. *Aust. Meteor. Mag., 31*, 171-177.

Hardtke, P. G. and J. Meincke, 1984: Kinematical interpretation of infrared surface patterns in the North Atlantic. *Oceanologica Acta, 7*, 373-378.

Hirsch, S. N., R. F. Kruckelberg, and F. H. Madden, 1971: The bispectral forest fire detection system. *Proc. of the 7th Intl. Symp. on Remote Sensing of the Environment, May 17-21*, ERIM, Univ. of Michigan, Ann Arbor, MI, 2253-2272.

———, 1964: Forest fire detection systems. *Proc. of the Western Forest Fire Research Council*, 3-5.

Hodge, W. T., 1973: Climatic data report-Southeastern snow storm, Feb. 8-11, 1973. *NOAA Technical Memorandum EDS NCC 2*, Dept. of Commerce, Washington, DC, 50 pp.

Hoffer, R. and C. Matzler, 1980: Investigations of snow parameters by radiometry in the 3- to 60-mm wavelength region. *J. Geophys. Res., 85*, 43-460.

Huff, F. A. and J. L. Vogel, 1978: Urban, topographic, and diurnal effects on rainfall in the St. Louis region. *J. Appl. Meteor., 17*, 565-577.

Hutchinson, G. E., 1957: *A Treatise on Limnology*. John Wiley and Sons, Inc., New York, NY, Vol. 1, 1015 pp.

Jarvinen, B. R. and C. G. Griffith, 1984: Forecasting rainfall in tropical cyclones using digitized infrared satellite data. *ERL Technical Paper*, Dept. of Commerce, Natl. Hurricane Center, Miami, FL, 12 pp.

Katzenberger, W. M. and J. A. Whelan, 1975: Currents of great Salt Lake, Utah. *Utah Geology, 2*, 103-107.

Kellogg, W. W., 1977: Effects of human activities on global climate—Part l. *WMO Bull., 26*, 229-240.

Kelts, K. and K. J. Hsu, 1978: Freshwater carbonate sedimentation. *Lakes: Chemistry, Geology, Physics*, A. Lerma (ed.), Springer-Verlag, New York, NY, 295-323.

Kienle, J. and G. E. Shaw, 1979: Plume dynamics, thermal energy and long-distance transport of volcanian eruption clouds from Augustine volcano, Alaska. *J. Volcanol. Geother. Res., 6*, 139-164.

Krueger, A. F., 1982: Geostationary satellite observations of the April 1979 Soufriere eruptions. *Science, 216*, 1108-1109.

Kukla, G. and J. Gavin, 1983: Recent fluctuations of Northern Hemisphere snow cover in autumn. *Proc. of the 8th Annual Climate Diagnostics Workshop, October 17-21*, Atmos. Envir. Serv., Downsview, Ontario, Canada, 289-296.

———, 1981: Snow cover and climate. *Glaciol. Data Report GD-11*, G. Kukla, A. Hecht, and D. Wiesnet (eds.), World Data Center A for Glaciology, Univ. of Colorado, Boulder, CO, 27-39.

——— and D. Robinson, 1981: Climatic value of operational snow and ice charts. *Glaciol. Data Report GD-11*, G. Kukla, A. Hecht, and D. Wiesnet (eds.), World Data Center A for Glaciology, Univ. of Colorado, Boulder, CO, 103-119.

——— and D. Robinson, 1979: Accuracy of snow and ice monitoring. *Glaciol. Data Report GD-5*, World Data Center A for Glaciology, Univ. of Colorado, Boulder, CO, 91-98.

Kunzi, K. F., S. Patil, and H. Rott, 1984: Snow-cover parameters retrieved from Nimbus-7 Scanning Multichannel Microwave Radiometer (SMMR). *Rev. Geophys. Space Phys., 22*, 452-467.

Landsberg, H. E., 1976: Concerning possible effects of air pollution on climate. *Bull. Amer. Meteor. Soc., 57*, 213-215.

———, 1970: Man-made climatic changes. *Science, 170*, 1265-1274.

———, 1956: *The Climate of Towns. Man's Role in Changing the Face of the Earth*, W. L. Thomas (ed.), Univ. of Chicago Press, Chicago, IL, 584-606.

Lange, A. M. T., M. C. Ingham, and C. A. Price, 1984: Distribution of maturing *Illex illecebrosus* relative to the shelf-slope water front of the northeastern United States. *Preprint, Sci. Council Mtg., N. W. Atl. Fish. Organiz.*, NAFO SCR 84/IX/109, Ser. No. N906, 18 pp.

Lauritson, L., G. J. Nelson, and F. W. Porto, 1970: Data extraction and calibration of TIROS-N/NOAA radiometers. *NOAA Technical Memorandum NESS 107*, Dept. of Commerce, Washington, DC, 44-66.

Laurs, R. M., 1971: Fishery-advisory information available to tropical Pacific tuna fleet via radio facsimile broadcast. *Comm. Fish. Rev., 33*, 40-42.

Legeckis, R. and J. W. Bane, Jr., 1983: Comparison of the TIROS-N satellite and aircraft measurements of Gulf Stream surface temperatures. *J. Geophys. Res., 88*, 4611-4616.

———, W. Pichel, and G. Nesterczuk, 1983: Equatorial long waves in geostationary satellite observations and in a multichannel sea surface temperature analysis. *Bull. Amer. Meteor. Soc., 64*, 133-139.

———, and A. L. Gordon, 1982: Satellite observations of theBrazil and Falkland currents 1975 to 1976 and 1978. *Deep Sea Res., 29*, 375-401.

———, and G. Cresswell, 1981: Satellite observations of sea-surface temperature fronts off the coast of western and southern Australia. *Deep Sea Res., 28A*, 297-306.

———, 1978: A survey of worldwide sea surface temperature fronts detected by environmental satellites. *J. Geophys. Res., 83*, 4501-4522.

——— and J. Pritchard, 1976: Algorithm for correcting the VHRR imagery for geometric distortions due to Earth curvature, Earth rotation, and spacecraft roll attitude errors. *NOAA Technical Memorandum NESS 77*, Dept. of Commerce, Washington, DC, 31 pp.

Lin, A., 1976: *A Survey of the Physical Limnology of Great Salt Lake*. Center for Water Resources Res., Utah Water Res. Lab., Logan, Utah, 82 pp.

———, P. Chang, and P. Sha, 1972: Some physio-chemical characteristics of the Great Salt Lake. *The Great Salt Lake and Utah's Water Resources. Proc. of the 1st Annual Conf. of the Utah Sect. of Amer. Water Res. Assoc.*, 45-65.

Lipes, R. G., 1982: Description of Seasat radiometer status and results. *J. Geophys. Res., 87*, 3385-3395.

———, R. L. Bernstein, V. J. Cardone, K. B. Katsaros, E. G. Njoku, A. S.Riley, D. B. Ross, C. T. Swift, and F. J. Wentz, 1979: Seasat scanning multichannel microwave radiometer: Results of the Gulf of Alaska workshop. *Science, 204*, 115-1417.

Liu, C. T., 1983: Tropical Pacific sea surface temperatures measured by Seasat microwave radiometer and by ships. *J. Geophys. Res., 88*, 1909-1914.

Martinec, J. and A. Rango, 1980: Discharge forecasts in mountain basins based on satellite snow cover mapping. *NASA Conf. Publ. 2116*, Natl. Aero. and Space Admin., Washington, DC, 223-238.

Matson, M., 1984: The 1982 El Chichon volcano eruptions—a satellite perspective. *J. Volcanol. and Geother. Res., 23*, 1-10.

——— and S. R. Schneider, 1984: Fire detection using the NOAA-series satellites. *NOAA Technical Report NESDIS-7*, Dept. of Commerce, Washington, DC, 34 pp.

——— and C. P. Berg, 1981: Satellite detection of seiches in great Salt Lake, Utah. *Water Resources Bull., 17*, 122-128.

——— and J. Dozier, 1981: Identification of subresolution high temperature sources using a thermal IR sensor. *Photogram. Eng. Remote Sens., 47*, 1311-1318.

——— and D. R. Wiesnet, 1981: New database for climate studies. *Nature, 289*, 451-456.

——— and R. Legeckis, 1980: Urban heat islands detected by satellite. *Bull. Amer. Meteor. Soc., 61*, 212.

———, E. P. McClain, D. F. McGinnis, Jr., and J. A. Pritchard, 1978: Satellite detection of urban heat islands. *Mon. Wea. Rev., 106*, 1725-1734.

Maul, G. A., 1984: *Introduction to Satellite Oceanography*. Martinis Nijoff Publ., Kluwer Academic Publ., Norwell, MA, 606 pp.

———, F. Williams, M. Roffer, and F. M. Sousa, 1984: Remotely sensed oceanographic patterns and variability of bluefin tuna catch in the Gulf of Mexico. *Oceanologica Acta, 7*, 469-479.

McClain, E. P., W. G. Pichel, and C. C. Walton, 1985: Comparative performance of AVHRR-based multichannel sea surface temperatures. *J. Geophys. Res., 90*, 11587-11601.

———, W. G. Pichel, C. C. Walton, Z. Ahmad, and J. Sutton, 1983: Multi- channel improvements to satellite-derived global sea surface temperatures. *Adv. Space Res., 2*, Pergamon Press, New York, NY, 43-47.

———, 1981: Multiple atmospheric-window techniques for satellite sea surface temperatures. Oceanog. from Space, J. F. R. Gower (ed.), *Marine Science, Vol. 13*, Plenum Press, New York, NY, 73-85.

———, 1980a: Passive radiometry of the ocean from space-An overview. *Bound. Layer Meteorol., 18*, 7-24.

———, 1980b: Environmental satellites. *McGraw-Hill Encyclopedia of Environmental Science*, McGraw-Hill,New York, NY, 15-30.

———, 1979: Satellite-derived Earth surface temperature, Chapter 3. J. S. Winston (ed.), Quantitative Meteorological Data from Satellites, *Technical Note 166, WMO-No. 531*, World Meteor. Organ., Geneva, Switzerland, 60-68.

———, 1973: Snow survey from Earth satellites-A technical review of methods. *WMO/IHD Report No. 19, WMO No. 353*, World Meteor. Organ. Int. Hydrol. Decade, Geneva, Switzerland, 42 pp.

McCormick, M. P., 1983: Aircraft and ground-based lidar measurements of El Chichon stratospheric aerosols. *Proc. of the 7th Annual Climate Diagnostics Workshop, October 18-22*, Natl. Ctr. for Atmos. Res., Boulder, CO, 386-389.

McGinnis, D. F., Jr., J. A. Pritchard, and D. R. Wiesnet, 1975: Determination of snow depth and snow extent from NOAA-2 satellite very high resolution radiometer data. *Water Resour. Res., 11*, 897-902.

McMillin, L. M. and D. S. Crosby, 1984: Theory and validation of the multiple window sea surface temperature technique. *J. Geophys. Res., 89*, 3655-3661.

Milman, A. S. and T. T. Wilheit, 1985: Sea surface temperatures from the scanning microwave radiometer on Nimbus-7. *J. Geophys. Res., 90*, 11631-11641.

Moore, G. K., 1978: Satellite surveillance of physical water-quality characteristics. *Proc. of the 12th International Symp. on Remote Sensing of the Environment, Manila, Phillippines, April 20-26, 1978*, Environ. Res. Inst. of MI, Ann Arbor, MI, 445-462.

Natl. Aero. and Space Admin./JPL, 1985: Satellite-derived sea surface temperature: Workshop III.*JPL Public. 85-63*, Natl. Aero. and Space Admin., Jet Propulsion Lab., Pasadena, CA, 183 pp.

National Weather Service, 1980: Great Lakes composite ice and surface temperature analyses. *Technical Procedures Bulletin Series No. 290*, Dept. of Commerce, NOAA, Washington, DC, 12 pp.

Njoku, E. G., J. M. Stacey, and F. T. Barath, 1980: The Seasat Scanning Multichannel Microwave Radiometer (SMMR): Instrument description and performance. *IEEE J. Oceanic Eng., 0E-5*, 100-115.

O'Brien, H. W. and R. H. Munis, 1975: Red and near-infrared spectral re- flectance of snow. *CRREL Research Report, 332*, Cold Regions Res. & Eng. Lab., U. S. Army Corps of Engineers, Hanover, NH, 18 pp.

Oke, T. R., G. B. Maxwell, and D. Yap, 1972: Comparison of urban/rural cooling rates at night. *Preprints, Conf. Urban Envir. and 2nd Conf. Biometeor., Philadelphia, PA*, Amer. Met. Soc., Boston, MA, 17-21.

Popham, R. W., 1968: Satellite applications to snow hydrology. *World Meteor. Organ. Int. Hydrol. Decade, WMO/IHD Report No. 7*, Geneva, Switzerland, 10 pp.

Prabhakara, C., G. Dalu, and V. G. Kunde, 1974: Estimation of sea surface temperature from remote sensing in the 11 to 13 μm window region. *J. Geophys. Res., 79*, 5039-5044.

Price, J. C., 1979: Assessment of the urban heat island effect through the use of satellite data. *Mon. Wea. Rev., 107*, 1554-1557.

Rango, A., 1981: Operational applications of satellite snow cover observations, executive summary. *NASA Tech. Paper 1822*, Washington, DC, 70 pp.

———, A. T. C. Chang, and J. L. Foster, 1979: The utilization of spaceborne microwave radiometers for monitoring snow pack properties. *Nordic Hydrology, 10*, 25-40.

———, 1975: Employment of satellite snowcover observations for improving seasonal runoff estimates. *NASA Special Publ. 391*, Natl. Aero. and Space Admin., Washington, DC, 157-174.

Rao, P. K., 1972: Remote sensing of urban "heat islands" from an environmental satellite. *Bull. Amer. Meteor. Soc., 53*, 647-648.

Robinson, M. K., R. A. Bauer, and E. H. Schroeder, 1979: Atlas of North Atlantic and Indian Ocean monthly mean temperatures and mean salinities of the surface layer. *NAVOCEANO Ref. Publ. 18*, Naval Oceanographic Office, Washington, DC, pp.

——— and R. A. Bauer, 1976: Atlas of North Pacific Ocean monthly mean temperature and mean salinities of the surface layer. *NAVOCEANO Ref. Publ. 2*, Naval Oceanographic Office, Washington, DC, 194 pp.

Robock, A. and M. Matson, 1983: Circumglobal transport of the El Chichon volcanic dust cloud. *Science, 221*, 195-197.

Rodgers, E. and H. Siddalingaisk, 1983: The utilization of Nimbus-7 SMMR measurements over land. *J. Clim. Appl. Meteor., 22*, 1753-1763.

Sawada, 1983: Analysis of eruption clouds by the 1981 eruptions of Alaid and Pagan volcanoes with GMS images. *Pap. Meteor. Geophy., 34*, 307-324.

Schanda, E., C. Matzler, and K. Kunzi, 1983: Microwave remote sensing of snow cover. *Intl. J. Remote Sensing, 4*, 149-158.

Schelske, C. L. and J. C. Roth, 1973: *Limnological Survey of Lakes Michigan, Superior, Huron, and Erie.* Great Lakes Res. Div., Univ. of Michigan, Ann Arbor, MI.

Schink, D., J. McCarthy, T. Joyce, G. Flierl, P. Wiebe, and D. Kester, 1982: Multidisciplinary program to study warm core rings. *EOS, 63*, 834-835.

Schneider, S. R., D. F. McGinnis, Jr., and G. Stephens, 1985: Monitoring Africa's Lake Chad basin with Landsat and NOAA satellite data. *Intl. J. of Remote Sensing, 6*, 59-73.

———, 1980: The NOAA/NESS program for operational snowcover mapping: Preparing for the 1980's. *NASA Conf. Publ. 2116*, Natl. Aero. and Space Adm., Washington, DC, 21-39.

——— and D. F. McGinnis, Jr., 1977: Spectral differences between VHRR and VISSR data and their impact on environmental studies. *Proc. of the 43rd Annual Meeting of the Amer. Soc. of Photogram., March 1, Washington, DC*, 470-480.

———, D. R. Wiesnet, and L. M. McMillin, 1976: River basin snow mapping at the National Environmental Satellite Service. *NOAA Technical Memorandum NESS 83*, Dept. of Commerce, Washington, DC, 19 pp.

Schwalb, A., 1978: The TIROS N/NOAA A-G satellite series. *NOAA Technical Memorandum NESS 95*, Dept. of Commerce, Washington, DC, 75 pp.

———, 1972: Modified version of the improved TIROS operational satellite (ITOS D-G). *NOAA Technical Memorandum NESS 35*, Dept. of Commerce, Washington, DC, 48 pp.

Scofield, R. A. and C. E. Weiss, 1977: A report on the Chesapeake Bay region nowcasting experiment. *NOAA Technical Memorandum NESS 94*, Dept. of Commerce, Washington, DC, 55 pp.

Shafer, B. A. and C. F. Leaf, 1980: Landsat derived snowcover as an input variable for snowmelt runoff forecasting in south central Colorado. *NASA Conf. Publ. 2116*, Natl. Aero. and Space Admin., Washington, DC, 151-170.

Simpson, J. and V. Wiggert, 1969: Models of precipitating cumulus towers. *Mon. Wea. Rev., 97*, 471-489.

Smith, W. L., P. K. Rao, R. Koffler, and W. K. Curtis, 1970: The determination of sea surface temperature from satellite high resolution infra-red radiation measurements. *Mon. Wea. Rev., 98*, 604-611.

Stephens, D. W. and D. M. Gillespie, 1976: Phytoplankton production in the great Salt Lake, Utah, and a lab study of algal response to enrichment. *Limnol. and Oceanogr., 21*, 74-87.

Strong, A. E., 1986: Monitoring El Niño using satellite based sea surface temperatures. *Ocean Air Interactions, 1* 11-28.

———, 1984: Monitoring El Chichon aerosol distribution using NOAA 7 satellite AVHRR sea surface temperature observations. *Geofis. Int., 23*, 129-141.

——— and E. P. McClain, 1984: Improved ocean surface temperatures from space-comparisons with drifting buoys. *Bull. Am. Meteor. Soc., 65*, 138-142.

——— and B. J. Eadie, 1978: Satellite observations of calcium carbonate precipitations in the Great Lakes. *Limnol. Oceanogr., 23*, 877-887.

——— and H. G. Stumpf, 1975: An evaluation of ERTS data for oceanographic uses through Great Lakes studies. *Final Rept. NASA Contract S-70246AG*, Natl. Aero. and Space Adm., GSFC, Greenbelt, MD, 20 pp.

———, 1974: Remote sensing of algal blooms by aircraft and satellite in Lake Erie and Utah Lake. *Remote Sens. of Envron., 3*, 99-107.

———, H. G. Stumpf, J. L. Hart, and J. A. Pritchard, 1974: Extensive summer upwelling on Lake Michigan during 1973 observed by NOAA 2 and ERTS 1 satellites. *Proc. of the 9th International Symp. on Remote Sensing of the Environment, Apr 15-19*, Environ. Res. Inst. of MI, Ann Arbor, MI, 923-932.

———, E. P. McClain, and D. F. McGinnis, Jr., 1971: Detection of thawing snow and ice packs through the combined use of visible and near- infrared measurements from Earth satellites. *Mon. Wea. Rev., 99*, 828-830.

Susskind, J. J. and D. Reuter, 1985: Retrieval of sea surface temperatures from HIRS/MSU. *J. Geophys. Res., 90*, 11602-11608.

———, J. Rosenfield, D. Reuter, and M. T. Chahine, 1984: Remote sensing of weather and climate parameters from HIRS2/MSU on TIROS N. *J. Geophy. Res., 89*, 4677-4697.

Swift, C. T., 1980: Passive microwave remote sensing of the ocean—A review. *Bound Layer Meteor., 18*, 24-54.

Tarpley, J. D., S. R. Schneider, E. J. Danaher, and G. I. Myers, 1980: An all digital approach to snowmapping using geostationary satellite data. *NASA Conf. Publ. 2116*, Natl. Aero. and Space Admin., Washington, DC, 267-279.

Thomas, G. E., 1983: Satellite measurements of the El Chichon stratospheric cloud. *Proc. of the 7th Annual Climate Diagnostics Workshop, October 17-22, 1982,* Natl. Cntr. for Atmos. Res., Boulder, CO, 390-393.

Thomas, R. L., A. L. Kemp, and C. F. Lewis, 1972: Report on the surficial sediment distribution of the Great Lakes Part 1, Lake Ontario. *Geol. Surv. Cen. Sci., Ser. 10*, 72-77.

Tiuri, M. and M. Hallikainen, 198l: Microwave emission characteristics of snow covered Earth surfaces by the Nimbus-7 satellite. *Eur. Microwave Conf. llth*, 233-238.

Todd, D. K. (ed.), 1970: Surface water. *The Water Encyclopedia*. Water Information Center, Port Washington, NY, 169-171.

Torrey, M. S., 1976: *Environmental Status of the Lake Michigan Region*. Argonne Natl. Lab, ANL/ES-40, V. 3. available from NTIS, Springfield, VA.

Tsuchiya, K., S. Yammamoto, C. Ishida, K. Takeda, Y. Yajima, and S. Morizono, 1982: The two frequency microwave radiometers over snow fields. *Proc. of the 1st Japan/U. S. Snow and Evapotranspiration Workshop, March 1982, Tokyo, Japan*.

Ulaby, F. T., and W. H. Stiles, 1980: The active and passive microwave response to snow parameters, 2, Water equivalent of dry snow. *J. Geophys. Res., 85*, 1045-l049.

Vastano, A. C. and R. L. Bernstein, 1984: Mesoscale features along the first Oyashio intrusion. *J. Geophy. Res., 89*, 587-596.

Walsh, J. E., 1984: Snow cover and atmospheric variability. *Amer. Sci., 72*, 50-57.

Walton, C. C., 1985: Satellite measurement of sea surface temperature in the presence of volcanic aerosol. *J. Climat. Appl. Meteor., 24*, 501-507.

———, 1980: Deriving sea surface temperature from TIROS-N data. *Remote Sensing of Atmospheres and Oceans*, A. Deepak, (Ed.), Academic Press, New York, 547-559.

———, R. L. Brower, and T. L. Signore, 1976: Satellite derived sea surface temperature from NOAA spacecraft. *NOAA Technical Memorandum NESS 78*, Dept. of Commerce, Washington, DC, 84 pp.

Weiler, R. R. and V. K. Chawla, 1968: The chemical composition of Lake Erie. *Proc. of the 11th Conf. on Great Lakes Res.*, Great Lakes Res. Div., Univ. of Michigan, Ann Arbor, MI, 593-608.

Weinreb, M. P. and M. L. Hill, 1980: Calculations of atmospheric radiances and brightness temperatures in infrared window channels of satellite radiometers. *NOAA Technical Report NESS 80*, Dept. of Commerce, Washington, DC, 43 pp.

Wentz, F. J., 1983: A model function for ocean microwave brightness temperatures. *J. Geophys. Res, 88*, 1892-1908.

White, D. and S. Rushforth, (Brigham Young Univ., Provo), 1973: (Personal communication).

White, W. S., (Mich. State Univ., East Lansing), 1974: Role of Calcium Carbonate in Lake Metabolism. Ph. D. Dissertation, 141 pp.

Wiesnet D. R., D. F. McGinnis, Jr. and J. A. Pritchard, 1974: Mapping of the 1973 Mississippi River floods by the NOAA-2 satellite. *Water Resources Bulletin, 10*, 1040-1049.

——— and D. F. McGinnis, Jr., 1973: Hydrologic applications of NOAA-2's Very High Resolution Radiometer. *Remote Sensing and Water Resources Management*, Amer. Water Resources Assoc., Urbana, IL, 179-190.

Wilheit, T. T., J. Greaves, J. Gatlin, D. Han, B. M. Krupp, A. S. Milman, and E. Chang, 1984: Retrieval of ocean surface parameters from the Scanning Multifrequency Microwave Radiometer (SMMR) on the Nimbus-7 satellite. *IEEE Trans. Geosci. Remote Sens., GE-22*, 133-143.

——— and A. T. C. Chang, 1980: An algorithm for retrieval of ocean surface and atmospheric parameters from the observations of the Scanning Multi-Channel Microwave Radiometer. *Radio Sci., 15*, 525-544.

Wilkerson, J. C., G. Born, and D. Lame, 1970: Seasat Gulf of Alaska 622-101, 208 pp.

Wilson, B. W., 1972: Seiches. *Advances in Hydroscience*, V. T. Chow (ed.), Academic Press, New York, NY, 1-94.

Zahl, P. A., 1967: Life in a dead sea—Great Salt Lake. *National Geographic, 32*, 252-263.

Contributors to Section VIII

Chang, Alfred T.	NASA Goddard Space Flight Center Greenbelt, MD 20771
Kniskern, Franklin E.	NOS/Joint Ice Center Washington, DC 20233
Grody, Norman C. Matson, Michael McClain, E. Paul McGinnis, David F. Jr. *Reinhardt, George Schneider, Stanley R. Strong, Alan E.	NOAA/NESDIS Washington, DC 20233

* (Moved to a new position)

Section IX

SATELLITES AND CLIMATE APPLICATIONS

National climate programs in many countries and an active World Climate Program led by the World Meteorological Organization testify to the current interest in climate from all aspects. Climate involves time scales long enough to aggregate the details of the weather and must deal with relatively large spatial scales as well. Satellites are ideal for providing the global perspective of the Earth that climate monitoring, analysis, and research require. Yet, until 1985, none of the instruments on the operational meteorological satellites were designed specifically for climate applications. Only a few experimental instruments carried on research satellites (especially Nimbus) could be directly linked to climate investigations. Nevertheless, great progress has been made in using the primarily meteorological satellite data for identifying climatic variations or solving some of the enigmas of climate.

These chapters deal with various aspects of climate applications. Chapter IX-1 details the progress made in estimating the radiation budget of the Earth. Chapters IX-2 and IX-3 examine the uses of satellite data for making inferences about clouds, aerosols, and ozone, the atmospheric constituents that influence the radiation budget of the atmosphere. Chapter IX-4 deals with real-time monitoring of the Earth climate.

Chapter IX-1

THE EARTH RADIATION BUDGET

IX-1.1 INTRODUCTION

Satellite observations of the radiation entering and leaving the Earth-atmosphere system and their use for studies of the Earth's climate are discussed in this chapter. Prior to the satellite era, the Earth Radiation Budget (ERB) was estimated by using radiative transfer calculations. Because of their importance to studies of climate, satellite measurements of the radiation budget were initiated early in the space observation program. The first successful measurements from space of the planetary radiation budget were made in 1959 from Explorer-7 (House, 1984). These measurements were limited in duration and were analyzed primarily with regard to synoptic weather patterns (Weinstein and Suomi, 1961). Radiation budget measurements (albedo and Outgoing Longwave Radiation [OLR]) continued with TIROS 2, 4, and 7. (Bandeen et al., 1961; Winston and Rao, 1962, 1963; Winston, 1967). These observations extended over longer periods, allowing for preliminary studies of seasonal and year to year variations.

Several Nimbus satellites carried Earth radiation budget instruments: the Medium Resolution Infrared Radiometer (MRIR) on Nimbus-3 (Raschke et al., 1973) and the Earth Radiation Budget instrument (ERB) on Nimbus-6 and 7 (Jacobowitz et al., 1979; 1984a; 1986). ERB was specifically designed to measure the planetary albedo and emitted flux over areas ranging from about 5° lat. × 5° long. to the full Earth disk, and to monitor the solar constant. The more advanced Earth Radiation Budget Experiment (ERBE) was developed by NASA in the late 1970s and early 1980s. ERBE instruments were flown onboard NOAA-9 and NOAA-10 and ERBS (Chapters I-2 and III-6) beginning in 1984. Data from these experiments are being analyzed.

Beginning in June 1974, estimates of the planetary albedo and OLR were obtained from scanning radiometers on NOAA operational spacecraft (Gruber, 1977; Gruber and Winston, 1978; Ohring et. al., 1984). The primary function of these radiometers was to obtain images of weather systems. Their most unique feature, however, was that they were made routinely, thus permitting the development of a truly climatic time series of nearly continuous observations. While they suffer deficiencies compared to the recent ERB and ERBE results, primarily because of the narrow spectral intervals used in the radiometers, recent research (Ohring and Gruber, 1983; Gruber and Jacobowitz, 1985) suggests that these data provide a reasonable view of the components of the radiation budget on large space and time scales. The discussions of the temporal and spatial variation of the planetary radiation budget in IX-1.3 are based on estimates derived from NOAA data acquired between June 1974 and February 1978 (with the OLR values revised by Gruber and Krueger, 1984), and albedo based on a solar constant of 1370.6 Wm^{-2}.

IX-1.2 SOLAR CONSTANT DETERMINATION FROM SATELLITE OBSERVATIONS

Earth orbiting satellites offer the means for obtaining quantitative measurements of solar irradiance impinging upon the Earth. Observations from the surface are difficult to use because they must be corrected for the effects of the intervening atmosphere. Solar irradiance observations have been made from satellites since the early 1960s, but the Nimbus-6 ERB experiment that started in July 1975 carried the first instrument capable of the accuracy and precision required for monitoring solar output. In November 1978, a remarkably significant improvement was made with the use of the self-calibrating cavity radiometer of the Nimbus-7 ERB experiment. Results of these observations are shown in Figure IX-1-1; the solar irradiance is adjusted to that for a mean Earth-Sun distance (solar constant) which takes out the variation introduced by the Earth's elliptical orbit.

The Nimbus-7 data for November 16, 1978 to March 31, 1986 (Figure IX-1-1) yield a solar constant of 1370.6 Wm^{-2} with a standard deviation of 0.73 Wm^{-2} (0.05% of the mean) (Hickey and Alton, 1983). Daily standard deviation values average around 0.29 Wm^{-2} (0.02% of the mean). Estimates for November 1981-July 1983 are preliminary.

In addition to the total solar measurements, solar spectral measurements have been made from satellites to study the various aspects of the physics of the Sun and its effects on the weather and climate of Earth.

IX-1.3 AVERAGE RADIATION BUDGET COMPONENTS

IX-1.3.1 Annual Cycle of Global Averages

Figure IX-1-2 shows the annual cycle of global means of albedo, OLR, absorbed solar radiation, and net radiation, based on the period June 1974-February 1978. The albedo shows a pronounced annual cycle with a maximum during Northern Hemisphere winter and a minimum during summer. The OLR also shows a pronounced annual variation with its maximum during Northern Hemisphere summer. This maximum is the result of a higher global mean temperature in July than in January because of the more extensive land areas of the Northern Hemisphere.

The amplitude of the annual cycle of absorbed solar radiation is about 5 Wm^{-2}, smaller than that of the

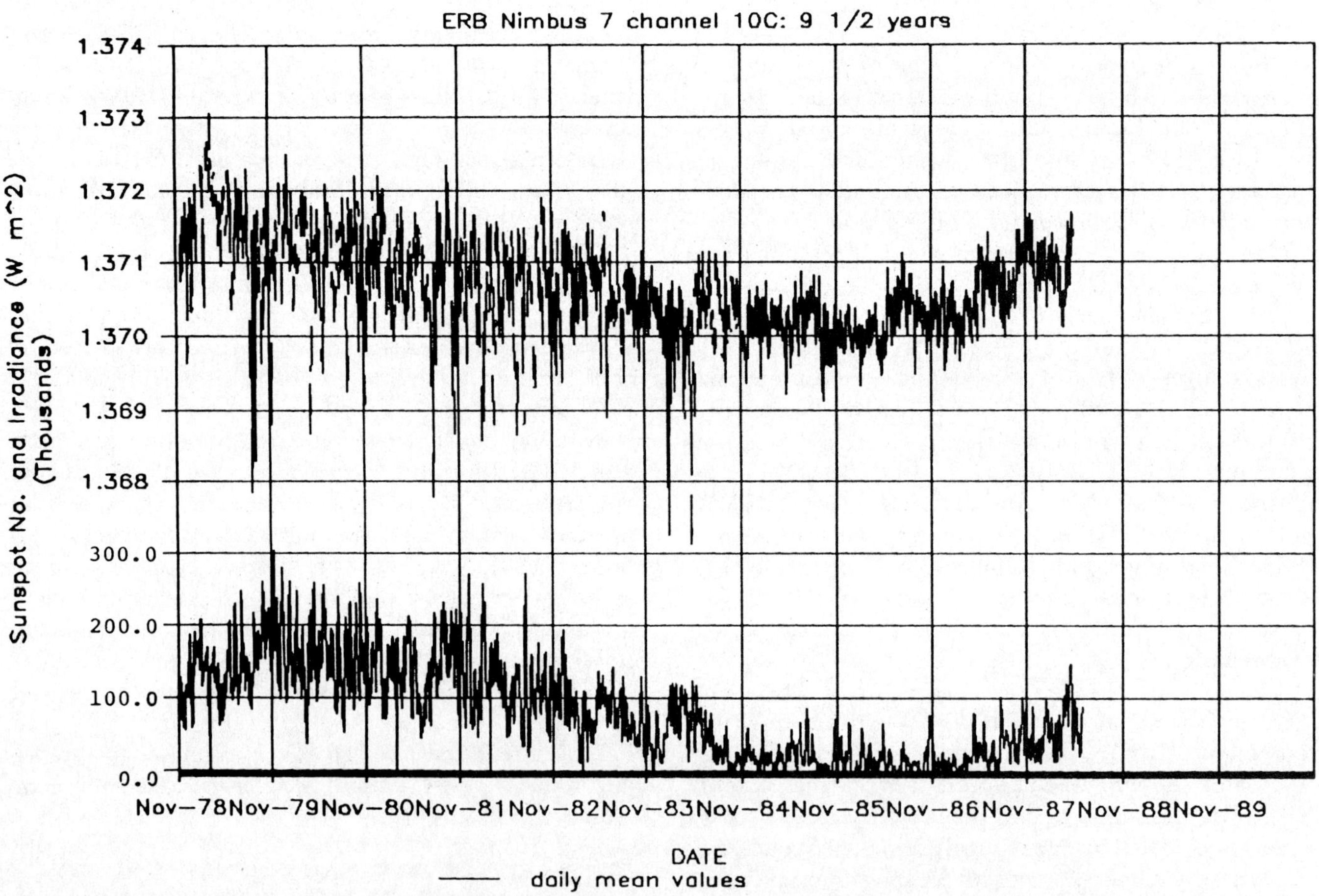

Figure IX-1-1. Daily mean solar irradiance from November 16, 1978 to April 30, 1988. Lower plot is sunspot number for the same period (Hickey et al., 1988).

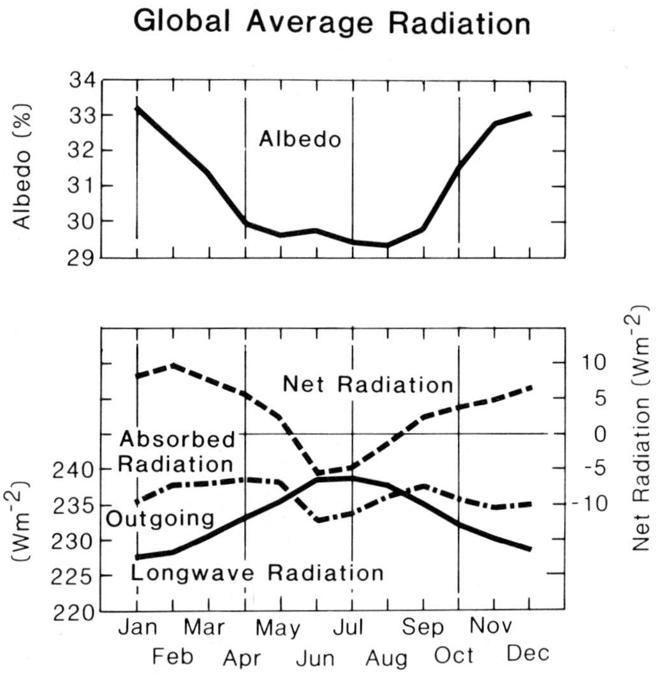

Figure IX-1-2. Annual cycle of albedo, outgoing longwave radiation, absorbed solar radiation, and net radiation based on 45-month SR data set (June 1974-February 1978).

OLR and much smaller than that implied by the variation in albedo. The annual variation of absorbed solar radiation has two minima; the one during Northern summer appears to be inconsistent with the minimum values of albedo at that time. This minimum albedo occurs when the Sun is farther from the Earth and the overall incoming solar radiation is less than during Northern Hemisphere winter. This difference in incoming solar energy ($\pm3.4\%$) is enough to dampen the amplitude of the absorbed solar radiation and to result in a phase variation in absorbed solar radiation that is more complex than implied by the relatively simple annual cycle of the albedo. As a result the amplitude and phase of the net radiation is essentially determined by the variations in OLR. The global annual average albedo is 31%, the absorbed solar energy 236 Wm^{-2}, and the OLR 233 Wm^{-2}. Thus, the net radiation exhibits a positive imbalance of 3 Wm^{-2}. Normally it is expected that the long-term average of net radiation should be near zero. Too much significance should not be attached to this surplus. It represents the small difference between two large numbers; a small error in either estimate of absorbed and/or the emitted radiation could account for this slight surplus. For example, a preliminary comparison with OLR data obtained for the month of April 1985 from the ERBE shows that the global flux obtained from the NOAA operational data is 3 Wm^{-2} less than the ERBE flux (Yang et al., 1988). The annual average comparison is not yet known.

IX-1.3.2 Average Seasonal Fields

The global distributions of albedo, OLR, and net radiation for the Northern Hemisphere winter and summer seasons[1] are shown in Figures IX-1-3 and IX-1-4. Each of these seasonal averages is based on scanning radiometer data obtained between June 1974 and February 1978 (Gruber and Winston, 1978). Each is shown on a Mercator (60°N to 60°S) and two polar stereographic projections (poleward of 50°N on the left and 50°S on the right).

Over the tropics and subtropics (30°N to 30°S), the broad areas of low OLR and high albedo over Indonesia, South America, and Africa are prominent, with a pronounced shift northward from winter to summer. These represent the active convective cloud regions over the three tropical continents. Zonally oriented minima of OLR and maxima of albedo, located mainly north of the Equator over the Atlantic, central and eastern Pacific, and the Indian Ocean, represent the convective cloudiness of the InterTropical Convergence Zone (ITCZ). The ITCZ moves only slightly northward from its winter to its summer positions, except over the Indian Ocean where in summer it merges with the extensive regions of low OLR and high albedo of the Asian summer monsoon.

The extensive regions of high OLR and low albedo north and south of the ITCZ represent the regions of little cloudiness associated with areas of low-level divergence and high pressure. These areas expand markedly northward over the subtropical North Pacific and North Atlantic in summer.

There are two distinct regions of the subtropics where high albedo and low OLR do not occur together. Over the desert regions (note especially the Sahara and Saudi Arabia) albedo values are as high as over cloudy regions, while the OLR is high because of the warm surface (particularly in summer). Over the eastern subtropical Pacific and Atlantic Oceans there are regions of high albedo and relatively high OLR associated with the frequent presence of low stratocumulus clouds, and relatively high temperatures.

The higher values of net radiation in the tropical and subtropical regions are skewed toward the regions of little cloudiness north of the Equator in summer and south of the Equator in winter. Aside from the increasing negative values that occur toward the winter (cold) hemisphere, the lowest values of net radiation

[1] All seasonal references henceforth are with respect to the Northern Hemisphere.

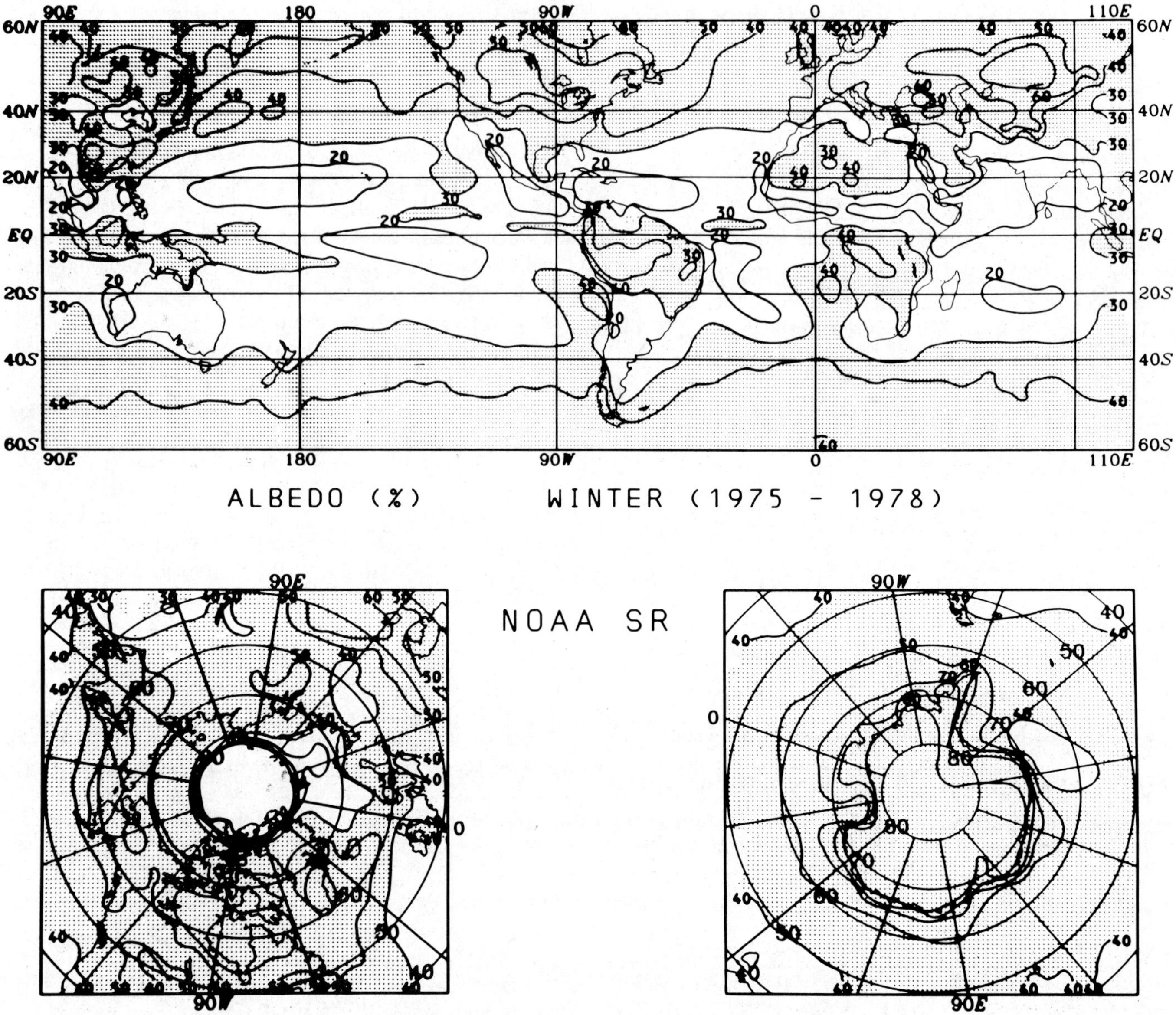

Figure IX-1-3a. Winter (December-February) albedo.

in the tropics and subtropics occur over deserts where the high values of both albedo and OLR result in values near zero. Other minima (e.g., less than 50 Wm^{-2}) over the tropics and subtropics occur in regions of cloudiness. The larger minima are found in regions of low level stratocumulus where both albedo and OLR are high. Minimal values also occur in some regions of major cloudiness such as South America, Africa, and Southeast Asia (in summer) where the effects of high albedo are dominant even though OLR values are relatively low.

In the zones 30° to 60°N and S, there are distinct differences between the Northern and Southern Hemispheres, particularly with respect to albedo. These are related to the differences in the distributions of continents and oceans. In the Northern Hemisphere, albedo values of more than 50% occur in winter over large continental regions dominated by snow cover. Although oceanic regions have somewhat lower values, they are still relatively high because of extensive cloudiness, particularly in the western and northern oceans. In summer most of the Northern Hemisphere continents have relatively low albedos while ocean albedos are generally higher. In the Southern Hemisphere, albedos are generally lower over continents than over oceans during both winter and summer, al-

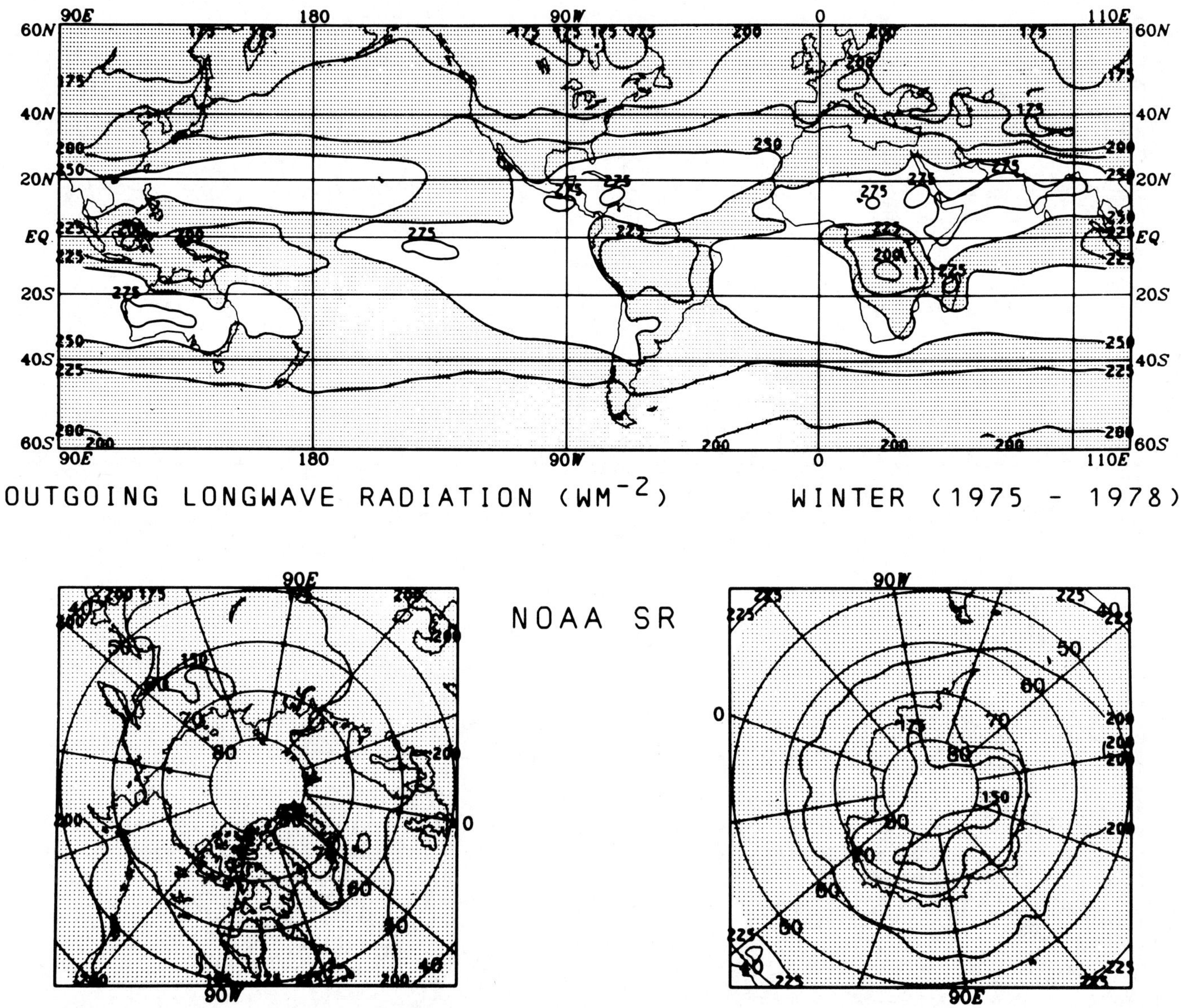

Figure IX-1-3b. Winter (December-February) outgoing longwave radiation.

though the principal continental perturbation occurs during Southern summer. The OLR pattern generally is more zonal in character than albedo in the middle latitudes during the winter of each hemisphere. The distinctly lower winter values of OLR over the continents, particularly in the Northern Hemisphere, are undoubtedly associated with low surface and air temperatures. In warm seasons, higher values of OLR generally predominate over the continents. The net radiation over mid-latitudes during the warm season is strongly influenced by the albedo pattern. In the cold season, the strong poleward gradient of net radiation is dominated by the gradient in available solar energy, although influences of the OLR distribution are also evident.

The polar regions are dominated in their summer seasons by the high albedos of Antarctica, and Greenland and the Arctic Ocean. Wintertime observations of albedo are possible closer to the North Pole than to the South Pole because of solar and satellite viewing geometry (Gruber, 1977). The OLR patterns over the poles are also strongly influenced by features of Antarctica and Greenland during both winter and summer. The Antarctic OLR values are quite low even during Southern summer and are similar to those over Greenland in winter. The lowest monthly mean OLR values observed anywhere in the world occur over Antarctica in Southern winter (100 to 125 Wm^{-2}).

The net radiation patterns of the polar caps are dominated by albedo during their summers and, be-

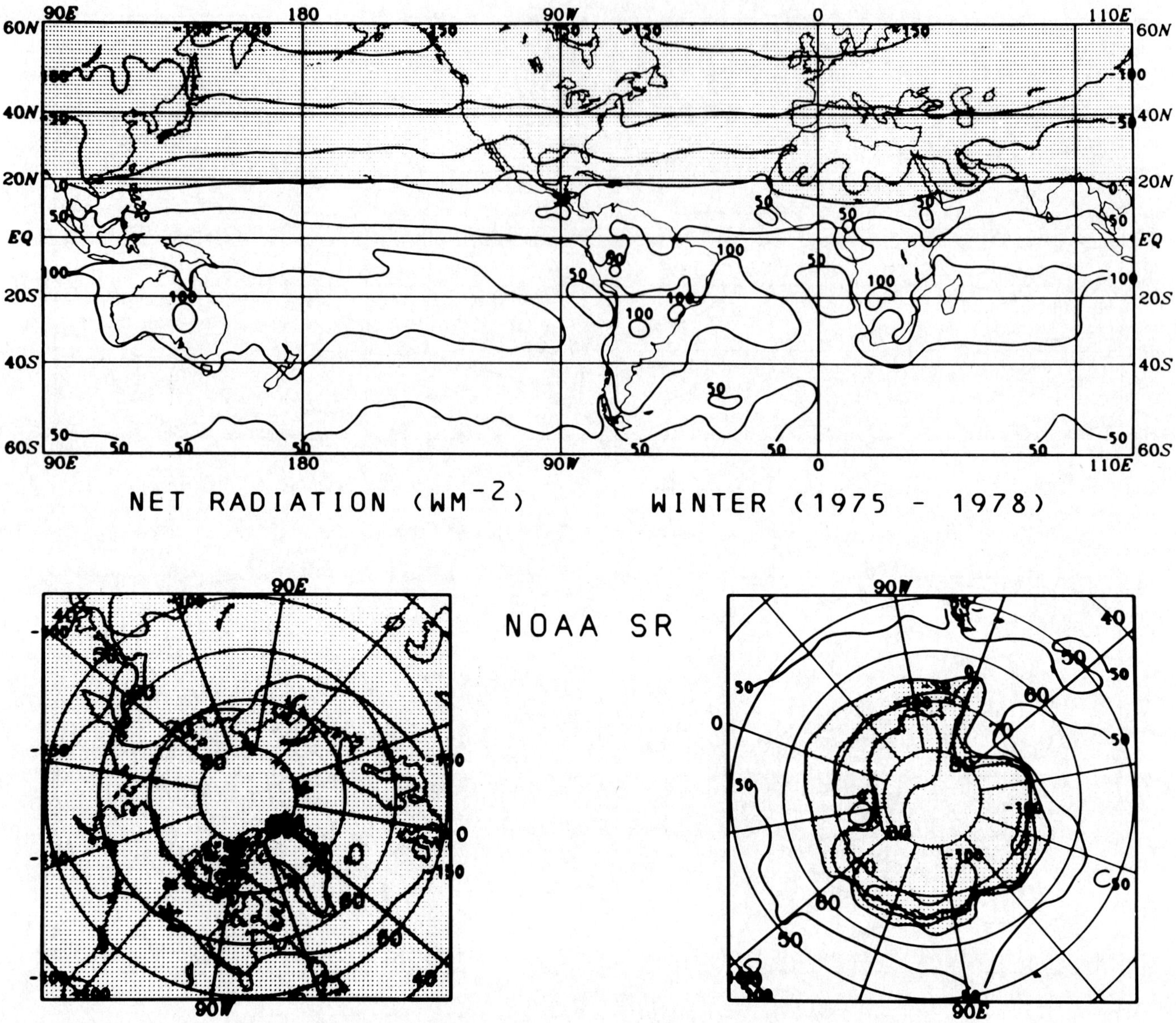

Figure IX-1-3c. Winter (December-February) net radiation.

cause there is so little insolation, by OLR during their winters.

IX-1.3.3 Annual Cycle of Components for Selected Climatic Regions

Monthly averages of NOAA scanning radiometer data can also be used to show the annual march of the Earth radiation budget for selected climatic regimes. These selections (Figures IX-1-5 through IX-1-9) are for some of the same regions chosen by Budyko (1974), who developed a climatology of the energy budget at the surface of the Earth. The monthly averages are for 3 or 4 years of data, based again on June 1974-February 1978 data (Gruber and Winston, 1978). The data are from the 2½° × 2½° grid points located closest to the climatic locations identified by Budyko. To assist in the interpretation of the observations, information on the annual cycle of cloudiness at these locations was obtained from Volumes 7, 10, and 12 of the "World Survey of Climatology" (Landsberg, 1972-77). In cases where no cloudiness data were available for a particular station, an attempt was made to find a nearby station (within 100 to 200 km) with cloud data, or to obtain the annual cycle of precipitation (as a surrogate measure of the annual variation of cloudiness) from "Cli-

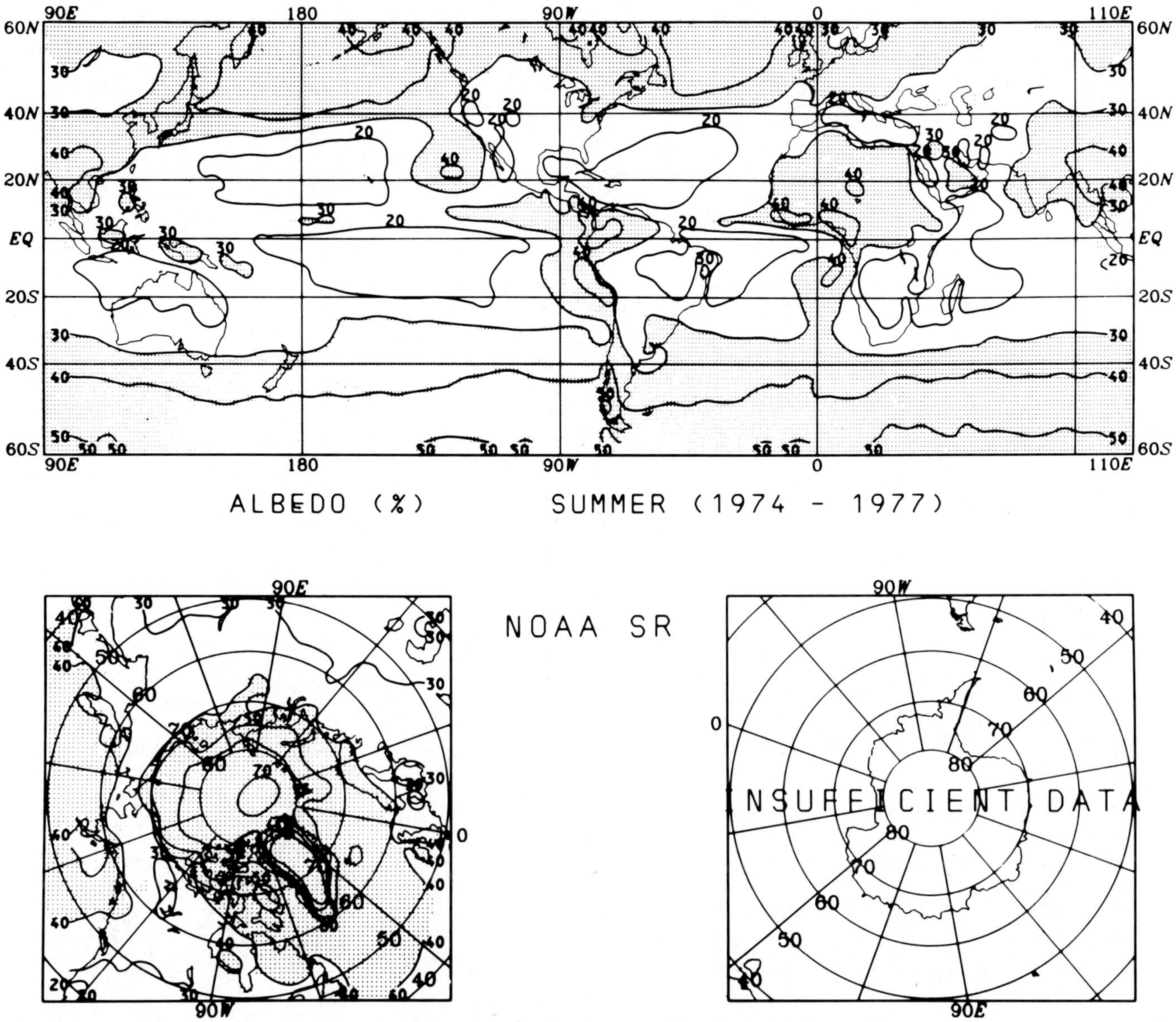

Figure IX-1-4a. Summer (June-August) albedo.

mates of the World" (U.S. Department of Commerce, 1969).

A. Equatorial Continental Climate—São Gabriel, Brazil (Figure IX-1-5). The net radiation curve shows two maxima (September and February) that are in qualitative agreement with the solar insolation cycle for an equatorial location. The second maximum occurs in February rather than in March, when the Sun is over the Equator, because the albedo curve shows a maximum in March. The longwave radiation curve shows a minimum in March, indicating a maximum of major cloudiness extending to high altitudes. This dominance of reflected solar radiation over emitted longwave radiation as it affects net radiation indicates that the albedo effect of clouds overwhelms their greenhouse effect, at least for a region where insolation is large.

The annual variation of net radiation is small, as expected for an Equatorial location. Of interest is the difference of 29 Wm^{-2} in net radiation between June and December—months with the same solar declination angle. The albedos are similar for the 2 months. A small part ($\sim$8 Wm^{-2}) of the net radiation difference can be explained by the longwave radiation difference between the 2 months; the largest part, however, is due to the difference in Earth-Sun distance between June and December.

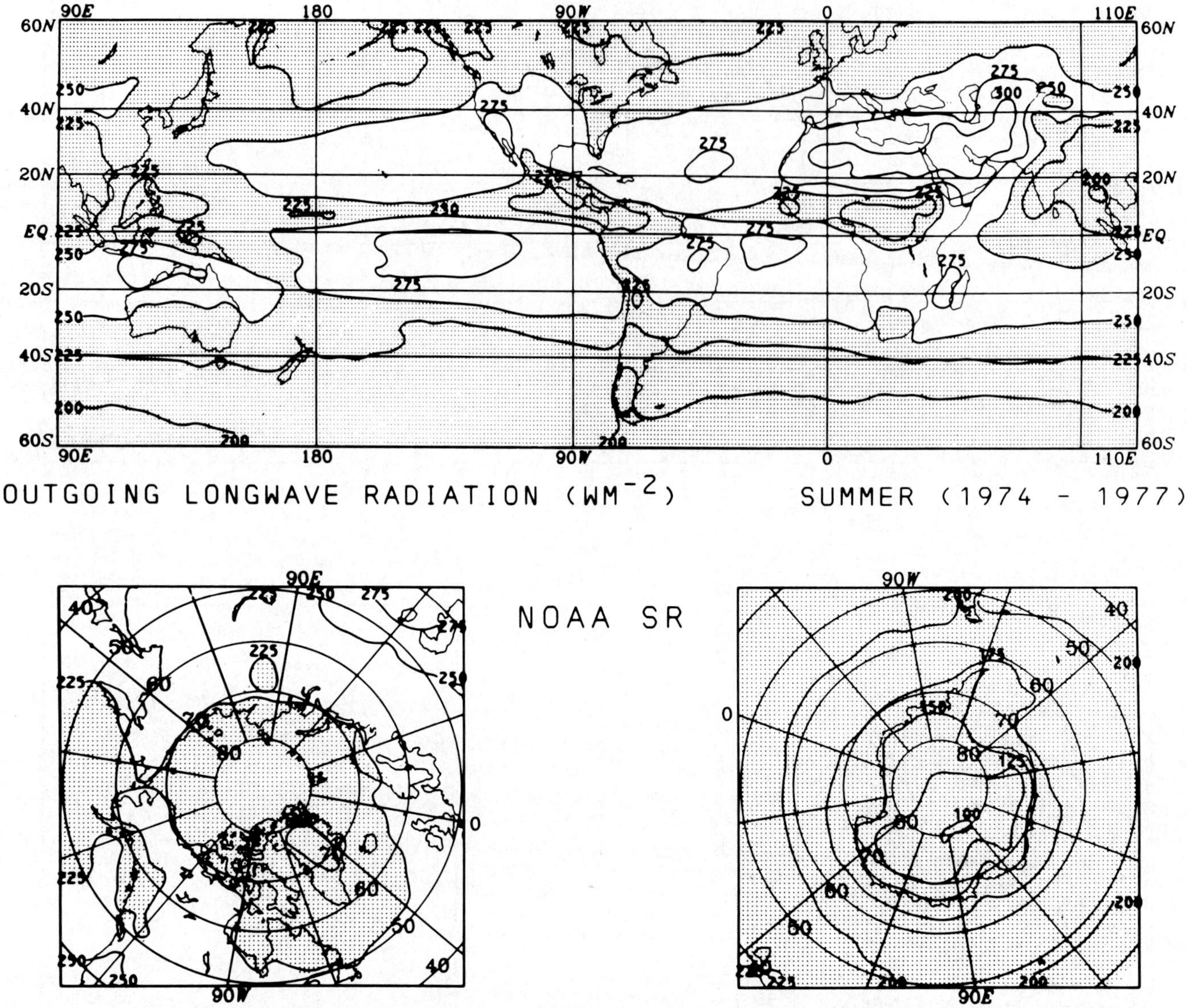

Figure IX-1-4b. Summer (June-August) outgoing longwave radiation.

B. Equatorial Monsoon Climate—Saigon, Vietnam (Figure IX-1-6). The effects of the monsoon circulation are most evident in the annual marches of albedo and longwave radiation. The late winter/early spring minimum in albedo and maximum in longwave radiation are associated with dry tropical air, with little cloudiness, that covers this area at this time. The radiation data suggest that the monsoon reaches its climax in September (maximum albedo, minimum longwave radiation); precipitation data confirm this observation.

C. Tropical Continental Climate—Aswan, Egypt (Figure IX-1-7). With little cloudiness or precipitation, the radiation budget of this desert area is controlled by the annual march of insolation. Both the net radiation and the longwave emission are at a maximum in summer and a minimum in winter. In contrast to the Equatorial regions, just described, the longwave emission curve here reflects the annual variation of temperature rather than that of cloudiness. The net radiation curve shows a deficit for most of the year. This is a result of the combination of clear skies, which permit escape of longwave surface radiation to space, and relatively high albedo (30 to 35%) despite the absence of cloudiness. For some other desert areas (e.g., Libya and Saudi Arabia), the net radiation is negative and less than -10 Wm^{-2} all year-round.

D. Subtropical Continental Climate—Aidin, U.S.S.R (Figure IX-1-8). The large annual cycle in cloudiness

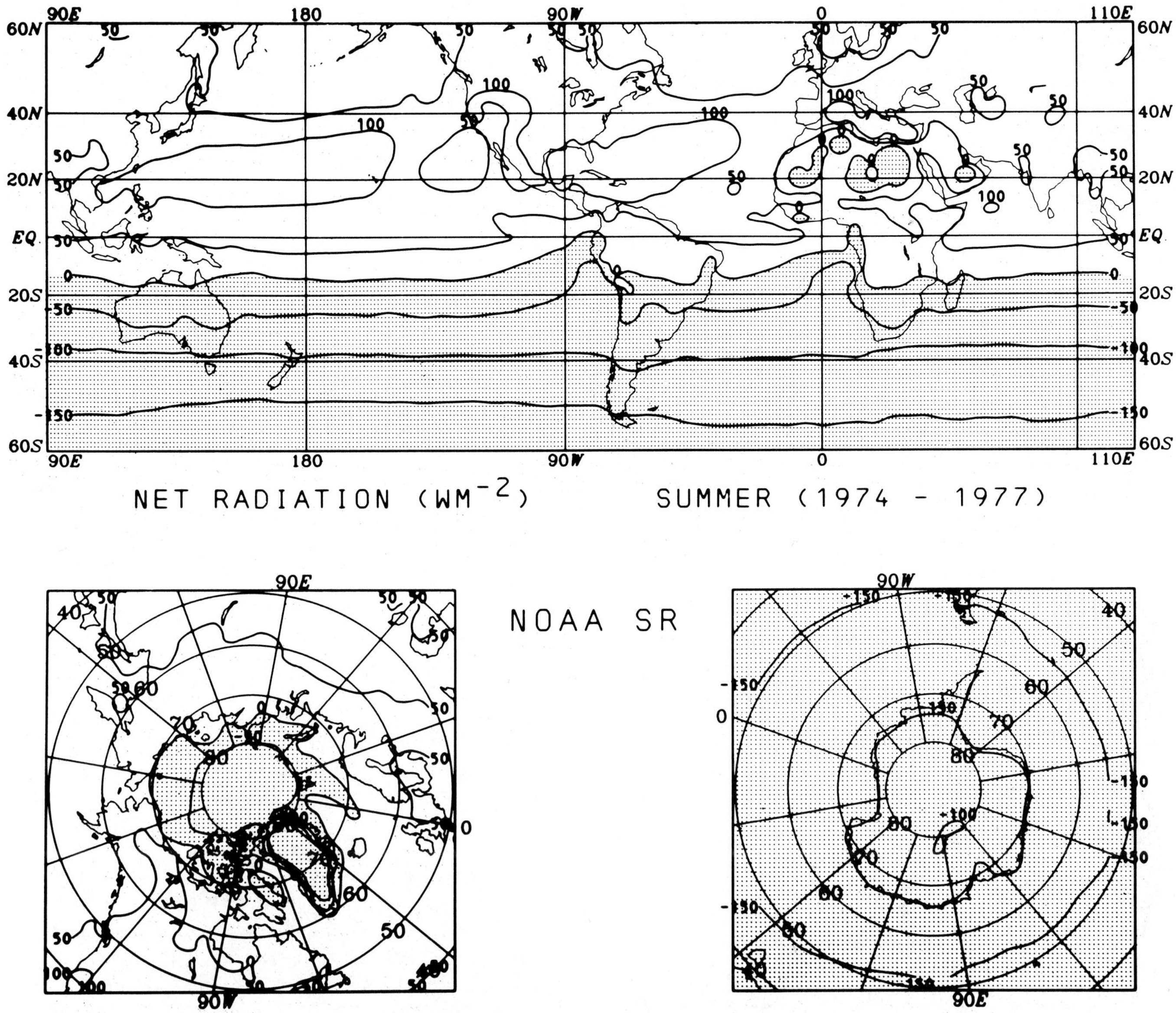

Figure IX-1-4c. Summer (June-August) net radiation.

with a maximum in winter is reflected in the albedo curve and in the longwave radiation curve. The longwave radiation is also influenced by the annual temperature cycle and the albedo may be influenced by winter snow; both are largely in phase with the annual variation of cloudiness.

E. Mid-latitude Continental Climate—Barnaul, U.S.S.R (Figure IX-1-9). The late winter peak in albedo (60%) is probably more influenced by surface snow cover than by cloudiness, which is at a minimum. Typical of mid-latitude continental areas, there is a very large annual range of longwave radiation (here over 70 Wm^{-2}) that reflects the large annual temperature range.

IX-1.4 SURFACE RADIATION BUDGET FROM SATELLITE OBSERVATIONS

Knowledge of the radiative balance at the surface, as well as at the top of the atmosphere, is important for all climate related studies. The surface radiation budget consists of the net (all wave) radiation and its components, the upwelling and downwelling shortwave (0.2-4.0 μm) and longwave (>4.0 μm) radiation. Calculations of these components require either direct observations of radiative fluxes or estimations based on observations of other atmospheric and surface variables. Satellite observations now can be used to monitor

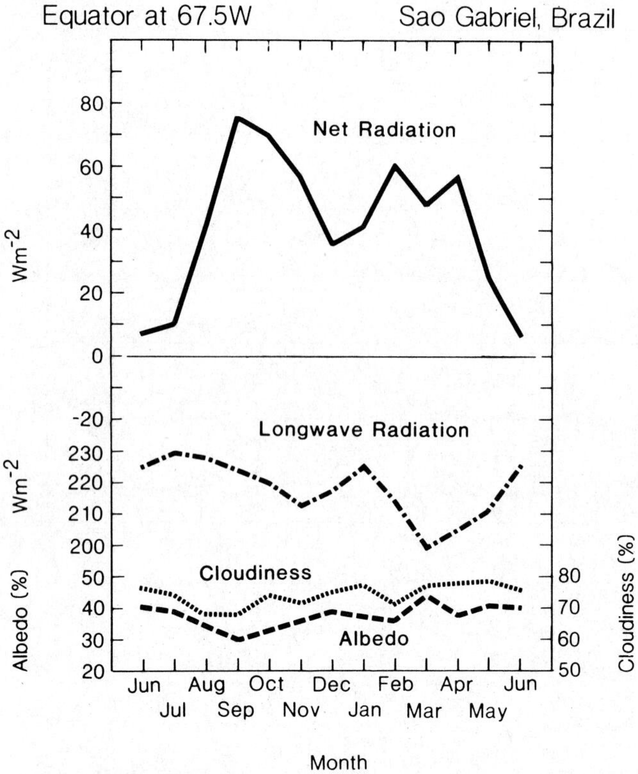

Figure IX-1-5. The annual variation of the Earth-atmosphere system radiation budget and cloudiness for an Equatorial continental location (Equator, 67.5°W—São Gabriel, Brazil).

the surface radiation budget over most of the globe. Since the satellite cannot measure surface energy budget components directly, empirical-theoretical methods are used to calculate these components.

Accuracy requirements for various space and time scales have been considered by the Committee on Space Research (COSPAR, 1978), World Climate Program-40 (1982), and World Climate Program-70 (1984). The accuracy requirements set forth by the World Climate Program-70 report specify ± 10 Wm^{-2} for surface net radiation with a scale of 250 km for monthly means, although accuracies of ± 20 Wm^{-2} would still be of some use.

IX-1.4.1 Downwelling Shortwave Radiation

Satellite radiometer measurements of shortwave radiation reflected by the Earth-atmosphere system contain much information about solar radiation reaching the surface. A measurement of solar energy reflected to space from the Earth-atmosphere immediately specifies the maximum amount of solar energy that can be absorbed at the surface. If the amount absorbed by the atmosphere can be estimated, then the satellite

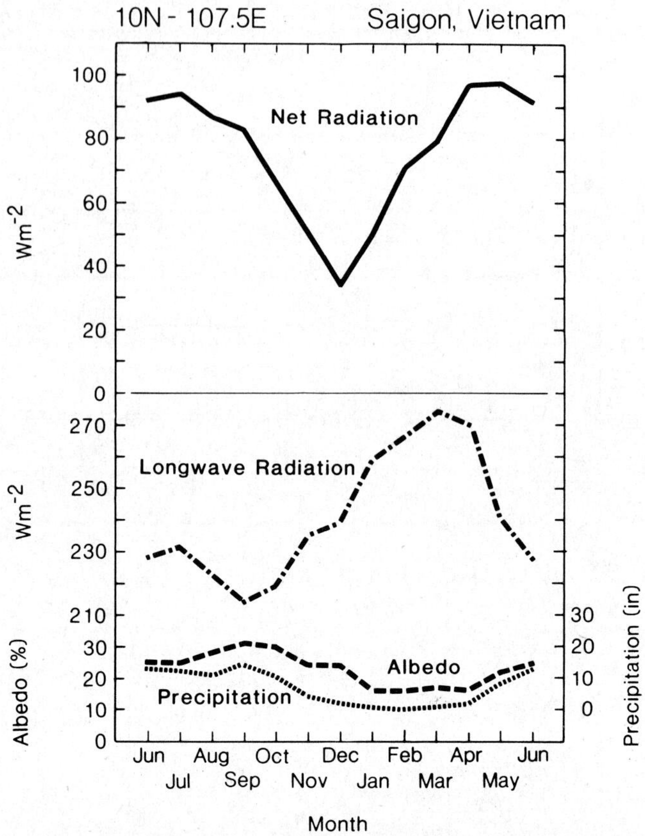

Figure IX-1-6. The annual variation of the Earth-atmosphere system radiation budget and precipitation for an Equatorial monsoon location (10°N, 107.5°E—Saigon, Vietnam).

measurement can be adjusted to yield an estimate of the shortwave radiation reaching the surface. If the satellite measurement is to be used to estimate the downwelling shortwave radiation, the albedo of the surface must also be known. This is especially important over regions of high reflectivity such as deserts and snow where it may be difficult to separate the albedo effects of cloudiness from the surface albedo.

Several models for using satellite shortwave observations and other information to estimate incident solar radiation have been developed (Tarpley, 1979; Gautier et al., 1980; Moser and Raschke, 1984; Pinker and Ewing, 1985). These models vary from the purely statistical to the mainly physical.

Tarpley (1979) used a statistical regression technique to obtain surface fluxes for various locations in the Great Plains from GOES shortwave VISSR data at a resolution of 50 × 50 km. In this model, cloud amount is estimated for a given location from the satellite visible counts. One of three regression equations is then used for each of three categories of cloud amount. For clear (or nearly clear) cases, the standard

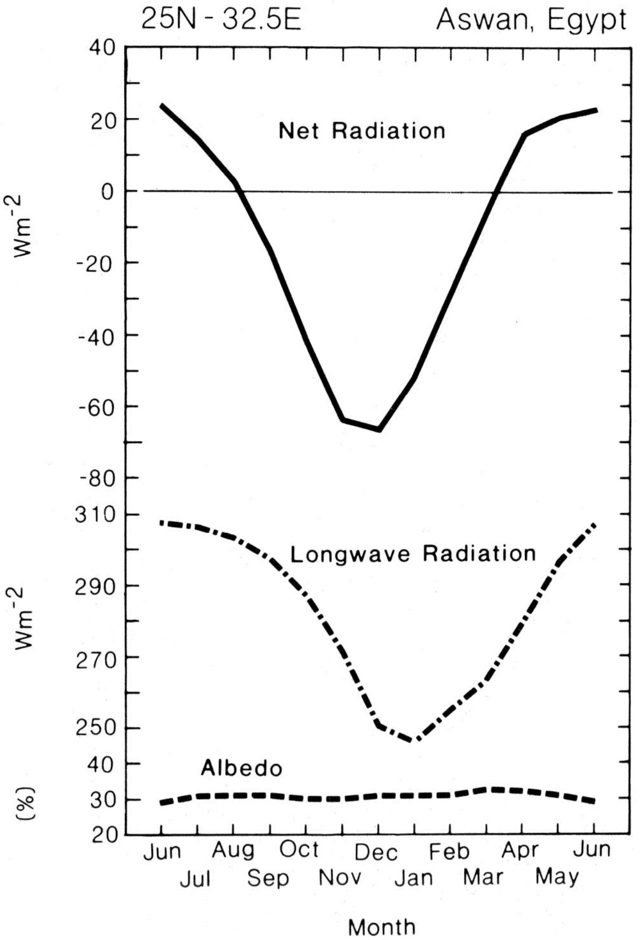

Figure IX-1-7. The annual variation of the Earth-atmosphere system radiation budget for a tropical continental location (25°N, 32.5°E—Aswan, Egypt).

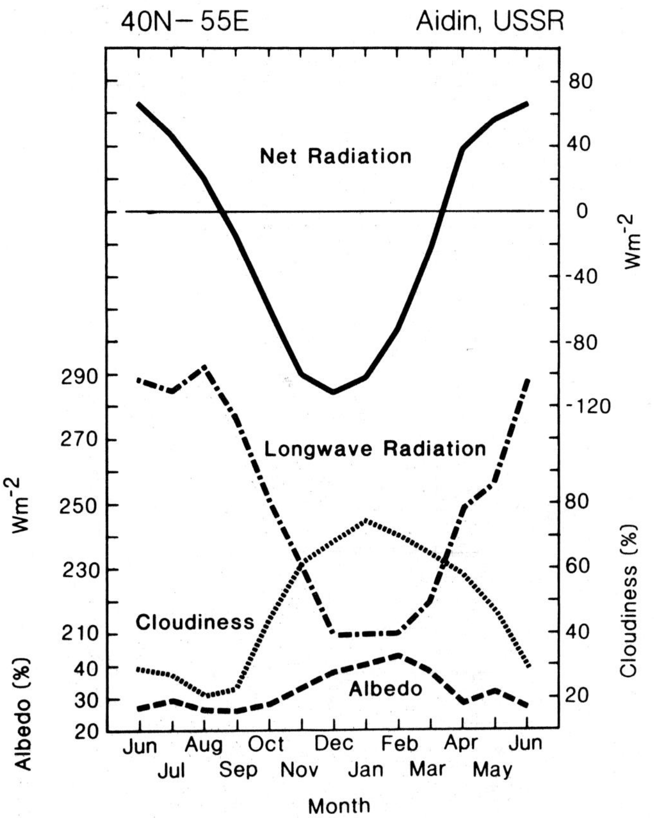

Figure IX-1-8. The annual variation of the Earth-atmosphere system radiation budget and cloudiness for a subtropical continental location (40°N, 55°E—Aidin, U.S.S.R).

error of estimate obtained using this regression technique was less than 10% of the mean hourly insolation. For partly cloudy skies, it increased to 30% and for completely cloudy skies to 50%. The standard error of estimate for total daily insolation was less than 10%.

Gautier et al. (1980) using a physical approach, obtained insolation from calibrated GOES data. The GOES data used were of high space resolution (2 km at 45° lat.) with a time resolution of 1 hour. The model considered most atmospheric effects, except for aerosols, and was independent of wavelength. Clear and cloudy cases were treated differently. The model was tested with coincident satellite and surface measurements for southern Canada. The standard error in estimated daily cumulative insolation amounted to 5% of the measured mean for clear days and 14% for cloudy days. The overall average standard error was less than 9% of the measured mean for all cases.

Moser and Raschke (1984) assumed that clouds perturb the "clear-sky" radiance field and that satellite observations of reflected solar radiation are indicators of clouds. The transmittance by clouds of solar radiation was calculated with the aid of a radiative transfer model. The approach was implemented over Europe with Meteosat observations for two months. A comparison with ground observations showed uncertainties of only 6% for monthly averages of daily sums.

Pinker and Ewing (1985) formulated a detailed and computationally efficient radiative transfer model for estimating surface solar insolation either hourly or daily. This more sophisticated representation of the physical processes in the atmosphere used as input geostationary visible measurements and other routine meteorological observations (e.g., precipitable water). The model was tested with daily cumulative values of insolation for 4 months; the results showed the mean value of the daily estimates to be 19.61 MJm^{-2}, compared to the mean measured value of 19.72 MJm^{-2}, a correlation coefficient between predicted and measured daily values of 0.94, and a standard error of estimate

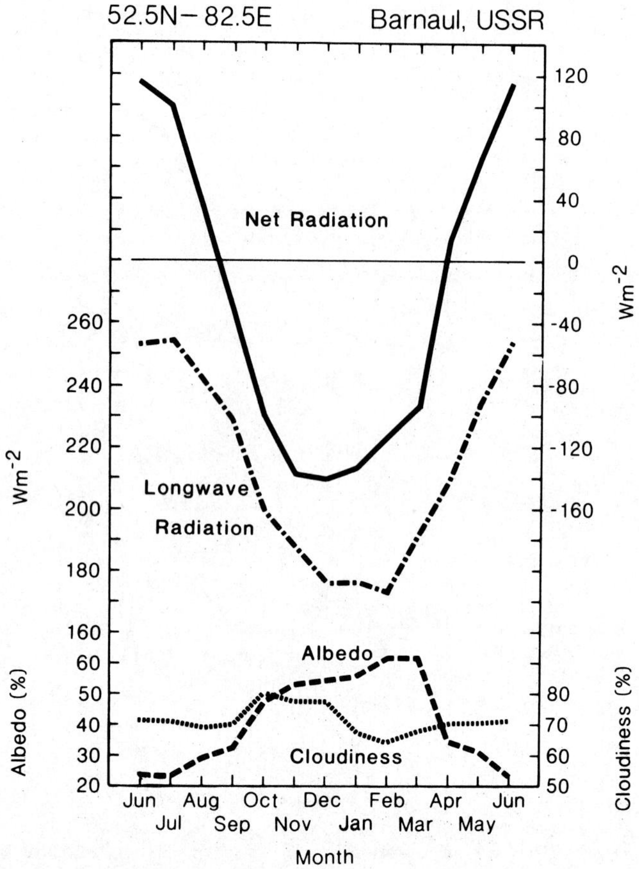

Figure IX-1-9. The annual variation of the Earth-atmosphere system radiation budget and cloudiness for a mid-latitude continental location (52.5°N, 82.5°E—Barnaul, U.S.S.R).

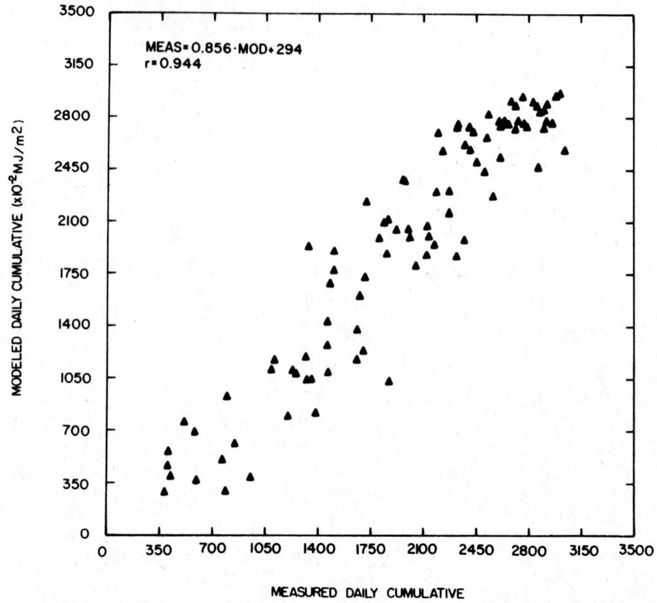

Figure IX-1-10. Scattergram of measured and satellite estimated values of daily cumulative insolation for Toronto, May-August 1982 (Pinker and Ewing, 1985).

of 2.47 MJm^{-2} or 12.5% of the mean observed value. The daily performance of the model for Toronto for the 4-month period is shown in Figure IX-1-10.

Standard errors of estimate were also calculated for averaging periods of from 1 to 10 days (Figure IX-1-11). Note that as the averaging period increased from 1 to 5 days, the standard error of estimate decreased rapidly to near 5%. A slower decrease, to near 3%, occurred as the averaging period approached 10 days.

IX-1.4.2 Upwelling Shortwave Radiation or Surface Albedo

In the absence of clouds and major aerosol concentrations, a shortwave satellite measurement is a useful approximation of the upwelling shortwave radiation, or the surface albedo.

Measurements from several polar orbiting and geostationary satellites are available for developing climatological datasets of surface reflectivity. Relevant sensors include: AVHRR on operational NOAA satellites, TM and MSS on Landsat; VISSR on GOES;

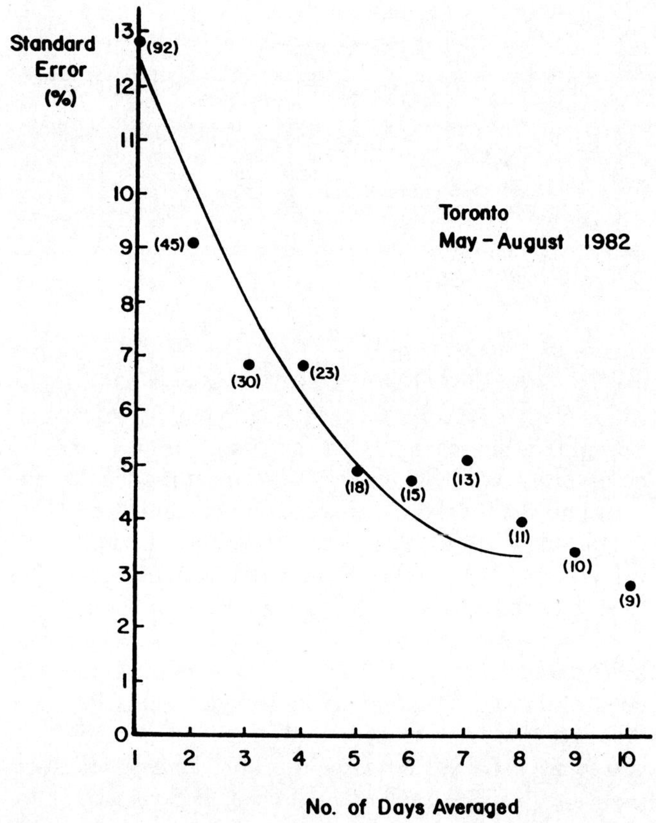

Figure IX-1-11. Standard errors of the satellite estimated global solar radiation for Toronto, as a function of the number of days averaged (Pinker and Ewing, 1985). The solid line represents a second order least square fit. Each point was weighted according to the number of cases it represents (in parentheses).

VIS channel on Meteosat; the broadband measurements from the Nimbus 6 and 7 ERB experiment; and ERBE. Important factors limit the usefulness of all these data sources: calibration deficiencies; differing filter functions and wavelength ranges; and orbit and sampling limitations. The ERBE data may provide the best estimates since they include measurement over the entire solar spectrum with both wide and narrow field of view instruments that are identical on different satellites. Also, observations taken at different local times provide diurnal sampling. Geostationary satellite observations, although limited to the visible part of the solar spectrum, have a special advantage since they provide frequent observations over the whole day.

In determining surface albedo from cloud-free satellite shortwave observations, the effects of the intervening atmosphere must be eliminated. These effects include the absorption and scattering of solar radiation (0.2-4.0 μm) by gases and aerosols. In principle, radiative transfer theory could be used to account for these effects over uniform terrain. A review of techniques available for calculating shortwave radiation can be found in Lenoble (1977). Modeling of the radiative transfer characteristics of most gases in spectral intervals relevant to surface reflectivity is relatively straightforward (McClatchey et al., 1972). Spatial and temporal variability of water vapor is one of the major problems. The treatment of aerosols is more complex because they not only vary in concentration in space and time, but also in shape, size distribution, and index of refraction (Braslau and Dave, 1973a; 1973b).

Numerous simulation studies have been conducted on atmospheric effects under selected atmospheric and surface conditions and for satellite instrument characteristics (Otterman and Fraser, 1976; Preuss and Geleyn, 1980; Otterman and Robinove, 1981). Based on more general simulations with the solar radiation absorption model of Lacis and Hansen (1974), Chen and Ohring (1984) obtained a simple linear relationship between planetary albedo and surface albedo. This relationship, applied to the minimum planetary albedo obtained over monthly periods by the NOAA scanning radiometer in the VIS range, appears to yield reasonable values for the broadband, cloud-free sky albedo for zonally averaged values.

Although, in principle, corrections can be made for atmospheric effects, the task is difficult in practice on a routine, large-scale basis. While precipitable water data could be obtained systematically from satellite soundings and routine meteorological analyses, aerosol information would be difficult to obtain.

It has long been recognized that the amount of energy reflected from the surface depends strongly on Sun elevation and on the direction of reflectance (Kondratyev, 1969; Salomonson and Marlatt, 1971; Kreibel, 1978; Pinker et al., 1980; Kimes, 1983). Satellite observations have confirmed this dependence. Taylor and Stowe (1984) have developed detailed angular reflectance models for eight uniform surface types (land; water; snow; ice; low, middle, and high water droplet clouds; and high ice clouds) using Nimbus-7 ERB data. Some of the angular reflectance properties of land, snow, and ice were sampled sparsely in the ERB data, because the noon Sun synchronous orbit of Nimbus-7 did not permit observations of many clear land areas at low solar elevations. Geostationary observations covering all solar zenith angles can be used to overcome this deficiency. Pinker and Razgaitis (1983) used the VIS channel of the GOES-E VISSR to estimate the diurnal variation of the surface albedo over three regions of the U.S. Great Plains.

Another factor that must be considered in interpreting reflectivity measurements is the strong wavelength dependence of reflected energy (Coulson et al., 1965; Kondratyev et al., 1981). In general, the 0.3 to 0.7 μm albedo is larger than the near-IR albedo for ice and snow, while the opposite is true for vegetation. Since most satellite instruments measure the reflected energy only in restricted spectral bands, surface radiation budget requirements cannot be met by most operational satellites.

A large number of site-specific studies of surface albedo and its variations have been made from a variety of satellite sensors. Courel et al. (1984) documented the evolution of the surface albedo of the Sahara-Sahel region during 1967-79 using hourly VIS data from Meteosat-1 and digital Landsat MSS data. Both Landsat and Meteosat observations yield albedos of 0.50 for the brightest spots of the Sahara.

Kaufman and Joseph (1982) developed a procedure for calculating surface albedo from satellite observations. The procedure is based on an analytical, approximate solution of the equations of radiative transfer and is primarily applicable in the 0.4 to 0.8 μm wavelength range for solar zenith angles from 10 to 60°, surface albedos between 0.03 and 0.50, and atmospheric optical thicknesses between 0.2 and 2. The procedure was used with a Landsat image of the African coast. Mekler and Joseph (1983) also described an empirical method for measuring the spectral surface albedo using Landsat imagery; the method permits routine construction of spectral albedo maps.

IX-1.4.3 Downwelling Longwave Radiation

A method for determining downward longwave radiation at the surface from satellite measurements (Darnell et al., 1983) uses data derived from HIRS/2

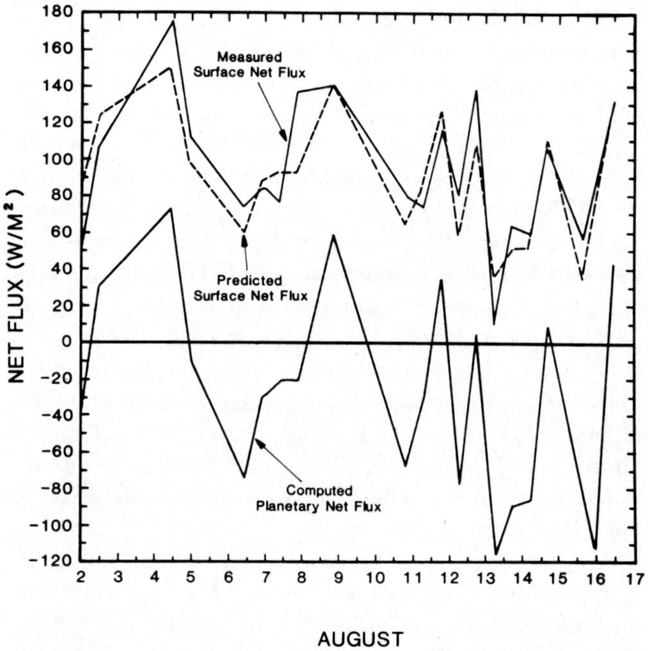

Figure IX-1-12. Daily averaged net radiation fluxes for August 1982, Ottawa Canada. Lower solid line: planetary net fluxes computed from GOES E data. Upper solid line: measured surface net fluxes. Upper dashed line: predicted surface net fluxes, generated from a regression equation based on the GOES computed planetary net fluxes, developed for Toronto, summer of 1982 (Pinker et al., 1985).

sounding inversions, which include cloud cover and height, and temperature and water vapor profiles. These satellite derived variables are used as input to an atmospheric radiative transfer model. Initial comparisons with measurements at the ground indicate accuracies of about 20 Wm^{-2}.

IX-1.4.4 Upwelling Longwave Radiation

Calculation of upwelling longwave radiation at the surface requires information on the temperature and emissivity of the surface. Emissivities of various types of surfaces over the Earth are generally well known, but the nature of the surface must be specified in space and time. Temperature determinations at the surface, for both ocean and land, using transparent (window) channels in the IR and in the MW portions of the spectrum are discussed in Chapters VI-2, VIII-4, and X-2. Essentially, satellite determinations of upwelling longwave radiation can be made wherever temperature of the Earth's surface is retrieved from satellite scanning radiometers or sounding instruments.

IX-1.4.5 Surface Net Longwave Radiation

Statistical models have been formulated for directly specifying the surface net IR radiation from the VAS instrument on the GOES satellites (Smith and Woolf, 1984). One advantage of VAS is the availability of hourly estimates. Under cloudy conditions, the radiation components are not directly observed, but under such conditions the net infrared flux (at least in the tropics) is generally small. The variation of net longwave radiation at the surface thus can be sampled satisfactorily from satellites over the tropics. Accuracies of the estimates made using this statistical model are about ±10 Wm^{-2} and ±20 Wm^{-2} for clear and cloudy conditions, respectively.

IX-1.4.6 Surface Net Radiation (All Wavelengths)

As the result of a study conducted over the central U.S., Pinker and Corio (1984) found that the net ra-

Table IX-1-1 *Regression statistics between time averaged values of measured and predicted planetary and surface net radiation flux for Toronto, Ottawa, and Quebec for May–August 1982, using different averaging periods (from 1 to 10 days). The days grouped were not necessarily consecutive (Pinker et al., 1985).*

	TORONTO				OTTAWA				QUEBEC			
NO. OF DAYS	NO. OF CASES	CORR.	STANDARD ERROR W/m²	%	NO. OF CASES	CORR.	STANDARD ERROR W/m²	%	NO. OF CASES	CORR.	STANDARD ERROR W/m²	%
1	67	0.924	17.8	14.5	67	0.914	19.6	16.0	63	0.932	16.2	15.9
2	33	0.921	13.4	10.8	33	0.937	13.1	10.7	31	0.937	11.9	11.6
3	22	0.926	11.9	9.6	22	0.940	10.8	8.8	21	0.966	8.6	8.4
4	16	0.913	11.8	9.5	16	0.949	9.7	7.9	15	0.921	9.7	9.3
5	13	0.867	12.0	9.7	13	0.955	7.9	6.4	12	0.956	7.3	7.0
6	11	0.922	10.8	8.7	11	0.964	7.2	5.9	10	0.980	4.5	4.4
7	9	0.901	11.6	9.3	9	0.944	9.8	7.9	9	0.981	4.8	4.7
8	8	0.880	11.7	9.4	8	0.965	6.8	5.5	7	0.939	5.9	5.5
9	7	0.865	11.4	9.1	7	0.946	8.3	6.7	7	0.978	5.6	5.5
10	6	0.874	11.3	9.0	6	0.950	6.9	5.5	6	0.978	4.8	4.7

diation and the outgoing longwave radiation at the top of the atmosphere are correlated with the net radiation at the surface ($r = 0.76$). The net radiation at the top of the atmosphere was derived from one VIS observation and two IR observations per day from NOAA-5 (Winston et al., 1979).

In a subsequent study, Pinker et al. (1985) re-examined the previously established relationship using daily totals of planetary net radiation computed from GOES-E data. Corresponding daily totals of surface net radiation were derived from the surface observations taken by the Canadian Environmental Service. Linear regression was performed between the computed values of planetary net radiation and measured surface net radiation. The derived relationships were then tested for their predictive value on independent data. Correspondence between predicted and observed values was very good (Figure IX-1-12). The regression statistics for this study are listed in Table IX-1-1. The correlation coefficients and standard errors of estimate are shown for averaging periods ranging from 1 to 10 days. The standard errors decrease substantially as the averaging period is increased, while the correlation coefficients generally remain in the same high range regardless of averaging. This method shows much promise for producing very useful climatological monitoring of the net surface radiation from geostationary satellites.

Chapter IX-2

CLOUDS AND TROPOSPHERIC AEROSOLS FROM SATELLITES

IX-2.1 INTRODUCTION

Interactions of clouds and aerosols with incoming solar and outgoing terrestrial radiation partly determine the Earth's energy balance and climate. Both clouds and tropospheric aerosols are definable as suspensions of particles in the atmosphere. The principal difference between them is the size of the particles. Aerosol particles typically range from 0.001 μm to 1 μm in radius, cloud particles from 1 μm to 100 μm. The larger sized cloud particles cause about 100 times greater absorption or scattering of radiation than do the aerosol particles. This explains, for example, why visibility is lower in fog (cloud) than in haze (aerosol). Aerosols show more of a decrease in absorption and scattering with increased wavelength than do clouds. Thus, clouds generally interact about equally with incoming solar radiation and emitted thermal radiation, whereas aerosols interact mainly in the shortwave (solar) portion of the spectrum. The actual effect of clouds and aerosols on the Earth's radiation balance and climate also depends on many other factors, such as chemical composition and concentrations, both vertically and horizontally.

Numerous studies have estimated the effects of clouds and aerosols on the Earth's climate (Rasool and Schneider, 1971; Schneider, 1972; Hansen et al., 1981; and Coakley et al., 1983). These studies indicate that clouds and aerosols can cause substantial warming or cooling of the Earth surface and atmosphere. These effects are comparable in magnitude with the effect of increasing CO_2. These results are only approximate, however, because the physical processes and the time and space variabilities of clouds and aerosols are not well represented in these models.

Satellite estimates of cloud and aerosol distributions in space and time can aid in improving the modeling of the effects of these quantities on climate variations. Also, regular monitoring of their changing distributions should be useful in understanding and anticipating climatic variations.

IX-2.2 TECHNIQUES FOR MEASURING CLOUDS FROM SATELLITES

With the coming of weather satellites, pictures of clouds taken from space became commonplace. It was clear from the beginning that the pictures contained much new information, but even today the quantitative characterization of clouds from satellite observations awaits complete solution. The earliest approach was to manually analyze the thousands of pictures from the satellites. These nephanalyses, although fairly representative of cloud patterns, basically involved much subjectivity and were tedious to produce (Clapp, 1964). The mass of data and the desire to remove subjectivity from the nephanalyses prompted efforts to produce automated, digital analyses of cloud pictures.

IX-2.2.1 Threshold Methods

The first automated technique used to analyze satellite cloud information was the "threshold method" (Arking, 1964), which is still in use. In its simplest form, the technique counts as "cloud covered" any image element for which the observed radiance differs by more than a certain amount (threshold value) from the radiance observed under cloudfree conditions. Any element with a smaller difference is considered "cloudfree." The fractional cloud cover is given by the ratio of the number of cloud covered elements to the total number of elements in the scene. Initially, a single threshold, based on VIS wavelengths, indicated cloud covered elements. Later, multiple thresholds categorized cloud cover into various fractional amounts based on the VIS radiances (Miller and Feddes, 1971). As IR scanning radiometers came into use, multiple IR thresholds were used to assign cloud covered elements to layers at different altitudes (Koffler et al., 1973; Stowe, 1984; Minnis and Harrison, 1984).

Figure IX-2-1 is an example of a multiple threshold IR technique used to classify Nimbus-6 Temperature-Humidity InfraRed (THIR) radiometer data for a (160 km)^2 scene into three cloud types (Stowe et al., 1978). The lack of well-defined demarcations between classifications, i.e., each bin does not contain a single mode from the histogram, limits the accuracy of this method. Since clouds generally are not divided into elements exactly the size of the image elements (typically 4 to 8 km on a side), their edges seldom coincide with the limits of the image elements. Common experience indicates that a substantial portion of global cloud cover is in the form of clouds smaller than the resolution of the observations. Consequently, errors in threshold derived cloud cover can be substantial when a large portion of the image elements are only partially cloud covered (Shenk and Solomonson, 1972).

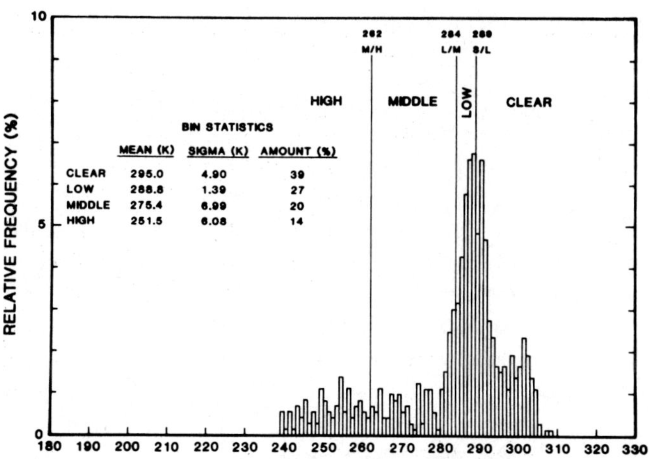

Figure IX-2-1. Application of multiple IR threshold to (7 km)^2 Nimbus-6 THIR 11 μm data for a (160 km)^2 region (13.75°S, 20.25°W, August 3, 1975).

Assessing errors in cloud cover derived from threshold methods has proven difficult. For (160-km)^2 areas, the error in cloud amount observed with (7-km)^2 IR image elements is about 10 to 20% (Stowe, 1984). If the errors are largely random, time averages, such as the monthly mean, will be unbiased and the errors will be considerably smaller (Minnis and Harrison, 1984; Stowe, 1984). Extensive error analyses carried on in the International Satellite Cloud Climatology Project (ISCCP) (Schiffer and Rossow, 1983) should determine whether errors using the threshold method do significantly decrease with substantial time averaging.

IX-2.2.2 Bispectral Methods

Joint use of VIS and IR window measurements from polar orbiter or geostationary radiometers provide better means of obtaining cloud cover and heights during daylight hours than the use of either alone. The bispectral method was initially introduced as a procedure to permit consideration of partially cloud covered scan spots (Reynolds and Vonder Haar, 1977).

Either radiation models or observations are used to estimate, for the VIS channel, the radiances expected for cloud-free and fully cloud covered scan spots. The observed VIS channel radiance for a scan spot can then be used to obtain the cloud cover. Using this cloud cover estimate, the IR radiance for the same scan spot can be used to estimate cloud top temperature and height.

Unfortunately, clouds exhibit considerable variability in the amount of sunlight they reflect. Figure IX-2-2 shows VIS reflectivities plotted against 11 μm radiances for a (250 km)^2 Pacific Ocean region that contains low level marine stratocumulus. Cloudfree 11 μm radiances are near 100 mW^{-2}sr^{-1}cm, but cloud layers emit at approximately 82 mWm^{-2}sr^{-1}cm. Emissions in the diagram near that of the layer are thought to originate from clouds that fill the FOV of the instrument. The large range of VIS reflectivities (20 to 50%) of these clouds can be attributed to variable amounts of liquid water within the clouds across the region (Platt, 1983). Because of this large range in reflectivity, cloud cover estimates based on observations of reflected sunlight are highly uncertain.

Nevertheless, the use of VIS and IR channels can provide information on cloud structure not available through use of a single channel (Desbois et al., 1982; Platt, 1983; Stowe et al., 1988). For example, Figure IX-2-3 shows VIS reflectivities plotted against 11 μm

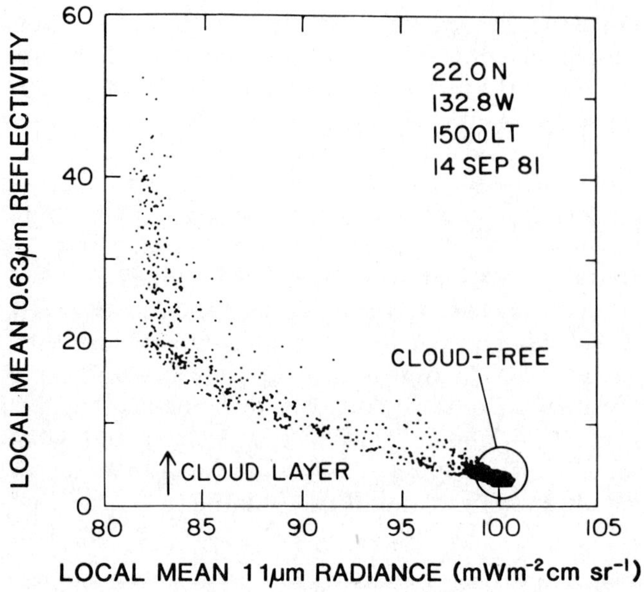

Figure IX-2-2. VIS reflectivity and 11 μm radiance for NOAA-7 AVHRR. Data are for a $(250\ \text{km})^2$ region of the Pacific Ocean off the coast of California. Each data point represents the mean radiance for an $(8\ \text{km})^2$ portion of the $(250\ \text{km})^2$ region formed by taking 2×2 arrays of the $(4\ \text{km})^2$ AVHRR GAC data. The pattern is typical of low level, single layered systems over the ocean.

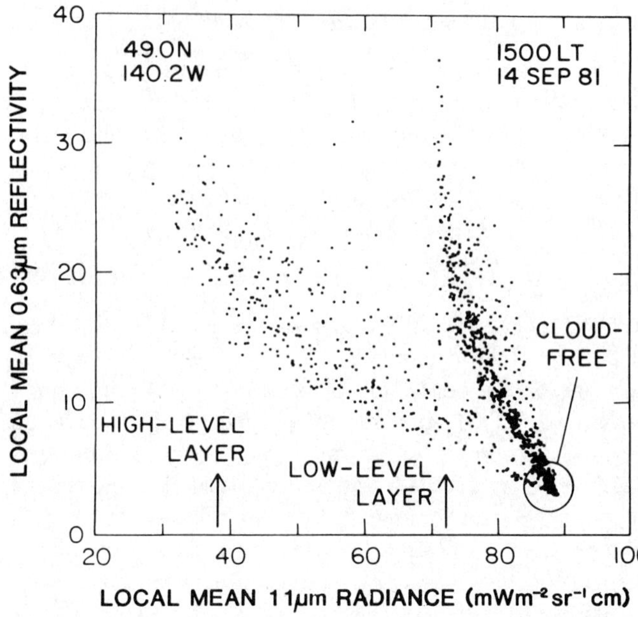

Figure IX-2-3. Same as Figure IX-2-2, but for two-layered system.

radiances, as in Figure IX-2-2, but for a two-layered cloud system. The two branches in the diagram clearly indicate the presence of two cloud layers.

IX-2.2.3 Spatial Coherence

Another approach to dealing with clouds that only partially fill a FOV is to recognize that a given cloud layer often spans a large region (200 to 1000 km). The varying radiances among the scan spots in such an area can be used to determine those associated with the cloudfree background and those of the cloud layer(s) in the scene (Coakley and Bretherton, 1982). Regions with low spatial variability and lower levels of thermal emission are associated with cloud layers; those with low spatial variability and high levels of thermal emission indicate a cloudfree background. Figure IX-2-4 shows images of TIROS-N AVHRR Global Area Coverage (GAC) data for a $(1000\text{-km})^2$ region over the Pacific Ocean. Figure IX-2-5 shows the IR data arranged in 16 arrays of $(4\text{-km})^2$ GAC pixels for each $(250\text{-km})^2$ area within the scene of Figure IX-2-4. For each $(8\text{-km})^2$ region the mean and standard deviation for the 4 pixels are plotted. The $(8\text{-km})^2$ regions with small variability in radiance produce the feet to the arches in each of the arrays in Figure IX-2-5. The left hand feet, with radiating temperatures near 282 K, represent the temperature of the top of the cloud layer, and the right hand feet near 293 K represent the temperatures of the cloudfree ocean background. The bodies of the arches in Figure IX-2-5 are due to $(8\text{-km})^2$ regions only partially covered by the clouds in the layer. Broken cloud cover generally results in changes in the fractional cloud cover from one imager scan spot to the next. Thus, the emitted radiance under broken cloud conditions exhibits considerable spatial variability even on the local $(8\text{-km})^2$ scale.

By extending this analysis to observations at both 3.7 and 11 μm, as is possible with AVHRR data at night, the fractional cover for two layered systems can be retrieved (Coakley, 1983).

X-2.2.4 Cirrus, Frontal Clouds, and Convective Cloud Clusters

The spatial coherence method is limited to layered, opaque cloud systems that are common features of the global cloud cover. Cirrus, frontal clouds, and convective cloud clusters are also common globally, but the spatial coherence method fails for such systems. With the possible exception of cloud top temperature for cirrus (McCleese and Wilson, 1976; Smith and Platt, 1978; Szejwach, 1982), no known method successfully characterizes the properties of these more complex cloud systems.

X-2.2.5 Sounder Retrievals

The methods for retrieving atmospheric temperature and moisture profiles from satellite sounders require

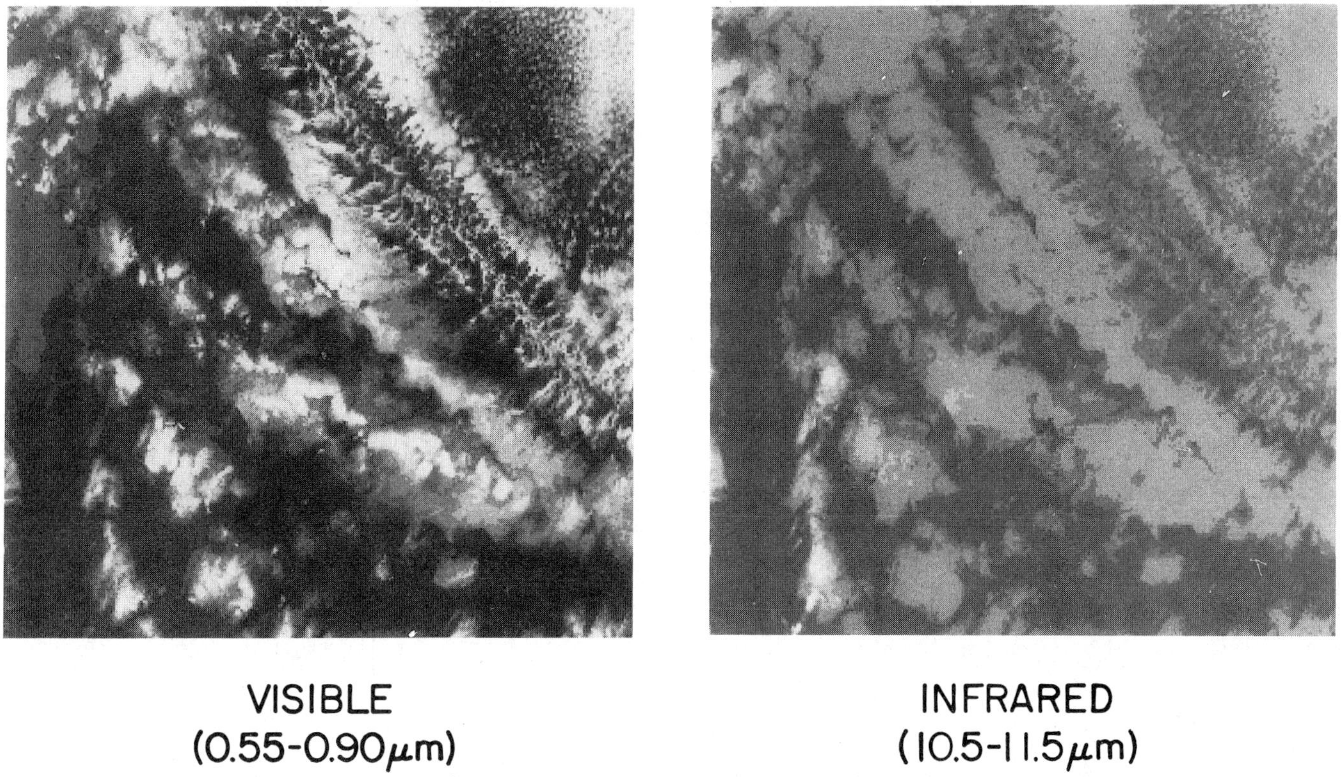

Figure IX-2-4. Images constructed from VIS reflectivities and 11 μm TIROS-N AVHRR radiances. The images depict clouds (lighter shades in both images) covering a (1000-km)^2 region of the Pacific Ocean (Coakley and Bretherton, 1982).

the elimination of cloudiness effects in cloud sensitive channels. Thus, cloud cover can be obtained as a by-product of the sounder retrieval methods (Smith, 1968; Smith et al., 1970; Chahine, 1974). The following concept is common to all sounder retrieval methods: radiances of cloud-free and cloud covered portions of adjacent FOV are assumed to be essentially constant. Owing to changes in cloud cover from one FOV to the next, however, the radiances of sequential FOV can be quite different. Using these radiance differences in at least two channels where the radiative properties of the clouds are the same (e.g., the 15 μm channels of the sounder instruments) permits determination of cloud cover amount and height of the cloud tops from the FOV radiances. Although this principle can be generalized to accommodate cases where multiple cloud layers are present (Chahine, 1982), it is rarely possible to do so. The retrieved cloud cover, assuming a single layer, is said to be an effective cloud cover at an effective altitude, i.e., that which reproduces the observed radiances (Susskind et al., 1984).

Cloud top heights obtained from satellite sounder data compare favorably with those obtained using *in situ* LIDAR and radiosonde observations (Smith and Platt, 1978). However, sounder data often fail to reveal the presence of low level clouds with emission temperatures near those of the surface (Wielicki and Coakley, 1981). Also, sounder data often fail to distinguish between clouds in distinctly different layers, even when the layers are moderately spaced in altitude and so have substantially different emission temperatures. Recent work with imagery data suggests that these problems could be reduced by decreasing instrument noise and improving both spatial resolution and sampling rate.

Despite their limitations, sounder data offer a valuable source of information on clouds. Because it is not necessary to assume that clouds fill the instrument FOV, sounder retrievals appear to be more generally applicable than threshold techniques. Furthermore, analyses of cloud cover at the variety of wavelengths afforded by the sounders (3 to 4 μm, 8 to 12 μm, and 15 μm) indicate the varying degrees to which clouds are transparent at these wavelengths (Chahine, 1982; Coakley, 1983).

IX-2.3 CLIMATIC APPLICATION OF SATELLITE CLOUD MEASUREMENTS

The Global Atmospheric Research Program (GARP/JOC, 1978; 1981) has identified five applica-

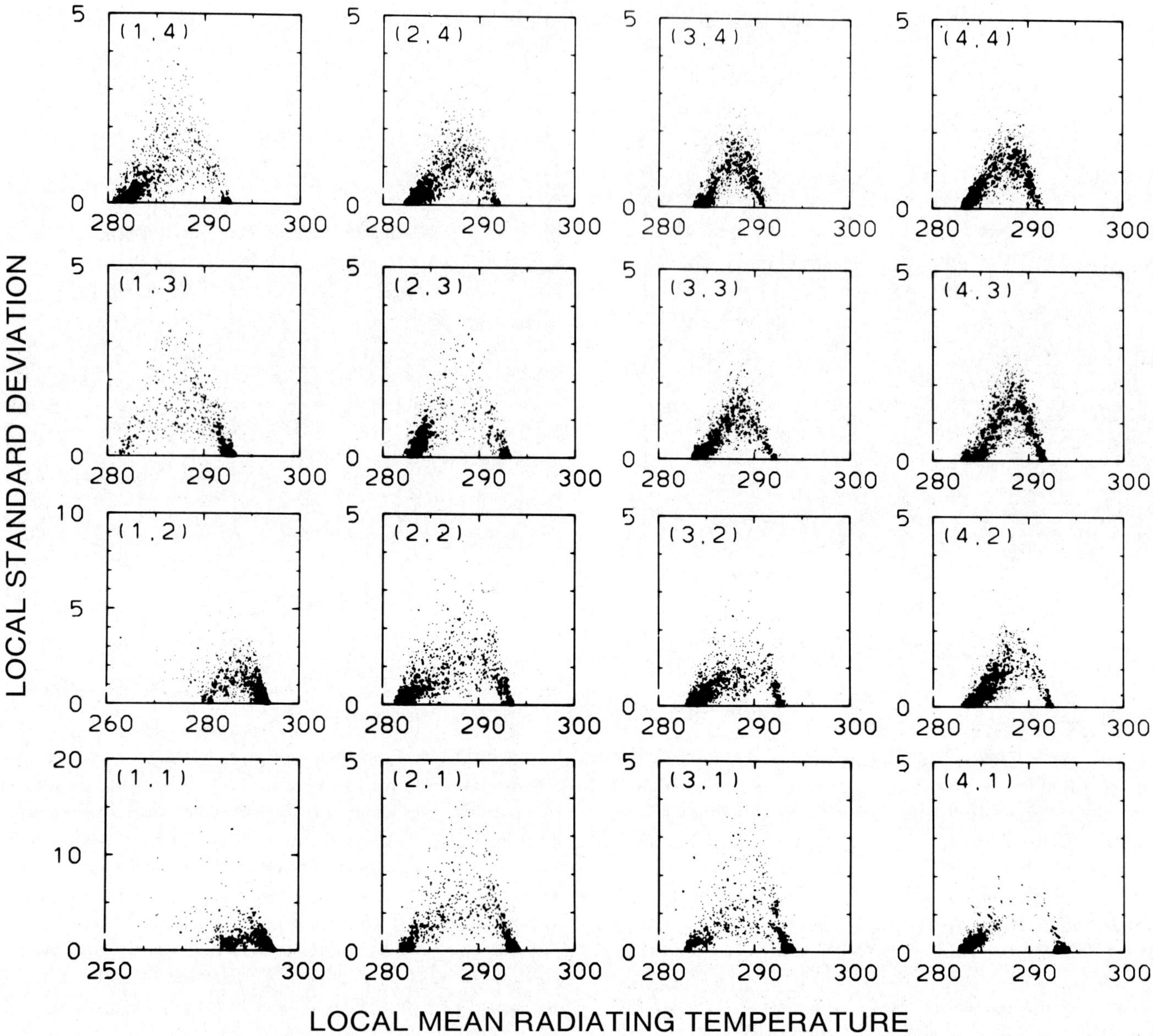

Figure IX-2-5. Spatial coherence analysis of the 11 μm radiances (transformed to brightness temperatures) imaged in Figure IX-2-4. Each panel represents a $(250\text{-km})^2$ portion of the $(1000\text{-km})^2$ region. The figure has the same orientation as the image. Each point in the figure gives the mean and standard deviation of emitted radiance for $(8\text{-km})^2$ portions of the associated $(250\text{-km})^2$ frame. The $(8\text{-km})^2$ regions are formed by taking 2×2 arrays of the $(4\text{-km})^2$ GAC data (Coakley and Bretherton, 1982).

tions for a global, multiyear climatology of cloudiness: (1) inclusion in and validation of climate models; (2) climate monitoring and diagnosis; (3) studies of the dependence of climate on clouds; (4) studies of cloudiness and the ERB; and (5) parameterization of cloud formation and radiative processes. The inadequacy of currently available cloud climatologies for satisfying these applications is recognized. A recent historical review of global cloud climatologies by Hughes (1984) concludes that: (1) there is at present no unique or generally accepted global cloud climatology; (2) climatologies vary widely in their representation of reality; and (3) climatologies do not allow accurate detection of year to year global cloud differences. The differences among cloud climatologies derived through 1980 are shown in Figure IX-2-6 for zonal average total cloud cover for January (Hughes, 1984). Positions of the maxima and minima are in agreement, but mag-

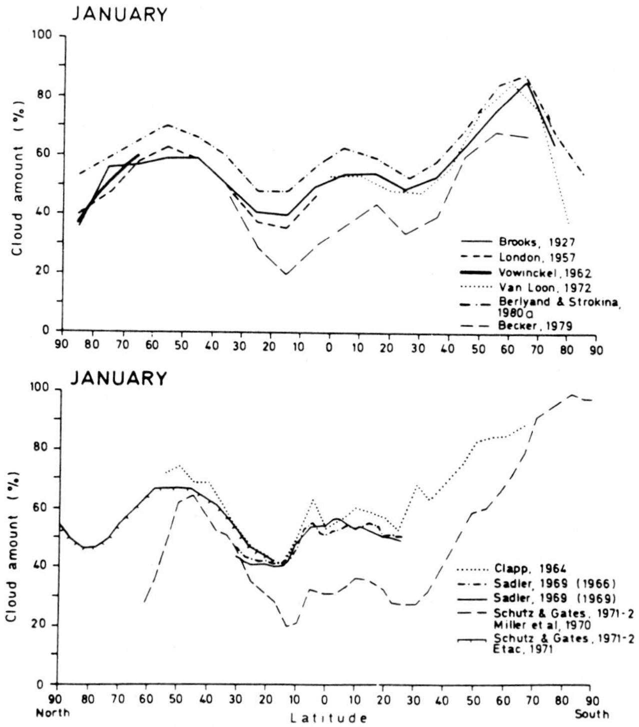

Figure IX-2-6. Comparison of meridional profiles of zonally averaged cloud amount for January (Hughes, 1984).

nitudes differ by as much as 30%. Regional differences can be even larger. One cause for the differences is that some climatologies are based on surface observations, some on satellite observations, and others on a combination of the two.

Estimates of zonal cloud cover for June 1979 from the Nimbus-7 program (Hwang et al., 1988), together with the climatologies of London (1957) and Berlyand et al. (1980), are shown in Figure IX-2-7. The most striking features of the satellite cloud estimates are: (1) the pronounced peak in cloudiness associated with the ITCZ; and (2) the greater cloud cover at local midnight than at local noon (about 4% globally averaged). Note that for the entire Northern Hemisphere, with the exception of the latitudes of the ITCZ, the satellite cloud cover estimates are less than those based on surface observations. In the ITCZ, the satellite observations exhibit much more cloud cover than the earlier estimates based on surface observations. Although most of the differences between zonally averaged satellite and surface cloud cover estimates may be attributable to inadequate sampling of surface data, particularly south of about 20°N, there are obviously differences resulting from the differing nature of the two types of observations. For example, surface observations yield higher estimates of cloud cover than sat-

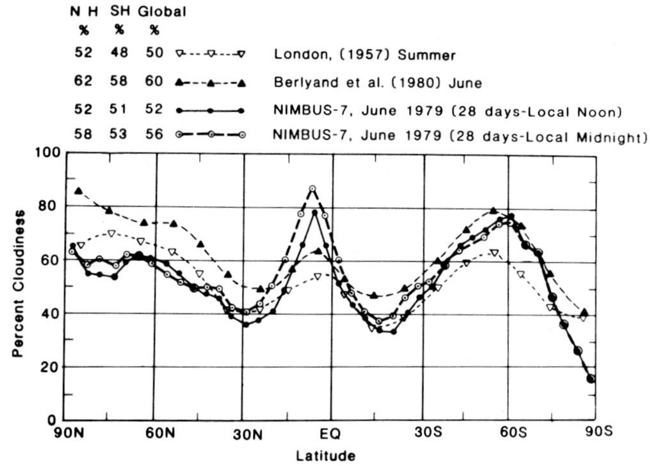

Figure IX-2-7. Comparison of initial results from Nimbus-7 derived cloud climatology with zonal mean cloudiness of London (1957) and Berlyand et al. (1980). The Nimbus-7 result is for June 1979 and is shown at local noon and midnight.

ellites when cloud coverage is broken to scattered because of the surface observer's different geometrical view of the sky.

IX-2.4 TECHNIQUES FOR MEASURING TROPOSPHERIC AEROSOLS FROM SATELLITES

Satellite instruments sensitive at UV, VIS, and near-IR wavelengths are used to detect the presence of aerosols in the troposphere. Passive techniques have been used to measure the radiance or polarization of reflected solar radiation (Hariharan, 1971; Rao et al., 1973; Santer et al., 1982; Prosch et al., 1982). Active techniques (LIDAR measurements) have been proposed, but not as yet implemented on satellites (Russell et al., 1983; Zuev, 1983).

Griggs (1975), and Carlson and Wendling (1977) showed through theoretical radiative transfer computations that the reflected solar radiance emerging vertically (satellite nadir direction) over an ocean surface increases as a nearly linear function of aerosol optical thickness for values ranging between 0.1 and 1. 0. The amount of increase (slope) depends on the wavelength, the solar illumination angle, and the aerosol size distribution and refractive indices. The slope is essentially independent of the vertical distribution of the aerosol particles.

Griggs (1977) compared simultaneous ground based measurements of aerosol optical thickness at a wavelength of 0.5 μm with Landsat-2 nadir radiances in three spectral intervals (Figure IX-2-8). A similar technique using AVHRR data from NOAA satellites

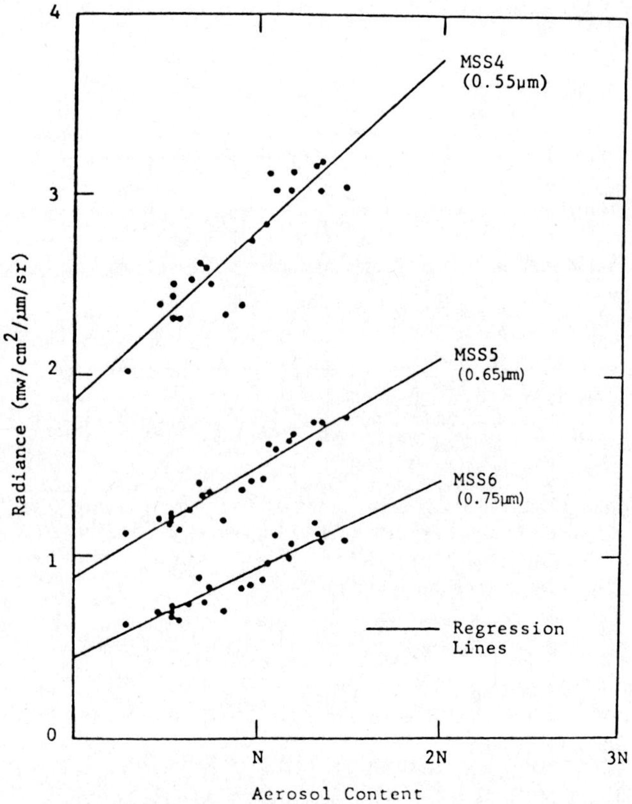

Figure IX-2-8. Observed relationship between nadir radiance and aerosol content for three channels of the MSS instrument on Landsat-2. *N* is equal to an optical thickness of 0.213 at a wavelength of 0.5 μm (Griggs, 1977).

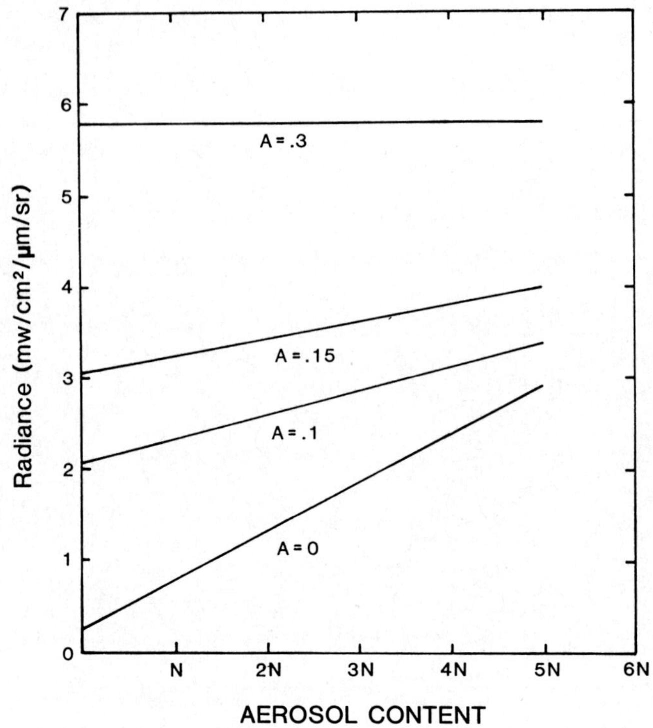

Figure IX-2-9. Theoretical relationship between nadir radiance and aerosol content for different Earth surface albedos at a wavelength of 0.75 μm (Griggs, 1981).

has since been developed (Griggs, 1983; Griggs and Stowe, 1984). This technique is restricted to low albedo surfaces such as the ocean, because the reflected radiation becomes insensitive to the amount of aerosol in the atmosphere as the surface albedo increases (Figure IX-2-9). Also, partial cloud cover within the FOV can be a serious source of error (Norton et al., 1980).

Theoretical computations indicate that it should be possible to infer an effective size distribution for the aerosol particles from simultaneous radiance measurements made at more than one wavelength (Griggs, 1983; Koepke and Quenzel, 1981). Figure IX-2-10 shows that the difference in radiance between two Landsat channels (0.65 μm and 0.75 μm) increases with increasing aerosol optical thickness, but by differing amounts depending on the Junge size distribution parameter, ν. For size distributions with relatively few large particles ($\nu = 4$), the difference between the radiances in the two channels varies more with increasing aerosol content than for distributions with relatively many large particles ($\nu = 2$). Modifications of the NOAA AVHRR instrument for the 1990s to make measurements at 0.63, 0.86 and 1.61 μm have been considered, partly to allow for more accurate and complete detection of aerosol properties (Ahmad and Sutton, 1988).

IX-2.5 CLIMATIC APPLICATIONS OF SATELLITE TROPOSPHERIC AEROSOL ESTIMATES

Several studies have illustrated potential applications of the measurements. Griggs (1983) has compared a large set of aerosol optical thickness estimates using Channel 1 of the NOAA-6 AVHRR with aerosol values derived from ground-based Sun photometers at ocean sites around the globe. The agreement between the two types of observations was excellent (Figure IX-2-11).

Aerosols affect the accuracy of other quantities remotely sensed by satellites (e.g., Stowe and Fleming, 1980; Strong, 1984). Griggs (1985) has developed an algorithm that corrects remotely sensed sea surface temperature for the effects of stratospheric aerosols from the El Chichón eruption, using the optical thickness measurements from AVHRR.

Carlson (1979) used radiance data from the NOAA-3 VHRR VIS channel to construct isopleths of aerosol optical thickness over the eastern Equatorial North Atlantic during seven days of a Saharan dust outbreak in the summer of 1974. Quantitative assessment of

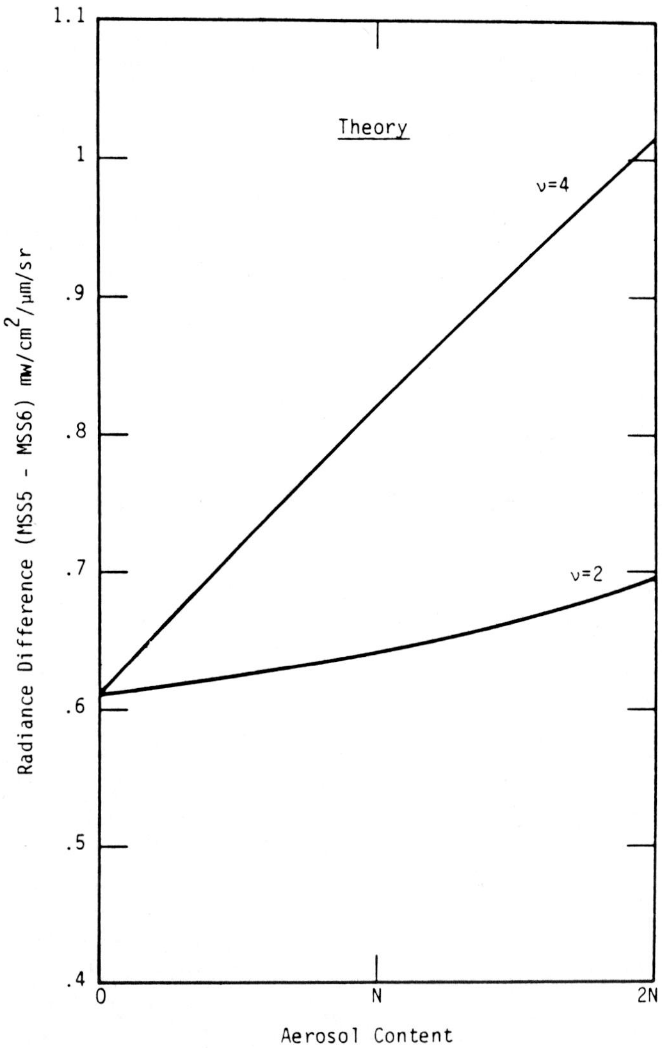

Figure IX-2-10. Theoretical difference in nadir radiance between MSS/5 (0.65 μm) and MSS/6 (0.75 μm) as a function of aerosol content for two different Junge size distributions: ν = 4, relatively small particles, and ν = 2, relatively large particles (courtesy M. Griggs).

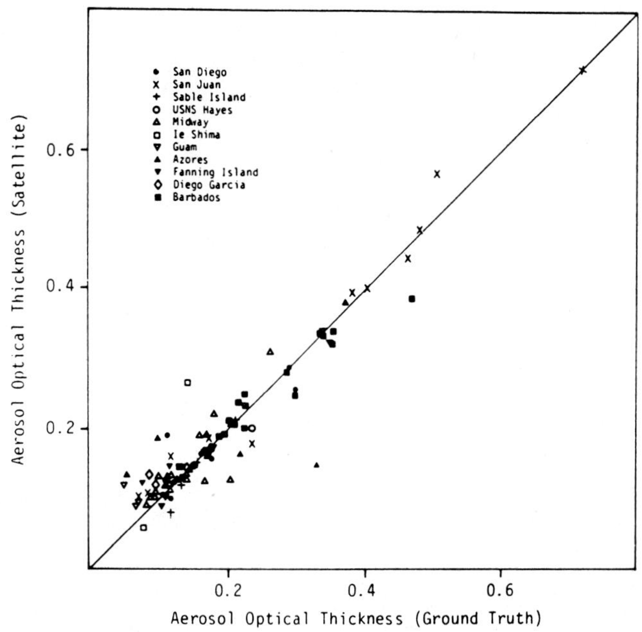

Figure IX-2-11. Comparison of NOAA-6 AVHRR and ground truth Sunphotometer estimates of aerosol content at eleven different geographical locations for solar zenith angle less than 70° (Griggs, 1985).

these data enabled him to relate the Saharan dust outbreaks to the dynamical patterns associated with the ITCZ and easterly waves.

Routine monitoring of aerosol optical thickness has begun at NOAA/NESDIS on an experimental basis (Rao et al., 1988). The most persistent aerosol feature observed has been the Saharan dust cloud in the tropical Atlantic. Other aerosol features that have been detected include: dust from other deserts of the world, air pollution related haze; and smoke from forest fires and slash burning of crops.

Construction of a global and multiple year climatology of tropospheric aerosol optical properties will be developed from these routine estimates from satellites. Such a climatology will make it possible to include the effects of aerosols in climate perturbation studies. The WMO Committee on Atmospheric Sciences (CAS) and the Radiation Commission of the International Association of Meteorology and Atmospheric Physics (IAMAP) have proposed an International Aerosol Climatology Project (IACP) with the objective of establishing a global aerosol data base for climate studies (WCP, 1983). The proposal recognizes the importance of satellites for observation of global variations of aerosol concentration.

Chapter IX-3

STRATOSPHERIC MONITORING

IX-3.1 INTRODUCTION

"Stratosphere" is used herein to refer to the middle atmosphere, i.e., the region between the top of the troposphere at 8 to 15 km and the base of the thermosphere at approximately 80 km. In this region, radiative, photochemical, and dynamical processes are closely related. The layer is also coupled with the atmospheric regions above and below it through exchanges of mass, momentum, energy, and chemical constituents.

Understanding the stratosphere is of critical climatic importance. It contains most of the atmospheric ozone that shields the biosphere from damaging UV radiation. It also influences the surface climate through radiative and dynamic linkages. Thus, it is necessary to understand its processes and to observe its long term changes and trends.

Satellite measurements provide considerable information on the climatic variations in temperature, ozone, and aerosols in the stratosphere.

IX-3.2 TEMPERATURE AND CIRCULATION

IX-3.2.1 Types of Measurements

Satellite estimates of stratospheric temperature profiles have been made continuously since April 1969 when the Satellite IR Spectrometer (SIRS/A) flew on Nimbus-3. Data have been gathered from a number of other instruments flown on satellites in succeeding years. Tables IX-3-1 and IX-3-2 (WMO, 1982) contain details on the various satellite temperature sounding instruments that have provided stratospheric information.

Until 1975, all measurements were made looking downward so that the radiation received came from a deep layer (Wark and Fleming, 1966). Measurements of radiation upwelling from a uniformly mixed constituent, such as carbon dioxide in bands near 15 μm, represent the mean blackbody temperature of a layer about 12- to 20-km thick. Chapters II-2, IV-2, and VII-9 describe the instrumentation and means for deriving soundings from these downward viewing instruments.

The first atmospheric limb-viewing instrument, the Limb Radiance Inversion Radiometer (LRIR), was flown on Nimbus-6. The Limb Infrared Monitor of the Stratosphere (LIMS), similar to LRIR, and the Stratospheric and Mesospheric Sounder (SAMS), also a limb viewing instrument, were flown on Nimbus-7. By scanning across the limb, atmospheric emission is measured as a function of altitude, which permits determination of the vertical distribution of temperature and trace constituents. From the tropopause to about 65 km (Gille and House, 1971), the narrow weighting functions (about 4 km) and frequent samples (1.5 km) of the instruments in the vertical allow higher vertical resolution and better depiction of atmospheric temperature structure than with downward viewing instruments. Horizontal resolution is not good, however, since 50% of the energy received at 10 mb comes from

Table IX-3-1 *NASA satellite measurements of atmospheric temperature profiles—measurement parameters (WMO, 1982)*

INSTRUMENT	SATELLITE	TIME PERIOD	LAT. RANGE (DEG.)	ALTITUDE RANGE	SPATIAL RESOLUTION			DUTY CYCLE	ESTIMATED MEASUREMENT UNCERTAINTY	COMMENTS
					VERTICAL	ALONG TRACK	CROSS TRACK			
IRIS	Nimbus-4	4/70–1/71	80 S to 80 N	10–35 km	~6 km	96 km	Orbit spacing	Continuous	1.5 K	Thermal emission measurements at noon and midnight local time.
SCR 6 Channel	Nimbus-4	4/70–12/72	80 S to 80 N	0–40 km	~7 km		Orbit spacing	Continuous	1 K	
SCR 16 Channel	Nimbus-5	12/72–6/76	80 S to 80 N	0.45 km	~7 km		Orbit spacing	Alternate days to alternate weeks	1 K	
LRIR	Nimbus-6	6/75–1/76	64 S to 84 N	15–65 km	3 km	12 km	Orbit spacing (~2500 km) (~1300 km by combining ascending and descending node data)	75%	1 K A 1 K P	Vertical resolution, accuracy, and precision are somewhat altitude dependent.
PMR	Nimbus-6	6/75–6/81	80 S to 80 N	40–90 km	~10 km		Orbit spacing	Continuous	1 K	
SAMS	Nimbus-7	10/78 to present	50 S to 70 N	15–90 km	10 km	100 km based on vertical scan	Orbit spacing Data interpolated to a grid 2.5° lat. × 10° long.	75% (3 days on, 1 day off)	3 K A 1 K P	Accuracy is a function of height. It is expected to improve somewhat with further processing.
LIMS	Nimbus-7	10/78 to 5/79	64 S to 84 N	10–70 km	~3 km	84 km	Orbit spacing (~2500 km) (~1300 km by combining ascending and descending node data)	75%	3 K A 1 K P	Effective duty cycle including ERB interference effects. Vertical resolution, accuracy, and precision are somewhat altitude dependent.

NOTE: A denotes accuracy
P denotes precision

Table IX-3-2 *NOAA satellite instruments used for derivation of atmospheric temperature profiles—measurement parameters (WMO, 1982)*

INSTRU-MENT	SATELLITE	TIME PERIOD	LAT. RANGE (DEG.)	ALTI-TUDE RANGE	SPATIAL RESOLUTION			DUTY CYCLE	ESTIMATED MEASURE-MENT UNCER-TAINTY	COMMENTS
					VERTI-CAL	ALONG TRACK	CROSS TRACK			
SIRS A	Nimbus-3	4/69–9/70	80 S to 80 N	0–30 km	5–10 km	300 km	400 km	Continuous	2–3 K	
SIRS B	Nimbus-4	4/70–10/72	80 S to 80 N	0.30 km	5-10 km	400 km	400 km	Continuous	2–3 K	
ITPR	Nimbus-5	12/72–6/74	82 S to 82 N	0–30 km	5–10 km	400 km	50 km	Intermittent	2–3 K A 1 K P	
VTPR	NOAA-2 NOAA-3 NOAA-4 NOAA-5	10/72 3/79	90 S to 90 N	0–30 km	4–10 km	600 km	500–700 km	Continuous	2–3 K A 1 K P	
HIRS	Nimbus-6	6/75–3/76	80 S to 80 N	0–40 km	5–10 km	200 km	250 km	Intermittent	2–3 K A 1 K P	
TOVS: SSU BSU (HIRS/2) MSU	TIROS-N NOAA-6 and subsequent spacecraft	3/79 to present	90 S to 90 N	0–50 km	5–10 km	210 km	220–500 km	Continuous	2–3 K A 1 K P	Two satellite system

NOTE: A denotes accuracy
P denotes precision

a 300-km horizontal path. In addition, the lifetime of these limb-scanning instruments has been limited.

IX-3.2.2 Operationally Produced Temperature and Circulation Data

Operational (downward viewing) sounders on NOAA spacecraft provide continuous monitoring of stratospheric temperature and circulation. NOAA/NMC maps for both Northern and Southern Hemispheres, at levels from 70- to 0.4-mb, are derived daily using vertical temperature estimates from the operational satellite sounders. Figures IX-3-1 and IX-3-2, based on NMC 2-mb maps during a Northern (1986-87) and a Southern (1987) Hemisphere season, illustrate the types of time variations that can be observed, largely by the use of satellite information. The time averaged temperature maps, shown in Figures IX-3-3 and IX-3-4 typify the patterns for midwinter Northern and Southern Hemisphere months. Note the lower temperatures over high latitudes in the Southern Hemisphere as compared with the Northern Hemisphere.

IX-3.2.3 Zonally Averaged Conditions

There are several published studies of multiyear data sets. Figure IX-3-5 shows north-south cross-sections of 4 year mean zonally averaged temperatures and winds from 1000 to 0.4-mb for Northern Hemisphere winter months. Individual months within this 4 year period exhibit a great deal of interannual variability (Geller et al., 1983; Hamilton, 1982).

Satellite data have added greatly to geographical coverage over oceans, above 30 km, and in the Southern Hemisphere. These data have confirmed many previously observed features, and have added an enormous amount of detail. Differences between Northern and Southern Hemispheres are particularly apparent. The Southern Hemisphere winter polar tropopause, at 190 K, is 15 K colder than its Northern Hemisphere counterpart, while the Southern Hemisphere summer polar tropopause, at 235 K, is 5 K warmer than the Northern Hemisphere summer tropopause. These differences in the two hemispheres are probably partially attributable to the variation in the Earth-Sun distance during the course of the year. Other interhemispherical differences can be observed at the stratopause (Labitzke, 1974; Barnett, 1974).

Hirota and Barnett (1977) have shown that mean temperatures near the winter pole at stratopause and mesopause levels are almost perfectly correlated in a negative sense, i.e., the mesopause is hottest during winter while the stratopause is hottest in summer. The latest measurements show that this is also true on time scales of a few days during warming events, when peaks or dips of temperature at the stratopause are mirrored by dips or peaks, respectively, near the mesopause. This, together with earlier findings by Fritz and Soules (1970) that winter warmings at extratropical northern latitudes were accompanied by global scale cooling in the tropics and in the Southern Hemisphere, clearly points to the important role of variations in meridional and vertical heat transport during warming events.

Mean zonal winds may be calculated geostrophically by using the meridional gradients in zonal mean temperatures. Leovy and Webster (1976), in comparing the Northern and Southern Hemisphere mean zonal winds during winters, found similar cross-sections in early winters, but in the Southern Hemisphere the strongest winds were 5° to 10° lat. closer to the pole.

IX-3.2.4 Planetary Scale Waves

Satellite observations have been very valuable for establishing the climatology and phenomenology of large scale waves, especially the propagation of Rossby waves in the mid-latitudes. Planetary scale waves are easily observed, because it is possible to study spatial fields and temporal variations of satellite sounder radiances without the necessity for absolute calibrations or for inversion of the radiances to obtain temperatures. Planetary waves are particularly suitable for study with satellite sounding measurements because of the large amplitudes of the temperature variations, large vertical wavelengths, and small phase motion between satellite observations.

Numerous studies have been made of stratospheric planetary waves using satellite observations. Fritz (1970), in an analysis of SIRS radiances, studied traveling and stationary waves in the stratosphere. He found that these waves have their greatest amplitudes in winter (as expected) and that the quasi-stationary waves were twice as strong in the Northern Hemisphere as in the Southern Hemisphere. The vertical structure of these waves has also been clarified. Hirota and Barnett (1977) showed that temperature waves can be of significant amplitude up to the mesopause, with the maximum amplitudes occurring as high as 65-km for wave numbers 1 and 2. Kohri (1981) found these maxima at 45 km during a less active winter. He confirmed Leovy and Webster's (1976) finding of large variations in wave amplitude, with maximum amplitudes occurring at 60° lat. in both hemispheres.

Hirota (1976) clearly showed that the axis of wavenumber 1 tilts westward with height, except in the tropics. This is consistent with the structure of vertically propagating planetary waves, and indicates pole-

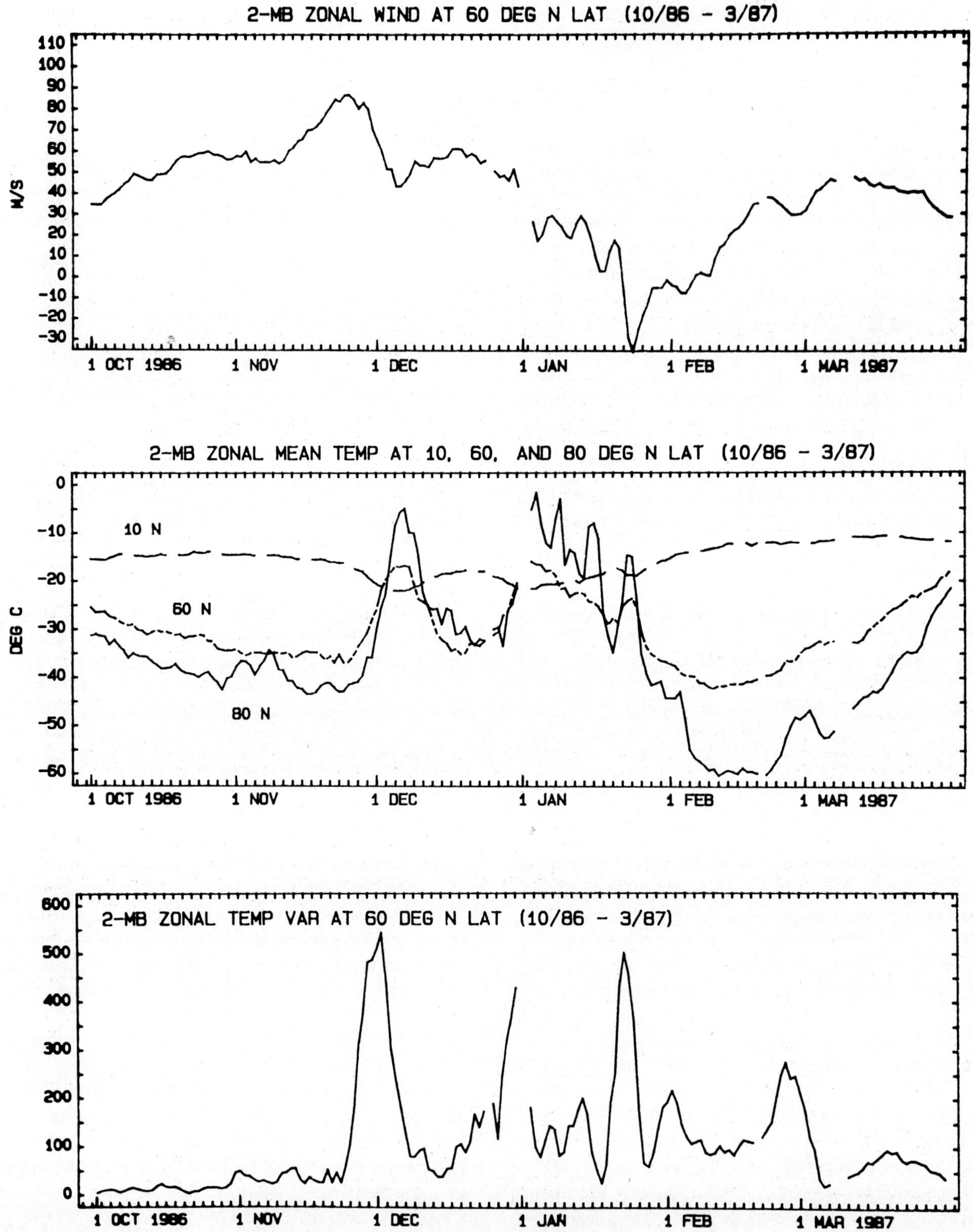

Figure IX-3-1. Time sequences of several stratospheric circulation parameters of the Northern Hemisphere, October 1986 through March 1987, from NMC analyses of TOVS data.

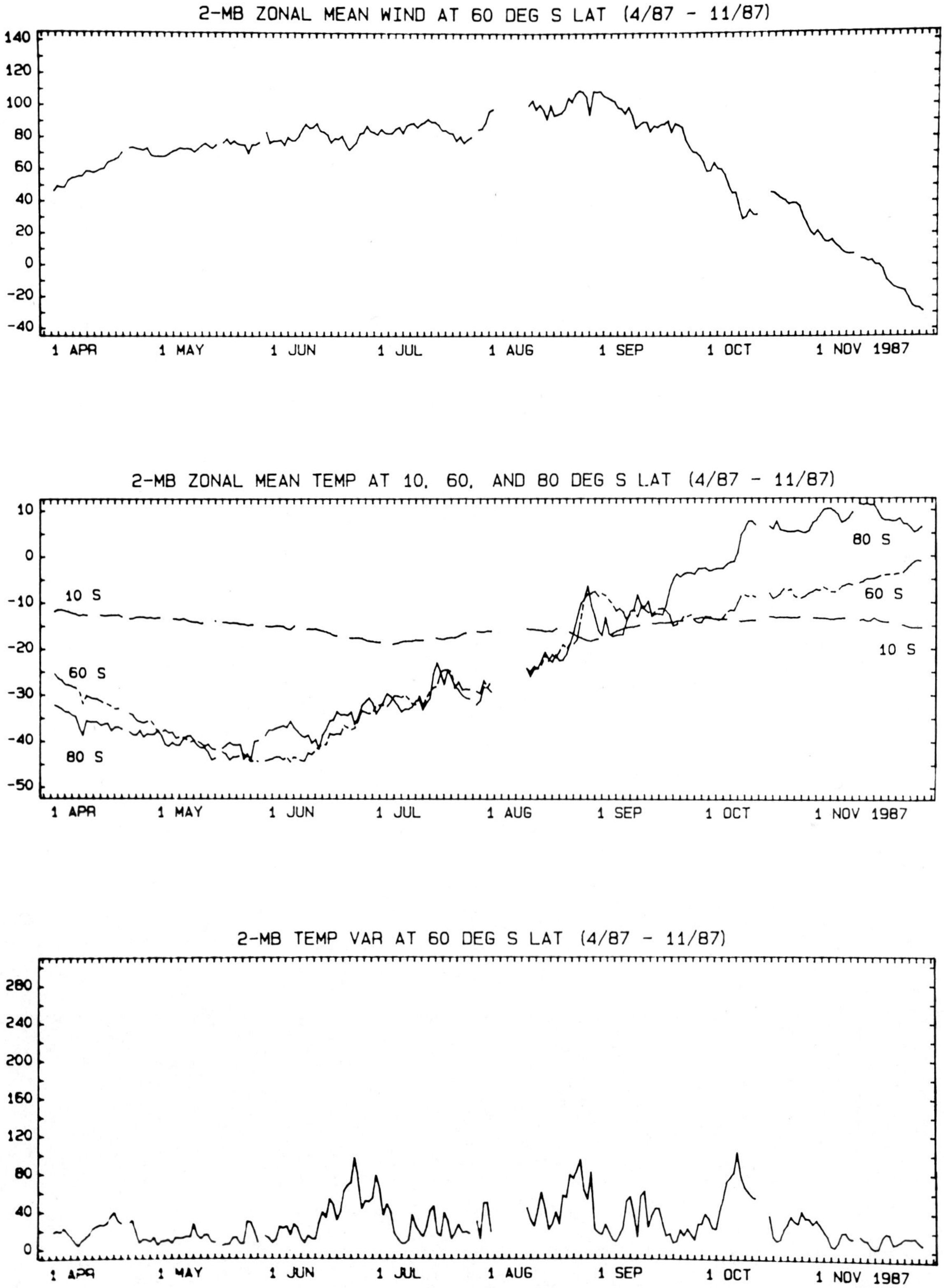

Figure IX-3-2. Time sequences of several stratospheric circulation parameters of the Southern Hemisphere, April through November 1987, from NMC analyses of TOVS data.

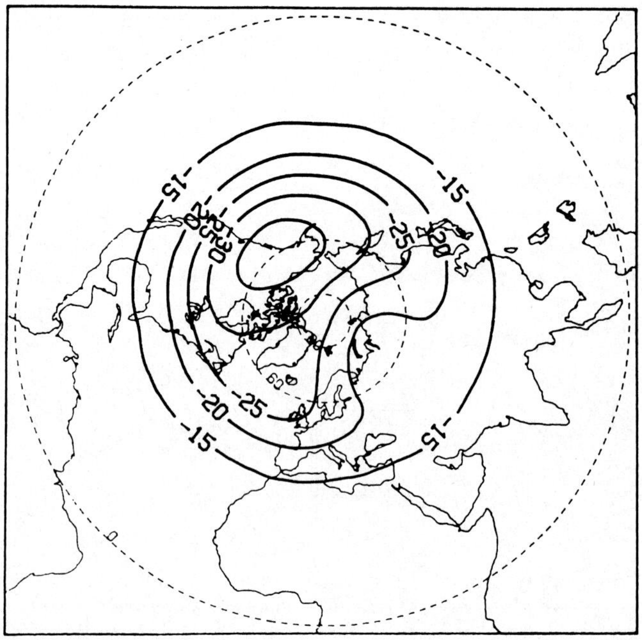

Figure IX-3-3. January mean 2-mb Northern Hemisphere temperature, 1979-86.

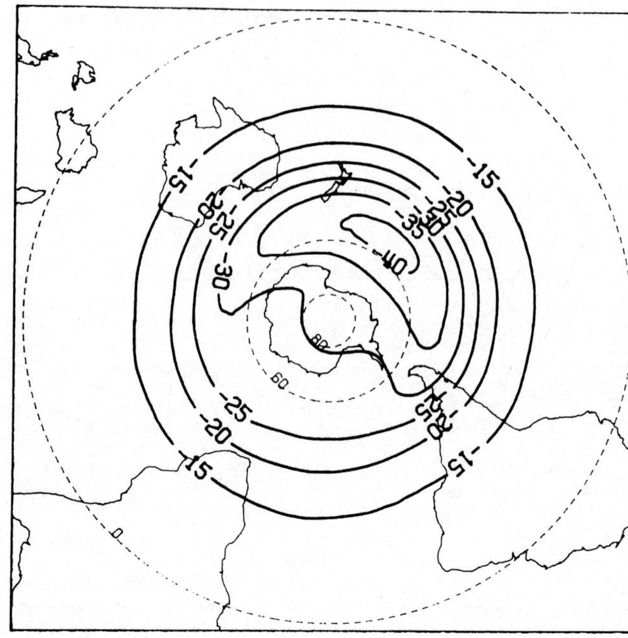

Figure IX-3-4. July mean 2-mb Southern Hemisphere temperature, 1978-85.

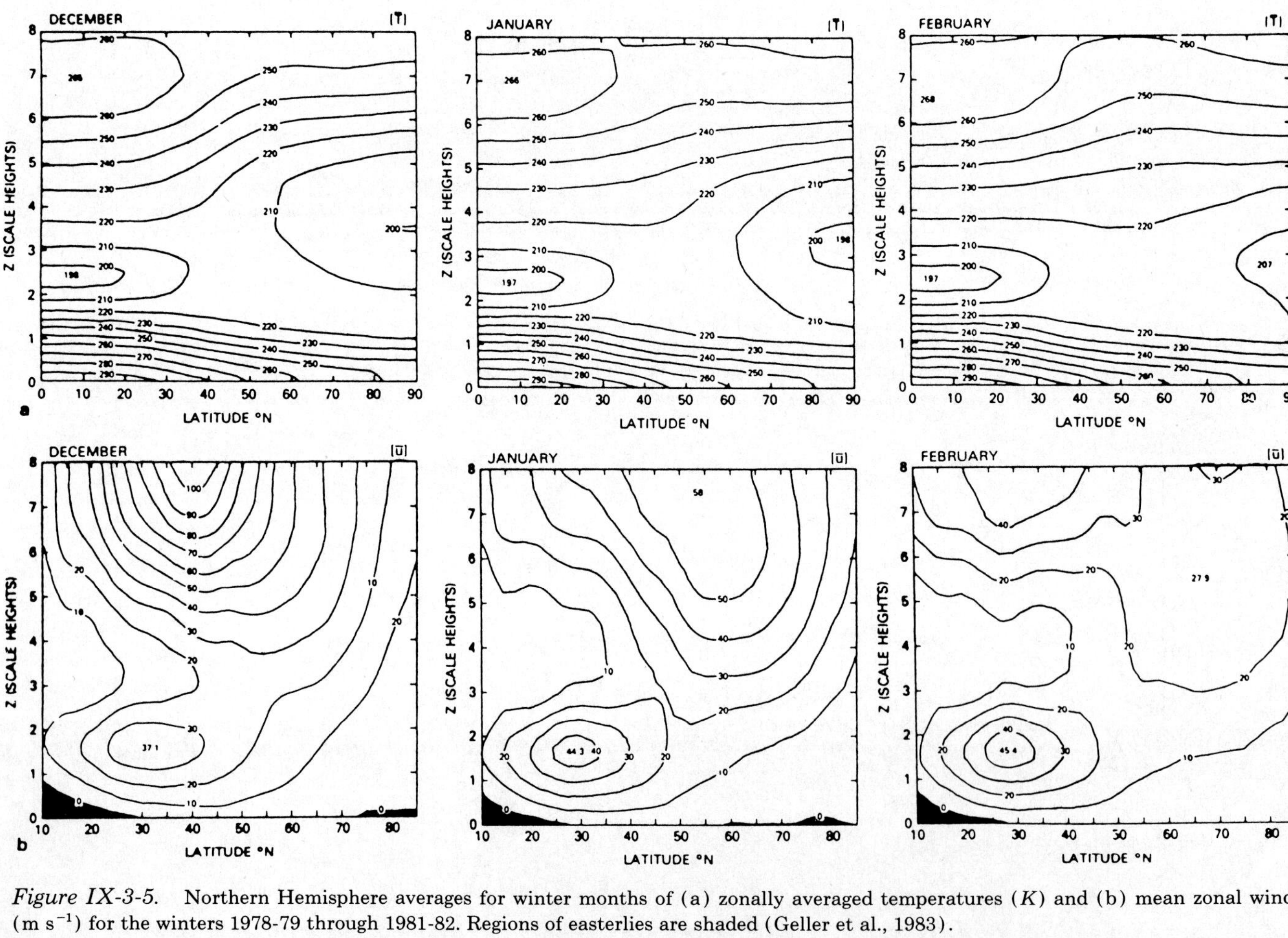

Figure IX-3-5. Northern Hemisphere averages for winter months of (a) zonally averaged temperatures (K) and (b) mean zonal winds (m s^{-1}) for the winters 1978-79 through 1981-82. Regions of easterlies are shaded (Geller et al., 1983).

ward transport of heat and associated upward vertical flux of energy. Horizontally, wavenumber 1 has a southwest-northeast tilt, which indicates a poleward transport of eddy momentum. Wavenumber 2 has similar, but less pronounced characteristics. Barnett (1975) has pointed out that these waves extend across the Equator in the equinoctial seasons, a feature previously unnoticed.

IX-3.2.5 Diagnostic Studies of Heat, Momentum, and Energy Budgets

With satellite data, the mean zonal and temporal structures of stratospheric disturbances can be defined and their slow, large-scale variations can be resolved. The general circulation in this region can be defined largely in terms of radiative forcing and dynamical interaction between large-scale, slowly varying waves and the mean zonal flow. At some point higher up in the middle atmosphere, small-scale eddies are believed to become as important as large-scale eddies in determining the mean zonal climatology.

These ideas can be tested by studying the heat, momentum, and energy budgets through studies of those phenomena resolved by synoptic analyses derived from satellite data. One particular procedure is to compare the mean meridional circulations obtained as residuals in the heat and momentum balances. Hartmann (1976) used Selective Chopper Radiometer (SCR) data to show that these two independently derived meridional circulations are qualitatively different in the upper stratosphere. Subsequent analyses of PMR data by Crane et al. (1980) and LRIR data by Gille et al. (1981) also showed large differences in the upper stratosphere and mesosphere between mean meridional circulations derived independently from the zonal mean heat and momentum balances. These differences are too large to be explained by errors on scales resolved by the satellite data and suggest that phenomena at other scales not resolved by the data are important. It is suggested that unresolved phenomena are already of major importance at 50-km. The inconsistencies in the large scale budgets can be rectified by postulating unresolved momentum sources or sinks which produce zonal flow decelerations with height. Estimates of the effect of momentum carried upward by unresolved gravity waves seem sufficient to account for these decelerations (Lindzen, 1981).

IX-3.3 SATELLITE OZONE MEASUREMENTS

For ozone, the focus is on two elements, the total integrated column amount and the vertical distribution. The former is of importance because of its influence on solar radiation reaching the Earth and its relationship to tropospheric weather patterns. The latter is significant because of its sensitivity to possible human influences and could serve as a means for providing early warnings of major changes in ozone.

Data on both of these elements are acquired by three types of measurements:

- atmospheric emission at 9.6 μm
- backscattered UV radiation
- solar occultation in the VIS Chappuis bands

Satellite ozone experiments are listed in Table IX-3-3, and the periods of operation are shown in Figures IX-3-6 and IX-3-7. The satellite ozone record be-

Table IX-3-3 *Satellite remote sensing instruments used for ozone derivation*

TOTAL OZONE			
IRIS	Infrared Interferometer Spectrometer	9.6 mm	Prabhakara et al., 1973
MFR	Multi-Filter Radiometer	9.6 mm	Lovill, 1974
TOVS	Tiros Operational Vertical Sounder	9.6 mm	Planet et al., 1984
BUV	Backscattered Ultraviolet	BUV	Mateer et al., 1971
SBUV	Solar Backscattered Ultraviolet	BUV	Heath et al., 1975
SBUV/2	Solar Backscattered Ultraviolet/2	BUV	Planet (personal communication)
TOMS	Total Ozone Mapping System	BUV	Heath et al., 1975
OZONE VERTICAL PROFILES			
BUV	Backscattered Ultraviolet (25–55 km)	BUV	London et al., 1977
BUV	Backscattered Ultraviolet (25–55 km)	BUV	Heath et al., 1973
SBUV	Solar Backscattered Ultraviolet (25–55 km)	BUV	Schneider et al., 1981
SBUV/2	Solar Backscattered Ultraviolet (25–55 km)	BUV	Planet (personal communication)
LRIR	Limb Radiance Inversion Radiometer (15–60 km)	9.6 mm	Gille and House, 1971
LIMS	Limb Infrared Monitor of the Stratosphere (16–60 km)	9.6 mm	Gille and Russell, 1984
SAGE	Stratospheric Aerosol and Gas Experiment (16–60 km)	Solar Occult	McCormick et al., 1984
SAGE/2	Stratospheric Aerosol and Gas Experiment (16–60 km)	Solar Occult	McCormick et al., 1984

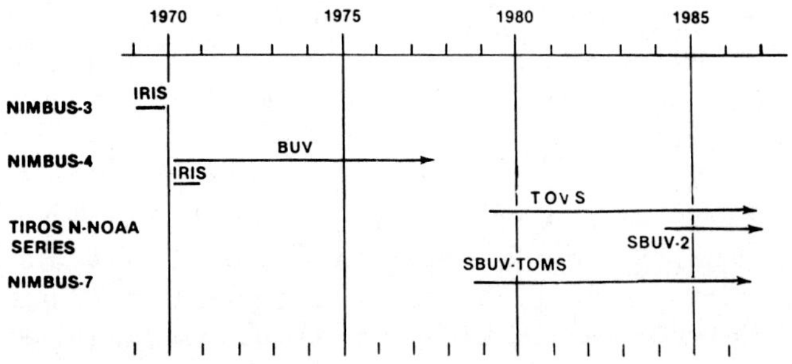

Figure IX-3-6. Schedule of satellites and measurements systems for total ozone.

gan in the late 1960s and will extend to the future with the incorporation of the SBUV/2 instrument on NOAA operational spacecraft.

A global climatology of total ozone has been derived from satellite data by Hilsenrath and Schlesinger (1981) and by Heath et al. (1982). Miller et al. (1979; 1982) list additional information gained from comparing satellite observations with ground-based coverage.

More recently, the emphasis has been on time variations of total atmospheric ozone over the globe. Figure IX-3-8 shows the monthly average total ozone (60°N to 60°S) obtained from the operational TOVS system. In the period from 1979 through 1981, the ozone values were generally higher than in 1982-83. In the winter of 1982-83, ozone values dropped to their lowest levels. This period of "ozone hole" is being examined in great detail and may be related to two events. The first is the volcanic eruption of El Chichón in April 1982 that spewed large amounts of material into the stratosphere (McCormick and Swissler, 1983); the second is the El Niño event of 1982-83 that was accompanied by large scale atmospheric circulation changes. Quiroz (1983)

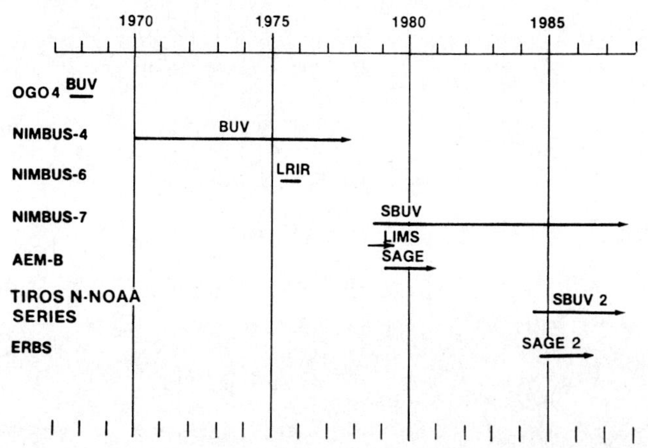

Figure IX-3-7. Schedule of satellites and measurement systems for ozone vertical profiles.

has shown the difficulty of separating the stratospheric temperature signals from these two events. As the record is updated it will be interesting to see if the ozone values return to their previous levels.

Comparisons of satellite and ground-based vertical profiles of ozone in the stratosphere have been presented by London et al. (1977) for OGO-4, Bhartia et al. (1984) for SBUV, McCormick et al. (1984) for SAGE, and Remsberg et al. (1984) for LIMS. Profile climatology has been presented by McPeters et al. (1984) and in map form by Nagatani and Miller (1984). Figure IX-3-9, an example of the latter, shows that strong longitudinal and latitudinal structures are present even in the monthly average.

Most important in observations of ozone at various levels will be the documentation of climatic variations and trends in ozone over seasons and years. As more is learned about these variations, it will become more feasible to separate natural variations from some of the man made alterations in the ozone distribution. The SBUV/2 instrument, now part of the NOAA operational satellite system (Figures IX-3-6 and IX-3-7), will provide routine monitoring of ozone so that further analysis of its variations can be carried out.

IX-3.4 STRATOSPHERIC AEROSOLS

The alterations of incoming solar radiation and outgoing longwave radiation by stratospheric aerosols (Chapter VIII-1) result in long term implications for climate change, especially after volcanic eruptions. The eruption of Tambora in 1815 in the Dutch East Indies was followed by what Stommel and Stommel (1979) called "the year without a summer" for Europe, New England, and other parts of the Earth. Krakatoa, in 1883, spewed ash to a height of 30 miles, plunging the surrounding region into darkness for 2½ days (Toon and Pollack, 1977). Mt. Agung in 1963 and El Chichón in 1982 were the latest eruptions to have significant impacts on the stratosphere. The full effect of volcanic eruptions on climate is still uncertain, as shown by Angell and Korshover (1984) who compared stratospheric and tropospheric temperatures after these most recent eruptions. The stratosphere showed a global warming after each eruption, but the troposphere showed a cooling after Mt. Agung and a warming after El Chichón. The warming after El Chichón probably was a result of the El Niño event that occurred at the same time. However, the effects of El Niño could not be separated from those of El Chichón.

Of recent public interest is the possibility of a nuclear winter following massive forest fires and firestorms set off by a large number of nuclear explosions. The dust

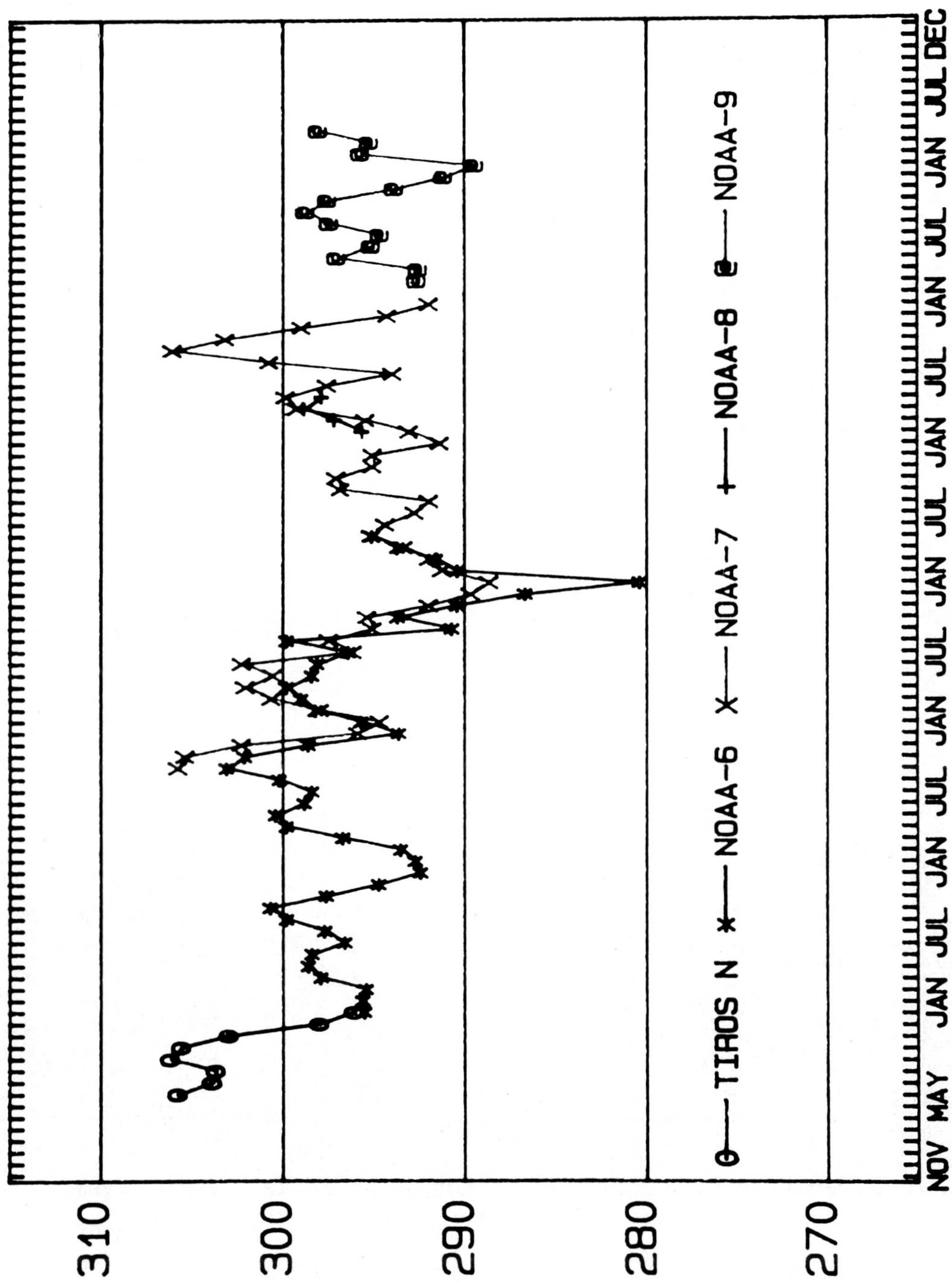

Figure IX-3-8. Time sequence of global (60°N to 60°S) monthly average total ozone as determined from TOVS system, 1979-1986.

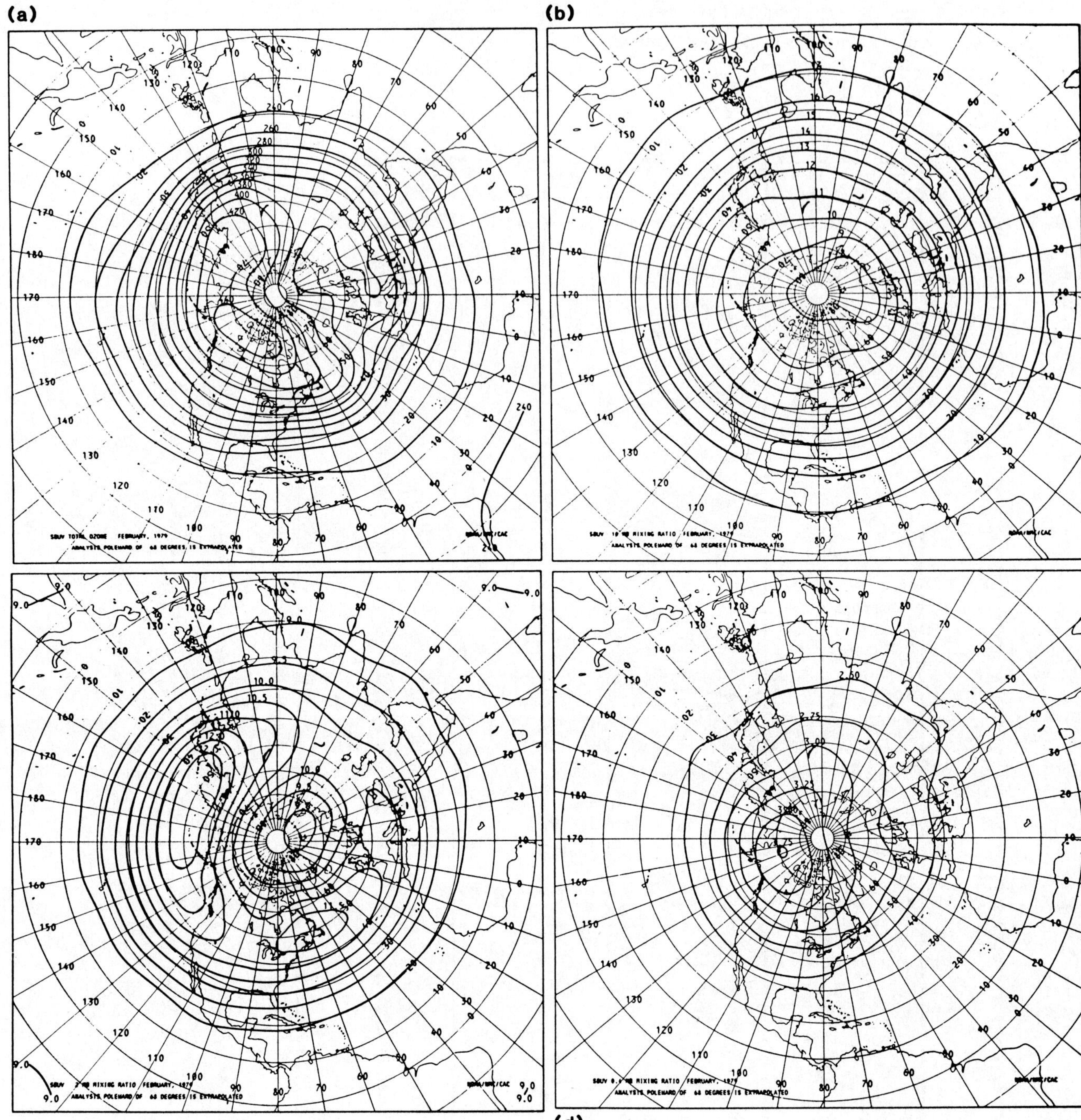

Figure IX-3-9. February 1979 monthly average patterns of (a) total ozone, (b) ozone mixing ratio ($\mu g \cdot g^{-1}$) at 10 mb, (c) 2 mb, and (d) 0.4 mb.

and soot created could remain suspended in the stratosphere, blocking out the Sun's rays, and lead to a nuclear winter (Turco et al., 1983; NAS, 1984).

Stratospheric aerosols have been the major focus of a series of solar occultation experiments by satellites since 1975 (McCormick et al., 1979). The first measurements were made by a hand held Sun photometer, the Stratospheric Aerosol Measurement (SAM) instrument, during the Apollo-Soyuz Test Project. These Apollo-Soyuz measurements of the solar intensity were made in a band centered at 0.83 μm (Pepin et al., 1977). Following this successful demonstration of the feasibility of such measurements, SAM/2 was flown on Nimbus-7 in 1978 and the SAGE (Chapter I-2) was flown on the Applications Explorer Mission satellite in 1979. The SAM/2 single channel Sun photometer measured the solar intensity at 1 μm for measurements of aerosol and molecular scattering. The SAGE radiometer yielded aerosol extinction profiles from channels at 0.45 μm and 1.0 μm and profiles of nitrogen dioxide and ozone from the 0.385 μm and 0.60 μm channels. Measurements from SAM and SAGE are taken at orbital sunrise and sunset (about 30 per day), but because of different orbits, SAM/2 takes measurements only at latitudes 64° to 80° (sunset in the Northern Hemisphere and sunrise in the Southern Hemisphere), while the SAGE instrument takes sunrise and sunset measurements between 79°S and 79°N. A schematic of a sunrise or sunset event is shown in Figure IX-3-10.

SAGE/2, launched from the space shuttle in 1984, measures solar intensity in seven bands between 0.385 μm and 1.02 μm. In addition to aerosol, ozone, and nitrogen dioxide measurements, SAGE/2 also measures WV at 0.94 μm.

A comparison of SAM/2 data with dustsonde and lidar measurements by Russell et al. (1981) showed that the extinction profiles agreed within measurement and conversion uncertainties. Figure IX-3-11 shows these comparisons and their respective standard deviation error bars.

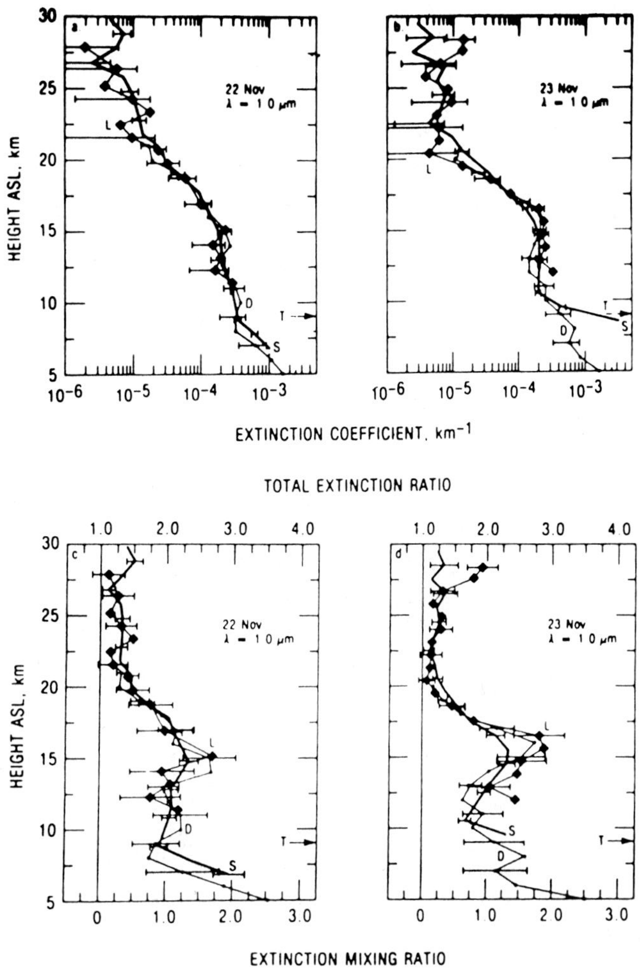

Figure IX-3-11. Comparison of extinction coefficients and ratios derived from SAM/2 (*S*), dustsonde (*D*), and lidar (*L*) measurements (Russell et al., 1981).

McCormick et al. (1981), in summarizing over 10,000 aerosol profiles in the first year of the SAM/2 operation, found an increased extinction in the winter hemisphere at 20 to 25 km which they attributed to more polar stratospheric clouds in winter. These polar stratospheric cloud sightings by SAM/2 led to a workshop on these clouds (Hamill and McMaster, 1984). One conclusion of this workshop was that while polar stratospheric clouds can cause an increase in cooling of several Wm^{-2}, neither the planetary radiation balance nor any dynamical process would be significantly affected. These clouds may, however, affect retrieval of other data by satellite instruments.

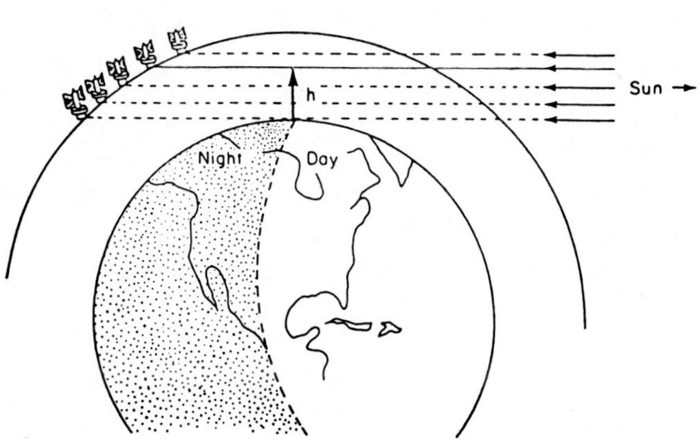

Figure IX-3-10. Geometry of Earth, Sun, and satellite during solar extinction measurements. Different layers of the atmosphere at tangent heights *h* are successively sampled during each satellite sunrise or sunset event (McCormick et al., 1979).

In April 1979, the Soufrière volcano on St. Vincent erupted several times. Although volcanic material reaching the stratosphere did not significantly affect temperature or climate, SAGE measured the volcanic debris. Figure IX-3-12 shows a normal sounding and the one affected by the volcanic debris. The total mass of the stratospheric ejecta from Soufrière measured by

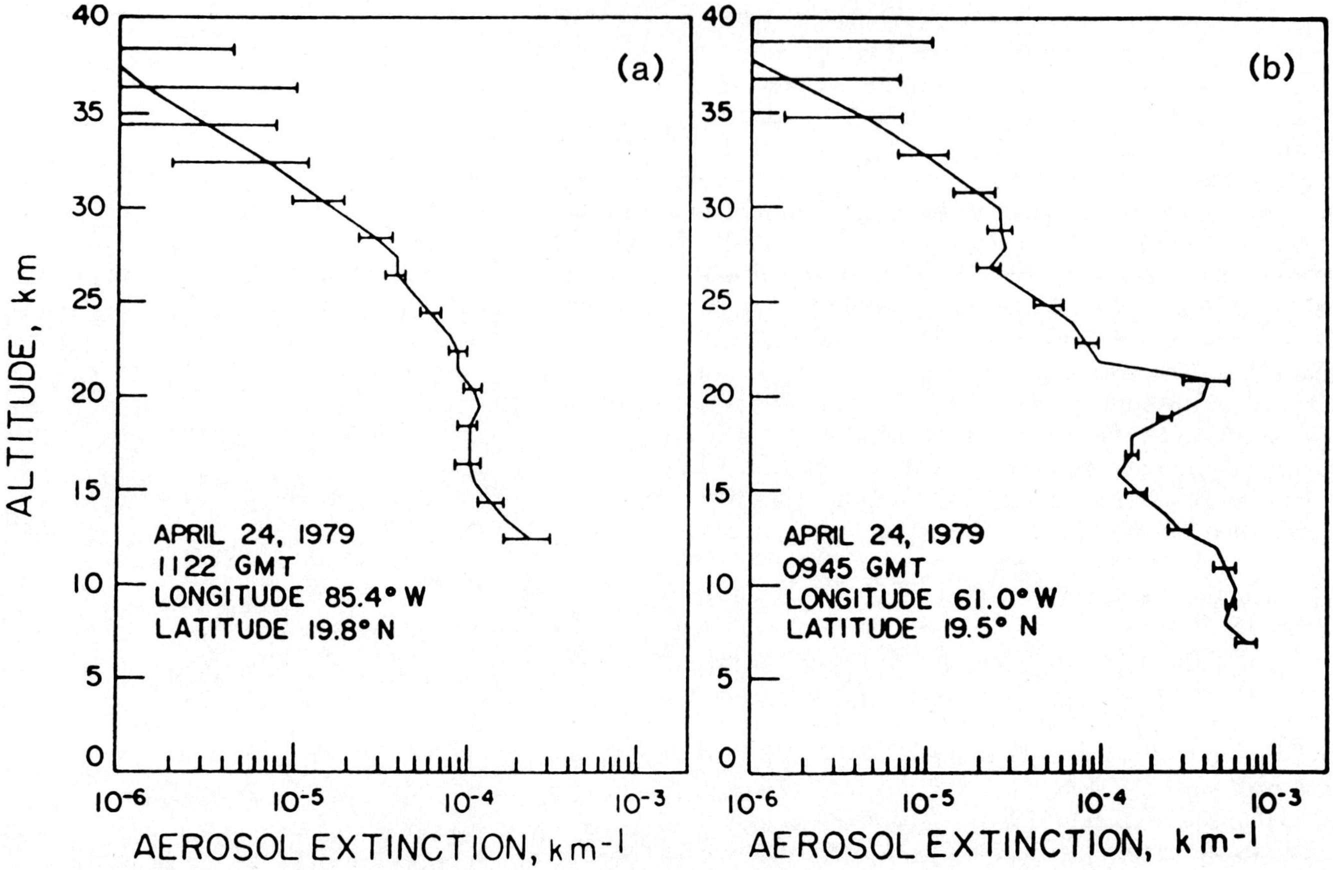

Figure IX-3-12. (a) Normal aerosol extinction profile as determined by SAGE satellite system; (b) enhanced aerosol profile (McCormick et al., 1981).

SAGE was less than 0.5% of the global stratospheric aerosol content. More significant was the El Chichón eruption in late March and early April, 1982. Unfortunately, neither SAGE nor SAM/2 sampled the main part of the stratospheric debris. However, McCormick et al. (1983) noted that SAM/2 measurements showed El Chichón debris did reach high latitudes and that most was confined to levels below 18 km. Using data from a lidar flight mission, McCormick and Swissler (1983) computed the stratospheric aerosol mass after the El Chichón eruption at 12 megatons compared to a normal background state of 0.5 megatons.

IX-3.5 STRATOSPHERIC TRACE CONSTITUENTS

Trace constituents other than aerosols are also measured by satellites. The LIMS experiment on Nimbus-7 obtained profiles of water vapor, nitrogen dioxide, and nitric acid in addition to ozone (Gille and Russell, 1984). At higher levels, the Solar Mesosphere Explorer (SME) takes measurements of ozone, water vapor, and nitrogen dioxide at mesospheric levels (Thomas et al., 1980). These trace constituents alter the radiation balance, change the ozone concentration, and can lead to long term changes in the stratosphere.

Chapter IX-4

USE OF SATELLITE DATA IN CLIMATE ANALYSIS

IX-4.1 INTRODUCTION

The Climate Analysis Center (CAC), a unit of the National Weather Service, is responsible for monitoring, analyzing, and predicting climate variations for periods from a few weeks to a few years. The monitoring and prediction functions are real time operations, while the analysis or diagnosis and improvement of monitoring and prediction methods are carried on in a developmental or research mode. The principal goals of monitoring and prediction are to identify, trace, and forecast important anomalies in various elements of the climate of the Earth-atmosphere system. Of major priority are those elements that directly or indirectly affect important socioeconomic functions. A good example of a major large-scale climate variation having major, widespread effects is the El Niño/Southern Oscillation (ENSO) phenomenon (Rasmusson and Wallace, 1983). ENSO episodes occur at irregular intervals of 2 to 10 years and last a year or two. They constitute the greater part of the interannual variability at many tropical locations and have been associated with substantial climatic anomalies outside the tropics as well. While the phenomenon is not yet understood well enough to be truly predictable, individual ENSO events in many respects follow a similar course (Rasmusson and Carpenter, 1982). Thus, it is of considerable importance for prediction and impact assessment to know the current and recent status of ENSO related climatic parameters.

Analysis of such large-scale climate variations requires a high quality data base that describes the behavior of the global climate system at time and space scales appropriate to the problem. At present, those scales are taken to be larger than about 2 weeks and about 6×10^4 km^2. Satellite data contribute in two important ways to the global CAC analysis effort. The space and time coverage of satellite observations is exceptionally complete in comparison to that from most other sources. The ability of satellites to observe areas where conventional data are unobtainable and to provide denser temporal and spatial sampling in rather inaccessible areas is very valuable. The other important benefit of satellites in the CAC analysis effort is that they often provide independent data sources that can be used to validate other data. Many of the data used by CAC are obtained from large area objective analyses such as the NMC Global Data Assimilation System (GDAS). Subtle biases in the GDAS, often more significant from the perspective of climate than of weather, can be difficult to discover. Fortunately, some forms of satellite data, such as outgoing longwave radiation (OLR), are not used in GDAS and so provide an independent check on anomalies in the routine analyses.

Satellite data also have some disadvantages. Data used for climatic applications rely to a considerable extent on the historical context of the data. Thus, satellite data with their short-term records (less than 25 years), sometimes discontinuous records, and the changing nature of instrumentation, are difficult to organize into consistent sets. Despite these problems, satellite data have proven especially useful in several aspects of climate analysis. The applications covered in this chapter are the determination and analysis of global snow cover and sea ice, analysis of sea surface temperature, and the determination of large scale precipitation.

IX-4.2 GLOBAL SNOW COVER AND SEA ICE EXTENT

IX-4.2.1 Introduction

Snow cover and sea ice extent have potential to influence global climate through their effects on surface roughness and on temperature and moisture near the surface. These effects alter the momentum and heat transfer in the lowest layers of the atmosphere and the upper levels of the ocean. The high reflectivities of snow and sea ice decrease the solar energy absorbed at the surface. On the other hand, both snow and sea ice have low thermal conductivities and can radiate at considerably different rates from the underlying land or water surfaces whose temperatures may be quite different.

While variations of snow cover and sea ice associated with geologic scale climate changes, such as the ice ages, are well documented, the relationships among the snow-ice-atmosphere-ocean components of the climate system are not so clear on shorter time scales. Recently, however, with the increasing availability of satellite estimates of snow cover and sea ice extent, new diagnostic studies have been carried out on monthly to seasonal time scales at spatial scales ranging from continental to global.

IX-4.2.2 Datasets

The weekly satellite snow cover analyses for the Northern Hemisphere prepared by the Satellite Analysis Branch of NESDIS and the monthly values derived from them are described in Chapter VIII-2. The monthly digitized data and the values of areal snow cover for North America, Eurasia, and the whole Northern Hemisphere have been especially useful in assessing both the month to month and year to year variations in snow cover. Since the dataset now covers a period of more than 20 years, it provides good estimates of climatic mean snow cover and its variations and trends.

Hand analyzed charts of sea ice extent, concentration, and thickness for the Arctic and the Antarctic have been produced weekly since January 1973 by the Joint (U.S. Navy-NOAA) Ice Center. These analyses are based on both VIS imagery and passive MW observations, and so are not restricted only to "cloud-free" observations. Weekly data are not currently being digitized as received, but monthly areal estimates are produced by CAC. The satellite record since 1973 permits computation of a base period mean that gives good estimates of the annual variation of ice in the polar oceans. The dataset also allows for examination of interannual variations over this period and on a current basis.

IX-4.2.3 Climate Studies of Satellite Observed Snow Cover and Sea Ice

Summaries of the variations in Northern Hemisphere snow cover over space and time have been made by Morse and Ropelewski (1983) and Matson et al. (1986). One example, the time variation of winter season snow cover over Eurasia, is shown in Figure VIII-2-5. An interesting aspect of these data is the apparent relationship between the mean Eurasian snow cover area in December-March and subsequent Indian summer (June-September) monsoon rainfall (Hahn and Shukla, 1976; Dickson, 1984). Figure IX-4-1 shows that an excess (deficiency) of snow cover over all of Eurasia during winter is frequently associated with a weaker (stronger) than normal monsoon over India the following summer. The physical mechanism for such a relationship is probably related to retarded or accelerated heating of the Asian continent in spring, but many aspects of the physical relationship are unclear. Also, it is not clear if this statistical relationship will remain stable in the future.

Satellite observations of sea ice show that the annual cycles for the Arctic and Antarctic are quite different (Figure IX-4-2). The annual variation of Arctic sea ice is limited because the Arctic Ocean is basically landlocked. Thus the annual range of fluctuation in the Arctic (maximum/minimum = 2.1) is much smaller than in the Antarctic (maximum/minimum = 5.0). Regional variations of Arctic sea ice may be more sensitive indicators of large-scale climate variations in regions such as the Davis Strait and the Greenland Sea where the ice is free to grow.

The annual cycle of Antarctic sea ice in the South Indian Ocean and Pacific Ocean sectors is more complex in structure than the total Antarctic sea ice. Sea ice in the Ross Sea and adjacent areas shows a tendency for two maxima in most years of the satellite record.

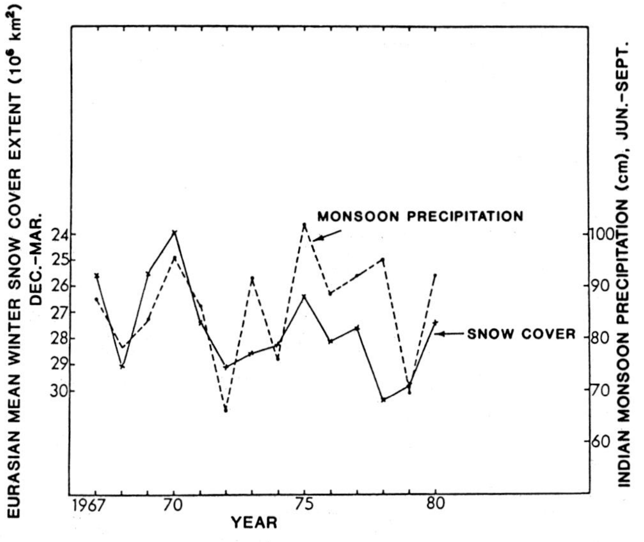

Figure IX-4-1. Eurasian winter (December-March) snow cover versus Indian monsoon (June-September) precipitation (Dickson, 1984).

It has been hypothesized (Ropelewski, 1983) that the secondary maximum of sea ice area in these sectors during October (the primary maximum occurs in July-August) is a response to wind forcing by the atmosphere.

IX-4.3 SEA SURFACE TEMPERATURE

Sea Surface Temperature (SST) plays a key role in the coupling between ocean and atmosphere. Namias (1959, 1976) argued convincingly that the evolution of SST anomalies is important for long range weather forecasting. Furthermore, recent studies of the ENSO (Rasmusson and Carpenter, 1982) have demonstrated the important climatic role of large-scale changes of tropical SST anomalies. Numerous modeling studies of the response of the atmosphere to SST anomalies suggest that SST comprise a significant source of air-sea interaction that may produce important tropical and extratropical climate anomalies (Shukla and Wallace, 1983).

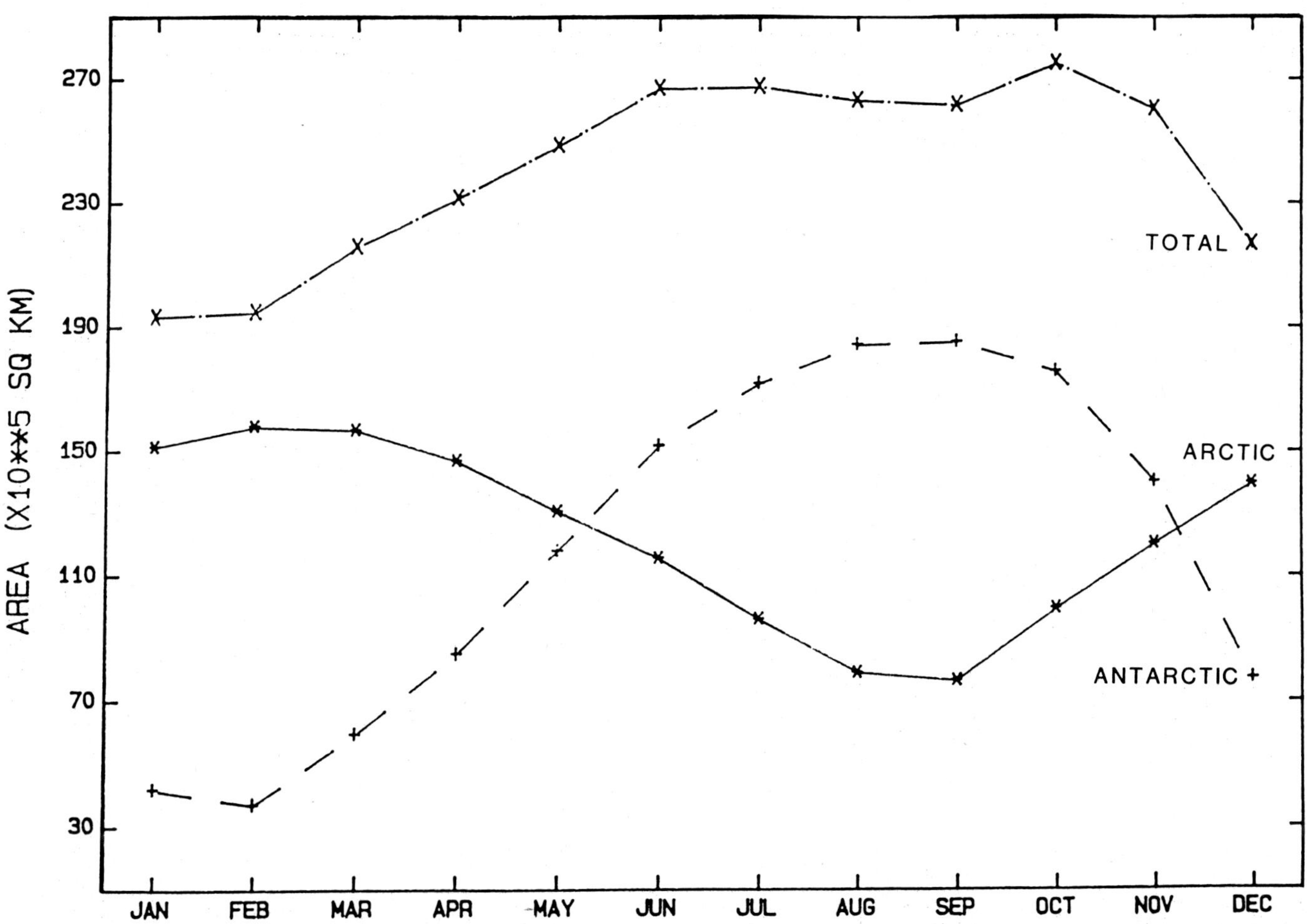

Figure IX-4-2. The annual cycle of Arctic, Antarctic, and total sea ice derived from the 11-year (1973-83) base period. Note that total sea ice tends to follow the Southern Hemisphere cycle, but with a much smaller relative change in amplitude through the year. The total global sea ice varies between 4% and 5% of the total area of the globe.

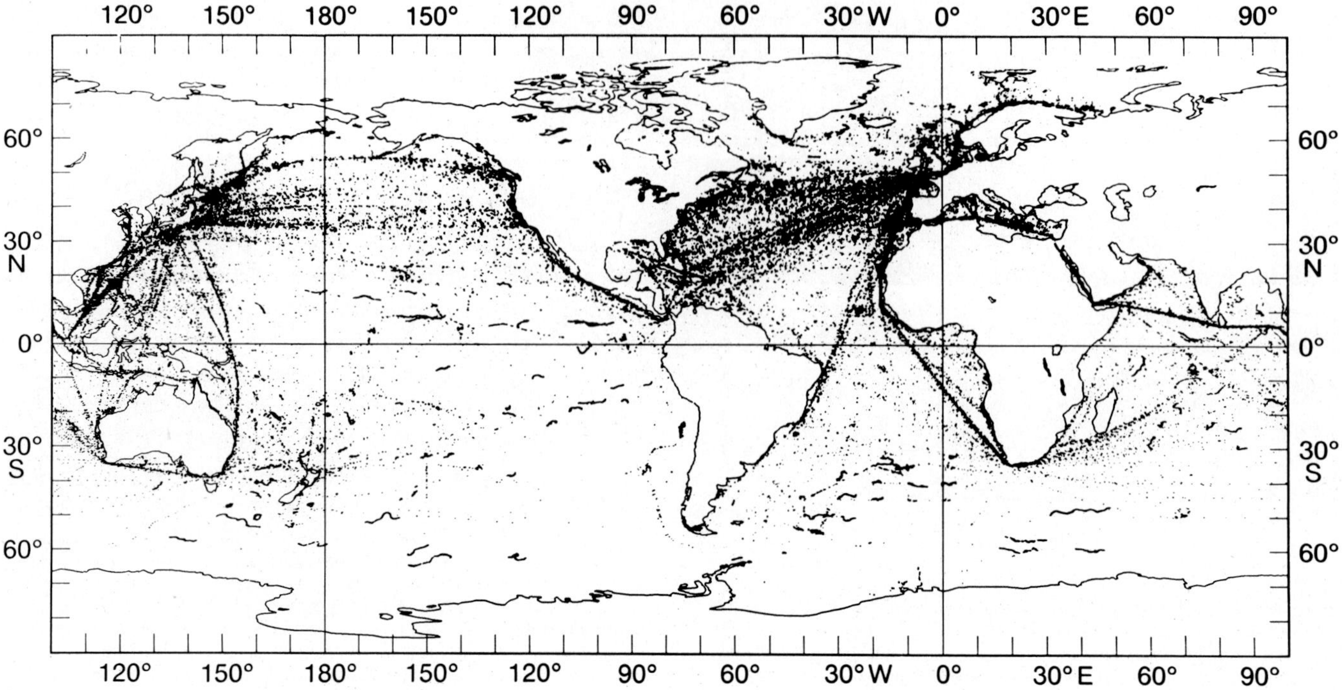

Figure IX-4-3. Distribution of *in situ* (ship and buoy) data received over the Global Telecommunication System for October 1986.

Ships and drifting and fixed buoys furnish approximately 100,000-120,000 SST observations per month. The distribution of observations received through the Global Telecommunications System during October 1986 is shown in Figure IX-4-3. These data are generally adequate to describe the SST patterns between 60°N and 30°S except over the central and eastern Pacific in the tropics and south of the Equator. Un-

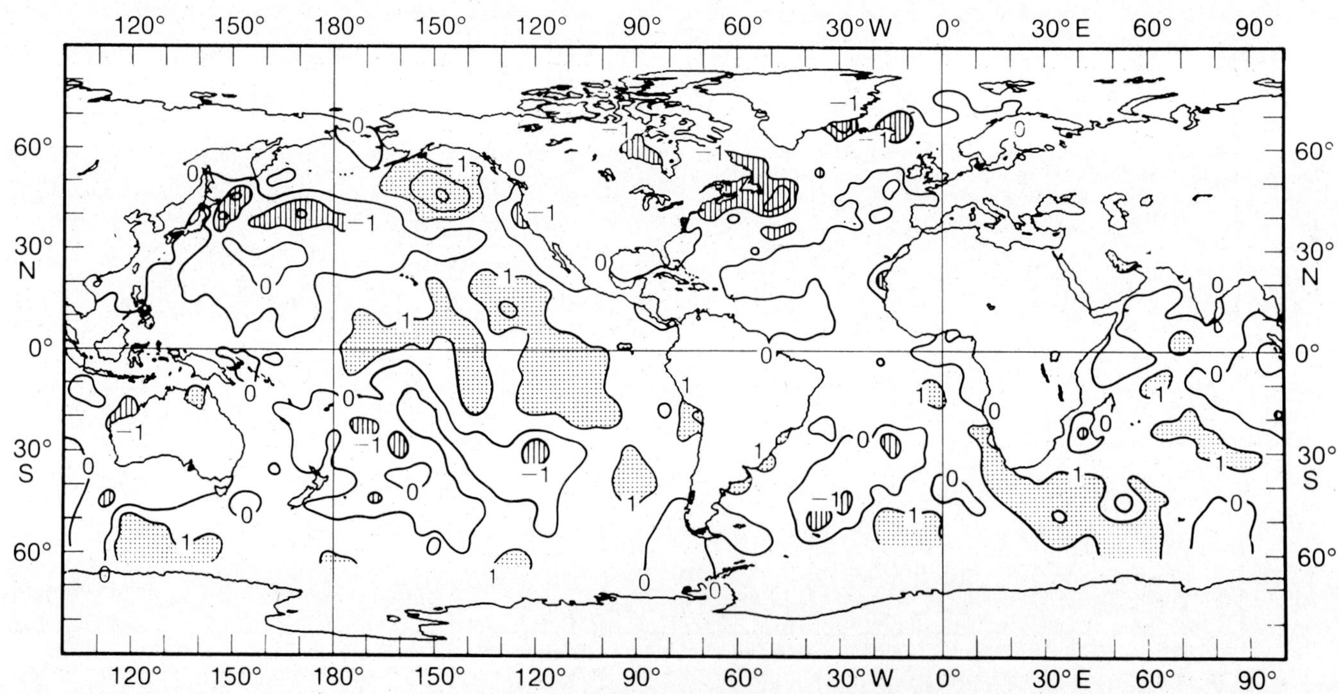

Figure IX-4-4. *In situ* sea surface temperature anomalies for October 1986, on a 2° lat.-long. grid. Anomalies are computed from the climatology of Reynolds (1988). Contour interval 0.5°C between ±°C; otherwise contour interval 1°C; values less than −1°C are shaded; values greater than 1°C are stippled.

fortunately the tropical Pacific is of extreme climatic importance because large-scale climate responses occur with positive anomalies during ENSO events and negative anomalies during "anti-ENSO" events. Individual ship observations are subject to large errors in both temperature and position. These *in situ* data must be screened using climatology and then analyzed with a nonlinear spatial filter based on median values (Rabiner et al., 1975). The use of medians results in the objective elimination of extreme values rather than smoothing their effects into surrounding grid values with weighted means. Use of this technique (Figure IX-4-4) eliminates smaller scale (less than 500-km) features. Although the larger scale details are generally accurate, much of the smaller scale variability portrayed over the tropical and South Pacific is unrealistic because of limited *in situ* data.

Satellite data can be used to significantly improve this analysis. Figure IX-4-5 shows the number of monthly satellite SST retrievals over the globe from 1982 to 1984. These retrievals are obtained from AVHRR data using the multichannel techniques described in Chapters VI-2 and VIII-4. Generally the satellite SST are much more numerous and more uniform in spacing than the *in situ* observations. The number of retrievals varies from 200,000 to almost 3 million per month. These variations result from satellite hardware problems, interpretation difficulties because of cloud cover, and, occasionally, high-level aerosols.

Figure IX-4-5. Number of global satellite AVHRR SST retrievals per month between 1982 and 1984.

Strong (1983) reported that aerosols introduced by the eruptions of El Chichón in March-April 1982 resulted in cold biases, at times exceeding 2°C, in the satellite SST retrievals. Figure IX-4-6 shows the satellite SST field for October 1986 (a month with a high number of retrievals).

Comparison of Figures IX-4-4 and IX-4-6 shows that the overall patterns are similar but with important dif-

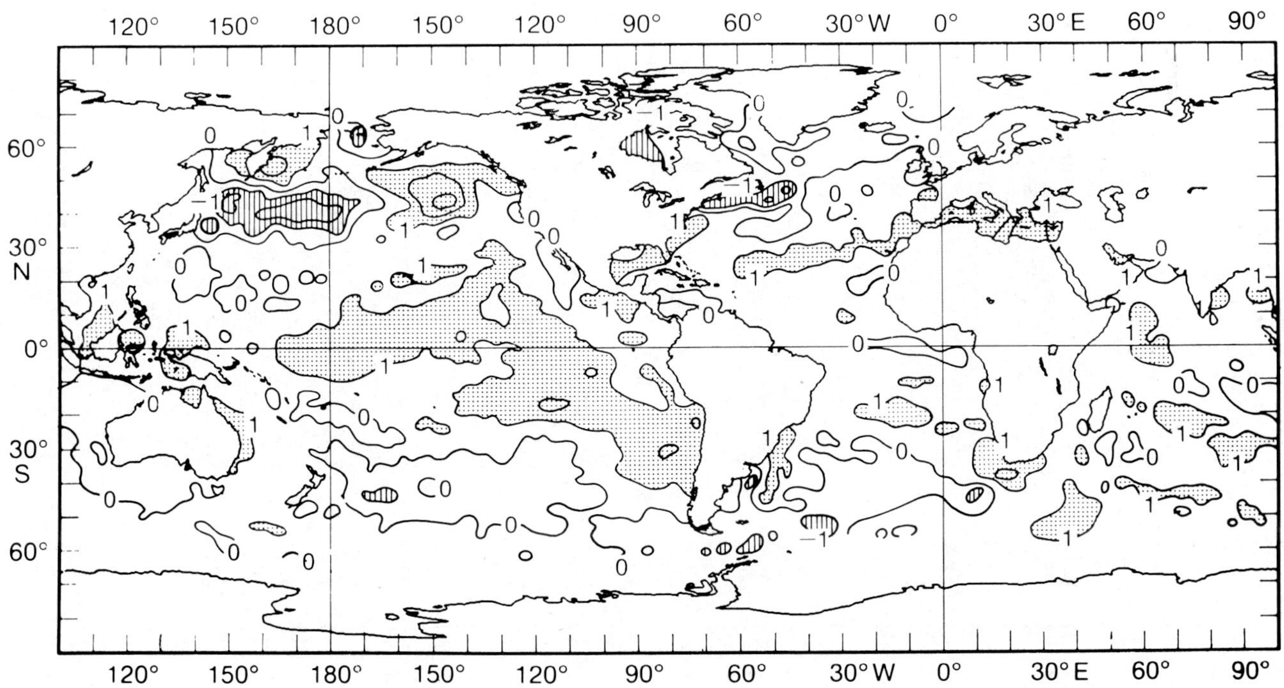

Figure IX-4-6. Satellite sea surface temperature anomalies for October 1986 (see legend of Figure IX-4-4).

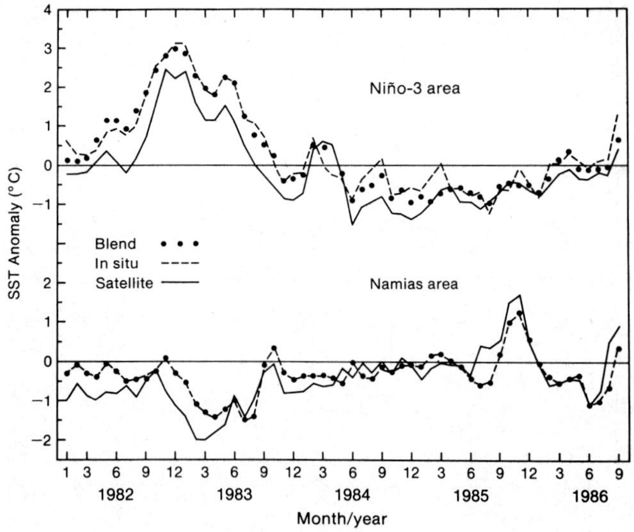

Figure IX-4-7. Time series of *in situ* and satellite and blended SST anomalies. The Pacific regions are: "Niño 3" (10°N-5°S, 150°W-90°W); "Namias" (30°N-50°N, 165°W).

ferences. The advantage of satellite data is clear in the large scale coherence over the tropical and South Pacific present in Figure IX-4-6 but not in Figure IX-4-4. There are large zonal differences in the Pacific and smaller scale differences where both types of data sources should be adequate.

Further illustrations of these differences are given in Figure IX-4-7, which shows time series of analyzed SST estimates from satellite and *in situ* observations for regions in the eastern equatorial Pacific (Niño 3) and the eastern mid-latitude North Pacific (Namias, 1976). Generally the satellite analysis is about 0.5°C lower than the *in situ* analysis. However, there are important reversals in these tendencies for several months. Figures IX-4-4, IX-4-6, and IX-4-7 demonstrate that the differences between the two analyses are dependent on time and location.

Both types of data have clear advantages. The *in situ* data offer direct measurement (i.e., groundtruth) with independence of errors in observations from different platforms; the satellite data offer expanded and more uniform data coverage. Combination of the two data sets is accomplished by a "blended" analysis method which uses *in situ* data to define "benchmark" temperature values in regions of frequent *in situ* observations, and uses the satellite data to define the "shape" of the field between these benchmarks. The resulting anomaly field is shown in Figure IX-4-8.

Comparison of the three analyses (Figures IX-4-4, IX-4-6, and IX-4-8) suggests that the blended analysis is an improvement over fields obtained solely from *in situ* data or satellite data. Large-scale biases or trends in the satellite data have little influence on the final solution. In our example, any satellite bias can affect

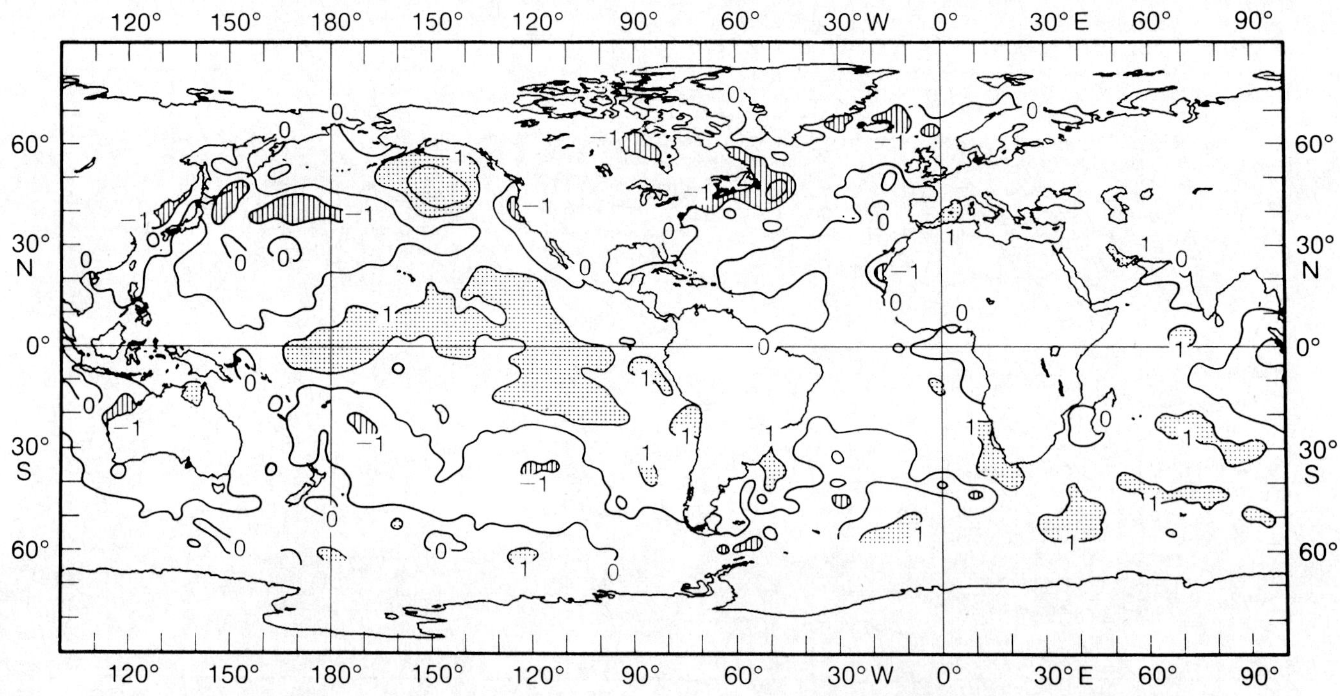

Figure IX-4-8. Blended (see text) sea surface temperature anomalies for October 1986 (see legend of Figure IX-4-4).

only the blended result south of 30°S where *in situ* data are sparse. Features in the satellite analysis, such as the coherent anomaly along the Equator in the Pacific, are maintained in the blended analysis although its size is slightly reduced by *in situ* "benchmarks." See Reynolds (1988) for further details.

IX-4.4 PRECIPITATION

IX-4.4.1 Introduction

Convective precipitation in the tropics, with its attendant release of latent heat, is one of the major forcing mechanisms of the general circulation. Large-scale anomalies in tropical precipitation are closely associated with global scale circulation anomalies (Rasmusson and Carpenter, 1982; Horel and Wallace, 1981). Careful monitoring of variations in tropical precipitation is of major importance in attempting to diagnose the global climate system. Knowledge of the actual precipitation over large areas is a major factor in recognizing climate events of societal significance. It is also potentially important for initialization of surface conditions in General Circulation Models (GCM) used to simulate the global climate (Mintz, 1981) and in analysis of divergence of the wind field used for numerical weather prediction (Julian, 1984).

Estimating average rainfall over large regions is challenging. The traditional method using rain gauges is generally inadequate for the small spatial scales of convective rainfall regimes that dominate large parts of the globe (Hudlow, 1979). A dense network of rain gauges is needed to accurately estimate areally averaged precipitation. Such networks in oceanic or remote land regions are not practical.

Remote sensing provides another method of measuring rainfall. Ground-based radars, now used routinely to detect precipitation and severe weather, produce good estimates of areally averaged rainfall (Hudlow and Patterson, 1979). As with rain gauges, their use is not possible over oceans or remote land areas.

The obvious solution is to use satellite-based sensors of either the active or passive type (Wilheit et al., 1977). An active operational system for large scale precipitation measurements is not yet practical. Thus, the use of passive sensors, in particular IR window measurements, appears to be the best approach at present to obtain precipitation estimates.

IX-4.4.2 Precipitation Inferred from Variations in Outgoing Longwave Radiation (OLR)

An objective estimation technique that has proven useful in climate analysis uses the OLR dataset from twice per day IR imagery (see Chapter IX-1). The OLR data have been available since June 1974 (March-December 1978 are missing). In the tropics, areas of low (high) OLR are strongly associated with high (low) concentrations of convective clouds. Heddinghaus and Krueger (1981) and Liebmann and Hartmann (1982) inferred variability in large-scale tropical rainfall from variations in OLR. Arkin (1984) compared variations in OLR with areally averaged rainfall from central tropical Pacific island stations. He found that about 40% of the variance in anomalous monthly mean rainfall for this region for the period June 1974-February 1978, could be explained using the anomalies in mean OLR. Lau and Chan (1983) used the number of days in a month with OLR <240 Wm^{-2} to estimate areally averaged tropical rainfall. Over large areas in the vicinity of the Indian subcontinent, Lau's quantity and the mean OLR are highly correlated, indicating that these methods are equivalent. In both cases, the estimates suffer from relatively sparse time sampling (2 per day) and from inadequate calibrations.

Anomalies in OLR have proven to be one of the most useful tools available for monitoring interannual variability in tropical convective precipitation. The OLR anomalies associated with the ENSO episode of 1982-83 were the largest observed until that time and were one of the clearest indications that a significant event was occurring. Diagnostic analyses and model simulations of this episode (Rasmusson and Wallace, 1983; Gill and Rasmusson, 1983; Arkin, 1983) successfully used OLR anomalies to infer anomalies in condensation heating in the tropics.

In Figures IX-4-9 through IX-4-11, Arkin's (1984) regression relationship between central equatorial Pacific island rainfall and OLR is used to estimate tropical rainfall anomalies during the episode. During June-August 1982, (Figure IX-4-9) rainfall was more than 200 mm above normal over much of the central tropical Pacific, mostly from 150°E to 180°. Positive anomalies of over 200 mm also extended into the eastern Pacific near 5°N. These anomaly patterns indicate a southward displacement and strengthening of the east Pacific ITCZ and a northeastward displacement of the South Pacific Convergence Zone (SPCZ). Both conditions are characteristic of ENSO episodes (Rasmusson and Carpenter, 1982; Arkin, 1982). Negative anomalies of more than 200 mm occurred over Indonesia and Australia.

By December 1982-February 1983, (Figure IX-4-10) an area of very strong positive anomalies near the Equator extended from just east of the dateline to about 100°W. Estimates of 1600 mm greater than normal were centered just south of the Equator near 140°W.

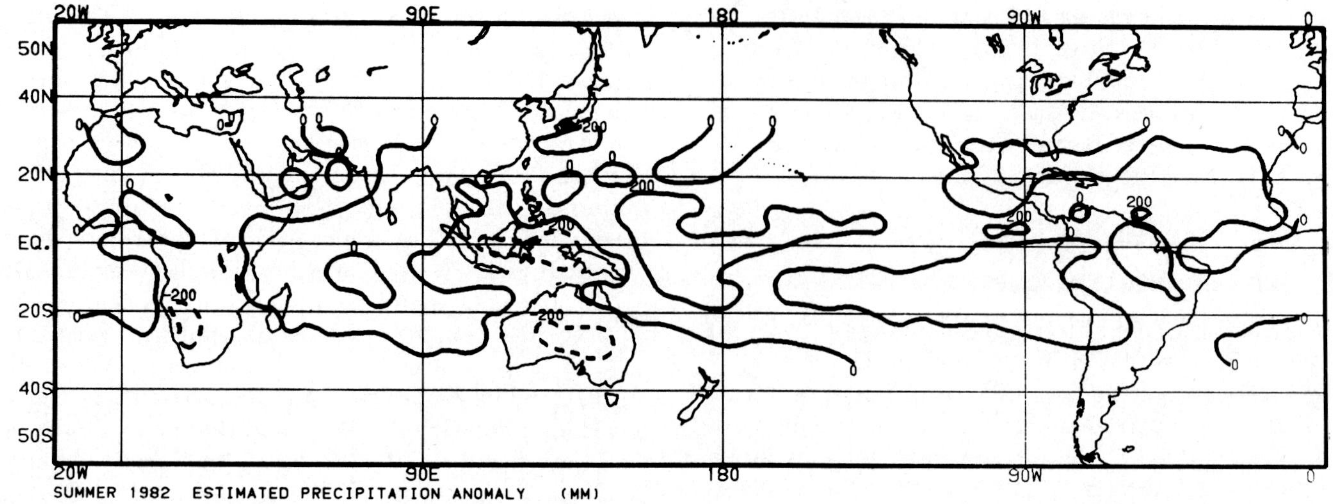

Figure IX-4-9. Precipitation anomaly (mm) estimates from anomalies in OLR for June-August 1982. Contour interval is 200 mm, with negative contours dashed.

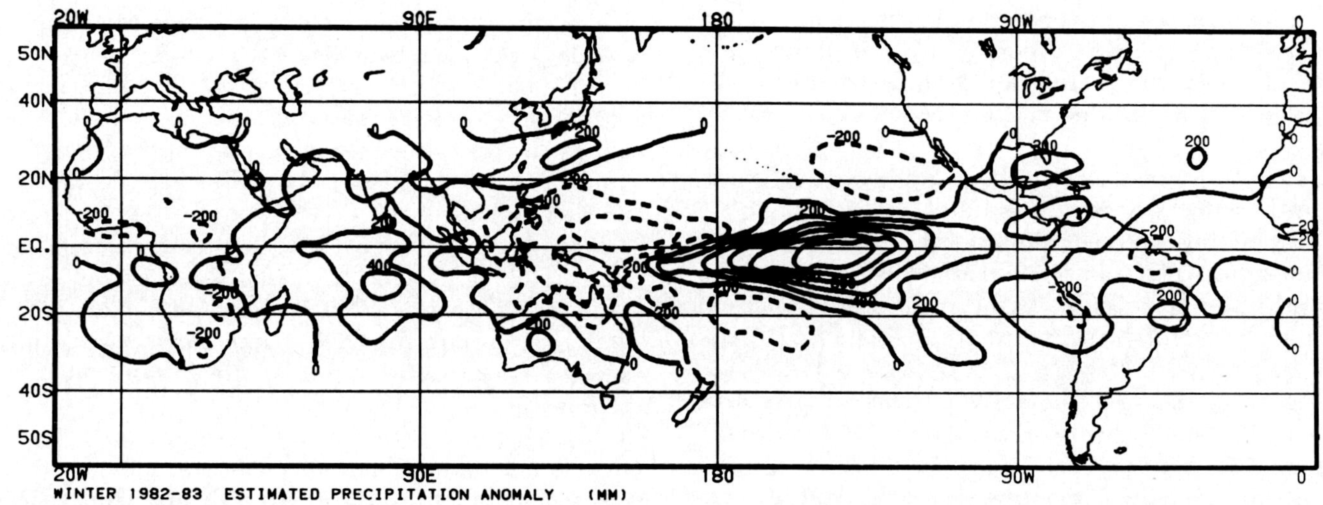

Figure IX-4-10. Precipitation anomaly (mm) estimates from anomalies in OLR for December 1982-February 1983 (see legend of Figure IX-4-9).

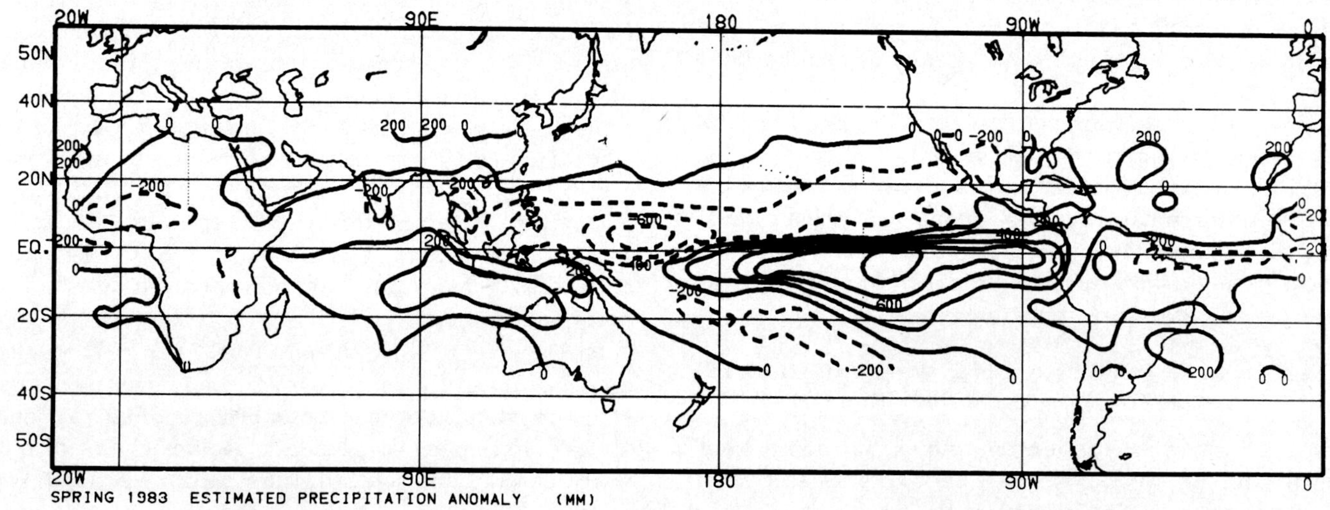

Figure IX-4-11. Precipitation anomaly (mm) estimates from anomalies in OLR for March-May 1983 (see legend of Figure IX-4-9).

Negative anomalies of as much as 600 mm occurred in several regions in the west Pacific. Dryer than normal conditions were indicated in eastern equatorial Brazil and the equatorial Atlantic.

During March-May 1983 (Figure IX-4-11), very large positive anomalies, up to 1000 mm, occurred between the dateline and South America centered near 5°S. Extensive negative anomalies extended from north of the Equator in the Western Pacific all the way across the North Pacific to Mexico. Prominent negative anomalies also occurred in the normal location of the SPCZ and along the Equator in the Atlantic.

These estimated rainfall anomalies contain substantial uncertainties. Even in the dependent data the correlation between station rainfall anomalies and OLR anomalies was only 0.65. It is likely, however, that the estimates, particularly those in the tropical Pacific, are representative of what actually occurred. It need hardly be pointed out that there would have been very much less information about the distribution of tropical rainfall anomalies during this episode if satellite data were unavailable.

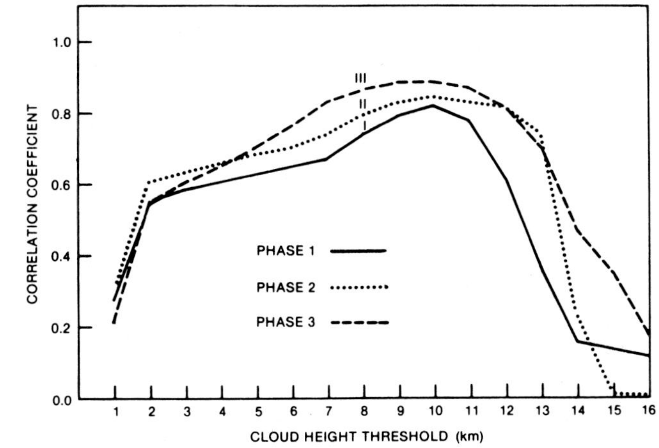

Figure IX-4-12. Correlation between 6-hourly rainfall accumulations over the GATE area and fraction of the area covered by cloud above various height thresholds (Arkin, 1979).

IX-4.4.3 The GOES Precipitation Index

Arkin (1979) and Richards and Arkin (1981) compared statistics of the IR measurements from GOES over various areas in the GARP Atlantic Tropical Ex-

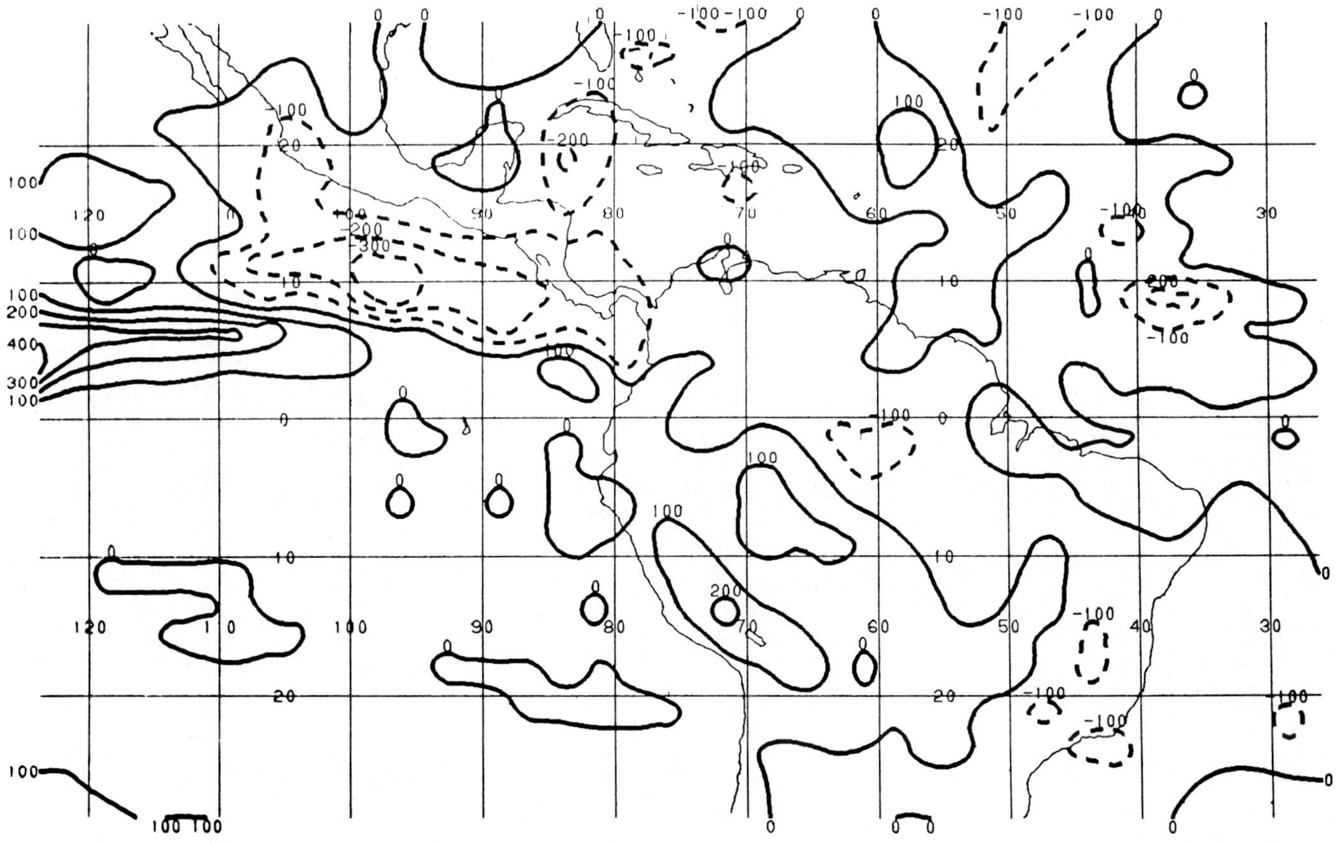

Figure IX-4-13. "Anomaly" (see text) in GPI for September-November 1982. Contours at intervals of 100 mm with solid contours for zero or positive values and dashed contours for negative values.

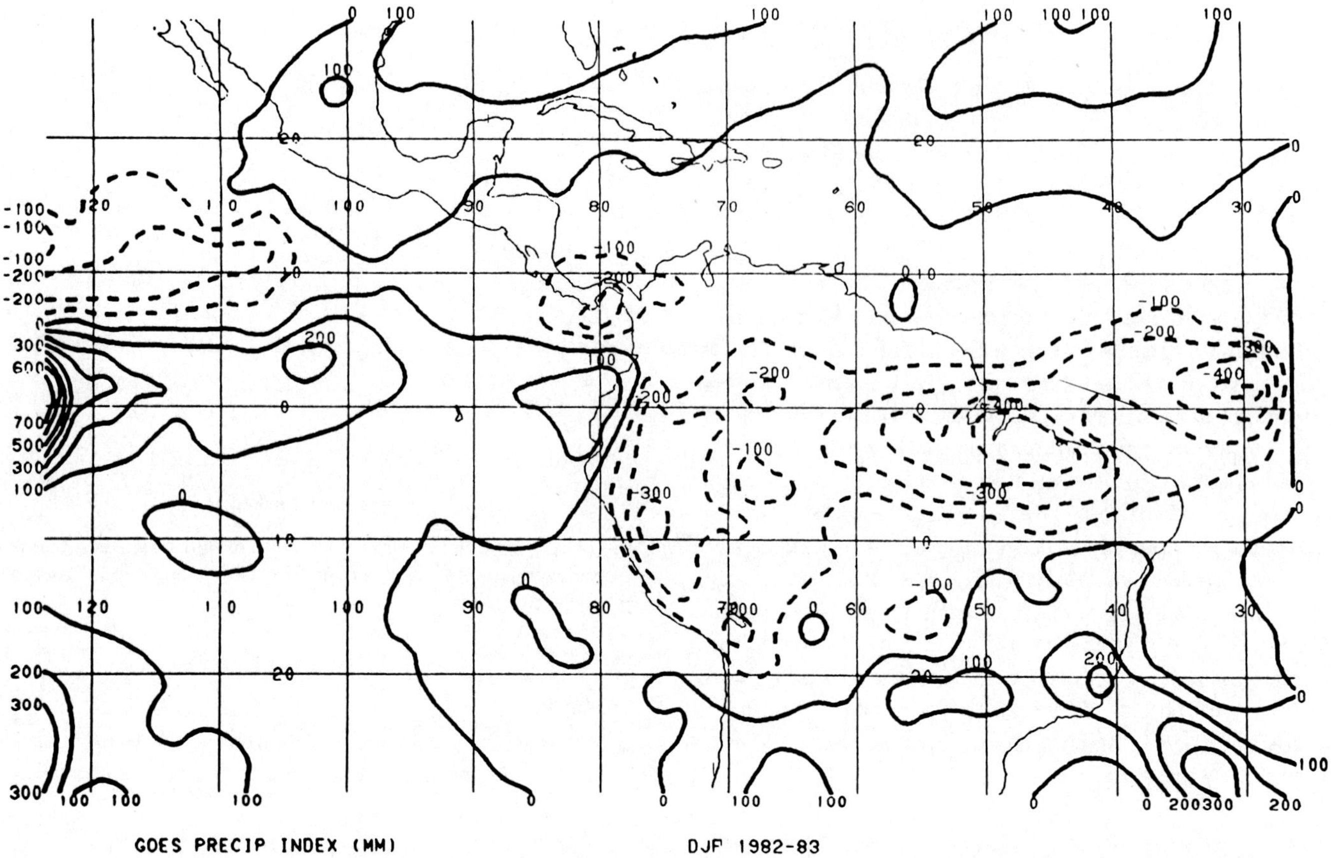

Figure IX-4-14. "Anomaly" (see text) in GPI for December 1982-February 1983 (see legend of Figure IX-4-13).

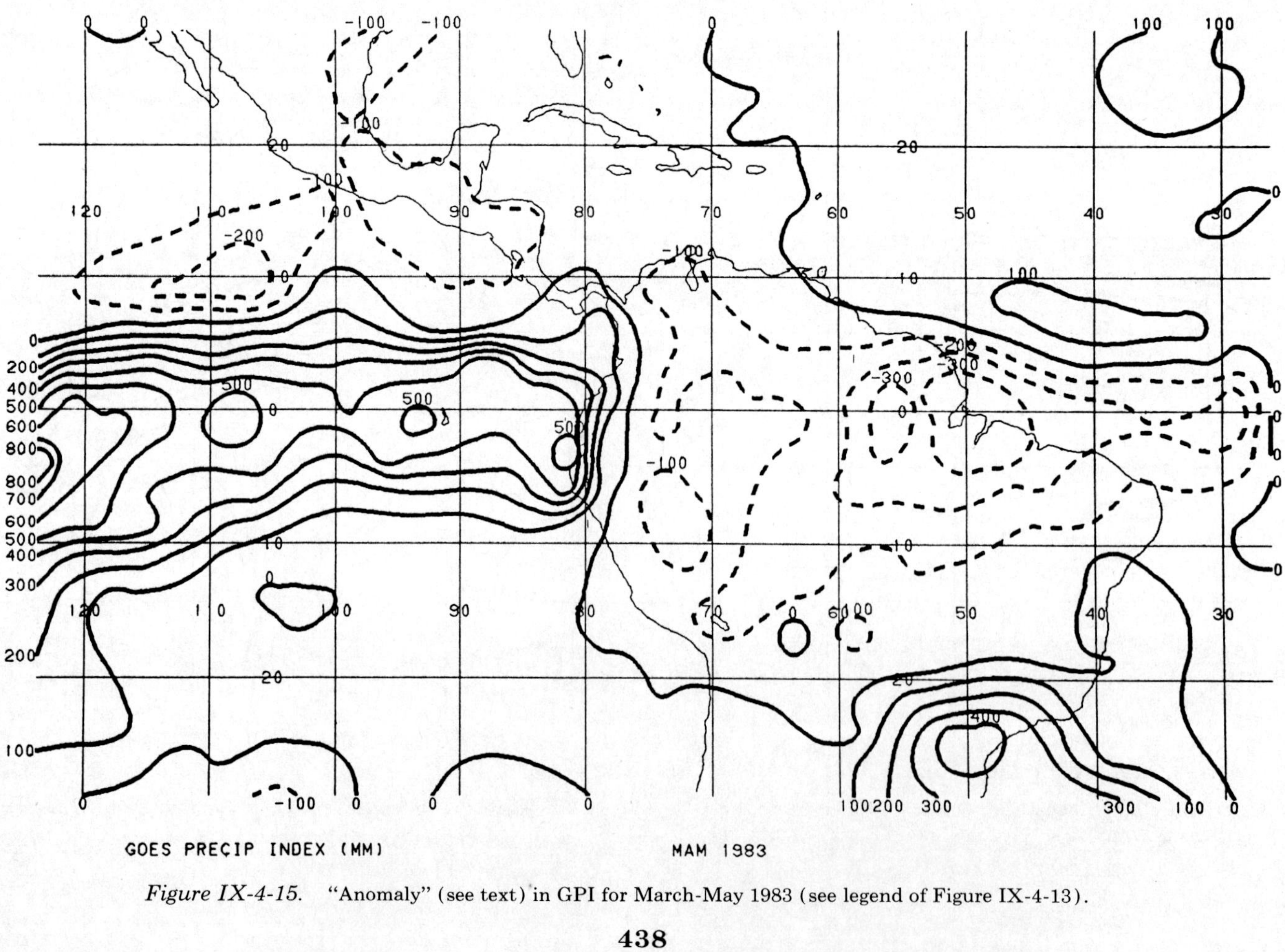

Figure IX-4-15. "Anomaly" (see text) in GPI for March-May 1983 (see legend of Figure IX-4-13).

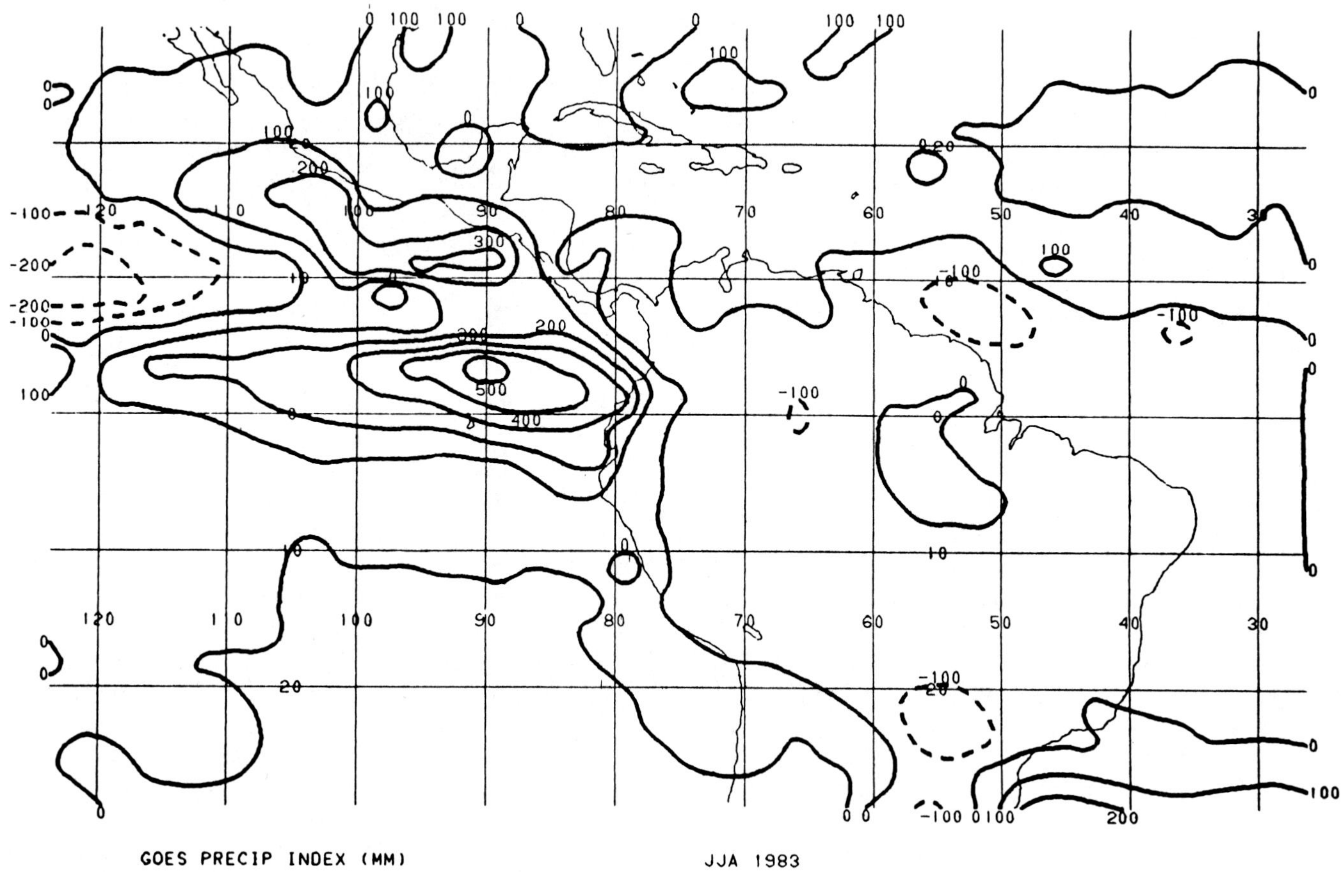

Figure IX-4-16. "Anomaly" (see text) in GPI for June-August 1983 (see legend of Figure IX-4-13).

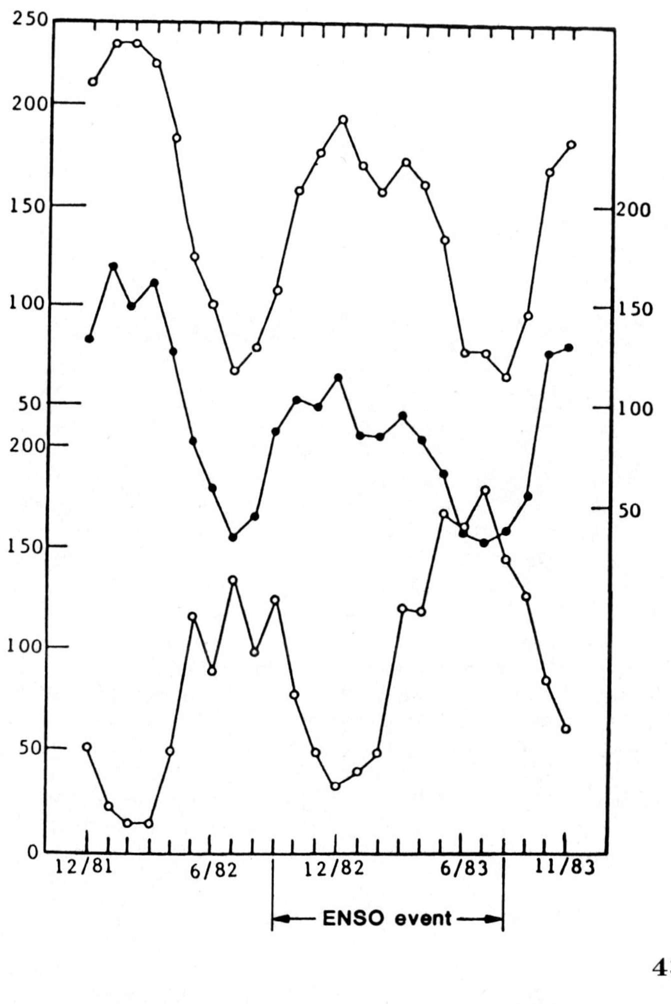

Figure IX-4-17. Monthly GOES Precipitation Index (mm) averaged over tropical South America (top), the Atlantic ITCZ (center), and the eastern Pacific ITCZ (bottom) from December 1981 to November 1983. Note that scales for the top and bottom plots are on the left margin while that for the center plot is on the right.

periment (GATE) arrays with digital radar estimates of rainfall. Arkin found that the 6 hour mean of the fraction of a 2° lat. × 2° long. box covered by cloud with Equivalent BlackBody Temperatures (EBBT) in a wide range (220-260 K) was highly correlated with rainfall during that period. Figure IX-4-12 shows that the variation of correlation with threshold for all three phases of the experiment was similar, with high correlation over a wide range of thresholds. Richards and Arkin (1981) showed that the correlation between rainfall and fractional coverage was poor for areas of 0.5° × 0.5°, but was as high as 0.8 for areas of 1.5° × 1.5° and 2.5° × 2.5°. Correlations increased somewhat with time averaging, but were quite high even for hourly estimates for the 2.5° × 2.5° box.

These studies indicated that large-scale area and time averages of convective precipitation could be estimated with a linear function of the time-mean fraction of properly sized areas covered by EBBT <235 K. Such estimates of precipitation from GOES IR data began in December 1981 and continue to the present. The GOES Precipitation Index (GPI) is given by the fractional coverage averaged over some time t (in hours) multiplied by $3t$. This attempt to use a relationship derived in the eastern tropical Atlantic to estimate precipitation in other regions assumes that the convective precipitation process is similar over all portions of the tropics and even over subtropical regions in the warmer part of the year. While there is little proof that such is in fact the case, Griffith et al. (1980) found that a similar relationship derived for convective rainfall in Florida could be used across the entire tropical Atlantic and eastern tropical South America. Further substantiation of the relationship between GOES fractional cloud cover and ground based rainfall measurements (rain gauge or radar) would be desirable.

To describe the interannual change in tropical rainfall in the eastern Pacific and South America, anomalies in GPI were computed for each season for December 1981-November 1983 in the form of seasonal values minus the 2-year seasonal mean. During September-November 1982 (Figure IX-4-13), the positive precipitation anomaly associated with the ENSO event was well west of South America, with an area of positive GPI "anomaly" extending from west of 125°W to about 100°W. Another area of above "normal" GPI extended along the Andes with a maximum near Lake Titicaca.

By December 1982-February 1983 (Figure IX-4-14), positive values extended along the Equator eastward to near 80°W, with an extensive area of negative values to the east. March-May 1983 (Figure IX-4-15) exhibited similar features, but the positive area in the eastern Pacific was of considerably larger magnitude. Several areas near the Equator in the eastern Pacific had values of GPI more than 1000 mm larger than during the same period in 1982. By June-August 1983 (Figure IX-4-16), large anomalies were found only in the eastern Pacific near and north of the Equator.

The mean GPI over large areas of the tropics exhibits a substantial annual cycle. The annual cycles prior to and during the 1982-83 ENSO event over South America, the Atlantic ITCZ, and the eastern Pacific ITCZ are shown in Figure IX-4-17. While the phase of the annual cycle was similar in each area throughout the period, peak values were roughly ⅓ less during the ENSO event in both the South American and Atlantic ITCZ areas, and about 50% greater in the eastern Pacific ITCZ.

BIBLIOGRAPHY AND REFERENCES FOR SECTION IX

Ahmad, Z. and J. Sutton, 1988: Multichannel estimation of aerosol properties over ocean from current and future AVHRR instruments. *Final Report*, NOAA/NESDIS Contract #NAS5-21680, Dept. of Commerce, Washington, DC, pp 80.

Angell, J. K. and J. Korshover, 1984: Comparison of tropospheric temperatures following Agung and El Chichon volcanic eruptions. *Mon. Wea. Rev., 112*, 1457-1463.

Arkin, P. A., (Univ. of Maryland), 1984: An examination of the southern oscillation in the upper tropospheric tropical and subtropical wind field. Ph. D. Dissertation, 240 pp.

——— and B. N. Meisner, 1984: Fluctuations in the large-scale cloudiness and precipitation regimes in the tropical eastern Pacific and South America associated with the 1982-1983 ENSO event. *Proc. of the 15th Conf. on Hurricanes and Tropical Meteorology, January 9-13, Miami, FL*, Amer. Meteor. Soc., Boston, MA, 10-13.

———, 1983: A diagnostic precipitation index from infrared satellite imagery. *Tropical Ocean-Atmosphere Newsletter, 17*, 5-7.

———, 1982: The relationship between interannual variability in the 200-mb tropical wind field and the Southern oscillation. *Mon. Wea. Rev., 110*, 1393-1404.

———, 1979: The relationship between fractional coverage of high cloud and rainfall accumulations during GATE over the B—scale array. *Mon. Wea. Rev., 107*, 1382-1387.

Arking, A. and J. Vemury, 1984: The Nimbus-7 ERB data set: A critical analysis. *J. Geophys. Res., 89*, 5089-5097.

———, 1964: Latitudinal distributions of cloud cover from TIROS III photographs. *Science, 143*, 569-572.

Atwater, M. A. and J. T. Ball, 1981: A surface solar radiation model for cloudy atmospheres. *Mon. Wea. Rev., 109*, 878-888.

Bandeen, W. R., R. A. Hanel, J. Licht, R. A. Stampfl, and W. G. Stroud, 1961: Infrared and reflected solar radiation measurements from the TIROS-2 meteorological satellite. *J. Geophys. Res., 66*, 3169-3185.

Barnett, J. J., 1975: Hemisphere coupling-evidence of a cross-equatorial planetary wave guide in the stratosphere. *Quart. J. Roy. Meteor. Soc., 101*, 835-845.

———, 1974: The mean meridional temperature behavior of the stratosphere derived from measurements by the selective chopper radiometer on Nimbus-4. *Quart. J. Roy. Meteor. Soc., 100*, 505-530.

Barrett, E. C., 1970: The estimation of monthly rainfall from satellite data. *Mon. Wea. Rev., 98*, 322-327.

Berlyand, T. G., L. A. Strokina, and L.Ye. Greshnikova, 1980: Zonal distribution of cloud cover over the earth. *Meteorology and Hydrology, 3*, 13-24.

Bhartia, P. K., K. F. Klenk, A. J. Fleig, C. G. Wellemeyer, and D. Gordon, 1984: Intercomparison of Nimbus-7 solar backscattered ultraviolet ozone profiles. *J. Geophys. Res., 89*, 5227-5238.

Braslau, N. and J. V. Dave, 1973a: Effect of aerosols on the transfer of solar energy through realistic model atmospheres, Part I: Non absorbing aerosols. *J. Appl. Meteor., 12*, 601.

——— and J. V. Dave, 1973b: Effect of aerosols on the transfer of solar energy through realistic atmospheres, Part II: Partly-absorbing aerosols. *J. Appl. Meteor., 12*, 616.

Briegleb, B. and V. Ramanathan, 1982: Spectral and diurnal variations in clear sky planetary albedo. *J. Appl. Meteor., 21*, 1160-1171.

Budyko, M. I., 1975: Mean air temperature as a function of the variations in solar radiation. *Meteorology and Hydrology, 10*, 1-10.

———, 1974: *Climate and Life*. Academic Press, New York, NY, 508 pp.

———, 1969: The effect of solar radiation variations on the climate of the earth. *Tellus 21*, 611-619.

Cadle, R. D. and G. W. Grams, 1975: Stratospheric aerosol particles and their optical properties. *Rev. Geophys. Space Phys., 13*, 475-501.

Campbell, G. G. and T. H. Vonder Haar, 1980a: An analysis of two years of Nimbus-6 Earth Radiation Budget observation: July 1975 to June 1977, *Atmos. Sci. Pap. No. 320*, Dept. of Atmos. Sci., Colorado State Univ., Fort Collins, CO, 83 pp.

——— and T. H. Vonder Haar, 1980b: Climatology of radiation budget measurements from satellites. *Atmos. Sci. Pap. No. 323*, Dept. of Atmos. Sci., Colorado State Univ., Fort Collins, CO, 74 pp.

Carlson, T. N., 1979: Atmospheric turbidity in Saharan dust outbreaks as determined by analyses of satellite brightness data. *Mon. Wea. Rev., 107*, 322-335.

——— and P. Wendling, 1977: Reflected radiance measured by NOAA 3 VHRR as a function of optical depth for Saharan dust. *J. Appl. Meteor., 16*, 1368-1371.

Cess, R. D., B. P. Briegleb, and M. S. Lian, 1982: Low-latitude cloudiness and climate feedback: Comparative estimates from satellite data. *J. Atmos. Sci., 39*, 53-59.

———, 1976: Climate change: An appraisal of atmospheric feedback mechanisms employing zonal climatology. *J. Atmos. Sci., 33*, 1831-1843.

Chahine, M. T., 1982: Remote sensing of cloud parameters. *J. Atmos. Sci., 39*, 159—170.

———, 1974: Remote sounding of cloudy atmospheres-I: The single cloud layer. *J. Atmos. Sci., 31*, 233-243.

Chen, T. S. and G. Ohring, 1984: On the relationship between clear sky planetary and surface albedos. *J. Atmos. Sci., 41*, 156-158.

———, G. Ohring, and H. Ganot, 1983: Some applications of satellite radiation observations to climate studies. *NOAA Technical Report NESDIS-4*, Dept. of Commerce, Washington, DC, 21 pp.

Clapp, P. F., 1964: Global cloud cover for seasons using TIROS nephanalysis. *Mon. Wea. Rev., 92*, 495-507.

Coakley, J. A., Jr. and D. G. Baldwin, 1984: Towards the objective analysis of clouds from satellite imagery data. *J. Clim. Appl. Meteor., 23*, 1065-1099.

———, 1983: Properties of multilayered cloud systems from satellite imagery. *J. Geophys. Res., 88*, 10818-10828.

———, R. D. Cess, and F. B. Yurevich, 1983: The effect of tropospheric aerosols on the Earth's radiation budget: A parameterization for climate models. *J. Atmos. Sci., 40*, 116-138.

——— and F. P. Bretherton, 1982: Cloud cover from high—resolution scanner data: Detecting and allowing for partially filled fields of view. *J. Geophys. Res., 87*, 4917-4932.

———, 1979: A study of climate sensitivity using a simple energy balance model. *J. Atmos. Sci., 36*, 260-269.

——— and B. A. Wielicki, 1979: Testing energy balance climate models. *J. Atmos. Sci., 36*, 2031-2039.

———, 1977: Feedback in vertical-column energy balance models. *J. Atmos. Sci., 34*, 465-470.

COSPAR, 1978: Toward an internationally coordinated Earth radiation budget satellite observing system: Scientific uses and systems considerations. *Report of the Alpach Workshop*, ICSU/COSPAR, 76 pp.

Coulson, K. L., G. M. Bouricius, and E. L. Gray, 1965: Optical reflection properties of natural surfaces. *J. Geophys. Res., 70*, 4601-4611.

Courel, M. F., R. S. Kandel, and S. I. Rasool, 1984: Surface albedo and the Sahel drought. *Nature, 307*, 528-531.

Crane, A. J., J. D. Haigh, J. A. Pyle, and C. F. Rogers, 1980: Mean meridional circulations of the stratosphere and mesosphere. *Pure Appl. Geophys., 118*, 307-328.

———, 1979: Annual and semiannual waves in the temperature of the mesosphere as deduced from Nimbus-6 PMR measurements. *Quart. J. Roy. Meteor. Soc., 105*, 509-520.

Crutcher, H. L. and J. M. Meserve, 1970: Selected-level height, temperatures, and dew-point temperatures for the Northern Hemisphere. U.S. Navy, *NAVAIR 50-IC-52*, Chief, Naval Operations, Washington, DC.

Darnell, W. L., S. K. Gupta, and W. F. Staylor, 1983: Downward longwave radiation at the surface from satellite measurements. *J. Climate Appl. Meteor., 22*, 1956-1960.

Davies, R., 1984: Reflected solar radiances from broken cloud scenes and the interpolation of scanner measurements. *J. Geophys. Res., 89*, 1259-1266.

Deirmendjian, D., 1969: *Electromagnetic Scattering on Spherical Polydispersions*, Elsevier Pub. Co., New York, NY, 290 pp.

Desbois, M., G. Seze, and G. Szejwach, 1982: Automatic classification of clouds on METEOSAT imagery: Application to high-level clouds. *J. Appl. Meteor., 21*, 401-412.

Dewey, K. F. and R. Heim, Jr., 1982: A digital archive of Northern Hemisphere snow cover, November 1966-December 1980. *Bull. Amer. Meteor. Soc., 63*, 1132-1141.

———, 1977: Daily maximum and minimum temperature forecasts and the influence of snow cover. *Mon. Wea. Rev., 105*, 1594-97.

Dickson, R., 1984: Eurasian snow cover versus Indian monsoon rain fall - an extension of the Hahn-Shukla results. *J. Clim. Appl. Meteor., 23*, 171-173.

Donner, L. J. and H. L. Kuo, 1984: Radiative forcing of stationary planetary waves. *J. Atmos. Sci., 41*, 2849-2868.

Ellis, J. S. and T. H. Vonder Haar, 1976: Zonal average earth radiation budget measurements from satellites for climate studies. *Atmos. Sci. Paper No. 240*, Dept. Atmos. Sci., Colorado State Univ., Fort Collins, CO, 74 pp.

Fraser, R., 1976: Satellite measurement of mass of Saharan dust in the atmosphere. *Appl. Opt., 15*, 2471-2479.

Fritz, S., 1970: Earth's radiation to space at 15 microns: Stratospheric temperature variations. *J. Appl. Meteor., 9*, 815-824.

——— and S. D. Soules, 1970: Large-scale temperature changes in the stratosphere observed from Nimbus-3. *J. Atmos. Sci., 27*, 1091-1097.

Garcia, O., 1981: A comparison of two satellite rainfall estimates for GATE. *J. Appl. Meteor., 20*, 430-438.

Gautier, C., G. Diak, and S. Masse, 1980: A simple physical model to estimate incident solar radiation at the surface from GOES satellite data. *J. Appl. Meteor., 19*, 1005-1012.

Geller, M. A., M. Wu, and M. E. Gelman, 1984: Troposphere-stratosphere monthly winter general circulation statistics for the Northern Hemisphere - interannual variations. *J. Atmos. Sci., 41*, 1726-1744.

———, M. Wu, and M. E. Gelman, 1983: Troposphere-stratosphere monthly winter general circulation statistics for the Northern Hemisphere - four year coverages. *J. Atmos. Sci., 40*, 1334-1352.

Gill, A. and E. M. Rasmusson, 1983: The 1982-83 climate anomaly in the equatorial Pacific. *Nature, 306*, 229-234.

Gille, J. C. and J. M. Russell III, 1984: The limb infrared monitor of the stratosphere: Experiment description, performance, and results. *J. Geophys. Res., 89*, 5125-5140.

———, R. R. Garcia, and W. J. Khori, 1981: Early winter meridional motions in the stratosphere and mesosphere determined from LRIR data. *3rd Conf. on Meteorology of the Upper Atmosphere, Jan. 20-22, San Diego, CA*, Amer. Meteor. Soc., Boston, MA.

——— and F. B. House, 1971: On the inversion of limb radiance measurements: Temperature and thickness. *J. Atmos. Sci., 28*, 1427-1442.

Global Atmospheric Research Program/JOC, 1981: *The International Satellite-Cloud Climatology Project*. Report of Workshop organized by IAMAP and COSPAR, Balaton-almodi, Hungary, 52 pp.

———, 1978: *JOC Study Conf. on Parameterization of Extended Cloudiness and Radiation for Climate Models*, Oxford, England, GARP Climate Dynamics Subprogram, WMO, Geneva, Switzerland, 37 pp.

Griffith, D. G., W. L. Woodley, J. S. Griffin, and S. C. Stromatt, 1980: Satellite-derived precipitation atlas for GATE. *NOAA Atlas Series*, Environ. Res. Lab., Boulder, CO, Superintendent of Documents No. 0-315-309, 280 pp.

Griggs, M., 1985: A method to correct satellite measurements of sea surface temperature for the effects of atmospheric aerosols. *J. Geophys. Res., 90* (D7), 12,951-12,959.

——— and L. L. Stowe, 1984: Measurements of aerosol optical parameters from satellites. *Proc. of the Intl. Radiation Symp., Perugia, Italy,* Deepak Publishing, Hampton, VA, 42-45.

———, 1984a: AVHRR aerosol groundtruth experiment. *NOAA Contract Final Report*, Contr.# NA-83-SAC-00106, Dept. of Commerce, Washington, DC, 36 pp.

———, 1984b: A method to correct AVHRR-derived sea surface temperatures for aerosol contamination. *NOAA Contract Final Report*, Contr.# NA-83-SAC-00076, Dept. of Commerce, Washington, DC, 53 pp.

———, 1983: Satellite measurements of tropospheric aerosols. *Adv. Space Res., 2*, 5, Pergamon Press, New York, NY, 109-118.

———, 1981: Satellite measurements of tropospheric aerosols. *NASA Contractor Rept. 3459*, Contr.# NAS1-15898, Natl. Aero. and Space Admin., Washington, DC, 58 pp.

———, 1977: Comment on "Relative atmospheric aerosol content from ERTS observations." *J. Geophys. Res., 82*, 4972.

———, 1975: Measurement of atmospheric aerosol optical thickness over water using ERTS-1 data. *J. Air Poll. Contr. Assoc., 25*, 622-626.

Gruber, A. and H. Jacobowitz, 1985: The longwave radiation estimated from NOAA polar orbiting satellites: An update and comparison with Nimbus-7 ERB results. *Adv. Space Res., 5,* Pergamon Press, New York, NY, 111-120.

——— and A. F. Krueger, 1984: The status of the NOAA outgoing longwave radiation data set. *Bull. Amer. Meteor. Soc., 65*, 958-962.

——— and J. S. Winston, 1978: Earth-atmosphere radiative heating based on NOAA scanning radiometer measurements. *Bull. Amer. Meteor. Soc., 59*, 1570-1573.

———, 1977: Determination of the Earth-atmosphere radiation budget from NOAA satellite data. *NOAA Technical Report, NESS 76*, Dept. of Commerce, Washington, DC, 28 pp.

———, 1972: Fluctuations in the positions of the ITCZ in the Atlantic and Pacific Oceans. *J. Atmos. Sci., 29*, 193-197.

Hahn, D. G. and J. Shukla, 1976: An apparent relationship between snow cover and Indian monsoon rainfall. *J. Atmos. Sci., 33*, 2461-62.

Hamill, P. and L. R. McMaster, 1984: Polar Stratospheric clouds: Their role in atmospheric processes. *NASA Conf. Pub. 2318*, NASA Workshop, Virginia Beach, VA, June 20-22, 1983, Natl. Aero. and Space Adm., GSFC, Washington, DC, 72 pp.

Hamilton, K., 1982: Some features of the climatology of the Northern Hemisphere stratosphere revealed by NMC upper atmosphere analyses. *J. Atmos. Sci., 39*, 2737-2749.

Hansen, J., D. Johnson, A. Lacis, S. Lebedoff, P. Lee, D. Rind, and G. Russell, 1981: Climatic impact of increasing atmospheric carbon dioxide. *Science, 213*, 957-966.

Hariharan, T. A., 1971: Atmospheric turbidity from polarization of outgoing visible radiation. *Pure Appl. Geophys., 86*, 198-208.

Hartman, D. L., H. H. Hendon, and R. A. Houze, Jr., 1984: Some implications of the mesoscale circulation in tropical cloud clusters for large-scale dynamics and climate. *J. Atmos. Sci., 41*, 113-121.

——— and D. A. Short, 1980: On the use of Earth radiation budget statistics for studies of clouds and climate. *J. Atmos. Sci., 37*, 1233-1250.

Hartman, D. T., 1976: The structure of the stratosphere in the Southern Hemisphere during late winter 1973 as observed by satellite. *J. Atmos. Sci., 33*, 1141-1154.

Heath, D. F., A. J. Fleig, A. J. Miller, T. G. Rogers, R. M. Nagatani, H. D. Bowman, V. G. Kaveeshwan, K. F. Klenk, P. K. Bhartia, and K. D. Lee, 1982: Ozone Climatology Series, Vol. 1: Atlas of Total Ozone, April 1970-December 1976. *NASA Reference Pub. 1098*, Natl. Aero. and Space Adm., Greenbelt, MD, 163 pp.

———, A. J. Krueger, H. A. Roeder, and B. D. Henderson, 1975: The solar backscatter ultraviolet and total ozone mapping spectrometer (SBUV/TOMS) for Nimbus-G. *Opt. Eng., 14*, 323.

———, C. T. Mateer, and A. J. Krueger, 1973: The Nimbus-4 backscatter ultraviolet (BUV) atmospheric ozone experiment—two years operation. *Pure Appl. Geophys., 106-108*, 1238-1253.

Heddinghaus, T. R. and A. F. Krueger, 1981: Annual and interannual variations in outgoing longwave radiation over the tropics. *Mon. Wea. Rev., 109*, 813-829.

Henderson-Sellers, A., 1980: Albedo changes: Surface surveillance from satellites. *Clim. Change, 2*, 275-281.

Hickey, J. R., B. M. Alton, H. L. Kyle, and D. Hoyt, 1988: Total solar irradiance measurements by ERB/Nimbus-7: A review of nine years. *Space Science Reviews*, Kluwer Academic Publishers, p. 151.

——— and B. M. Alton, 1983: Extraterrestrial solar irradiance results from the ERB experiment of Nimbus-7. *Proc. of the 5th Conf. on Atmos. Radiation, June 6-9, Baltimore, MD*, Amer. Meteor. Soc., Boston, MA, 444-447.

———, L. L. Stowe, H. Jacobowitz, P. Pellegrino, R. H. Maschoff, F. House, and T. H. Vonder Haar, 1980: Initial solar irradiance determinations from Nimbus-7 cavity radiometer measurements. *Science, 208*, 281-283.

Hilsenrath, E. and B. M. Schlesinger, 1981: Total ozone seasonal and interannual variations devised from the 7-year Nimbus-4 data set. *J. Geophys. Res., 86*, 12087-12096.

Hirota, I. and J. J. Barnett, 1977: Planetary waves in the winter mesosphere-preliminary analysis of Nimbus-6 PMR results. *Quart. J. Roy. Meteor. Soc., 103*, 487-498.

———, 1976: Seasonal variation of planetary waves in the stratosphere observed by the Nimbus-5 SCR. *Quart. J. Roy. Meteor. Soc., 102,* 757-770.

Horel, J. D. and J. M. Wallace, 1981: Planetary-scale atmospheric phenomena associated with the southern oscillation. *Monthly Weather Review, 109*, 813-829.

House, F., 1984: Observing the Earth radiation budget from satellites: Past, present, and a look to the future. *Advances in Space Research, 5*, 89-98.

Hudlow, M. D., 1979: Mean rainfall patterns for the three phases of GATE. *J. Appl. Meteor., 18*, 1656-1669.

———, and V. L. Patterson, 1979: GATE Radar Rainfall Atlas. NOAA Special Report, Environmental Data and Information Service, Washington, DC, 155 pp.

Hughes, N. A., 1984: Global cloud climatologies: A historical review. *J. Climate Appl. Meteor., 23*, 724-751.

Hwang, P. H., L. L. Stowe, H. Y. M. Yeh, H. L. Kyle and the Nimbus-7 Cloud Processing Team, 1988: The Nimbus-7 global cloud climatology. *Bull. Am. Meteor. Soc., 69*, 743-752.

Jacobowitz, H., H. V. Soule, H. L. Kyle, F. B. House, and the Nimbus-7 ERB Experiment Team, 1984a: The Earth Radiation Budget (ERB) experiment— An overview. *J. Geophys. Res., 89*, 5021-5038.

———, R. J. Tighe, and the Nimbus-7 ERB Experiment Team, 1984b: The Earth Radiation Budget derived from the Nimbus-7 ERB experiment. *J. Geophys. Res., 89*, 4997-5010.

———, W. L. Smith, H. B. Howell, F. W. Nagle, and J. R. Hickey, 1979: The first 18 months of planetary radiation budget measurements from the Nimbus-6 ERB experiment. *J. Atmos. Sci., 36*, 501-507.

———, L. L. Stowe, and J. R. Hickey, 1978: The Earth radiation budget (ERB) experiment. *The Nimbus-7 Users Guide*, Natl. Aero. and Space Adm., GSFC, Greenbelt, MD, 33-69.

Johnston, J., 1963: *Econometric Methods*. McGraw Hill, New York, NY, 300 pp.

Julian, P. R., 1984: Objective analysis in the tropics: A proposed scheme. *Mon. Wea. Rev., 112*, 1752-1767.

Junge, C. E., C. W. Chagnon, and J. E. Manson, 1961: Stratospheric aerosols. *J. Meteor., 18*, 81-108.

Kaufman, Y. J. and J. H. Joseph, 1982: Determination of surface albedos and aerosol extinction characteristics from satellite imagery. *J. Geophys. Res., 87*, C2, 1287-1299.

Kidwell, K. B., 1981: NOAA polar orbiter data (TIROS-N and NOAA-6) users guide. Dept. of Commerce, NESDIS, Washington, DC, 146 pp.

Kilonsky, B. J. and C. S. Ramage, 1976: A technique for estimating tropical open-ocean rainfall from satellite observations. *J. Appl. Meteor., 15*, 972-975.

Kimes, D. S., 1983: Dynamics of directional reflectance factor distributions for vegetation canopies. *Appl. Opt., 22*, 1364-1372.

Koepke, P. and H. Quenzel, 1981: Turbidity of the atmosphere determined from satellite: Calculation of optimum wavelength. *J. Geophys. Res., 86*, 9801-9805.

——— and H. Quenzel, 1979: Turbidity of the atmosphere from satellites: Calculation of optimum viewing geometry. *J. Geophys. Res., 84*, 7847-7856.

Koffler, R. A., A. G. DeCotiis, and P. K. Rao, 1973: A procedure for estimating cloud amount and height from satellite infrared radiation data. *Mon. Wea. Rev., 101*, 240-243.

Kohri, W. J., (Drexel Univ., Philadelphia, PA), 1981: LRIR observations of the structure and propagation of the stationary planetary waves in the Northern Hemisphere during December 1975. Ph. D. Dissertation, 355 pp.

Kondratyev, K.Ya., V. I. Korzov, V. V. Mukhenberg, and L. N. Dyachenko, 1981: The shortwave albedo and surface emissivity. *Joint Scientific Committee Study Conf. on Land Surface Processes in Atmospheric General Circulation Models, (World Climate Research Program)*, Natl. Aero. Space and Adm., GSFC, Greenbelt, MD, 463 pp.

———, L. N. Dyachenko, and M. P. Piatovskaya, 1974: On the relationship between the Earth-atmosphere system albedo and the earth's surface albedo. *Earth Survey Problems*, Akademic-Verlag Berlin, 473 pp.

———, 1969: *Radiation in the Atmosphere*. Academic Press, New York, NY.

Kowalik, W. S., S. E. Marsh, and R. J. P. Lyon, 1982: A relation between LANDSAT digital numbers, surface reflectance and the cosine of the solar zenith angle. *Remote Sens. Environ., 12*, 39-55.

Kriebel, K. T., 1978: Measured spectral bidirectional reflection properties of four vegetated surfaces. *Appl. Opt., 17*, 253-259.

Labitzke, K., 1974: The temperature in the upper stratosphere: Differences between hemispheres. *J. Geophys. Res., 79*, 2171-2175.

Lacis, A. A. and J. E. Hansen, 1974: A parameterization for the absorption of solar radiation in the Earth's atmosphere. *J. Atmos. Sci., 31*, 118-133.

Landsberg, H. E., 1972-77: *World Survey of Climatology*. Elsevier Scientific Pub. Co., Amsterdam, Vols. 7, 10, and 12.

Lau, K. M. and P. H. Chan, 1983: Short-term climate variability and atmospheric teleconnections from satellite-observed outgoing longwave radiation. Part I: Simultaneous relationships. *J. Atmos. Sci., 40*, 2735-2750.

Lemke, P., E. W. Trinkl, and K. Hasselman, 1980: Stochastic dynamic analysis of polar sea ice variability. *J. Phys. Oceanogr., 10*, 2100-2120.

Lenoble, J., 1977: *Standard Procedures to Compute Atmospheric Radiative Transfer in a Scattering Atmosphere, Numerical Comparison of the Methods*. Intl. Assoc. of Meteor. and Aeronomy Radiation Commission, 125 pp.

Leovy, C. B. and P. J. Webster, 1976: Stratospheric long waves: Comparison of thermal structure in the Northern and Southern Hemispheres. *J. Atmos. Sci., 33*, 1624-1638.

Lian, M. S. and R. D. Cess, 1977: Energy balance climate models: A reappraisal of ice-albedo feedback. *J. Atmos. Sci., 34*, 1058-1062.

Liebmann, B. and D. L. Hartmann, 1982: Interannual variations of outgoing IR associated with tropical circulation changes during 1974-78. *J. Atmos. Sci., 39*, 1153-1162.

Lindzen, R. S., 1981: Turbulence and stress owing to gravity wave and tidal breakdown. *J. Geophys. Res., 86*, 9707-9714.

London, J., J. E. Frederick, and G. P. Anderson, 1977: Satellite observations of the global distribution of stratospheric ozone. *J. Geophys. Res., 82*, 2543-2556.

———, 1957: A study of the atmospheric heat balance. *Final Report*, Contract No. AF19(122)-165, New York Univ., New York, NY, 99 pp.

Lovill, J. E.. 1974: A comparison of the Southern and Northern Hemisphere general circulation characteristics as determined by satellite ozone data. *Proc. of the Int. Conf. on Struct. Compos. and Gen. Cir. of the Upper and Lower Atmos. and Possible Anthropogenic Perturbations, Jan. 14-25, 1974, Melbourne, Australia*, 340-359.

Manabe, S. and R. T. Wetherald, 1975: The effects of doubling the

CO_2 concentration on the climate of a general circulation model. *J. Atmos. Sci., 32*, 3-15.

——— and R. T. Wetherald, 1967: Thermal equilibrium of the atmosphere with a given distribution of relative humidity. *J. Atmos. Sci., 24*, 241-259.

Mateer, C. L., D. F. Heath, and A. J. Krueger, 1971: Estimation of total ozone amount from satellite measurements of backscattered ultraviolet Earth radiance. *J. Atmos. Sci., 28*, 1307-1311.

Matson, M., C. F. Ropelewski, and M. S. Varnadore, 1986: An atlas of satellite-derived Northern Hemispheric snow cover frequency. *NOAA Atlas*, Dept. of Commerce, NESDIS, Washington, DC, 75 pp.

McClain, E. P., W. G. Pichel, and C. C. Walton, 1985: Comparative performance of AVHRR-based multichannel sea surface temperatures. *J. Geophys. Res., 90*, 11587-11601.

McClatchey, R. A., R. W. Fenn, J. E. A. Selby, F. E. Volz, and J. S. Garing, 1972: *Optical properties of the atmosphere*. AFCRL-72-0497, Air Force Cambridge Res. Lab., Hanscom Field, Bedford, MA, 115 pp.

McCleese, D. J. and L. S. Wilson, 1976: Cloud top heights from temperature sounding instruments. *Quart. J. Roy. Meteor. Soc., 102*, 781-790.

McCormick, M. P., T. J. Swissler, E. Hilsenrath, A. J. Krueger, and M. T. Osborn, 1984: Satellite and correlative measurements of stratospheric ozone: Comparisons of measurements made by SAGE, ECC balloons, chemiluminescent and optical rocketsondes. *J. Geophys. Res., 89*, 5315-5320.

——— and T. J. Swissler, 1983: Stratospheric aerosol mass and latitudinal distribution of the El Chichon eruption cloud for October 1982. *Geophys. Res. Lett., 10*, 877-880.

———, C. R. Trepte, G. S. Kent, 1983: Spatial changes in the stratospheric aerosol associated with the north polar vortex. *GPL, 10*, 941-944.

———, G. S. Kent, G. K. Yue, and D. M. Cunnold, 1982: Stratospheric aerosol effects from Soufriere volcano as measured by the SAGE satellite system. *Science, 216*, 1115-1118.

———, W. P. Chu, G. W. Grams, P. Hamill, B. M. Herman, L. R. McMaster, T. J. Pepin, P. B. Russell, H. M. Steele, and T. J. Swissler, 1981: High-latitude stratospheric aerosols measured by the SAM II satellite system in 1978 and 1979. *Science, 214*, 328-331.

———, P. Hamill, T. J. Pepin, W. P. Chu, T. J. Swissler, and L. R. McMaster, 1979: Satellite studies of the stratospheric aerosol. *Bull. Amer. Meteor. Soc., 60*, 1038-1046.

McPeters, R. D., D. F. Heath, and P. K. Bhartia, 1984: Average ozone profiles for 1979 from the Nimbus-7 SBUV instrument. *J. Geophys. Res., 89*, 5199-5214.

Mekler, Y. and J. H. Joseph, 1983: Direct determination of surface albedos from satellite imagery. *J. Clim. Appl. Meteor., 22*, 530-536.

———, H. Quenzel, G. Ohring, and I. Markus, 1977: Relative atmospheric aerosol content from ERTS observations. *J. Geophys. Res., 82*, 967-970.

Miller, A. J., R. M. Nagatani, T. G. Rogers, A. J. Fleig, and D. F. Heath, 1982: Total ozone variations 1970-74 using backscattered ultraviolet (BUV) and ground-based observations. *J. Appl. Meteor., 21*, 621-630.

———, B. Kortz, and D. F. Heath, 1979: Comparison of backscatter ultraviolet (BUV) and ground-based total ozone fields for December 1970. *Pure Appl. Geophys., 117*, 355-360.

Miller, D. B. and R. G. Feddes, 1971: *Global Atlas of Relative Cloud Cover, 1967-70*. United States Air Force/Dept. of Commerce, Washington, DC, 237 pp.

Minnis, P. and E. F. Harrison, 1984: Diurnal variability of regional cloud and clear-sky radiative parameters derived from GOES data. Part I: Analysis method; Part II: November 1978 cloud distributions; Part III: November 1978 radiative parameters. *J. Clim. Appl. Meteor., 23*, 993-1051.

Mintz, Y., 1981: A brief review of the present status of global precipitation measurements. *Precipitation Measurements from Space—Workshop Report,* Natl. Aero. and Space Adm., GSFC, Greenbelt, MD, D1-D4.

Morse, B. and C. F. Ropelewski, 1983: Spatial and temporal distribution of Northern Hemisphere snow cover. *NOAA Technical Report NESDIS-6*, Dept. of Commerce, Washington, DC, 31 pp.

Moser, W. and E. Raschke, 1984: Incident solar radiation over Europe estimated from METEOSAT data. *J. Clim. Appl. Meteor., 23*, 166-170.

Naber, P. S. and J. A. Weinman, 1984: The angular distribution of infrared radiances emerging from broken fields of cumulus. *J. Geophys. Res., 89*, 1249-1258.

Nagatani, R. M. and A. J. Miller, 1984: Stratospheric ozone changes during the first year of SBUV observations. *J. Geophys. Res., 89*, 5191-5198.

Namias, J., 1976: Negative ocean-air feedback systems over the north Pacific in the transition from warm to cold seasons. *Mon. Wea. Rev., 9*, 1107-1121.

———, 1959: Recent seasonal interactions over the north Pacific waters and the overlying atmospheric circulations. *J. Geophys. Res., 64*, 631-646.

National Academy of Sciences, 1984: *The effects on the atmosphere of a major nuclear explosion*. National Academy Press, Washington, DC, 193 pp.

Norton. C. C., F. R. Mosher, B. Hinton, D. W. Martin, D. Santek, and W. Kohlow, 1980: A model for calculating aerosol turbidity over the oceans from geostationary satellite data. *J. Appl. Meteor., 89*, 633-644.

Oerlemans, J. and H. M. Van den Dool, 1978: Energy balance climate models: Stability experiments with a refined albedo and updated coefficients for infrared emission. *J. Atmos. Sci., 35*, 371-381.

Ohring, G., A. Gruber, and R. G. Ellingson, 1984: Satellite determinations of the relationship betwen total longwave radiation flux and infrared window radiance. *J. Clim. Appl. Meteor., 23*, 416-425.

———, 1983: Recent observational studies of the effect of clouds on the radiation budget of the earth-atmosphere system. *Abstracts, Intl. Association of Meteorological and Atmospheric Physics General Assembly, Hamburg, Germany*, p. 474.

——— and A. Gruber, 1983: Satellite radiation observations and climate theory. *Adv. in Geophys., 25*, 237-304.

———, P. F. Clapp, T. R. Heddinghaus, and A. F. Krueger, 1981: The quasi-global distribution of the sensitivity of the Earth-atmosphere radiation budget to clouds. *J. Atmos. Sci., 38*, 2539-2541.

——— and P. Clapp, 1980: The effect of changes in cloud amount on the net radiation at the top of the atmosphere. *J. Atmos. Sci., 37*, 447-454.

——— and J. Mariano, 1964: Changes in the amount of cloudiness and the average surface temperature of the Earth. *J. Atmos. Sci., 21*, 448-450.

Oort, A. H. and E. M. Rasmusson, 1971: Atmospheric circulation statistics. *NOAA Professional Paper No. 5*, Dept. of Commerce, Washington, DC, 323 pp.

Otterman, J. and C. J. Robinove, 1981: Effects of the atmosphere on the detection of surface changes from Landsat multispectral scanner data. *Intl. J. Remote Sens. 2*, 351-360.

——— and R. S. Fraser, 1976: Earth atmosphere system and surface reflectivities in arid regions from LANDSAT MSS measurements. *Remote Sens. Environ., 5*, 247-266.

Pepin, T. J., M. P. McCormick, W. P. Chu, F. Simon, T. J. Swissler,

R. R. Adams, K. H. Crumbly, and W. H. Fuller, Jr., 1977: Stratospheric aerosol measurements. *NASA SP-421*, Natl. Aero. and Space Adm., Washington, DC, 127-136.

Pinker, R. T. and J. A. Ewing, 1985: Modeling surface solar radiation: Model formulation and validation. *J. Clim. Appl. Meteor., 24*, 389-401.

———, J. A. Ewing, and J. D. Tarpley, 1985: The relationship between the planetary and surface net radiation. *J. Clim. Appl. Meteor., 24*, 1262-1268.

——— and L. A. Corio, 1984: Surface radiation budget from satellites over the Great Plains. *Mon. Wea. Rev., 112*, 203-215.

——— and J. A. Razgaitis, 1983: Estimating the solar zenith dependence of clear-sky planetary albedo for land surfaces from the GOES satellite. *J. Geophys. Res., 88*, C10, 6007-6011.

———, O. E. Thompson, and T. F. Eck, 1980: The albedo of a tropical evergreen forest. *Quart. J. Roy. Meteorol. Soc., 106*, 551-558.

Planet, W. G., D. S. Crosby, J. H. Lienesch, and M. L. Hill, 1984: Determination of total ozone amount from TIROS radiance measurements. *J. Clim. Appl. Meteor., 23*, 308-316.

Plass, G. N. and G. W. Kattawar, 1972: Effect of aerosol variation on radiance in the Earth's atmosphere-ocean system. *Appl. Opt., 11*, 1598-1604.

Platt, C. M. R., 1983: On the bi-spectral method for cloud parameter determination from satellite VISSR data: Separating broken cloud and semitransparent cloud. *J. Clim. Appl. Meteor., 22*, 429-439.

——— and G. L. Stephens, 1980: The interpretation of remotely sensed high cloud emittances. *J. Atmos. Sci., 37*, 2314-2322.

Prabhakara, C., E. B. Rodgers, and V. V. Salomonson, 1973: Remote sensing of the global distribution of total ozone and the inferred upper tropospheric circulation from Nimbus IRIS experiments 1. *Pure Appl. Geophys., 106-108*, 1226-1237.

Preuss, H. J. and J. F. Geleyn, 1980: Surface albedos derived from satellite data and their impact on forecast models. *Arch. Met. Geoph. Biokl., Ser. A, 29*, 345-356.

Prosch, T., D. Henning, and E. Raschke, 1982: Studies on polarization of light over water surfaces. *Ann. Meteor., 18*, 113-115.

Quiroz, R. S., 1983: The isolation of stratospheric temperature change due to the El Chichon volcanic eruption from non-volcanic signals. *J. Geophys. Res., 88*, 6773-6780.

Rabiner, L. R., M. R. Sambar, and C. E. Schmidt, 1975: Applications of a non-linear smoothing algorithm to speech processing. *Proc. of the IEEE Trans. Acoust. Speech Signal, ASSP-23*, 552-557.

Ramanathan, V., L. B. Collis, and R. E. Boughner, 1976: Sensitivity of surface temperature and atmospheric temperature to perturbations in the stratospheric concentration of ozone and nitrogen dioxide. *J. Atmos. Sci., 33*, 1092-1112.

Rao, C. R. N., L. L. Stowe, E. P. McClain, and J. Saper, 1988: Development and application of aerosol remote sensing with AVHRR data from the NOAA satellites. *Aerosols in Climate*, Deepak Publishing, Hampton, VA, 69-80.

———, T. Takashima, and R. B. Toolin, 1973: Measurements and interpretation of the polarization of radiation emerging from the atmosphere at an altitude of 28 km over south-western New Mexico (USA). *Quart. J. Royal Meteor. Soc., 99*, 294-302.

Raschke, E., T. H. Vonder Haar, M. Pasternak, and W. R. Bandeen, 1973: The radiation balance of the Earth-atmosphere system from Nimbus 3 radiation measurements. *NASA Technical Note D-7249*, Natl. Aero. and Space Adm., Washington, DC, 73pp.

Rasmusson, E. M. and J. M. Wallace, 1983: Meteorological aspects of the El Niño/Southern Oscillation. *Science, 222*, 1195-1202.

——— and T. H. Carpenter, 1982: Variations in tropical sea surface temperature and surface wind fields associated with the Southern Oscillation/El Niño. *Mon. Wea. Rev., 110*, 354-384.

Rasool, S. I. and H. J. Bolle, 1984: ISLSCP: International satellite land surface climatology project. *Bull. Amer. Met. Soc., 65*, 143-144.

——— and S. H. Schneider, 1971: Atmospheric carbon dioxide and aerosols: effects of large increases on global climate. *Science, 173*, 138-141.

Remsberg, E. E., J. M. Russell, J. C. Gille, L. L. Gordley, P. L. Bailey, W. G. Planet, and J. E. Harries, 1984: The validation of Nimbus-7 LIMS measurements of ozone. *J. Geophys. Res., 89*, 5161-5178.

Reynolds, D. W. and T. H. Vonder Harr, 1977: A bi-spectral method for cloud parameter determination. *Mon. Wea. Rev., 105*, 446-457.

Reynolds, R. W., 1988: A real-time global sea surface temperature analysis. *J. Clim., 1*, 75-86.

———, 1982: A monthly averaged climatology of sea surface temperatures. *NOAA Technical Report NWS 31*, Dept. of Commerce, Washington, DC, 35 pp.

Richards, F. and P. A. Arkin, 1981: On the relationship between satellite observed cloud cover and precipitation. *Mon. Wea. Rev., 109*, 1081-1093.

Richardson, A. J., 1982: Relating Landsat digital count values to ground reflectance for optically thin atmospheric conditions. *Appl. Opt., 21*, 1457-1464.

Riehl, H., 1954: *Tropical Meteorology*. McGraw-Hill, New York, NY, 392 pp.

Robock, A. and M. Matson, 1983: The circumpolar transport of the El Chichon volcanic dust cloud. *Science, 221*, 195-197.

———, 1980: The seasonal cycle of snow cover, sea ice and surface albedo. *Mon. Wea. Rev., 108*, 267-285.

Rodgers, C. D., 1984: Coordinated study of the behavior of the middle atmosphere in winter (PMP-1). *Handbook for MAP, 12*, Scientific Committee on Solar-Terrestrial Physics, Urbana, IL, 154 pp.

Rogers, J. C. and H. van Loon, 1979: The seesaw in winter temperatures between Greenland and northern Europe. Part II: Some oceanic and atmospheric effects in middle and high latitudes. *Mon. Wea. Rev., 107*, 509-519.

Ropelewski, C. F., 1983: Spatial and temporal variations in Antarctic sea-ice (1973-82). *J. Clim. Appl. Meteor., 22*, 470-473.

Russell, P. B., B. M. Morley, and E. V. Browell, 1983: Spaceborne lidar measurement accuracy: Simulation of aerosol, cloud, molecular density, and temperature retrievals. *Adv. Space Res.,2,* Pergamon Press, New York, NY, 49-53.

———, T. J. Swissler, M. P. McCormick, W. P. Chu, J. M. Livingston, and T. J. Pepin, 1981: Satellite and correlative measurements of the stratospheric aerosol I: An optical model for data conversions. *J. Atm. Sci, 38*, 1279-1294.

Sadler, J. C., 1968: *Average cloudiness in the tropics from satellite observations*. East-West Center Press, Honolulu, HI, 22 pp.

Salomonson, V. V. and W. E. Marlatt, 1971: Airborne measurements of reflected solar radiation. *Remote Sens. Environ., 2*, 1.

Santer, R., M. Herman, and J. Lenoble, 1982: Retrieval of aerosol optical characteristics from polarization measurements of reflected solar radiation above the oceans. *Adv. Space Res., 2*, Pergamon Press, New York, NY, 65-70.

Schiffer, R. A. and W. B. Rossow, 1983: The international satellite cloud climatology project: The first project of the world climate research programme. *Bull. Amer. Meteor. Soc., 64*, 779-784.

Schneider, S. H. and C. Mass, 1975: Volcanic dust, sunspots, and temperature trends. *Science, 190*, 741-746.

———, 1972: Cloudiness as a global climatic feedback mechanism: The effects on the radiation balance and surface temperature of variations in cloudiness. *J. Atmos. Sci., 29*, 1413-1422.

Schneider, W. H., P. K. Bhartia, K. F. Klenk, and C. L. Mateer, 1981: An optimum statistical techniques for ozone profile retrieval from backscattered UV radiances. *Proc. of the 4th Conf. on Atmos.*

Radiation, June 16-18, Toronto, Ontario, Canada, Amer. Meteor. Society, Boston, MA, 33-37.

Shenk, W. E. and V. V. Salomonson, 1972: A simulation study exploring the effects of sensor spatial resolution on estimates of cloud cover from satellites. *J. Appl. Meteor., 12*, 214-220.

Shukla, J. and J. M. Wallace, 1983: Numerical simulation of the atmospheric response to equatorial Pacific sea surface temperature anomalies. *J. Atmos. Sci., 40*, 1613-1629.

Simpson, J. J. and C. A. Paulson, 1980: Small-scale sea surface temperature structure. *J. Phys. Ocean., 10*, 399-410.

Smith, W. L. and H. M. Woolf, 1984: Geostationary satellite sounder (VAS) observations of longwave radiation flux. *WMO World Climate Program Report WCP 70*, Report of an International Meeting of Experts in Satellite Systems to Measure the Earth's Radiation Budget Parameters and Climate Change Signals, August 20-September 2, 1983, Igls, Austria, 36 pp.

——— and C. M. R. Platt, 1978: Comparison of satellite deduced cloud heights with indications from radiosonde and ground-based laser measurements. *J. Appl. Meteor., 17*, 1796-1802.

———, 1976: The use of eigenvectors of statistical covariance matrices for interpreting satellite sounding radiometer observations. *J. Atmos. Sci., 33*, 1127-1140.

———, H. M. Woolf, and W. J. Jacob, 1970: A regression method for obtaining real-time temperature and geopotential height profiles from satellite spectrometer measurements and its application to Nimbus 3 SIRS observations. *Mon. Wea. Rev., 98*, 582-603.

———, 1968: An improved method for calculating tropospheric temperature and moisture from satellite radiometer measurements. *Mon. Wea. Rev., 96*, 387-396.

Stommel, H. and E. Stommel, 1979: The year without a summer. *Sci. Amer., 240*, (6), 176-186.

Stowe, L. L., C. G. Wellemeyer, T. F. Eck, H. Y. M. Yeh, and the Nimbus-7 Cloud Data Processing Team, 1988: Nimbus-7 global cloud climatology. Part I: algorithms and validation. *J. of Clim., 1*, 445-470.

———, 1984: Evaluation of NIMBUS-7 THIR/CLE and Air Force three-dimensional nephanalysis estimates of cloud amount. *J. Geophys. Res., 89*, 5370-5380.

——— and H. E. Fleming, 1980: The error in satellite retrieved temperature profiles due to the effects of atmospheric aerosol particles. *Remote Sens. Envir., 9*, 57-64.

———, M. Chen, H. Jacobowitz, and I. Ruff, 1978: Classification of clouds for the Nimbus Satellite ERB Experiment using THIR data. *Proc. of the 3rd Conf. on Atmospheric Radiation, June 28-30, Davis, CA*, Amer. Meteor. Soc., Boston, MA, 103-106.

Strong, A. E., 1984: Monitoring El Chichon aerosol distribution using NOAA-7 satellite AVHRR sea surface temperature observations. *Geofis. Int., 23*, 129-142.

———, 1983: Satellite-derived sea surface temperature errors due to El Chichon aerosol cloud. *Tropical Ocean-Atmosphere Newsletter, No. 18*, 14-15.

Susskind, J., J. Rosenfield, D. Reuter, and M. T. Chahine, 1984: Remote sensing of weather and climate parameters from HIRS2/MSU on TIROS N. *J. Geophys. Res., 89*, 4677-4697.

Swift, C. T., 1980: Passive microwave remote sensing of the ocean—a review. *Bound. Layer Meteor., 18*, 25-54.

Szejwach, G., 1982: Determination of semitransparent cirrus cloud temperature from infrared radiances: Application to Meteosat. *J. Appl. Meteor., 21*, 384-393.

Taljaard, J. J., H. Van Loon, H. L. Crutcher, and R. L. Jenne, 1969: Climate of the upper air: Southern Hemisphere, Vol. I. Temperatures, dew points, and heights at selected pressure levels. U. S. Navy *NAVAIR 50-IC-55*, Chief, Naval Operations, Washington, DC.

Tarpley, J. D., S. R. Schneider, and R. L. Money, 1984: Global vegetation indices from the NOAA-7 meteorological satellite. *J. Clim. Appl. Meteor, 23*, 491-494.

———, 1979: Estimating incident solar radiation at the surface from geostationary satellite data. *J. Appl. Meteor., 18*, 1172-1181.

Taylor, V. R. and L. L. Stowe, 1984: Reflectance characteristics of uniform earth and cloud surfaces derived from Nimbus-7 ERB. *J. Geophys. Res., 89*, D4, 4987-4996.

Thomas, G. E., C. A. Barth, E. R. Hansen, C. W. Hord, G. M. Lawrence, G. H. Mount, G. J. Rottman, D. W. Rusch, A. I. Stewart, R. J. Thomas, J. London, P. L. Bailey, P. J. Crutzen, R. E. Dickinson, J. C. Gille, S. C. Liu, J. F. Noxon, and C. B. Farmer, 1980: Scientific objectives of the solar mesosophere explorer mission. *PAGEOPH, 118*, 591-615.

Toon, O. B. and J. B. Pollack, 1977: Volcanoes and climate. *Natur. Hist.*, 13-26.

Tucker, C. J., 1978: A comparison of satellite sensor bands for vegetation monitoring. *Photo. Eng. Remote Sens., 44*, 1369-1380.

Turco, R. P., O. B. Toon, T. P. Ackerman, J. B. Pollack, and C. Sagan, 1983: Nuclear winter: Global consequences of multiple nuclear explosions. *Science, 222*, 1283-1292.

U.S. Department of Commerce, 1969: *Climates of the World*. Superintendent of Documents, U.S. Government Printing Office, Washington, DC, 28 pp.

Vonder Haar, T. H. and V. E. Suomi, 1971: Measurements of the Earth's radiation budget from satellites during a five-year period. Part 1: Extended time and space means. *J. Atmos. Sci., 28*, 305-314.

Wagner, A. J., 1973: The influence of average snow depth on monthly mean temperature anomaly. *Mon. Wea. Rev., 101*, 624-626.

Walsh, J. E., 1984: Snow cover and atmospheric variability. *Amer. Scientist, 72*, 50-57.

———, D. R. Tucek, and M. E. Peterson, 1982: Seasonal snow cover and short-term climatic fluctuations over the United States. *Mon. Wea. Rev., 110*, 1474-85.

——— and C. M. Johnson, 1979: An analysis of Arctic Sea ice fluctuations 1953-77. *J. Phys. Oceanogr., 9*, 580-591.

Wark, D. Q. and M. E. Fleming, 1966: Indirect measurements of atmospheric temperature profiles from satellites: 1. Introduction. *Mon. Wea. Rev., 94*, 351-362.

Warren, S. G. and S. H. Schneider, 1979: Seasonal simulation as a test for uncertainties in the parameterizations of a Budyko-Sellers zonal climate model. *J. Atmos. Sci., 36*, 1377-1391.

Wielicki, B. A. and J. A. Coakley, Jr., 1981: Cloud retrieval using infrared sounder data: Error analysis. *J. Appl. Meteor., 20*, 157-169.

Weinstein, M. and V. Suomi, 1961: Analysis of satellite infrared radiation measurements on a synoptic scale. *Mon. Wea. Rev., 89*, 419-428.

Wetherald, R. T. and S. Manabe, 1975: The effect of changing the solar constant on the climate of a general circulation model. *J. Atmos. Sci., 32*, 2044-2059.

Wilheit, T. T., A. T. Chang, M. S. V. Rao, E. V. Rodgers, and J. S. Theon, 1977: A satellite technique for quantitatively mapping rainfall rates over the ocean. *J. Appl. Meteor., 16*, 551-560.

Winston, J. S., A. Gruber, T. I. Gray, M. S. Varnadore, C. L. Earnest, and L. P. Manello, 1979: Earth atmospheric radiation budget derived from NOAA satellite data June 1974-February 1978, Vol. 1., U.S. Department of Commerce, NOAA, NESS, Washington, DC.

———, 1971: The annual course of zonal mean albedo as derived from ESSA-3 and 5 digitized picture data. *Mon. Wea. Rev., 99*, 818—827.

———, 1967: Planetary scale characteristics of monthly mean long—wave radiation and albedo and some year to year variations. *Mon. Wea. Rev., 95*, 235-256.

——— and P. K. Rao, 1963: Temporal and spatial variations in the planetary scale outgoing long-wave radiation as derived from TIROS II measurements. *Mon. Wea. Rev., 9l*, 641-657.

——— and P. K. Rao, 1962: Preliminary study of planetary scale outgoing long-wave radiation as derived from TIROS-2 measurements. *Mon. Wea. Rev., 90*, 307-310.

World Climate Program—70, 1984: Report of an international meeting of experts on satellite systems to measure the earth's radiation budget parameters and climate change signals (Igls, Austria, 29 August-2 September 1983), 36 pp.

World Climate Program, 1983: Experts meeting on aerosols and their climatic effects. Report on workshop organized by WMO-CAS and Radiation Commission of Intl. Assoc. of Meteor. and Aeronomy, *WCP-55*, WMO, December, 1983, 107 pp.

World Climate Program—40, 1982: Radiation budget of the Earth and its atmosphere. K. Ya. Kondratyev and E. Raschke with contributions from H. J. Preuss and J. Schmetz.

World Meteorological Organization, 1982: *WMO Global Ozone Research and Monitoring Project WMO Report No. 11*. The stratosphere 1981 theory and measurements. World Meteorological Organization, Geneva, Switzerland, 2-29.

Yang, S. K., A. Gruber, B. Liebmann, and A. J. Miller, 1988: Intercomparison of outgoing longwave radiation obtained from ERBE and NOAA operation satellite: A preliminary result. *Proc. of the 3rd Conf. on Satellite Meteor. and Ocean., February 1-5, Anaheim, CA*, Amer. Meteor. Soc., Boston, MA, 138-141.

Zuev, V. E., 1983: Laser sounding of aerosols using airborne and space facilities. *Adv. Space Res., 20,* Pergamon Press, New York, NY, 39-47.

Contributors to Section IX

Coakley, James A., Jr.	Department of Atmospheric Sciences Oregon State University Corvallis, OR 97331
Pinker, Rachel T.	University of Maryland College Park, MD 20742
Arkin, Phillip A. Gelman, Melvyn E. Miller, Alvin J. Nagatani, Ronald M. Reynolds, Richard W. Ropelewski, Chester F.	NOAA/NWS Washington, DC 20233
Gruber, Arnold Ohring, George Stowe, Larry L.	NOAA/NESDIS Washington, DC 20233

Section X

USE OF WEATHER SATELLITE DATA IN AGRICULTURAL APPLICATIONS

Weather satellite data offer a unique source of information for many agricultural applications. Methods for assessing and mapping the condition of vegetation using satellite data are given in Chapter X-1. A vegetation index is defined and its usefulness is discussed in terms of AVHRR sensor data from NOAA satellites. Applications involve both graphic and visual displays of the digital data. Vegetation dynamics are displayed on color maps that are useful in drought and flood assessment, and wildfire management.

Chapter X-2 treats the determination of solar radiation, maximum/minimum temperatures, and precipitation estimates from satellite data. All of these elements are related to plant growth. Sensor data from polar orbiting and geosynchronous satellites are used in physical and statistical models for determining surface insolation. Surface temperatures are obtained from AVHRR IR data, and shelter temperatures are inferred from TOVS data.

Chapter X-1

METHODS OF VEGETATION MONITORING AND THEIR APPLICATIONS

X-1.1 INTRODUCTION

The first use of satellites for monitoring agricultural production was the Large Area Crop Inventory Experiment (LACIE) by the U.S. Department of Agriculture (USDA), NASA, and NOAA from 1975 through 1978. The technical approach in LACIE was to use Landsat data together with conventional weather data to estimate wheat production for the U.S. and the U.S.S.R. One conclusion from LACIE was that crop yield models required more weather data and that weather satellites were the most likely source of such data (Hill et al., 1980). About the same time as LACIE, the Earth Satellite Corporation began using satellite data to estimate precipitation and insolation for a commercial agricultural assessment project. A second joint program among USDA, NASA, and NOAA, the Agricultural Resource Inventory Survey Through Aerospace Remote Sensing (AgRISTARS), began in 1980 and ran through 1985. In this program the uses of operational environmental satellite data for agricultural purposes were developed.

Polar orbiting NOAA and Landsat satellites are useful for agricultural applications. Among their attributes are: (1) regular, repetitive observation patterns; (2) large-scale synoptic views of the Earth surface from space; and (3) the availability of a permanent information archive for establishing baseline data. Characteristics of the two polar orbiting satellite systems are compared in Table X-1-1 and NOAA. ATN is also discussed in Chapter III-1.

X-1.2 SATELLITE CAPABILITIES FOR VEGETATION MONITORING

The assessment of vegetation by multispectral remote sensing is based on the inherent reflectance characteristics of vegetation and the differential reflectance discernible in the VIS and the near-IR bands. Some typical reflectance curves are shown in Figure X-1-1.

The strong absorption of light by chlorophyll in green leaves results in reflectances of less than 20% in the 0.5 to 0.7 μm spectral interval (VIS green and red). These values increase sharply to near 60% in the 0.7 to 1.3 μm range (Hoffer and Johannsen, 1969). These reflectance characteristics have been used widely, especially with Landsat data, to classify land cover type, to estimate crop acreage, and to detect plant stress (Myers, 1983; Tucker, 1978, 1979).

The assessment of vegetation from a polar orbiting satellite is affected by the Sun-target-sensor geometry, the varying amount of cloud cover, the frequency of observation, the inherent resolution of the remote sensing system, the contribution of bare surfaces to the scene reflectance, and the interpretation of cause and effect relationships in the evaluation of the data.

Polar orbiting satellites, such as NOAA and Landsat, operate at altitudes of 725 to 920 km. Thus, reflectance

Table X-1-1 *Selected characteristics of the NOAA AVHRR and Landsat MSS Systems**

CHARACTERISTICS	NOAA AVHRR	LANDSAT MSS**
Inclination of orbit	99.092°	98.91°–99.1°
Height above surface	850 km	916.6 km
Number of orbits/day	14.2	14
Times of coverage at Equator (LST) for NOAA-7, 9, and 11	0230 descending 1430 ascending	0930
Orbital period	102 min	103.3 min
Latitudinal coverage	90°N–90°S	82°N–82°S
Cycle duration	1 day	18 days
Ground coverage	3000 km	185 km
Field of view (FOV)	±56°	±5.78°
Instantaneous field of view (IFOV)	1.39–1.51 mrad	0.086 mrad
Ground resolution (nadir)	1.1 km	79 m
Ground resolution (maximum off-nadir)	2.4 km along track 6.9 km across track	79.5 m along track 80 m across track
Number of channels	One VIS One near IR One middle IR Two thermal IR	Two VIS Two near IR

* Adapted from Townshend and Tucker (1984).

** Landsat-4 and 5 have lower orbits (725 km) and a cycle duration of 16 days but overall the MSS has similar properties to previous sensors in the series. Landsat-4 and 5 also carry the 7 band, 30 m IFOV Thematic Mapper sensor.

characteristics of the surface can be altered as the reflected energy traverses the whole atmosphere before being detected by the satellite sensor. These atmospheric effects are increased significantly in off-nadir viewing, as is the case with the NOAA AVHRR sensor. The ±56° scan angle of the AVHRR produces atmospheric optical thicknesses ranging from 1.0 at the nadir to 2.2 at the extremities (Holben and Fraser, 1984). To deal with the problems of extreme look angles, some researchers consider only the central part of the swath. A significant effect of such usage is to reduce the effective coverage for some applications (Tucker et al., 1984; Miller and Moore, 1983; Townshend and Tucker, 1984).

The use of VIS and reflective IR wavelengths for viewing surface features is highly dependent upon cloud conditions. Clouds effectively block out the surface radiance and most of the information content is lost. Cloud elements smaller than the pixel resolution and thin cirrus clouds are most difficult to detect (Tucker et al., 1984).

The relative frequency of observation becomes increasingly important as it relates to the growth stage of vegetation. For the Landsat system, the repeat cycle is once every 16 days with one satellite or 8 days with two satellites (providing clouds do not interfere). The NOAA polar orbiting satellite provides the distinct advantage of daily observation.

Resolution of the remote sensing system, or Instan-

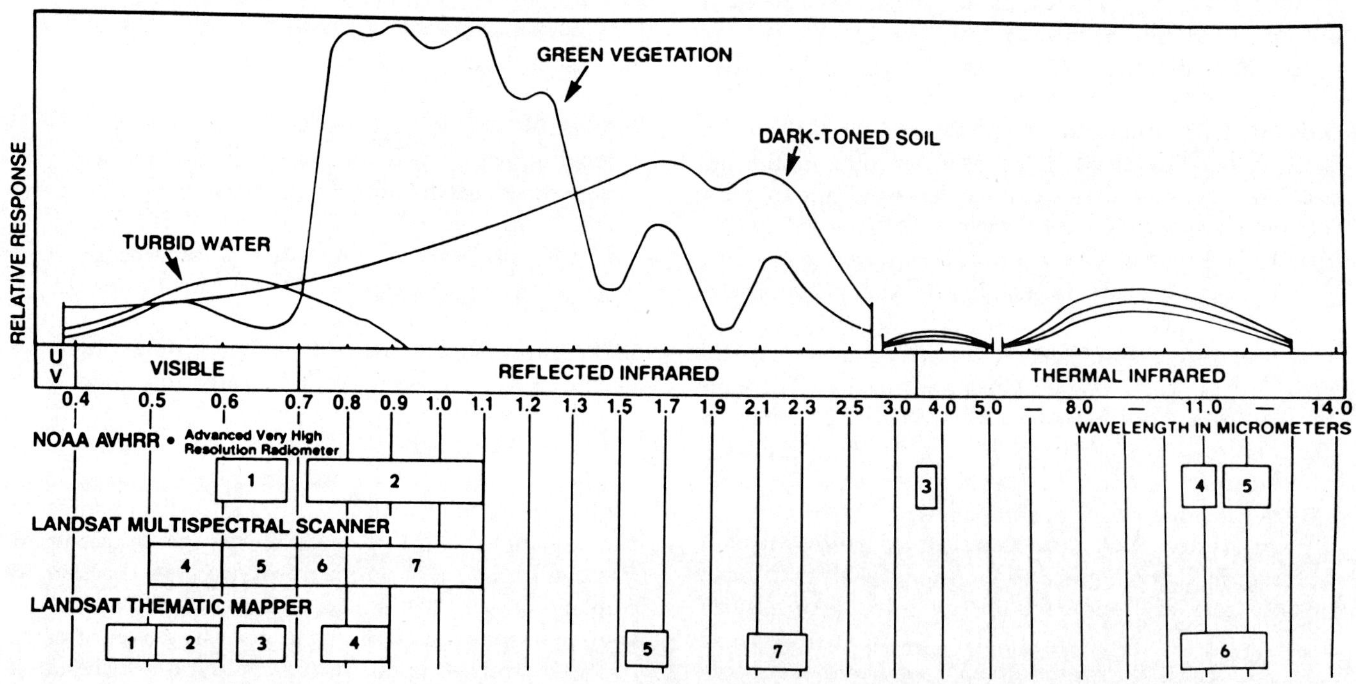

Figure X-1-1. The spectral sensitivity of three typical surfaces as related to bands sensed by three multispectral systems (courtesy of EROS Data Center).

taneous Field of View (IFOV), ranges from 80 m for the Landsat MSS (30 m for the thematic mapper) to 1,100 m for the AVHRR. Since the signal received by the sensor is the integrated value over the entire IFOV, the contributions of variable constituents of the area viewed must be considered carefully in evaluation of the signal. Obviously, the higher resolution of the Landsat observations is very advantageous.

The information from a remote sensor consists of a series of digital values from each part of the sensed spectrum. These signals vary according to the presence or absence of vegetation, the vegetation density, the opacity of the atmosphere, the moisture content of the background soil, etc. These data do not in and of themselves, however, provide information on the causes of the observed conditions. Diminished values in reflected IR may indicate that vegetation is under stress. The causes, however, usually must be determined by other means (e.g., direct field observations).

X-1.3 USE OF THE AVHRR DATASET

X-1.3.1 Spatial Resolution

The highest resolution AVHRR data (1.1 km at nadir) are available only as real-time High Resolution Picture Transmission (HRPT) data and in the recorded Local Area Coverage (LAC) dataset available for only 10 min per orbit (Kidwell, 1984). Otherwise, data for all parts of the globe are available in the Global Area Coverage (GAC) dataset, but at the reduced resolution of approximately 4 km. Samples of these GAC data are mapped to polar stereographic projections of the Northern and Southern Hemispheres. This is called the PSG product. Although these data still have the GAC 4-km resolution, they are used to represent areas varying from about 15 × 15 km at the equator to 30 × 30 km at the poles (Tarpley et al., 1984).

X-1.3.2 Time Series

Because of the daily observations from the AVHRR, time series analyses of vegetative conditions, locally, regionally, and globally, have significant potential. Time plots of vegetation index values are used weekly for assessment purposes. These plots of weekly values compare the variation of vegetation indices from the current growing season with those from each of the two previous years for specific agroclimatic regions over all agricultural areas of the world. The indices in the time series plots, however, are not error-free, as factors other than vegetation conditions can influence the values. Experience with the satellite vegetation indices and a knowledge of agrometeorology in each region are required for skillful interpretation of these assessment products (van Dijk et al., 1987).

X-1.3.3 Cloud Screening

According to Tarpley et al. (1984), on any single day about half of the Earth surface is obscured by clouds. The mid-afternoon overpass time of the currently used NOAA satellites compounds the problem because of the frequent occurrence of small cumulus clouds at that time of day.

Several techniques to reduce cloud contamination are being investigated. One is to filter the data in the mapping process so that the highest vegetation index value encountered for an array location is mapped. The disadvantage of this method is that under completely clear conditions it selects the greenest area within an array location and causes the map to be less representative of typical vegetation conditions. Another technique that is less effective as a cloud filter, but produces a random sampling of vegetation, is to set a threshold on the VIS brightness (Channel 1) to discriminate clouds. With both methods the data are mapped so that the vegetation index in an array location consists of a value computed from a *single* AVHRR pixel (no averaging).

To further reduce cloud contamination, 7-day maximum vegetation index composites are produced from the daily arrays. For each composite period, data are retained at each array location for the day with the largest normalized difference vegetation index. This greatly reduces cloud effects in the composite, except for areas which were cloudy for all seven days. Maximum value compositing greatly improves the product because some atmospheric and scan angle effects, which act to reduce the computed vegetation index, are automatically excluded.

Another cloud screening technique has been described by Tucker et al. (1984). This technique uses the Channel 5 thermal band of the AVHRR to detect the presence of clouds through the effective blackbody temperature. Although it may not be sensitive enough to detect very small low clouds, this technique is considered valuable for detecting high cirrus clouds with very low reflectivity.

X-1.4 ASSESSMENT PRODUCTS AND THEIR APPLICATIONS

Assessment of the latest global agricultural conditions has traditionally used rainfall and temperature information for each region and compared current values with historical normal values. A variety of other information is also considered, such as reports from field agricultural services, the news media, and diplomats. In no case is the quality and quantity of timely information for foreign areas equal to that available domestically. For some foreign regions, timely information can be very meager. Thus, satellite derived vegetation estimates, even though of limited accuracy, can augment available information about foreign agricultural conditions.

Satellite derived vegetation indices are related to the amount and condition of vegetation viewed, so the most useful satellite products are time-series plots of vegetation indices over regions of interest. The time-series plot of vegetation index in a region for the current year is compared with that for a previous year or years. Such a comparison can reveal significant trends in vegetation reflectance despite variable conditions in the atmosphere. If conditions prevailing in the baseline year are known, even generally, the current year's environmental conditions can be evaluated by reference to the baseline year. This is illustrated below in connection with African drought.

Comparison of indices to assess crop stress works best in regions where there are a reasonable number of cloud-free days during the crop seasons. Such regions include most of the major wheat, corn, and soybean areas of the globe. In regions where cloudiness is prevalent, such as over Central America and Southeast Asia, comparisons with historical plots are difficult. However, a good deal of useful information is still available from the satellite data. In particular, the satellite VIS and IR data can be used in their more traditional meteorological roles of detecting year-to-year variations in major cloudiness and precipitation.

X-1.4.1 Assessment Products

Assessment products currently available include time-series plots of vegetation indices for selected areas and maps of vegetation conditions based upon a derived color-coding system.

Various mathematical combinations of VIS and near-IR channel values (AVHRR Channels 1 and 2; Landsat MSS Channels 5 and 7) have been found to be sensitive indicators of the presence of green vegetation and are referred to as vegetation indices (Tucker et al., 1984).

Two vegetation indices have been routinely calculated from AVHRR Channels 1 and 2 since April 1982. A simple Vegetation Index (VI) and a Normalized Difference Vegetation Index (NDVI) are computed as follows:

$$\mathrm{VI} = \mathrm{Ch}_2 - \mathrm{Ch}_1 \qquad \text{(X-1-1)}$$

$$\mathrm{NDVI} = \frac{\mathrm{Ch}_2 - \mathrm{Ch}_1}{\mathrm{Ch}_2 + \mathrm{Ch}_1} \qquad \text{(X-1-2)}$$

where Ch_1 and Ch_2 are calibrated data from Channels 1 (VIS) and 2 (near-IR) of the AVHRR. The normalized difference vegetation index is preferred for global vegetation monitoring because it partially compensates for changing illumination conditions, surface slope, and viewing aspect. Clouds, water, and snow reflect more in the VIS than in the near IR, so for these features, the NDVI is negative. Rock and bare soil reflect similarly in these bands, so their vegetation indices are near zero. For vegetation, NDVI ranges from 0.1 to 0.6; the higher values are associated with greater density and "greenness" of the plant canopy. Atmospheric effects such as scattering by dust and aerosols, Rayleigh scattering, and subpixel-sized clouds all act to increase Ch_1 with respect to Ch_2 and to reduce the computed vegetation indices. The reduction is greater at high scan angles. Figure X-1-2 is an example of a weekly "greenness" image from NOAA 7.

AVHRR and Landsat MSS vegetation indices are in close agreement over the same regions (Gray and McCrary, 1981); high correlations ($r = 0.86$) between NOAA-6 AVHRR data and Landsat MSS data were found over an area along the Brazil-Argentina border. The correlations were surprisingly high, considering that: the Landsat data were collected over a 3-day period (March 15-17, 1980); the AVHRR data were collected on a single day (March 16, 1980); the Landsat data were obtained some 2 hours later than NOAA-6; and that Landsat MSS Band 7 and NOAA-6 AVHRR Channel 2 differed somewhat in spectral characteristics. Both Landsat and NOAA AVHRR data have been used to monitor land use changes in the Lake Chad basin (Schneider et al., 1985). The NOAA vegetation index was found to be effective for monitoring vegetation patterns within this highly changeable region. Comparisons of crop yields estimated from AVHRR data with USDA-reported crop yields for districts in Texas, Oklahoma, and Kansas showed high correlations ($r = 0.87$) (Barnett and Thompson, 1983).

Maps portraying satellite estimates of vegetation conditions are compiled in two general forms. One simply displays a vegetation index on a map, as in Figure X-1-2, with a single color or gray level assigned for each range of values. The other is a color display in which the shades of color are directly related to the values of a normalized satellite vegetation index. Initially a False Color InfraRed (FCIR) system was used and more recently the Ambroziak Color Coordinate System (ACCS) was developed as an improvement upon the FCIR system. For crop monitoring, the FCIR display is basically a two-color display of one VIS and one near-IR channel. The Earth's surface can look quite different when observed in a VIS channel as compared with the near-IR. The FCIR display uses red and cyan (blue-green), with each color's brightness in

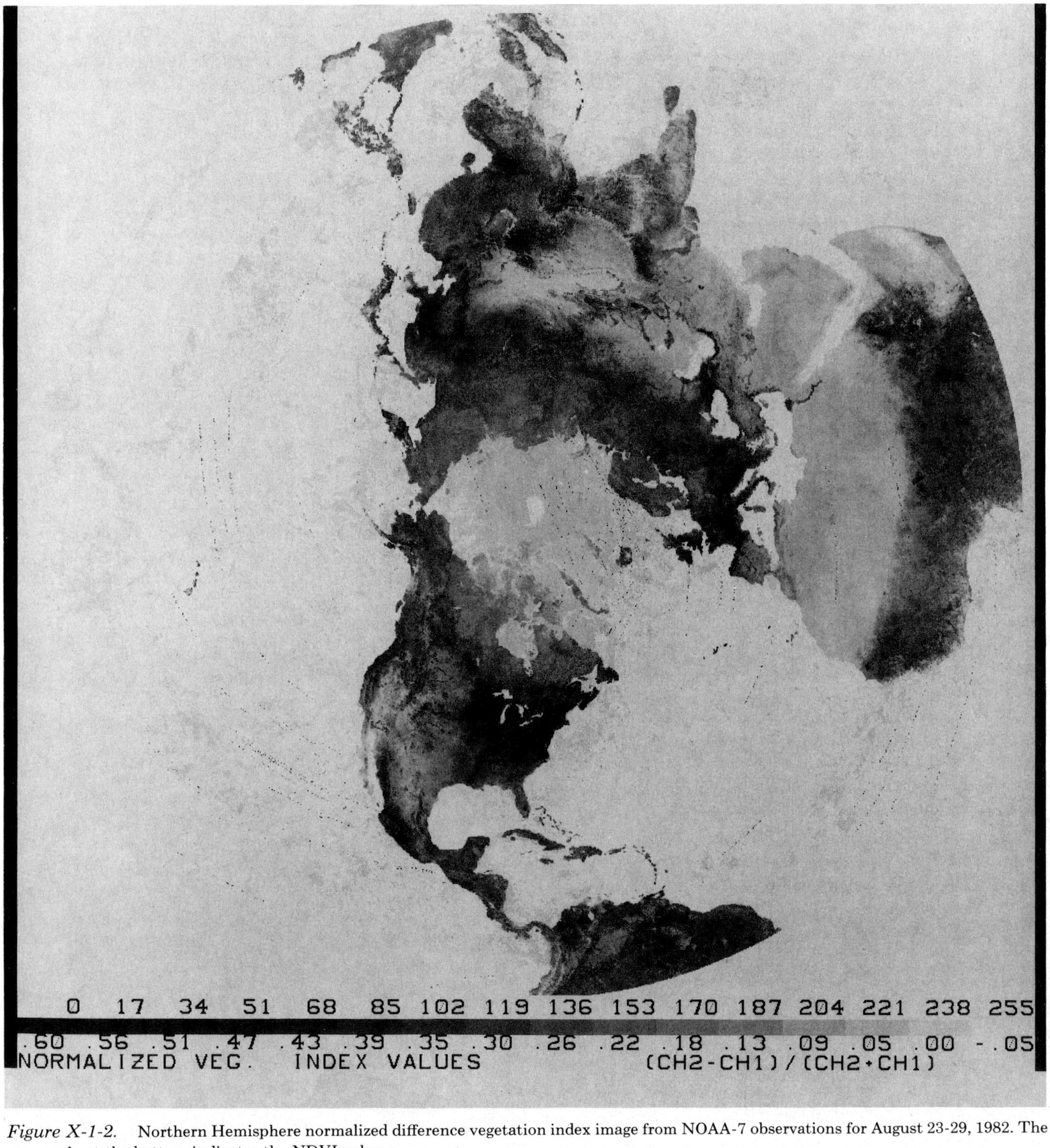

Figure X-1-2. Northern Hemisphere normalized difference vegetation index image from NOAA-7 observations for August 23-29, 1982. The gray scale at the bottom indicates the NDVI value.

a television display directly related to "normalized" vegetation index values. Positive values (increasing biomass) are in shades of red, and negative values (water) in shades of cyan. Where the two are equal the color is white or gray. Shades of color (or degree of saturation) are exactly equal to the index values. For example, an index of 0.60 appears as a pink that is 60% of saturated red.

Since changes in shades (or percent saturation) of red are equal to changes in the vegetation index, it is easy to make qualitative comparisons of the state of the vegetation on the ground—the redder the image, the more the vegetation. This system has been in use for some time for depicting the conditions of the vegetation, but it has a serious flaw. The subtle changes in saturation may be sufficient to detect patterns, but they are too subtle for making visual comparisons between points that are not adjacent.

The ACCS replaces the poorly resolved percent of saturation of a single-color with hue. It is much easier to see the difference between red and orange (hue) than between red and reddish pink (shading or saturation). The first ACCS assigned green to an index value of 1, red to zero, and blue to −1. The result is a fairly natural looking scene with green vegetation, dark blue water, red and brown soils, and yellow stressed vegetation. Clouds are bright red, however, an annoying feature for the analyst. To cure this problem, the red signal is reduced to zero for the very high VIS brightness values of clouds. This change produces an image of more natural appearance (Ambroziak, 1984).

The ACCS proved to be extremely useful in monitoring the status of tropical vegetation during the Northern Hemisphere summer of 1984. Estimates produced just from the images agreed closely with assessments produced from all other data sources. Patterns of drought and plentiful rainfall appeared in finer detail than was possible by use of station data alone.

X-1.4.2 Drought Assessment

A. Africa. The 1983-84 droughts in Africa showed that regularly available and frequent satellite information can be used for assessment of crop conditions and production potential. The potential is greatest for early warning of food shortages, especially in areas where information is scarce (van Dijk et al., 1984; Johnson et al., 1987). In Sudan, where current meteorological data are unavailable, the vegetation index for 1984 was lower than for 1983 (Figure X-1-3). The agricultural area of Sudan was divided into smaller assessment areas of 1° lat. × 2° long. (Figure X-1-4). Time plots of the vegetation index for 1983 and 1984 for four of these assessment areas are illustrated in Figure X-1-5. In area 4, a semi-arid region with average annual precipitation of 150-600-mm, the extremely low

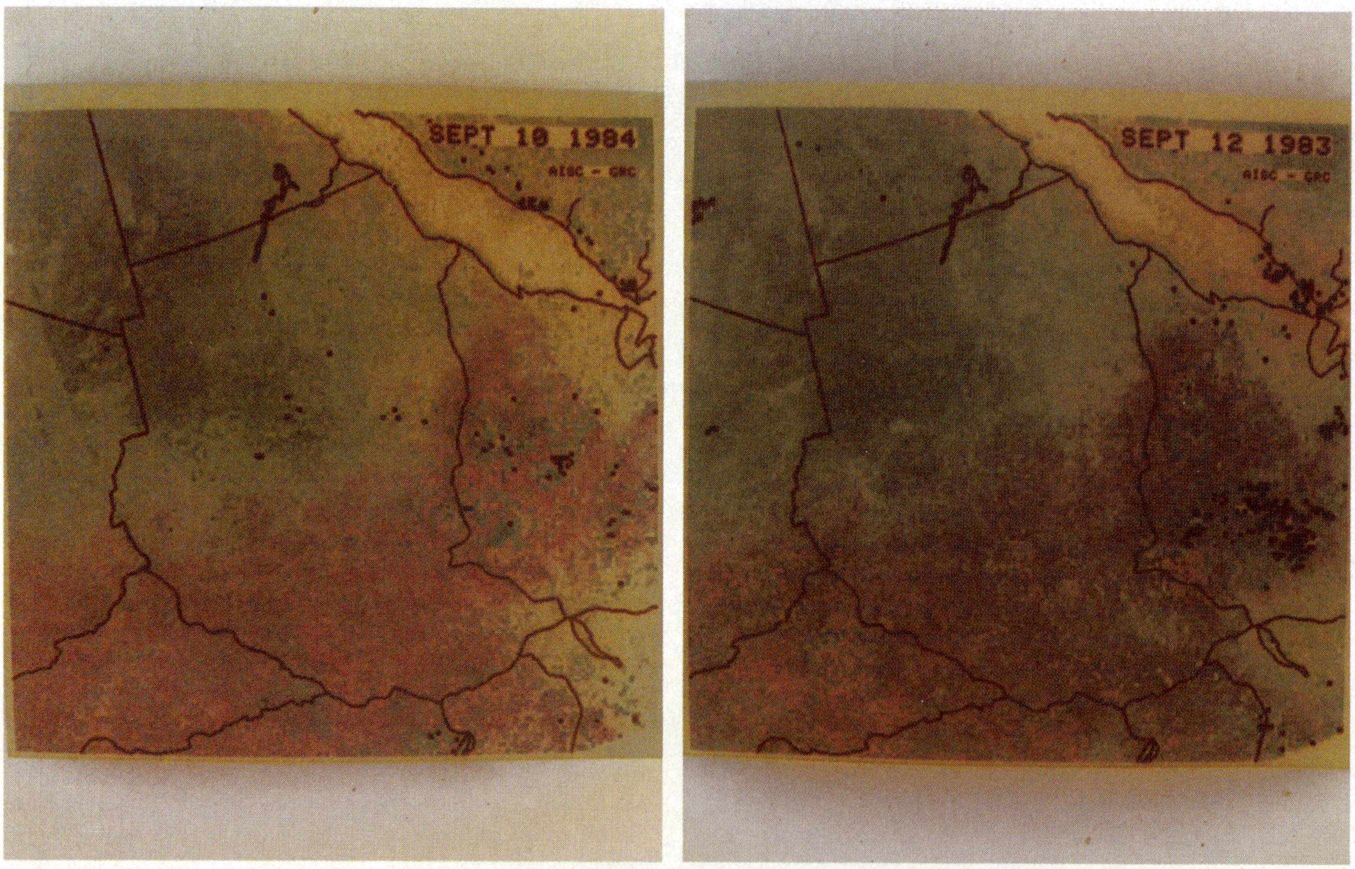

Figure X-1-3. ACCS images of Sudan for September 1983 and 1984 (van Dijk, 1986).

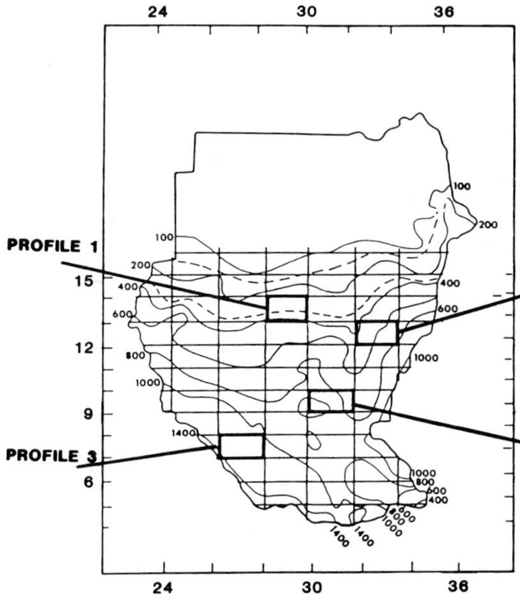

Figure X-1-4. Selected 1° lat. × 2° long. assessment areas for the construction of vegetation indices. Heavy outlines show areas for which detailed analyses of time plots for 1983 and 1984 were made (see text). Solid and dashed isohyets of average annual precipitation (mm) are also shown (van Dijk, 1986).

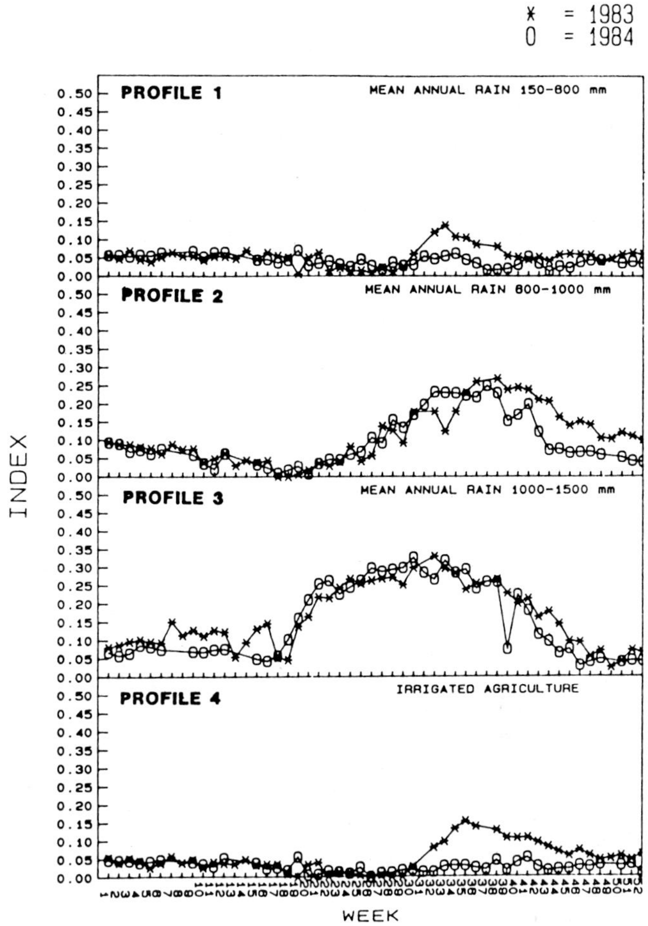

Figure X-1-5. Time plots of vegetation indices for four areas in Sudan (see Figure X-1-4) for 1983 and 1984 (van Dijk, 1986).

vegetation index values of 1984 for this region were essentially at the desert level. The 1984 growing season conditions were very unfavorable for sorghum, millet, maize, groundnut, and sesame crops.

In area 1, a rainfed farming region with average annual precipitation of 600 to 1000 mm, the vegetation index for the first part of the 1984 crop season was higher than 1983 and the middle part about the same as 1983. During October, however, the vegetation index for 1984 was considerably lower than in 1983. Late September and October are the harvest months for this region. The vegetation index alone cannot explain whether the 1984 drop was due to early harvesting, unfavorable crop conditions, or a combination of the two. More information is needed to resolve this question.

In area 3, an intensive agricultural region with average annual precipitation of 1000 to 1500 mm, the vegetation index was about the same for 1983 and 1984, indicating about the same crop growing conditions for both years.

In area 2, an irrigated farming region along the Blue and White Nile Rivers, the vegetation index was much lower for 1984 than 1983, indicating unfavorable crop growth conditions with lower yields.

Overall results of a more detailed study of all areas of Sudan for 1984 showed that the unfavorable crop conditions were very localized and mainly in the drier areas (Figure X-1-6). Because the intensive agricultural area showed the same vegetation index values for 1983 and 1984, the reduction in total yield for the whole country was not as dramatic as expected.

B. The United States Midwest. In July, August, and September of 1983, this area was affected by drought and heat stress that had a devastating effect on the crops (Figure X-1-7). Two areas were selected for study using the satellite vegetation index (Figure X-1-8). One area, *MO*, is primarily forested in the Missouri Ozarks. A second area, *NMO*, is a corn and soybean producing area straddling the corners of Missouri, Iowa, Kansas, and Nebraska. According to Figure X-1-7, area *MO* was suffering only from heat stress, while area *NMO* was affected by both heat and drought. Time plots of the vegetation indices of both areas for May-July in 1982 and 1983 are compared in Figure X-1-9.

In area *MO*, the vegetation index for 1983 was greater than that of 1982 at the end of May. After mid-July, however, the effects of heat stress on the forest vegetation apparently resulted in a lower vegetation index for 1983 than for 1982.

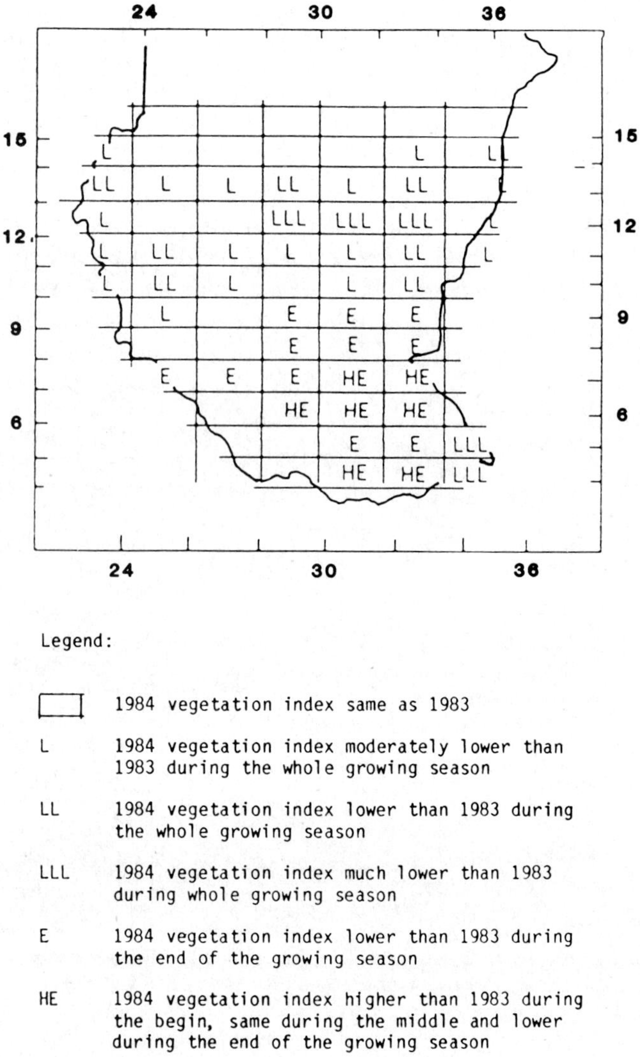

Figure X-1-6. Vegetation stress map of Sudan calculated for 1° lat. × 2° long. areas (van Dijk, 1986).

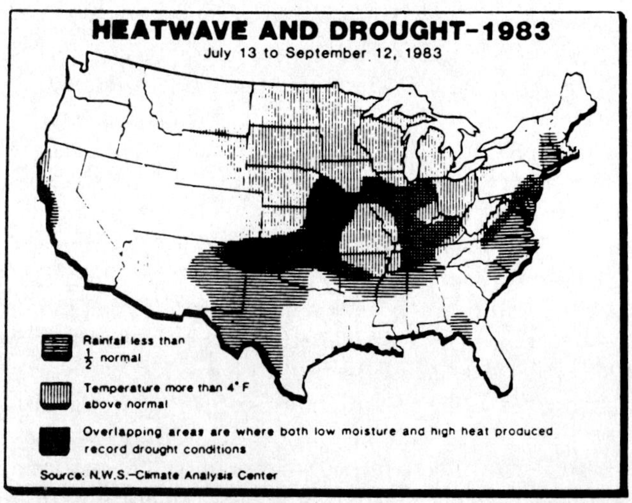

Figure X-1-7. Heatwave and drought status, 1983.

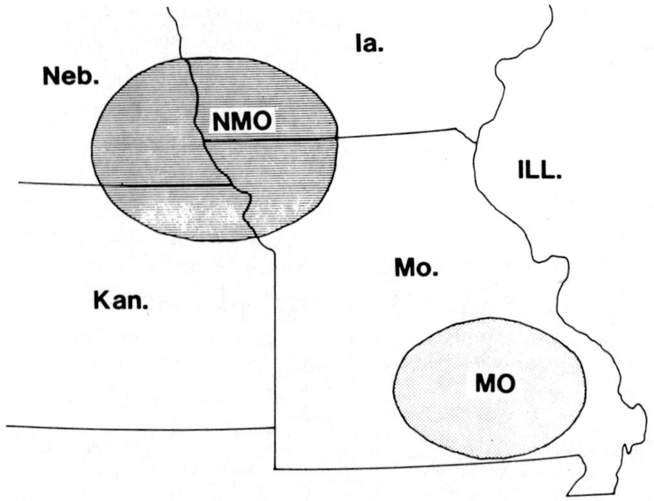

Figure X-1-8. Study area locations.

In area *NMO,* vegetation indices were nearly the same for both years during late spring and early summer. In 1982 the vegetation index continued to increase into July and the corn yield was good. Less increase in vegetation index was seen in July 1983. The corn yield in 1983 was much lower than in 1982. The heat and drought stress affected the vegetation index of both annual field crops and perennial forest. The vegetation index was an overall stress indicator. The difference in the vegetation index was greater for the corn and soybean area in 1982 and 1983 than for the forest (van Dijk et al., 1984).

X-1.4.3 Flood Assessment

Floods annually destroy or threaten significant quantities of agricultural production. Information on duration and extent of flooding may be derived from satellite data. Assessments of damage and possible impact on production may be made and ameliorating measures suggested.

Rohde et al. (1976) used Landsat imagery to map the 1975 spring flood on the Red River of North Dakota. Allison and Schmugge (1979) used Nimbus-5 data in the analysis of floods in eastern Australia. Helfert et al. (1981) used TIROS N data to evaluate flooding along the Ouachita and Mississippi drainage basins in northern Louisiana and southern Arkansas in May 1979. They found that by using a simple vegetation index, the ready identification and crude classification of land, water, and cloud cover regions were possible.

X-1.4.4 Wildfire Management Program

In a joint project, the U.S. Geological Survey EROS Data Center, the Bureau of Land Management (BLM), and NOAA evaluated the use of AVHRR data in BLM's

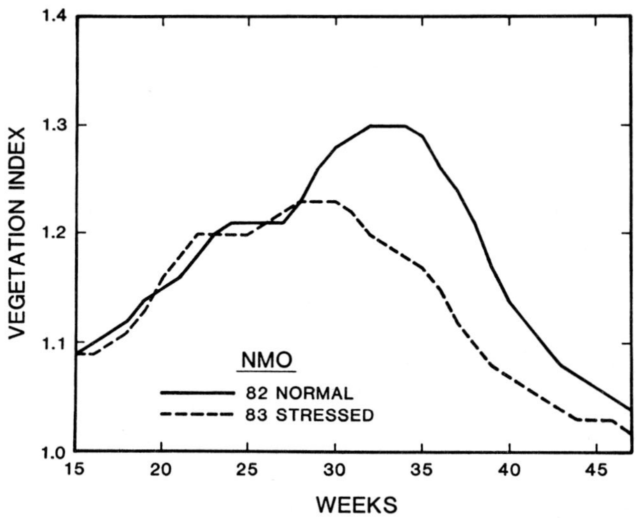

Figure X-1-9a. Vegetation indices for area *NMO* for 1982 and 1983.

national wildfire management program. For 5 days in March and April 1982, the normalized vegetation index was calculated for three study sites in northwestern Arizona. These calculations were made at 1-km resolution from AVHRR LAC data within 1 week of the observations.

It was found that the vegetation index corresponded to the fire fuels measured on the ground by field personnel. Thus, the index could be used to identify areas highly susceptible to wildfires in 1982. It was concluded that the information content of the AVHRR vegetation index provided district fire management personnel with the capability of: (1) detecting, locating, and ranking areas according to the buildup of fire fuels and risk of

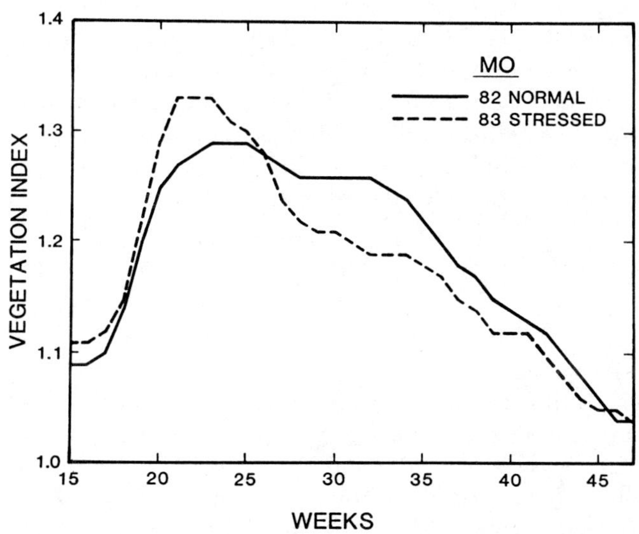

Figure X-1-9b. Vegetation indices for area *MO* for 1982 and 1983.

wildfires; (2) estimating time of senescence of the herbaceous vegetation for use as an indicator of when fire problems will increase; and (3) identifying regions of higher risk so fire equipment can be strategically located to maximize its utility (Miller and Moore, 1983).

X-1.5 AREAS REQUIRING RESEARCH

Higher resolution than that provided by the PSG product is essential in certain areas badly affected by cloudiness, such as Central America and Southeast Asia, and in regions such as the Caribbean. Here, LAC data at full 1.1-km-resolution are needed. The much greater amount of data to be processed must be offset by careful selection of regions and clear observation dates.

Vast improvement in PSG data usefulness and reliability would result from improvements in cloud screening. These are likely to include use of the thermal IR channels along with Channels 1 and 2. Care must be exercised, however, to be sure that improvements in cloud screening do not make future PSG data totally incompatible with archived PSG data.

Attempts should be made to account for, or at least circumvent, some of the factors that cause observation to observation scatter in the AVHRR data. Known factors are amount of cloudiness in the scene, look angle, and amount of water vapor in the atmosphere. Considerable improvements in the PSG data would result from understanding and reducing the effects of these factors.

Since the major value of the PSG data lies in being able to compare vegetation amount in the same geographical area from one year to another, any factor that changes the AVHRR data from year to year is serious. One such factor is the orbit drift which causes the satellite to cross a given point about half an hour later in 1984 than in 1983. Solar illumination was reduced in 1984 relative to 1983 and the Channel 1 and Channel 2 values were reduced proportionally. This particular effect can be minimized in the data by considering a ratio of Channel 2 to Channel 1 as a vegetation index.

Variation in illumination also occurred when NOAA-7 was replaced by a new satellite. If data from an early morning pass must be compared to that from a mid-afternoon pass, inconsistencies are anticipated.

Currently AVHRR PSG products are available about 2 weeks after they are acquired. GAC or LAC products may be available somewhat sooner, but require greater processing time due to their resolution. This is a vast improvement over Landsat data availability, but is often too late to make quasi-real-time evaluations of probable effects of droughts, floods, etc. A shorter processing time would be desirable.

Chapter X-2

SOLAR RADIATION, MAX/MIN TEMPERATURE, AND DAILY PRECIPITATION[1]

X-2.1 INTRODUCTION

The most important and variable determinants of year-to-year changes in agricultural production are weather and climate anomalies. Unfavorable fluctuations of precipitation and/or temperature can cause crop failures, delay planting and harvesting, and increase disease and insect infestations. Too little snow cover exposes crops to winterkill; hot, dry winds completely desiccate plants within 24 to 48 hours; and continuously overcast skies deprive plants of radiation required for photosynthesis.

Weather information is used together with other sources of information, such as newspaper reports and knowledge of agricultural practices, to monitor agricultural conditions. Crop production forecasters rely primarily on routine weather observations. Forecast accuracies are reduced when observations are sparse, unreliable, or delayed too long. In many countries meteorological observations are inadequate for agricultural monitoring. Other quantities important to agricultural monitoring, but not routinely measured at the ground, are solar radiation, canopy temperature, and vegetation indices.

[1] Primary Contributor: J. Dan Tarpley

Methodologies have been developed to produce quantitative data for agricultural monitoring from operational satellite data. Estimates of precipitation, insolation, and surface temperature from satellite data are not directly comparable with ground measurements because of spatial averaging and other problems inherent in their interpretation. In many cases these differences are advantageous: a single satellite estimate of average conditions over a large area may be more representative than a single point measurement within the area. Current agricultural models, originally developed with conventional meteorological data, can now use satellite data to form more comprehensive estimates of conditions, particularly where conventional data are sparse or undependable.

X-2.2 SOLAR RADIATION ESTIMATES

Incident solar radiation is the primary source of energy for growing vegetation. Crop development and yield are direct functions of photosynthesis associated with the amount of accumulated solar energy. Solar radiation also plays a key role in evapotranspiration. VIS observations from satellites provide an excellent source of information about the amount of solar radiation reaching the plant canopy. The methods of estimating the solar radiation reaching the surface from satellite observations have been discussed in IX-1.4.1. Results of an experiment using geostationary visible data in a physical model of Pinker and Ewing (1985) were illustrated in Figures IX-1-10 and IX-1-11.

The procedures used by NESDIS for estimating solar radiation on a real-time basis have been statistical. These methods require coincident satellite and ground (pyranometer) observations to develop the coefficients in the regression equation. The first such regression equation was developed by Tarpley et al. (1978) using coincident GOES and ground based pyranometer data over the Great Plains in the summer of 1977. This equation was used to make surface solar radiation estimates on an operational basis from July 1980 through February 1983 over agriculturally important areas of the U.S. An improved methodology with a revised regression equation was developed by Justus et al. (1986) from coincident GOES and pyranometer data for July through December 1980 from the Great Plains to the East Coast. This new method was implemented operationally in late February 1983.

The method produces estimates of daily total insolation; these are based on hourly estimates made from GOES data between 0800 and 1600 LST, with interpolation used toward both sunrise and sunset and for any other missing hourly values. The original resolution of the data is 8 km at the subpoint. Averages of 5 × 5 arrays are obtained for each 1° lat. × 1° long. grid point. Figure X-2-1 shows a comparison of such daily estimates with coincident pyranometer measurements for the July-December 1980 dependent dataset.

An evaluation of independent, routinely produced GOES insolation estimates was made for the period March-September 1983 (Sullivan et al., 1984). Comparisons were made between the satellite derived daily total insolation and measurements by 23 pyranometer stations. Each ground observation was compared against the satellite estimate at the nearest grid point. Typically, the RMS errors of the satellite estimate against ground observations were between 45 and 75, and the bias was no greater than ±25 ly. Other comparisons have been made with pyranometer data from Argentina (Espoz and Brizuela, 1983), the Western U.S., and a dense network in Ohio. All show errors in the daily estimate of between 10 and 15% of the daily mean.

A pyranometer station typical of the midwest is in Canfield, Ohio. A time plot of the difference between ground observed and satellite derived insolation at the nearest gridpoint is shown in Figure X-2-2. The statistics for different time groupings and for weekly averaged insolation are listed in Table X-2-1. Differences

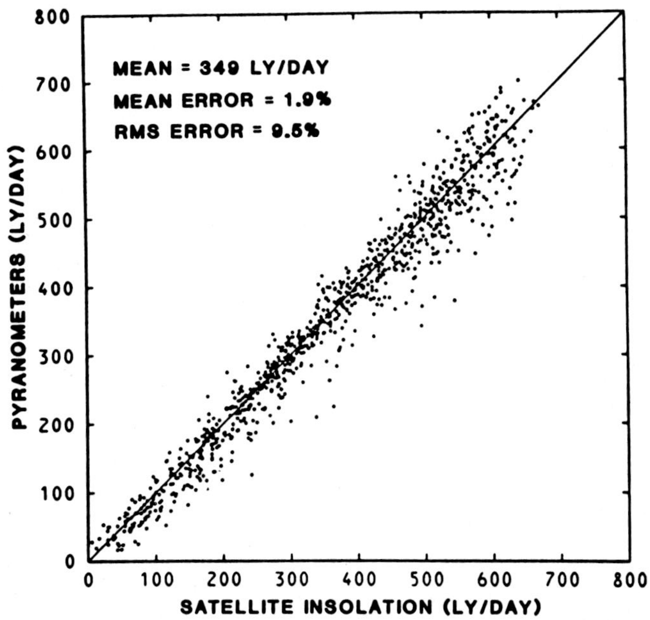

Figure X-2-1. Observed daily total insolation vs. satellite estimate for dependent dataset. The mean error and RMS error are expressed as percentages of the daily mean.

were significantly larger in the spring than in the summer. The large springtime errors were very likely associated with a greater number of synoptic-scale weather systems common to the midwest in spring, while summer was characterized by stable weather conditions and a higher proportion of clear weather. The comparison of weekly averages shows better agreement because most of the daily error is random.

A problem with the operational insolation algorithm is mistaking snow cover for clouds. When the ground is snow covered, the reflectivity is high and the insolation equation produces low estimates of insolation as though the scene were cloudy. Using the current GOES sensors, there is no way to automatically distinguish snow from clouds.

X-2.3 SURFACE TEMPERATURE ESTIMATES

Air temperature is significantly related to crop development and conditions. Current operational crop and soil moisture models require daily maximum and minimum shelter temperatures and dew point temperatures. These are used because they were the only temperature data available to agricultural model developers. The high spatial variability of shelter temperature contributes to crop monitoring errors where observations are sparse and unrepresentative of a large region. Because satellite data essentially provide areally averaged information, a single satellite estimate may

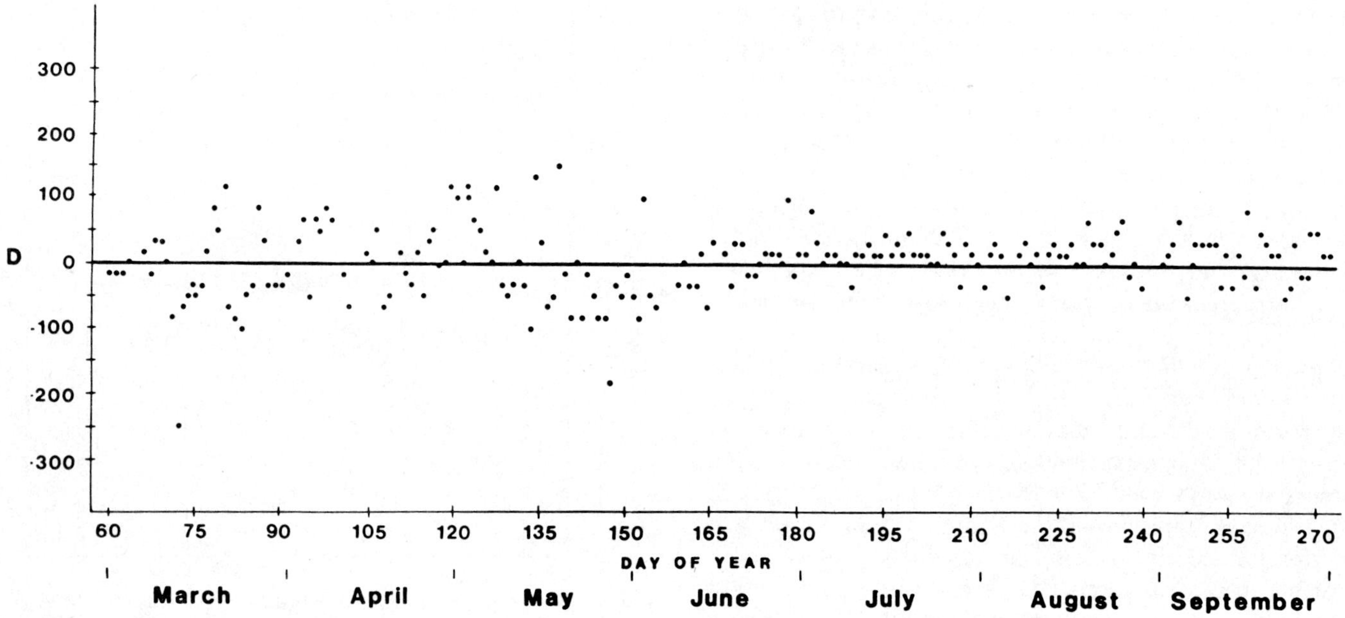

Figure X-2-2. Daily difference between satellite insolation estimates and ground observations (ly/day) for the 1983 growing season at Canfield, Ohio.

be more representative of a large area than one or two conventional measurements within the area.

Canopy (or skin) temperature may be more directly related to growth and evapotranspiration than shelter temperature. The difference between the two is a measure of crop stress (Idso et al., 1977; Jackson et al., 1981). The temperature of the radiating surface is also related to evapotranspiration, soil moisture, and surface energy balance (Carlson et al., 1981; Price, 1982a). The ability to observe canopy (or skin) temperature directly is an advantage of satellite observations. Canopy and skin temperatures are equivalent when the sensor FOV is filled with vegetation. Bare soil, water, and differing plant communities within a FOV complicate the relation of skin to canopy temperature.

Since most crop models were developed from shelter temperature, satellite derived surface temperatures have generally been expressed as equivalent shelter temperatures.

X-2.3.1 Canopy Temperature Estimation

Satellite observations in the thermal IR window (10 to 12 μm) are used to obtain estimates of canopy or skin temperature (Chapters VI-2, VIII-1, VIII-4, and IX-1). The possibility of using the AVHRR split window technique, discussed in VIII-4, to measure land surface temperature has been investigated by Price (1984). He found that surface temperature can be calculated with an RMS error of 2 to 3°C for a vegetated land surface. The errors result primarily from imperfect knowledge of the water vapor and variations in surface

Table X-2-1 *Statistics of differences between daily and weekly insolation observed by pyranometer and estimated from satellite for Canfield, Ohio, March–September 1983*

	DAILY				WEEKLY
TIME INTERVAL	MARCH–APRIL	MAY–JULY	AUGUST–SEPTEMBER	MARCH–SEPTEMBER	MARCH–SEPTEMBER
RMS (ly/day)	62.2	53.3	33.8	51.3	26.1
Correlation coef.	0.92	0.95	0.97	0.95	0.98
Largest negative diff. (ly/day)	−258	−182	−58	−258	−71
Largest positive diff. (ly/day)	123	145	85	145	49
Number of days	54	88	57	199	30

emissivity as a function of wavelength. Price used emissivity values characteristic of vegetation; emissivity is more variable for bare soil or rocks.

The diurnal temperature range of the land surface is influenced by the amount of moisture in the surface layer of the soil. The tendency of soils to resist temperature change in response to an applied heat source is called thermal inertia, which is dependent upon soil type and soil moisture content. The temperature response of soil to the diurnally varying solar insolation (i.e., the soil's diurnal heat capacity) is a function of the soil thermal inertia. Theoretical arguments relating soil moisture availability, evapotranspiration, and surface energy balance to observations of diurnal surface temperature range have been presented by Price (1980, 1982a, 1982b) and Carlson et al. (1981). Their results were applied to data from the Heat Capacity Mapping Mission (HCMM) satellite. HCMM case studies showed that surface moisture differences are readily detectable in areas where there are large contrasts between amounts of soil moisture, e.g., urban areas and irrigated valleys within deserts.

Hourly IR observations from GOES as a source of thermal inertia information using the HCMM techniques have been studied with similar results (Carlson et al., 1984). The shortcomings of GOES data are their low resolution (4 × 7 km) and the inability to correct the data for atmospheric moisture. The lack of atmospheric correction may not be important since the quantity of interest is the diurnal range of surface temperature.

While the studies cited indicate that AVHRR data can provide accurate estimates of surface temperature, routine operational application is difficult because of the cloud problem. Over oceans, the detection and elimination of cloud-contaminated pixels are relatively easy to achieve because of the uniformity of the sea surface and the conservative nature of the sea surface temperature field. Over land, the extreme heterogeneity of the surface and the rapidly changing surface temperature make automatic cloud filtering of AVHRR data very difficult. A solution to this problem would be for analysts to identify and eliminate cloud-contaminated scenes interactively, but this procedure is labor intensive and costly.

Satellite surface temperature estimates have been used to delineate areas of freezing for frost warnings and for monitoring freezing events that can affect food production. GOES IR data facilitate freeze forecasting by following the diurnal progression of the freeze line. Accurate freeze forecasts permit growers to protect crops only when there is a significant freeze threat. Such a forecast system is in operation at the University of Florida (Bartholic and Sutherland, 1978; Martsolf, 1983a; 1983b). Freeze events occur in Florida only after cold frontal passages so the atmosphere is dry and GOES imagery can be converted to skin temperature with little error caused by atmospheric moisture.

X-2.3.2 Shelter Temperature

The TIROS Operational Vertical Sounder (TOVS) system on NOAA polar orbiters has been used in methods for estimating shelter temperatures. The polar orbiters provide the global coverage required for crop monitoring. Orbital passes near 0230 and 1430 local time are at relatively convenient times for determining daily temperatue extremes.

A simple linear regression approach has been developed to estimate shelter temperature from satellite soundings (Davis and Tarpley, 1983). Sets of satellite soundings and shelter temperatures, coincident in space and time, were used to develop regression coefficients. The hourly surface observations used included stations within a 1.5° lat. radius of the sounding location at the closest hours before and after the time of the satellite sounding. Data included cloud amount, temperature, dewpoint, and wind velocity.

Figure X-2-3 shows the average standard deviation of the difference between satellite retrieval shelter temperatures and ground observations for April 1982 through September 1983 (independent data) over the eastern two-thirds of the U.S. The seasonal variation of the difference between satellite and ground estimates is striking. During winter months the satellite estimates are less accurate, probably because of greater horizontal temperature gradients and large surface temperature inversions characteristic of winter conditions. The nighttime satellite estimates show a bias of between 1° and 2°C (satellite too warm) in winter because of strong inversions over cold ground.

There have been several case studies of errors in

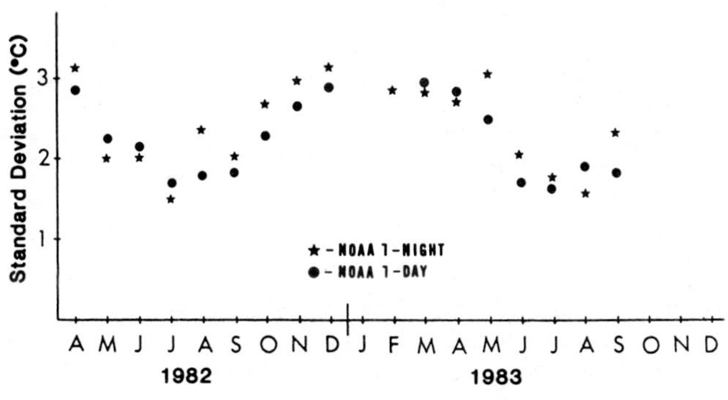

Figure X-2-3. Monthly mean standard deviation of the difference between satellite estimates of shelter temperature and groundtruth.

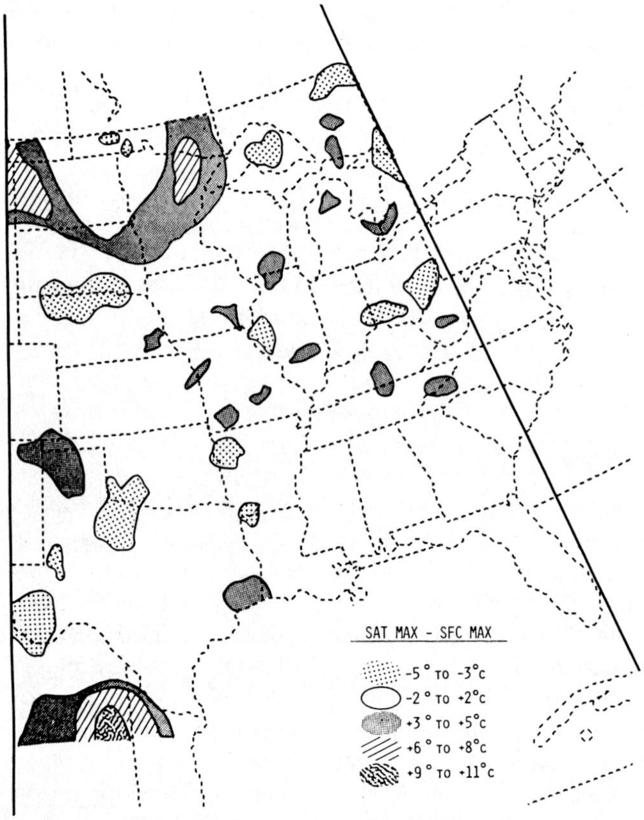

Figure X-2-4. Difference between analyses of satellite estimates and surface observations of daily maximum temperature for September 8, 1984. Differences greater than ±2°C are shaded.

maximum and minimum temperature fields estimated from satellite data. Figure X-2-4 shows the field of errors for maximum shelter temperatures estimated from TOVS data on a single day, September 8, 1984. Over most of the area, the difference was between ±2°C. Large errors occurred over the Northern Plains and northern Mexico. In both cases the satellite estimate was 6 to 10°C warmer than the observed maximum and both resulted from satellite temperatures in warm areas being spread by the analysis procedure into colder areas. If the density of satellite retrievals had been greater, this would have been less of a problem.

X-2.4 DAILY PRECIPITATION ESTIMATES

Crop yield prediction models require daily precipitation reports throughout the year. Precipitation outside the growing season determines the soil moisture available to the crops at the beginning of the growing season. Precipitation amount and its distribution during the growing season critically affect the vigor and productivity of the crops.

In many areas of the world, synoptic weather stations are too widely spaced to represent the distribution of rainfall adequately at resolutions needed for agricultural monitoring. Significant rainfall between stations is unreported. Operationally available VIS and IR satellite data, however, can fill in the likely precipitation distribution where synoptic data are sparse. The techniques described in Chapters VII-6 and IX-4 can be useful in estimating precipitation amounts for agricultural assessment. Where geostationary observations are available, complete diurnal coverage of major precipitation, particularly the highly convective type, can be rather fully assessed. Accuracies of simple, objective estimates such as the GOES precipitation index (discussed in Chapter IX-4) naturally tend to be more accurate when summed in time and averaged over moderately sized areas. Cumulative precipitation amounts, of course, are generally most useful for vegetative growth, so such simple satellite precipitation indices can be very useful in crop yield production.

BIBLIOGRAPHY AND REFERENCES FOR SECTION X

Allison, L. J. and T. J. Schmugge, 1979: A hydrological analysis of East Australianfloods using Nimbus-5 electrically scanning radiometer data. *Bull. Amer. Meteor. Soc., 60*, 1414-1427.

Ambroziak, R. A., 1984: A new method of incorporating meteorological satellite data into global crop monitoring. *Proc. of the 18th Intl. Symp. Remote Sens. Environ., Oct. 1-5, 1984, Paris, France*, Environmental Res. Inst. of Michigan, Ann Arbor, MI, 9 pp.

Atwater, M. A. and J. T. Ball, 1978: Intraregional variations of solar radiation in the eastern United States. *J. Appl. Meteor., 17*, 1116-1125.

——— and P. S. Brown, Jr., 1974: Numerical computations of the latitudinal variation of solar radiation for an atmosphere of varying opacity. *J. Appl. Meteor., 13*, 289-297.

Barnett, T. L. and D. R. Thompson, 1983: Large area relation of Landsat MSS and NOAA-6 AVHRR spectral data to wheat yields. *Remote Sens. Environ., 13*, 277-290.

Barrett, E. C. and D. W. Martin, 1981: *The Use of Satellite Data in Rainfall Monitoring*. Academic Press, New York, NY, 329 pp.

Bartholic, J. F. and R. A. Sutherland, 1978: Future freeze forecasting. *Proc. Fla. State Hort. Soc., 91*, Tallahassee, FL, 334-336.

Brakke, T. W. and E. T. Kanemasu, 1981: Insolation estimates from satellite measurements of reflected radiation. *Remote Sens. Environ. 11*, 157-167.

Carlson, T. N., F. G. Rose, and E. M. Perry, 1984: Regional-scale estimates of surface moisture availability from GOES infrared satellite measurements. *Agron. J., 76*, 972-979.

———, J. K. Dodd, S. G. Benjamin, and J. N. Cooper, 1981: Satellite estimation of the surface energy balance, moisture availability, and thermal inertia. *J. Appl. Meteor., 20*, 67-87.

Davies, J. A., W. Schertzer, and M. Nunez, 1975: Estimating global solar radiation. *Bound.-Layer Meteor., 9*, 33-52.

Davis, P. A., and J. D. Tarpley, 1983: Estimation of shelter temperature from operational satellite sounder data. *J. Climate Appl. Meteor., 22*, 369-376.

Espoz, C. and A. Brizuela, 1983: Application of remote sensing and agrometeorological methods for crop assessment in the Pampa Humeda. *FAO Fellowship Program Report, ARG 81/002.*

Gatlin, J. A., R. J. Sullivan, and C. J. Tucker, 1984: Considerations ofand improvements to large-scale vegetation monitoring. *IEEE Trans. on Geoscience and Remote Sensing, 22*, 496-502.

Gautier, C., G. Diak, and S. Masse, 1980: A simple physical model to estimate incident solar radiation at the surface from GOES satellite data. *J. Appl. Meteor., 19*, 1005-1012.

Gray, T. I. and D. G. McCrary, 1981: Meteorological satellite data — A tool to describe the health of the world's agriculture. *AgRISTARS, Rep. EW-NI-04042*, Johnson Space Center, Houston, TX, 7 pp.

Griffith, C. R., W. L. Woodley, P. G. Gruber, D. W. Martin, G. Stout, and D. N. Sikdar, 1978: Rain estimation from geosynchronous satellite imagery—Visible and infrared studies. *Mon. Wea. Rev., 106*, 1153-1171.

Hanson, K. J., 1971: Studies of cloud and satellite parameterization of solar irradiation at the Earth's surface. *Proc. of the Miami Workshop on Remote Sensing*, Dept. of Commerce, Washington, DC, 133-148.

———, T. H. Vonder Haar, and V. E. Suomi, 1967: Reflection of sunlight to space and absorption by the Earth and atmosphere over the United States during spring 1962. *Mon. Wea. Rev., 95*, 354-362.

Hay, J. E. and K. J. Hanson, 1978: A satellite-based methodology for determining solar irradiance at the ocean surface during GATE. *Bull. Amer. Meteor. Soc., 59*, 1549.

Heitkemper, L. J., 1985: An interactive method for estimating rainfall from satellites. *Intl. Conf. on Interactive Information and Processing Systems for Meteorology, Oceanography, and Hydrology, January 7-11, 1985, Los Angeles, CA*, Amer. Meteor. Soc., Boston, MA, 271-278.

Helfert, M. R., D. G. McCrary, and T. I. Gray, 1981: Utilization of meteorological satellite imagery for world-wide environmental monitoring the Lower Mississippi River flood of 1979—Case 1. *AgRISTARS, Rep. EW-N1-04104*, Johnson Space Center, Houston, TX, 26 pp.

Hill, J. D., N. Strommen, C. Sakamoto, and S. LeDuc, 1980: LACIE - An application of meteorology for United States and foreign wheat assessment. *J. Appl. Meteor., 19*, 22-34.

Hoffer, R. M. and C. J. Johannsen, 1969: *Remote Sensing in Ecology*, P. L. Johnson (ed.), Univer. of Georgia Press, Athens, GA, 244 pp.

Holben, B. and R. S. Fraser, 1984: Red and near-infrared sensor response to off-nadir viewing. *Intl. J. Remote Sensing, 5*, 145-160.

Idso, S. B., R. D. Jackson, and R. J. Reginato, 1977: Remote sensing of crop yields. *Science, 196*, 19-25.

Jackson, R. D., S. B. Idso, R. J. Reginato, and P. J. Pinter, Jr., 1981: *Water Resources Res., 12*, 1133-1138.

Johnson, G. E., A. van Dijk, and C. M. Sakamoto, 1987: The use of AVHRR data in operational agricultural assessment in Africa. *Geocarto. Intl., 1*, 41-60.

Justus, C. G., M. V. Paris, and J. D. Tarpley, 1986: Satellite measured insolation in the United States, Mexico, and South America. *Remote Sens. of Environ., 20*, 57-83.

Kidwell, K. B., 1984: NOAA Polar Orbiter Data (Tiros-N, NOAA-6, NOAA-7, and NOAA-8) Users Guide. Dept. of Commerce, Washington, DC, 144 pp.

Klink, J. C. and K. J. Dollhopf, 1986: An evaluation of satellite-based insolation estimates for Ohio. *J. Climate Appl. Meteor., 25*, 1741-1751.

Liou, K. N., 1976: On the absorption, reflection and transmission of solar radiation in cloudy atmospheres. *J. Atmos. Sci., 33*, 798-805.

Martsolf, J. D., 1983a: Florida frost information and dissemination system. *Proc. of the Conf. on Cooperative Climate Seminar, Tallahassee, FL*, USDS, NOAA, National Climate Program Office, Washington, DC, 209-225.

———, 1983b: Satellite thermal maps provide detailed views and comparison of freezes. *The Citrus Industry, 64*, 24-31.

McClain, E. P., 1983: Multichannel improvements to satellite-derived global sea surface temperatures. *Adv. Space Res., 2*, 43-47.

McMillin, L. M. and D. S. Crosby, 1984: Theory and validation of the multiple window sea surface temperature technique. *J. Geophys. Res., 89*, 3655-3661.

Miller, W. A. and D. G. Moore, 1983: Time-series vegetation monitoring with NOAA satellite data. *Proc. of the Intl. Conf. Renewable Resource Inv. for Monitoring Changes and Trends*, Oregon State Univer., Corvallis, OR, 86-89.

Myers, T. P. and R. F. Dale, 1983: Predicting daily insolation with hourly cloud height and coverage. *J. Climate Appl. Meteor., 22*, 537-545.

Myers, V. I., 1983: *Manual of Remote Sensing*, Vol. II. Chap. 33, Robert N. Colwell (ed.), Amer. Soc. of Photogram., Falls Church, VA, 2111-2228.

Price, J. C., 1984: Land surface temperature measurements from the split window channels of the NOAA-7 Advanced Very High Resolution Radiometer. *J. Geophys. Res., 89,* 7231-7237.

———, 1982a: On the use of satellite data to infer surface fluxes at meteorological scales. *J. Appl. Meteor., 21,* 1111-1122.

———, 1982b: Estimation of regional scale evapotranspiration through analysis of satellite thermal infrared data. *IEEE Trans. Geosci. Remote Sens., GED-D2D0,* 286-292.

———, 1980: The potential of remotely sensed thermal infrared data to infer surface soil moisture and evaporation. *Water Resour. Res., 16,* 787-795.

Raphael, C. and J. E. Hay, 1984: An assessment of models which use satellite data to estimate solar irradiance at the Earth's surface. *J. Climate Appl. Meteor., 23,* 832-844.

Revfeim, K. J. A., 1978: A simple procedure for estimating global daily radiation on any surface. *J. Appl. Meteor., 17,* 1126-1131.

Rohde, W. G., J. V. Taranik, and C. A. Nelson, 1976: Inventory and mapping of flood inundation using interactional digital image analysis techniques. *Proc. of the 2nd Pecora Mem. Symp.,* Amer. Soc. of Photogram., Falls Church, VA, 131-143.

Schneider, S. R., D. F. McGinnis, Jr., and G. Stephens, 1985: Monitoring Africa's Lake Chad basin with LANDSAT and NOAA satellite data. *Intl. J. of Remote Sens., 6*(1), 59-73.

———, D. F. McGinnis, Jr., and J. A. Gatlin, 1981: Use of NOAA/AVHRR visible and near-infrared data for land remote sensing. *NOAA Technical Report NESS 84,* Dept. of Commerce, Washington, DC, 48 pp.

Sullivan, G. H., V. French, S. K. LeDuc, J. L. Sebaugh, and W. W. Wilson, 1984: Evaluation of satellite-derived estimates of solar radiation. Statistical Reporting Service, USDA, Washington, DC.

Tarpley, J. D., S. R. Schneider, and R. L. Money, 1984: Global vegetation indices from the NOAA-7 meteorological satellite. *J. Climate Appl. Meteor., 23,* 491-494.

———, 1979: Estimating incident solar radiation at the surface from geostationary satellite data. *J. Appl. Meteor., 18,* 1172-1181.

———, S. R. Schneider, J. E. Bragg, and M. P. Waters, III, 1978: Satellite data set for solar incoming radiation studies. *NOAA Technical Memorandum, NESS 96,* Dept. of Commerce, Washington, DC, 36 pp.

Townshend, J. R. G. and C. J. Tucker, 1984: Objective assessment of Advanced Very High Resolution Radiometer data for land cover mapping. *Intl. J. of Remote Sens., 5,* 497-504.

Tucker, C. J., J. A. Gatlin, and S. R. Schneider, 1984: Monitoring vegetation in the Nile delta with NOAA-6 and NOAA-7 AVHRR imagery. *Photo. Eng. Remote Sens., 50,* 53-61.

———, 1979: Red and photographic infrared linear combinations for monitoring vegetation. *Remote Sens. Env., 8,* 127-150.

———, 1978: A comparison of sensor bands for vegetation monitoring. *Photo. Eng. Remote Sens., 44,* 1369-1380.

van Dijk, A., S. L. Callis, C. M. Sakamoto, and W. L. Decker, 1987: Smoothing vegetation index profiles: An alternative method for reducing radiometric disturbance in NOAA/AVHRR data. *Photo. Eng. Remote Sens., 53,* 1059-1067.

——— (Univ. of Missouri, Columbia, MO), 1986: A crop condition and crop yield estimation method based on NOAA/AVHRR satellite data. Ph.D. Dissertation, 139-146.

———, D. McCrary, S. LeDuc, and M. Lyon, 1984: AVHRR training set. (unpublished manuscript), 9 pp.

Vonder Haar, T. H. and J. S. Ellis, 1975: Solar energy microclimate as determined from satellite observations. *Proc. Soc. Photo-Opt. Instrum. Eng., 68,* 18-22.

Whitney, L. F., 1982: A statistical approach to rainfall estimation using satellite and conventional data. *NOAA Technical Report NESS 89,* Dept. of Commerce, Washington, DC, 50 pp.

Contributors to Section X

Hadeen, Kenneth D. — Director, National Climatic Data Center
NESDIS, NOAA Federal Building
Asheville, NC 28801

Johnson, Gary E. — NOAA/NESDIS, Office of Research and Applications
Satellite Applications Laboratory
Climatic Applications Branch
Federal Building
608 East Cherry Street
Columbia, MO 65201

LeDuc, Sharon K. — NOAA/ERL Atmospheric Science Modeling Division
Atmospheric Research & Exposure Assessment
EPA
MD 80
Research Triangle Park
Raleigh, NC 27711

Tarpley, J. Dan — NOAA/NESDIS,
Washington, DC 20233

Section XI

FUTURE SATELLITE SYSTEMS AND APPLICATIONS

The previous sections described the evolution of the current satellite systems and their applications in various scientific disciplines. It spanned a period of over 25 years and enabled us to understand the environment we live in. This section provides information on future satellite systems and their capabilities. In order to be realistic and imaginative, the expected systems for the next 25 years are examined. Many of them are already in advanced planning stages and concept studies have been initiated. Chapter XI-1 describes the polar platform—a massive system that can accommodate many sensor systems to satisfy the needs of several scientific disciplines. Chapter XI-2 contains some of the issues related to the satellite systems that need a complete examination and evaluation.

Chapter XI-1

THE POLAR ORBITING PLATFORM ERA[1]

XI-1.1 INTRODUCTION

The preceding pages have taken the reader from TIROS-I in 1960 to NOAA-K, L, and M and GOES-I to M in the 1990s. As of early 1989, NOAA plans to continue with the NOAA-K, L, M series and even extend it to include NOAA-N. The successive polar orbitor satellite series, NOAA-O, P, and Q, will be a free-flying afternoon series compatible with the forthcoming polar orbiting platforms. The NOAA O, P, Q series and the platforms are expected to include all or some of the new instruments described in this chapter. The Space Station / Polar Orbiting Platform Program is intended to serve the world as a source of satellite-sensed data for decades.

NASA and the European Space Agency (ESA) each plan to launch Polar Orbiting Platforms. The first NASA platform (Figure XI-1-1) will include NASA research instruments plus some NOAA prototype operational instruments (Figure XI-1-2). The ESA series of platforms will carry NOAA operational instruments plus ESA research instruments. There may be a third platform (NASA) carrying only a research payload. Japan has also announced its intention to fly a research platform. The scenario of projected polar platforms is given in Table XI-1-1. Table XI-1-2 lists the projected NOAA instruments which are potential instrument payloads for the NASA and ESA platforms. The payloads for the NASA and ESA platforms will differ because of the differences in mission, payload capability, and Equator crossing time of the two platforms.[2]

XI-1.2 THE POLAR ORBITING PLATFORMS—NOAA PAYLOADS

The Advanced Medium Resolution Imaging Radiometer (AMRIR) imager and Advanced Microwave Sounding Unit (AMSU) sounder are critical to NOAA's forecast and monitoring mission. The ozone sensor, the Global Ozone Monitoring Radiometer (GOMR), will only be on an afternoon platform because it requires more solar illumination than is available in the morning. The Altimeter (ALT) and Operational Scatterometer (OPSCATT) are other possible NOAA contributions to the NASA or ESA platforms.

XI-1.3 NOAA'S NEW POLAR ORBITING PLATFORM ERA INSTRUMENTS

The AMRIR, which combines AVHRR and HIRS capabilities, is seen as an 11-channel radiometer (VIS and IR) with a nominal resolution of 800 m at nadir. It will scan from horizon to horizon, cross track, and will continue to provide direct broadcast transmissions in both high resolution and reduced resolution modes.

The GOMR consists of two separate instruments. The GOMR Limb Sounder (GLS) is an infrared limb

[1] Primary Contributor: Bruce H. Needham

[2] Payloads for the proposed platforms have not been fully determined.

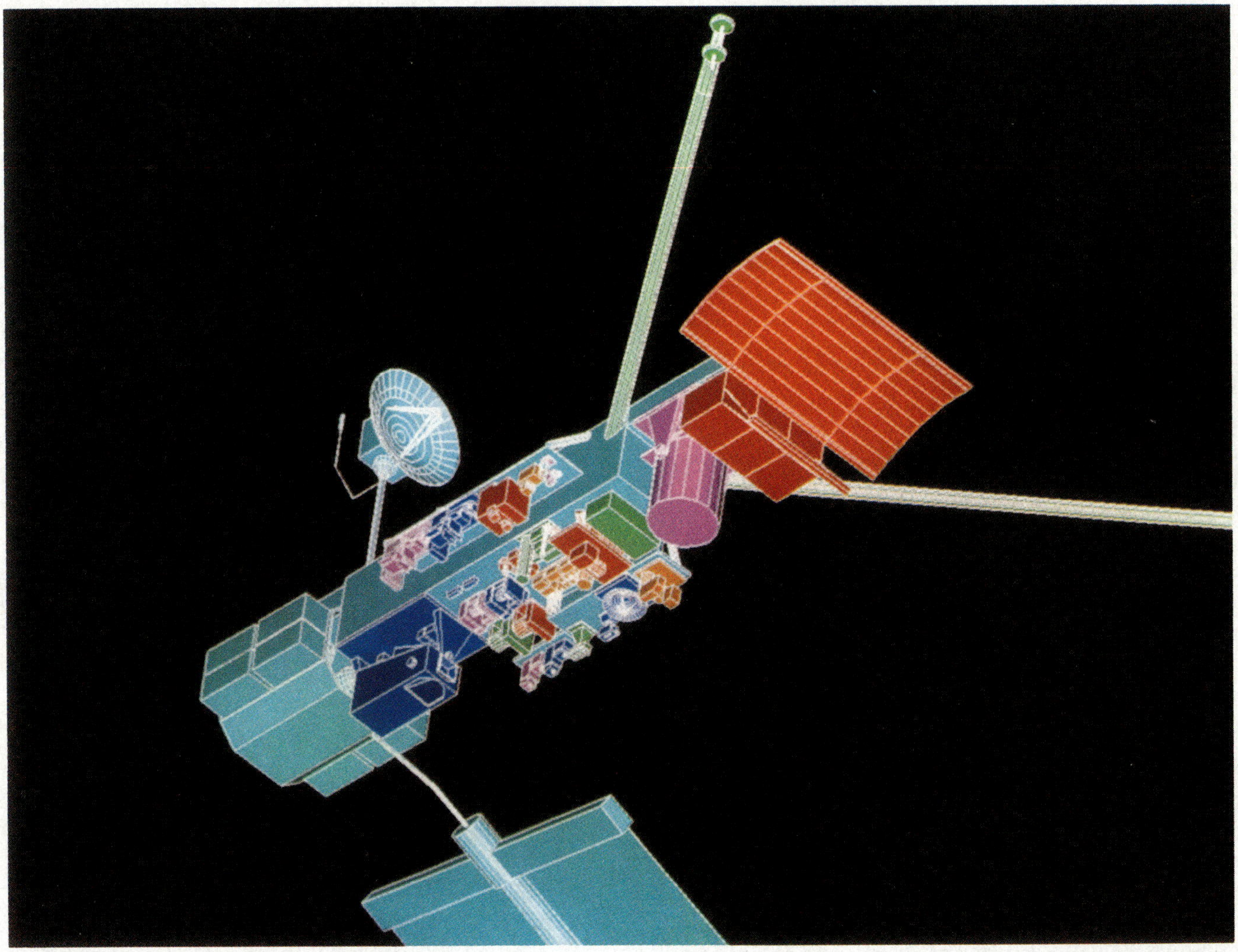

Figure XI-1-1. Tentative NASA platform.

sounder for acquiring detailed vertical soundings. It contains 10 channels between 6 and 15 μm, scans vertically on the limb, and provides about a 2-km vertical resolution. The second portion of the GOMR is the GOMR Nadir Sounder (GNS) which scans cross track for measuring total ozone in the air column. The GNS has six channels between 0.305 and 0.360 μm.

The Earth Radiation Budget Instrument (ERBI), a successor to Earth Radiation Budget Experiment (ERBE) sensors on NOAA-9 and 10, consists of two units: a nonscanning unit with five channels ranging from 0.2 to 50 μm, and a scanning unit with 3 channels from 0.12 to 5 μm. A Conical Scanning Radiometer (CSR) from the Federal Republic of Germany may replace the ERBI scanner on the ESA platform.

The ALT is a continuously operating active microwave system for determining ocean surface height, wave height, surface currents, wind velocity, sea ice boundaries, and other geophysical data. It will operate in the Kuband (13.6 GHz). It is a nadir viewing instrument with a spatial resolution of about 24.2 km (depending on sea state) and a vertical resolution of about 3.5 cm.

Table XI-1-1 *Baseline polar platform scenarios*

PLATFORM	NASA #1	NASA #2	ESA #1	JAPAN #1
Altitude	824 km	824 km	824 km	800 km
Equator Crossing (Ascending, LST)	1330	1330	0930	1100
Instruments	Research & Operational	Research	Research & Operational	Research
Estimated Launch Date	1995	1996–1997	1997	1998
Estimated Weight (Payload)	4000 kg	3500 kg	2900 kg	2100 kg

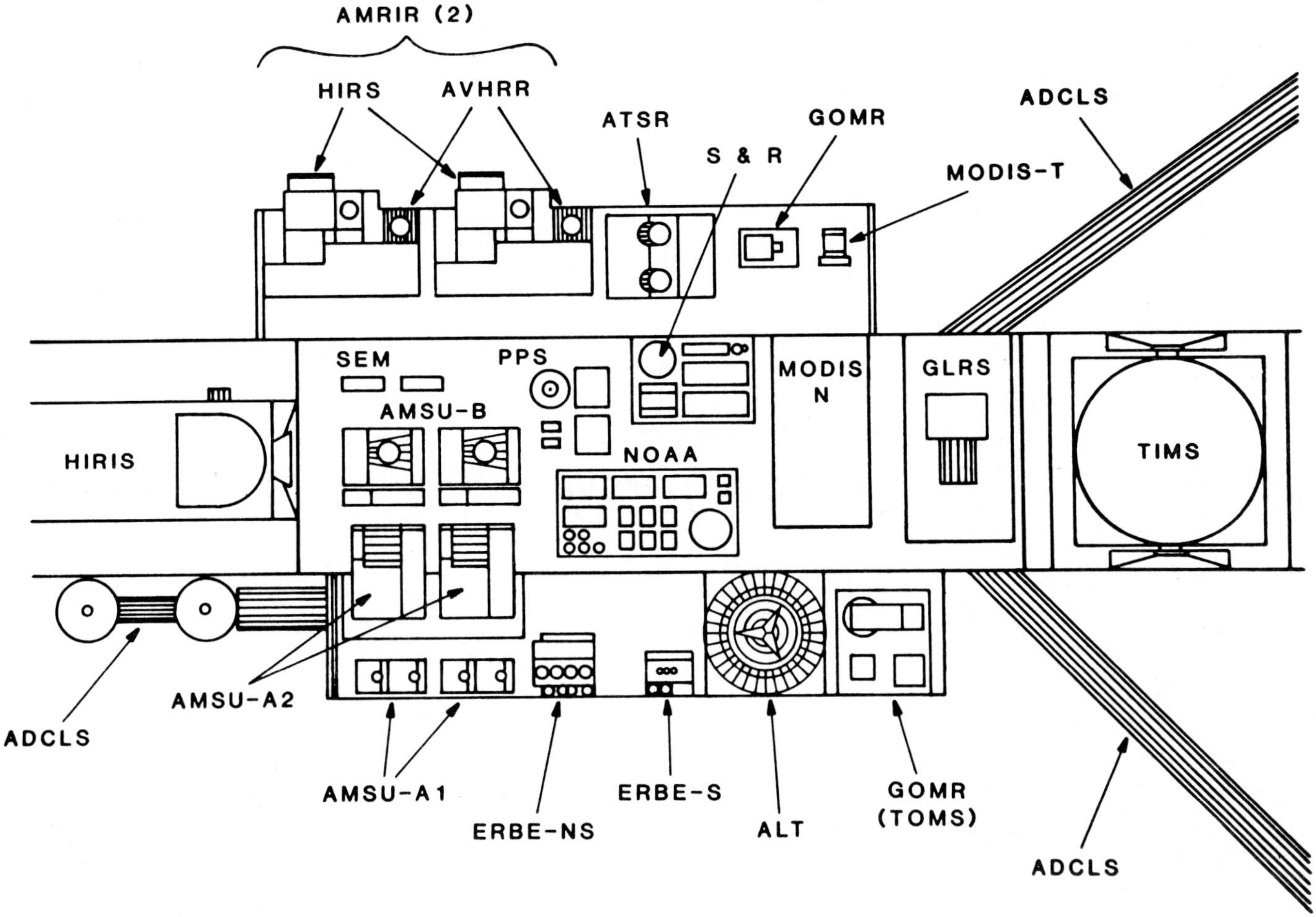

Figure XI-1-2. NASA Platform 1-nadir view with potential instrument payload.

Table XI-1-2 *Proposed NOAA operational instruments for the polar platforms*

AMRIR	Advanced Medium Resolution Imaging Radiometer
GOMR	Global Ozone Monitoring Radiometer
AMSU/A, B	Advanced Microwave Sounding Units A & B
ERBI	Earth Radiation Budget Instrument
SEM	Space Environment Monitor
ASEM	Augmented Space Environment Monitor
ALT	Altimeter
OPSCATT	Operational Scatterometer
Argos-2	Data Collection and Location System
S & R	Search and Rescue
CCDH	Command, Communications, and Data Handling Equipment

The OPSCATT is an active microwave system for providing wind vector data from which global vector surface wind fields can be inferred. It operates at 13.995 GHz, using six stick antennas, four with single and two with dual polarization. Spatial resolution is 25 km, with a swath width of 120 to 700 km on either side of nadir.

Portions of an Augmented Space Environment Monitor (ASEM) will be carried either on the NOAA free flyer or on the NASA Platform. The ASEM consists of the basic Space Environment Monitor (SEM) plus:

- An Atmospheric X-Ray Imager (AXI) to monitor the intensity and ranging from 1 to 100 keV at a spatial resolution of 20 km and an energy resolution of 3 keV. The instrument views the nadir with a field of view of ±60°.
- A Three-Axis Magnetometer (TAM) to measure the *in situ* vector magnetic field.
- A Local Electric Field Instrument (LEFI) to monitor the ionospheric electric field vector imposed by the magnetosphere.
- A RAdio Beacon (RAB) to permit determination of the total electron content of the ionosphere along the line of sight between the beacon and receiver.

These new instruments are expected to be on-board the NASA and ESA platforms tentatively scheduled for launch in 1996 and 1997. If the schedule can be adhered to, there will be a smooth transition from NOAA N (the last of the current NOAA satellites) to the long mission, Polar Orbiting Satellites (NOAA-O,P,Q) and the Polar Orbiting Platforms.

Chapter XI-2

ISSUES FOR THE FUTURE[1]

> There is an old meteorological doggerel: "More data, more data, from pole to equator, Measuring everything, everywhere, all of the time."

Environmental satellites have brought close to reality the fulfillment of this weatherman's dream. The availability of these platforms and their vast arrays of measurements raise unresolved issues, however, that will have significant bearing upon the long-term nature of their usage in the future.

Sensor performance is one case in point. In the continuing quest for improved instruments, is there a practical point of diminishing returns beyond which it is not worth the technical effort, cost, and data processing to wring out the additional data that would result? If there is such a point in sensor resolution, for example, what is it?

From the standpoint of the user community's ability to use the large amounts of data becoming available, is there also a point of diminishing returns? As the community becomes increasingly awash in imagery and bits, keeping its head above water is more important than ever if data are to be utilized quickly, accurately, and at reasonable cost. Are too many data being archived? Are too many duplicative? Are enough thrown away?

Are we also trying to do too much with single observing platforms? Environmental satellites, once "limited-mission" vehicles, have become "multi-mission," if not "do-all" spacecraft. The more they become so, the bigger and more complicated they turn out to be. Designed by engineers, they are turned over to technicians to operate. As complexity increases, the skills and capacities of operating personnel are stretched to the limit.

These are some of the technical issues to be decided. There are also political ones, in particular privatization and convergence (the merging of systems operated by separate organizations). In 1983, the U.S. Government considered privatizing, that is, selling to industry the satellites, ground facilities, and operations of the NOAA and GOES satellites. A Request for Proposals was prepared, but Congress stepped in to prohibit the divestiture from taking place. This effort to commercialize the satellite operations of the National Oceanic and Atmospheric Administration thus came to an end. The idea persists, however.

The fact that the U.S. Air Force operates a Defense Meteorological Satellite Program, consisting of polar orbiting spacecraft, and that NOAA also operates a polar orbiting system has led to a long-standing issue as to how closely the two systems should be merged. The systems are similar, but not identical in makeup or mission. Possible options include construction and launch of polar orbiters for NOAA by the Air Force, using a common design, and complete incorporation and operation of the NOAA system by the Air Force.

Readers will quickly recognize the many kinds of

[1] Primary Contributor: Gordon Vaeth

effects that privatization and/or convergence could have upon the nature and availability of future civil type environmental satellite services supplied by the U.S.

Finally there is the issue, larger than any other, of the ability of those who receive environmental data from space to use those data purposefully, particularly for the protection of life and property. Satellite imagery showing the onset of hazardous weather is hardly more than ornamental if those on the ground cannot translate it into effective warnings, precautions, population evacuations, and the like. Until an adequate response to satellite observations can be made, the potential of orbiting environmental platforms can never be more than partly realized. Ensuring such a response in developing countries is probably the most difficult and critical of all the tasks ahead.

Contributors to Section XI

Needham, Bruce H.	NOAA/NESDIS
Vaeth, J. Gordon	Washington, DC 20233

ACRONYMS AND ABBREVIATIONS

ABI — National Weather Service station identifier for Abilene, Texas
ACCS — Ambroziak Color Coordinate System
A/D — Analogue to Digital
ADB — Active DataBase
ADCLS — Advanced Data Collection and Location System
ADESS — Automatic Data Editing and Switching System
AEG — Atlantic Environmental Group, NOAA
AEM-A — Applications Explorer Mission-A
AEM-B — Applications Explorer Mission-B
AFGWC — Air Force Global Weather Central at Offutt AFB, Nebraska
AFOS — Automation of Field Operations and Services
AFSC — Air Force System Command Space Division, Fairchild AFB, WA
AGC — Automatic Gain Control
AgRISTARS — Agricultural Resource Inventory Survey Through Aerospace Remote Sensing
AIDJEX — Arctic Ice Dynamics Joint Experiment
AIRS — Atmospheric Infrared Sounder
AL — Assigned Level
AKM — Apogee Kick Motor
ALT — Altimeter
AM — Amplitude Modulation
AMF — Apogee Motor Firing
AMRIR — Advanced Medium Resolution Imaging Radiometer
AMSR — Advanced Microwave Scanning Radiometer
AMSU — Advanced Microwave Sounding Unit
AOC — Argos Operations Committee
AOCS — Attitude and Orbit Control Subsystem
AOIPS — Atmospheric and Oceanographic Information Processing System
AOL — Airborne Oceanographic Lidar
APL — Applied Physics Laboratory of the Johns Hopkins University
APT — Automatic Picture Transmission
Argos — Data Collection and Location System (France)
ASDAR — Aircraft to Satellite Data Relay
ASEM — Augmented Space Environment Monitor
ATLID — Atmospheric Lidar
ATN — Advanced TIROS-N Spacecraft
ATS — Applications Technology Satellite
ATSR — Along Track Scanning Radiometer
AVCS — Advanced Vidicon Camera System
AVHRR — Advanced Very High-Resolution Radiometer
AWIPS — Advanced Weather Interactive Processing System
AWS — Air Weather Service
AWS — Automatic Weather Stations
AXI — Atmospheric X-Ray Imager
BB — Alphabetical Designation Given to an Image Enhancement Curve
BDT — Bulk Data Transfer
BECO — Booster Engine Cutoff
BESEX — Joint U.S./U.S.S.R. Bering Sea Experiment
Bhaskara-1 — India's Experimental Geostationary Satellite
BLM — Bureau of Land Management
bpi — Bits per inch
BPJ — Booster Package Jettison
bps — Bits per second
BUV — Backscatter Ultraviolet
C-BAND — Communications Band
C-MAN — Coastal Marine Automated Network Station
CA — Cloud Analysis
CAC — Climate Analysis Center
CAPPI — Constant Altitude PPI radar
CAS — Commission on Atmospheric Sciences (WMO)
CASID — Climate Air-Sea Interactive Drifter
CB — Alphabetical Designation Given to an Enhancement Curve
CCC — Central Cold Cover
CCD — Charged Coupled Devices
CCF — Central Computer Facility
CCI — Charged Coupled Imager
CCIR — Consultative Committee on International Radio
CCR — Cloud Cover Radiometer
CCT — Computer-Compatible Tape
CDA — Command and Data Acquisition
CDAS — Command and Data Acquisition Station
CDDF — Central Data Dissemination Facility
CDDF — Central Data Distribution Facility
CDF — Coded Digital Fax
CDHF — Central Data Handling Facility
CDROM — Compact Disk Read-Only Memory
CE — Cold Eddy
CEOS — Committee on Earth Observation Satellites
CEOS/WGD — Committee on Earth Observation Satellites/Working Group on Data Management
CGMS — Coordination of Geostationary Meteorological Satellites
CHARM — Coastal Habitat Assessment, Research, and Mensuration Program
CHL-*a* — Chlorophyll *a*
CLAES — Cryogenic Limb Array Etalon Spectrometer
CLS — Contingency Landing Site

CMV	Cloud Motion Vectors
CNES	Centre National d'Études Spatiales, France
CO_2	Carbon Dioxide
CODAR	Coastal Ocean Dynamics Applications Radar
COE	U.S. Army Corps of Engineers
Cosmos	Satellite for Either Meteorological or Oceanic Purposes (U.S.S.R.)
COSPAR	Committee on Space Research
COSPAS	Space System for Search of Distressed Vessels (U.S.S.R.)
CR	Correlation Radiometer
CROHMS	Columbia River Operational Hydrometeorological Management System
CRS	AFSC Command Readout Stations at Loring AFB in Maine and Washington
CRT	Cathode-Ray Tube
CSR	Conical Scanning Radiometer
CTH	Cloud Top Height
CTW	Cloud Track Wind
CW	Continuous Wave
CZCS	Coastal Zone Color Scanner
DACS	Data Access and Command Subsystem
DATTS	Data Acquisition Telecommand and Tracking Station Located at Michelstadt, Germany
dB	Decibel
DBMS	DataBase Management System
DBS	Direct Broadcast Satellite
DBS	Direct Broadcast System
DCLS	Data Collection and Location System
DCP	Data Collection Platform
DCPR	Data Collection Platform Relay
DCPR	Data Collection Platform Report
DCS	Data Collection System
DIAL	Differential Absorption Lidar
DIGS	Delta Inertial Guidance System
DJF	December-January-February
DMA	Defense Mapping Agency
DMSP	Defense Meteorological Satellite Program
DOD	Department of Defense
DPC	Data Processing Center
DPSS	Data Processing Subsystem
DPU	Data Processing Unit
DRCC	Data Referencing and Conditioning Centre
DRT	Data Relay Transponder
DRU	Data Recovery Unit
DS	Dwell Sounding
DSB	Direct Sounder Broadcast
DSD	Data Set Directory
EAS	Electronic Animation System
EBBT	Equivalent Blackbody Temperature
ECMWF	European Centre for Medium Range Weather Forecasts (U.K.)
EDA	Electronically Despun Antenna
EDIS	Environmental Data and Information Service
EIR	Enhanced Infrared Imagery
EIRP	Effective Irradiated Power
EM	Electromagnetic
EMR	Electromagnetic Radiation
EMA	Electrodynamics of the Middle Atmosphere
ENSO	El Niño/Southern Oscillation
EOLE	French Satellite Designed to Collect Data from Free-Floating Balloon
EOS	Earth Observing System
EOSDIS	EOS Data and Information System
EPS	Energetic Particles Sensor
ERB	Earth Radiation Budget
ERBE	Earth Radiation Budget Experiment
ERBE-NS	ERBE Nonscanner Instrument
ERBE-S	ERBE Scanner Instrument
ERBI	Earth Radiation Budget Instrument
ERBS	Earth Radiation Budget Satellite
ERIM	Environmental Research Institute of Michigan
ERL	Environmental Research Laboratory, NOAA
EROS	Earth Resources Observation System
ERS-1	Earth Resources Satellite, European Space Agency
ERTS	Earth Resources Technology Satellite
ESA	European Space Agency
ESC	Earth Satellite Corporation
ESMR	Electrically Scanned Microwave Radiometer
ESOC	European Satellite Operations Center, Darmstadt, Germany
ESRO	European Space Research Organization
ESSA	Environmental Science Services Administration
ESSA-1	Environmental Science Services Administration Satellite (e.g., number 1)
EUMETSAT	European Organization for the Exploration of Meteorological Satellites
FACC	Ford Aerospace & Communications Corporation
FANAS	Forecast for Ascending Node for Automatic Satellites Message—GTS
FAX	Facsimile
FCIR	False Color Infrared System
FGGE	First GARP Global Experiment (FGGE)
FM	Frequency Modulation
FNOC	Navy Fleet Numerical Operations Center, Monterey, California
FNOC	Navy Fleet Oceanographic Center
FOV	Field Of View
FT	File Transfer
FY	First Year
FZ	Fracture Zone

G/T	Gain per Temperature
GAC	Global Area Coverage
GARP	Global Atmospheric Research Program
GATE	GARP Atlantic Tropical Experiment
GCM	General Circulation Model
GDAS	Global Data Assimilation System
GDR	German Democratic Republic
Ge	Germanium
GEM	Goddard Earth Model
GEM10	Goddard Earth Model 10
GEO	Geosynchronous Earth Orbiter
GEOS	Geodetic Observing Satellite
Geosat	Geodesy Satellite (Navy)
GERTS	General Electric Radio Tracking System
GFOV	Geometrical Field of View
GHz	Gigahertz
GLAS	Goddard Laboratory for Atmospheric Studies
GLRS	Geodynamics Laser Ranging System
GLS	GOMR Limb Sounder
GMI	Gray-McCrary Index of Vegetation
GMS	Geostationary Meteorological Satellite, Japan
GMSS	Geostationary Meteorological Satellite System (Japan)
GMSSA	GMS Standard Atmosphere
GNS	GOMR Nadir Sounder
GOASEX	Gulf of Alaska Seasat Experiment
GOES	Geostationary Operational Environmental Satellite (NOAA)
GOES-Next	Geostationary Satellite Being Planned
GOES-Tap	Land-Line Distribution Network for GOES Data
GOMR	Global Ozone Monitoring Radiometer
GOMS	Geosynchronous Orbit for Meteorological Satellite System (U.S.S.R.)
GOSSTCOMP	Global Operational Sea Surface Temperature Computation
GPI	GOES Precipitation Index
GS	Gulf Stream
GSFC	Goddard Space Flight Center, NASA
GTS	Global Telecommunications System of the WMO
GVHRR	Geosynchronous Very High-Resolution Radiometer
GWE	Global Weather Experiment
HALOE	Halogen Occultation Experiment
HCMM	Heat Capacity Mapping Mission
HEPAD	High Energy Proton and Alpha Detector
HF	High Frequency, Between 3 and 30 MHz
HgCdTe	Mercury Cadmium Telluride
HIF	Historical Instrument File
Himawari	(Sunflower) Japanese GMS Satellite
HIRIS	High-Resolution Imaging Spectrometer
HIRS	High-Resolution Infrared Sounder
HMMR	High-Resolution Multifrequency Microwave Radiometer
HRDI	High-Resolution Doppler Interferometer
HRFAX	High-Resolution Fascimile
HRIR	High-Resolution Infrared Radiometer
HRIS	High-Resolution Imaging Spectrometer
HRPT	High-Resolution Picture Transmission
HRV	High-Resolution Visible Imager
IACP	International Aerosol Climatology Project
IAMAP	International Association of Meteorology and Atmospheric Physics
ICEX	Ice and Climate Experiment
ICSU	International Council of Scientific Unions
IDCPS	International DCP System
IF	Intermediate Frequency
IFFA	Interactive Flash Flood Analyzer
IFOV	Instantaneous Field of View
IFREMER	Institut Francais de Recherche pour l' Exploitation de la Mer, France
IGY	International Geophysical Year
IGFOV	Instantaneous Geometric Field of View
IHS	Intensity, Hue, Saturation
Insat	Indian National Satellite System
InSb	Indium Antimonide
IPOMS	International Polar Orbiting Meteorological Satellites
IR	Infrared
IRIS	Infrared Interferometer Spectrometer
IRLS	Interrogation, Recording, and Location System
IRVA	Image Rectifying Video Analyzer
ISAMS	Improved Stratospheric and Mesospheric Sounder
ISCCP	International Satellite Cloud Climatology Project
ITCZ	Intertropical Convergence Zone
ITIR	Intermediate Thermal Infrared Radiometer
ITOS	Improved TIROS Operational System (ITOS Series Included ITOS-1 and NOAA-1 to 5)
ITOS-D	One of the ITOS Series, Prior to Launch
ITPR	Infrared Temperature Profile Radiometer
ITT	International Telephone and Telegraph Company
JANUS	Preliminary Program to Develop Concept for Rocket Launched Satellite
JASIN	Joint Air-Sea Interaction Program
JJA	June-July-August
JMA	Japan Meteorological Agency
JOC	Joint Organizing Committee
JPL	Jet Propulsion Laboratory
JTWC	Joint Typhoon Warning Center
keV	Kiloelectron Volt
kHz	Kilohertz
kW	Kilowatt
LAC	Local Area Coverage
LACIE	Large Area Crop Inventory Experiment
Landsat	U.S. Earth Resources Satellite

LARAR	Large Antenna Real Aperture Radar
LASA	Laser Atmosphere Sounder and Altimeter
LASER	Light Amplification by Stimulated Emission of Radiation
LASERFAX	Laser Facsimile system
LAWS	Laser Atmospheric Wind Sounder
LBF	Level of Best Fit
LEFI	Local Electric Field Instrument
LIDAR	Light Detection and Ranging
LIMS	Limb Infrared Monitor of the Stratosphere
L/O	Liftoff
LRFAX	Low Resolution Facsimile
LRIR	Limb Radiance Infrared Radiometer
LST	Local Solar Time
LST	Local Standard Time
LTE	Local Thermodynamic Equilibrium
LUT	Local User Terminal
MACH	Mensuration and Assessment of Coastal Habitat
MAG	Magnetometer
MAM	March-April-May
MAREPS	Marine Reports
MB	Alphabetical Designation Given to an Image Enhancement Curve
Mbps	Megabits per Second
MCC	Mesoscale Convective Complex
MCC	Mission Control Center
MCF	Master Control Facility
McIDAS	Man-Computer Interactive Data Access System
MCO	Mostly Covered Cloud Field
MCSST	Multichannel Sea Surface Temperature
MDD	Meteorological Data Dissemination
MDUC	Meteorological Data Utilization Center at New Delhi, India
MDUS	Medium Scale Data Utilization Station
MEPED	Medium Energy Proton and Electron Detector
MERIS	Medium Resolution Imaging Spectrometer
Meteor	System of Meteorological Satellites, U.S.S.R.
Meteosat	Geostationary Satellite (EUMETSAT)
METROMEX	Metropolitan Meteorological Experiment, St. Louis
MeV	Mega Electron Volt
MFR	Multifilter Radiometer
MGCS	Meteosat Ground Computer System
MHz	Megahertz
MIEC	Meteorological Information Extraction Centre
MIRP	Manipulated Information Rate Processor
MIZEX-EAST	Marginal Ice Zone Experiment in the Greenland Sea
MIZEX-WEST	Marginal Ice Zone Experiment in the Bering Sea
MLS	Microwave Limb Sounder
MMIPS	Man-Machine Interactive Processing System
MO	Designation for Forested Area Located in the Missouri Ozarks
MOCC	Meteosat Operations Control Centre
MODIS	Moderate Resolution Imaging Spectrometer
MOP	Meteosat Operational Program
Molniya	Communications Satellite (U.S.S.R.)
MPD	Magnetospheric Particle Detector
mrad	Milliradian
MRIR	Medium Resolution Infrared Radiometer
MRS	Main Ranging Station
MSC	Meteorological Satellite Center, Japan
MSC	Mission Satellite Center
MSI	Multispectral Imaging
MSP	Micrometeorological Sensor Processor
MSS	Multispectral Scanner
MST	Moscow Standard Time
MSU	Microwave Sounding Unit
MSU-E	Multizonal Scanning Unit with Electronic Scanning
MSU-SK	Multizonal Scanning Unit with Mechanical Scanner
MU	Microwave Unit
MW	Megawatt
MW	Microwave
MY	Multiyear
NAS9070	Mainframe Computer, NOAA Facility
NASA	National Aeronautics and Space Administration
NASA-ERL	National Aeronautics and Space Administration-Environmental Resources Laboratory
NASA/GSFC	National Aeronautics and Space Administration/Goddard Space Flight Center
NASDA	National Space Development Agency of Japan
NBS	National Bureau of Standards
NCAR	National Center for Atmospheric Research, Boulder, Colorado
NCDC	National Climatic Data Center
NDBC	NOAA Data Buoy Center
NDVI	Normalized Difference Vegetation Index
NEFC	Northeast Fisheries Center
NEMS	Nimbus-Experiment Microwave Spectrometer
NESC	National Environmental Satellite Center
NESDIS	National Environmental Satellite, Data, and Information Service
NESS	National Environmental Satellite Service
NEΔT	Noise Equivalent Temperature Difference
NFJ	Noise Fairing Jettison
NFMOA	National Fish Meal and Oil Association
NHC	National Hurricane Center
NHRL	National Hurricane Research Laboratory

Nimbus	NASA Research and Development Satellite
NMAS	NWS/NOAA Marine Advisory Service
NMC	National Meteorological Center, Washington, DC
NMFS	National Marine Fisheries Service
NMO	Corn and Soybean Producing Area Where the States of Missouri, Iowa, Kansas, and Nebraska Meet
NOAA	National Oceanic and Atmospheric Administration
NOAA-2	Polar Orbiting Satellite Operated by NOAA (e.g., number 2)
NORSEX	Norwegian Remote Sensing Experiment
NORSEX	Norwegian Sea Experiment
NoSASS	Without Seasat-A Satellite Scatterometer (SASS) data
NoSat	Without Satellite Data
NPOP	NASA Polar Orbiting Platform
NRCS	Normalized Radar Backscattering Cross Section
NRL	Naval Research Laboratory
NROSS	Navy Remote Ocean Sensing System
NRZ1	Non-Return to Zero-Change on One
NRZL	Non-Return to Zero Level
NSSFC	National Severe Storms Forecast Center, Kansas City, Missouri
NVI	Normalized Vegetation Index
NWP	Numerical Weather Prediction
NWS	National Weather Service
OBS	Observations
OCI	Ocean Color Instrument
OGO-4	Orbiting Geophysical Observatory Satellite
OLR	Outgoing Longwave Radiation
OLS	Operational Linescan System
OPSCATT	Operational Scatterometer
ORSTOM	French Office de la Recherche Scientifique et Technique Outre Mer, France
OSDPD	Office of Satellite Data Processing and Distribution at NESDIS
OSU	Oregon State University
PC	Personal Computer
PCM	Phase Code Modulation
PCM	Pulse Code Modulation
PDB	Primary DataBase
PDUS	Primary Data User Station
PEM	Particle Environment Monitor
PI	Photo Interpretation
PI	Principal Investigator
PMC	Pressure Modulated Cell
PMF	Product Master File
PMR	Pressure Modulated Radiometer
PMT	Photo Multiplier Tube
POES	Polar Orbiting Environmental Satellite
POOES	Polar Orbiting Operational Environmental Satellite
PPI	Plan Position Indicator
PPS-PPD	Precision Positioning System-Precision Orbit Determination System
PRL	Polar Research Laboratories
PROFS	Program for Regional Observing and Forecasting Services
PRT	Platinum Resistance Thermometer
PSG	Polar Stereographic
PTT	Platform Transmitter Terminals
PUF	Product User File
r	Correlation Coefficient
R&D	Research and Development
R/V	NOAA Research Vessels (e.g., Oceanographer)
RAB	Radio Beacon
RADAR	Radio Detection And Ranging
RAM	Random Access Memory
RAWARC	Radar Warning and Coordination
RBV	Return Beam Vidicon Camera
RCA	Radio Corporation of America
RCC	Rescue Coordination Center
RCS	Reaction Control Subsystem
RF	Radio Frequency
RFC	River Forecast Center
RGON	Remote Geophysical Observing Network
RH	Relative Humidity
RHI	Range Height Indicator
RHS	Right Hand Side
RISOP	Rapid Interval Scan Operations
RJE	Remote Job Entry
RMS	Root Mean Square
RMSD	Root Mean Square Differences
rpm	Revolutions per Minute
S-band	Super High Frequency, 1.55 to 5.20 GHz
S/N	Signal to Noise Ratio
SAGE	Stratospheric Aerosol and Gas Experiment
SAM	Stratospheric Aerosol Measurement
SAMS	Stratospheric and Mesospheric Sounder
SAR	Search And Rescue
SAR	Synthetic Aperture Radar
SAR	Sarcasso Water
SAR-C	Synthetic Aperature Radar C-band
SARSAT	Search and Rescue Satellite-Aided Tracking
SASS	SEASAT-A Satellite Scatterometer
SATEMP	Satellite Temperatures, WMO Distributed Bulletin
SATOB	Satellite Observation Synopsis, WMO Distributed Bulletin
SBUV	Solar Backscatter Ultraviolet Radiometer
SCAMS	Scanning Microwave Spectrometer
SCATT	Scatterometer
SCMR	Surface Composition Mapping Radiometer
SCR	Surface Contour Radar
SCR	Selective Chopper Radiometer
SDB	System DataBase
SDDS	Satellite Data Distribution System

SDSD Satellite Data Service Division, NESDIS
SDUS Secondary Data User Station
SDUS Small Scale Data Utilization Station
SE Standard Error
SEAMAP Southeast Assessment Monitoring and Prediction
SEAS Shipboard Environmental Data Acquisition System
Seasat An Oceanographic Research Satellite (U.S.)
SECO Sustainer Engine Cutoff
SEFC Southeast Fisheries Center
SEM Solar Environment Monitor
SEM Space Environment Monitor
SEP NWS Office Identifier for Stephenville, TX
SESC Space Environment Services Center, Boulder, CO
SFDF Satellite Field Distribution Facility
SFSS Satellite Field Services Station
SHF Super High Frequency
SHW Shelf Water
SIO Scripps Institution of Oceanography
SIP Soundings Improvement Project
SIR Shuttle Imaging Radar
SIR-B' Shuttle Imaging Radar-B' Mission
SIRS Satellite IR Spectrometer
Skylab A U.S. Manned Space Station
SLAR Side Looking Airborne Radar
SLW Slope Water
SME Solar Mesosphere Explorer
SMMR Scanning Multichannel Microwave Radiometer
SMM ACRIM Solar Maximum Mission Active Cavity Radiometer Irradiance Monitor
SMS Synchronous Meteorological Satellite
SMS GOES Synchronous Meteorological Satellites/ Geostationary Operational Environmental Satellite
SOCC Satellite Operations Control Center
SOLSTICE Solar-Stellar Intercomparison Experiment
SOP Special Observing Period
SPCZ South Pacific Convergence Zone
SPM Solar Proton Monitor
SPOT System Probatoire d'Observation de la Terre, Satellite operated by Spot Image, France
SR Scanning Radiometer
SSCC Spin-Scan Cloudcover Camera
SSEC Space Science and Engineering Center, University of Wisconsin
SSMI Special Sensor Microwave Imager
SSMT Special Sensor Microrwave Temperature
SST Sea Surface Temperature
SSU Stratospheric Sounding Unit
STS Space Transportation System (NASA)
SURSAT Canadian Surveillance Satellite Experiment
SUSIM Solar Ultraviolet Spectral Irradiance Monitor
SWH Significant Wave Height
SWIS Satellite Weather Information System
SWOP Stereo Wave Observation Project
SZA Solar or Satellite Zenith Angles
T-Number Stage of Tropical Cyclone Intensity Based Upon Typical Long Term Growth Rate Observed in Satellite Imagery
TACC Tracking and Control Center
TACS Tracking and Control Station
TAM Three Axis Magnetometer
TARS Turn Around Ranging Station
TBM Terabit Memory System
TCLBS Tropical Constant Level Balloon System
TED Total Energy Detector
TGS Tri-Glycine Sulphate
THIR Temperature-Humidity Infrared Radiometer
TIMS Thermal Infrared Multispectral Scanner
TIP TIROS Information Processor
TIROS-1 Television and Infrared Observation Satellite (e. g., number 1)
TM Thematic Mapper
TOGA Tropical Ocean Global Atmosphere
TOMS Total Ozone Mapping Spectrometer
Topex Topography Experiment (Oceanic)
TOS TIROS Operational System
TOVS TIROS Operational Vertical Sounder
TPT Television Picture Terminal
TRIFAR Three-Frequency Airborne Radar
TSO Time-Sharing Option
TV Television
T1 First Stage of Tropical Storm Intensity or Growth
UARS Upper Atmosphere Research Satellite
UDA Ultra High Frequency Data Collection System Antenna
UHF Ultra High Frequency
USAPC United States Argos Processing Center
USB Unified S-Band Receiver
USCG U.S. Coast Guard
USDA U.S. Department of Agriculture
USGS U.S. Geological Survey
UTC Universal Time Coordinated
UTH Upper Tropospheric Humidity
UV Ultraviolet
VAS VISSR Atmospheric Sounder
VDM VISSR Digital Multiplexer
VECO Vernier Engine Cutoff
VHF Very High Frequency
VHRR Very High Resolution Radiometer
VHS Video Home System
VHSC Video Home System Cassette
VI Vegetation Index
VIRGS VISSR Image Registration and Gridding System
VIRR Visible Infrared Radiometer

VIS	Visible
VISSR	Visible Infrared Spin Scan Radiometer
VRA	VHF Real-Time Antenna
VTPR	Vertical Temperature Profile Radiometer
WBDCS	Wideband Data Collection System
WE	Warm Eddy
WEFAX	Weather Facsimile
WINDCO	Wind Computation Operation
WINDII	Wind Imaging Interferometer
WMO	World Meteorological Organization of the United Nations
WOCE	World Ocean Climate Experiment
WORM	Write Once, Read Many Times
WSFO	Weather Service Forecast Office
WV	Water Vapor
WWW	World Weather Watch
XBT	Expendable Bathythermograph
ZA	Alphabetical Designator Given to an Image Enhancement Curve
ZTS	Zoom Transfer Scope

GLOSSARY[1]

ABSORPTION BAND—A range of wavelengths in the electromagnetic spectrum within which radiant energy is absorbed by a substance.

ACTIVE SYSTEM—A remote sensing system that transmits its own electromagnetic emanations at an object and then records the energy reflected or refracted back to the sensor. Radar, acoustic soundings, and leaders are examples of active systems.

ALBEDO—The ratio of the amount of electromagnetic radiation reflected by a body to the amount incident upon it, often expressed as a percentage.

ALGORITHM—1) A fixed step-by-step procedure to accomplish a given result; usually a simplified procedure for solving a complex problem; also a full statement of a finite number of steps. 2) A computer-oriented procedure for resolving a problem.

ANALOGUE—A form of data display in which values are shown graphically. Also a form of computing in which values are represented by directly measurable quantities, such as voltages or resistances. Analogue computing methods contrast with digital methods in which values are treated numerically.

ANIMATION SYSTEM—A system that displays sequential images/data in a moving depiction (movie).

ARC CLOUD—A narrow curved line of cumulus clouds forming along a thunderstorm outflow boundary similar to a mini cold front. Arc clouds are frequently seen in both visible and infrared satellite imagery.

ASCENDING NODE—The northbound Equator crossing of a satellite; given in degrees longitude, date, and time (UTC) for any given orbit or pass.

ATMOSPHERIC CORRECTIONS—Algorithms employed to account for absorption or reflection characteristics of atmospheric gases.

ATMOSPHERIC WINDOWS—Those wavelength ranges in which radiation can pass through the atmosphere with relatively little attenuation. In the visible and infrared portions of the spectrum, there are windows at approximately 0.3 to 2.5, 3.0 to 4.0, 4.2 to 5.0, and 7.0 to 15.0 μm.

ATMOSPHERIC SOUNDINGS—Measurement of atmospheric parameters, such as temperature and water vapor, for various height or pressure levels.

ATTENUATION—The scattering and absorption of radiant energy by clouds or the atmosphere.

BACKSCATTERING—The scattering of radiant energy into the hemisphere of space bounded by a plane normal to the direction of the incident radiation and lying on the same side as the incident ray; the opposite of forward scatter.

BAND—1) A selection of wavelengths. 2) Frequency band. 3) Absorption band. 4) A group of tracks on a magnetic drum. 5) a range of radar frequencies, such as X-band, Q-band, etc.

BAUD—data rate in communications expressed as bits per second.

BIOMASS—The total mass of living matter (vegetation) in a specific geographic area.

BLACKBODY—An ideal emitter which radiates energy at the maximum possible rate per unit area at each wavelength for any given temperature. A blackbody also absorbs all the radiant energy incident upon it. No actual substance behaves as a true blackbody.

BLACKBODY RADIATION—The electromagnetic radiation emitted by an ideal blackbody; it is the theoretical maximum amount of radiant energy of all wavelengths that can be emitted by a body at a given temperature. The spectral distribution of blackbody radiation is described by Planck's law and related radiation laws.

BRIGHTNESS—1) The attribute of visual perception in accordance with which an area appears to emit more or less light. 2) Luminance. 3) The luminous flux emitted or reflected per unit projected area per unit solid angle.

BRIGHTNESS TEMPERATURE—1) The temperature of a blackbody radiating the same amount of energy per unit area at the wavelengths under consideration as the object being viewed. Also called effective temperature. 2) The apparent temperature of a non-blackbody determined by measurement with an optical pyrometer or radiometer. (See EQUIVALENT BLACKBODY TEMPERATURE.)

BUSLOAD—The spacecraft excluding the instruments and sensors making up the payload.

CALIBRATION—The act or process of comparing certain specific measureants from an instrument to specific values of the physical parameters being measured using a known source. (See VALIDATION.)

CHANNEL—Used to refer to specific wavelengths of on-board satellite sensors.

CHLOROPHYLL *A* — One of the main pigments present in the principal groups of phytoplankton.

CLOSED CELL STRATOCUMULUS—A term often used to describe oceanic stratocumulus viewed in satellite imagery. With warm air aloft, the clouds are spread laterally within the temperature inversion. Small clear zones form where debris from adjacent cells converge, and descend, giving the appearance of large cells with small clear borders.

CLOUD BAND—A nearly continuous cloud formation with a distinct long axis, a length-to-width ratio of at least 4:1, and a width greater than 1° lat.

CLOUD ELEMENT—The smallest cloud form that can be resolved in satellite imagery.

COMMA CLOUD SYSTEM—A cloud system with a shape resembling the comma punctuation mark. Comma patterns vary in size from small convective complexes to large storm systems. The comma shape results from differential rotation of the cloud border, but is further influenced by adjoining patterns of upward and downward motions.

[1] Some of the entries are courtesy of the American Society of Photogrammetry and Remote Sensing: Colwell, R. N., 1983: *Manual of Remote Sensing,* Vol. I. ASPRS, Falls Church, VA.

COMMA HEAD—The rounded portion of the cloud comma pattern to the left of the axis of maximum wind speed. (See COMMA CLOUD SYSTEM.) This portion shows the most rotation and frequently contains an overcast of middle and high clouds. This region often produces most of the precipitation.

COMMA TAIL—The portion of the comma cloud that is to the right of and nearly parallel to the axis of maximum winds. (See COMMA CLOUD SYSTEM.) The tail consists of generally low- and mid-level (or lower based) clouds. It tends to be convectively active during the afternoon and evening.

COMPUTER COMPATIBLE TAPES—Tapes containing digital data. These are standard magnetic tapes, 19 cm wide, in 9-track or 7-track format.

CONTRAST—Use of opposing elements such as shades of gray or colors to highlight differences in a scene. (See ENHANCEMENT.)

CONTRAST STRETCHING—Improving the contrast of images by digital processing. The original range of digital values is expanded to utilize the full contrast range of the recording film or display device.

COUNTS—Digitized output of a sensor.

CONVENTIONAL DATA—Data collected by ordinary means such as thermometers, anemometers, radiosondes, etc., as contrasted with data collected by a remote-sensing system. (See GROUNDTRUTH.)

DATA ACQUISITION SYSTEM—The collection of devices and media that measures and records physical variables prior to input to the data processing system.

DATA GRANULE—The smallest aggregation of data items that is independently managed (e.g., described, inventoried, retrieved). This, for example, may be a swath or scene for scanner data, or for time series data, a fixed-interval segment (e.g., an hour, an orbit). The granule may, but need not, correspond to a physical segment on a data volume (e.g., a file on a tape) or to an entire volume (e.g., a tape).

DATASET—A logically meaningful grouping or collection of similar or related data. Data having all of the same characteristics (source or class of source, processing level and algorithms, resolution, etc.), but different independent variable ranges, are normally considered part of a single data set.

DATASET CATALOG—Next level down from DATASET DIRECTORY; detailed information about whole datasets; typically specific to a discipline, data center, and/or other facility.

DATASET DIRECTORY—A set of descriptions of a large number of datasets containing high-level information suitable for making an initial determination of the potential usefulness of a dataset for some application. Information on the location of more detailed descriptions and/or the dataset itself will be found in this directory.

DEFORMATION ZONE—A neutral point (col) in the atmosphere's motion, relative to itself, where air parcels undergo contraction in one direction, and elongation in the perpendicular direction. The narrow zone along the axis of elongation is called the deformation zone. Gradients of conservative properties increase in deformation zones as parcels of air approach, but do not cross the axis of elongation.

DEFORMATION ZONE CLOUD SYSTEM—Clouds in deformation zones undergo elongation in one direction and contraction in the perpendicular direction. The shape of these clouds depends upon the synoptic pattern and the amount of moisture and convergence. Often, clouds evaporate on one side of an elongation axis, producing a smooth cloud boundary.

DEGRADATION—The lessening of quality of data or images because of any optical, electronic, or mechanical distortions (noise) in the data collection or image forming systems.

DESCENDING NODE—The southbound Equator crossing of a satellite; given in degrees longitude, date, and time for any given orbit or pass.

DIELECTRIC CONSTANT—Electrical property of matter that influences radar returns; also referred to as complex dielectric constant.

DIGITAL—The presentation or recordation of data or information in numerical format.

DOPPLER EFFECT—A change in the observed frequency of electromagnetic radiation or other waves caused by relative motion between the source and the observer.

DRIFT—The horizontal track of an object (e.g., clouds) caused by the wind or fluid motion.

DUAL-FREQUENCY RADAR—A radar, or pair of radars, operating at different frequencies but synchronized to view the same target.

DVORAK TECHNIQUE—A cloud pattern analysis procedure that estimates the current intensity of tropical cyclones by using environmental satellite imagery. The technique was developed by Vern Dvorak, NOAA/NESDIS, in the early 1970s.

EARTH SYNCHRONOUS ORBIT—An orbit in which the motion of the satellite is synchronous with the rotation of the Earth so that the position of the satellite appears to remain stationary from the Earth in space and time. (See GEOSTATIONARY SATELLITE.)

ELECTROMAGNETIC RADIATION—Energy propagated through space or through material media in the form of an advancing interaction between electric and magnetic fields. The term radiation, alone, is commonly used for this type of energy. Also called electromagnetic energy.

ELEMENT—The smallest definable object of interest in a scene. It is a single item in a collection, population, or sample.

EMISSIVITY —The ratio of the radiation given off by a surface to the radiation given off by a blackbody at the same temperature; a blackbody has an emissivity of 1, other objects between zero and 1.

ENHANCED INFRARED IMAGERY—Enhancement is the process by which the contrast between cloud and surface features in infrared imagery is increased to simplify interpretation. This is done by displaying the available gray shades through a smaller temperature range. Depending on what combinations of temperature thresholds/segments are used, nearly any combination of low-, middle-, and high-level cloud features can be highlighted.

ENHANCED VISIBLE IMAGERY—Similar to enhanced

infrared imagery except that various gray shades are assigned to specific ranges of brightness in the visible. Enhanced VIS imagery is useful in providing better early morning views of fog/stratus and aiding in prediction of fog/stratus dissipation times.

ENHANCEMENT CURVE—A particular combination of temperature ranges assigned to specific gray shades to provide contouring and to highlight specific features such as tropical storms, convective activity, stratus/fog, sea surface temperatures, etc.

ENHANCED "*V*"—A cloud top signature sometimes seen in enhanced IR imagery in which the coldest cloud tops form a "*V*" shape with the base of the "*V*" on the upwind side of a convective storm and warmer cloud tops in the center of the "*V*." Storms which show this cloud top feature are often associated with severe weather.

EPHEMERAL DATA—Data that: 1) Help to characterize the conditions under which the remote sensing data were collected. 2) May be used to calibrate the sensor data prior to analysis. 3) Include such information as the positioning and spectral stability of sensors, Sun angle, platform attitude, etc.

EPHEMERIS—Position of a satellite in space as a function of time. Ephemeris data are used for gridding satellite imagery. Since ephemeris gridding is based solely on the predicted position of the satellite, it is susceptible to errors from vehicle pitch, orbital eccentricity, and the oblateness of the Earth.

EQUATORIAL CROSSING TIME—The local solar time of a satellite's passage over the Equator. Orbital numbers are assigned at the ascending crossing point, and are advanced by one when the spacecraft (polar) completes a full revolution about the poles and again reaches the Equator.

EQUIVALENT BLACKBODY TEMPERATURE—(Also known as BRIGHTNESS TEMPERATURE.) The temperature of a blackbody that would give the same energy output per unit area at the wavelength(s) under consideration, as that of the object being viewed.

EXTRATROPICAL—Usually used to describe weather phenomena in regions outside of the tropics (poleward of 23°N or 23°S).

FACSIMILE—A graphic display of information produced by the electrical transmission of data. It is displayed as dots on lines of constant or varying intensities.

FARINFRARED—The preferred term for the wavelengths in the infrared region greater than ~ 30 μm.

FILTER—1) Any material which, by absorption or reflection, selectively modifies the radiation transmitted through an optical system. 2) To remove a certain component or components of radiation, usually by means of a filter, although other devices may be used.

FIELD OF VIEW—See INSTANTANEOUS FIELD OF VIEW.

G/T—An engineering parameter where G is the gain of the receiving and T is the total system noise temperature. G/T is the antenna gain in decibels minus the noise temperature in dB.

GEOSTATIONARY SATELLITE—A satellite in a west-to-east orbit at an altitude of 35,800 km (22,300 statute miles) above the Equator. At this altitude it circles the axis of the Earth once in 24 hours making its speed in orbit synchronous with the Earth's rotation. It remains essentially stationary over a given geographical point above the Equator. GOES, Meteosat, and GMS are examples of geostationary enviromental satellites. (See SYNCHRONOUS SATELLITE.)

GRAVITY WAVE—Wave disturbance in which buoyancy acts as the restoring force on parcels displaced from hydrostatic equilibrium. Frequently seen in satellite imagery as a wave-like cloud formation within vertical temperature inversions.

GRAY BODY—A radiating surface whose radiation has essentially the same spectral energy distribution as that of a blackbody at the same temperature, but whose emissive power is less. Its absorptivity is nonselective.

GRAY SCALE—A calibrated sequence of gray tones ranging from black to white.

GRIDDING—The procedure of locating environmental satellite imagery on a certain geographic reference frame, i.e., placing latitude and longitude information on a satellite image. GOES and GMS are gridded by computer. NOAA analogue data are gridded manually.

GROUNDTRUTH—Term coined for data and information obtained *in situ* within the atmosphere or at or below the Earth's surface to aid in interpretation of remotely sensed data.

HYDROGRAPHIC—Referring to the study of waters (oceans, lakes, and rivers) including physical characteristics (oceanography, limnology) and elements affecting safe navigation.

HYPOXIA—A deficiency in the amount of oxygen reaching an organism.

IMAGE—1) The counterpart of an object produced by the reflection or refraction of light when focused by a lens or mirror. 2) The recorded representation (commonly as a photo-image) of an object produced by optical, electro-optical, optical mechanical, or electronic means.

IMAGE ENHANCEMENT—Any one of a group of operations that improve the detectability of targets or categories. These operations include, but are not limited to, contrast improvement, edge enhancement, spatial filtering, noise suppression, image smoothing, and image sharpening.

INFRARED IMAGERY—Satellite imagery sensed in portions of the 3 to 13 μm wavelength region of the electromagnetic spectrum. Usually refers to the thermal IR, particularly the 11 μm window unless otherwise stated.

INGESTING—Entering or inputting data, algorithms, etc., into a computer-based system for further processing, analysis, or evaluation.

INSOLATION—Incident solar energy

INSTANTANEOUS FIELD OF VIEW—The spot size or instantaneous geographic coverage of a satellite sensor. The narrow field of view designed into detectors, particularly scanning radiometer systems, allows for electromagnetic radiation from only a small area to be recorded at any one instant.

INTERACTIVE IMAGE PROCESSING—The use of an operator or analyst at a computer or analytical console to provide the means of assessing, preprocessing, extracting features, classifying, identifying, and displaying the original or processed imagery for subjective evaluations and further interactions.

INVENTORY—Information about data granules; usually used to identify or locate the needed dataset granules, given the specification of the independent variable range(s); may contain information extracted from the data granules (e.g., percent cloud cover) as well as information to enable ordering (e.g., Volume ID, file names, etc.) Inventory also includes information on the availabilty of browse data sets to enable the selection of a particular browse data.

MAPPED DATA—Earth located data that are displayed on a specific map projection, such as Mercator, polar sterographic, etc.

MARSHALL-PALMER DISTRIBUTION—An empirical expression developed by Marshall and Palmer relating number of drops (within diameter intervals) per unit volume for various rainfall rates.

MESOSCALE—Describes events that occur on a local scale (10 to 100 km) and are of short duration, typically zero to 3 hours. Thunderstorms, tornadoes, and flash floods are events usually described as mesoscale.

MESOSCALE CONVECTIVE COMPLEX—Often called the MCC, is the older designation for the MESOSCALE CONVECTIVE SYSTEM.

MESOSCALE CONVECTIVE SYSTEM—A well-organized cluster of thunderstorms with cloud top temperatures of −32°C or lower over at least 100,000 km^2 and temperatures of −52°C or lower over at least 50,000 km^2 for at least six hours.

METSAT—Meteorological satellite, such as the DMSP, NOAA, GOES, Meteosat, and GMS satellites. Metsat refers to weather (and environmental) satellites in general, not to a specific one.

MICROWAVE—Electromagnetic radiation having wavelengths between approximately 1 mm and 1 m, or 0.3 to 300 GHz in frequency, bounded on the short wavelength side by the far infrared (at 1 mm) and on the long wavelength side by very high-frequency radio waves. Passive systems operating at these wavelengths are sometimes called microwave systems. Active systems are called radar, although the literal definition of radar requires a distance-measuring capability not always included in active systems. The exact limits of the microwave region are not defined.

MIDDLE INFRARED—The preferred term for the wavelengths in theinfrared region extending from about 3 μm to around 30 μm.

MIE SCATTERING—Scattering of radiation by particles of size equal to or larger than the wavelength of the incident radiation. The resulting scattering is a complicated function of geometry, particle size, and the optical properties of the particle.

MULTICHANNEL SYSTEM—Usually used for scanning systems capable of observing and recording several channels of data simultaneously, preferably through the same aperture.

MULTIPLEXER(ING)—A system that combines several individual data communications lines or channels that are not used to full capacity into a single line that is more efficient.

MULTISPECTRAL—Generally used for remote sensing in two or more spectral bands, such as visible and infrared.

MULTISPECTRAL (LINE) SCANNER—A remote sensing device that operates on the same principle as the infrared scanner, except that it is capable of recording data in the ultraviolet and visible portions of the spectrum as well as the infrared.

MUNSELL COLOR CHART—A radiometrically traceable color system based on the Commission International de l'Éclairage (CIE).

NADIR—Satellite subpoint, or the point on the Earth's surface that is centered directly below the satellite. This is also called the point of zero nadir angle for a satellite in Earth orbit.

NAVIGATION—Refers to locating remotely sensed data on the Earth. Algorithms to accomplish this employ altitude and orbit of the spacecraft and attributes of the sensor.

NEAR INFRARED—The preferred term for the shorter wavelengths in the infrared region extending from about 0.75 μm (visible red) to around 3 μm. The term usually emphasizes the radiation reflected from plant materials, which peaks around 0.85 μm. It is sometimes called solar infrared, since it has most significance during the daylight hours.

NEPHANALYSIS—A graphic synoptic depiction of types and amounts of cloudiness from satellite or other observations.

NODAL INCREMENT—Degrees longitude between successive polar orbiter ascending nodes. It is the amount of turning of the Earth under the satellite, measured in degrees longitude, that takes place during one nodal period.

NODAL PERIOD—The time elapsing between successive passages of a satellite through successive ascending nodes.

NODE—One of the two points of intersection of the orbit of a satellite with the plane of the Equator of the Earth. The ascending node of Equator crossing refers to that point on the plane of the Equator at which the satellite crosses from the Southern to the Northern Hemisphere. The descending node of Equator crossing denotes that point at which the satellite crosses the plane of the Equator from the Northern to the Southern Hemisphere.

OPAQUE—Impervious to radiation at the wavelength being discussed. Not transparent or translucent.

OPEN CELL CUMULUS—Cumulus clouds that form as polygons or ellipses and tend to have clear centers and cloud edges. Open cell cumulus are found mainly over oceans in areas of strong cold air advection and strong low-level winds.

ORBIT—The path which a satellite follows in its motion through space, relative to the Earth.

ORBIT NUMBER—Refers to a particular circuit path beginning at the satellite's ascending node. The number from launch to the first ascending node is designated as zero.

OUTFLOW BOUNDARY—A surface boundary left by the horizontal spreading of cool air associated with a thunderstorm. The boundary may be of only recent origin, or may have existed as a distinct boundary for over a day. Outflow boundaries may intersect each other or other lines (fronts, low-level jets) and "focus" new convection. (See ARC CLOUD.)

OVERLAP—The area common to two successive photos or scan swaths along the same or adjacent flight or orbital strip; the amount of overlap is expressed as a percentage of photo or scanned area.

PASSIVE SYSTEM—A sensing system that detects or measures radiation emitted by the target. It is distinguished from ACTIVE SYSTEM.

PAYLOAD—The environmental sensors and instruments carried on the spacecraft.

PELAGIC—Of or pertaining to the seas or oceans.

PHOTOSYNTHETIC PIGMENTS—Any of a group of light-absorbing pigments.

PHYTOPLANKTON—Floating and drifting plant life of the sea.

PICTURE—Representation of a scene by a photographic positive print or transparency, made from a negative, produced by the direct action of visible light or electromagnetic radiation (EMR) outside the visible part of the spectrum and converted into visible EMR by an optical-mechanical or wholly electronic scanner.

PIXEL—Picture Element. The smallest element (spectral or spatial) sensed by a satellite sensor. A row of pixels forms a scan line; a series of scan lines creates an image.

POLARIZATION—The direction of vibration of the electrical field vector of electromagnetic radiation.

POLAR ORBITING SATELLITE—A satellite whose orbit lies in a plane passing through the center of the Earth and is inclined to the equatorial plane so that the subpoint track traverses polar latitudes on every orbit.

PRECESSION RATE—The fixed space angular motion of the orbital line of nodes; positive to the east, negative to the west. Precession rate for a sunsynchronous orbit is −0.986° per day, or about 360° per year. The net effect is that the satellite's orbital plane rotates slowly around the Earth at the same rate and direction that the Earth rotates around the Sun; hence Sun-synchronous.

PRINCIPAL POINT—The point on the Earth where the satellite sensor is focused at any time during the orbit. If the sensor vertical axis is perpendicular to the Earth's surface, the principal point coincides with the subpoint.

QUANTITATIVE PRECIPITATION ESTIMATION—The estimation of rainfall or rainfall rates based on satellite data.

RADIO FREQUENCY—The frequency of radio energy, specifically the frequency of a given radio carrier wave.

RADIOMETER—An instrument for quantitatively measuring the intensity of electromagnetic radiation (radiant energy) in some band of wavelengths in a part of the electromagnetic spectrum. Usually used with a modifier, such as an infrared radiometer or a microwave radiometer.

RADIOSONDE—A balloon-borne instrument designed to measure atmospheric temperature, pressure, and humidity. This instrument transmits the data on a radio frequency throughout its ascent.

RAPID INTERVAL SCAN—The mode of data gathering by GOES whereby an image is produced at 5- to 15-minute intervals instead of the normal 30-minute interval. Rapid interval scan is used primarily for research and to observe rapidly developing severe weather situations.

RAWINSONDE—A RADIOSONDE that is tracked by a radar or radio direction finder to measure wind speed and direction.

RAYLEIGH SCATTERING—The wavelength-dependent scattering of electromagnetic radiation by spherical particles in the atmosphere much smaller (about one-tenth or less) than the wavelengths scattered.

REAL TIME—Time in which reporting/recording of events is simultaneous with the events. Real time of satellite data is the time at which the data are received. It is distinguished from archive retrieval of the data.

RESOLUTION—Size of the smallest object or smallest physical unit (e.g., temperature) that can be sensed in the satellite data.

S1 SKILL SCORE—Forecast verification criteria determined by

$$\frac{100 \sum |e_G|}{\sum |G_L|}$$

where e_G is the error of the forecast height difference and G_L is the observed or forecast height difference between grid points, whichever is larger. The lower the score, the more useful the forecast.

SCAN LINE—The narrow strip on the ground that is swept by the instantaneous field of view of a detector in a scanner system.

SCANNER—1) Any device that scans, and can thus produces an image. (See SCANNING RADIOMETER.) 2) A radar set incorporating a rotatable antenna, or radiator element, motor drives, mounting, etc. for directing a searching radar beam through space and imparting target information to an indicator.

SCANNING RADIOMETER—A radiometer, which by the use of a rotating or oscillating plane mirror, can scan a path, usually normal, to the movement of the radiometer.

SCOFIELD/OLIVER TECHNIQUE—A manual technique designed to estimate thunderstorm rainfall using consecutive GOES enhanced infrared imagery. (See QUANTITATIVE PRECIPITATION ESTIMATION.)

SEICHE—1) A standing wave oscillation found in a bounded water body that continues after the initiating force has terminated. 2) Oscillation of a fluid in response to a disturbing force having the same frequency as the naturally occuring frequency of that fluid, e.g., tides.

SENSOR—Any device that gathers energy, electromagnetic radiation or other, converts it into a signal, and presents it in a form suitable for obtaining information about the environment.

SHORTWAVE RADIATION—Radiation in the visible and near-visible portion of the electromagnetic spectrum (0.4 to 1.0 μm in wavelength).

SIDELOOKING RADAR—An all weather, day/night remote sensor which is particularly effective in imaging large areas of terrain. It is an active sensor, generating and transmitting its own energy. The return energy received is used to produce a photo-like picture of the ground. Also referred to as side-looking airborne radar: SLAR.

SIGNAL—The effect (e.g., pulse of electromagnetic energy) conveyed over a communications path or system. Signals are received by the sensor from the scene and converted to another form for transmission to the processing system.

SIGNATURE—Any characteristic or series of characteristics by which a material, phenomenon, or change may be recognized in an image or data set.

SIGNATURE ANALYSIS TECHNIQUES—Techniques that use the variation in the spectral reflectance or emittance of objects as a method of identifying the objects.

SOUNDINGS—See ATMOSPHERIC SOUNDINGS

SPECTRAL BAND—An interval in the electromagnetic spectrum defined by two wavelengths, frequencies, or wavenumbers.

SPECTRAL CHANNEL—A portion of the electromagnetic spectrum delineated by specific radiant energy wavelengths. As related to environmental satellites, sensors respond to incident energy in specific spectral bands.

SPECTRAL INTERVAL—The width, generally expressed in wavelength or frequency of a particular portion of the electromagnetic spectrum. A given sensor is designed to measure or be sensitive to energy received at the satellite from that part of the spectrum. Also termed SPECTRAL CHANNEL or SPECTRAL BAND.

SPECULAR REFLECTION—The reflectance of electromagnetic energy without scattering or diffusion, as from a surface that is smooth in relation to the wavelengths of incident energy. Also called mirror reflection.

SPLIT WINDOW—1) Segmentation of a remote-sensing window channel into parts, i.e., smaller wavelength increments. 2) Frequently used to describe a technique used to compute satellite-observed sea surface temperatures.

STRETCHED GRIDDED DATA—Data from an orbit that is stretched to eliminate most of the panoramic distortion caused by the Earth's curvature and scanning geometry of the satellite sensors.

STRETCHED VISSR DATA—GOES VISSR image data that have been stretched in time by the temporary storage and retransmission of those data at the CDA by the use of the SDB.

STEREO(STEREOSCOPIC)IMAGE—That mental impression of a three-dimensional object which results from stereoscopic vision (stereo viewing).

SUBPOINT—The point on the Earth centered directly below a satellite. This is also called the point of zero nadir angle for a satellite in Earth orbit. See NADIR.

SUBTRACK—The locus of subpoints on the Earth made by a satellite as it moves through its orbit.

SUN-SYNCHRONOUS ORBIT—An orbit which precesses 360° during the course of the year permitting a satellite in this orbit to obtain views of a given geographical area, at the same local times each day. A Sun-synchronous satellite always crosses the Equator on the ascending node at the same local solar time. (See PRECESSION RATE.)

SUN GLINT—When the sun illuminates a water surface, a portion of the shortwave radiation is reflected to space. This reflection appears in visible imagery as a light area on an otherwise dark water area. When the surface is perfectly smooth, it becomes SPECULAR REFLECTION.

SYNCHRONOUS SATELLITE—A satellite orbiting the Earth which makes one revolution in 24 hours synchronous with the Earth's rotation. (See GEOSTATIONARY SATELLITE.)

SYNOPTIC—Referring to an overall view. In meteorology referring to data obtained simultaneously over a wide area to describe a nearly instantaneous picture of the state of the atmosphere.

SYNTHETIC APERTURE RADAR—An active microwave sensor that uses the amplitude of the return signal to determine surface roughness and the Doppler shift to determine position.

T-NUMBER—A numerical designator, assigned to satellite observed tropical storm cloud patterns, that estimates the strength of the storm. (See DVORAK TECHNIQUE.)

TELEMETRY—The process of measuring a quantity or quantities, transmitting the measured value to a distant station, and there interpreting, indicating, or recording the quantities measured. Typically, spacecraft telemetry includes parameters such as spacecraft temperature, power supply, and orbital information which are transmitted to Earth to monitor spacecraft status.

TERMINATOR—The great circle of transition from daylight to darkness on the Earth's surface. The terminator is detectable in environmental satellite visible imagery as a gradual change from easily visible cloud features to a nearly black image.

TEXTURE—In a photo image, the frequency of change and arrangement of tones, often due to highlights and shadows created by the irregular surface being viewed.

THERMAL BAND—A general term for middle-infrared wavelengths where the signal is dependent on the temperature of the emitting surface. It is used to describe the atmospheric windows 8 to 14 μm and 3 to 6 μm.

THERMOCLINIC—A steep vertical temperature gradient in a layer of water which is appreciably greater than gradients above or below it.

THERMAL INFRARED—The preferred term for the middle wavelength range of the IR region, extending roughly from 3 μm at the end of the near infrared, to about 15 or 20 μm, where the far infrared begins. In practice the limits represent the envelope of energy emitted by the Earth behaving as a gray body with a surface temperature around 290 K (27°C). (See GRAY SCALE.)

TONE—Each distinguishable shade of gray from white to black on an image.

ULTRAVIOLET RADIATION—Electromagnetic radiation of shorter wavelength than visible radiation but longer than X-rays; roughly, radiation in the wavelength interval between 10^{-3} and 0.4 μm.

UPWELLING—1) Rising of subsurface water to the surface. It is common along coast lines when the wind blows parallel to the coast and the surface water is displaced away from the coast and replaced by water from below. 2) Radiation moving from the surface upward.

VALIDATION—Comparison of a measurement from an instrument with an alternative measurement of a parameter by a different instrument or technique of acceptable reliability.

VIDICON CAMERA SYSTEM—A storage-type electronically scanned photoconductive television camera tube, which often has a response to radiations beyond the limits of the visible region. Particularly useful in early space applications, since no film was required.

VISIBLE IMAGERY—(Also known as VISUAL IMAGERY.) Environmental satellite imagery sensed in the 0.4 to 0.75 μm wavelength region of the electromagnetic spectrum.

VORTEX—In its most general use, any flow possessing vorticity. More often the term refers to a flow with closed streamlines. Frequently used to describe a semicircular or circular cyclonic cloud pattern.

VORTICITY—A vector measure of the local atmospheric rotation.

WARM-CORE RINGS—Anticyclonic rings/circulations that form throughout the slope water region between the Continental Shelf and the Gulf Stream.

WATER VAPOR IMAGERY—A display of data taken in one of the water vapor channels (e.g., 6.7 or 7.3 μm). Atmospheric water vapor absorbs outgoing radiation in these regions, resulting in a decreased temperature being sensed by the satellite.

WINDOW—A band of the electromagnetic spectrum which offers maximum transmission and minimal attenuation through a particular medium with the use of a specific sensor. (See ATMOSPHERIC WINDOWS.)

ZENITH—The point in the celestial sphere that is exactly overhead; opposed to nadir.

AMERICAN METEOROLOGICAL SOCIETY

INDEX

Page numbers in italics indicate figures; page numbers followed by t indicate tabular material.